Progress in the Science of Functional Dyes

Yousuke Ooyama · Shigeyuki Yagi
Editors

Progress in the Science of Functional Dyes

Springer

Editors
Yousuke Ooyama
Hiroshima University
Higashi-Hiroshima, Hiroshima, Japan

Shigeyuki Yagi
Osaka Prefecture University
Sakai, Osaka, Japan

ISBN 978-981-33-4391-7 ISBN 978-981-33-4392-4 (eBook)
https://doi.org/10.1007/978-981-33-4392-4

This Springer imprint is published by the registered company Springer Nature Singapore Pte Ltd.
The registered company address is: 152 Beach Road, #21-01/04 Gateway East, Singapore 189721, Singapore

Preface

The dawn of organic dye (color) chemistry will undoubtedly be Sir W. H. Perkin's serendipitous discovery of the first synthetic organic dye *Mauve* (Mauveine) in 1856, and has led to the first synthetic indigo in 1878 by J. F. W. Adolf von Baeyer, Nobel Laureate in Chemistry in 1905. Nearly 100 years after the first synthetic indigo, the traditional organic dyes as dyestuffs and pigments have developed into "Functional Dyes" which is a term coined by Japanese dye-chemists T. Kitao and Z. Yoshida et al. in the late 1970s, defined as attractive materials generating a new function that is changing the color and luminescence, recording information, or converting energy (light to electricity or vice versa, etc.) by external stimuli including light, temperature, pressure, pH, electric and magnetic fields, atmospheric gas, etc. Indeed, the first *International Symposium on Functional Dyes* was successfully held in Osaka, Japan, in 1989, which resulted in major breakthroughs in functional dye chemistry and technology. The symposium was followed by two major symposia, *East Asia Symposium on Functional Dyes and Advanced Materials* (EAS) and *International Symposium on Functional π-Electron Systems* (F-π). As the result of considerable theoretical, synthetic, and physical researches, functional dyes have actually been applied as semiconductors, emitters in optoelectronics devices, photosensitizers, optical and luminescent sensors for biological and environmental monitoring systems, and therapeutic agents for medical purposes, and so on.

Today, Sustainable Development Goals (SDGs) has been adopted as the 2030 agenda for sustainable development by all United Nations Member States in 2015. It provides a shared blueprint for peace and prosperity for people and the planet, now and into the future. As we can see, functional dyes have attracted growing interest as one of the most promising materials contributing to SDGs. Judging from the world situation, we sense just the dawning of a new era for functional dye chemistry. Consequently, this prompted us to write a new book regarding functional dyes passing on the wisdom and expertise to the next generation.

"Color Chemistry" (published by Wiley-VCH, the first, second, and third editions in 1987, 1991, and 2003, respectively) edited by Heinrich Zollinger, who was an inspiring teacher and researcher of Dyestuff and Textile Chemistry, has been a very successful textbook for synthesis, properties, and applications of organic dyes and pigments. Klaus Hunger edited "Industrial Dyes" (published by Wiley-VCH in 2003), which is a reference book comprising the principal classes of industrially produced dyes and their toxicological, ecological, and legal survey. Thus, the aim of this book "Progress in the Science of Functional Dyes" is to give readers cutting-edge information on functional dye chemistry evolved from "Color Chemistry" and "Industrial Dyes". This book is written by Japanese senior and young dye-researchers, although we understand that there are a wide variety of opinions from the reader in this regard.

This book actually covers a wide range of relevant topics to functional dyes, from synthesis and functionalities to applications. Making a survey of recent progresses in functional dye chemistry provides the opportunities not only to understand the structure–property relationships of a variety of functional dyes but also to know how they are applied to practical uses from electronic devices to biochemical analyses. From classic dyes such as cyanines, squaraines, porphyrins, phthalocyanines, and so on to the newest functional π-conjugation systems, various types of functional dyes are taken up in the book, especially focusing on their state-of-the-art topics and future aspects. Indeed, Part I is the synthesis and properties of cyanine dyes, squaraine dyes, porphyrins, phthalocyanines, BODIPY dyes, and their analogues as important structures for functional dyes. Part II deals with crystal polymorphism and dye aggregation, photochromism, two-photon absorption properties, solid-state photoluminescence, and circular polarized luminescence as the unique advanced properties of functional dyes. In Part III, various applications of functional dyes such as chemosensors and bioimaging, white-light emissive materials based on supramolecular approach, photodynamic therapy, photoenergy conversion (dye-sensitized solar cells), photoenergy conversion (organic photovoltaic), organic light-emitting diode are introduced. Therefore, interested readers will benefit from the scientific context where organic dyes and pigments are comprehensively expounded from the chemical aspects.

We thank the authors and publishers for copyrighted materials reproduced with their permission. Thanks are also due to Dr. Shinichi Koizumi, Senior Editor of Chemical Division, Springer, who lead us to this project, as well as Ms. Taeko Sato, who carefully edited the manuscript and Ms. Kowsalya Raghunathan for the project coordination.

Wish for the end of the Novel Coronavirus Disease 2019 (COVID-19).

Hiroshima, Japan
Osaka, Japan
July 2020

Yousuke Ooyama
Shigeyuki Yagi

Contents

Part I Important Structures for Functional Dyes

1 **Polymethine Dyes** 3
Masaki Matsui

2 **Squaraine Dyes** 21
Takeshi Maeda

3 **Porphyrins: Syntheses and Properties** 49
Jun-ichiro Setsune

4 **Phthalocyanine and Related Analogues** 85
Soji Shimizu

5 **BODIPY Dyes and Their Analogues** 119
Yasuhiro Kubota

Part II Properties of Functional Dyes

6 **Molecular and Crystal Structures of Polymorphic Organic Dyes and Coloured Organic Compounds** 223
Shinya Matsumoto and Jiyong Hwang

7 **Photochromism** 263
Seiya Kobatake

8 **Red and Near-IR Fluorescent Two-Photon Absorption Dyes** 283
Tsutomu Ishi-i

9 **Molecular Designs for Solid-State Luminescent Properties and Recent Progresses on the Development of Functional Luminescent Solid Materials** 309
Kazuo Tanaka, Masayuki Gon, and Yoshiki Chujo

10 **Circularly Polarized Luminescence (CPL) Based on Planar Chiral [2.2]Paracyclophane** 343
Yasuhiro Morisaki

Part III Applications of Functional Dyes

11 Fluorescent Chemosensors 377
Yosuke Niko and Shigeru Watanabe

12 White-Light Emissive Materials Based on Supramolecular Approach 409
Yuji Kubo

13 Photodynamic Therapy 445
Keiichi Imato and Yousuke Ooyama

14 Photoenergy Conversion (Dye-Sensitized Solar Cells) 469
Yousuke Ooyama

15 π-Conjugated Polymers Incorporating Naphthalene-Based Nitrogen-Containing Heteroaromatics for Organic Photovoltaics 541
Itaru Osaka

16 Luminescent Materials for Organic Light-Emitting Diodes 561
Shigeyuki Yagi

Part I
Important Structures for Functional Dyes

Chapter 1
Polymethine Dyes

Masaki Matsui

Abstract Polymethine dyes consist of heteroaromatic end and linkage groups. They are classified into cyanines, merocyanines, and oxonols. Their UV–Vis and near-infrared (NIR) absorption band are affected by the heteroaromatics and linkage groups. The change in the absorption band of polymethine dyes by aggregate formation has been explained by the molecular exciton coupling theory. Polymethine dyes have been used as data recording materials. Polymethine dyes can be used as sensitizers in dye-sensitized solar cells. Though usual dyes are solid, liquid trimethine dyes have been reported. The fluorescence intensity of the liquid trimethine dyes is drastically enhanced in liquid nitrogen.

Keywords Synthesis · UV–vis absorption band · Data recording materials · Sensitizers

1.1 Introduction

In the nineteenth century, it was strongly desired to prepare "Quinine" which acts as an antimalarial drug. It was known that an antimalarial compound "Cinchonine" was contained in the sap of Cinchona officinalis. An English chemist, G. Williams, isolated magnificent blue material from "Cinchonine" in 1856. Later, the structure was proved to be the mixture of mono- and trimethine dyes **1.1** and **1.2** as shown in Fig. 1.1. This material was named cyanine (*cyanos* = blue). It is well known that W. H. Perkin discovered Mauve or Mauveine in 1856. However, the discovery of cyanine is hardly known, probably due to the fact that cyanine was very weak upon light irradiation to become a commercial dye for coloration purpose.

In 1873, a German chemist, H. W. Vogel, found that cyanine dyes could enhance the photosensitivity in photography. As World War I broke out in 1914, it was desired to develop the studies on the sensitizers in the field of photography. An English chemist, W. H. Miles, further developed the studies on cyanine dyes.

M. Matsui (✉)
Gifu University, Gifu, Japan
e-mail: matsuimasaki_family@nifty.com

Y. Ooyama and S. Yagi (eds.), *Progress in the Science of Functional Dyes*,
https://doi.org/10.1007/978-981-33-4392-4_1

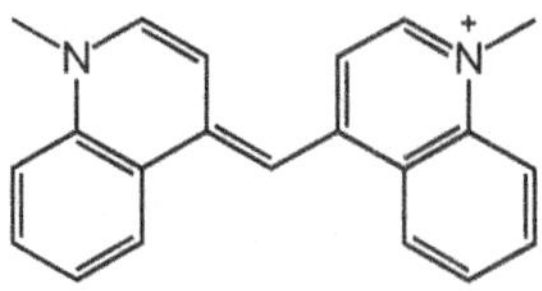

1.1

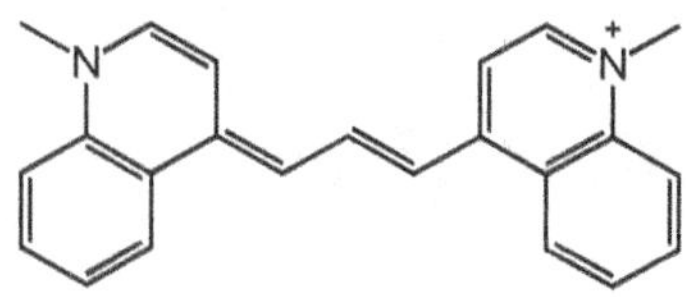

1.2

Fig. 1.1 William's cyanines

Thus, the first practical application of cyanine dyes comes from the spectral sensitizers for silver halide photography. Now, polymethine dyes have been used as the recording materials in optical disks and as probes in biological systems. They can be also used as the sensitizers in dye-sensitized solar cells (DSSCs).

There are excellent reviews for cyanines (Hishiki 1974; Yasui 1987; Mishra et al. 2000; Tatikolov 2012). In this chapter, the bases of polymethine dyes such as synthesis and physical properties and their topics are described.

1.2 Bases of Polymethine Dyes

1.2.1 Classification of Polymethine Dyes

The term "polymethine dyes" was proposed by König (1922). Figure 1.2 shows three basic structures of polymethine dyes. In these structures, heteroatoms are charged and are linked by a conjugated chain with an odd number of carbon atoms. When two heteroatoms are nitrogen, they are cationic. Though most cyanine dyes are symmetrical, the unsymmetrical derivatives are also known. They are subdivided according to the number of methine groups such as mono-, tri-, penta-, and heptamethines.

Cationic

(1-1)

Neutral

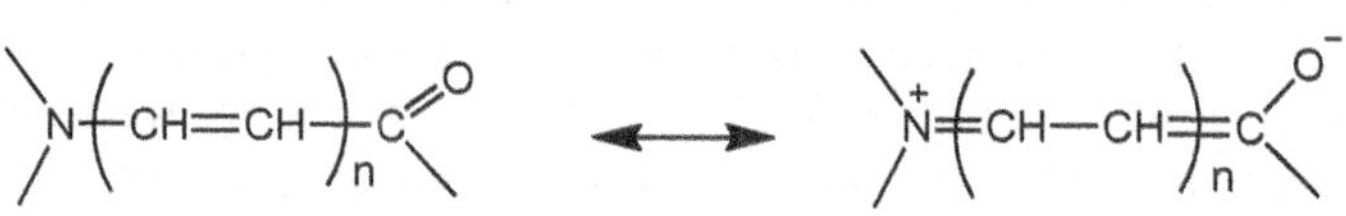

(1-2)

Anionic

(1-3)

Fig. 1.2 Basic structures of polymethine dyes

Here, it should be noted that the methine group belongs only to the linkage group. The methine group contained in the heteroaromatic moiety is not included. It is confusing that, in the long history of cyanine dyes, the terms "carbo-, dicarbo-, and tricarbocyanines", which correspond the number of n in Figs. 1.1, 1.2, 1.3, and 1.4, have been also used.

When one heteroatom is nitrogen and another one is oxygen, they are neutral and possibly expressed as a twitter ionic structure. Merocyanine dyes are classified in this group.

When two heteroatoms are oxygen, they are anionic. Oxonol dyes belong to this group.

The general structure of polymethine dye **1.3** is shown Fig. 1.3. The structure of polymethine dyes consists of two hetero end aromatics and a linkage group between them. The examples of hetero end aromatics are shown in groups A and B. Group A shows the hetero end aromatics for cyanine and merocyanine dyes. The use of same hetero end aromatics in group A at both sides produces the symmetrical cyanine dyes and that of different ones, unsymmetrical derivatives. Group B indicates the hetero end aromatics for oxonol dyes. The use of hetero end aromatics in group A and B in each terminal affords merocyanine dyes.

One of the unique points of polymethine dyes is the linkage group. The C–C bond length in the linkage group of benzothizolyl cyanine dyes was measured to be 1.41 Å by single X-ray crystallography. This length is between those of double (1.34 Å) and single (1.54 Å) C–C bonds, indicating that electrons are delocalized in the methine linkage. The polymethine chain is generally linked in the α- or γ-position to the nitrogen of the heterocyclic group in cyanines. Squarylium dyes are the special type of the linkage. Recently, squarylium cyanine dyes have attracted much attention due to their bathochromic and intense absorption and emission as functional dyes. Croconium cyanine dyes are much more bathochromic than the squarylium derivatives. The linkage group in polymethine dyes is easily substituted to get desired properties such as UV–Vis and NIR absorption and fluorescence bands.

Thus, the polymethine dyes consist of the combination of hetero end aromatics and linkage groups. Furthermore, the aza analogues also exist. Therefore, an infinite number of polymethine dyes are considerable.

1.2.2 Synthesis

Examples for the synthesis of polymethine dyes are indicated in Fig. 1.4. A quaternary salt having an acidic methyl group such as **1.4**, which is in equilibrium with the corresponding enamine **1.5** in solution, reacts with an aromatic compound having a good leaving group such as **1.6** to give a monomethine dye **1.7** (1–4).

Unsymmetrical monomethine dyes **1.10** and **1.13** are also obtained by similar reactions (1–5 and 1–6).

A symmetrical *meso*-substituted trimethine dye **1.15** is prepared from two molar amounts of **1.8** with 1,1,1-triethoxyalkanes **1.14** (1–7).

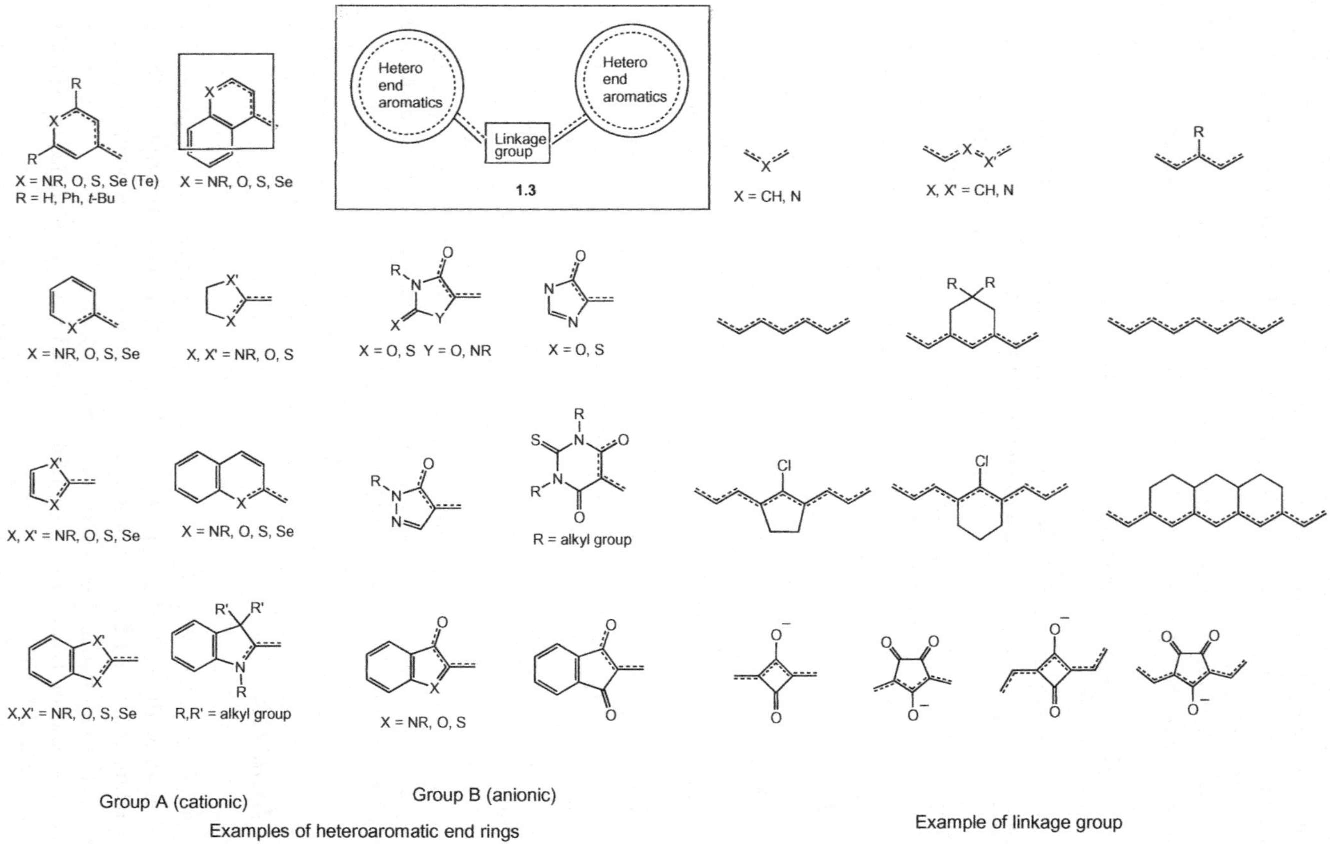

Fig. 1.3 Structure of polymethine dyes

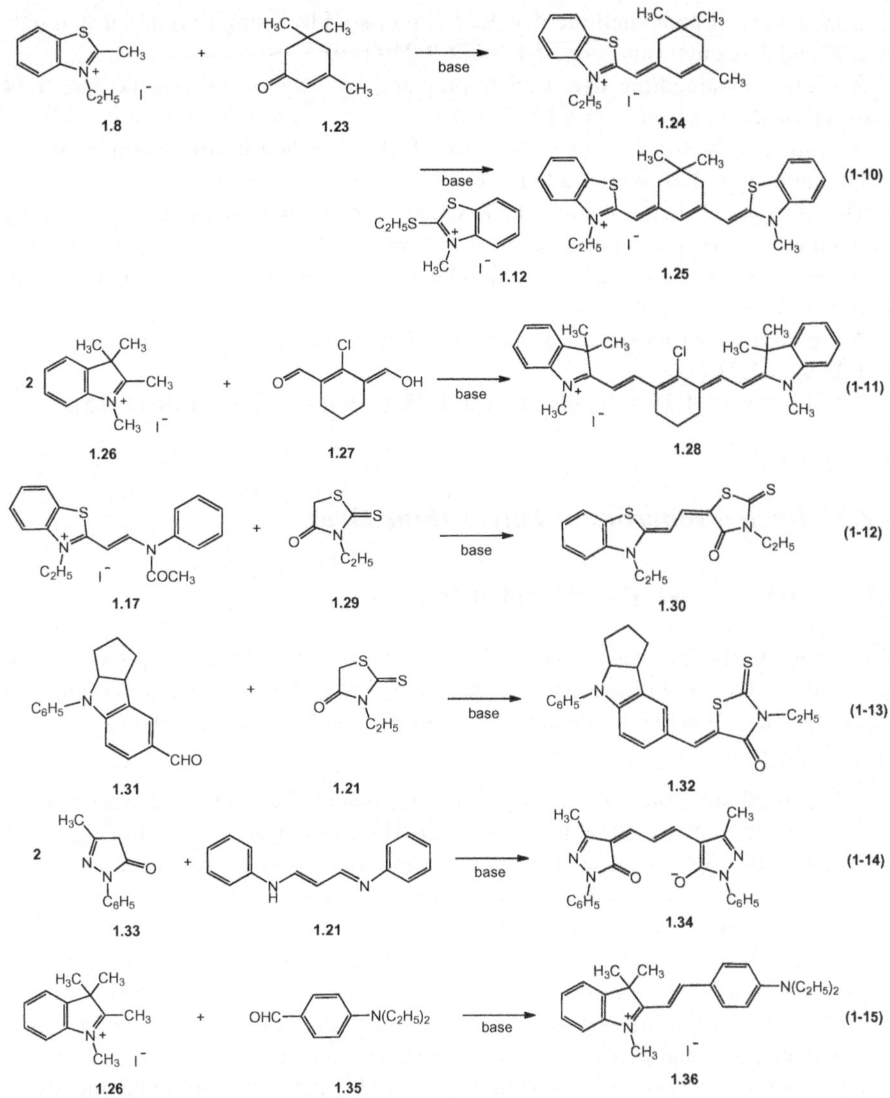

Fig. 1.4 Examples for the synthesis of polymethine dyes (1). Examples for the synthesis of polymethine dyes (2)

An unsymmetrical trimethine dye **1.19** is obtained by way of a hemicyanine **1.17**, which is produced by the reaction of **1.18** and diphenylformamidine **1.16** produced by the reaction of two molar amounts of aniline with triethyl orthoformate. The hemicyanine **1.17** is acetylated to give **1.18**, which is more active than **1.17**. Then, compound **1.18** reacts with **1.11** to give **1.19** (1–8).

A symmetrical pentamethine dye **1.22** is prepared by using two molar amounts of **1.20** and 3-phenyliminopropenylaniline **1.21** (1–9).

A cyclic pentamethine dye **1.25** is prepared by way of an intermediate **1.24** followed by the reaction with **1.12** (1–10).

A cyclic symmetrical heptamethine dye **1.28** is formed by the reaction of two molar amounts of **1.26** with **1.27** (1–11).

The acetylated hemicyanine **1.17** reacts with active methylene-containing rhodamine **1.29** to give a merocyanine dye **1.30** (1–12).

A merocyanine dye **1.32** is provided by the reaction of an aldehyde **1.31** with rhodanine **1.29** (1–13).

A symmetrical oxonol dye **1.34** is prepared by the reaction of two molar amounts of **1.33** with **1.21** (1–14).

The reaction of **1.26** with an aldehyde **1.35** gives a styryl dye **1.36** (1–15).

1.2.3 Basic Properties of Polymethine Dyes

1.2.3.1 UV–Vis and NIR Absorption Spectra

The change in the absorption spectra is shown in Fig. 1.5. The absorption band of polymethine dyes are influenced by (1) the number of carbon atoms in the conjugated linkage group, (2) the kind of end heteroaromatics, and (3) the kind of substituent at the linkage group:

(1) The usual structures of most cyanine and oxonol dyes are symmetrical. They are an alternative chromophoric system. The absorption maximum (λ_{max}) and molar absorption coefficient (ε) of cyanine dyes, whose structure is shown in **1.37**, become more bathochromic and more intense with extension of the π-conjugation system until heptamethine ($n = 3$). Though the λ_{max} of nona- ($n = 4$) and undecamethine ($n = 5$) derivatives is bathochromic, the absorption band becomes broad, which comes from *cis–trans* isomerization in the linkage group. Cyclization of the linkage group is carried out to inhibit the isomerization. When one olefinic bond increases in the conjugated linkage group, the λ_{max} causes a bathochromic shift about 100 nm. For example, in the case of indolenium dyes, the λ_{max} is observed at 545 nm for trimethine ($n = 1$), 636 nm for pentamethine ($n = 2$), and 745 nm for heptamethine ($n = 3$) derivatives, respectively. The λ_{max} of oxonol dyes **1.38** also becomes more bathochromic with increasing chain length due to extension of π-conjugation system.

(2) The λ_{max} also depends on the kinds of end heteroaromatics. Benzoxazole, indolenine, benzothiazole, and lepidine have been used as heteroaromatics. For example, in the case of trimethine dye **1.37** ($n = 1$), the λ_{max} is in the following order of the heteroaromatics: benzoxazole (485 nm) < indolenine (545 nm) < benzothiazole (557 nm) < lepidine (605 nm). The λ_{max} almost depends on the

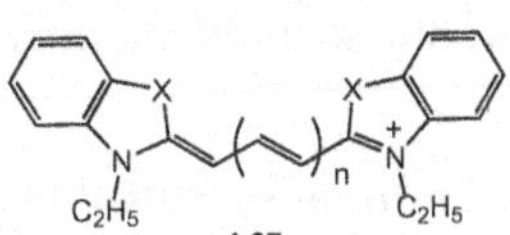

1.37

X	n	λ_{max} / nm
$C(CH_3)_2$	1	545
$C(CH_3)_2$	2	636
$C(CH_3)_2$	3	745
O	0	372
O	1	485
O	2	580
S	0	422
S	1	557
S	2	650
S	3	762
(CH=CH)	0	522
(CH=CH)	1	605
(CH=CH)	2	708
Se	0	429
Se	1	570
Se	2	660
Se	3	770

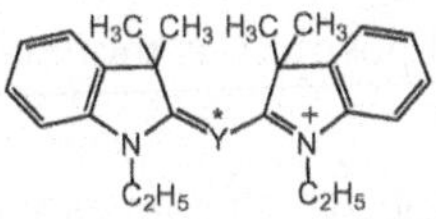

1.39

Y	λ_{max} / nm
CH	422
N	375

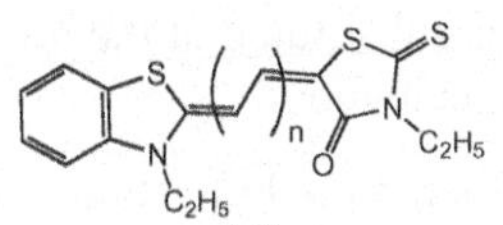

1.42

n	λ_{max} / nm
0	432
1	528
2	605
3	635

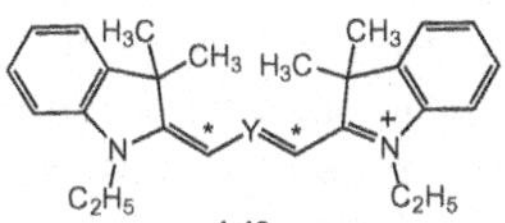

1.40

Y	λ_{max} / nm
CH	553
N	602

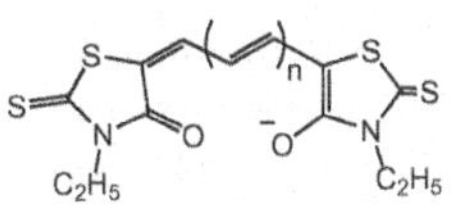

1.38

n	λ_{max} / nm
0	542
1	613
2	714

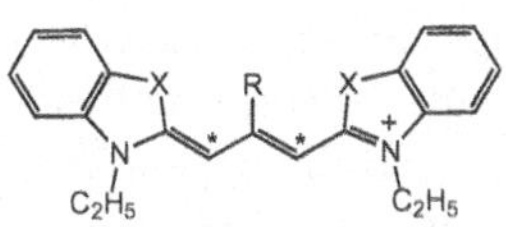

1.41

X	R	λ_{max} / nm
O	H	483
O	CH_3	491
S	H	557
S	CH_3	543
Se	H	570
Se	CH3	553

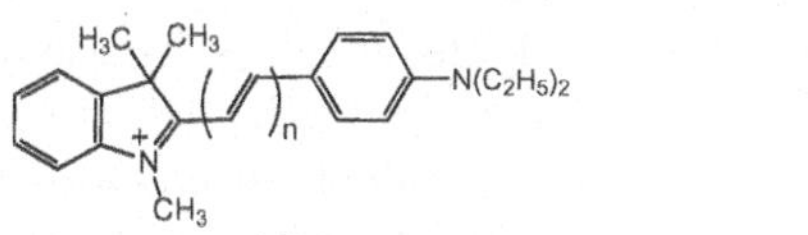

1.43

n	λ_{max} / nm
1	525
2	568

Fig. 1.5 Absorption spectra of polymethine dyes. **a** In methanol

basicity of the heteroaromatics. The larger the basicity of heteroaromatic ring, the longer the λ_{max}.

(3) When the methine group is substituted with a nitrogen atom, the λ_{max} also changes. In the symmetrical cyanine dyes, starred marks (*) are put at the even number of carbon atoms starting from the terminal nitrogen atoms. When the electron-donating group is introduced into the starred position or the electron-withdrawing group at the unstarred position, the λ_{max} causes a bathochromic shift. In the case of monomethine dye **1.39**, the aza analogue, in which an electron-negative nitrogen atom is substituted at the starred position, is more hypsochromic than the methine analogue. Meanwhile, in the case of trimethine dye **1.40**, the aza analogue, in which an electro-negative nitrogen atom

is substituted at the unstarred position, is more bathochromic than the methine analogue.

When a methyl group is introduced at the unstarred *meso*-position in trimethine dye **1.41**, the λ_{max} depends on the kind of heteroaromatics. In the case of benzoxazolyl derivative, the λ_{max} causes a bathochromic shift due to the introduction of the electron-donating methyl group at the unstarred position. However, in the case of benzothiazolyl and benzoselenolyl derivatives, the λ_{max} is hypsochromic due to steric effect between the heteroatoms and the methyl group.

In the cases of merocyanine **1.42** and styryl dyes **1.43**, the λ_{max} causes bathochromic shift when the π-conjugation system is extended. As merocyanine and styryl dyes are a push–pull chromophoric system, as stronger the push moiety and as stronger the pull moiety, the more bathochromic the λ_{max}.

The ε value of cyanine dyes is usually larger than 100,000 $dm^3\ mol^{-1}\ cm^{-1}$ due to wide planarity. Some cyanine, merocyanine, and styryl dyes are fluorescent.

1.2.3.2 Self-Aggregation

The self-association of dye molecules is a unique and important phenomenon of polymethine dyes. The aggregates can show characteristic changes in the absorption band. Comparing with the monomer absorption band, the bathochromically shifted one is called J-band (J for Jelley) and the hypsochromically shifted one H-band (H for hypsochromic). The absorption bands formed by aggregates are explained by the molecular exciton coupling theory (Kasha et al. 1965).

In diluted solutions, as there are no intermolecular interactions between dye molecules, the excited molecules are just deactivated to the ground state. Meanwhile, in concentrated solutions or in the crystalline form, the hole in HOMO can electrostatically have interactions with the electrons in LUMO. As a result, an exciton, a pair of a hole and an electron, is formed to delocalize the excited state (excitonic state).

The energy levels of an arranged dimer in the isolated two molecules have been proposed by Kasya (Kasha et al. 1965). The excitonic state of the dye aggregate splits into two levels E' and E'' through the interaction of transition dipoles as shown in Fig. 1.6. One dipole phase relation is toward the same direction and another one in the opposite direction. In the case of parallel transition dipoles, the E' level, in which the dipole phase relation is canceled, is forbidden to cause a blue shift, while in the case of in-line transition dipoles, the E'' level, in which the dipole phase relation is canceled, is forbidden to cause a red shift. In the case of oblique transition dipoles, Davydov splitting corresponding to E' and E'' levels is observed.

When dye molecules stack up like a sandwich, corresponding to parallel transition dipoles, the hypsochromic shift is observed. When dye molecules are arranged side by side, corresponding to in-line transition dipoles, the bathochromic absorption shift is observed. When the angle of slippage, the angle between the line of center of a column of dye molecules and the long axis of any one of the parallel molecules,

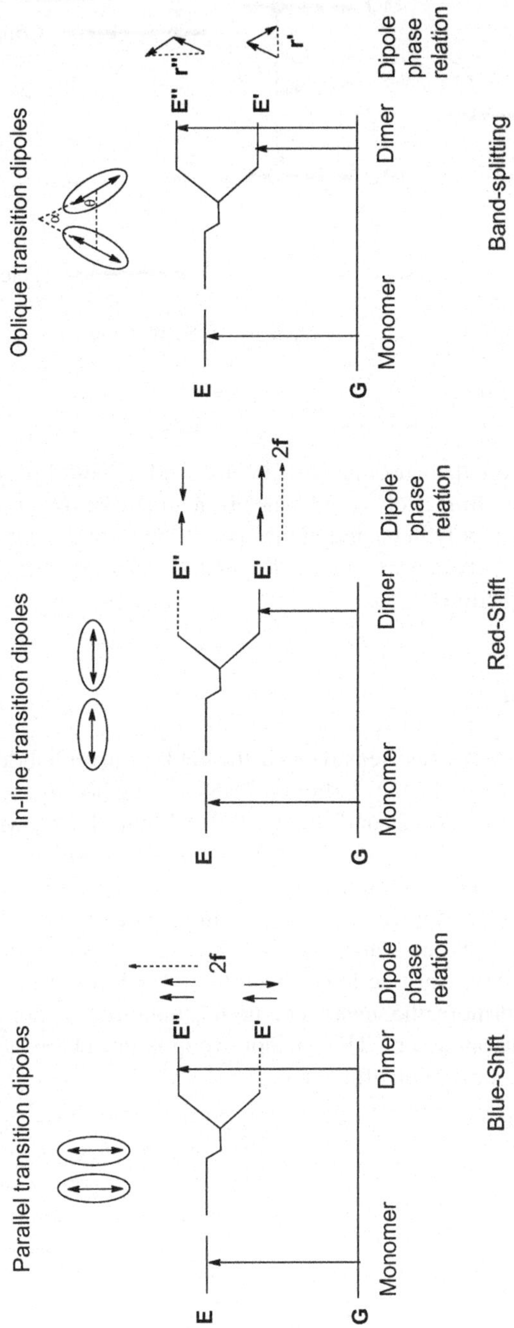

Fig. 1.6 Schematic explanation for absorption shift

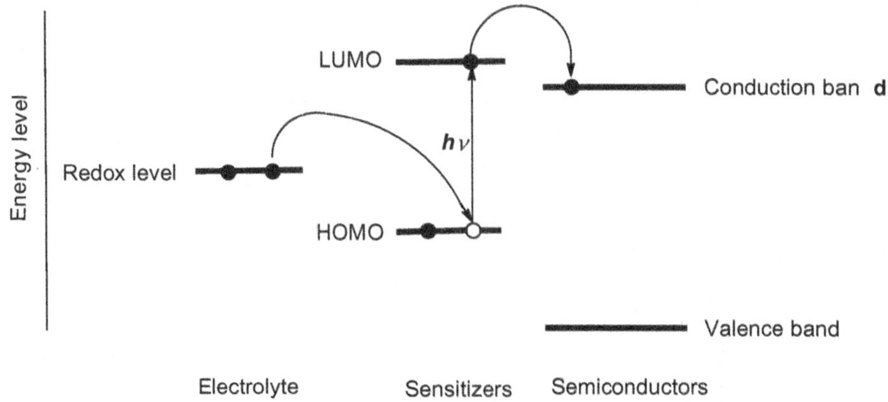

Fig. 1.7 Principle of sensitizers

is larger than 54.7°, the hypsochromic shift is observed. When this is smaller than 54.7°, the bathochromic shift is observed. The arrangement of dye molecules in the concentrated solution, in polymers, and in the crystalline form depends on the intermolecular interactions between the molecules such as π/π-interactions, hydrogen bonding, and *n*/π interactions.

1.2.3.3 Sensitization

Many polymethine dyes can act as sensitizers in the fields of silver halide photography and photoconductors. Recently, the studies on DSSCs using organic sensitizers have attracted much attention (O'Regan and Grätzel 1991). Figure 1.7 shows the principle of the sensitizers in DSSCs. When the LUMO level of the sensitizers is higher than that of the conduction band of semiconductors such as TiO_2 and ZnO, the excited electron can be thermodynamically injected into the conduction band. Then, silver halides and semiconductors are sensitized. In the case of DSSCs, in addition, the HOMO level of sensitizers must be lower than the redox level of the electrolyte to accept electrons. Furthermore, the sensitizers should contain an anchor group to show an affinity for the semiconductors. This anchor group is introduced at the acceptor moiety to enhance the conversion efficiency.

1.3 Topics in Polymethine Dyes

1.3.1 Data Recording Materials

In this field, many kinds of materials such as azo complex dyes, phthalocyanines, naphthoquinone dyes, cyanines, and oxonols have been proposed. Most important properties for the materials are (1) the absorption band for laser light, (2) smooth melting or sublimation upon laser irradiation, (3) good solubility in organic solvents, and (4) good durability.

As the writing and reproducing laser wavelength for CD-R is 750–830 nm, pentamethine dyes are most important. Pentamethine dyes can act as singlet oxygen sensitizers to decompose themselves. Therefore, a singlet oxygen quencher must be added to show good durability. Therefore, pentamethine dyes are used in the presence of a singlet oxygen quencher. A recording layer consisting of a pentamethine dye **1.44** and a singlet oxygen quencher **1.45** has been proposed (Yanagisawa et al. 1993).

A material **1.46** which has a singlet oxygen quencher as a counter anion has been reported (Namba et al. 1996).

Later, new counter anion quenchers **1.47**, which show good solubility in organic solvents, were proposed (Namba et al. 1999).

As the recording and reproducing wavelength for DVD-R is 550–620 nm, indolenine and benzindolenine trimethine dyes are suitable for this purpose. Symmetrical derivatives **1.48** and **1.49** and unsymmetrical trimethine dyes **1.50** have been proposed as the materials.

It is known that when tetracyanonaphthoquinone (TCNQ) was added to polymethine dyes, photostability is improved. However, TCNQ is not soluble in organic solvents. Therefore, dye **1.51**, in which TCNQ is contained as a part of the counter anion, has been proposed (Meng et al. 2001).

An oxonol dye **1.52**, which does not contain heavy metals, hase been proposed. This compound can produce gaseous CO_2 by decomposition. Bipyridyl salts effectively quench the singlet state of the oxonol dye to inhibit the singlet oxygen formation (Mikoshiba et al. 2006). The structures of these dyes are shown in Fig. 1.8.

1.3.2 Sensitizers for DSSCs

In 2003, a merocyanine dye called "indoline dye" **1.53** (**D149**) was reported to act as a highly efficient sensitizer in DSSCs (Horiuchi et al. 2003). "Indoline" means the hydrogenated form of indole. In 2008, the octyl derivative **1.54** (**D205**) was reported to show higher conversion efficiency than **D149** on TiO_2 due to improved open-circuit photovoltage (V_{oc}) (Ito et al. 2008)). The octyl group can prevent approaching I_3^- species in an electrolyte. Then, the electron transfer from the semiconductor surface to I_3^- species decreased to improve V_{oc}. Later, an indoline dye **1.55** was reported to

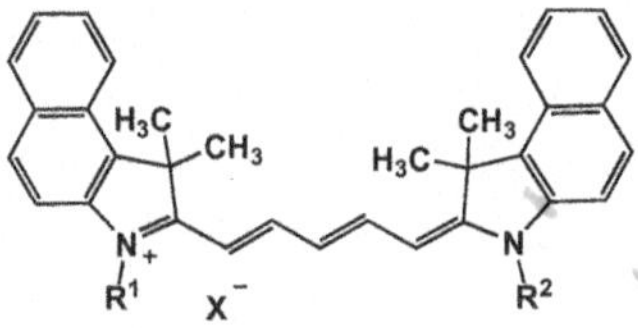

1.44

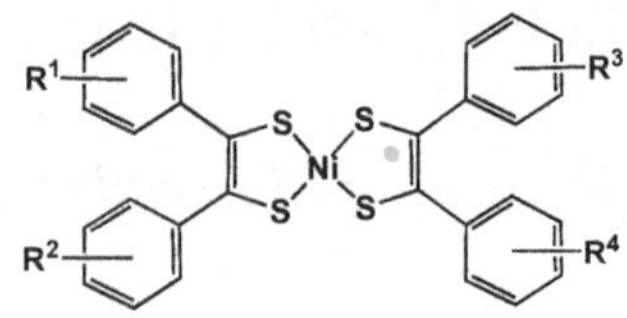

1.45

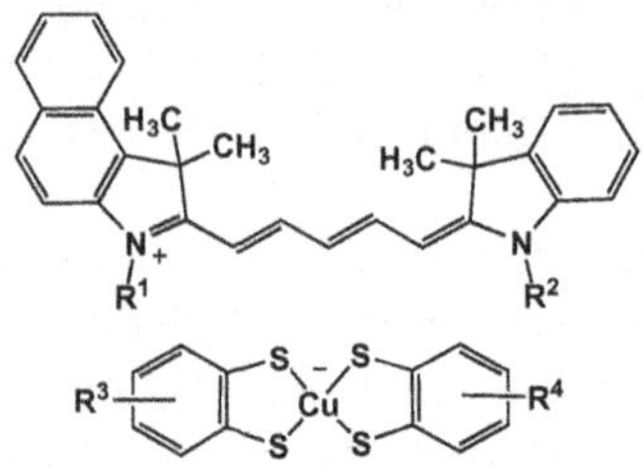

1.46

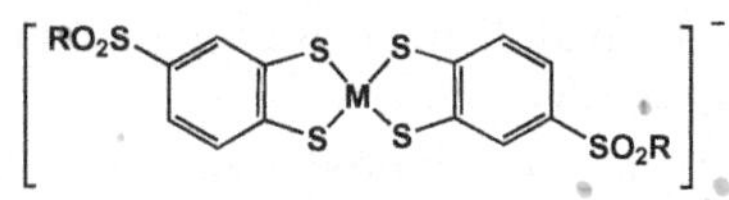

1.47

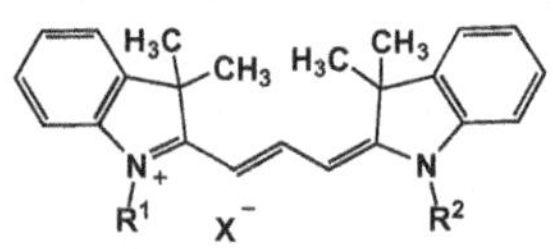

1.48

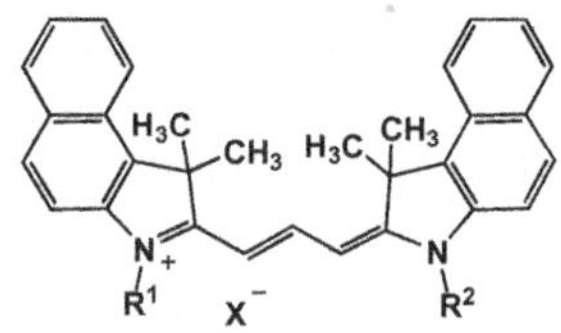

1.49

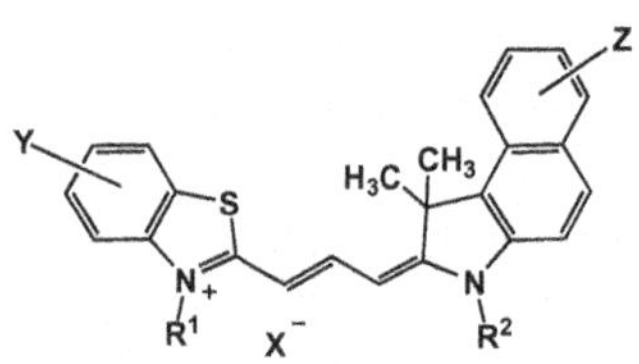

1.50

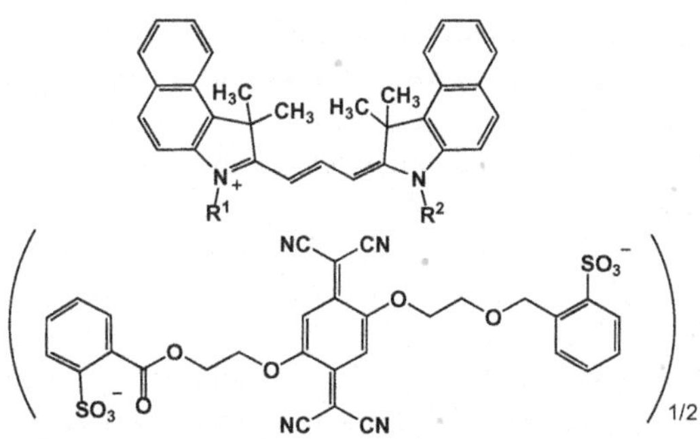

1.51

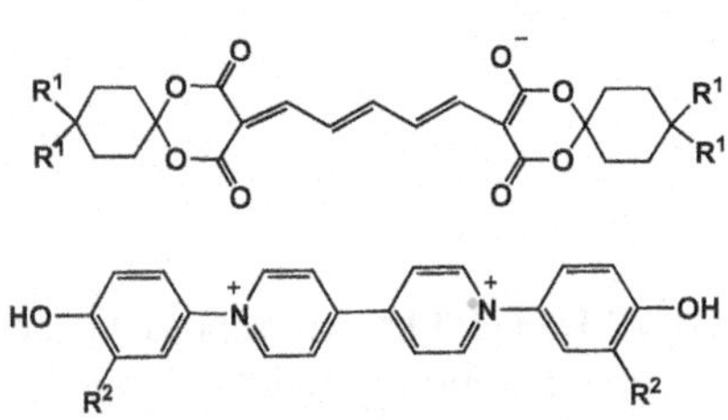

1.52

Fig. 1.8 Data recording polymethine dyes

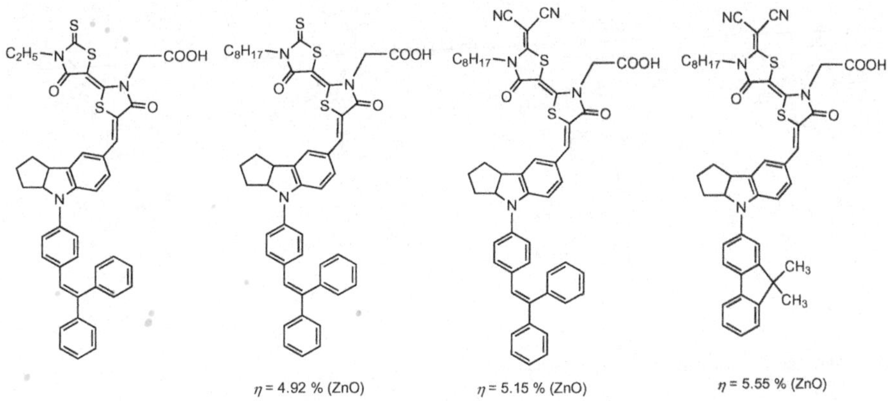

1.53 (D149) **1.54 (D205)** **1.55** **1.56**

Fig. 1.9 Development of indoline dyes in DSSCs

show higher conversion efficiency than **D205** due to the improvement of short-circuit photocurrent (J_{sc}) on ZnO that comes from a more bathochromic absorption band (Higashijima et al. 2011). Finally, a dye **1.56**, in which more bulky fluorenyl moiety is substituted, exhibited higher conversion efficiency than **1.55** due to the inhibition of aggregate formation of dye molecule on ZnO (Higashijima et al. 2012) as shown in Fig. 1.9.

It is necessary for the sensitizers to have an anchor group which shows an affinity for the semiconductor substrate. Generally, a carboxylic acid group is used as the anchor group to form the complexes with semiconductors. The adsorbed dye is usually extracted with polar solvents such as dimethylacetamide. The degradation of DSSCs comes from desorption of the sensitizers from the semiconductor substrate. When there are two anchor groups showing higher acidity in the sensitizer, the stability of the solar cell is improved. The acidity of α-cyanoacrylic acid is higher than that of rhodamine acetic acid. The indoline dye **1.60** showed the best stability among **1.57**, **1.58**, **1.59**, and **1.60.** The structures are depicted in Fig. 1.10. Actually, **1.60** is not extracted by dimethylacetamide (Matsui et al. 2016a).

1.3.3 Liquid Polymethine Dyes

Though most dyes are usually solid, liquid dyes were found. Not only liquid coumarins (Park et al. 2009) and azo dyes (Biradar et al. 2016), but also liquid styryl and cyanines dyes have been reported.

In the case of styryl dyes **1.61**, indicated in Fig. 1.11, the melting point changes depending on the counter anion X and substituents R^1 and R^2. Especially, the counter anion X drastically affects the melting point. Though bromide or iodide is usually used as the counter anion, their melting point is high. When

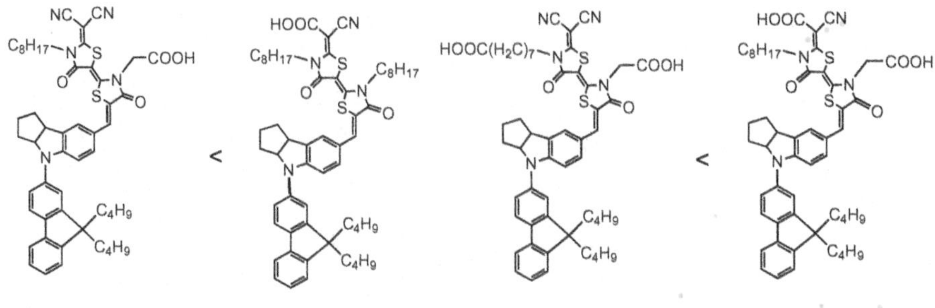

1.57 1.58 1.59 1.60

Fig. 1.10 Stable indoline dyes in DSSCs

Fig. 1.11 Liquid styryl dyes

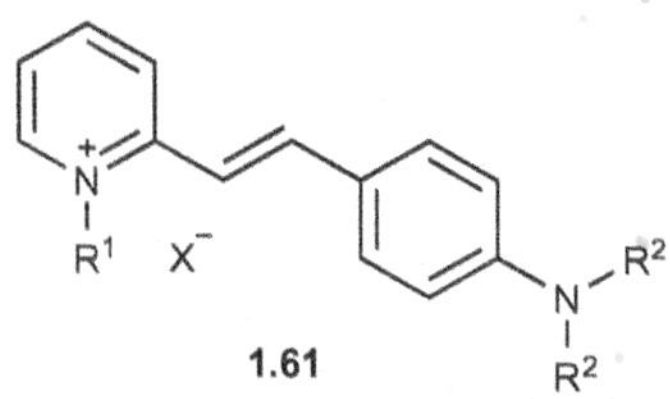

1.61

R^1	R^2	X	mp / °C
C_4H_9	C_2H_5	Br	249-250
C_4H_9	C_2H_5	$(CF_3SO_2)_2N$	121-122
C_4H_9	C_2H_5	$(C_4F_9SO_2)_2N$	81.6
$C_{12}H_{25}$	$C_{10}H_{21}$	$(C_4F_9SO_2)_2N$	27.2
$(CH_2CH_2O)_3CH_3$	$(CH_2CH_2O)_3CH_3$	$(C_4F_9SO_2)_2N$	-4.0

bis[perfluorobutylsulfonyl]imide was used as the counter anion, the melting point becomes low, being 81.6 °C. This result comes from steric effects of the counter anion to prevent intermolecular interactions between the dye molecules (Matsui et al. 2014a). Furthermore, the substitution of R^1 and R^2 with medium alkyl groups such as decyl and dodecyl groups lowered the melting point. When the R^1 and R^2 moieties are substituted with oxyethylene groups, the styryl derivative dye is viscous liquid at room temperature having a glass transient temperature (T_g) at −4.0 °C (Matsui et al. 2014b).

In the case of trimethine dyes, the melting point becomes low by changing the counter anion and substituent R. The liquid dyes **1.62**, shown in Fig. 1.12, are viscous compounds. When the fluorescence spectrum of the liquid trimethine dyes was measured in liquid nitrogen in neat form, the fluorescence intensity of liquid derivatives drastically increased whereas that of the corresponding solid derivatives slightly increased or were similar (Matsui et al. 2016b).

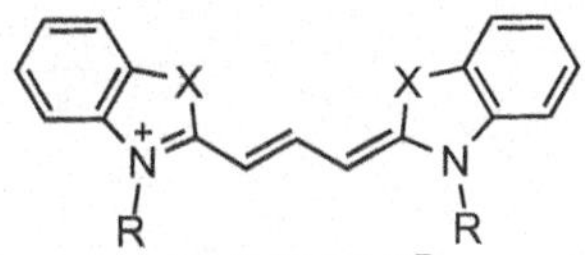

$(C_4F_9SO_2)_2N^-$

1.62

X	R	T_g / °C
$C(CH_3)_2$	$C_{12}H_{25}$	-18
S	$(CH_2CH_2O)_3CH_3$	-23
O	$(CH_2CH_2O)_3CH_3$	-22

Fig. 1.12 Liquid trimethine dyes

1.3.4 Fluorescence of Polymethine Dyes

Polymethine dyes are fluorescent in solution. Many polymethine dyes have been used as probes. Recently, the fluorescence quantum yield (Φ_f) of pentamethine dye was improved by annelating the linkage group. The Φ_f of **1.64** and **1.65** were measured to be 0.69 and 0.55, respectively, whereas that of **1.63** was 0.15 (Michie et al. 2017) as shown in Fig. 1.13.

Though most polymethine dyes are fluorescent in solution, the fluorescence intensity decreases in the crystalline form. In the case of indolenium trimethine dyes **1.66**, the dipropyl derivative having a bis(trifluoromethylsulfonyl)imide anion exhibited the highest Φ_f in the crystalline form as indicated in Fig. 1.14. The X-ray structure

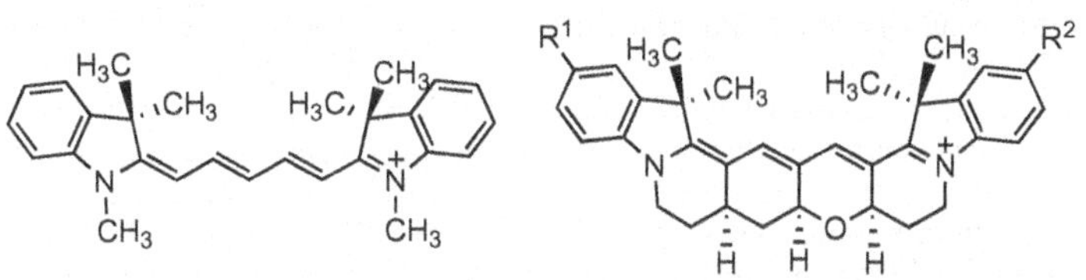

1.63

1.64: $R^1 = R^2 = H$
1.65: $R^1 = SO_3^-$, $R^2 = COOH$

R	λ_{max}	ε	λ_{em}	Φ_f	τ / ns
1.63[a]	638	214,000	657	0.15	0.7
1.64[a]	662	206,000	677	0.69	2.5
1.65	670[b]	199,000[b]	683[b]	0.55[b]	1.7[c]

Fig. 1.13 Enhancement of fluorescence intensity of pentamethine dye. **a** In methanol, **b** in pH 7.4 PBS, **c** in H_2O

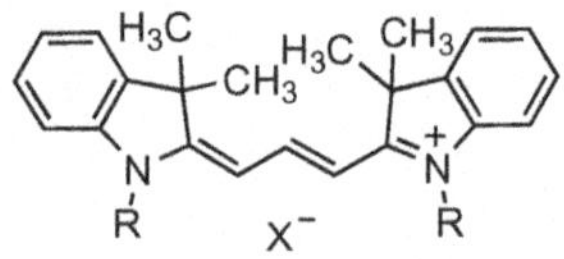

1.66

R	X	In toluene		Solid state (crystalline)	
		F_{max} / nm	Φ_f	F_{max} / nm	Φ_f
CH_3	$(CF_3SO_2)_2N$	573	0.07	648	0.03
C_3H_7	I	578	0.04	677	0.03
C_3H_7	$(CF_3SO_2)_2N$	574	0.12	657	0.31
C_3H_7	$(C_4F_9SO_2)_2N$	574	0.12	648	0.02

Fig. 1.14 Solid-state fluorescence of trimethine dyes

of this compound indicates an isolated monomer-type packing that comes from the anion and propyl groups, whereas the other derivatives show the isolated dimer-type packing or consequent π/π stacking (Matsui et al. 2015).

References

Biradar S, Kasugai R, Kanoh H, Nagao H, Kubota Y, Funabiki K, Shiro M, Matsui M (2016) Liquid azo dyes. Dyes Pigm. 125:249

Higashijima S, Miura H, Fujita T, Kubota Y, Funabiki K, Yoshida T, Matsui M (2011) Highly efficient new indoline dye having strong electron-withdrawing group for zinc oxide dye-sensitized solar cell. Tetrahedron 67:6289

Higashijima S, Inoue Y, Miura H, Kubota Y, Funabiki K, Yoshida T, Matsui M (2012) Organic dyes containing fluorene-substituted indoline core for zinc oxide dye-sensitized solar cell. RSC Adv. 2:2721

Hishiki Y (1974) Recent progress in cyanine dyes syntheses. Yuki Gosei Kyokai Shi 32:971

Horiuchi T, Miura H, Uchida S (2003) Highly-efficient metal-free organic dyes for dye-sensitized solar cells. Chem Commun 3036

Ito S, Miura H, Uchida S, Takata M, Sumioka K, Liska P, Comte P, Péchy P, Grätzel M (2008) High-conversion-efficiency organic dye-sensitized solar cells with a novel indoline dye. Chem Commun 5194

Kasha M, Rawls HR, El-Bayoumi MA (1965) The exciton model in molecular spectoscopy. Pure Appl Chem 11:371

König W (1922) Chemistry of the quinocyanines. Constitution of the pinacyanoles. Ber 55:3293

Matsui M, Kubota Y, Funabiki K (2014a) Low melting-point 2-pyridinium styryl dyes. J Jpn Soc Colour Mater 87:3

Matsui M, Kubota Y, Funabiki K (2014b) Liquid 2-pyridinium styryl dyes having oxaalkyl units. J Jpn Soc Colour Mater 87:187

Matsui M, Ando S, Fukushima M, Shibata T, Kubota Y, Funabiki K (2015) Fluorescence properties of indolenium carbocyanine dyes in solid state. Tetrahedron 71:3728

Matsui M, Tanaka N, Kubota Y, Funabiki K, Jin J, Higashijima S, Miura H, Manseki K (2016a) Long-term stability of novel double rhodanine indoline dyes having one and two anchor carboxyl group(s) in dye-sensitized solar cells. RSC Adv 6:33111

Matsui M, Yamamoto T, Kubota Y, Funabiki K (2016b) Survey, fluorescence spectra, and solubility of liquid cyanine dyes. New J Chem 40:10187

Meng FS, Yang SI, Tian H, Su JH, Chen KC (2001) CN1311184A

Michie MS, Götz R, Franke C, Bowler M, Kumari N, Magidson V, Levitus M, Loncarek J, Sauer M, Schnermann MJ (2017) Cyanine conformational restraint in the far-red range. J Am Chem Soc 139:12406

Mikoshiba H, Inagaki Y, Morishima S, Satou N, Warishi K, Akiba M, Motoki M, Ichijima S, Shibata M, Hashimoto H, Kubo H (2006) Fujifilm Research & Development 51:55

Mishra A, Behera RK, Behera PK, Mishra BK, Behara GB (2000) Cyanines during the 1990s: a review. Chem Rev 100:1973

Namba T, Shinkai M, Inoue T, Kitagata S (1996) US5574715

Namba T, Kitagata S, Shinkai S, Suzuki M, Kimura S, Hirako K (1999) US5958650

O'Regan B, Grätzel M (1991) A low-cost, high-efficiency solar cell based on dye-sensitized colloidal TiO_2 films. Nature 353:737

Park S-Y, Kubota Y, Funabiki K, Matsui M (2009) Survey of liquid coumarin dyes and their fluorescence properties. Chem Lett 38:162

Tatikolov AS (2012) Polymethine dyes as spectral-fluorescent probes for biomacromolecules. J Photochem Photobiol C 13:55

Yanagisawa S, Sakai T, Chyman T, Araki Y, Matsui F (1993) US5204220

Yasui S (1987) Syntheses and characteristics of functional polymethine dyes. J Jpn Soc Colour Mater 60:212

Chapter 2
Squaraine Dyes

Takeshi Maeda

Abstract Squaraine dyes which are composed of a cyclobutenedione core with aromatic or heterocyclic components show sharp and intense electronic absorption in the areas of visible to near-infrared regions and often fluorescence emission. These prominent optical properties arouse our interest in various fields of applications using the dye. In order to respond to the diverse demands of squaraine dyes in application fields, considerable effort have been made in the past decades to design and synthesize symmetrical and unsymmetrical squaraine dyes by means of classical condensation reaction of squaric acid moiety with electron-rich compounds. A novel approach using transition-metal catalyzed cross-coupling is developed to construct squaraine chromophores. This approach allows not only to attach desired functionalities on peripheral parts of chromophores but also to synthesize oligomeric and polymeric squaraine dyes. In addition to the molecular level of study, the supramolecular architectures have been constructed by non-covalent interaction between the dye molecules. This section gives an overview of the recent advances in syntheses and structures of squaraine dye with particular attention to the novel synthetic protocol.

Keywords Condensation · Transition metal-catalyzed cross-coupling · Squaric acid · Squaraine

2.1 General Remarks

Since squaric acid (3,4-dihydroxycyclobut-3-ten-1,2-dione), a kind of oxocarbonic acid, began to be used for research and development in the middle of the 1960s, a variety of new organic compounds was synthesized from squaric acid (West and West 1980). Among them, squaraine dye that is obtained by the dehydration condensation reaction between squaric acid, and electron-rich compounds have been actively studied due to its excellent optical property. Treibs et al. first reported that squaraine dye can be obtained by the condensation reaction between squaric acid and pyrrole

T. Maeda (✉)
Department of Applied Chemistry, Graduate School of Engineering, Osaka Prefecture Univesity, Naka-ku, Sakai 599-8531, Japan
e-mail: tmaeda@chem.osakafu-u.ac.jp

Y. Ooyama and S. Yagi (eds.), *Progress in the Science of Functional Dyes*,
https://doi.org/10.1007/978-981-33-4392-4_2

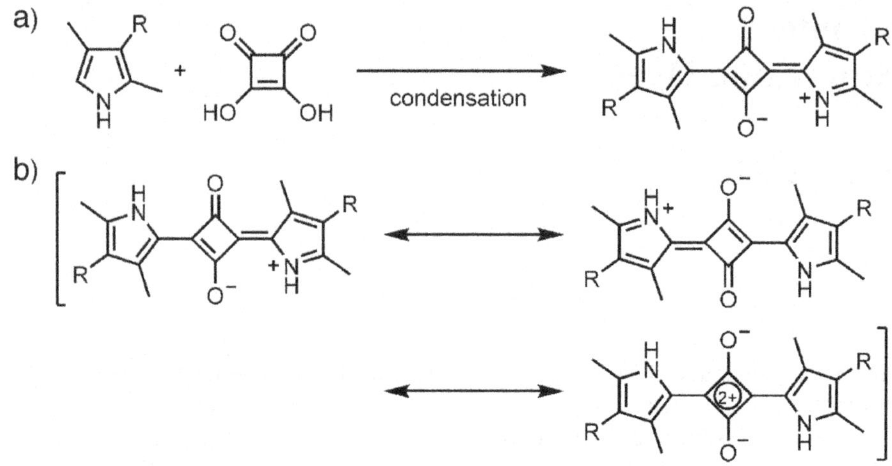

Fig. 2.1 Synthetic scheme of the first reported squaraine dye (**a**) and its resonance structures (**b**)

derivatives or phloroglucinol (Fig. 2.1a) (Treibs and Jakob 1965). Since squaraine dye can be expressed as an equally contributing resonance form like cyanine dye, it is considered to be an analogue of cyanine dye (Fig. 2.1b).

Unlike cationic cyanine dye, squaraine dye is neutral, so it is expressed as a zwitterion structure. Since squaraine dye is composed of an electron-donating aromatic ring (heterocycle) and electron-accepting squaric acid residue, it has a donner-accepter-donner structure. Therefore, a symmetric squaraine dye is treated as a quadrupole from the viewpoint of electron distribution, and it is polarized in its ground state. Like cyanine dye, squaraine dye exhibits very strong electronic absorption with a narrow half-width in the long wavelength region. Owing to the excellent stability stemming from its electrically neutral structure, optical property and structural diversity, squaraine dye is widely used in various fields such as a charge generator in a xerographic photoreceptor (Law 1993,1992), solar cells (Chen et al. 2015), nonlinear optical materials (Chen et al. 1994), fluorescence sensing materials (Das et al. 1994) and photodynamic therapy (Ramaiah et al. 1997). If squaraine dye demands high performance in various application fields, it is required to synthesize it by designing the molecular structure that is specified to the respective application field. In response to this request, many novel squaraine dyes that can be applicable to the respective field have recently been reported. In the present paper, recently reported new squaraines are reviewed focusing on their synthesis methods and structures.

2.2 Electronic Absorption and Fluorescence Properties

Squaraine dyes exhibited a prominent electronic absorption band in the range from the far-red to near-infrared region. These dyes offer the attraction of high molar extinction coefficients reaching -10^5 L mol^{-1} cm^{-1}. Figure 2.2 shows the electronic absorption and fluorescence spectra of a typical example of squaraine dye with indolenine components. The dye exhibits narrow and intense absorption ($\varepsilon = 341{,}000$) with the maximum at 643 nm. As can be seen in the absorption spectrum, the vibronic shoulder was frequently observed in the high energy side of intense absorption bands for squaraine dyes.

Generally, the electronic transition from the ground state (S_0) to the first excited state (S_1) corresponding to the lower energy absorption band is mainly composed of HOMO to LUMO excitation. Figure 2.3 shows the isodensity plot of HOMO and LUMO of a squaraine dye with indolenine components. The HOMO is spread over the whole π-system of the central four-membered ring including oxygen atoms and indolenine parts. The LUMO is mainly located on the polymethine chain. The theoretical calculation has revealed that a charge transfer from the electron-donating aromatic ring to the squaric acid unit scarcely happens in the $S_0 \rightarrow S_1$ electronic transition, but the charge transfer occurs mainly from the oxygen atom in the squaric acid residue to the four-membered ring at the center (Bigelow and Freund 1986). Many squaraine dyes show fluorescence emission with small Stokes shifts. This indicates the small difference between the ground state and excited state dipole moments and the electronic structures of these dyes.

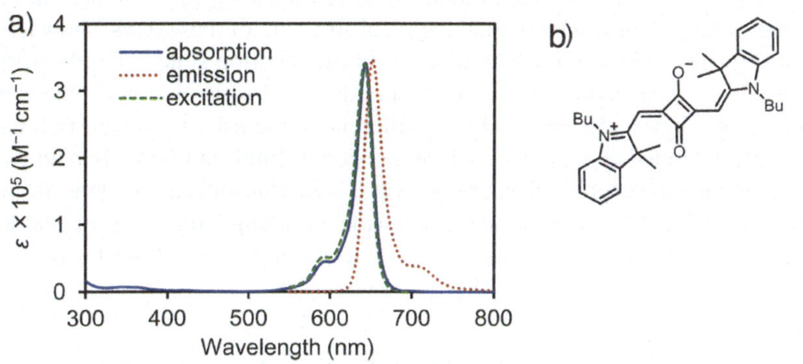

Fig. 2.2 Electronic absorption and fluorescence spectra of symmetrical squaraine dye with indolenine component (in toluene) (**a**), its chemical formula (**b**)

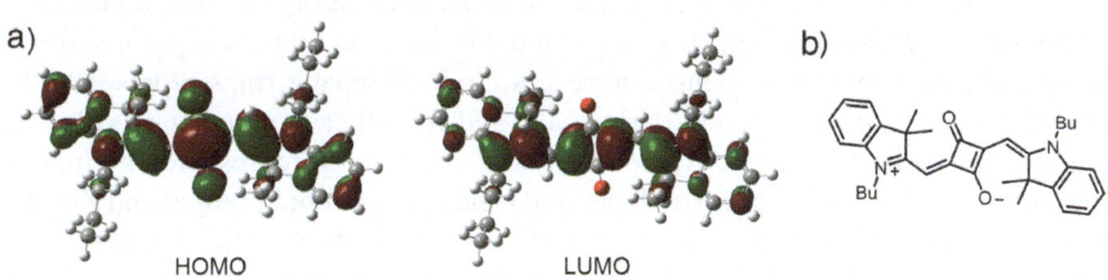

Fig. 2.3 Isodensity plots of HOMO and LUMO for indolenine-based squaraine dyes obtained by DFT calculation at the B3LYP/6-31G(d) level of theory (**a**) and its chemical formula (**b**)

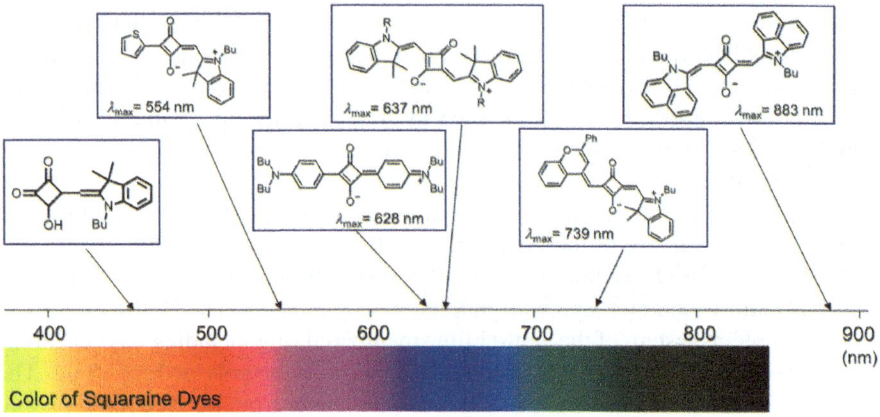

Fig. 2.4 A diagram of the transition energy (wavelength) of maximum absorption bands for squaraine dyes with various electron-rich components incorporated on 1,3-position of squaric acid residue

The transition energy for its maximum absorption depends on the structure of the aromatic ring and the heterocycle that are substituted at the 1,3-positions of the squaric acid (Fig. 2.4). The symmetrical squaraine dyes with *N,N*-dialkylaniline and indolenine components, which are the typical structure of these dyes, show a prominent absorption band around 620–650 nm in the solution state. The absorption of a squaraine dye consisting of thiophene and the semi-squaraine was observed at a higher energy region (554 nm). On the other hand, the use of electron-rich components with extended π-systems such as benzopyrylium and benzo[*c,d*]indolenine makes the transition energy of squaraine dyes lower. This indicates that the absorption band of squaraine dye can be adjusted in a wide wavelength region from visible light to near-infrared light by selecting the substituent components at the 1,3-positions.

2.3 Synthesis of Squaraine Dyes by Condensation Reaction

Symmetric squaraine dyes are generally obtained by the condensation reaction between one-equivalent squaric acid and two-equivalent aromatic amine/phenol or heterocyclic components. These dyes are roughly classified into two groups; one is the dye obtained by the condensation reaction with aromatic rings represented by pyrrole and *N,N*-dialkyl aniline derivatives, and the other is that obtained by the condensation reaction with activated methylene compounds represented by indolenine and benzothiazole (Fig. 2.5) (Maahs and Hegengerg 1966; Sprenger and Ziegenbein 1968; Ziegenbein and Sprenger 1966; Shaw et al. 2018). The squaraine dyes obtained by the condensation reaction have been summarized in some review articles (Law 1993; Schmidt and West 1980; Sreejith et al. 2008a; Beverina and Salice 1207; McEwen and Wallace 2009; Yagi and Nakazumi 2008).

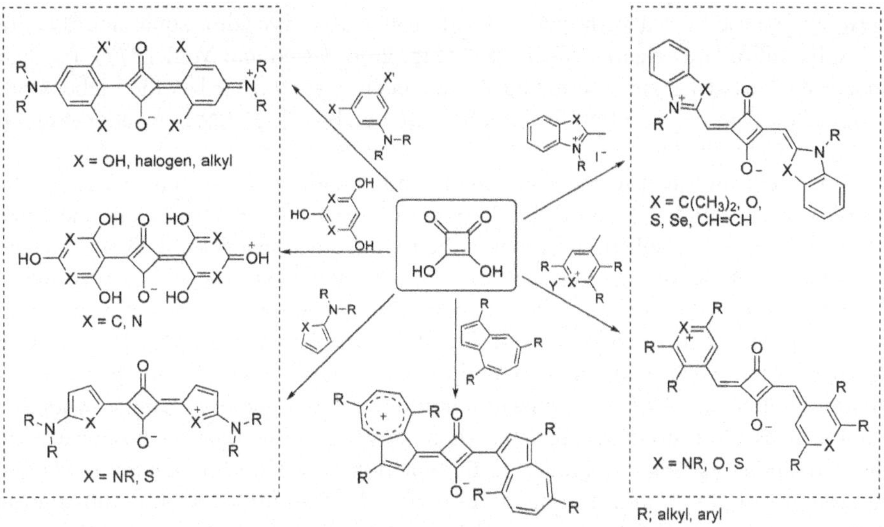

Fig. 2.5 Representative squaraine dyes with symmetrical structures

In the condensation reaction between an electron-rich arene and squaric acid, semi-squaraine is produced as an intermediate (Fig. 2.6a). At the succeeding second stage of the condensation reaction, arene can attack the 2- and 3-positions of the semi-squaraine. However, the 1,2-substitutions are scarcely produced because squaraine dye of the 1,3-substitution is obtained regioselectively (Law and Bailey 1986). On the other hand, in the Friedel-Crafts reaction between arene and 3,4-dichloro-3-cyclobutene-1,2-dione that is acid chloride of squaric acid obtained by the reaction

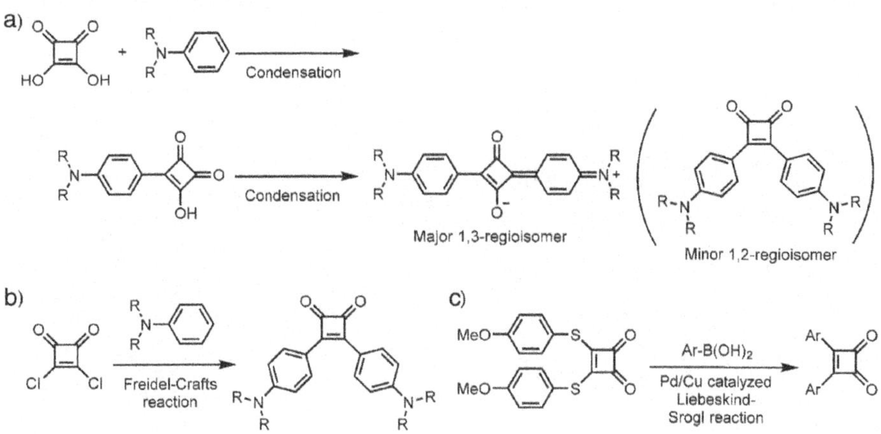

Fig. 2.6 Squaraine derivatives obtained by condensation (**a**), Friedel-Crafts reaction (**b**), and Liebeskind-Srogl reaction (**c**)

between squaric acid and thionyl chloride (or oxalyl chloride), squaraine dyes of 1,2-substitution are selectively obtained (Fig. 2.6b) (Ried and Vogl 1977). Further, squaraine dyes of 1,2-substitution have been recently synthesized by the Liebeskind-Srogl cross-coupling reaction using palladium catalyst (Fig. 2.6c) (Aguilar-Aguilar and Pena-Cabrera 2007).

When activated methylene compounds and squaric acid are reacted under the azeotropic condition, 1,3-substituted squaraine dyes are selectively obtained like arene. However, 1,2-substitution that is the regioisomer seems to be also produced by this reaction. Beverina et al. have clarified that the squaraine of 1,2-substitution is regioselectively produced by the condensation reaction between ethyl ester of squaric acid and benzothiazolium salt (Fig. 2.7a) (Ronchi et al. 2011). In the reaction between squaric acid and benzothiazolium salt, squaric acid of 1,3-substitution is generally obtained. As for the phenomenon that regioselectivity changes when the squaric acid ester is used, the present authors are further investigating the cause using the corresponding semi-squaraine. In the reaction between semi-squaraine (**1**) and benzothiazolium salt, an isomeric mixture of 1, 2-substitution and 1, 3-substitution is produced (Fig. 2.7b). On the other hand, when ethyl ester of semi-squaraine (**2**) is used, squaraine dye of 1, 2-substitution is regioselectively obtained (Fig. 2.7c). The difference in the two reactions can be explained associated with the electron density at the position of squaric acid. In the case of **1**, the reaction becomes less susceptible to nucleophilic attack, because the electron density at the 2-position in the skeleton of the squaric acid is increased by the deprotonation. Thus, the reaction

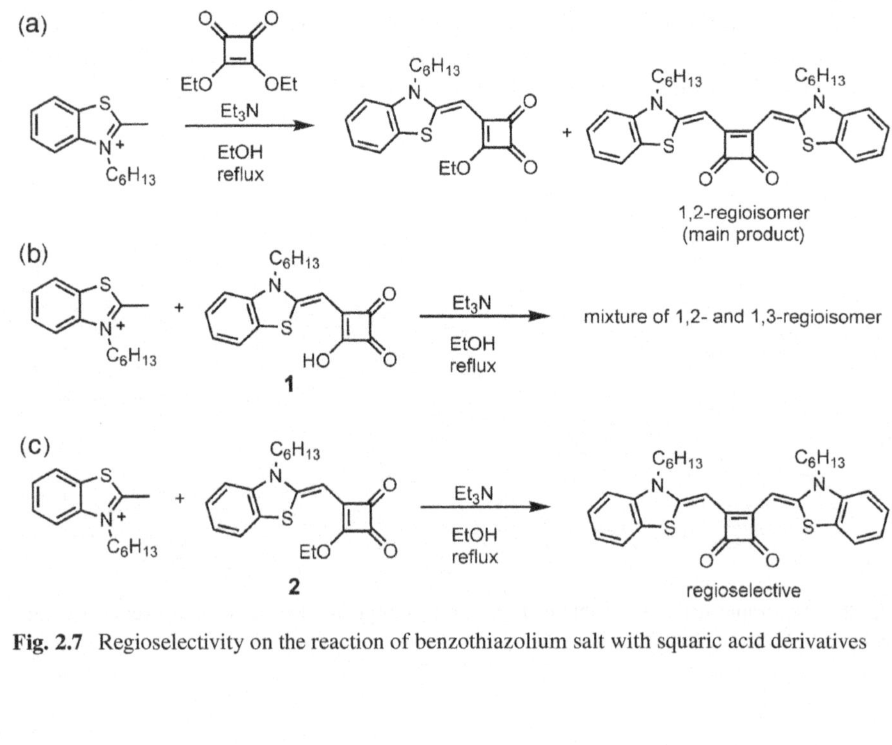

Fig. 2.7 Regioselectivity on the reaction of benzothiazolium salt with squaric acid derivatives

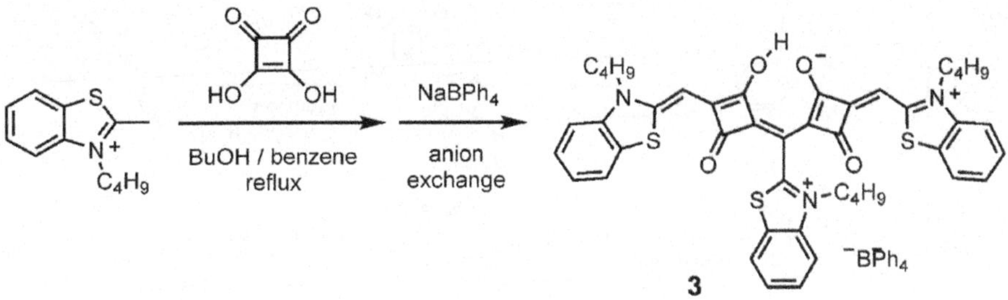

Fig. 2.8 A squaraine dye having two semi-squaraine components bridged on methine carbon

at the 3-position of the squaric acid occurs. On the other hand, the electron density at the 3-position of the squaric acid of **2** increases by the resonance effect of the benzothiazole. Therefore, it is considered that the 2-position of the squaric acid residue with lower electron density comes under nucleophilic attack.

Generally, squaraine dyes can be obtained by heating one-equivalent squaric acid and two-equivalent nucleophilic agent in the mixed solvent of alcohol and benzene/toluene. In the reaction with activated methylene compounds, symmetric squaraine dyes are obtained through corresponding enamine derivatives by using a small amount of quinoline. In these reactions, it has been confirmed that small amounts of dyes that have an absorption band in the near-infrared region were produced. From the X-ray crystal structure analysis, Nakazumi et al. clarified that this byproduct with a near-infrared absorption band is composed of two semi-squaraine parts bridged by methine carbons, and it is bis-squaraine dye **3** that is planar due to the intramolecular hydrogen bonds formed among the parts of squaric acid (Nakazumi et al. 1994). They also clarified that the dye **3** in which semi-squaraine derivatives are condensed can be obtained at high yield by eliminating base catalysts that had been used in the conventional reaction systems (Fig. 2.8).

It is assumed that dye **3** is obtained by the reaction of the squaraine dye product with semi-squaric acid derivatives as intermediates.

Symmetric squaraine dyes have been described above, but it should be noted that there have been lots of reports on asymmetric squaraine dyes. Since the introduction of squaraine dye with asymmetric structure greatly expands the diversity of the molecular structure, it would enable more detailed designing of the optical and electrochemical properties including the absorption wavelength region. The asymmetric squaraine dyes are synthesized by a method: semi-squaraine is first synthesized by condensing 1:1 of squaric acid and electron-rich aromatic rings (heterocycles), and then the different aromatic rings (heterocycles) are further condensed (Keil and Hartmann 2001). Since the reaction using semi-squaraine and equimolar squaric acid generates the mixture of 1, 3-substitute of squaraine dyes and semi-squaraine, it is not appropriate as a condition to synthesize semi-squaraine. The reaction between one-equivalent nucleophile and diester derivatives of squaric acid (or acid chloride of squaric acid) produces the corresponding semi-squaraine derivatives at high yield in many cases. Therefore, this reaction is efficient for the stepwise synthesis

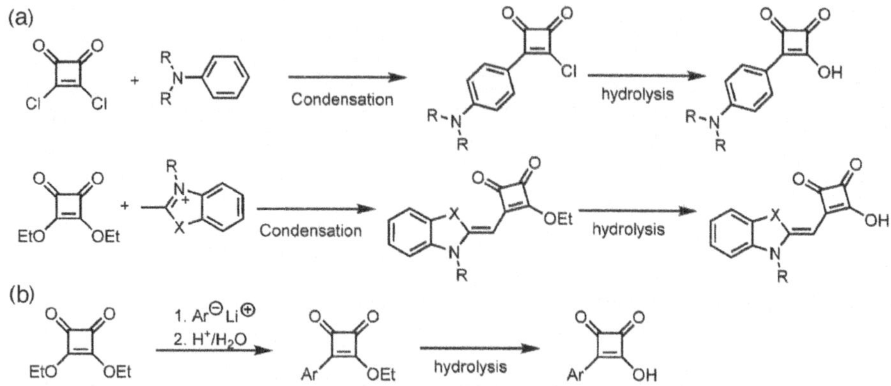

Fig. 2.9 Selective synthesis of semi-squaraine derivatives

of asymmetric squaraine dyes (Fig. 2.9a) (Terpetschnig and Lakowicz 1993). Also, the other synthesis method was reported where a kind of arene is first lithiated by a lithium agent, then it is reacted with diester derivatives of squaric acid as nucleophile (Fig. 2.9b) (Liebeskind et al. 1988). By hydrolyzing the ester and acid chloride intermediates obtained in these reactions, it is expected to synthesize asymmetric squaraine as semi-squaraines.

2.4 Squaraine Dyes Synthesized by Pd-Catalyzed Cross-Coupling

Squaric acid reacts with highly activated nucleophiles such as *N,N*-dialkyl aniline derivatives, phenol and active methylene compounds, resulting in the production of the desired squaraine dyes. However, it is natural that squaric acid does not react with compounds with low nucleophilicity. Nonetheless, in order to respond to the diverse needs of squaraine dyes in application fields, it is required to develop methods to synthesize squaraine replacing the classical condensation reaction. Liebeskind et al. reported that stannyl cyclobutenedione derivative **4** obtained by the reaction between 3,4-diisopropyl squarate and silyl stannyl compounds reacts with various kinds of haloarene in the presence of Pd catalyst, resulting in the production of the corresponding semi-squaraine derivatives (Fig. 2.10) (Liebeskind and Fengl 1990).

By using this method, bis-squaraine dyes whose two squaraine skeletons are connected by various aromatic rings can be synthesized. Bis-semi-squaraine **5** can be obtained in the following procedure; various kinds of diiodoarene are first reacted with stannyl cyclobutenedione derivative **4** with palladium/copper catalyst, and then the reaction product is hydrolyzed under acidic condition (Yagi et al. 1417). Subsequently, bis-squaraine dye **6** can be obtained by the reaction between **5** and heterocyclic quaternary salt with active methyl groups (Fig. 2.11). For **6b** and **6c** in

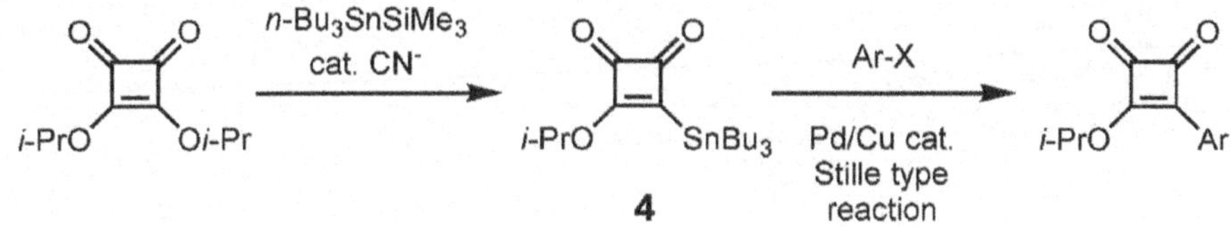

Fig. 2.10 Synthesis of semi-squaraines using stannyl cyclobutenedione derivative as nucleophilic equivalent of squaric acid

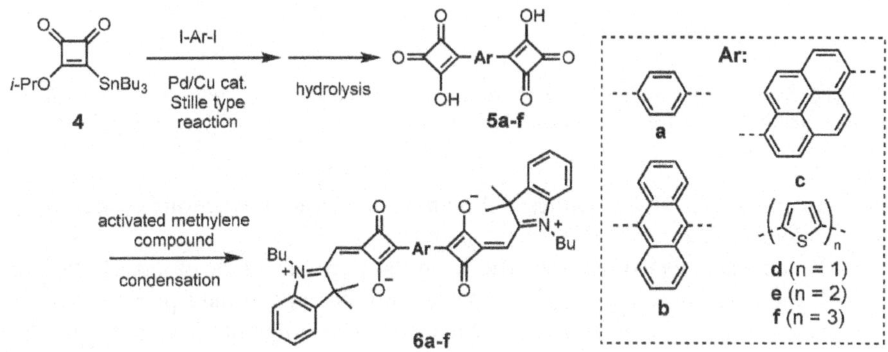

Fig. 2.11 Bis-squaraine dyes with various aromatic linkers

which anthracene and pyrene are used at the connecting parts, the absorption peak is shifted to longer wavelength region by 15–54 nm compared with that of **6a** with phenylene skeleton (Yagi et al. 2008; Nakazumi et al. 2005). In the case of bis-squaraine dye **6d** in which thiophene with high electron-donating property is used at the connecting parts, the absorption maximum was observed at 785 nm, which was the longest among the series of compounds. It was concluded that the reason for the longest wavelength is that the electron transition by intramolecular charge transfer is promoted by the existence of thiophene rings. For **6e** and **6f** where bithiophene and terthiophene are used at the connecting parts, the absorption was observed at a shorter wavelength region than that of **6d** with thiophene skeleton. It is considered that the reason is that the overlapping of the pz orbitals decreases by the rotation of the carbon-carbon bonds, resulting in the hindrance of the expansion of the π-conjugation system. These studies clearly showed that the absorption can be shifted to the longer side by incorporating the skeleton of the π-conjugation system between two semi-squaraine molecules. Thus, it can be said that they proposed an effective method to synthesize squaraine dyes with a near-infrared absorption band.

It was reported that a new type of dye where squaraine chromophores are linearly connected can be synthesized by using stannyl cyclobutenedione derivative **4** as nucleophilic equivalent with palladium/copper catalyst (Fig. 2.12) (Yagi et al. 2011). The squaraine dye **8** where a cyclobutene skeleton is introduced in the molecular end can be synthesized by the palladium/copper catalyzed cross-coupling reaction with **4** followed by hydrolysis. Finally, dye **9** where semi-squaraine skeleton is connected

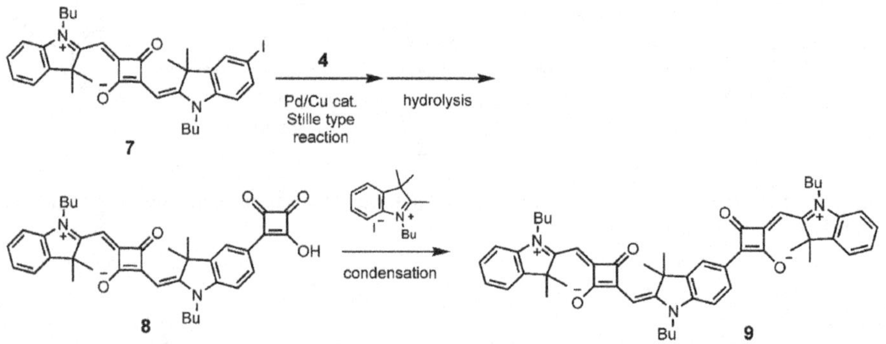

Fig. 2.12 Synthesis of a squaraine-based dye where semi-squaraine skeleton is connected with the conventional squaraine dye

with squaraine dye can be synthesized by the condensation reaction between dye **8** and indolenium salt.

In the same manner using symmetric squaraine dye with iodine group at both ends, dye **12** where semi-squaraines are connected at both ends of the squaraine dye can be obtained (Fig. 2.13). In addition, when a similar reactions are repeated, dye **13** in which five cyclobutene skeletons are linearly connected can be obtained. Although the absorption maximum of symmetric squaraine dye as a raw material is observed at 642 nm, the absorption bands for dyes **9**, **12** and **13** were observed at 763 nm,

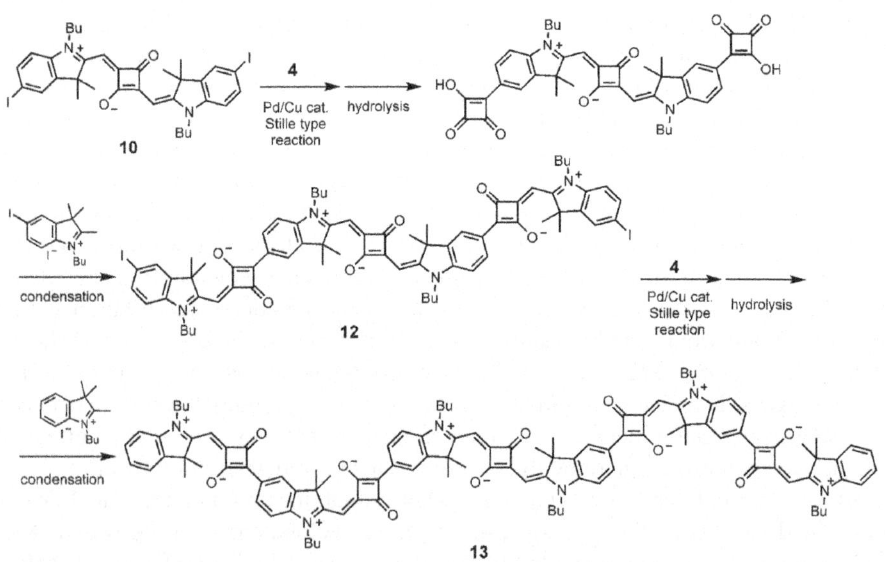

Fig. 2.13 Synthesis of oligomeric squaraine dye by the Pd-catalyzed cross-coupling reaction

862 nm and 940 nm, respectively. This result indicates that the electronic absorption remarkably shifts to the longer side by connecting squaraine chromophores. From these results, it was revealed that if the catalytic cross-coupling using stannyl cyclobutenedione derivatives as a substrate is utilized, cyclobutenedione skeleton can be introduced into the position that has been difficult to introduce by conventional condensation reactions.

2.5 Functionalization of Squaraine Dyes

Since squaraine dye has a rigid π-conjugation system, its solubility itself is low. When alkyl groups are introduced in aromatic rings and heterocyclic components of squaraine dye, its solubility would be secured. Furthermore, in order to make squaraine dye be soluble or dispersed in an aqueous solvent, hydrophilic groups such as carboxylic acid are generally introduced. Hydrophilic squaraine dye is used as a fluorescent labeling agent in the analysis of protein and biological tissues and as an ion sensor. Unsymmetrical squaraine dye **14** with carboxylic group at the *N*-alkyl ends of the aniline skeleton forms an aggregate in an aqueous solvent. For this dye, it was observed that the absorption band shifts to the higher energy side due to the exciton coupling leading the remarkable fluorescence quenching (Fig. 2.14). When human serum albumin (HSA) is added to this solution, the squaraine dye forms a 1:1 complex with the HSA, and the absorption band at the lower energy region and the fluorescence emission originating from individual chromophore are observed (Nakazumi et al. 2003; Welder et al. 2003). In this way, it is considered that squaraine dye, which uses the changes in the optical property based on the aggregation-dissociation as a signal, is a labeling agent of non-covalent bond type which does not need chemical modification. Thus, it enables rapid analysis of protein. Based on this principle, near-infrared absorbing squaraine dye was also found, which similarly amplifies the fluorescence in near-infrared region by using squaraine dye composed of expanded π-conjugation components (Nakazumi et al. 2005). Furthermore, focusing on a phenomenon in which boronic acid group works as a hydrophilic

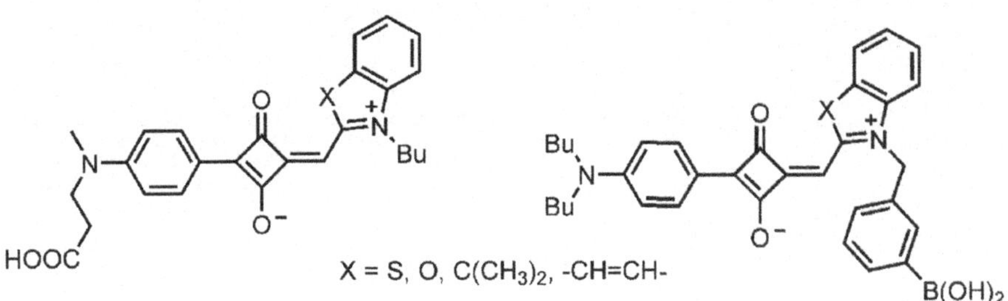

14 **15**

Fig. 2.14 Squaraine dyes bearing polar functionality for fluorescent labeling agents group by forming borate salt and forms complex with 1,2-diol, squaraine dye with boronic acid group **15** has been developed. It was shown that this dye can be applied to the analysis of bacteria and sialic acid, utilizing signal changes in electronic absorption/fluorescence properties caused by the formation of aggregates in aqueous solvent and the extinction of chromophore aggregates due to the interaction with 1,2-diol in the sugar skeleton (Saito et al. 2012; Ouchi et al. 2015).

As mentioned above, hydrophobic squaraine chromophores are not completely soluble but dispersed in aqueous solvents and form dye aggregates. The low solubility in water is the obstacle for the biocompatibility needed for bioimaging. To manage the solubility problem, vesicles and micelles consisting of surfactants such as phospholipids were used with squaraine dyes bearing hydrophilic peripheral substituents (Fig. 2.15). Unsymmetrical squaraine dyes with carboxy groups on the end of *N*-alkyl chains (**16**) are incorporated as a guest into phospholipid (dimyristoylphosphatidylcholine) vesicles (Chen et al. 1995). In addition to the carboxy group, an ethylene glycol chain is also useful to provide hydrophilicity. Pyrrole- and aniline-based squaraine dyes with ethylene glycol units (**17, 18**) are synthesized and encapsulated into the micelles. The micelle-encapsulated dyes are used for two-photon fluorescence imaging and cyanide ion sensing (Ahn et al. 2012; Liu et al. 2017).

Other than these examples, squaraine dyes are widely used as molecular sensors that change the optical property by the interaction with the targets such as metal ions at their recognition sites (Ajayaghosh 2005; Sun et al. 2016). Squaraines inherently have two distinct coordination sites at the aniline nitrogen and the oxygen from cyclobutenedione moiety. Julolidine-based squaraine dye **19** shows an absorption band at 660 nm in acetonitrile (Fig. 2.16). Upon the addition of H^+, the absorption band disappears and broad absorption at 690 nm is simultaneously developed. This bathochromic shift is attributed to the protonation of the oxygen atom of squaric acid residue. In contrast, the dye **20** (n = 1) bearing azacrown moiety shows a hypsochromic shift of absorption upon the addition of H^+. This is due to the protonation at the nitrogen atom which was stabilized by the hydrogen bonding interaction in the azacrown ring. Dyes **20** and **21** also show color changes with the addition of metal ions. Dye **20** (n = 3) with large azacrown macrocycles can sense Ba^{2+}

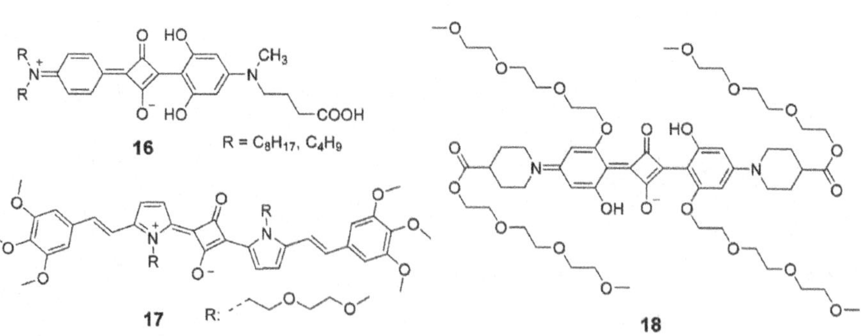

Fig. 2.15 Squaraines with peripheral hydrophilic groups encapsulated in micelles and vesicles

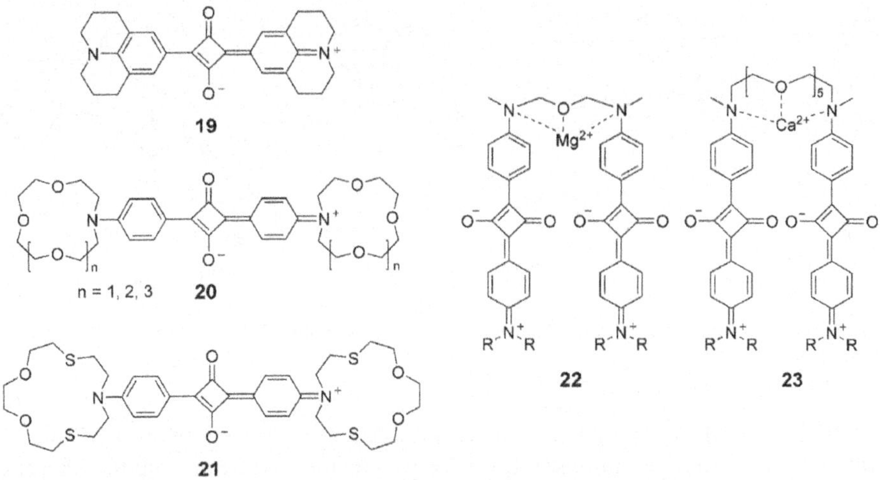

Fig. 2.16 Squaraine dyes with coordination sites for cation sensing

among the alkali (Li^+, Na^+, K^+) and alkaline earth metals (Mg^{2+}, Ca^{2+}, Ba^{2+}). The result of selective sensing is in accordance with the ability of aza18crown6 to bind Ba^{2+}. Thus, squaraine dyes can be utilized for colorimetric sensing triggered by the interaction between ions and peripheral coordination sites. Furthermore, dye **21** can selectively sense Hg^{2+} among other metal ions (Fe^{3+}, Cu^{2+}, Ni^{2+}, etc.) because of the high affinity of sulfur atoms and Hg^{2+} (Ros-Lis et al. 2008).

Ajayaghosh et al. reported the bichromophoric squaraines for colorimetric sensing. Dyes **22** and **23** interact with Mg^{2+} and Ca^{2+} at the ether linker between two chromophores. The coordination of ions causes the folding of chromophores to form H-aggregates. The absorption spectral changes occurred due to the exciton coupling in the H-aggregates. The signaling event of squaraines arises not only from perturbation of electronic structure caused by the interaction with analytes, but also from the interaction between chromophores directed by the coordination of ions (Arunkumar et al. 2005a; Ajayaghosh et al. 2002).

Various substituents can be introduced on the aromatic rings and heterocycles of squaraine dye. The optical and electrochemical properties of squaraine dye depend on the aromatic ring (heterocycle). On the other hand, although a functional group located around the chromophore gives perturbation of the electronic transition as an auxochrome, the effect of the functional group on the electronic absorption property of the squaraine chromophore is in most cases limited (Law 1987). Due to the excellent ability in light absorption, squaraine dye is widely used in various fields of applications by introducing functional groups. Nazeeruddin et al. have clarified that far-infrared absorbing squaraine dye **24** composed of indolenine with carboxylic group has high ability as a sensitizing dye in dye-sensitized solar cells using titanium oxide (Fig. 2.17).

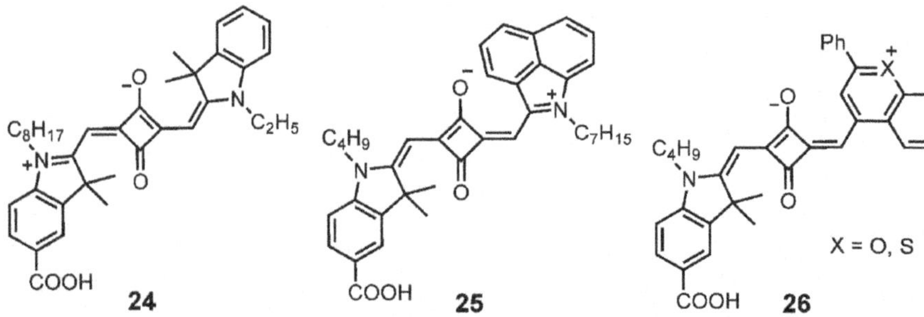

Fig. 2.17 Far-red and near-infrared absorbing squaraine dyes with carboxy group for dye-sensitized solar cells

Since the carboxyl group introduced in a squaraine chromophore becomes an adsorption site on titanium oxide, the effective electron injection from the photoexcited dye to the titanium oxide becomes possible (Yom et al. 2007). In addition, unsymmetrical squaraine dyes **25**, **26** composed of heterocycles with expanded π-skeleton and carboxy-indolenine components are also applied to dye-sensitized solar cells (Maeda et al. 2011a). These dyes exhibited intense near-infrared absorption, thanks to the expanded π-skeleton. Upon the use in dye-sensitized solar cells, they contribute to high photoelectric conversion efficiency in the near-infrared region.

As mentioned above, if squaraine chromophores are linearly connected by catalytic cross-coupling using stannyl cyclobutenedione derivative as the substrate, it is possible to synthesize a dye with high absorption ability in the near-infrared region. Utilizing this method, near-infrared-absorbing squaraine dye **27** in which carboxylic acid is introduced at the molecular end was synthesized, which is applied to the sensitizing dye for dye-sensitized solar cells (Fig. 2.18) (Maeda et al. 2011b). Since the squaraine dye **28** with three cyclobutene skeletons exhibits a photosensitization effect at the near-infrared region over 800 nm (Maeda et al. 2013).

A dye with two squaraine chromophores taking carbazole skeleton as the central skeleton has been synthesized using the Stile-type reaction followed by a condensation reaction (Fig. 2.19) (Maeda et al. 2018). It was clarified that these dyes exhibit a split absorption band originating from the intramolecular exciton interaction between two chromophores in the molecule. The dye-sensitized solar cells using these dyes exhibited high light harvesting capability in a wide range due to the split of the absorption band originating from the exciton interaction. In this way, new squaraine dyes with various aromatic ring components have been developed by using the cross-coupling reaction between haloarene and stannyl cyclobutenedione derivatives. These synthesis methods enable the tuning of the light harvesting ability by precisely adjusting the expansion of the π-conjugation system and the intramolecular exciton interaction.

Although a squaraine dye has a structure in which electron-rich π-conjugation compounds are condensed at the 1,3- or 1,2-positions of the cyclobutenedione moiety, it is also possible to introduce an electron-accepting substituent like the dicyanovinylene group at the 2-position of the squaric acid (Tatarets et al. 2005). Würthner

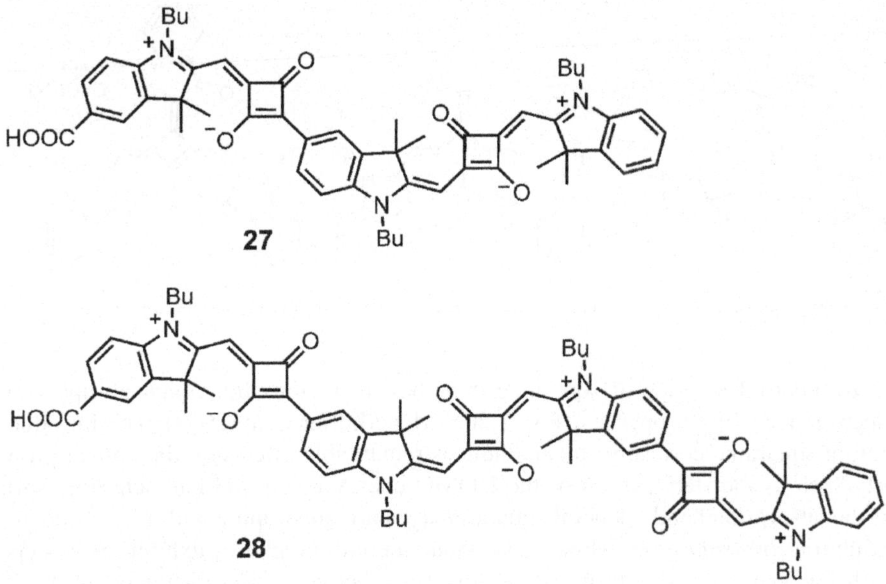

Fig. 2.18 Near-infrared absorbing squaraine dyes with linearly π-extended structures used in dye-sensitized solar cells

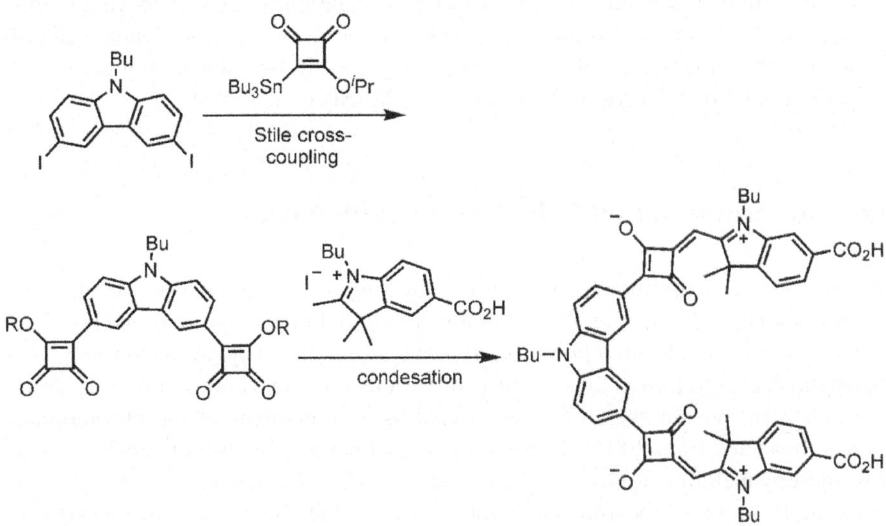

Fig. 2.19 Exciton-coupled squaraine dyes for dye-sensitized solar cells

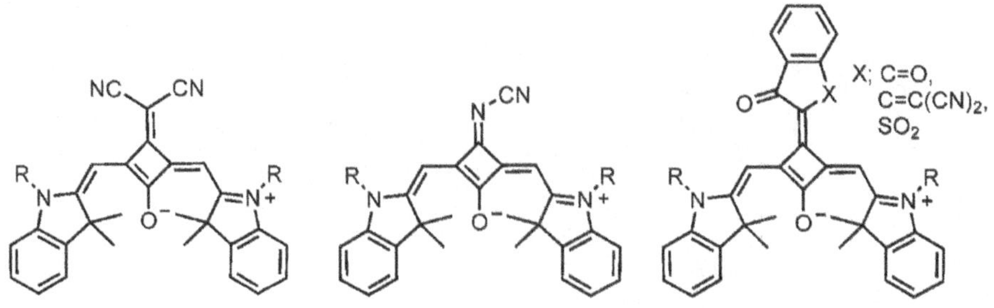

Fig. 2.20 Squaraine dyes with acceptor units on the 2-position of cyclobutenedione moiety

et al. reported a series of squaraine dyes having various functional groups with electron-accepting properties (Fig. 2.20) (Mayerhöffer et al. 2013). The introduction of squaraine core substituents induces remarkable effects on the optical property such as the stabilization of the HOMO energy levels and bathochromic shift of the absorption band. General squaraine dyes are quadrupole with C_{2h} symmetry exhibiting *transoid*-type, while these squaraine chromophores exhibit *cisoid*-type with C_{2v} symmetry due to the steric hindrance between the substituents at the 2-position and the aromatic ring (heterocycle) components. Therefore, this squaraine dye becomes a dipole, and affects the packing structure of molecules in a crystal or a solid thin film. Focusing on these characteristics, squaraine dyes with dicyanomethylene groups are applied to organic field-effect transistors, electron donor materials in organic thin-film solar cells and sensitizing dyes in dye-sensitizing solar cells (Gsänger et al. 2014; Mayerhöffer et al. 2009; Maeda et al. 1303).

2.6 Squaraine-Based Polymers and Oligomers

In order to provide prominent absorption in the long wavelength region and elucidate the mechanism of exciton interaction among chromophores, oligomers and polymers that have squaraine chromophores as a repeat unit have been synthesized (Fig. 2.21). Ajayaghosh et al. reported the synthesis of various polymers composed of squaraine dye in which pyrrole skeleton (**29**) is used as a component of the chromophore (Ajayaghosh and Eldo 2001). These dyes are synthesized by the condensation reaction, thereby including regioisomeric units in the main chain, because of the condensation at 1,3- and 1,2-position of squaric acid residue. Havinga et al. reported the π-conjugated polymers consisting of squaraine chromophores (**30**) (Havinga et al. 1993). These polymers are also synthesized by the polycondensation of squaric acid and homoditopic activated methylene compounds. Since the condensation at 1,2-position of squaric acid is hardly supressed in polymerization, 1,2- and 1,3-substituted structures should be included in the polymer chain. Apart from the ununiformed polymer structures, these polymers show prominent absorption beyond 900 nm and have an extremely low band gap (−1.3 eV). A series of π-conjugated polymer based

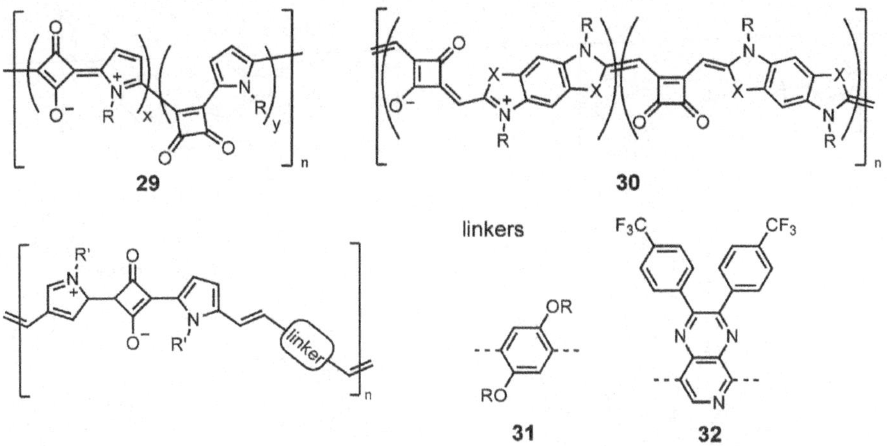

Fig. 2.21 Polycondensation products of squaric acid and pyrrole (29), homoditopic activated methylene derivatives (30), and bispyrrole moieties with linker components (31, 32)

on pyrrole-based squaraine chromophores (**31**, **32**) is also synthesized by the polycondensation of squaric acid with bispyrrole monomers with aromatic linkers (Eldo and Ajayaghosh 2002; Shi et al. 2011a). These dyes also exhibited broad and intense absorption in the near-infrared region. The onset of the low energy absorption bands is observed beyond 1100 nm.

Polymers composed of squaraine chromophores have been synthesized by the catalytic cross-coupling using squaraine with halogen substituents as a monomer (Fig. 2.22). From squaraine dyes composed of indolenine having a bromo group at the 5-position, the corresponding polymer **33** was obtained by the Yamamoto coupling using nickel catalyst. The maximum absorption of this polymer is bathochromically shifted in comparison to its monomeric counterpart and observed at 738 nm in chloroform. It was clarified that the cause of the red-shifted absorption was due to the exciton coupling among the squaraine chromophores (Völker et al. 1098).

Further, polymer **34** was obtained by the Suzuki-Miyaura cross-coupling of halogenated squaraine and phenylene diboronic acid ester as co-monomer (Maeda et al. 2590). Unlike polymer **33**, bathochromic shift caused by the exciton coupling is scarcely observed for **34**. This indicates that the phenylene linkers isolate chromophores and prevent exciton coupling. These polymers have been applied to the donor materials in organic thin-film solar cells, and exhibited the photoelectric conversion ability in the range from far-red to near-infrared region. Polysquaraine bearing dicyanovinylene core-substituents **35** are also synthesized by the same strategy to obtain polymer **33**. The combination of spectroscopic analysis and calculations suggests that the polymer can adopt the helical H-aggregated conformation and the zigzag conformation. Interestingly, a cyclic trimer **36** is isolated in addition to polymer **35** (Völker and Lambert 2012).

Since halogenated squaraine dye becomes a substrate in catalytic coupling reaction, squaraine chromophores can be introduced in various molecular skeletons.

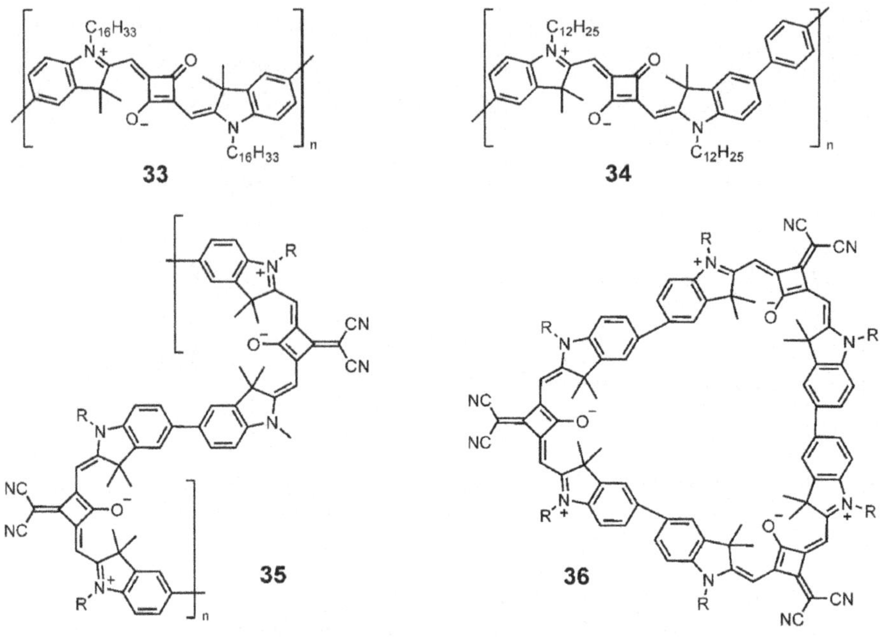

Fig. 2.22 Polysquaraines synthesized by Pd-catalyzed cross-coupling polymerization

Scherer et al. synthesize the oligomeric squaraine dye through the combination of cross-coupling and condensation. The intermediate **37** bearing active methylene terminals was synthesized by the Stille cross-coupling between brominated squaraine dyes and indolenine derivatives with trialkyl-stannyl thiophene skeleton. Also, the squaraine oligomer **38** with thiophene linkers was synthesized by the condensation reaction (Fig. 2.23) (Scherer et al. 2002).

2.7 Miscellaneous Containing Squaraine Chromophores

Two-photon absorption (2PA) defined as the simultaneous absorption of two photons of the same or different energy by a molecule enables a large variety of improved and novel technological capabilities such as microfabrication, high-resolution fluorescence microscopy, non-destructive imaging of biological tissues, and photodynamic therapy (He et al. 2008). The important factor to gain a large 2PA cross section is found to be a donor-acceptor-donor conjugated system. Therefore, squaraine chromophores have a potential for 2PA dyes due to the intrinsic donor-acceptor-donor structure. Actually, squaraine-based compounds **32** and **36** show a large 2PA cross section. Marder et al. reported squaraine dyes bearing extended π-systems **39** and bearing zinc porphyrin components **40** (Fig. 2.24) (Chung et al. 2006; Odom et al. 2009). Lambert et al. also demonstrated the branched squaraine chromophores for

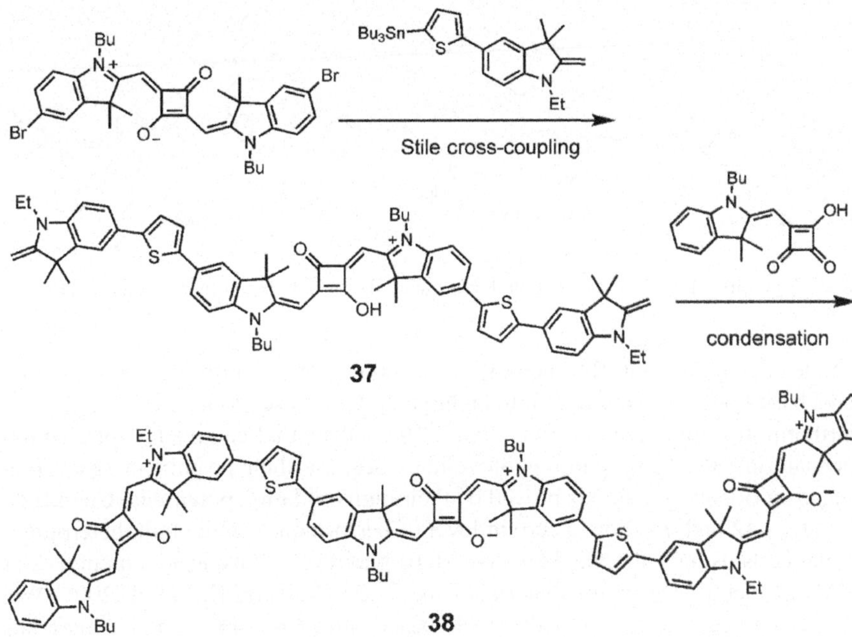

Fig. 2.23 Oligomeric squaraine dyes with thiophene linkers obtained by the combination of Stile cross-coupling and condensation reaction

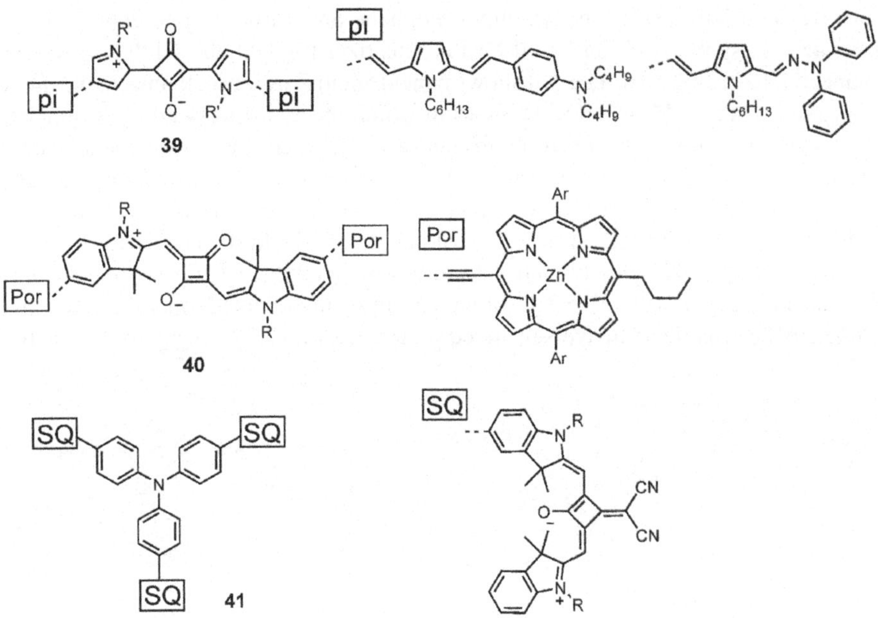

Fig. 2.24 Squaraine-based multiple chromophoric systems for two-photon absorbing materials

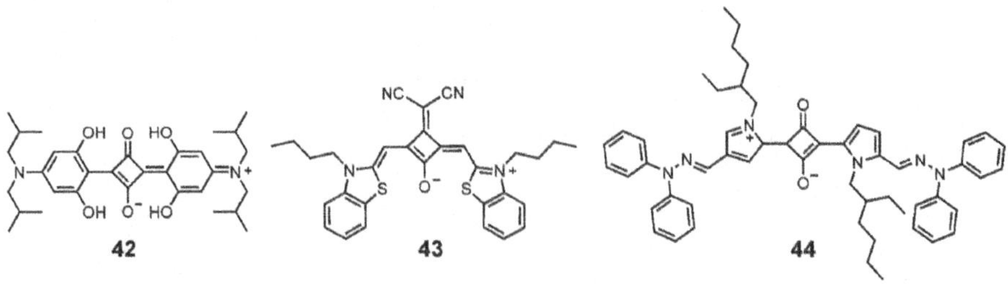

Fig. 2.25 Squaraine dyes used as electron donor materials for bulk heterojunction solar cells

2PA molecules (Ceymann Rosspeintner et al. 2006). These multiple chromophoric systems based on squaraines exhibited a large 2PA cross section.

Furthermore, squaraine dyes were one of the substantial candidates for electron donor materials used in organic photovoltaics because their prominent absorption and electron-donating nature is proved by their high oxidation potentials (Chen et al. 2015). Dyes **42** and **43** show a certain level of performance in the bulk heterojunction solar cells using fullerene derivative, (6,6)-phenyl C_{61} butyric acid methyl ester (PCBM) as electron acceptor materials (Fig. 2.25) (Mayerhöffer et al. 2009; Wei et al. 2011). Pagani et al. reported pyrrole-based squaraines **44** for donor materials (Silvestri et al. 2008).

The intense absorption in the far-red and near-infrared region of squaraine is also useful for the sensitizers of dye-sensitized solar cells. Unsymmetrical squaraine sensitizer consisting of triphenylamine-thiophene and carboxy indolenine components **45** was synthesized and used for the sensitizer for TiO_2-based dye-sensitized solar cells (Li et al. 2010). The cell shows panchromatic response, thanks to the broad absorption band of **45** on TiO_2. Dyes **46** in which the cyanoacryl group is attached to the squaraine chromophores through rigid π-conjugated spacers were synthesized by catalytic cross-coupling reaction as a key reaction using the corresponding halogenated squaraine dye as substrate. These dyes have high performance as sensitizers in dye-sensitized solar cells (Fig. 2.26) (Shi et al. 2011b; Jradi et al. 2015).

Squaraine dye **47** with branch structure was synthesized by the Sonogashira coupling between triarylamine with ethynyl groups and iodinated squaraines. It is also applied to the sensitizers in dye-sensitized solar cells (Fig. 2.27) (Nguyen et al. 2016).

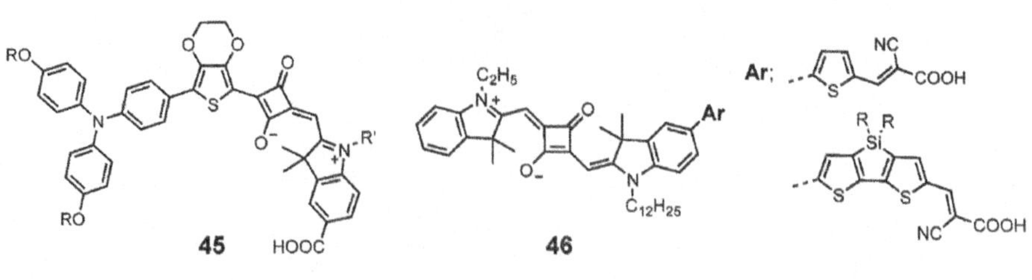

Fig. 2.26 Squaraines with peripheral donor components designed for dye-sensitized solar cells

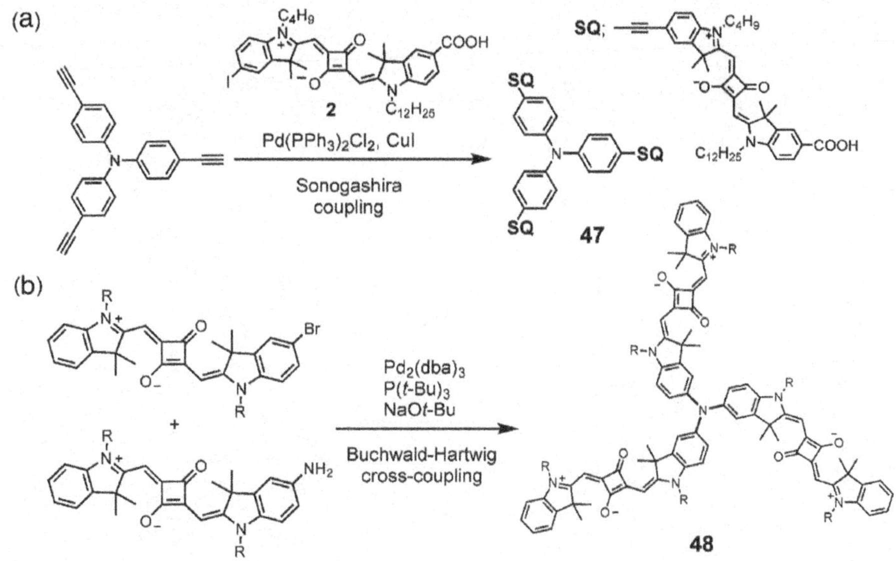

Fig. 2.27 Branched squaraine dyes prepared by Pd-catalyzed coupling reactions

Lambert et al. successfully synthesized the squaraine dye **48** which has a branch structure connected by amino groups by the Buchwald-Hartwig coupling using amino-functionalized squaraines. They clarified the effect of the exciton coupling on the branch structure (Ceymamm Balkenhohl et al. 2016).

2.8 Supramolecular Assembly of Squaraine Dyes

Squaraine dyes, especially *N,N*-dialkylaniline-based squaraines, easily react with nucleophiles such as thiol. The intense absorption band is diminished because the chromophore was broken by the nucleophilic attack. Their high reactivity is utilized for the colorimetric sensing of thiols (Fig. 2.28a) (Sreejith et al. 2008b). On the other hand, Smith et al. demonstrated the stabilization of squaraine chromophores through the supramolecular approach. They utilize the rotaxane structure in which squaraines are encapsulated in macrocycles (Fig. 2.28b). The rotaxane **49** can be synthesized through a clipping strategy that is the macrocyclization using acid chloride in the presence of squaraine dyes bearing bulky peripheral substituents. The rotaxane structure contributes to not only the protection of their reactive site but also to the suppression of aggregation in poor solvents. These benefits provide a chance to use it for bioimaging (Arunkumar et al. 2005b).

Squaraines have a rigid π-conjugated system thereby have a tendency to aggregate in poor solvents as well as the solid state. For the construction of a uniform

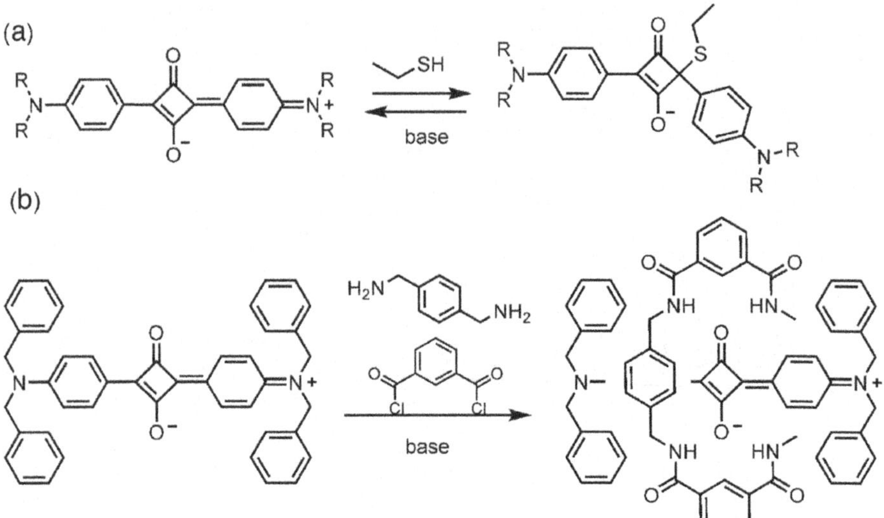

Fig. 2.28 Reaction of the anilino-squaraine with ethanethiol (**a**) and the squaraine-based rotaxane

supramolecular assembly based on squaraines, it is important to optimize chromophore structures and the alkyl chains. Würthner et al. demonstrated the cooperative self-assembly of pyrrole-based squaraine dye **51** (Fig. 2.29a, b) (Ajayaghosh et al. 2008). The dye exhibited an intense absorption band at 664 nm in CH_2Cl_2. In less polar toluene and methylcyclohexane, hypsochromic shift of its absorption was observed. This indicated the aggregation with face-to-face π-stacks (H-type aggregation). The dye self-assembles into extended fiber-like aggregates which are stabilized by dispersive interactions and hydrogen bonds. Bichromophoric dye **50** can aggregate into a spherical and extended micellar structure in the presence of Ca^{2+} (Fig. 2.29c, d) (Mayerhöffer and Würthner 2012).

2.9 Conclusions

In this section, the methods to synthesize squaraine dyes from the classical methods by the condensation reactions to the catalytic cross-coupling reactions are explained. This section also described squaraine derivatives in which various functional groups are introduced, and groups of compounds whose branch structure and macromolecular structure are composed of squaraine skeletons. The squaraine chromophores exhibit specific sharp electronic absorption, and the absorption bands become broad or split based on the exciton coupling due to the approach of the chromophores. Therefore, when considering the application of squaraine dyes, it is required to take

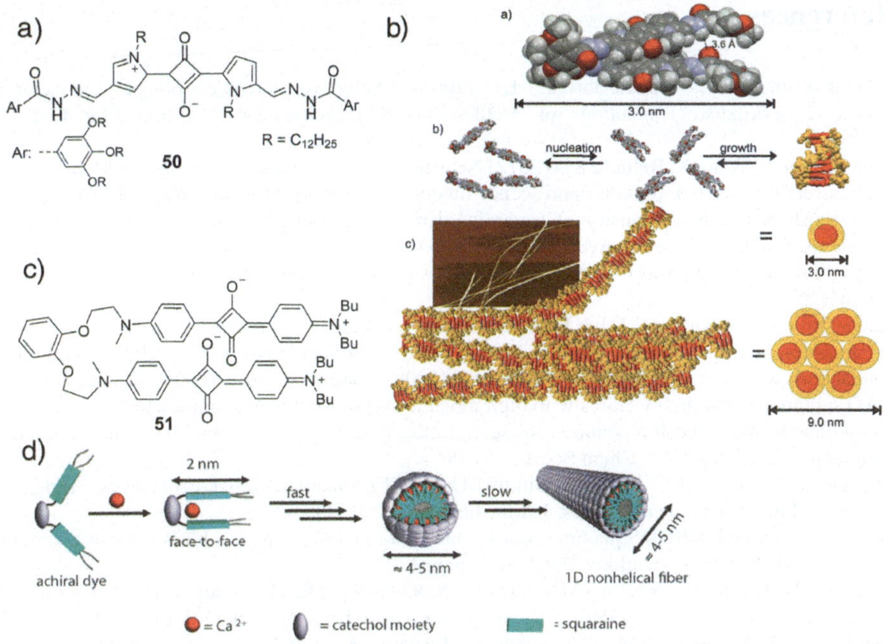

Fig. 2.29 Uniform supramolecular architectures formed by the self-assembly of squaraine dyes. **a** A structure of dyes 50 aggregated to the fibril-like architecture. **b** The dimer model (**a**), A schematic illustration of the aggregation process starting from monomeric species via the nucleation step to the formation of extended aggregate strands (**b**), the model of aggregate bundles and an AFM image of the aggregates (**c**). Reproduced from Ref. 77 with permission from The Royal Society of Chemistry. **C** A structure of bichromophoric squaraines 51. **d** Schematic representation of the process for Ca2+ induced self-assemble of 51. Reproduced from Ref. 78 with permission from The Royal Society of Chemistry

an aggregate formation of dyes into consideration in molecular designing. The advantages of squaraine chromophores are the expression of the strong absorption in the long wavelength region and the diversity of the structural components bonded to the cyclobutene skeletons. For the reason of the difficulty in the synthesis and the lack of stability, the number of near-infrared absorbing organic dyes is limited compared with that of visible-light absorbing ones. It is expected that the need for near-infrared absorbing dyes will more and more increase in various fields such as sensing, optoelectronics, and organic electronics. Since squaraine dye has a potential that can meet these demands owing to its stability and structural diversity, its further development is expected.

References

Aguilar-Aguilar A, Pena-Cabrera E (2007) Selective cross-couplings. sequential stille–liebeskind/srogl reactions of 3-chloro-4-arylthiocyclobutene-1,2-dione. Org Lett 9, 4163

Ahn HY, Yao S, Wang X, Belfield KD (2012) Near-infrared-emitting squaraine dyes with high 2PA cross-sections for multiphoton fluorescence imaging. ACS Appl Mater Interfaces 4:2847

Ajayaghosh A (2005) Chemistry of squaraine-derived materials: near-IR dyes, low band gap systems, and cation sensors. Acc Chem Res 38:449

Ajayaghosh A, Eldo J (2001) A novel approach toward low optical band gap polysquaraines. Org Lett 3:2595

Ajayaghosh A, Arunkumar E, Daub J (2002) A highly specific Ca^{2+}-Ion sensor: signaling by exciton interaction in a rigid–flexible–rigid bichromophoric "h" foldamerangew. Chem Int Ed 41:1766

Ajayaghosh A, Chithra P, Varghese R, Divya KP (2008) Controlled self-assembly of squaraines to 1D supramolecular architectures with high molar absorptivity. Chem Commun 969

Arunkumar E, Ajayaghosh A, Daub J (2005a) Selective calcium ion sensing with a bichromophoric squaraine foldamer. J Am Chem Soc 127:3156

Arunkumar E, Forbes CC, Noll BC, Smith BD (2005b) Squaraine-derived rotaxanes: sterically protected fluorescent near-IR dyes. J Am Chem Soc 127, 3288

Beverina L, Salice P (2010) Squaraine compounds: tailored design and synthesis towards a variety of material science applications. Eur J Org Chem 1207

Bigelow RW, Freund H (1986) An MNDO and CNDO / S(S + DES CI) study on the structural and electronic properties of a model squaraine dye and related cyanine. Chem Phys 107, 159

Ceymamm H, Balkenhohl M, Schmiedel A, Holzapfel M, Lambert C (2016) Localised and delocalised excitons in star-like squaraine homo- and heterotrimers. Phys Chem Chem Phys 18:2646

Ceymann H, Rosspeintner A, Schreck MH, Muetzel C, Stoy A, Vauthey E, Lambert C (2016) Cooperative enhancement versus additivity of two-photon-absorption cross sections in linear and branched squaraine superchromophores. Phys Chem Chem Phys 18:16404

Chen CT, Marder SR, Cheng LT (1994) Molecular first hyperpolarizabilities of a new class of asymmetric squaraine dyes. J Chem Soc Chem Commun 259

Chen H, Law KY, Perlstein J, Whitten DG (1995) Amphiphilic squaraine dye aggregates: evidence for a cyclic chiral structure as a general supramolecular structure for aggregates of dyes and aromatic molecules. J Am Chem Soc 117:7257

Chen G, Sasabe H, Igarashi T, Hong Z, Kido J (2015) Squaraine dyes for organic photovoltaic cells. J Mater Chem A 3:14517 and references therein

Chung S-J, Zheng S, Odani T, Beverina L, Fu J, Padilha LA, Biesso A, Hales JM, Zhan X, Schmidt K, Ye A, Zojer E, Barlow S, Hagan DJ, Van Stryland EW, Yi Y, Shuai Z, Pagani GA, Bredas J-L, Perry JW, Marder SR (2006) Extended squaraine dyes with large two-photon absorption cross-sections. J Am Chem Soc 128:14444

Das S, Thomas KG, Thomas KJ, Kamat PV, George MV (1994) Photochemistry of squaraine syes: 8. photophysical properties of crown ether squaraine fluoroionophores and their metal ion complexes. J Phys Chem 98, 9291

Eldo J, Ajayaghosh A (2002) New low band gap polymers: control of optical and electronic properties in near infrared absorbing π-conjugated polysquaraines. Chem Mater 14:410

Gsänger M, Kirchner E, Stolte M, Burschka C, Stepanenko V, Pflaum J, Würthner F (2014) High-performance organic thin-film transistors of j-stacked squaraine dyes. J Am Chem Soc 136:2351

Havinga EE, ten Hoeve W, Wynberg H (1993) Alternate donor-acceptor small-band-gap semiconducting polymers; polysquaraines and polycroconaines. Synth Met 55:299

He GS, Tan L-S, Zheng Q, Prasad PN (2008) Multiphoton absorbing materials: molecular designs, characterizations, and applications. Chem Rev 108:1245–1330

Jradi FM, Kang X, O'Neil D, Pajares G, Getmanenkp YA, Szymanski P, Parker TC, El-Sayed MA, Marder SR (2015) Near-infrared asymmetrical squaraine sensitizers for highly efficient dye

sensitized solar cells: the effect of π-bridges and anchoring groups on solar cell performance. Chem Mater 27:2480
Keil D, Hartmann H (2001) Synthesis and characterization of a new class of unsymmetrical squaraine dyes. Dyes Pigms 49:161
Law K-Y (1993) Organic photoconductive materials: recent trends and developments. Chem Rev 93:449
Law K-Y (1987) Squaraine chemistry: effects of structural changes on the absorption and multiple fluorescence emission of bis[4-(dimethylamino)phenyl]squaraine and its derivatives. J Phys Chem 91:5184
Law KY (1992) Squaraine chemistry: design, synthesis and xerographic properties of a highly sensitive unsymmetrical fluorinated squaraine. Chem Mater 4, 605
Law KY, Bailey FC (1986) Sqararaine chemistry: Synthesis of bis(4-dimethylaminophenyl)squaraince from dialkyl squarates. Mechanism and scope of the synthesis. Can J Chem 64, 2267
Liebeskind LS, Fengl RW (1990) 3-stannylcyclobutenediones as nucleophilic cyclobutenedione equivalents. synthesis of substituted cyclobutenediones and cyclobutenedione monoacetals and the beneficial effect of catalytic copper iodide on the stille reaction. J Org Chem 55, 5359
Liebeskind LS, Fengl RW, Wirtz KR, Shawe TT (1988) An improved method for the synthesis of substituted cyclobutenediones. J Org Chem 53:2482
Liu T, Liu X, Valencia MA, Sui B, Zhang Y, Belfield KD (2017) Far-red-emitting TEG-substituted squaraine dye: synthesis, optical properties, and selective detection of cyanide in aqueous solution. Eur J Org Chem 3957
Li J-Y, Chen C-Y, Lee C-P, Chen S-C, Lin T-H, Tsai H-H, Ho K-C, Wu C-G (2010) Unsymmetrical squaraines incorporating the thiophene unit for panchromatic dye-sensitized solar cells. Org Lett 12:5454
Maahs G, Hegengerg P (1966) Syntheses and derivatives of squaric acid. Angew Chem Int Ed 5:888
Maeda T, Shima N, Tsukamoto T, Yagi S, Nakazumi H (2011a) Unsymmetrical squarylium dyes with π-extended heterocyclic components and their application to organic dye-sensitized solar cells. Synth Met 161:2481
Maeda T, Hamamura Y, Miyanaga K, Shima N, Yagi S, Nakazumi H (2011b) Near-infrared absorbing squarylium dyes with linearly extended π-conjugated structure for dye-sensitized solar cell applications. Org Lett 13:5994
Maeda T, Tsukamoto T, Seto A, Yagi S, Nakazumi H (2012) Synthesis and characterization of squaraine-based conjugated polymers with phenylene linkers for bulk heterojunction solar cells. Macromol Chem Phys 213, 2590
Maeda T, Arikawa S, Nakao H, Yagi S, Nakazumi H (2013) Linearly π-extended squaraine dyes enable the spectral response of dye-sensitized solar cells in the NIR region over 800 nm. New J Chem 37:701
Maeda T, Mineta S, Fujiwara H, Nakao H, Yagi S, Nakazumi H (2013) Conformational effect of symmetrical squaraine dyes on the performance of dye-sensitized solar cells. J Mater Chem A 1, 1303
Maeda T, Nguyen TV, Kuwano Y, Chen X, Miyanaga K, Nakazumi H, Yagi S, Soman S, Ajayaghosh A (2018) Intramolecular exciton-coupled squaraine dyes for dye-sensitized solar cells. J Phys Chem C 122:21745
Mayerhöffer U, Deing K, Gruß H, Braunschweig K, Meerholz F. Würthner (2009) Outstanding short-circuit currents in BHJ solar cells based on NIR-absorbing acceptor-substituted squaraines. Angew Chem Int Ed 48, 8776
Mayerhöffer U, Würthner F (2012) Cooperative self-assembly of squaraine dyes. Chem Sci 3:1215
Mayerhöffer U, Gsänger M, Stolte M, Fimmel B, Würthner F (2013) Synthesis and molecular properties of acceptor-substituted squaraine dyes. Chem Eur J 19:218
McEwen JJ, Wallace KJ (2009) Squaraine dyes in molecular recognition and self-assembly. Chem Commun 6339

Nakazumi H, Natsukawa K, Nakai K, Isagawa K (1994) Synthesis and structure of new cationic squarylium dyes. Angew Chem Int Ed 33:1001

Nakazumi H, Colyer CL, Kaihara K, Yagi S, Hyodo Y (2003) Red luminescent squarylium dyes for noncovalent HSA labeling. Chem Lett 32:804

Nakazumi H, Ohta T, Etoh H, Uno T, Colyer CL, Hyodo H, Yagi S (2005) Near-infrared luminescent bis-squaraine dyes linked by a thiophene or pyrene spacer for noncovalent protein labeling. Synth Met 153:33

Nguyen TV, Maeda T, Nakazumi H, Yagi S (2016) Linear and tripodal squaraine sensitizers with triphenylamine donor components for dye-sensitized solar cells. Chem Lett 45:291

Odom S, Webster S, Padilha L, Peceli D, Hu H, Nootz G, Chung S, Ohira S, Matichak J, Przhonska O (2009) Synthesis and two-photon spectrum of a Bis(Porphyrin)-substituted squaraine. J Am Chem Soc 131:7510–7511

Ouchi K, Colyer CL, Sebaiy M, Zhou Z, Maeda T, Nakazumi H, Shibukawa M, Saito S (2015) Molecular design of boronic acid-functionalized squarylium cyanine dyes for multiple discriminant analysis of sialic acid in biological samples: selectivity toward monosaccharides controlled by different alkyl side chain lengths. Anal Chem 87:1933

Ramaiah D, Joy A, Chandrasekhar N, Eldho NV, Das S, George MV (1997) Halogenated squaraine dyes as potential photochemotherapeutic agents: synthesis and study of photophysical properties and quantum efficiencies of singlet oxygen generation. Photochem Photobiol 65, 783

Ried W, Vogl M (1977) Reaktionen von cyclobutendionen, XLV: reaktion von cyclobutendionen mit fluoren. Liebigs Ann Chem 101–105

Ronchi E, Ruffo R, Rizzato S, Albinati A, Beverina L, Pagani G (2011) A, regioselective synthesis of 1,2- versus 1,3-squaraines. Org Lett 13:3166

Ros-Lis JV, Martínez-Máñez R, Sancenón F, Soto J, Spieles M, Rurack K (2008) Squaraines as reporter units: insights into their photophysics, protonation, and metal-ion coordination behaviour. Chem Eur J 14:10101

Saito S, Massie TL, Maeda T, Nakazumi H, Colyer CL (2012) On-column labeling of gram-positive bacteria with a boronic acid functionalized squarylium cyanine dye for analysis by polymer-enhanced capillary transient isotachophoresis. Anal Chem 84:2452

Scherer D, Dörfler R, Feldner A, Vogtmann T, Schwoerer M, Lawrentz U, Grahn W, Lambert C (2002) Two-photon states in squaraine monomers and oligomers. Chem Phys 279:179

Schmidt AH (1980) The chemistry of squaraines. In: West R (ed) Oxocarbons. Academic Press, New York, p 185

Shaw SK, Schreiber CL, Roland FM, Battles PM, Brennan SP, Padanilam SJ, Smith BD (2018) High expression of integrin $\alpha v\beta 3$ enables uptake of targeted fluorescent probes into ovarian cancer cells and tumors. Bioorg Med Chem 26:2085

Shi Q, Chen W-Q, Xiang J, Duan X-M, Zhan X (2011a) A low-bandgap conjugated polymer based on squaraine with strong two-photon absorption. Macromolecules 44:3759

Shi Y, Hill RBM, Yum J-H, Dualeh A, Barlow S, Grätzel M, Marder SR, Nazeeruddin MK (2011b) A high-efficiency panchromatic squaraine sensitizer for dye-sensitized solar cells. Angew Chem Int Ed 50:6619

Silvestri F, Irwin MD, Beverina L, Facchetti A, Pagani GA, Marks TJ (2008) Efficient squaraine-based solution processable bulk-heterojunction solar cells. J Am Chem Soc 130:17640–17641

Sprenger HE, Ziegenbein W (1968) Cyclobutenediylium dyes. Angew Chem Int Ed 7:530

Sreejith S, Carol P, Chithra P, Ajayaghosh A (2008a) Squaraine dyes: a mine of molecular materials. J Mater Chem 18:264

Sreejith S, Divya K, Ajayaghosh A (2008b) A near-infrared squaraine dye as a latent ratiometric fluorophore for the detection of aminothiol content in blood plasma. Angew Chem Int Ed 47:7883

Sun W, Guo S, Hu C, Fan J, Peng X (2016) Recent development of chemosensors based on cyanine platforms. Chem Rev 116:7768

Tatarets AL, Fedyunyaeva IA, Terpetschnig E, Patsenker LD (2005) Synthesis of novel squaraine dyes and their intermediates. Dyes Pigms 64:125

Terpetschnig E, Lakowicz JR (1993) Synthesis and characterization of unsymmetrical squaraines: a new class of cyanine dyes. Dye Pigms 21:227

Treibs A, Jakob K (1965) Cyclotrimethine dyes derived from squaric acid. Angew Chem Int Ed 4:694

Völker SF, Uemura S, Limpins M, Mingebach M, Deibel C, Dyakonov V, Lambert C (2010) Polymeric squaraine dyes as electron donors in bulk heterojunction solar cells. Macromol Chem Phys 211, 1098

Völker SF, Lambert C (2012) Exciton coupling effects in polymeric cis-indolenine squaraine dyes. Chem Mater 24:2541

Wei G, Wang S, Sun K, Thompson ME, Forrest SR (2011) Solvent-annealed crystalline squaraine: PC70BM (1:6) solar cells. Adv Energy Meter 1:184

Welder F, Paul B, Nakazumi H, Yagi S, Colyer CL (2003) Symmetric and asymmetric squarylium dyes as noncovalent protein labels: a study by fluorimetry and capillary electrophoresis. J Chromatog B 793:93

West R (1980) Histry of The Oxocarbons. In: West R (ed) Oxocarbons. Academic Press, New York, p 1

Yagi S, Nakazumi H (2008) Squarylium dyes and related compounds. Top Heterocycl Chem 14:133

Yagi S, Murayama S, Hyodo Y, Fujie Y, Hirose M, Nakazumi H (2002) Synthesis and light absorption/emission properties of novel bis-squaraine dyes with extensively conjugated π-electron systems. J Chem Soc, Perkin Trans 1, 1417

Yagi S, Ohta T, Akagi N, Nakazumi H (2008) The synthesis and optical properties of bis-squarylium dyes bearing arene and thiophene spacers. Dyes Pigms 77:525

Yagi S, Nakasaku Y, Maeda T, Nakazumi H, Sakurai Y (2011) Synthesis and near-infrared absorption properties of linearly π-extended squarylium oligomers. Dyes Pigms 90:211

Yom JH, Walter P, Huber S, Rentsch D, Geiger T, Nüesch F, De Angelis F, Grätzel M, Nazeeruddin MK (2007) Efficient far red sensitization of nanocrystalline TiO_2 films by an unsymmetrical squaraine dye. J Am Chem Soc 129, 10320

Ziegenbein W, Sprenger HE (1966) Condensation products of squaric acid and azulenic hydrocarbons. Angew Chem Int Ed 5:893

Chapter 3
Porphyrins: Syntheses and Properties

Jun-ichiro Setsune

Abstract It is well known that *meso*-tetraarylporphyrins are obtained readily by condensation reactions between pyrroles and aromatic aldehydes. The 2+2 type condensation of aryldipyrrylmethanes and the second aromatic aldehyde is useful for synthesizing porphyrins with mixed *meso*-aryl groups, and further transformations of the porphyrin periphery are performed conveniently by using organometallic methodology. Some *meso*-aryl substituted porphyrins are designed for supramolecular application and photosensitizing effect. Porphyrins substituted at pyrrole β-positions are obtainable by using Barton–Zard synthesis, and these compounds are further converted into benzoporphyrins that are structurally similar to phthalocyanines and play an important role in the development of a variety of photo-functional materials. Porphyrins more heavily π-extended than benzoporphyrin are focused in the last part of this chapter. Multiple porphyrin cores are directly bridged by π-units or aromatic substituents at the porphyrin periphery are forced to come into π-conjugation with the porphyrin π-system by Scholl reactions, leading to strong infrared absorption bands at longer wavelength beyond 1000 nm, very small HOMO–LUMO gap, and very high two-photon absorption efficiency.

3.1 Introduction

Porphyrins are fully π-conjugated macrocycles where four pyrrole rings are connected by sp^2-hybridized carbon atoms. The macrocyclic π-conjugation and the dianionic tetradentate coordination site provide basis for a wide variety of applications. Photophysical, electrochemical, and coordination properties of porphyrins have been drawing great interest in a variety of scientific fields and these properties are tunable by the abundant methodology for structure modification. Comprehensive reviews on the current porphyrin studies have been published as multi-volume reference books (Kadish et al. 2000, 2010; Dolphin 1978). This chapter describes functionalized porphyrins with focusing on their synthetic methodology and their

J. Setsune (✉)
Department of Chemistry, Graduate School of Science, Kobe University, Kobe, Japan
e-mail: setsunej@kobe-u.ac.jp

Y. Ooyama and S. Yagi (eds.), *Progress in the Science of Functional Dyes*,
https://doi.org/10.1007/978-981-33-4392-4_3

properties caused by structural diversity. Although some aspects of application to material science and medicine will be mentioned here, they are arbitrarily selected just for examples. A large number of porphyrin analogues (expanded, contracted, and isomeric porphyrins) have been developed these days and their unusual structure and properties have been adding great insights into the porphyrin chemistry. These subjects have recently been reviewed (Sessler et al. 2017), and not included here.

3.2 *meso*-Tetra-substituted Porphyrins

3.2.1 Condensation of Pyrrole and Aldehyde

meso-Tetraarylporphyrins are produced by simple procedures known as a Rothemund method and a Longo-Adler method (Rothemund 1935; Adler et al. 1967). After heating a mixture of pyrrole and aromatic aldehyde in acetic acid or propionic acid under aerobic conditions, the reaction mixture was left overnight at room temperature to generate crystalline A_4-type porphyrins with D_{4h} symmetry usually in the yields more or less 20%. Since rotation of *meso*-aryl groups with *ortho*-substituents is limited, conformational isomers called atropisomers are separated and their stability is dependent on the bulkiness of the *ortho*-substituents. Collman's group constructed an O_2-binding heme model by using an αααα-isomer of *meso*-tetra(*o*-aminophenyl)porphyrin **2** that was obtained by condensation of pyrrole and *ortho*-nitrobenzaldehyde followed by $SnCl_2$ reduction (Collman et al. 1975). **2** is regarded as a scaffold for molecular architecture of unique stereochemistry (Fig. 3.1). Four possible atropisomers (αααα, ααββ, αααβ, αβαβ) occur at equilibrium in a 1:2:4:1 statistical ratio, but Lindsey found that the αααα isomer with all four amino groups on the same face of the porphyrin plane was obtained in 66% yield after heating the benzene solution of the equilibrium mixture in the presence of silica gel due to

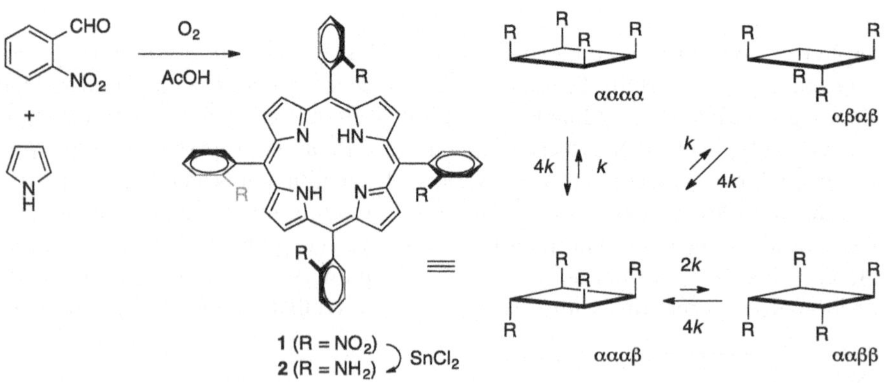

Fig. 3.1 Atropisomerism of *meso*-tetraarylporphyrins

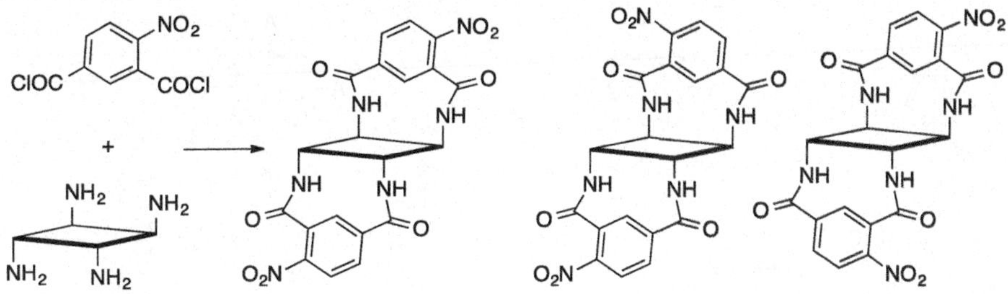

Fig. 3.2 Chiral porphyrins derived from *meso*-tetraarylporphyrins

the preferential adsorption of the target isomer (Lindsey 1980). On the other hand, Nishino reported that the αβαβ atropisomer of *meso*-tetra(*o*-nitrophenyl)porphyrin **1** could be enriched in refluxing toluene (Nishino et al. 1992). Therefore, equilibration in toluene at reflux followed by reduction and chromatographic separation produced 52% yield of the pure αβαβ atropisomer of **2**.

Unsymmetrical bridges were introduced into the ααββ-isomer of *meso*-tetra(*o*-aminophenyl)porphyrin **2** to give a 1:1 mixture of *meso*-**3** and *chiral*-**3** in the studies on molecular recognition by Ogoshi, Kuroda, and coworkers (Fig. 3.2) (Kuroda et al. 1995). The *chiral*-**3** was separated into a C_2-symmetric enantiomer pair by HPLC on a chiral phase and the binding constants of their Zn complexes toward amino acid esters were dependent on the chirality of the enantiomeric porphyrins with sevenfold difference.

Sugiura, Sakata, and coworkers separated A_3B-type porphyrin **4** and A_2B_2-type porphyrin **5** in 12% and 3% yield, respectively, from a mixture of porphyrin products in the mixed condensation of pyrrole with two different aromatic aldehydes under Longo–Adler reaction conditions (Fig. 3.3) (Sugiura et al. 1999). The ester groups of **4** and **5** were converted into formyl groups and then subjected to the mixed condensation with pyrrole again to give the pentameric porphyrin **6** in 25% yield using Lindsey's protocol for hindered porphyrins. That is, a one-flask two-step procedure of acid-catalyzed condensation at room temperature in highly diluted CH_2Cl_2 solution followed by DDQ oxidation of the porphyrinogen intermediate (**7** in Fig. 3.4) (Lindsey et al. 1987). The pentamer **6** was utilized as a precursor to the dendritic molecular architecture composed of 21 porphyrin nuclei and 64 benzene units through the condensation with pyrrole under Lindsey conditions. The final product was obtained in 44% yield and its molecular image was actually observed by STM as a square shape of 65 Å along the sides that is consistent with the theory 65.7 Å.

Mixed condensations are an easy way to functional porphyrins such as used for chirality sensors. Nakanishi and Berova had pointed out that the exciton coupling of two porphyrins in a skewed conformation gives a strong CD couplet at the Soret band region at ca. 400 nm and its sign is diagnostic of the absolute configuration of the optically active guests that was bound cooperatively to the two porphyrin units (Huang

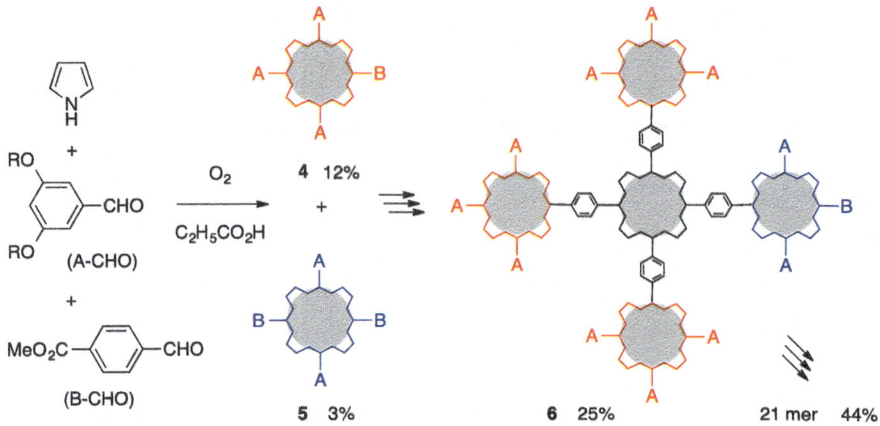

Fig. 3.3 Synthesis of square-shaped porphyrin oligomers

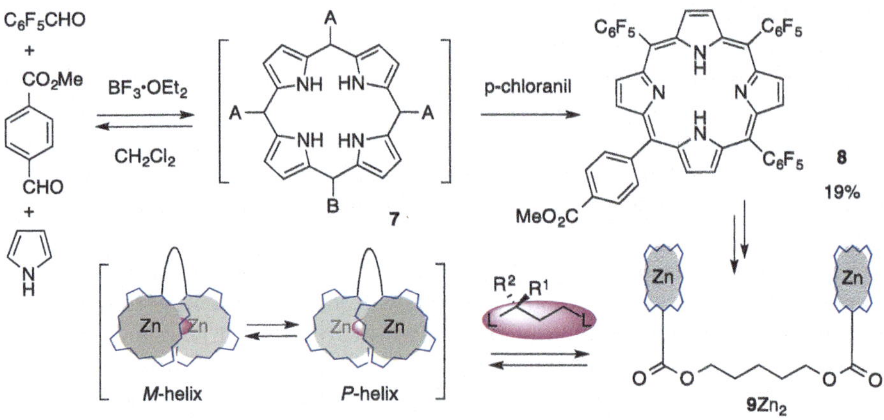

Fig. 3.4 Direct synthesis of A₃B-type *meso*-tetraarylporphyrin and its application to diporphyrin

et al. 1998). Their molecular design of the diporphyrin chiral sensor was modified by Borhan and coworkers. The mixed condensation of pentafluorobenzaldehyde, 4-carbomethoxybenzaldehyde, and pyrrole in the molar ratio of 3:1:4 according to Lindsey's protocol afforded 19% yield of A_3B-type porphyrin **8** after chromatographic separation of a mixture of porphyrin products with statistical distribution of *meso*-aryl substitution pattern (Fig. 3.4) (Li et al. 2008). **8** was converted to Zn porphyrin dimer **9**Zn_2 which was used for chirality sensing of optically active guest molecules such as diamines, aminoalcohols, and dialcohols. *meso*-Pentafluorophenyl groups increased Lewis acidity of Zn(II) and the guest binding was dependent on the chain length of the linker. Binding of the asymmetric guest molecules to **9**Zn_2 induces face-to-face orientation of two porphyrin rings with their electronic transition moments in a skewed orientation. Thus, chirality of the guest molecule is translated into the helicity of the host molecule.

3.2.2 Lindsey Synthesis Using Dipyrrylmethanes

The above-mentioned Lindsey protocol of acid-catalyzed condensation followed by oxidation is applicable to selective formation of *trans*-A_2B_2-type porphyrin **11** from aryldipyrrylmethanes **10** and the second aromatic aldehyde (Fig. 3.5) (Ravikanth et al. 1998). But it should be remembered that decomposition of **10** and a porphyrinogen intermediate to pyrrole and aromatic aldehyde is also acid-catalyzed to result in scrambling of the *meso*-aryl substitution pattern in the porphyrin products. Lindsey also reported an alternative route to **11** under milder reaction conditions taking advantage of more reactive intermediate, aryldipyrrylmethane-2,9-dicarbinol **13′** that is effectively obtained by sequential reactions of **10** with aroyl chloride and $NaBH_4$ (Zaidi et al. 2006). This synthetic method using aryldipyrrylmethane and aryldipyrrylmethane-2,9-dicarbinol was applied to A_3B-type porphyrins and AB_2C-type porphyrins. Since aryldipyrrylmethane-2,9-dicarbinol with three different aryl groups **13** was prepared by way of monoacylation of **10** with 2-pyridyl thioester under low temperature, ABCD-type porphyrins **14** were also prepared (Rao et al. 2000; Lindsey 2010).

Dimeric AB_2C-type porphyrin **15** that is known as a gable porphyrin was designed so as to make supramolecular cyclic assembly of six units (Fig. 3.6) (Takahashi and Kobuke 2005). Kobuke and coworkers obtained 2% yield of **15** through

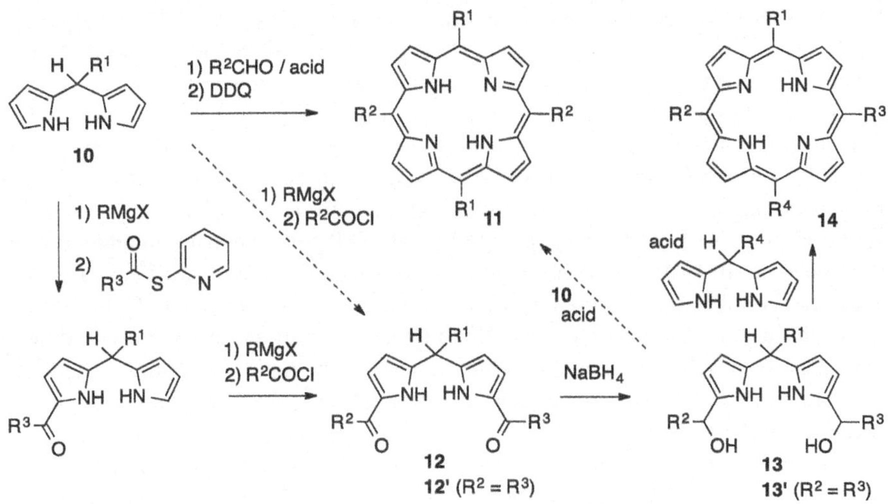

Fig. 3.5 Synthetic routes to *meso*-tetraarylporphyrins of various substitution patterns

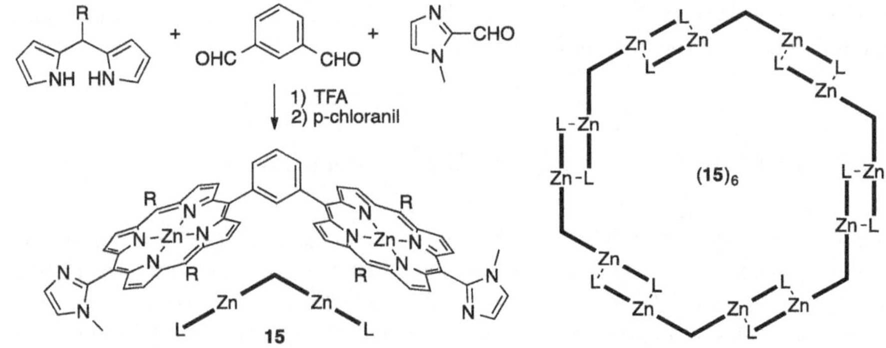

Fig. 3.6 Synthesis of a gable porphyrin for modeling a light-harvesting complex

TFA-catalyzed mixed condensation of 5-alkyldipyrrylmethane, 1-methylimidazole-2-aldehyde, and isophthalaldehyde. Since preparation yields in the single porphyrin ring-forming reaction which is usually less than 20%, double porphyrin ring-forming reactions would be very low and separation of the target porphyrin is problematic. Reversible coordination of imidazole to Zn allowed reorganization into the cyclic oligomers of 12 porphyrin units. The dimeric porphyrin units of a slipped cofacial orientation constituting the macro ring assembly $(\mathbf{15})_6$ showed close analogy to the dimeric chlorophyll units assembled in the light-harvesting complexes of photosynthetic bacteria.

Aida and coworkers reported A_2B_2-type porphyrin **16** in 13% synthetic yield and A_3B-type porphyrin as a byproduct in 4% yield through acid-catalyzed condensation of 5-(2,6-dimethoxyphenyl)-2,3,7,8-tetramethyldipyrrylmethane and benzaldehyde (Fig. 3.7) (Mizuno et al. 2000). The fully substituted porphyrin **16** undergoes a saddle shape distortion of the porphyrin plane in order to relieve steric crowding in its periphery, which results in the introduction of chirality. Although these enantiomeric forms, (*M*, *M*)- and (*P*, *P*)-form, interconvert rapidly by macrocyclic inversion at room temperature, this conformational change slowed down when they bind two molecules of carboxylic acids. If optically active guests were bound, either one of the diastereomeric pair is favored. For example, a strong CD signal in the negative sign was generated in the visible region (450–500 nm) when (*S*)-mandelic acid was added to **16** to lead to very high (>98%) diastereoselectivity. Thus, absolute configuration of optically active acids can be determined by the CD Cotton effect due to the porphyrin chromophore. When these diastereomerically biased adducts were dissolved in acetic acid, the chiral guests were replaced by acetic acids to generate enantiomerically biased adducts, the optical purity of which lasts in the timescale of days.

In addition to this chiral memory phenomenon, photoresponsive change in optical purity of the (*S*)-mandelic acid adduct of **16** was reported in on–off cycles of irradiation of the Soret band. That is, light irradiation caused racemization and the favored original diastereomeric adduct was reassembled in the dark. Crystals generated from

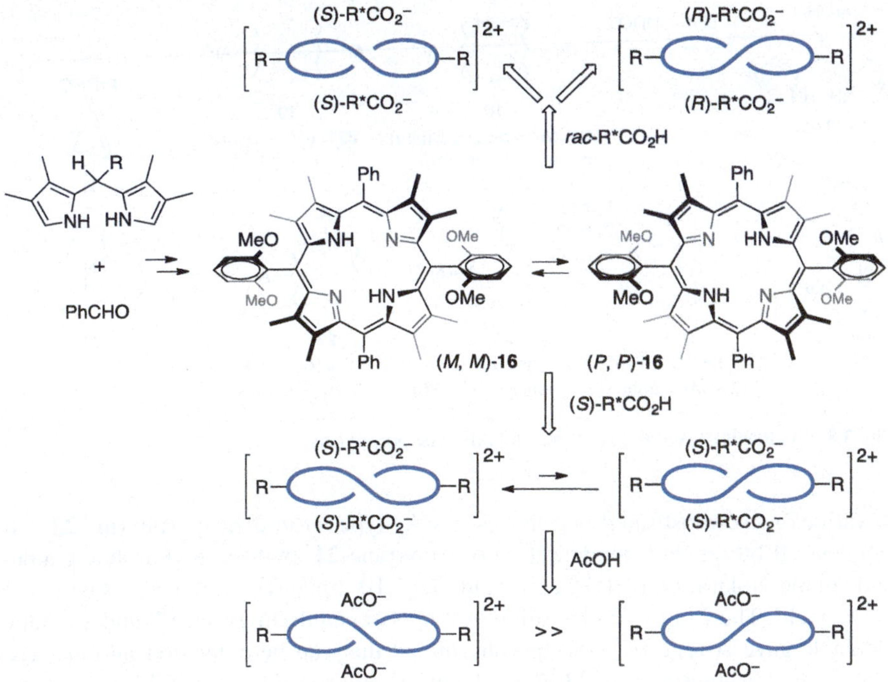

Fig. 3.7 Chirality sensing and optical resolution of carboxylic acids by using *trans*-A_2B_2-type *meso*-tetraarylporphyrin of D_2 symmetry

a solution of a mixture of **16** and racemic mandelic acid derivatives (1:2.1) were found to be a conglomerate leading to spontaneous optical resolution of the guests (Mizuno et al. 2006).

3.2.3 *Porphyrins with Mixed* meso-*Substituents*

A variety of porphyrins of various *meso*-substitution patterns can be derivatized from readily available simple porphyrins by introducing *meso*-substituents and their further transformation as demonstrated by Senge and coworkers (Fig. 3.8) (Senge 2011). *trans*-A_2-type *meso*-diarylporphyrins **18** are synthesized conveniently by Lindsey protocol of reacting an equimolar mixture of dipyrrylmethane **17** and aromatic aldehyde with TFA (0.1 equiv) in CH_2Cl_2 followed by DDQ (2 equiv) oxidation. The preparation yields of this 2+2 condensation sometimes amount to ~50% when aromatic aldehydes are not sterically demanding (Senge et al. 2010). Mixed condensation of **17** affords *trans*-AB-type *meso*-diarylporphyrins **19**, for example, 5-*p*-tolyl-15-*o*-anisylporphyrin was prepared in 22% yield (Fig. 3.8) (Senge et al. 2011). Simple procedures for **18** and **19** in relatively good yields provide a practical approach

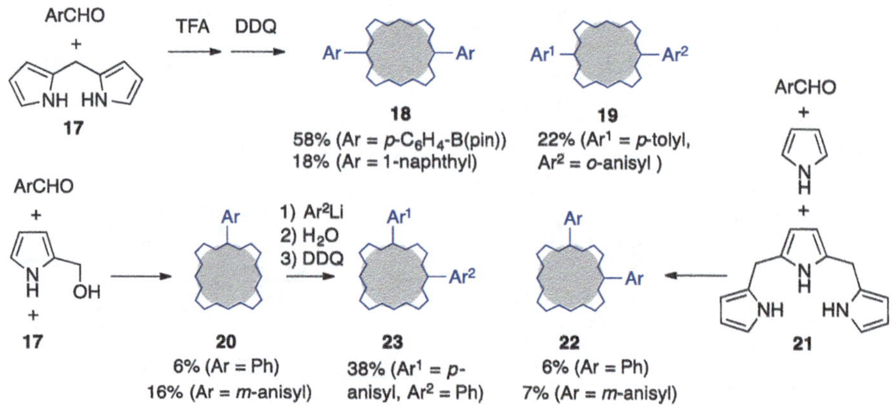

Fig. 3.8 Synthesis of A-type, A_2-type, and AB-type porphyrins

to various *meso*-substituted porphyrins. *cis*-A_2-type *meso*-diarylporphyrins **22** was synthesized by the 3+1 condensation of tripyrrane **21**, pyrrole, and aromatic aldehyde in the molar ratio of 1:1:2, for example, 5,10-diphenylporphyrin was prepared in 6% yield. The 2+1+1 condensation of 2-hydroxymethylpyrrole, **17**, and aromatic aldehyde gave A-type *meso*-arylporphyrins **20** that can be converted into *cis*-AB-type *meso*-diarylporphyrins **23** by nucleophilic addition of organolithium reagents (Ryppa et al. 2005).

Porphyrin *meso*-positions undergo ordinary electrophilic substitution reactions such as formylation, nitration, halogenation, and so on. Furthermore, various substituents can be introduced into the *meso*-position of porphyrin by means of transition metal-catalyzed cross-coupling reactions (Fig. 3.9) (Senge 2011; Ryan et al. 2011). *trans*-AB-type *meso*-diarylporphyrins **19** was readily brominated to

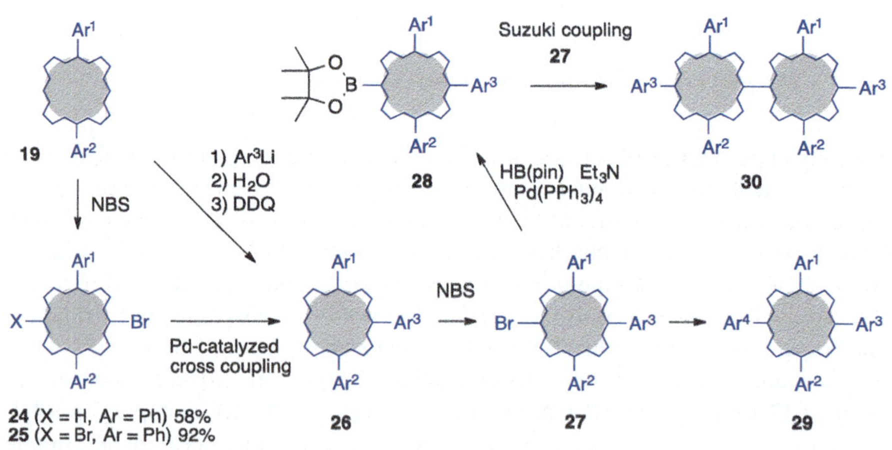

Fig. 3.9 Post-introduction of *meso*-substituents into porphyrin

afford monobromo compound **24** and dibromo compound **25** in good yields. These porphyrins can be subjected to various Pd- or Ni-catalyzed cross-coupling reactions to introduce (hetero)aromatic units by Suzuki coupling and Stille coupling, alkynyl units by Sonogashira coupling, vinyl units by Mizorogi-Heck reaction, and amines by Buchwald–Hartwig reaction. ABC-type *meso*-triarylporphyrins **26** are obtained in good yields in this way from **24,** and these porphyrins were prepared directly from **19** in good yields through the S_NAr reaction. It is noteworthy that alkyllithium reagents work well to introduce a *meso*-alkyl substituent that is relatively difficult to achieve by metal-catalyzed cross-coupling reactions. The iterative procedures starting from **26** afforded ABCD-type *meso*-tetraarylporphyrins **29**. Dioxaborolanylporphyrin **28** was readily prepared by Pd-catalyzed borylation of bromoporphyrin **27** with pinacolborane (HB(pin)), and it was employed as a coupling partner in the Suzuki coupling, for example, symmetrical *meso,meso*-linked diporphyrin **30** was synthesized.

3.2.4 meso-*Substituted Porphyrins as Photosensitizers*

Figure 3.10 illustrates synthesis of a photosensitizer, YD2-*o*-C8, that showed remarkable solar-to-electric power conversion efficiency (PCE) as used in the dye-sensitized solar cell (DSSC) known as a Grätzel cell (Yella et al. 2011). Since the 11.9% PCE under standard air mass 1.5 G illumination exceeds that of DSSC based on the ruthenium sensitizers, push–pull type porphyrin structures with electron-donating substituents and electron-withdrawing substituents arranged at the periphery are extensively studied. These structures are believed to enhance

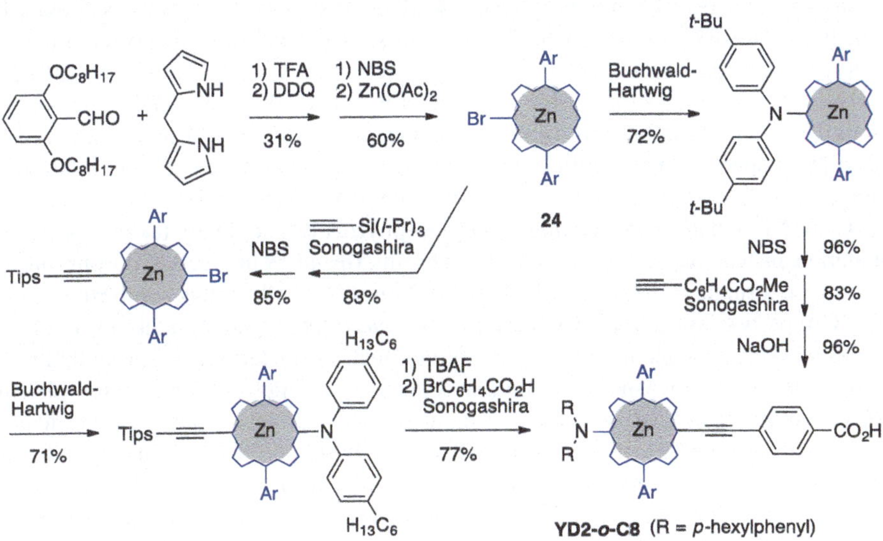

Fig. 3.10 Synthesis of *trans*-A_2BC-type push–pull porphyrins

Table 3.1 Power conversion efficiency (PCE) of Grätzel cell with porphyrin photosesitizers

Photosesitizer	PCE (%)	Ar	R	x	y
YD2-*o*-C8	11.9	2,6-di-octyloxy-phenyl	hexyl	0	1
GY50	12.8			1	1
GY21	2.5			1	0
SM315	13.0		2,4-di-hexyloxy-phenyl	1	1
SM371	12.0			0	1
YD2	8.4	3,5-di-*t*-butylphenyl	hexyl	0	1
ZnPBAT	>YD2	–	–	1	–
ZnPBA	YD2>	–	–	0	–

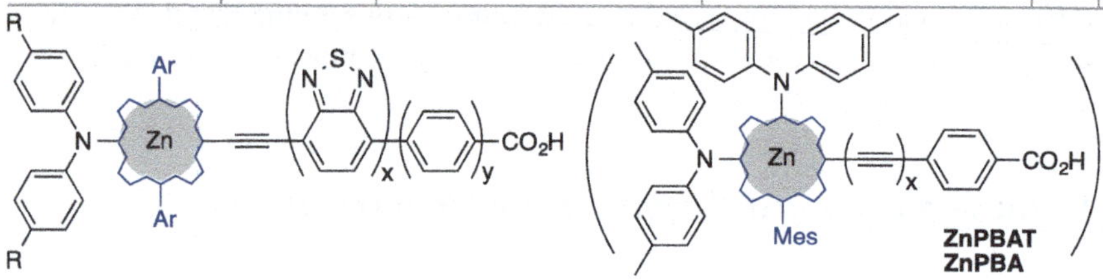

light-harvesting property that is originated from the intense absorption bands of porphyrins extending from visible to NIR region. Furthermore, introduction of long-chain alkoxy groups was suggested to help suppressing charge recombination process. The Grätzel's group prepared the *trans*-A_2BC type porphyrin, YD2-*o*-C8, as a photosensitizer through Sonogashira coupling, Buchwald-Hartwig amination, and the second Sonogashira coupling, in sequence starting from 5,15-di(2,6-dioctyloxyphenyl)-10-bromoporphyrin **24**. Senge and coworkers recently proposed Buchwald-Hartwig amination of **24** as the first step leading to a shorter route to this type of porphyrins (Fig. 3.10) (Meindl et al. 2017). The PCE value of 11.9% of YD2-*o*-C8 was greater than 8.4% of the reference dye YD2 under the same conditions using $Co^{II/III}$ tris(bipyridyl) complex as electrolyte (Table 3.1), and their UV–vis absorption spectra have a tailing up to 700 nm with absorption maxima at 442 nm (log ε 5.3), 576 nm (log ε 4.1), and 638 nm (log ε 4.4) in CH_2Cl_2.

If *cis*-A_2- and *cis*-AB-type *meso*-diarylporphyrins **22** and **23** are taken as starting materials instead of *trans*-*meso*-diarylporphyrins **18** and **19** in Fig. 3.8, porphyrins of additional structural diversity are obtained. Figure 3.11 illustrates synthesis of *cis*-A_2BC-type porphyrin (ZnPBAT) having two *meso*-amino substituents by Imahori and coworkers through Sonogashira coupling, Buchwald–Hartwig double amination, and the second Sonogashira coupling, in sequence starting from 5-mesitylporphyrin **20** (Kurotobi et al. 2013). They investigated on the effect of the asymmetrically enhanced push–pull electronic structure on the DSSC performance and found that introduction of the second amino group caused improvement of the light-harvesting efficiency in visible region and 10% increase in the power conversion efficiency when ZnPBAT was compared with YD2 that is substituted with single diarylamino group

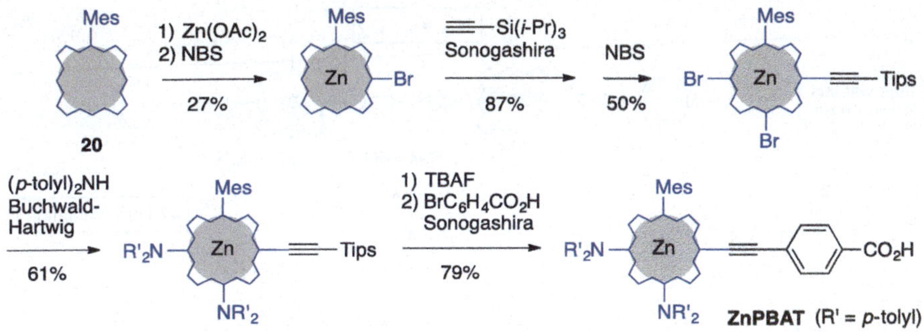

Fig. 3.11 *cis*-A_2BC-type push–pull porphyrin of strong intramolecular dipole moment

at the 5-*meso* position and a 4-carboxyphenylethynyl group at the 15-*meso* position (Table 3.1) (Bessho et al. 2010).

The Grätzel's group reported photosensitizers structurally related to YD2-*o*-C8. Insertion of a 2,1,3-benzothiadiazole unit between the benzoic acid anchoring group and ethyne linker improved PCE performance in the case of GY50 (Yella et al. 2014). Since 2,1,3-benzothiadiazole is strongly electron-withdrawing group, enhanced push–pull effect of GY50 caused red-shift of the absorption bands with more coverage of the visible to NIR wavelength region. Significantly reduced PCE in GY21 indicated that the benzene spacer between the 2,1,3-benzothiadiazole and carboxylic acid plays a key role in suppressing back electron transfer. When the electron-donating effect was strengthened by introducing 2,4-dialkoxyphenyl group at the 5-diarylamino substituent in the case of SM315, 13% PCE was achieved (Mathew et al. 2014). Relationship between photosensitizer structure and PCE is summarized in Table 3.1.

Water-soluble porphyrin, as a photosensitizer, transfers excitation energy to generate singlet molecular oxygen $^1O_2$ that is cytotoxically leading to cell death in photodynamic therapy (PDT) of cancer (Sternberg and Dolphin 1998). Anderson reported A_2BC-type porphyrin **31** that was designed for high intracelluar uptake (Fig. 3.12) (Kuimova et al. 2009). The butadiyne-linked diporphyrin structure of **31** causes conformational change between a planar form and a less stable twisted form through the rotation around the butadiyne axis. The ratio of these conformers was found to depend on the viscosity of the media that was detected by their distinctive fluorescence at 780 nm of the planar form and at 710 nm of the twisted form. The content of the twisted form was found to increase as increase in viscosity, this is because the rotation around the butadiyne axis becomes slower in viscous media. This fluorescent ratiometric molecular rotor provides insight into the diffusion-mediated cellular processes. They found that the diporphyrin **31** incorporated into live cells caused significant increase in the viscosity of the intracellular environment upon irradiation of light. This was interpreted in terms of the crosslinking reactions induced by singlet oxygen generated by the photosensitizing effect of **31**.

Porphyrin-related compounds have been widely studied in the field of PDT due to their intense absorption at long-wavelength visible region that leads to $^1O_2$ generation

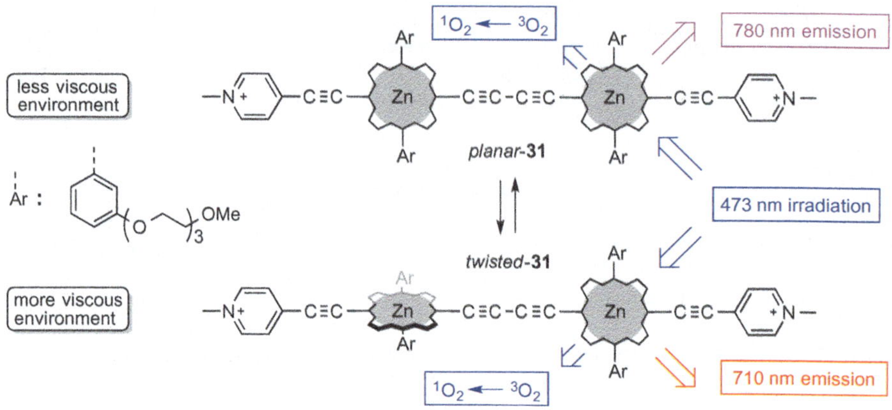

Fig. 3.12 Butadiyne-linked diporphyrin as a fluorescent ratiometric molecular rotor

by way of the triplet excited state due to effective intersystem crossing (Sternberg and Dolphin 1998). Important improvements of the photobiological effect depend on photosensitizers excited by photoirradiation at NIR region (700–1300 nm), because the NIR light can penetrate deeply into the biological tissues. The two-photon absorbing photosensitizers can be photo-excited by intense laser pulses at NIR region and the pinpoint focus of the laser beam enables treatment in the deep area of tissues. Therefore, porphyrin compounds of high two-photon absorption (2PA) efficiency have been extensively studied. The 2PA efficiency is estimated by 2PA cross section (σ) in the unit of GM as the ordinary one-photon absorption depends on the molecular extinction coefficient (ε). Push–pull porphyrin structure with extended π-conjugation is thought to have great magnitude of 2PA cross section as well as in the case of DSSC. Although simple monomeric porphyrins exhibit very small 2PA cross section less than 50 GM with femto-second pulses, the diporphyrin **31** showed a very high 2PA cross section of 17,000 GM with femto-second pulses at 916 nm and its 2PA PDT effect was demonstrated in vivo (Collins et al. 2008).

In the template-directed synthesis of cyclic porphyrin oligomers developed by Anderson and coworkers (Bols and Anderson 2018), 5,15-diarylporphyrin **18** as a starting material was brominated and then subjected to Pd-catalyzed Sonogashira coupling with trialkylsilylacetylene to give 5,15-diaryl-10,20-diethynylporphyrin **32** after desilylation (Fig. 3.13) (Sprafke et al. 2011). In the presence of properly designed template, Glacer oxidative dimerization of the ethyne units was successfully accomplished to give π-conjugated macro ring. The cyclic hexamer of Zn porphyrin *cyclo*-(**32**)$_6$ with butadiyne spacers was formed in 21% yield as a 1:1 complex with hexapyridine template $\mathbf{T}_6$. The fluorescence spectrum of *cyclo*-(**32**)$_6$ •$\mathbf{T}_6$ showed NIR bands ranging from 900 to 1300 nm, and the excited state delocalized over the whole macro ring was generated within less than 0.5 ps after the light absorption.

The dimeric porphyrin **33** worked well in this Glacer homocoupling reaction to afford the same macrocycle in better yield (Fig. 3.14). Although a small amount

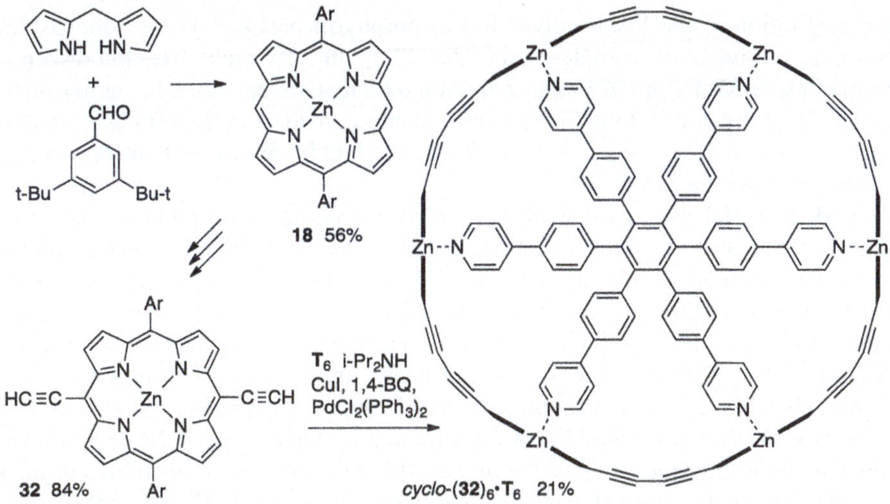

Fig. 3.13 Template-directed synthesis of π-conjugated cyclohexamer of Zn porphyrin

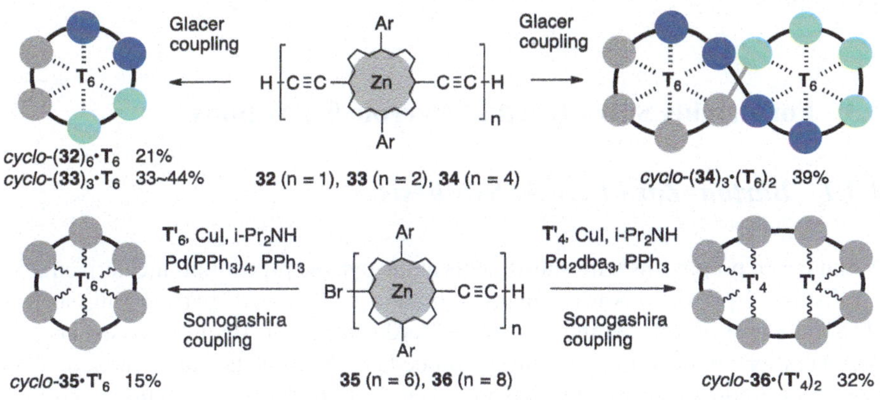

Fig. 3.14 Template-directed synthesis of π-conjugated cyclooligomers of Zn porphyrin

of the cyclododecamer was formed as a byproduct in these cyclization reactions, the Glacer reaction of the butadiyne-linked linear tetramer **34** in the presence of $\mathbf{T_6}$ as a mismatch combination for the *cyclo*-(**32**)$_6$ generated, in 39% yield, a cyclic porphyrin dodecamer *cyclo*-(**34**)$_3$ that includes two template molecules inside the cavity (Kondratuk et al. 2014). A smaller hexapyridine template $\mathbf{T'_6}$ having a hexaethynylbenzene core instead of the hexaphenylbenzene core of the $\mathbf{T_6}$ template was used in the cyclization of ethyne-linked linear oligomers of porphyrin (Rickhaus et al. 2017). The linear ethyne-linked porphyrin hexamer **35** was subjected to Sonogashira coupling reaction to give a 1:1 complex, *cyclo*-**35•**$\mathbf{T'_6}$, in 15% yield. A tetrapyridine template $\mathbf{T'_4}$ has a tetraethynylbenzene core and worked well in the cyclization of the linear ethyne-linked porphyrin octamer **36** by Sonogashira coupling to give a 1:2 complex, *cyclo*-**36•(T'$_4$)$_2$**, in 32% yield. The fluorescence band of *cyclo*-**35•T'$_6$** at 1073 nm is red-shifted relative to that of the linear hexamer, *linear*-**35**, at 909 nm and the fluorescence quantum yield of *cyclo*-**35•T'$_6$** is smaller by the factor of 1000, which is indicative of the highly delocalized singlet excited state of the cyclohexamer.

Anderson, Rebane, and coworkers studied two-photon absorption efficiency of their π-conjugated oligomers (Drobizhev et al. 2006). The 2PA cross section values (σ) of the linear Zn porphyrin oligomers increased by 450 times upon going from the monomer **32**$(SiR_3)_2$ to the dimer **33**$(SiR_3)_2$, but only by 2.4 times from the dimer **33**$(SiR_3)_2$ to the tetramer **34**$(SiR_3)_2$. The σ value increased gradually from 22,000 GM to 37,000 GM when the porphyrin units increased from 4 to 8 in a series of butadiyne-linked linear oligomers end-capped with trialkylsilyl groups. The 2PA cross section of *cyclo*-(**32**)$_6$ •**T$_6$** determined by femto-second 2P fluorescence intensity with laser pulses at 1000 nm was 151,000 GM that is much greater than 23,000 GM for the linear hexameric porphyrin oligomer of **32** end-capped with trialkylsilyl groups (Mikhaylov et al. 2016). This great σ value is responsible for the effective π-conjugation over the whole ring that is promoted by the complexation of the template.

3.3 Porphyrins Substituted at Pyrrole β-Positions

3.3.1 Barton–Zard Pyrrole Synthesis

A number of porphyrins with substituents at the pyrrole β-positions are derived from β-substituted pyrroles. Among many synthetic methods for pyrrole ring such as classical Knorr pyrrole synthesis, Barton–Zard synthesis is frequently used because of generality, applicability, and simple procedure. Nitro olefins are generated from nitroaldol products (acetoxynitroalkanes **37**) in situ under basic reaction conditions of Barton–Zard synthesis and then undergo [3+2]-cycloaddition with isocyanoacetate to give 3,4-disubstituted pyrrole-2-carboxylate **38** (Fig. 3.15) (Barton and Zard 1985). The ester function at one of the pyrrole α-positions is regarded as a protecting group, and it can be easily removed or utilized as a *meso*-carbon of porphyrin structure leading to versatile application. A related methodology, van Leusen pyrrole synthesis, using tosylmethyl isocyanide (TosMIC) and Michael acceptor is useful in the synthesis of 3,4-disubstituted pyrroles **39** having electron-withdrawing groups (EWG) at the β-position (van Leusen et al. 1992). 2-Stannylpyrrole **40** was also obtainable in good yields by one-pot procedure, and it is useful in the Stille coupling reactions leading to various building blocks for porphyrin analogues (Dijkstra et al. 1998).

Naturally occurring porphyrins have alkyl substituents at all the pyrrole β-positions. These 2,3,7,8,12,13,17,18-octaalkylporphyrins are different from

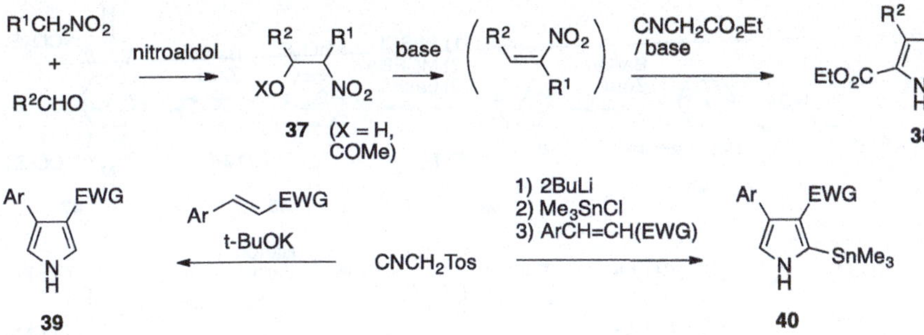

Fig. 3.15 Barton–Zard pyrrole synthesis and van Lausen pyrrole synthesis

5,10,15,20-tetraarylporphyrins in their stereochemistry, electronic structure, and reaction behavior. 2-Carboethoxy-3,4-diethylpyrrole **38** was readily decarboxylated and then subjected to the ordinary porphyrin synthesis using formaldehyde to give octaethylporphyrin (OEP) **41** in 30% yield (Sessler et al. 1992). Ponomarev and coworkers reported that Vilsmeier formylation of OEPCu(II) followed by $NaBH_4$ reduction produced *meso*-(dimethylaminomethyl)OEPCu(II) **42** in good yield. When **42** was allowed to react with MeI, ethylene-bridged diporphyrin **43** was obtained as a bisCu(II) complex (Fig. 3.16) (Borovkov et al. 1999). Binding optically active guest molecule by cooperative metal–ligand coordination bondings forces the bisZn(II) complex to take the face-to-face and skewed orientation of their porphyrin rings. Supramolecular chirogenesis of the bisZn(II) complex was extensively studied in the chirality sensing by Borovkov, Inoue, and coworkers (Borovkov et al. 2004).

Ono, Uno, and coworkers used a Diels–Alder adduct **44** as a nitro olefin equivalent in the Barton–Zard pyrrole synthesis to give ethanoisoindole **45** (Fig. 3.17) (Uno et al. 2000). Phenylsulfenyl chloride addition, MCPBA oxidation, and HCl elimination proceeded in the transformation from **45** to **46**. This reaction sequence is of great importance because simple alkenes are almost quantitatively converted into substrates for Barton–Zard pyrrole synthesis. Diels–Alder adduct **48** between cyclobutadiene and dimethyl acetylenedicarboxylate was also converted to the phenylsulfonyl olefin **49** (Ito et al. 1998, 2001). The phenylsulfonyl olefins **46** and **49** are reactive as well as nitro olefin to give the corresponding pyrroles **47** and **50** in

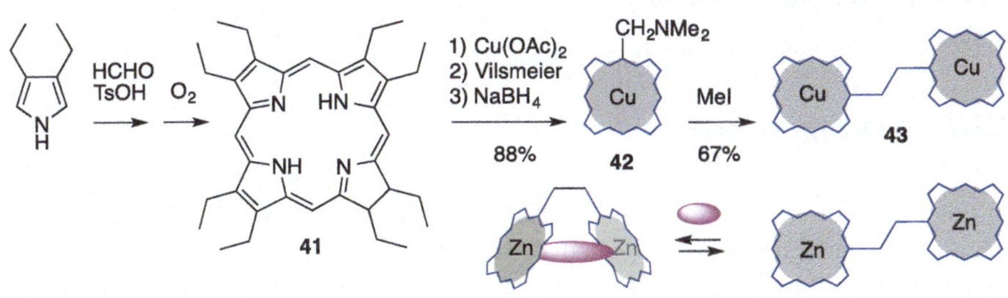

Fig. 3.16 Octaethylporphyrin and ethylene-bridged porphyrin

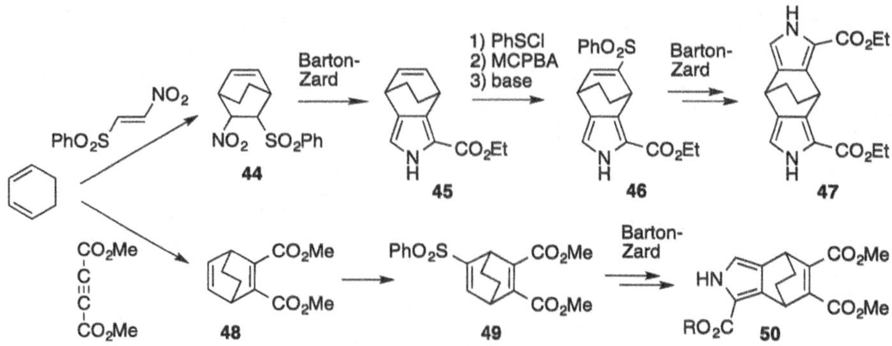

Fig. 3.17 Barton–Zard synthesis of [2,2,2]-octadiene-fused pyrroles

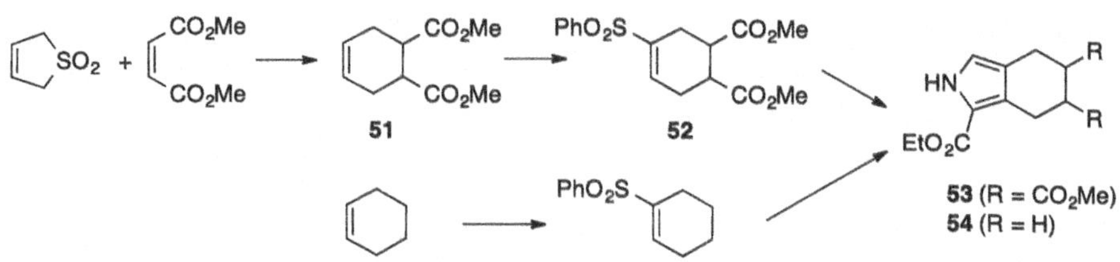

Fig. 3.18 Barton–Zard pyrrole synthesis using cyclohexene

more than 80% yield. These pyrroles were designed to liberate ethylene on pyrolysis by retro Diels–Alder reaction to generate benzo units in the final step of porphyrin synthesis.

Sulfolene as an inexpensive butadiene precursor was conveniently employed by Finikova, Vinogradov, and coworkers in the Diels–Alder reaction to give dicarbomethoxycyclohexene **51**. This Diels–Alder adduct was then converted to phenylsulfonyl cyclohexene **52** and subsequently to the tetrahydroisoindole **53** by Barton–Zard reactions (Finikova et al. 2001). When cyclohexene was taken as a starting material, the α-free pyrrole **54** was also obtained in good yield (Fig. 3.18) (Finikova et al. 2004). These cyclohexene-fused pyrrole units are readily converted into the benzene-fused pyrrole units by oxidation under mild reaction conditions especially after porphyrin ring formation.

3.3.2 *Benzoporphyrins*

Tetrabenzoporphyrins where benzene rings are fused at the pyrrole β-positions are chemically stable, and their extended π-conjugation stabilizes their cationic states, which changes the basicity and redox potentials. In addition to these features, the

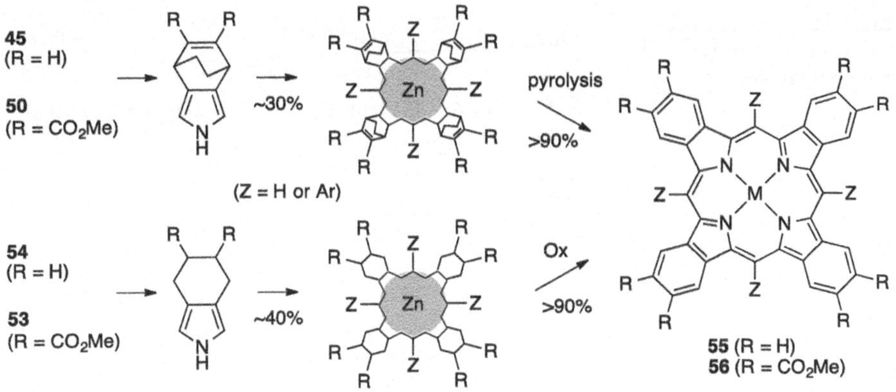

Fig. 3.19 Synthesis of benzoporphyrins

red-shifted UV–vis absorption spectra of benzoporphyrins are suitable as optoelectronic materials. The above-mentioned pyrroles **45**, **50**, **53**, and **54** are important building blocks for benzoporphyrins. Since the highly planar benzoporphyrins tend to form a face-to-face stacked array leading to insoluble materials, purification is usually difficult. The precursor porphyrins substituted with aliphatic groups at the pyrrole β-positions were prepared in 30–40% yield from the corresponding pyrroles by condensation with formaldehyde and aryl aldehydes. These porphyrins can be dissolved in organic solvents and circumvent problems associated with low solubility of benzoporphyrins (Fig. 3.19) (Carvalho et al. 2013). High temperature at ~200 °C is needed to induce retro Diels–Alder reaction in the pyrolysis as the final step to tetrabenzoporphyrins **55** and **56** starting from **45** and **50** (Ito et al. 1998, 2001). These porphyrins were alternatively obtained in good yields under much milder reaction conditions by oxidative aromatization starting from **53** and **54**, but oxidizing agents like metal salt, DDQ, or acid are needed (Finikova et al. 2001, 2004). The retro Diels–Alder method allows preparation of device assemblies of benzoporphyrins by introducing precursor porphyrins into the device assemblies before pyrolysis.

In the UV–vis spectrum of *meso*-unsubstituted benzoporphyrin **56** (Z = H), the Soret band at 454 nm and the Q band at 648 nm are red-shifted by 49 nm and 84 nm, respectively, relative to the bicyclo[2.2.2]octadiene-fused precursor (Table 3.2). In particular, this Q band of **56** (Z = H) is very strong with its absorption coefficient being in the order of 10^5 M^{-1} cm^{-1} (Ito et al. 2001). UV–vis absorption bands of *meso*-tetraphenylbenzoporphyrin **56** (Z = Ph) are further red-shifted by ~30 nm relative to **56** (Z = H).

Uno and coworkers prepared bicyclo[2.2.2]octadiene-fused dimeric porphyrin **57** in 21% yield using double 3+1 condensation of tripyrrane and the dipyrrole derived from **47** (Fig. 3.20). Subsequent retro Diels–Alder reaction afforded benzene-fused dimeric porphyrin **58** (Uno et al. 2007). The strong Soret band of **58** (M = 2Zn) at 475 nm is significantly red-shifted from those (399 nm and 414 nm) of **57** (M = 2Zn) (Table 3.2). The Q band of **58** at 638 nm is as intense as the Soret band and

Table 3.2 Selected major UV–vis absorption bands of benzoporphyrins

Porphyrin	λ_{max} nm (log ε)		solvent
	Soret bands	Q bands	
55 (M = Zn, Z = Ph)	463 (6.44)	609 (5.09), 652 (5.68)	$CHCl_3$
56 (M = Zn, Z = Ph)	485 (5.65)	621 (4.35), 667 (4.85)	$CHCl_3$
56 (M = Zn, Z = H)	454 (5.62)	597 (4.28), 648 (5.06)	$CHCl_3$
57 (M = 2Zn)	399 (5.65), 414 (5.65)	533 (4.40), 574 (4.74)	$CHCl_3$
58 (M = 2Zn)	394 (4.98), 475 (5.36)	621 (4.78), 638 (5.32)	C_5H_5N
60 (M = 5Zn)	404 (5.82), 423 (5.61)	534 (4.83), 573 (4.90)	$CHCl_3$
61 (M = 5Zn)	415 (5.56), 498 (5.32)	721 (4.80), 763 (5.61)	C_5H_5N

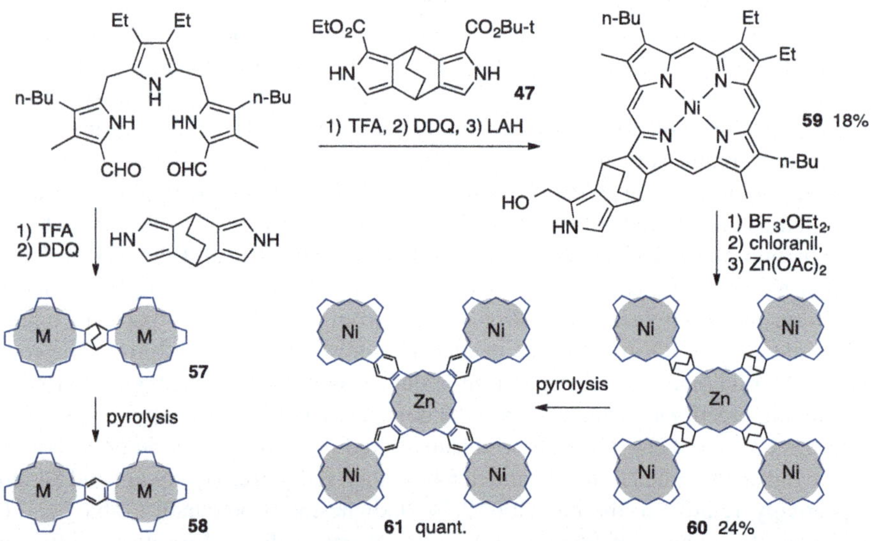

Fig. 3.20 Benzene-fused multiporphyrins

red-shifted by 64 nm in comparison with weak Q bands of **57** at 533 nm, and 574 nm. The single 3+1 condensation of tripyrrane and dipyrrole **47** followed by reduction of the ester group gave bicyclo[2.2.2]octadiene-fused porphyrin **59** in 18% yield. The pyrrole-2-carbinol part of **59** reacted under ordinary reaction conditions of porphyrin synthesis to give 24% yield of the pentameric porphyrin **60** where four porphyrin rings are connected to the pyrrole β-positions of the central porphyrin ring. Pyrolysis of **60** proceeded almost quantitatively to generate cruciform porphyrin **61** that has overall π-conjugation among 5 porphyrin units (Uoyama et al. 2010).

The UV–vis spectrum of **61** (M = 5Zn) showed strong Soret bands at 415 nm and 498 nm. These Soret bands are red-shifted from those (404 nm and 423 nm) of **60** (M = 5Zn) (Table 3.2). The Q band of **61** (M = 5Zn) at 763 nm is as intense as the Soret band and far red-shifted in comparison with weak Q bands of **60** (M = 5Zn) at 534 nm, and 573 nm and also with the Q band (638 nm) of the benzene-fused dimeric Zn porphyrin **58** (M = 2Zn), whereas the UV–vis spectra of bicyclo[2.2.2]octadiene-fused oligomeric Zn porphyrins **57** and **60** are not so different from that of monomeric Zn porphyrin.

Two-photon absorption cross section provides information on how far the π-conjugation is expanding in the molecular systems. The 2PA cross section values of the benzene-fused oligomeric Zn porphyrin **58** (M = 2Zn) and **61** (M = 5Zn) measured by z-scan method were reported as 3000 GM (λ_{ex} 1275 nm) and 3900 GM (λ_{ex} 1500 nm), respectively (Uoyama et al. 2010). These values are much larger than those (<100 GM) of monomeric porphyrins, but they are less than those (10^4–10^5 GM) of *meso,meso*-butadiyne-linked multiporphyrins like **33**$(SiR_3)_2$ and **34**$(SiR_3)_2$. It was suggested that electronic interaction between porphyrin π-systems by way of the pyrrole β-positions is not so effective as by way of the *meso*-positions.

Smith and coworkers prepared the cruciform porphyrin having four *meso*-tetraphenylporphyrin units instead of *meso*-unsubstituted porphyrin units of **61** in a simple procedure (Jaquinod et al. 1998). The phenylsulfonyl group was introduced into sulfolene itself through phenylsulfenyl chloride addition, MCPBA oxidation, and HCl elimination and then subjected to the Barton–Zard pyrrole synthesis. Pyrolysis of the resulting sulfolene-fused pyrrole **62** at 240 °C in the presence of *meso*-tetraphenylporphyrin (TPP) produced pyrroloporphyrin **64** in ca. 30% yield after DDQ oxidation of the initially formed pyrrolochlorin **63** (Fig. 3.21). The pyrroloporphyrin **64** was reduced to the corresponding carbinol that is an analogue of **59** and then converted into the cruciform porphyrin after DDQ oxidation.

Smith and coworkers also used *meso*-tetraphenylporphyrin (TPP) as a starting material in the synthesis of directly β-fused porphyrin trimer **66** (Fig. 3.22) (Paolesse et al. 2000). Readily available mononitroTPPNi(II) was regarded as a substrate for the Barton–Zard synthesis and it actually gave pyrroloporphyrin **65** after decarboxylation. It is noteworthy that β-nitroporphyrins react in a similar

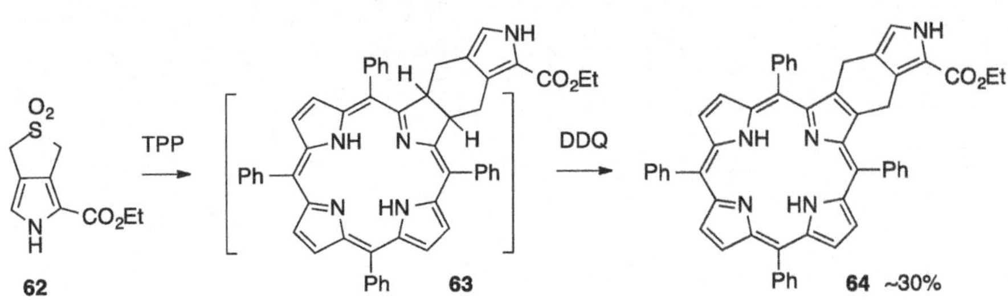

Fig. 3.21 TPP-derived building block of cruciform porphyrin

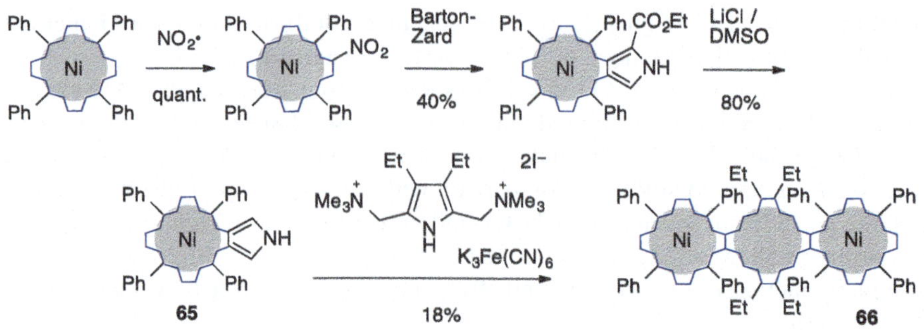

Fig. 3.22 Directly linked trimeric porphyrin

fashion to nitroalkene. Bisquarternary ammonium salt derived from 2,5-bis(N,N-dimethylaminomethyl)pyrrole was allowed to react with **65** under neutral conditions in the presence of oxidizing agent to afford 18% yield of the target porphyrin **66**. This β-fused trimer showed intense UV–vis bands at 486 nm (log $\varepsilon = 5.3$) and 716 nm (log $\varepsilon = 5.2$).

Kräutler and coworkers obtained sulfolene-fused porphyrin **67** in 44% yield by ordinary condensation reaction of arylaldehyde and solfolene-fused pyrrole that is derived from **62** (Fig. 3.23). Pyrolysis of **67** at 140 °C in the presence of 1,4-benzoquinone induced [4+2] cycloaddition and subsequent DDQ oxidation and metal insertion generated tetrabenzoporphyrin **68** with further π-conjugation to the quinone moieties (Banala et al. 2009). Fusion of 1,4-naphthoquinone instead of benzene at the pyrrole β-positions of *meso*-tetraarylporphyrin caused red-shifts of the Soret band by ~90 nm and of the Q band by ~70 nm in their UV–vis spectra (the major absorptions at 555 nm and 725 nm) of **68** in chlorinated solvents. **68** is a black-colored porphyrin with the lowest absorption coefficient (8700 M^{-1} cm^{-1}) over the whole UV–vis range (230–770 nm) seen as an absorption valley at 634 nm.

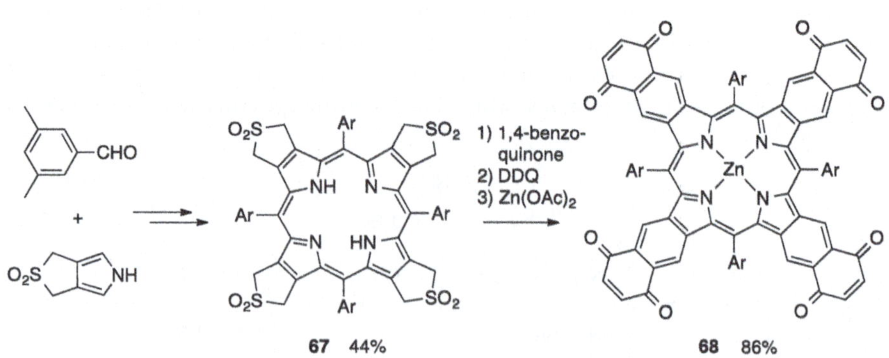

Fig. 3.23 A black-colored 1,4-naphthoquinone-fused porphyrin

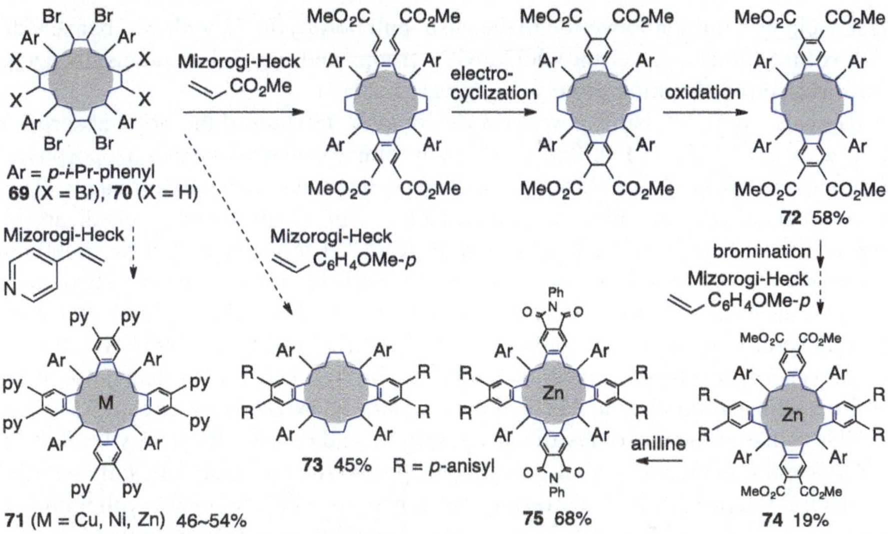

Fig. 3.24 Water-soluble porphyrin and push–pull porphyrin

Organometallic synthetic methods were applied in synthesizing benzoporhyrins by way of brominated *meso*-tetraarylporphyrins **69** and **70** that were obtained by using N-bromosuccinimide (NBS). Tetrabromination occurs regioselectively at the opposite pyrrole rings of the *meso*-tetraarylporphyrin free base in ~60% yield and octabromination of the Ni complexes proceeds in more than 80% yield (Chumakov et al. 2009). Water-soluble porphyrins are useful especially in the biomedical application, and *meso*-tetraarylporphyrins where four *meso*-aryl groups are pyridinium or aryl sulfonate have been frequently studied. Highly water-soluble porphyrins with eight hydrophilic groups at all the pyrrole β-positions have been synthesized recently through one-pot three-step reactions of Cu, Ni, and Zn complexes of **69** and vinylpyridines using Pd catalyst in 46–54% yields (Fig. 3.24) (Jiang et al. 2012). These metal complexes of *meso*-tetraaryltetrabenzoporphyrin **71** (M = Zn) can be protonated at the eight pyridyl units, which caused red-shifts by ~30 nm of the Soret band to 525 nm and the Q band to 708 nm. The Q band at 684 nm of the protonated complex of **71** (M = Ni) showed unusually high intensity as a monomeric metalloporphyrin. This octacation of **71** could be dissolved in distilled water with concentration more than 15 mM. When the tetrabromoporphyrin **70** was allowed to react with methyl acrylate under similar reaction conditions, the one-pot reactions of Mizorogi-Heck coupling, 6π-electrocyclization, and oxidative aromatization occurred to give the dibenzoporphyrin **72** with four methyl ester units in 58% yield. **72** was converted into the Zn(II) tetrabromodibenzoporphyrin in 42% yield and then subjected to the Pd-catalyzed coupling reaction with 4-methoxystyrene under the same Mizorogi-Heck conditions. The Zn(II) tetrabenzoporphyrin **74** formed in

45% yield was further converted to the push–pull porphyrin **75** with two benzenedicarboximide units as electron-withdrawing groups and two 3,4-di-*p*-anisylbenzene units as electron-donating groups (Kumar et al. 2018).

The push–pull tetrabenzoporphyrin Zn complex **75** showed the Soret absorption band at 507 nm that is red-shifted by 51 nm in comparison with the dibenzoporphyrin Zn complex of **73** with two 3,4-di-*p*-anisylbenzo units. It is also red-shifted by 30 nm in comparison with the dibenzoporphyrin Zn complex with two benzenedicarboximide unit derived from **72**. The Q band of **75** at 721 nm appeared at by ca. 80 nm longer wavelength than those push-only or pull-only dibenzoporphyrins. Theoretical study on the electronic structure indicated that the HOMO–LUMO gaps of the push-only and pull-only dibenzoporphyrins are not so different although both HOMO and LUMO energy levels of the push-only dibenzoporphyrin are higher than those of the pull-only dibenzoporphyrin. It is of interest that the HOMO energy level of **75** is very similar to that of the push-only dibenzoporphyrin and the LUMO energy level of **75** is very similar to that of the pull-only dibenzoporphyrin, which significantly reduced the HOMO–LUMO gap. This result was consistent with the cyclic voltammetric analysis; 1.75 V for **75** is by 0.25 V smaller than those dibenzoporphyrins.

3.3.3 Reactions of Porphyrins at Pyrrole β-Positions

Since *meso*-tetraarylporphyrins are readily available, robust, and soluble in organic solvents, great effort has so far been directed to structural modification at pyrrole β-positions. Bringmann and coworkers investigated Pd-catalyzed borylation of 2-bromo-*meso*-tetraphenylporphyrin **76** and obtained β-dioxaborolanylporphyrin **77** in 70% yield using toluene–water two phase system (Fig. 3.25) (Bringmann et al. 2008). If this reaction was run in DMF, carbopalladation on the adjacent *meso*-phenyl group occurred to give almost exclusive formation of indene-fused porphyrin **78** without participation of bis(pinacolato)diboron ($B_2(pin)_2$) in the reaction. A closely related cyclopentadiene-fused porphyrins **80** was reported by Osuka's group through Pd-catalyzed reaction between *meso*-bromoporphyrin **79** and alkynes (Sahoo et al. 2006). The Ni and Cu complexes **80** showed split Soret bands; one at 380–390 nm and the other at 480–490 nm. The additional π-conjugation in **80** (M = Ni) decreased electrochemical HOMO–LUMO gap ($\Delta E = 1.66$ V) relative to that ($\Delta E = 2.29$ V) of *trans*-A_2 type diarylporphyrin Ni complex. Once β-borylated porphyrin **77** was available, it was subjected to the Pd-catalyzed Suzuki–Miyaura cross coupling with **76** leading to β,β-linked diporphyrin **81** in 73% yield. Some bimetallic complexes of **81** could be resolved into enantiomers by HPLC on a chiral phase. The rotational stability at the β,β-axis depends on the metal but racemization of the optically resolved bisZn complex was not seen at room temperature.

Osuka, Shinokubo, and coworkers reported that the pyrrole β-position adjacent to the unsubstituted *meso*-position could be directly borylated (Hata et al. 2005). When the $[Ir(COD)OMe]_2$-catalyzed direct borylation of aromatic compounds by $B_2(pin)_2$ was applied to 5,15-diarylporphyrin and 5,10,15-triarylporphyrin, the

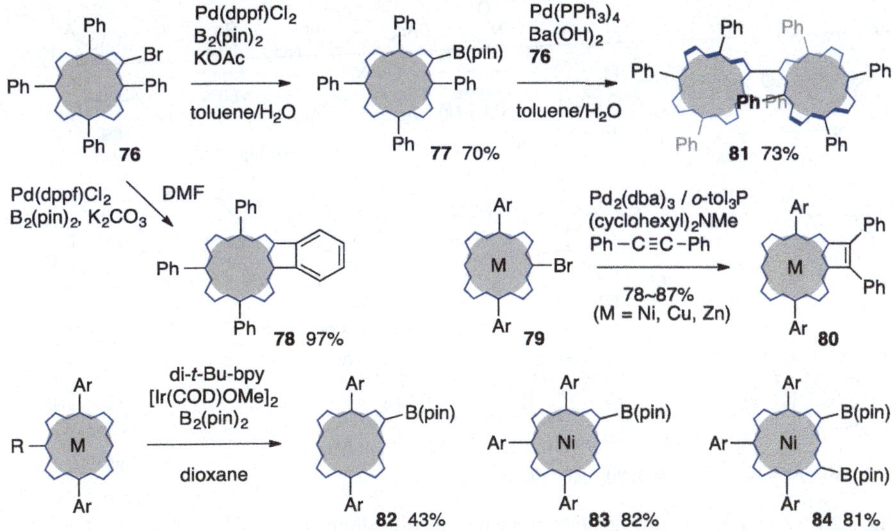

Fig. 3.25 Borylation of porphyrins

3,3,4,4-tetramethyldioxaborolanyl group was introduced selectively through the C-H activation at the pyrrole β-position that is sterically less congested than the *meso*-position. By changing the $B_2(pin)_2$/porphyrin ratio, monoborylated and diborylated porphyrins (**82**, **83**, **84**) were obtained in 43–82% yield. These borylated porphyrins were conveniently employed for various Pd-catalyzed cross-coupling reactions.

The pyrrole β-position of *meso*-tetraarylporphyrins is functionalized by oxidation reactions (Fukui et al. 2017). Brückner, Rettig, and Dolphin reported that OsO_4-mediated *cis*-dihydroxylation of $TPPH_2$ and TPPNi(II) gave 2,3-dihydroxy-*meso*-tetraarylchlorin **87** in 49% and 72% yield, respectively (Fig. 3.26) (Brückner et al. 1998). These compounds were further oxidized by DDQ to generate 2,3-dioxochlorins **88** in 73 and 65% yield (Daniell et al. 2003). **88** was alternatively synthesized by Crossley and coworkers by oxidizing β-amino- and β-hydroxyporphyrin **85** and **86** that were prepared conveniently from β-nitroporphyrin (Crossley and King 1984; Crossley et al. 1991). The Crossley's group utilized this 2,3-dioxochlorin **88** as a key compound in the development of various porphyrin oligomers of extended π-conjugation. The dimeric porphyrin **90** ($n = 0$) that was prepared by condensation of **88** with 1,2,4,5-tetraaminobenzene was converted to the 12,12′,13,13′-tetraoxobischlorin and then reacted with 1,2,4,5-tetraaminobenzene and **88** at both ends. Thus formed tetrameric porphyrin **90** ($n = 2$) showed electrochemical HOMO–LUMO gap of 0.8 eV (Crossley and Burn 1991). Brückner and coworkers obtained a pyrrole-modified porphyrin called indaphyrin **91** from 2,3-dihydroxy-*meso*-tetraarylchlorin **87** under acidic aerobic conditions (McCarthy et al. 2004). That is, a secochlorin bisaldehyde intermediate formed by the oxidative

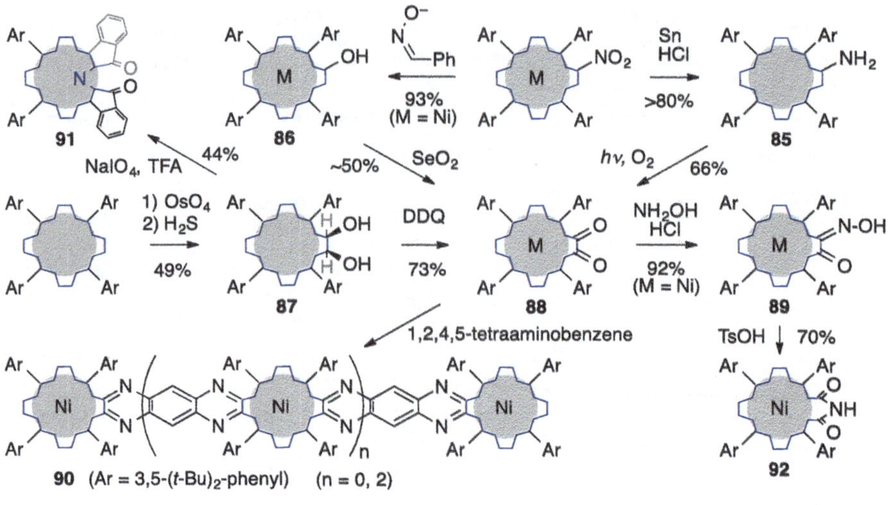

Fig. 3.26 Oxidative functionalization at the pyrrole β-position

C–C bond cleavage of the vicinal diol part of **87** with sodium periodate undergoes Friedel–Crafts type aromatic substitution at the *meso*-phenyl *ortho*-position promoted by trifluoroacetic acid (TFA).

Indaphyrin **91** and its metal complexes have a helical conformation in order to relieve electronic repulsion of carbonyl oxygens to one another and each resolved enantiomer was quite stable with a racemization barrier greater than 113 kJ/mol at 25 °C. UV–Vis spectrum of **91** and its metal complexes show broad absorptions in the region between 400 and 600 nm. Free base **91** has split Soret-like bands at 419 nm and 554 nm with $\varepsilon = {\sim}4 \times 10^4$ (Samankumara et al. 2015; Götz et al. 2015). α-Keto oximes **89** formed from metal complexes **88** (M = Ni, Pd, Pt) of 2,3-dioxochlorins readily underwent Beckmann rearrangement with *p*-toluenesulfonic acid (TsOH), in which case the pyrrole-3-oxo-4-oxime part was converted into 6-membered pyrazine imide in the product **92** (Akhigbe and Brückner 2013).

3.4 π-Extended Porphyrins

3.4.1 *Electrophilic Substitution with* meso-*Aryl Groups*

π-Extension of porphyrin core causes great influence on the absorption bands and redox potentials, which extends the applicability of porphyrin compounds to various scientific fields. Porphyrins with *meso*-acetylene groups (appeared in 3.1.4), benzoporphyrin derivatives (appeared in 3.2.2), and linear oligomers **66** and **90** are examples of this category. π-Conjugation between *meso*-aryl groups and porphyrin core could be enabled by simple procedure as already mentioned in the compounds **78**, **80**, and **91**.

Callot and coworkers used Friedel–Crafts acylation as a key reaction in connecting *meso*-aryl *ortho*-positions and pyrrole β-positions (Fig. 3.27) (Richeter et al. 2003). *cis*-A_2B_2 type *meso*-tetraarylporphyrin **93** was obtained in 5.1% yield from the porphyrin products mixture in the mixed condensation of pyrrole, *o*-methoxycarbonylbenzaldehyde, and 3,5-di-*t*-butylbenzaldehyde in the ratio of 2:1:1. The ester group of **93** was converted into acid chloride and then subjected to Friedel–Crafts reaction conditions. The major product **94** obtained in 44% had bistetralone-fused porphyrin structure. The Soret band at 516 nm (log $\varepsilon = 4.9$) and the Q band at 738 nm (log $\varepsilon = 4.1$), 826 nm (log $\varepsilon = 4.1$) in the UV–vis spectrum were remarkably red-shifted. A mixture of 2,12- and 2,13-dibromoTPPH_2 **95** obtained by regioselective dibromination of TPPH_2 was reacted with CuCN to give di-cyanoTPPCu **96** effectively by nucleophilic aromatic substitution. Hydrolysis of the cyano group followed by a similar Friedel–Crafts procedure afforded differently bistetralone-fused porphyrins **97** and **98** in 27% and 30% yield, respectively, after chromatographic separation.

Brückner's group also showed that if α-keto oxime free base **89** was reacted with TsOH, the Beckmann rearrangement leading to **92** was suppressed and the adjacent *meso*-phenyl group participated to generate a quinoline ring leading to mono- and bisquinoline-fused porphyrins **100** and **103** (Fig. 3.28) (Akhigbe et al. 2015). These pyrrole-modified porphyrins were alternatively synthesized in better yields through DDQ oxidation of the oxime moiety of **89** and **100** by way of mono- and bisquinoline N-oxide **99** and **101**. These quinoline-fused porphyrin free bases show Q bands in the range of 730–780 nm. Ruppert and coworkers reported that the bisquinoline-fused

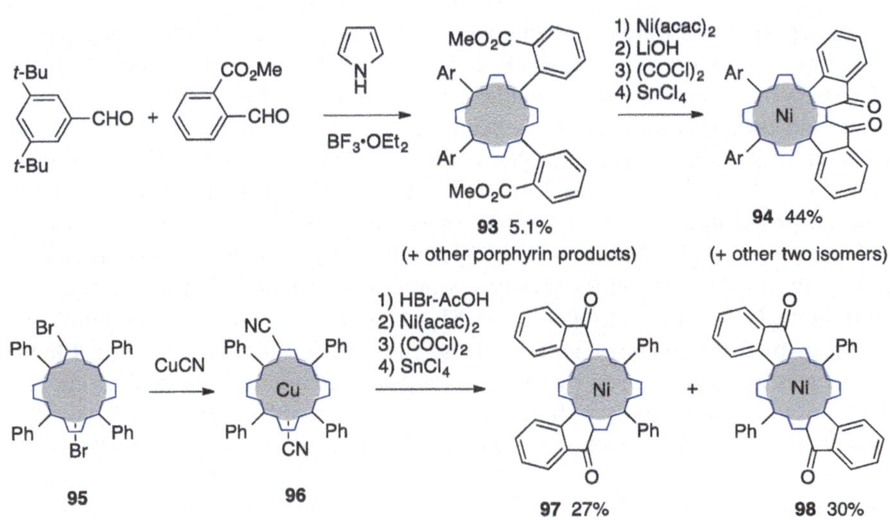

Fig. 3.27 Friedel–Crafts acylation between the *meso*-aryl *o*-position and the pyrrole β-position

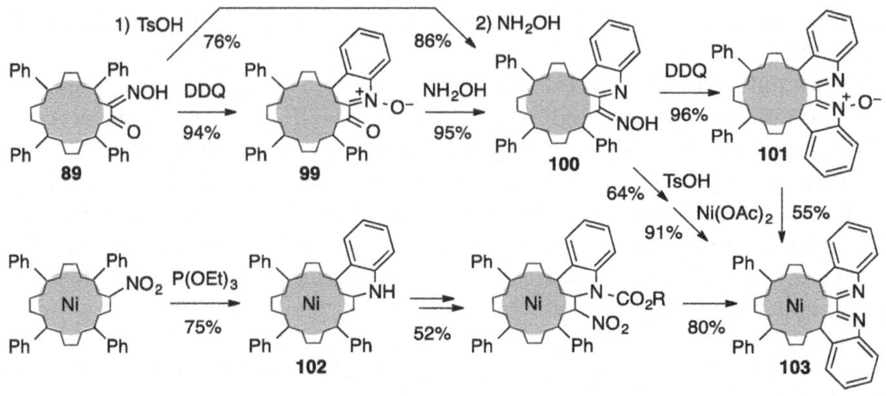

Fig. 3.28 Pyrrole-modified porphyrins with quinoline-fusion

porphyrin **103** was regarded to have an external bidentate coordination site and can be obtained in a shorter synthetic route starting from readily available β-nitroporphyrin (Jeandon and Ruppert 2011). *meso*-Phenyl C-H insertion of nitrene generated from nitro group with triethyl phosphite afforded dihydroquinoline-fused porphyrin **102** effectively. Protection of the NH group and nitration of **102** followed by just heating produced **103** in good yield.

3.4.2 Oxidative Coupling of β-Aminoporphyrins

β-Aminoporphyrins were readily prepared and showed interesting reactions that are owing to their nature of being easily oxidized. β-Amino-*meso*-tetraarylporphyrins **85** were conveniently applied to synthesizing dimeric porphyrins (Fig. 3.29). Bringmann and coworkers intended to obtain β,β-linked diporphyrin **104** by Ullmann coupling of 2-amino-3-bromoporphyrin **105**, but the product obtained in 78% yield turned out to be pyrazine-fused diporphyrin (M = Ni) **106** (Bruhn et al. 2014). The β,β-linked diporphyrin **104** was formed from **85** (M = Ni) in 94% yield by oxidizing with $AgPF_6$. **104** is intrinsically chiral diporphyrin, and its conformational change by rotation around the β,β-axis is inhibited even at 100 °C. In comparison with the bisNi complex of diporphyrin **81**, stereochemical stability is significantly increased by the presence of amino groups. The pyrazine-fused diporphyrin **106** (M = Zn) was also synthesized by Mandoj, Paolesse, and coworkers by reacting 2,3-diamino-*meso*-tetraphenylporphyrin **107** with 2,3-dioxochlorin **88** or diethyl oxalate (Mandoj et al. 2013). Shinokubo and coworkers recently investigated DDQ oxidation of β-aminoporphyrins **85** (M = Ni) with bulky *meso*-tetraaryl groups and found formation of the pyrazine-fused diporphyrin **108** in high yields (Ito et al. 2015). Although *meso*-tetraphenyl derivatives **106** take C_{2h} symmetric conformation, *meso*-tetramesityl derivative **108** has twisted π-conjugation system of D_2 symmetry. The

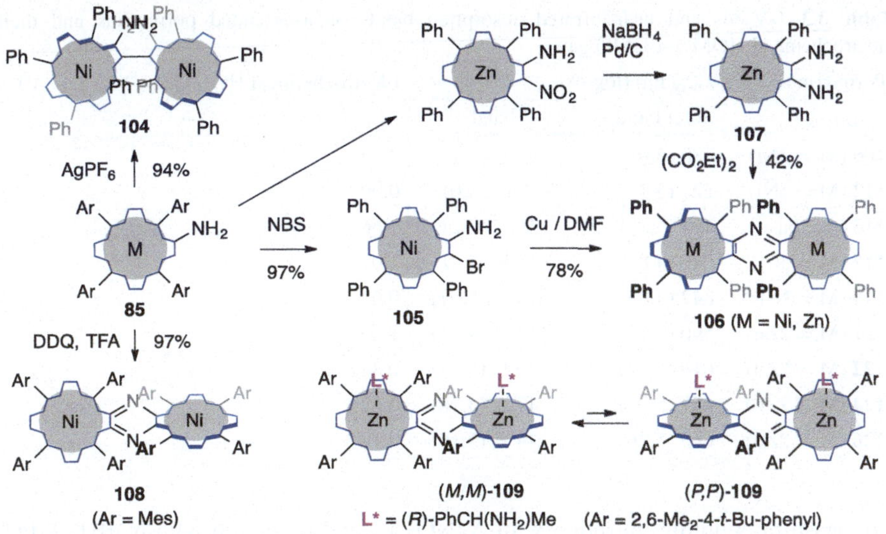

Fig. 3.29 Oxidative dimerization of aminoporphyrins

bisZn complex **109** is interchanging rapidly between a (*M*, *M*)-form and a (*P*, *P*)-form and binding (*R*)-1-phenylethylamine induced helical sense bias in favor of the (*M*, *M*)-form with 65:35 diastereoselectivity (Ito et al. 2016).

Fujimoto and Osuka demonstrated that π-conjugation between two porphyrin cores is much more enhanced by 1,5-naphthyridine-fusion in comparison with the pyrazine-fusion seen in **106** (Fujimoto and Osuka 2018). The 1,5-naphthyridine-fused diporphyrin **112** was obtained in 84% yield by Pt(II)-catalyzed cyclization of bisNi(II) complex of β-to-β ethynylene-bridged *meso*-aminoporphyrin dimer **111** followed by oxidation with PbO_2 (Fig. 3.30). Ni(II) complex **111** was prepared starting from β-borylporphyrin **83** by NIS-CuI iodination, (Fujimoto et al. 2014)

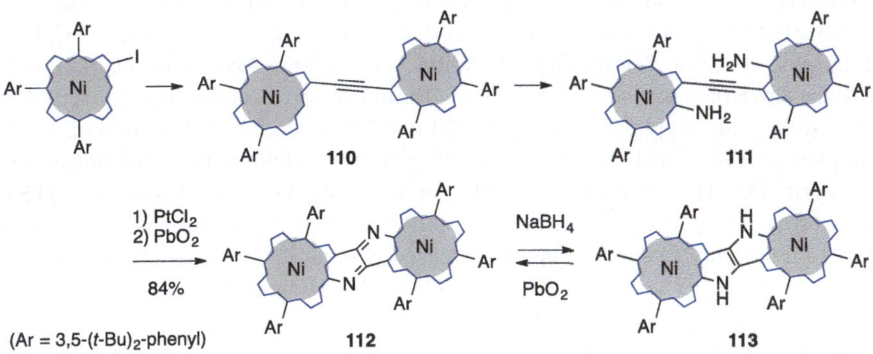

Fig. 3.30 1,5-Naphthyridine-fused diporphyrin

Table 3.3 UV–vis and near-infrared absorption bands of π-extended porphyrins and their electrochemical HOMO–LUMO gap

Porphyrin	λ_{max} nm (log ε)		Electrochemical HOMO–LUMO gap (V)
	Soret band	Q band	
106 (M = 2Ni)	480 (5.0)	619 (4.1)	2.16
112 (M = 2Ni)	521 (5.1)	1011 (5.0)	0.99
115 (free base)	474 (5.30)	698 (4.82)	1.66
117 (M = Ni)	648 (4.8)	1136 (4.7)	1.01
119 (M = Ni)	842 (4.85)	1417 (5.07)	0.61
121 (M = 2Zn)	590 (5.19)	1141 (4.71)	1.09
122 (M = 2Zn)	618 (5.24)	1323 (5.05)	0.84
124 (M = 2Zn)	657 (5.05)	1322 (5.16)	0.62
126 (M = 2Zn)	665 (5.26)	1495 (5.16)	0.77

Sonogashira coupling, nitration with $AgNO_2$-I_2, and reduction with $NaBH_4$-Pd/C. The UV–vis major band of **106** (M = Zn) appears at 492 nm (log ε = 5.06) and their Q bands are at shorter wavelength region than 650 nm. Its electrochemical HOMO–LUMO gap is 2.16 V that is only slightly smaller than 2.24 V of TPPZn. In contrast, the HOMO–LUMO gap (0.99 V) of **112** measured by cyclic voltammetry is remarkably smaller than the reduced form **113** (1.8 V). The major absorption bands of **112** at 521 nm (log ε = 5.1) and 1011 nm (log ε = 5.0) are far red-shifted in comparison with those (496 and 663 nm) of **113** (Table 3.3).

3.4.3 Oxidative Fusion of Porphyrin Periphery

Oxidative coupling of aromatic rings with dehydrogenative C–C bond formation is a versatile method for extending π-conjugation (Grzybowski et al. 2013). This Scholl type oxidation has been applied to porphyrin compounds (Lewtaka and Gryko 2012). Stepien and coworkers introduced phenanthrene units at the periphery of porphyrin using Scholl type oxidation (Fig. 3.31) (Mysłiwiec et al. 2012). 3,4-Diarylpyrrole that was prepared by Barton–Zard reaction was converted to octakis(3,4-dialkoxyphenyl)porphyrin **114** in 62% yield. Scholl oxidation of the Zn complex of **114** with $FeCl_3$ proceeded to give 97% yield of the tetraphenanthro-porphyrin **115**. This efficient transformation to highly π-extended porphyrin **115** is noteworthy. The π-extension caused red-shift of the Soret band from 427 to 474 nm, and the maximum Q band intensity increased from 4.38 (log ε at 522 nm) to 4.90 (log ε at 650 nm). The electrochemical HOMO–LUMO gap decreased from 2.27 to 1.66 V upon going from **114** to **115**.

More effective π-extension is enabled by oxidative aromatic coupling between *meso*-aryl groups and porphyrin core (Fig. 3.32). Osuka and coworkers obtained

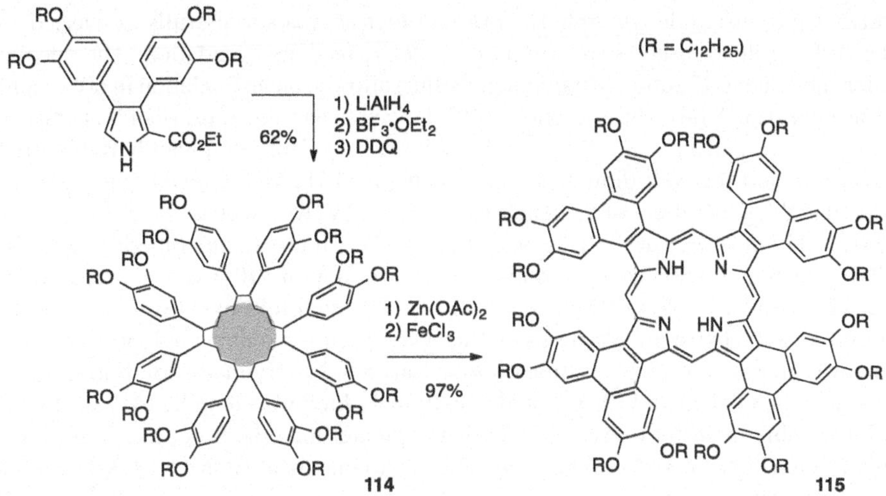

Fig. 3.31 Scholl oxidation to tetraphenanthroporphyrin

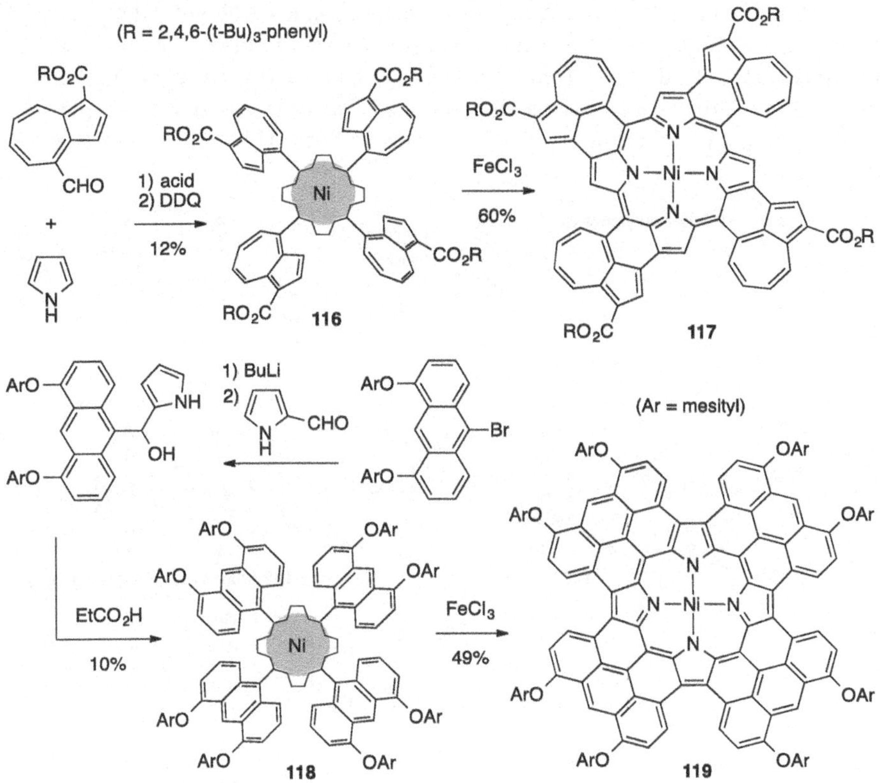

Fig. 3.32 Quadruply azulene- and anthracene-fused porphyrins

Ni(II) *meso*-tetraazulenylporphyrin **116** and then it was successfully converted to the fully azulene-fused porphyrin **117** by $FeCl_3$-mediated oxidation. The special ester substituents (2,4,6-tri-*t*-butylphenyl ester) allowed clean oxidation in 60% yield (Kurotobi et al. 2006). The Ni complex **117** showed strong absorption bands at 684 nm (log $\varepsilon = 4.8$) and 1136 nm (log $\varepsilon = 4.7$) (Table 3.3). This significantly red-shifted spectrum, and the well diminished electrochemical HOMO–LUMO gap (1.01 V) illustrate the great effect of azulene fusion. The 2PA cross section value ($\sigma = 7170$ GM) of **117** measured at 1380 nm excitation was very high as a monomeric porphyrin (Pawlicki et al. 2009). The quadruply fused Ni porphyrin **119** was also prepared by Anderson and coworkers (Davis et al. 2011). The eightfold oxidative ring closure occurred at the pyrrole-β-positions of Ni(II) *meso*-tetra-9-anthracenylporphyrin **118** that was prepared by condensation of 9-anthracenyl-2-pyrrylmethanol. It is remarkable that the Q band of **119** appeared at extremely long wavelength (1417 nm) with a remarkable intensity (log $\varepsilon = 5.07$) that is greater than the Soret-like absorption at 842 nm (log $\varepsilon = 4.85$) (Table 3.3). The electrochemical HOMO–LUMO gap of 0.61 V was very small.

Scholl type oxidation was applied to introduce direct triple bridges between two porphyrin cores of *meso*-triarylporphyrins (Fig. 3.33). Thompson and coworkers oxidized 5,15-diaryl-10-pyrenylporphyrin Zn(II) complex **120** with DDQ-$Sc(OTf)_3$ to give *β,β*-, *meso,meso*-, *β',β'*-triply fused diporphyrin **121** (Diev et al. 2010). This diporphyrin was further oxidized with $FeCl_3$ to cause oxidative coupling between *meso*-pyrenyl groups and pyrrole β-positions. The fully fused Zn complex **122** showed the major absorptions at 618 nm (log $\varepsilon = 5.24$) and 1323 nm (log $\varepsilon =$

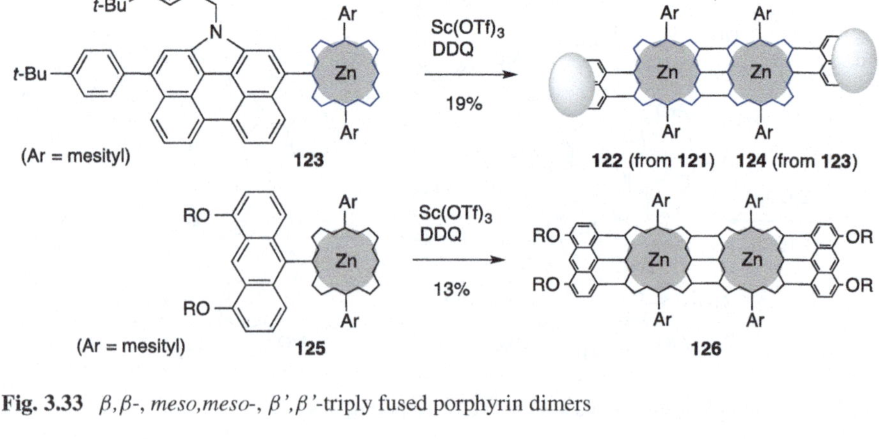

Fig. 3.33 *β,β*-, *meso,meso*-, *β',β'*-triply fused porphyrin dimers

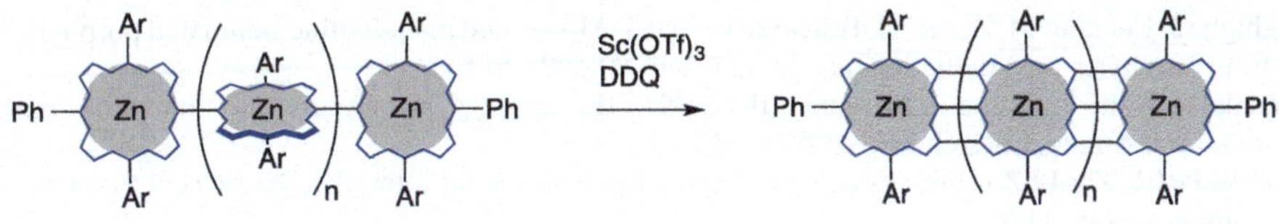

Fig. 3.34 *β*,*β*-, *meso*,*meso*-, *β'*,*β'*-triply fused porphyrin dimers

5.05). This NIR band of **122** is red-shifted by 182 nm and the electrochemical HOMO–LUMO gap (0.84 V) of **122** decreased by 0.25 V in comparison with those of **121** (Table 3.3). Kim and coworkers studied this oxidation chemistry of *meso*-triarylporphyrin **123** having a N-annulated perylene group instead of a pyrenyl group of **120** (Luo et al. 2015). The fully fused diporphyrin **124** showed the major absorptions at 657 nm (log $\varepsilon = 5.05$) and 1322 nm (log $\varepsilon = 5.16$) and a HOMO–LUMO gap of 0.62 V. Anderson's group also reported similar oxidative fusion in the case of *meso*-triarylporphyrin **125** having a 9-anthracenyl group (Davis et al. 2010). The fully fused diporphyrin **126** showed intense absorptions at 665 nm (log $\varepsilon = 5.26$) and 1495 nm (log $\varepsilon = 5.16$) and a HOMO–LUMO gap of 0.77 V.

Ag^+-mediated *meso-meso* coupling of porphyrins and subsequent oxidative fusion leading to *β*,*β*-, *meso*,*meso*-, *β'*,*β'*-triply fused multiporphyrins **128** with DDQ-$Sc(OTf)_3$ were originally developed by Osuka's group (Fig. 3.34) (Tsuda and Osuka 2001). Dimeric, trimeric, and tetrameric porphyrins show NIR absorption bands from 1100 to 1600 nm, and their 2PA cross section values (σ) were measured; 11,900 GM (λ_{ex} 1200 nm) for the dimer, 18,500 GM (λ_{ex} 2300 nm) for the trimer, and 41,200 GM (λ_{ex} 2300 nm) for the tetramer (Ahn et al. 2006; Nakamura et al. 2008).

As seen in many examples noted above, porphyrins show basic reaction behaviors characteristic of aromatic compounds, but unusual reaction behaviors were also observed probably due to macrocycle π-conjugation that stabilizes a radical state. Various organometallic transformations for modification of porphyrin structures have caused great developments in the porphyrin chemistry these days. It is well illustrated by the fact that photophysical and electrochemical properties can be fine-tuned to allow versatile applications as functional materials.

References

Adler AD, Longo FR, Finarelli JD, Goldmacher J, Assour J, Korsakoff L (1967) A simplified synthesis for meso-tetraphenylporphine. J Org Chem 32:476

Ahn TK, Kim KS, Kim DY, Noh SB, Aratani N, Ikeda C, Osuka A, Kim D (2006) Relationship between two-photon absorption and the π-conjugation pathway in porphyrin arrays through dihedral angle control. J Am Chem Soc 128:1700–1704

Akhigbe J, Brückner C (2013) Expansion of a pyrrole in meso-tetra-phenylporphyrin to a pyrazine imide moiety-using a beckmann rearrangement. Eur J Org Chem 3876–3884

Akhigbe J, Luciano M, Zeller M, Brückner C (2015) Mono- and bisquinoline-annulated porphyrins from porphyrin β,β'-dione oximes. J Org Chem 80:499–511

Banala S, Rühl T, Wurst K, Kräutler B (2009) "Blackening" porphyrins by conjugation with quinones. Angew Chem Int Ed 48:599–603

Barton DHR, Zard SZ (1985) A new synthesis of pyrroles from nitroalkenes. J Chem Soc Chem Commun 1098–1100

Bessho T, Zakeeruddin SM, Yeh C-Y, Diau EW-G, Grätzel M (2010) Highly efficient mesoscopic dye-sensitized solar cells based on donor-acceptor-substituted porphyrins. Angew Chem Int Ed 49:6646–6649

Bols PS, Anderson HL (2018) Template-directed synthesis of molecular nanorings and cages. Acc Chem Res 51:2083–2092

Borovkov VV, Lintuluoto JM, Inoue Y (1999) Synthesis of Zn-, Mn-, and Fe-containing mono- and heterometallated ethanediyl-bridged porphyrin dimers. Helv Chim Acta 82:919–934

Borovkov VV, Hembury GA, Inoue Y (2004) Origin, control, and application of supramolecular chirogenesis in bisporphyrin-based systems. Acc Chem Res 37:449–459

Bringmann G, Götz DCG, Gulder TAM, Gehrke TH, Bruhn T, Kupfer T, Radacki K, Braunschweig H, Heckmann A, Lambert C (2008) Axially chiral β,β'-bisporphyrins: synthesis and configurational stability tuned by the central metals. J Am Chem Soc 130:17812–17825

Brückner C, Rettig SJ, Dolphin D (1998) Formation of a meso-tetraphenylsecochlorin and a homoporphyrin with a twist. J Org Chem 63:2094–2098

Bruhn T, Witterauf F, Götz DCG, Grimmer CT, Würtemberger M, Radius U, Bringmann G (2014) C,C- and N,C-coupled dimers of 2-aminotetraphenylporphyrins: regiocontrolled synthesis, spectroscopic properties, and quantum-chemical calculations. Chem Eur J 20:3998–4006

Carvalho CMB, Brocksom TJ, de Oliveira KT (2013) Tetrabenzoporphyrins: synthetic developments and applications, Chem Soc Rev 42:3302–3317

Chumakov DE, Khoroshutin AV, Anisimov AV, Kobrakov KI (2009) Bromination of porphyrins (Review). Chem Heterocycl Comp 45:259–283

Collins HA, Khurana M, Moriyama EH, Mariampillai A, Dahlstedt E, Balaz M, Kuimova MK, Drobizhev M, Yang VXD, Phillips D, Rebane A, Wilson BC, Anderson HL (2008) Blood-vessel closure using photosensitizers engineered for two-photon excitation. Nat Photon 2:420–424

Collman JP, Gagne RR, Reed CA, Halbert TR, Lang G, Robinson WT (1975) Picket fence porphyrins. Synthetic models for oxygen binding hemoproteins. J Am Chem Soc 97:1427–1439

Crossley MJ, Burn PL (1991) An approach to porphyrin-based molecular wires: synthesis of a bis(porphyrin)tetraone and its conversion to a linearly conjugated tetrakisporphyrin system. J Chem Soc Chem Commun 1569–1571

Crossley MJ, King LG (1984) Novel heterocyclic systems from selective oxidation at the β-pyrrolic position of porphyrins. J Chem Soc Chem Commun 920–922

Crossley MJ, Burn PL, Langford SJ, Pyke SM, Stark AG (1991) A new method for the synthesis of porphyrin-α-diones that is applicable to the synthesis of trans-annular extended porphyrin systems. J Chem Soc Chem Commun 1567–1568

Daniell HW, Williams SC, Jenkins HA, Brückner C (2003) Oxidation of meso-tetraphenyl-2,3-dihydroxychlorin: simplified synthesis of β,β'-dioxochlorins. Tetrahedron Lett 44:4045–4049

Davis NKS, Thompson AL, Anderson HL (2010) Bis-anthracene fused porphyrins: synthesis, crystal structure, and near-IR absorption. Org Lett 12:2124–2127

Davis NKS, Thompson AL, Anderson HL (2011) A porphyrin fused to four anthracenes. J Am Chem Soc 133:30–31

Diev VV, Hanson K, Zimmerman JD, Forrest SR, Thompson ME (2010) Fused pyrene–diporphyrins: shifting near-infrared absorption to 1.5 μm and beyond. Angew Chem Int Ed 49:5523–5526

Dijkstra HP, ten Have R, van Leusen AM (1998) A direct synthesis of 2-(trimethylstannyl)pyrroles from Michael acceptors and stannylated tosylmethyl isocyanide. J Org Chem 63:5332–5338

Dolphin D (ed) (1978) The porphyrins. Academic Press, New York

Drobizhev M, Stepanenko Y, Rebane A, Wilson CJ, Screen TEO, Anderson HL (2006) Strong cooperative enhancement of two-photon absorption in double-strand conjugated porphyrin ladder arrays. J Am Chem Soc 128:12432−12433

Finikova O, Cheprakov A, Beletskaya I, Vinogradov S (2001) An expedient synthesis of substituted tetraaryltetrabenzoporphyrins. Chem Commun 261–262

Finikova O, Cheprakov A, Beletskaya I, Carroll P, Vinogradov S (2004) Novel versatile synthesis of substituted tetrabenzoporphyrins. J Org Chem 69:522–535

Fujimoto K, Osuka A (2018) A 1,5-naphthyridine-fused porphyrin dimer: intense nir absorption and facile redox interconversion with its reduced congener. Chem Eur J 24:6530–6533

Fujimoto K, Yorimitsu H, Osuka A (2014) Facile preparation of β-haloporphyrins as useful precursors of β-substituted porphyrins. Org Lett 16:972−975

Fukui N, Fujimoto K, Yorimitsu H, Osuka A (2017) Embedding heteroatoms: an effective approach to create porphyrin-based functional materials. Dalton Trans 46:13322–13341

Götz DCG, Gehrold AC, Dorazio SJ, Daddario P, Samankumara L, Bringmann G, Brückner C, Bruhn T (2015) Indaphyrins and indachlorins: optical and chiroptical properties of a family of helimeric porphyrinoids. Eur J Org Chem 3913–3922

Grzybowski M, Skonieczny K, Butenschçn H, Gryko DT (2013) Comparison of oxidative aromatic coupling and the Scholl reaction. Angew Chem Int Ed 52:9900–9930

Hata H, Shinokubo H, Osuka A (2005) Highly regioselective Ir-Catalyzed β-borylation of porphyrins via C−H bond activation and construction of β−β-linked diporphyrin. J Am Chem Soc 127:8264–8265

Huang X, Rickman BH, Borhan B, Berova N, Nakanishi K (1998) Zinc porphyrin tweezer in host−guest complexation: determination of absolute configurations of diamines, amino acids, and amino alcohols by circular dichroism. J Am Chem Soc 120:6185–6186

Ito S, Murashima T, Uno H, Ono N (1998) A new synthesis of benzoporphyrins using 4,7-dihydro-4,7-ethano-2H-isoindole as a synthon of isoindole. Chem Commun 1661–1662

Ito S, Uno H, Murashima T, Ono N (2001) Synthesis of benzoporphyrins functionalized with octaester groups. Tetrahedron Lett 42:45–47

Ito S, Hiroto S, Lee S, Son M, Hisaki I, Yoshida T, Kim D, Kobayashi N, Shinokubo H (2015) Synthesis of highly twisted and fully π-conjugated porphyrinic oligomers. J Am Chem Soc 137:142–145

Ito S, Hiroto S, Ousaka N, Yashima E, Shinokubo H (2016) Control of conformation and chirality of nonplanar π-conjugated diporphyrins using substituents and axial ligands. Chem Asian J 11:936–942

Jaquinod L, Siri O, Khoury RG (1998) Linear fused oligoporphyrins: potential molecular wires with enhanced electronic communication between bridged metal ions. Chem Commun 1261–1262

Jeandon C, Ruppert R (2011) A porphyrin with two coordination sites: the biquinoline ligand as a new potential external chelate. Eur J Org Chem 4098−4102

Jiang L, Zaenglein RA, Engle JT, Mittal C, Hartley CS, Ziegler CJ, Wang H (2012) Water-soluble ionic benzoporphyrins. Chem Commun 48:6927–6929

Kadish KM, Smith KM, Guilard R (eds) (2000) The porphyrin handbook. Academic Press, San Diego

Kadish KM, Smith KM, Guilard R (eds) (2010) Handbook of porphyrin science. World Scientific, Singapore

Kondratuk DV, Sprafke JK, O'Sullivan MC, Perdigao LMA, Saywell A, Malfois M, O'Shea JN, Beton PH, Thompson AL, Anderson HL (2014) Vernier-templated synthesis, crystal structure, and supramolecular chemistry of a 12-porphyrin nanoring. Chem Eur J 20:12826–12834

Kuimova MK, Botchway SW, Parker AW, Balaz M, Collins HA, Anderson HL, Suhling K, Ogilby PR (2009) Imaging intracellular viscosity of a single cell during photoinduced cell death. Nat Chem 1:69–73

Kumar S, Jiang X, Shan W, Jinadasa RGW, Kadish KM, Wang H (2018) β-Functionalized trans-A2B2 push–pull tetrabenzoporphyrins. Chem Commun 54:5303–5306

Kuroda Y, Kato Y, Higashioji T, Hasegawa J, Kawanami S, Takahashi M, Shiraishi N, Tanabe K, Ogoshi H (1995) Chiral amino acid recognition by a porphyrin-based artificial receptor. J Am Chem Soc 117:10950–10958

Kurotobi K, Kim KS, Noh SB, Kim D, Osuka A (2006) A quadruply azulene-fused porphyrin with intense near-IR absorption and a large two-photon absorption cross section. Angew Chem Int Ed 45:3944–3947

Kurotobi K, Toude Y, Kawamoto K, Fujimori Y, Ito S, Chabera P, Sundström V, Imahori H (2013) Highly asymmetrical porphyrins with enhanced push-pull character for dye-sensitized solar cells. Chem Eur J 19:17075–17081

Lewtaka JP, Gryko DT (2012) Synthesis of π-extended porphyrins via intramolecular oxidative coupling. Chem Commun 48:10069–10086

Lindsey JS (2010) Synthetic routes to meso-patterned porphyrins. Acc Chem Res 43:300–311

Lindsey J (1980) Increased yield of a desired isomer by equilibriums displacement on binding to silica gel, applied to meso-tetrakis(o-aminophenyl)porphyrin. J Org Chem 45:5215

Lindsey JS, Schreiman IC, Hsu HC, Kearney PC, Marguerettaz AM (1987) Rothemund and Adler-Longo reactions revisited: synthesis of tetraphenylporphyrins under equilibrium conditions. J Org Chem 52:827–836

Li X, Tanasova M, Vasileiou C, Borhan B (2008) Fluorinated porphyrin tweezer: a powerful reporter of absolute configuration for erythro and threo diols, amino alcohols, and diamines. J Am Chem Soc 130:1885–1893

Luo J, Lee S, Son M, Zheng B, Huang K-W, Qi Q, Zeng W, Li G, Kim D, Wu J (2015) N-annulated perylene-substituted and fused porphyrin dimers with intense near-infrared one-photon and two-photon absorption. Chem Eur J 21:3708–3715

Mandoj F, Nardis S, Pudi R, Lvova L, Fronczek FR, Smith KM, Prodi L, Genovese D, Paolesse R (2013) β-pyrazino-fused tetrarylporphyrins. Dyes Pigm 99:136–143

Mathew S, Yella A, Gao P, Humphry-Baker R, Curchod BFE, Ashari-Astani N, Tavernelli I, Rothlisberger U, Nazeeruddin MdK, Grätzel M (2014) Dye-sensitized solar cells with 13% efficiency achieved through the molecular engineering of porphyrin sensitizers. Nat Chem 6:242–247

McCarthy JR, Hyland MA, Brückner C (2004) Synthesis of indaphyrins: meso-tetraarylsecochlorin-based porphyrinoids containing direct o-phenyl-to-β-linkages. Org Biomol Chem 2:1484–1491

Meindl A, Plunkett S, Ryan AA, Flanagan KJ, Callaghan S, Senge MO (2017) Comparative synthetic strategies for the generation of 5,10- and 5,15-substituted push-pull porphyrins. Eur J Org Chem 3565–3583

Mikhaylov A, Kondratuk DV, Cnossen A, Anderson HL, Drobizhev M, Rebane A (2016) Cooperative enhancement of two-photon absorption in self-assembled zinc-porphyrin nanostructures. J Phys Chem C 120:11663−11670

Mizuno Y, Aida T, Yamaguchi K (2000) Chirality-memory molecule: crystallographic and spectroscopic studies on dynamic molecular recognition events by fully substituted chiral porphyrins. J Am Chem Soc 122:5278–5285

Mizuno Y, Alam MdA, Tsuda A, Kinbara K, Yamaguchi K, Aida T (2006) Hermaphroditic chirality of a D 2-symmetric saddle-shaped porphyrin in multicomponent spontaneous optical resolution: inclusion cocrystals with double-helical porphyrin arrays. Angew Chem Int Ed 45:3786–3790

Mysłiwiec D, Donnio B, Chmielewski PJ, Heinrich B, Stepien M (2012) Peripherally fused porphyrins via the Scholl reaction: synthesis, self-assembly, and mesomorphism. J Am Chem Soc 134:4822–4833

Nakamura Y, Jang SY, Tanaka T, Aratani N, Lim JM, Kim KS, Kim D, Osuka A (2008) Two-dimensionally extended porphyrin tapes: synthesis and shape-dependent two-photon absorption properties. Chem Eur J 14:8279–8289

Nishino N, Kobata K, Mihara H, Fujimoto T (1992) Efficient preparation of αβαβ-atropisomer of meso-tetra(o-aminophenyl)porphyrin. Chem Lett 1991–1994

Paolesse R, Jaquinod L, Sala FD, Nurco DJ, Prodi L, Montalti M, Natale CD, D'Amico A, Carlo AD, Lugli P, Smith KM (2000) β-Fused oligoporphyrins: a novel approach to a new type of extended aromatic system. J Am Chem Soc 122:11295–11302

Pawlicki M, Collins HA, Denning RG, Anderson HL (2009) Two-photon absorption and the design of two-photon dyes. Angew Chem Int Ed 48:3244–3266

Rao PD, Dhanalekshmi S, Littler BJ, Lindsey JS (2000) Rational syntheses of porphyrins bearing up to four different meso substituents. J Org Chem 65:7323–7344

Ravikanth M, Strachan J-P, Li F, Lindsey JS (1998) Trans-substituted porphyrin building blocks bearing iodo and ethynyl groups for applications in bioorganic and materials chemistry. Tetrahedron 54:7721–7734

Richeter S, Jeandon C, Kyritsakas N, Ruppert R, Callot HJ (2003) Preparation of six isomeric bis-acylporphyrins with chromophores reaching the near-infrared via intramolecular Friedel–Crafts reaction. J Org Chem 68:9200–9208

Rickhaus M, Jentzsch AV, Tejerina L, Grübner I, Jirasek M, Claridge TDW, Anderson HL (2017) Single-acetylene linked porphyrin nanorings. J Am Chem Soc 139:16502–16505

Rothemund P (1935) Formation of porphyrins from pyrrole and aldehydes. J Am Chem Soc 57:2010–2011

Ryan A, Gehrold A, Perusitti R, Pintea M, Fazekas M, Locos OB, Blaikie F, Senge MO (2011) Porphyrin dimers and arrays. Eur J Org Chem 5817–5844

Ryppa C, Senge MO, Hatscher SS, Kleinpeter E, Wacker P, Schilde U, Wiehe A (2005) Synthesis of mono- and disubstituted porphyrins: A- and 5,10-A2-type systems. Chem Eur J 11:3427–3442

Sahoo AK, Mori S, Shinokubo H, Osuka A (2006) Facile peripheral functionalization of porphyrins by Pd-catalyzed [3+2] annulation with alkynes. Angew Chem Int Ed 45:7972–7975

Samankumara LP, Dorazio SJ, Akhigbe J, Li R, Nimthong-Roldán A, Zeller M, Brückner C (2015) Indachlorins: nonplanar indanone-annulated chlorin analogues with panchromatic absorption spectra between 300 and 900 nm. Chem Eur J 21:11118–11128

Senge MO (2011) Stirring the porphyrin alphabet soup-functionalization reactions for porphyrins. Chem Commun 47:1943–1960

Senge MO, Shaker YM, Pintea M, Ryppa C, Hatscher SS, Ryan A, Sergeeva Y (2010) Synthesis of meso-substituted ABCD-type porphyrins by functionalization reactions. Eur J Org Chem 237–258.

Senge MO, Fazekas M, Pintea M, Zawadzka M, Blau WJ (2011) 5,15-A2B2-and 5,15-A2BC-type porphyrins with donor and acceptor groups for use in nonlinear optics and photodynamic therapy. Eur J Org Chem 5797–5816

Sessler JL, Mozaffari A, Johnson MR (1992) 3,4-Diethylpyrrole and 2,3,7,8,12,13,17,18-octaethylporphyrin [Pyrrole, 3,4-diethyl and 21H,23H-porphine, 2,3,7,8,12,13,17,18-octaethyl-]. Org Syn 70:68–72

Sessler JL, Gross Z, Furuta H (eds) (2017) Expanded, contracted, and isomeric porphyrins, a thematic issue. Chem Rev 117:2201–3881

Sprafke JK, Kondratuk DV, Wykes M, Thompson AL, Hoffmann M, Drevinskas R, Chen W-H, Yong CK, Kärnbratt J, Bullock JE, Malfois M, Wasielewski MR, Albinsson B, Herz LM, Zigmantas D, Beljonne D, Anderson HL (2011) Belt-shaped π-systems: relating geometry to electronic structure in a six-porphyrin nanoring. J Am Chem Soc 133:17262–17273

Sternberg ED, Dolphin D (1998) Porphyrin-based photosensitizers for use in photodynamic therapy. Tetrahedron 54:4151–4202

Sugiura K, Tanaka H, Matsumoto T, Kawai T, Sakata Y (1999) A Mandala-patterned Bandanna-shaped porphyrin oligomer, $C_{1244}H_{1350}N_{84}Ni_{20}O_{88}$, having a unique size and geometry. Chem. Lett. 1193–1194

Takahashi R, Kobuke Y (2005) Hexameric and pentameric slipped-cofacial dimers: toward an artificial light-harvesting complex. J Org Chem 70:2745–2753

Tsuda A, Osuka A (2001) Fully conjugated porphyrin tapes with electronic absorption bands that reach into infrared. Science 293:79–82

Uno H, Ito S, Wada M, Watanabe H, Nagai M, Hayashi A, Murashima T, Ono N (2000) Synthesis and structures of pyrroles fused with rigid bicyclic ring systems at β-positions. J Chem Soc Perkin Trans 1:4347–4355

Uno H, Nakamoto K-i, Kuroki K, Fujimoto A, Ono N (2007) Synthesis of porphyrin dimers fused with a benzene unit. Chem Eur J 13:5773–5784

Uoyama H, Kim KS, Kuroki K, Shin J-Y, Nagata T, Okujima T, Yamada H, Ono N, Kim D, Uno H (2010) Highly pure synthesis, spectral assignments, and two-photon properties of cruciform porphyrin pentamers fused with benzene units. Chem Eur J 16:4063–4074

van Leusen D, van Echten E, van Leusen AM (1992) Chemistry of sulfonylmethyl isocyanides. 37. Synthesis of 3,4-disubstituted pyrroles bearing substituents of electron-withdrawing and/or electron-donating nature. J Org Chem 57:2245–2249

Yella A, Lee H-W, Tsao HN, Yi C, Chandiran AK, Nazeeruddin MK, Diau EW-G, Yeh C-Y, Zakeeruddin SM, Grätzel M (2011) Porphyrin-sensitized solar cells with cobalt (II/III)–based redox electrolyte exceed 12 percent efficiency. Science 334:629–634

Yella A, Mai C-L, Zakeeruddin SM, Chang S-N, Hsieh C-H, Yeh C-Y, Angew MG (2014) Molecular engineering of push-pull porphyrin dyes for highly efficient dye-sensitized solar cells: the role of benzene spacers. Chem Int Ed 53:2973–2977

Zaidi SHH, Fico RM Jr, Lindsey JS (2006) Investigation of streamlined syntheses of porphyrins bearing distinct meso substituents. Org Process Res Dev 10:118–134

Chapter 4
Phthalocyanine and Related Analogues

Soji Shimizu

Abstract Phthalocyanine (Pc) has been one of the most important industrial dyes exhibiting lasting blue or green color and used in various research fields including not only organic electronics but also biomedical field. The aim of this chapter is to overview the chemistry of Pc and its ring contracted analogue, subphthalocyanine (SubPc), with an emphasis on their syntheses and optical properties from a viewpoint of molecular symmetry–property relationships. A basic theoretical description of the absorption properties of Pc and SubPc, which is referred to as Gouterman's four orbital theory, is introduced. In addition, magnetic circular dichroism (MCD) spectroscopy, a powerful spectroscopic tool to analyze the absorption spectra of Pc, SubPc, and related analogues, is briefly described. Finally, recent advances in the synthesis of new Pc-based NIR chromophores will also be covered.

Keywords Phthalocyanine · Subphthalocyanine · Functional dyes · Near-infrared absorption · Magnetic circular dichroism (MCD)

4.1 Introduction

The first synthesis of phthalocyanine (Pc, Fig. 4.1) traces back to 1907 when Braun and Tcherniac obtained an insoluble blue pigment in a reaction of *o*-cyanobenzamide at high temperature (Braun and Tcherniac 1907). A quarter-century after this accidental discovery, Linstead et al. characterized the structure of the free base of unsubstituted Pc (H_2Pc) and investigated its optical properties in their pioneering research (Ficken et al. 1934). Meanwhile, Robertson determined the crystal structure of Pc by X-ray diffraction analysis (Robertson and Woodward 1935). Owing to the robust blue and green color, Pc was found its initial use as a coloring staff known as "Phthalocyanine Blue BN" and "Phthalocyanine Green G". From the late twentieth century,

S. Shimizu (✉)
Department of Chemistry and Biochemistry, Graduate School of Engineering, Kyushu University, Fukuoka 819-0395, Japan
e-mail: ssoji@cstf.kyushu-u.ac.jp

Center for Molecular Systems (CMS), Kyushu University, Fukuoka 819-0395, Japan

Y. Ooyama and S. Yagi (eds.), *Progress in the Science of Functional Dyes*,
https://doi.org/10.1007/978-981-33-4392-4_4

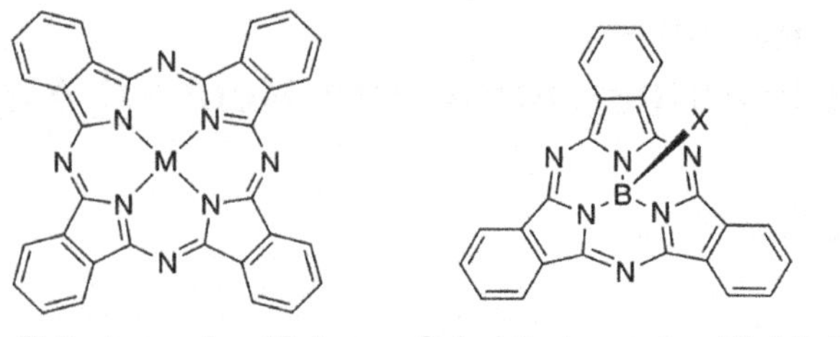

Fig. 4.1 Chemical structures of metal complex of Pc and SubPc. M represents the central metals or main group elements

the application of Pc has widely spread into the industrial and biomedical fields, such as catalysts, deodorants, optical disks, photodynamic therapy, semiconductors, solar cells, nonlinear optics, and so forth (Kadish et al. 2003; Leznoff and Lever 1989; McKeown 1998; Kobayashi and Fukuda 2006).

Almost 60 years after the discovery of Pc, subphthalocyanine (SubPc, Fig. 4.1) was accidentally synthesized by Meller and Ossko during their synthetic investigation on a boron complex of Pc using boron trichloride as a template (Meller and Ossko 1972). Owing to the trigonal pyramid coordination geometry of boron, a cyclotrimerization reaction of *o*-phthalonitrile provided SubPc. Since then, SubPc has been a sole contracted analogue of Pc, and only the boron complex has been known. In the compound name of SubPc, "*sub*" refers to Pc-like macrocycle containing three isoindole or pyrrole subunits (Sessler et al. 2017). Due to the contracted structure with a smaller 14π-electron conjugation than the 18π-electron conjugation of Pc, SubPc generally exhibits a vivid pink-to-purple color (Claessens et al. 2014). Another unique feature of SubPc is its bowl-shaped structure, which enables concave-convex π-π stacking interaction to form supramolecular architectures such as self-assembled one-dimensional columns and co-crystals with curved π-molecules (Shimizu et al. 2011; Sánchez-Molina et al. 2013; Konarev et al. 2015; Rhoda et al. 2016). SubPc has recently been studied in practical applications such as nonlinear optics (Sastre et al. 1996), organic field-effect transistors (Renshaw et al. 2010), organic light-emitting diodes (Morse et al. 2011), and organic photovoltaic cells (Mutolo et al. 2006; Duan et al. 2016).

Prominent properties of Pc and SubPc, such as intense absorption in the visible region, electrochemical properties, and ligand properties, arise from their macrocyclic 18π- and 14π-electron conjugation comprising imino-nitrogen-bridged four and three isoindole rings, respectively.

Figure 4.2 depicts the absorption spectrum of nickel complex of tetra-*tert*-butyl-substituted Pc in $CHCl_3$. The sharp, intense absorption around 670 nm arising from an *x/y* polarized π–π* transition is called Q band, whereas the broad bands in the higher energy region (300–400 nm) is called B band, which is often referred to as Soret band. SubPc exhibits similar Soret and Q bands in a shorter wavelength region than those of Pc due to its contracted conjugation system (Fig. 4.2). Although Pc shares the same origin of the Soret and Q bands with its structural analogue called porphyrin, Pc and porphyrin exhibit completely different absorption spectral profiles, as shown in Fig. 4.2. Gouterman's four-orbital theory described in Sect. 4.3.1 well

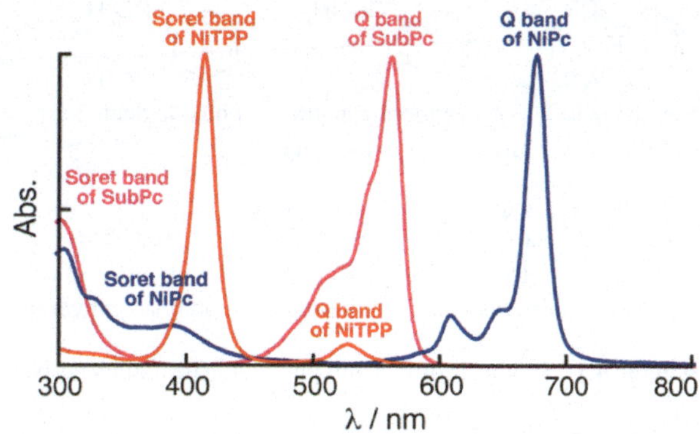

Fig. 4.2 UV/vis absorption spectra of porphyrin, Pc, and SubPc (nickel complexes of *meso*-tetraphenyl-substituted porphyrin (red) and tetra-*tert*-butyl-substituted phthalocyanine (blue), and subphthalocyanine with an axial chlorine ligand (pink))

explains a forbidden nature of the Q band of porphyrin and partially allowed nature of that of Pc (Gouterman 1959). Owing to the sharp, intense Q band in the far-red region, Pc has attracted attention as a potential structure to create near-infrared (NIR) chromophores.

Since the recent development in the Pc and SubPc synthesis has been reported elsewhere, this chapter focuses on the basic chemistry of Pc and SubPc, such as their conventional synthesis and structure–optical property relationship with an emphasis on how Gouterman's four-orbital theory can be applied to explain the absorption spectra of various kinds of Pc and SubPc analogues.

In this chapter, magnetic circular dichroism (MCD) spectroscopy is also introduced. MCD spectroscopy is a powerful spectroscopic technique to provide information about degeneracy and non-degeneracy of the excited states based on MCD signals called *Faraday* A, B, and C terms (Kobayashi et al. 2012; Mack et al. 2007). Michl et al. disclosed the relationship between MCD signal patterns of π-conjugated macrocycles and their frontier energy diagrams (Michl 1978).

In the last part of this chapter, the recent progress in the synthesis of Pc and related analogues exhibiting unique antiaromatic character and NIR absorption is covered.

4.2 Synthesis of Phthalocyanine and Subphthalocyanine

4.2.1 Synthesis of Symmetric H_2Pcs

H_2Pc and metal complexes of phthalocyanine (MPcs) can be synthesized from phthalonitrile or its derivatives, such as 1,3-diiminoisoindoline, *o*-cyanobenzamide, phthalimide, phthalic anhydride, and phthalic acid (Scheme 4.1) (Sharman et al. 2003). Since the Pc and MPc synthesis is comprehensively summarized elsewhere (Leznoff Lever 1989; McKeown et al. 2003), four fundamental synthetic methods, (1) strong base method, (2) lithium method, (3) 1,3-diiminoisoindoline method, and (4) electrosynthetic method, are introduced in this section.

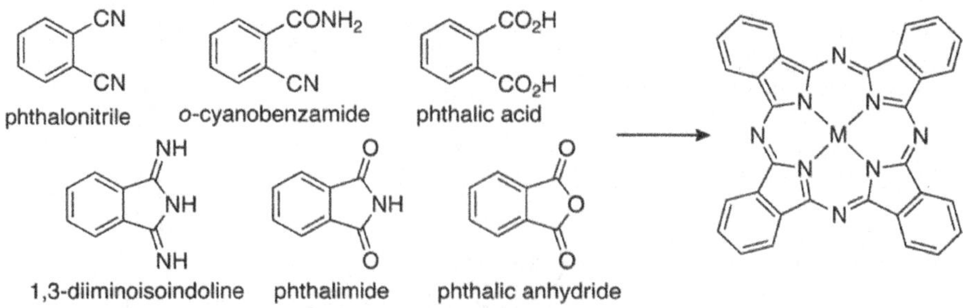

Scheme 4.1 Synthesis of Pc from phthalonitrile and its derivatives

(1) Strong base method (Scheme 4.2): Phthalonitrile (2.6 g, 20 mmol) and 1,8-diazabicyclo[5,4,0]undec-7-ene (DBU) or 1,5-diazabicyclo[4,3,0]non-5-ene (DBN) (20 mmol) were refluxed in 50 mL of ethanol (or propanol) for 18–24 h to provide H_2Pc in approximately 20% and 40% yields for the use of DBU and DBN, respectively. Instead of DBU and DBN, a basic solvent such as *N,N*-dimethyl-2-aminoethanol (DMAE) can also be used for mild and clean reactions. Because the strong base method was developed by Tomoda et al., this method is also called "Tomoda method" (Tomoda et al. 1980).

(2) Lithium method (Scheme 4.3): In this method, lithium alkoxide, formed in situ from lithium and alcohol with an alkyl chain length from C_3 to C_8, is used for cyclotetramerization of phthalonitrile. The central lithium ion is readily removed upon treatment with acid to provide H_2Pc. The practical reaction conditions are described as follows.

After lithium (20 mg) is dissolved in 6 mL of dry 1-pentanol under heat, 1,4-dibutylthio-2,3-dicyanobenzene (0.40 g, 1.3 mmol) is added, and the resulting mixture is refluxed for 1 h. After cooled down to room temperature, a methanol solution (100 mL) containing a few drops of conc. HCl is poured into the reaction mixture. A blue precipitate is filtered and purified by silica gel column chromatography. After recrystallization from chloroform and methanol, octakis-butylthio-substituted H_2Pc is obtained in 38% yield (0.15 g).

(3) 1,3-Diiminoisoindoline method (Scheme 4.4): 1,3-Diiminoisoindoline (20 g, 0.14 mol) suspended in 100 mL of DMAE is refluxed for 7 h. The resulting mixture is filtered, while it is hot, and the residue is washed with ethanol and acetone and dried to provide H_2Pc in 85% yield (15 g) (Brach et al. 1970).

Scheme 4.2 Strong base method

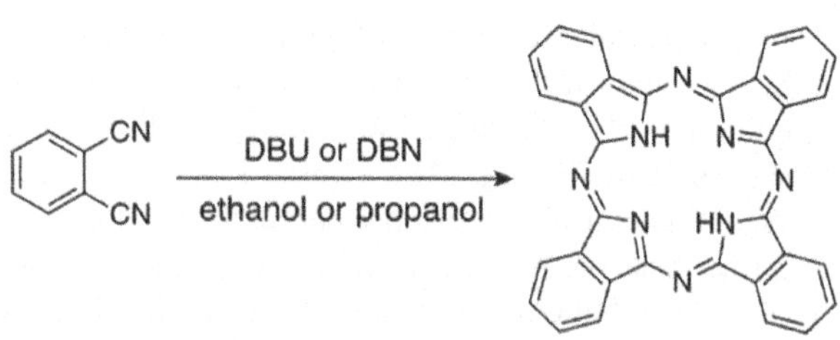

Scheme 4.3 Lithium method

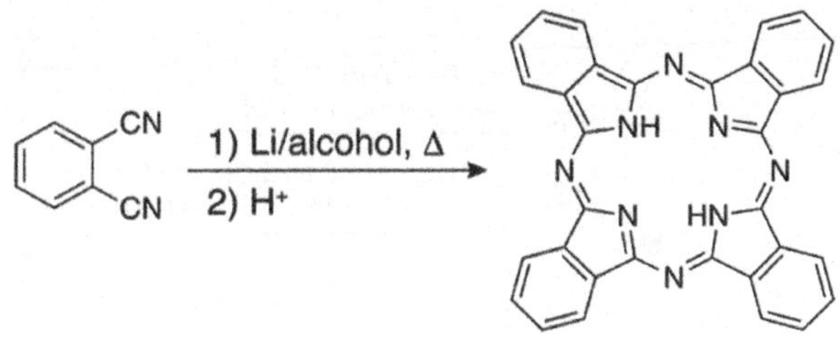

Scheme 4.4 1,3-Diiminoisoindoline method

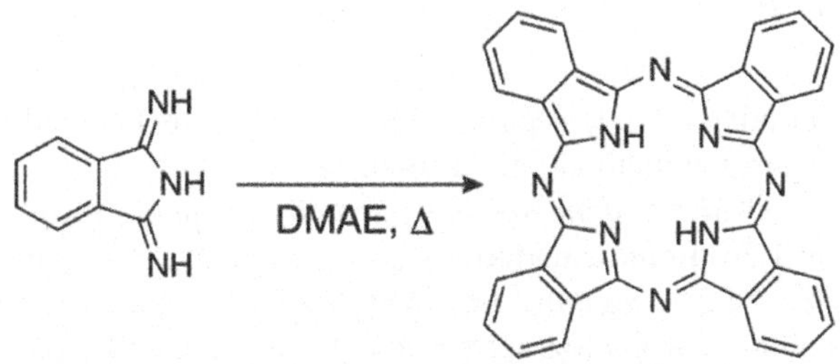

Scheme 4.5 Electrosynthetic method

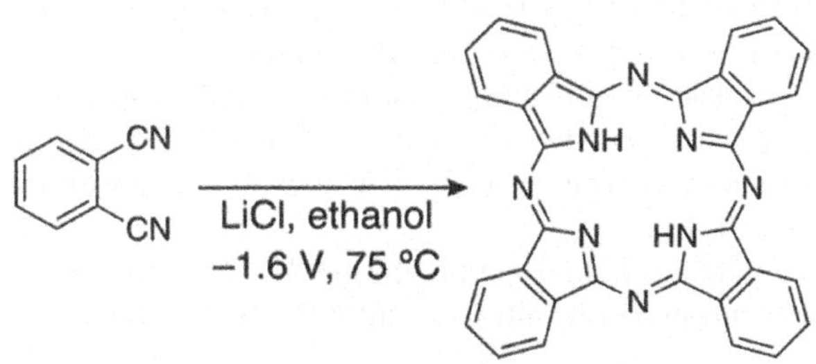

(4) Electrosynthetic method (Scheme 4.5): A dry ethanol solution (100 mL) containing lithium chloride (3.0 g, 0.071 mol) is added to an electrochemical cell. Phthalonitrile is added to the anode compartment until the concentration becomes 30 mM. Electrolysis is conducted at −1.6 V and 75 °C. The resulting mixture is poured into 100 mL of 0.2 M H_2SO_4 and stirred for a while. The precipitate is filtered off and washed with water and acetone. H_2Pc is then extracted by Soxhlet extractor with acetone. The yield is approximately 60–70% (Uosaki and Ueda 1992).

4.2.2 *Synthesis of Low-Symmetry Pcs*

Low-symmetry Pcs (A_3B-type Pc as shown in Scheme 4.6) are synthesized from a mixed-condensation reaction of two kinds of phthalonitriles or 1,3-diiminoisoindolines with different substituents. In this section, three methods,

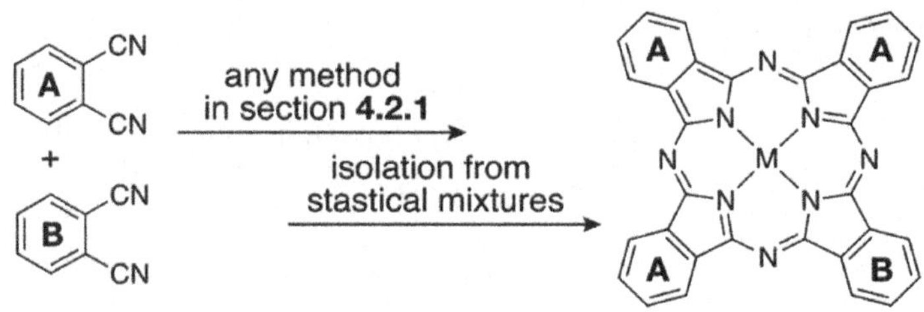

Scheme 4.6 Mixed-condensation method. **A** and **B** represent benzo-rings with different substituents

(1) mixed-condensation method, (2) polymer-bound method, and (3) SubPc ring-expansion method, are introduced.

(1) Mixed-condensation method (Scheme 4.6): Two kinds of phthalonitriles or 1,3-diiminoisoindolines are reacted in one of the four methods described in Sect. 4.2.1, and the obtained statistical mixture of compounds is separated by column chromatography, such as silica gel and size-exclusion column chromatography. Examples of this kind of reaction can be found elsewhere. The key reaction parameters of this method are *well mixing of two precursors by an agate mortar, concentrations (amount of solvent), temperature, and template.* Trial and error during optimization of reaction conditions is indispensable.

(2) Polymer-bound method (Scheme 4.7): To prevent a tedious procedure to isolate A_3B-type Pc, a polymer-bound method was developed by Leznoff et al. (Hall et al. 1982).

Polymer-bound trityloxyhexanol (7.4 g) and 4-nitrophthalonitrile (7.5 g, 43 mmol) are suspended in nitrobenzene (75 mL) containing Adogen® 464 (1.5 g), and the

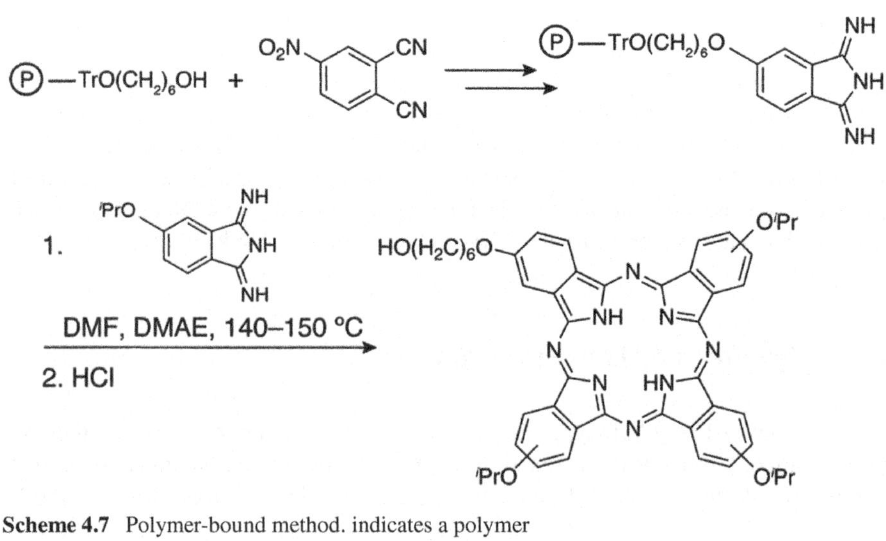

Scheme 4.7 Polymer-bound method. indicates a polymer

suspension is vigorously stirred at 40–50 °C under nitrogen until 4-nitrophthalonitrile is completely dissolved. To the solution, 25% aqueous solution of potassium hydroxide (75 mL) is added. After stirring at 60 °C for 22 h, the mixture is cooled to room temperature. The obtained polymer is washed with water, methanol, and THF and extracted by Soxhlet extractor to provide phthalonitrile-appended polymer (7.3 g).

Sodium (100 mg, 4.4 mmol) is dissolved in dry methanol (10 mL), and the polymer (2.0 g) and dry THF (15 mL) is added. To the solution, ammonia gas is bubbled at room temperature for 1 h, and then under reflux for 4 h. The mixture is cooled to room temperature and left overnight. The resulting polymer is filtered and washed with dry methanol, THF, methanol, and diethyl ether to provide 1,3-diiminoisoindoline-appended polymer (2.0 g).

The 1,3-diiminoisoindoline-appended polymer (1.5 g) is suspended in a mixed solvent of DMF and DMAE (1:1 (v/v)) containing 5-isopropoxy-1,3-diiminoisoindoline (1.2 g). The resulting mixture is heated at 140–150 °C and vigorously stirred under nitrogen for 20 h. After cooled to room temperature, the mixture is poured into water (80 mL). A green solid residue is filtered, dried, and extracted with Soxhlet extractor using dichloromethane for 18 h. Polymer-bound H_2Pc (1.6 g) is then obtained as a deep blue solid residue.

To the obtained polymer-bound H_2Pc (1.6 g) suspended in dry 1,4-dioxane (40 mL), dry HCl gas is bubbled until the concentration of HCl in the solution becomes approximately 0.3 N. The resulting mixture is stirred for 51 h at room temperature. The polymer residue is filtered, washed with a little amount of methanol and diethyl ether, and extracted with Soxhlet extractor using dichloromethane until the extract becomes colorless. After removal of the solvent from the combined solution, the fraction is purified by alumina gel column chromatography using methanol and ethyl acetate as an eluent by changing the ratio from 15:85 to 60:40 (v/v). A_3B-type H_2Pc is obtained in 24% yield (0.16 g) from the starting polymer-bound phthalonitrile.

(3) SubPc ring-expansion method (Scheme 4.8): Kobayashi et al. revealed ring-opening reactivity of SubPc in the presence of 1,3-diiminoisoindoline to provide a A_3B-type Pc. This method largely depends on the relative reactivities of SubPc and 1,3-diiminoisoindoline precursors. A combination of 1,3-diiminoisoindoline with electron-donating substituents and SubPc with electron-withdrawing substituents normally affords high selectivity of the A_3B-type Pc and small amounts of undesired byproducts, such as B_4-type Pc. Practical reaction conditions are described as follows (Musluoglu et al. 1992).

A DMSO/1-chloronaphthelene solution (2:1 (v/v)) of SubPc with an axial chlorine ligand (0.48 g, 1.1 mmol) and 15-crown-5-fused 1,3-diiminoisoindoline (2.6 g, 7.8 mmol) is heated at 80–90 °C for 12 h. A dark green precipitate formed upon addition of ethanol after cooled to room temperature is filtered and washed with hot water, ethanol, and diethyl ether to provide mono-15-crown-5-fused H_2Pc in 45% yield (0.35 g).

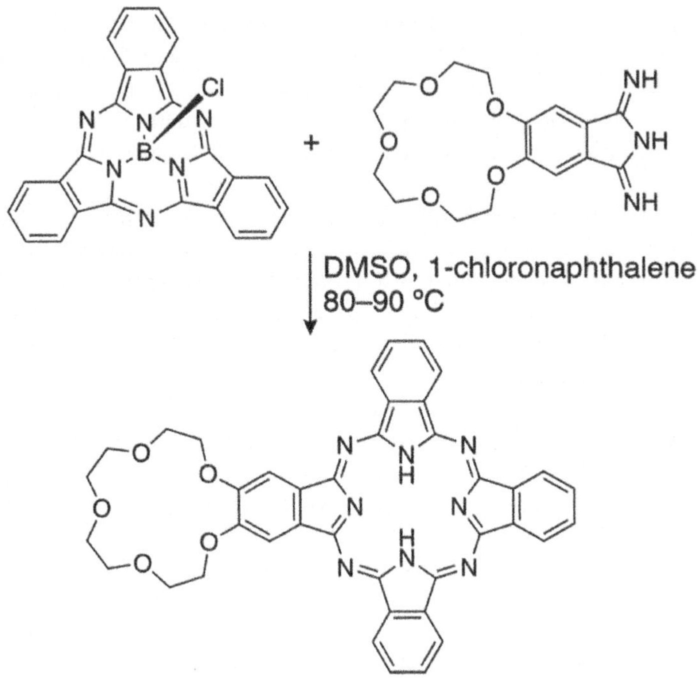

Scheme 4.8 SubPc ring-expansion method

4.2.3 *Synthesis of SubPc*

SubPc is synthesized from phthalonitrile and BCl_3 or BBr_3 (Scheme 4.9). Commercially available 1 M solution of BCl_3 is frequently used. A standard reaction procedure for the SubPc synthesis is described as follows (Claessens et al. 2003).

A 1 M *p*-xylene solution of BCl_3 (2 mL) is added to dry phthalonitrile (2 mmol) under an argon atmosphere. The mixture is stirred under reflux for 20 min. After evaporation of the solvent, the residual solid is extracted with toluene (100 mL). The solution is evaporated, and the resulting solid is washed with methanol (50 mL) and hexane (50 mL) to provide SubPc with an axial chlorine ligand in 82% yield (240 mg). In some cases, further purification by recrystallization or column chromatography is necessary.

Scheme 4.9 Synthesis of SubPc

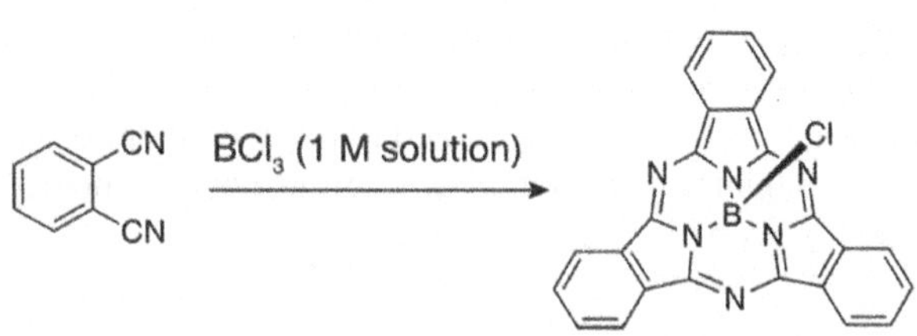

4.3 UV/vis Absorption Properties of Pc and SubPc

4.3.1 Gouterman's Four-Orbital Model and Theoretical Description of Absorption Properties of Pc and SubPc

A theoretical framework for the description of absorption properties of porphyrin was initially constructed to explain the energy and intensity differences between the Soret and Q bands of porphyrin. Then the established framework, which is known to date as "Gouterman's four-orbital model," was used to describe the absorption properties of Pc (Gouterman 1959; Fukuda and Kobayashi 2010).

In the middle of the twentieth century, Simpson attempted to explain the absorption spectra of porphyrins using 18-membered cyclic polyene having 18π-electrons (Simpson 1949). Although this model successfully described the energy differences between the Soret and Q bands, completely forbidden Q bands in this model failed to explain the fact that the Q bands of porphyrin can be observed as weak, but distinct bands. Meanwhile, on the basis of π-molecular orbitals (MOs) of porphyrin obtained by the LCAO technique, Platt et al. tried to describe the absorption properties (Longuet-Higgins et al. 1950). The energy difference between the Soret and Q bands were reproduced by their model because the degeneracy of the HOMO in Simpson's model is lifted by lowering the symmetry from the cyclic polyene to the D_{4h} metalloporphyrin skeleton. However, it was unable to explain the observed differences in intensities between the Soret and Q bands. In 1959, Gouterman demonstrated how Simpson's model is related to Platt's LCAO MOs by applying configurational interactions to one-electron excited states consisting of transitions from the HOMO (a_{2u}) and HOMO–1 (a_{1u}) to degenerate LUMO (e_g) (Fig. 4.3) (Gouterman 1959) The degree of configuration largely depends on the relative energy differences between the two one-electron excited states. In the case of porphyrin, the HOMO and HOMO–1 (or the next HOMO of porphyrin π-skeleton when metal- or substituent-centered MOs are inserted in between) is nearly degenerate. As a result, two excited states strongly interact with each other to produce the Soret and Q bands as sum and difference of two configurations, respectively. Therefore, the Soret band becomes allowed and more intense compared with the forbidden Q band.

In the case of Pc, the a_{2u} orbital is stabilized compared with the a_{1u} orbital by more electronegative nitrogen than carbon because the a_{2u} orbital has large MO coefficients (electron density) on the *meso*-positions (Fig. 4.3). The lifted degeneracy of the HOMO causes smaller configurational interactions, resulting in more intense Q bands of Pc than those of porphyrin. The major contribution of the HOMO–LUMO transitions to the Q bands in the case of Pc also indicates that the intensity and energy of the Q bands are significantly perturbed by molecular symmetry and peripheral substituents. Considering that the Q band of Pc generally appears around 650 nm, Pc is one of the potential structures to create NIR chromophores.

The absorption properties of SubPc can be explained in a similar manner to Pc because SubPc holds degenerate LUMO due to the three-fold molecular symmetry. Therefore, SubPc exhibits an intense Q band around 550 nm, which is blue-shifted

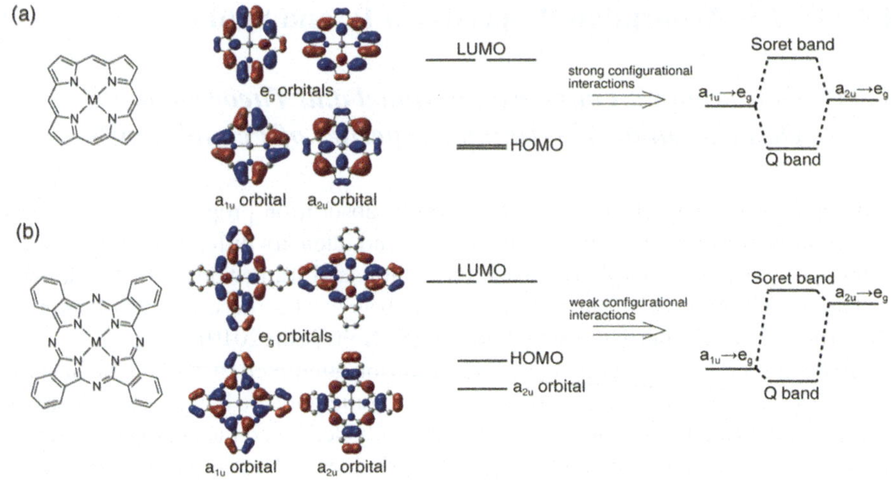

Fig. 4.3 Frontier molecular orbital diagrams of **a** porphyrin and **b** Pc and configurational interactions

from that of Pc because of the contracted 14π-electron conjugation of SubPc (Fig. 4.2).

4.3.2 MCD Spectroscopy in Pc Chemistry

MCD spectroscopy is a powerful spectroscopic technique to give an insight into the electronic structures of Pc, SubPc, and related analogues (Kobayashi et al. 2012). MCD spectroscopy studies optical activity under the magnetic fields known as the Faraday effect. Therefore, different from circular dichroism (CD), an MCD signal is observed even when a molecule lacks chirality. MCD spectra are recorded by mounting a magnet into a sample compartment of a normal CD spectrometer, in which the magnetic field is aligned parallel to light propagation.

MCD spectra contain a sum of characteristic bands called Faraday *A*, *B*, and *C* terms. The Faraday *A* term arises from the Zeeman splitting of a degenerate excited state under a magnetic field. The Faraday *B* term is caused by the magnetic field induced mixing of transitions as a second-order effect and observed for magnetically coupled non-degenerate transitions. The Faraday *C* term is observed for molecules having a degenerate ground state. Because the Faraday *C* term arises from the Boltzmann population distribution, the intensity of the Faraday *C* term is temperature-dependent. Due to the non-degenerate ground state, the MCD spectra of Pc, SubPc, and their analogues mainly consist of Faraday *A* or *B* terms (Fig. 4.4). The Faraday *A* term is observed as a derivative-shaped signal corresponding to the absorption band with an inflection point at the absorption maximum, whereas the Faraday *B* term is a band with absorption-like spectral profile irrespective of its sign. Faraday *A* terms are

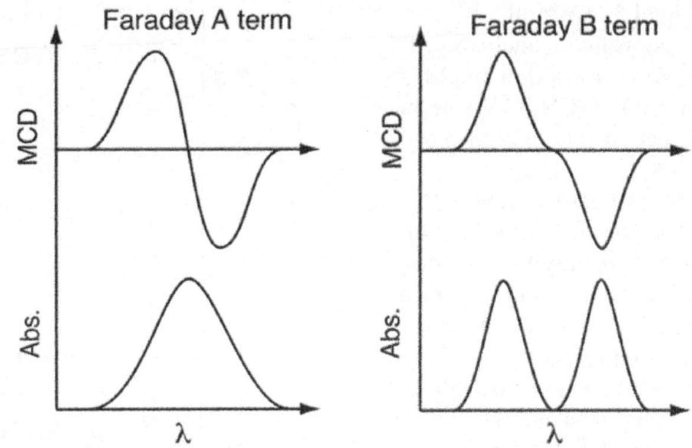

Fig. 4.4 Representation of MCD Faraday *A* and *B* terms

observed for symmetrically substituted MPc and SubPc because of their degenerate LUMO. When the degeneracy of the LUMO is lifted by lowering the symmetry, a couple of Faraday *B* terms is observed.

The relationship between the frontier MO diagram and MCD sign sequence was comprehensively investigated by Michl et al. They introduced a perimeter model to explain MCD sign sequences based on the magnitudes of the orbital splitting energies, ΔHOMO and ΔLUMO which denote the energy gap between the HOMO and HOMO–1 (or the next HOMO of the chromophore) and that between the LUMO and LUMO+1, respectively (Michl 1978). When ΔHOMO is greater than ΔLUMO, a minus-to-plus sign sequence in ascending energy is observed for Faraday *A* term or coupled Faraday *B* terms corresponding to both the Soret and Q bands. The sign sequence becomes opposite (plus-to-minus) when ΔLUMO is greater than ΔHOMO. Therefore, the MCD signal pattern and sign sequence in the Soret and Q band regions can provide information about the degeneracy and relative energy differences of the frontier orbitals, which are perturbed by several factors, such as substituents and molecular symmetries.

A band deconvolution analysis of UV/vis and MCD spectral data set using the same band parameters, such as band shapes, bandwidths, and centers of the bands, enables accurate band fitting compared with band deconvolution only based on UV/vis spectral data, which cannot avoid arbitrariness. The SIMPFIT program developed by Stillman et al. can perform this kind of simultaneous band deconvolution analysis (Mack and Stillman 2001). Figure 4.5 shows the deconvolution results of the Q band region of a zinc complex of Pc with an axial cyano ligand ((CN^-)ZnPc) (Mack and Stillman 1995).

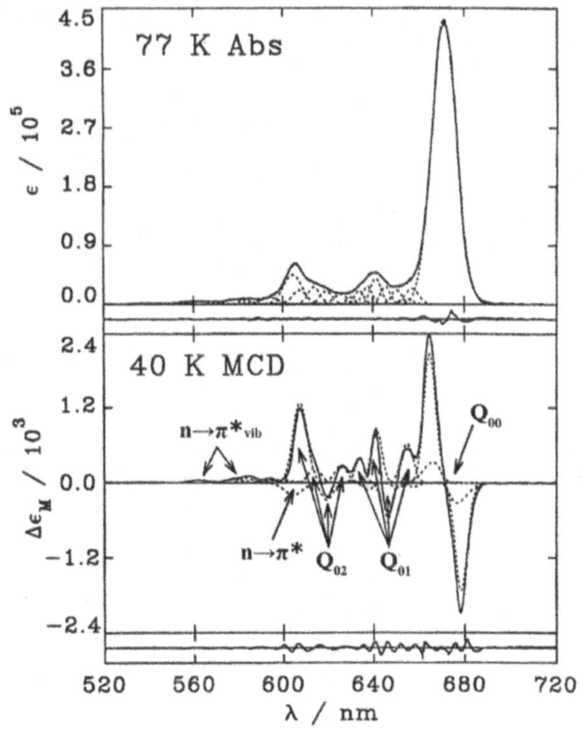

Fig. 4.5 Spectral deconvolution analysis of the UV/vis absorption and MCD spectra of $(CN^-)ZnPc$ in the Q band region recorded at 77 K for the UV/vis absorption spectrum and 40 K for the MCD spectrum in 5:2 dimethylformamide–dimethylacetamide solution. Solid, dashed, and dotted lines represent the experimental and calculated spectral data and the deconvoluted *B* terms, respectively. Redrawn with permission from ref. (Mack and Stillman 1995) Copyright 1995 ACS

4.3.3 UV/vis Absorption and MCD Spectra of Perturbed Pc and SubPc

In this section, examples of UV/vis absorption and MCD spectra of Pc and SubPc exhibiting perturbed optical properties by substituents and molecular symmetries are introduced.

Due to the degenerate LUMO arising from the D_{4h} symmetric structure, MPcs bearing less-perturbing substituents, such as alkyl substituents, exhibit a single Q band as shown in Fig. 4.2. Upon lowering the molecular symmetry from the D_{4h} symmetry of MPc to the D_{2h} symmetry of H_2Pc due to the diagonally arranged NH protons on the isoindole rings, H_2Pc exhibits split Q bands because of the lifted degeneracy of the LUMO. As mentioned in Sect. 4.3.1, the Q bands of Pc mainly consist of the transitions from the HOMO to the degenerate LUMO in the case of MPc or to the LUMO and LUMO+1 in the case of H_2Pc. Therefore, changes in the energy levels of these frontier orbitals can be observed as changes in the Q band spectral profile. As shown in Fig. 4.6, upon deprotonation of unsubstituted H_2Pc by addition of base, the split Q bands are merged into a single band due to symmetrization from the D_{2h} symmetry of H_2Pc to the D_{4h} symmetry of the deprotonated form (Pc^{2-}) (Ledson and Twigg 1975).

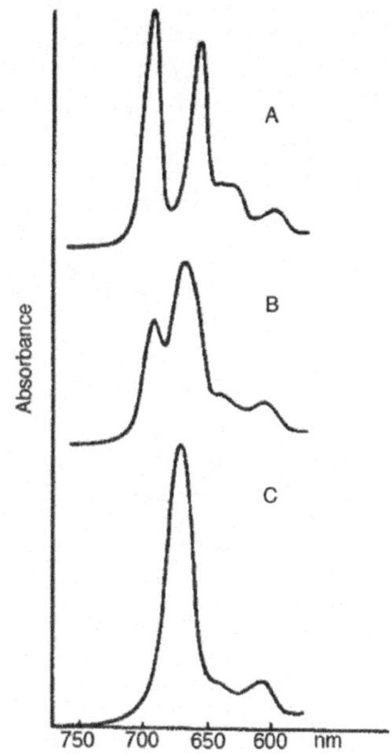

Fig. 4.6 Absorption spectral profiles of the Q band. (A) H_2Pc in benzotrichloride and (B, C) after the addition of NaOH/EtOH solution. Redrawn with permission from ref. (Ledson and Twigg 1975). Copyright 1975 Elsevier

The relationship between the molecular symmetry of Pc and the Q band spectral profile was investigated in detail by Kobayashi et al (Mack and Kobayashi 2011). They disclosed changes in the Q band spectral profiles for a series of low-symmetry Pcs, in which benzo-rings were added to or abstracted from the structure of Pc.

In the case of symmetric abstraction and addition of benzo-rings, which produces the structures of tetraazaporphyrin (TAP), naphthalocyanine (Nc), and anthracocyanine (Ac), the split Q bands of their free base forms red-shifted with decrease of the Q band splitting energy (H_2TAP: 619 and 551 nm, H_2Pc: 698 and 664 nm, H_2Nc: 784 nm and H_2Ac: 858 nm) upon increasing the number of peripheral benzo-rings (Fig. 4.7) (Kobayashi et al. 2004). Corresponding to the split Q bands of H_2TAP and H_2Pc, the Faraday *B* terms with trough and peak in ascending energy were observed. In contrast, despite the D_{2h} molecular symmetry of H_2Nc and H_2Ac, which do not possess degenerate term according to group theory, Faraday *A* term-like MCD signals were observed. This MCD signal is referred to as pseudo-Faraday *A* term, which is seen when two Faraday *B* terms accidentally appear in close energy (Kaito et al. 1977).

The observed trend in the red-shifts and decrease of the Q band splitting energies were explained in terms of changes in the frontier MO diagrams estimated based on electrochemical measurements and theoretical calculations. Figure 4.8 shows the partial frontier MO diagrams calculated by the ZINDO/s method. Upon addition of benzo-rings from H_2TAP to H_2Ac, the HOMO is more significantly destabilized than the LUMO and LUMO+1 because of the large MO coefficients of the HOMO on the periphery of the benzo-rings. The energy gap between the LUMO and LUMO+1

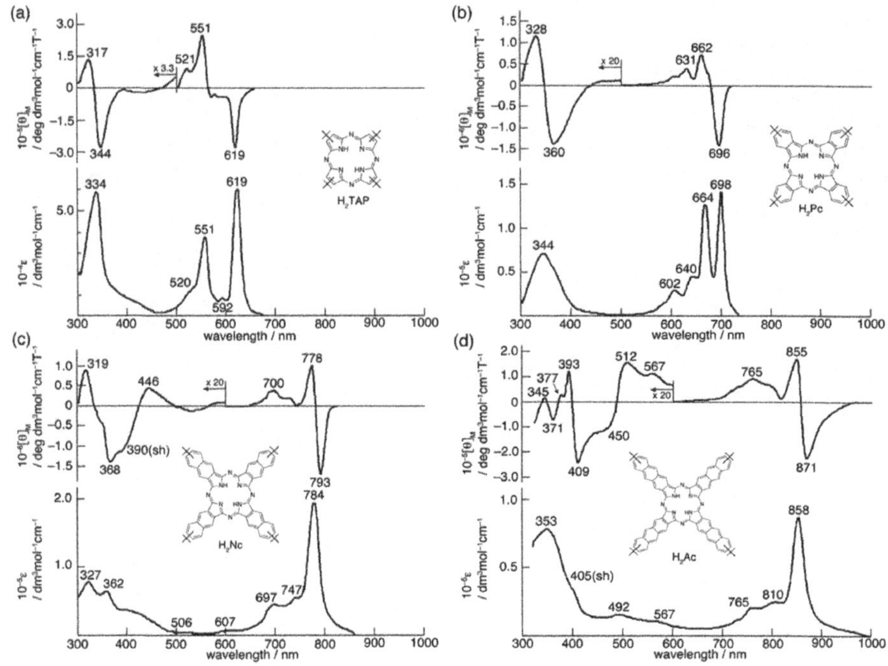

Fig. 4.7 UV/vis absorption (bottom) and MCD (top) spectra of **a** H_2TAP, **b** H_2Pc, **c** H_2Nc, and **d** H_2Ac in pyridine

Fig. 4.8 Partial MO diagram of H_2TAP, H_2Pc, H_2Nc, and H_2Ac calculated by the ZINDO/s method

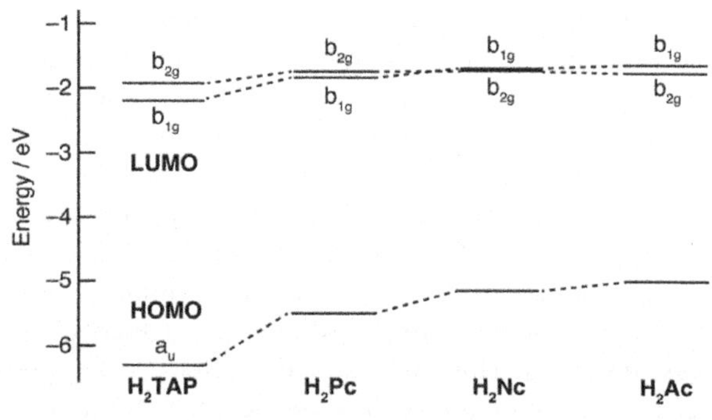

also decreases. These calculation results well reproduced the observed changes in the absorption spectra.

Upon addition of benzo-rings to the structure of TAP, structures of low-symmetry Pcs can be generated. Kobayashi et al. synthesized a series of compounds (**0Zn**–**4Zn** in Fig. 4.9) from phthalonitrile and maleonitrile by the lithium method (Miwa et al. 2004).

A single Q band of **0Zn** (ZnTAP) and **4Zn** (ZnPc) was observed at 637 and 677 nm, respectively. Corresponding to this Q band, the MCD spectra exhibited a

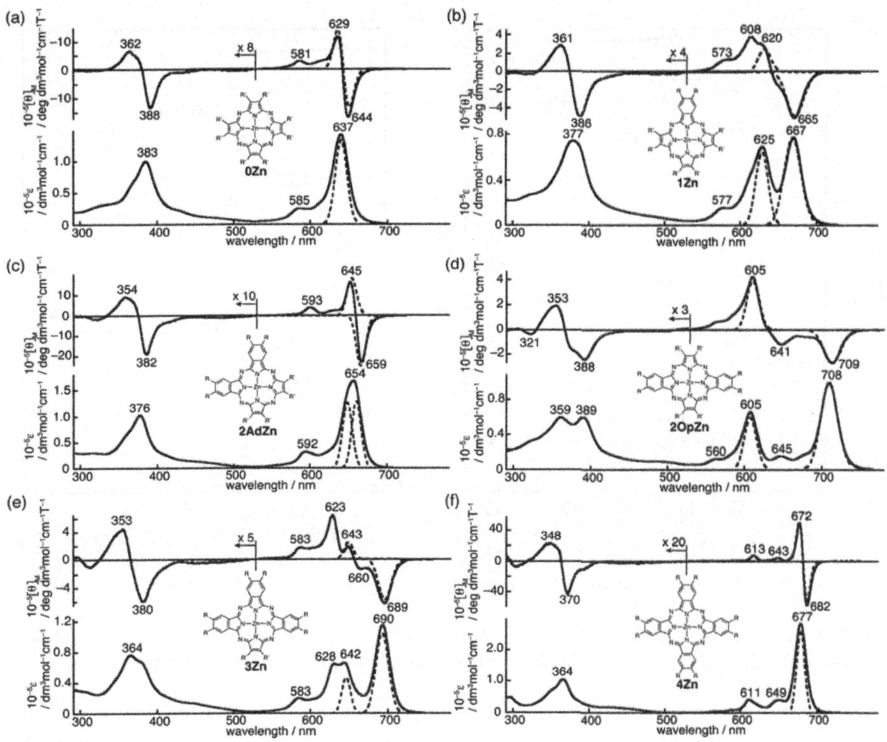

Fig. 4.9 UV/vis absorption (bottom) and MCD (top) spectra of a series of low-symmetry Pcs **a 0Zn**, **b 1Zn**, **c 2AdZn**, **d 2OpZn**, **e 3Zn**, and **f 4Zn** (R = octyloxy and R' = phenyl) in toluene containing 1×10^{-2} M pyridine. The split Q_x and Q_y bands evaluated by the band deconvolution analysis are also shown by dashed lines

Faraday *A* term. The Q band split into Q_x and Q_y bands upon lowering the molecular symmetry from the D_{4h} symmetry of **0Zn** and **4Zn** to the C_{2v} symmetry of **1Zn** (667 and 625 nm) and **3Zn** (690 and 642 nm) and the D_{2h} symmetry of **2OpZn** (708 and 605 nm). The MCD spectra and band deconvolution analysis unambiguously distinguished the split Q bands and their vibronic contributions. Among these low-symmetric TAPs, **2OpZn** exhibited the largest splitting energy, whereas the Q band of **2AdZn** was observed as a broad single band at 654 nm. As the band deconvolution analysis is shown by a dashed line in Fig. 4.9, two split Q bands are close in energy in the case of **2AdZn**. Therefore, the Faraday *B* terms of **2AdZn** were observed as a pseudo-Faraday *A* term (Kaito et al. 1977). Band deconvolution analysis also revealed a decrease in the splitting energy of the Q bands in order of **2OpZn** (2380 cm^{-1}) > **3Zn** (1040 cm^{-1}) > **1Zn** (960 cm^{-1}) > **2AdZn** (280 cm^{-1}).

The calculated frontier MO diagrams and redox potentials estimated by the cyclic voltammetry are consistent with the observed changes in the absorption and MCD spectra (Fig. 4.10). The HOMO is constantly destabilized upon increasing the number of attached benzo-rings, whereas the energy levels and degeneracy of the LUMO

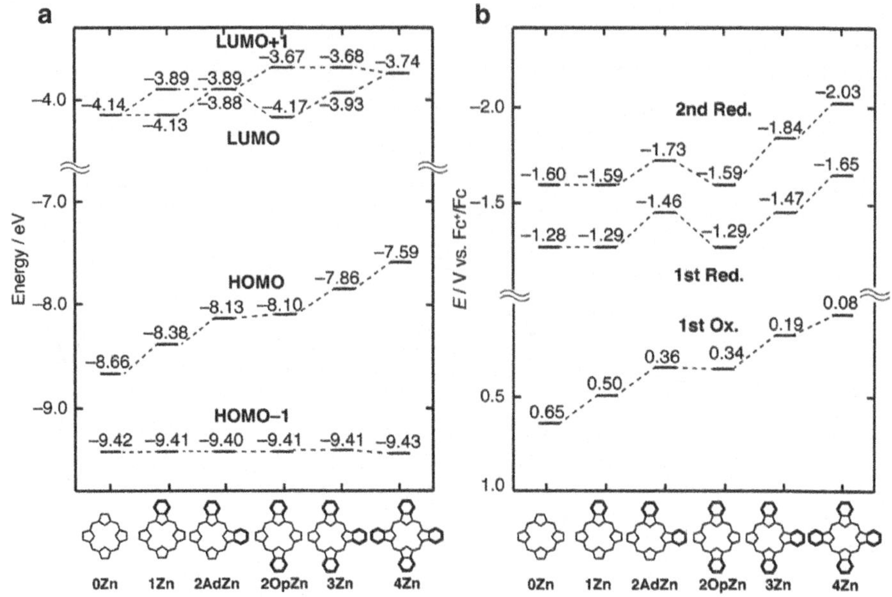

Fig. 4.10 **a** Calculated energies of the four frontier orbitals and **b** electrochemical data recorded at a scan rate of 50 mV s^{-1} in *o*-DCB containing 25 mM pyridine and 0.1 M TBAP

are affected by the positions of the benzo-rings. These changes can be rationally explained in terms of the distribution patterns of the frontier MO coefficients. As with MPc (a_{1u} orbital in Fig. 4.3), the HOMO of a metal complex of TAP (MTAP) is delocalized on the whole molecule so that the extent of destabilization is independent of the positions of the benzo-rings but dependent on the number of benzo-rings. However, since the LUMO of MTAP is localized along the x and y molecular axis (e_{gx} and e_{gy} orbitals, Fig. 4.11), the degeneracy of the LUMO is largely affected by the positions of the benzo-rings. In the case of **2OpZn**, in which the second benzo-ring is attached to the y axis, the LUMO+1 is further destabilized, whereas the LUMO remains unchanged. In contrast, the LUMO of **2AdZn** is destabilized to the same energy of the LUMO+1 due to the introduction of the second benzo-ring to the x axis.

When the isoindole rings are partially replaced with saturated β,β-sp^3-hybridized pyrroline rings, a similar relationship between molecular symmetry and Q band splitting energies is observed (Fig. 4.12) (Fukuda et al. 2004).

In contrast to Pc, the optical properties of TAP are significantly affected by peripheral substituents so that positional isomers exhibit different absorption spectra. Shimizu and Kobayashi synthesized a series of positional isomers of push–pull TAP from a reaction of 1,1,2-tricyano-2-*tert*-butylaminoethylene (Shimizu et al. 2014). Reflecting the arrangement of the push–pull substituents, the positional isomers (C_{4h}, D_{2h}, C_{2v}, and C_s) exhibited characteristic Q bands (Fig. 4.13). The observed red-shifts were also dependent on the overall chromophore symmetry. The frontier MO

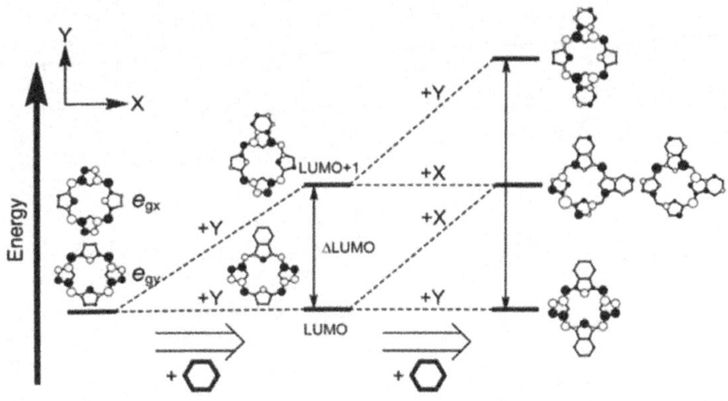

Fig. 4.11 Effects of the fused benzo-rings on the LUMO and LUMO+1 energies. Reprint with permission from ref. (Mack and Kobayashi 2011) Copyright 2011 ACS

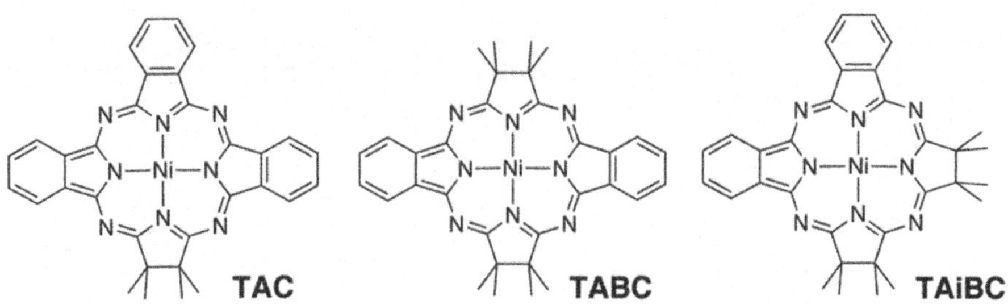

Fig. 4.12 Structures of β,β-sp^3-hybridized phthalocyanine analogues. **TAC**, **TABC**, and **TAiBC** denote tetraazachlorin, tetraazabacteriochlorin, and tetraazaisobacteriochlorin, respectively

analyses and TDDFT calculations revealed the Q band splitting patterns. Because of the trade-off of stabilization and destabilization caused by the pull (cyano) and push (*tert*-butylamino) substituents, C_{4h} and C_{2v} isomers possess degenerate and nearly degenerate LUMO, respectively (Fig. 4.14).

The relationship between the chromophore symmetry and absorption properties observed for Pc and TAP can be similarly applied to SubPc and related analogues. Push–pull substitution causes two types of positional isomers with C_3 and C_1 symmetries in the case of subporphyrazine (SubPz), which is a benzo-ring-abstracted analogue of SubPc (Scheme 4.10) (Liang et al. 2014). Due to the instability, the C_1 isomer (**MeOPh-C_1**) was obtained only when 1,1,2-tricyano-2-*p*-methoxyphenylethylene was used as a precursor. Reactions of other precursors in that study (tricyanoethylenes with *p*-tolyl and *p*-trifluoromethylphenyl substituents) exclusively provided the C_3 isomers (**Tol-C_3** and **CF_3Ph-C_3** in Scheme 4.10). An increase of the donor ability in order of *p*-trifluoromethylphenyl, tolyl, and *p*-methoxyphenyl caused constant red-shifts of the Q band and fluorescence spectra with a linear correlation to the Hammett σ_p parameters (Fig. 4.15). In contrast to the drastic changes in the Q band energies of the push–pull TAPs in response to

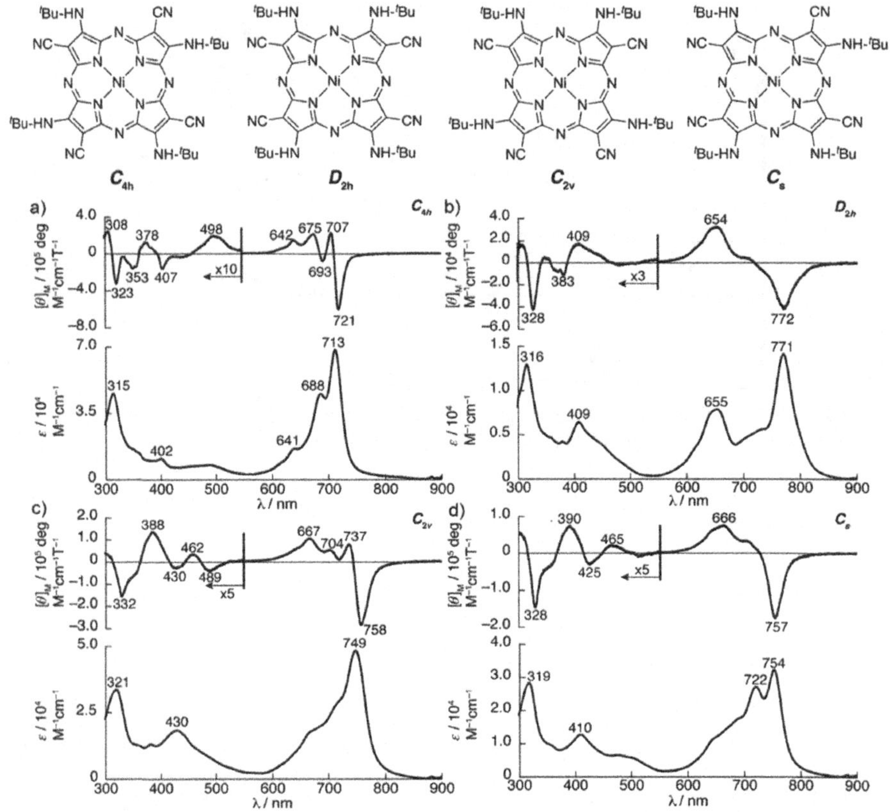

Fig. 4.13 Structures of the positional isomers of push–pull TAPs. UV/vis absorption (bottom) and MCD (top) spectra of **a** C_{4h}, **b** D_{2h}, **c** C_{2v}, and **d** C_s isomers in $CHCl_3$, except for D_{2h} isomer in pyridine. Redrawn with permission from ref. (Shimizu et al. 2014) Copyright 2014 Wiley

the substitution patterns, the chromophore symmetries of **MeOPh-C_3** and **MeOPh-C_1** have a marginal effect, but rather significantly cause changes in the intensity of the charge-transfer band around 500 nm. All C_3 isomers exhibited a derivative-shaped Faraday *A* term corresponding to the Q band, whereas **MeOPh-C_1** showed a pseudo-Faraday *A* term (Kaito et al. 1977).

In addition to the changes in the absorption spectral profiles, the asymmetric substitution of SubPz endows inherent chirality. This kind of chirality is regarded as bowl chirality. The enantiomers of **Tol-C_3** were successfully separated. The mirror-imaged CD spectra with similar spectral profiles to that of the absorption spectrum of **Tol-C_3** irrespective of their signs were observed (Fig. 4.15).

Upon symmetric extension of SubPz with benzo-rings at the β,β'-positions, which leads to the structures of SubPc and subnaphthalocyanine (SubNc), the Q band constantly red-shifts, as with the same trend of the red-shifts seen for TAP, Pc,

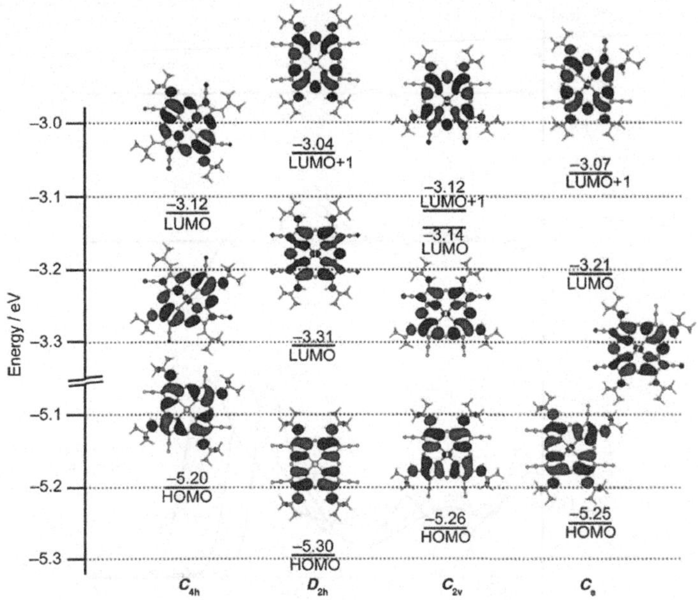

Fig. 4.14 Partial frontier MO diagrams of the positional isomers of push–pull TAPs (B3LYP/6-31G(d)). Redrawn with permission from ref. (Shimizu et al. 2014) Copyright 2014 Wiley

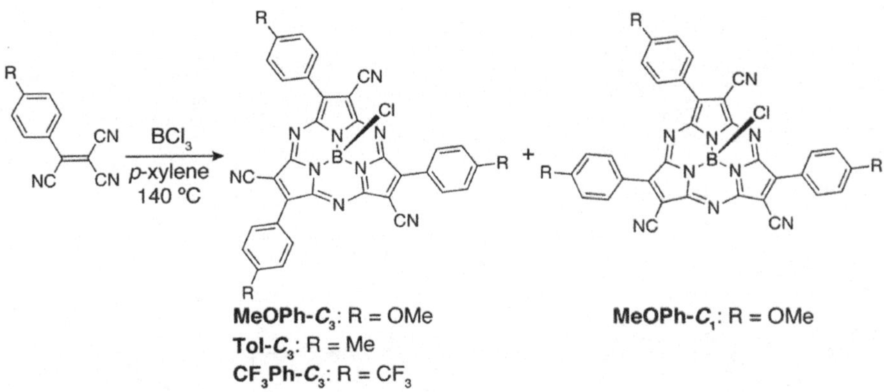

Scheme 4.10 Synthesis of push–pull SubPz

and Nc (Claessens et al. 2014). In contrast, the peripheral extension at the α,β-positions of SubPc has a smaller effect on the red-shift of absorption but endows the inherent chirality to the molecule (Shimizu et al. 2011). Two diastereomers of 1,2-subnaphthalocyanine (1,2-SubNc) exhibited similar UV/vis absorption, fluorescence, and MCD spectra (Fig. 4.16). However, the CD intensity of the C_1 isomer is one third of that of the C_3 isomer. On the basis of the band deconvolution analysis

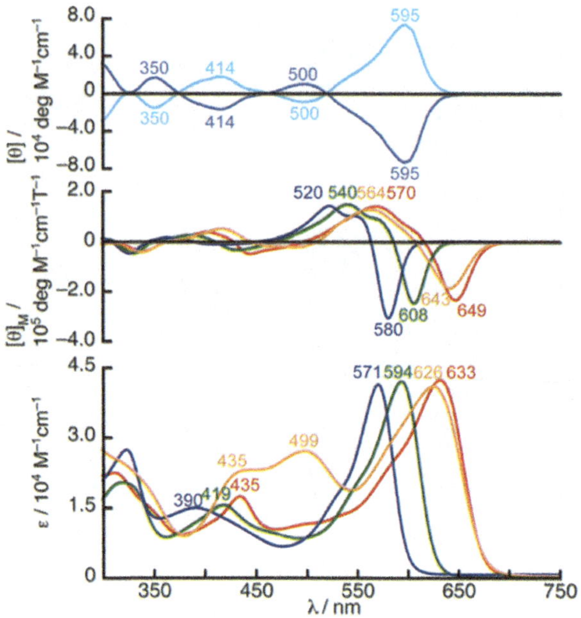

Fig. 4.15 UV/vis absorption (bottom) and MCD (middle) spectra of **MeOPh-C_3** (red), **Tol-C_3** (green), **CF_3Ph-C_3** (blue), and **MeOPh-C_1** (orange) in toluene, and CD spectra of enantiomers of **Tol-C_3** in toluene (top). Reprint with permission from ref. (Liang et al. 2014) Copyright 2014 RSC

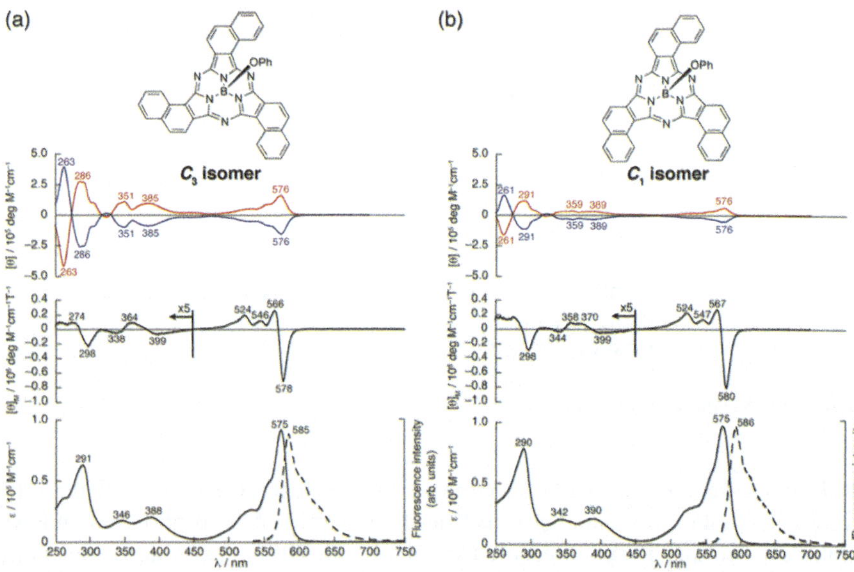

Fig. 4.16 UV/vis absorption (bottom), CD (top), and MCD (middle) spectra of the **a** C_3 isomer and **b** C_1 isomer of 1,2-SubNc in $CHCl_3$. Fluorescence spectra in $CHCl_3$ are shown by a dashed line (bottom). Redrawn with permission from ref. (Shimizu et al. 2011) Copyright 2011 ACS

using UV/vis absorption, CD, and MCD spectra, Shimizu and Kobayashi experimentally estimated the oscillator and rotational strengths and ascribed the observed changes in the CD intensities to smaller magnetic dipole moments of the C_1 isomer due to the oppositely arranged naphthalene ring. Since SubPc and its derivatives exhibit fluorescence, the inherent chirality of low-symmetry SubPcs can be used to generate circularly polarized luminescent materials.

4.4 Recent Examples of Pc Analogues with Unique Optical Properties

In this section, recent examples of Pc analogues with unique properties are introduced, focusing on NIR absorption and antiaromaticity. Sections 4.4.1 and 4.4.2 describe the NIR absorbing phosphorous(v) Pc and contracted and expanded Pc analogues, respectively.

4.4.1 Phosphorous(v) Pc as an NIR Chromophore

As described in the previous sections, the Q band of Pc in the far-red region (ca. 650 nm) is affected by several perturbations, such as molecular symmetry and substituents. It is, therefore, rational to target the NIR absorption using Pc as a fundamental chromophore skeleton. Synthetic investigation toward this goal has been intensively conducted from the early stage of Pc chemistry. In most cases, peripheral extension of the conjugated systems by fusing aromatic ring units or by oligomerization via peripheral benzene rings was attempted. Although the red-shifts of the Q band were achieved based on this strategy, such NIR-absorbing Pc analogues were found to be rather unstable due to the inevitably highly lying HOMO. Push-pull substitution also destabilizes the HOMO energy level because of the predominant push effect compared with the pull effect. Therefore, to create stable NIR chromophores, it is important to attain a narrow HOMO–LUMO gap by effectively controlling the HOMO and LUMO energy levels.

Recently, Furuyama and Kobayashi reported a rational molecular design to achieve such systems (Furuyama and Kobayashi 2017). As shown in Fig. 4.3., the HOMO (a_{1u} orbital) has large electron density on the α-positions of the benzo-rings, whereas large MO coefficients can be found on the central nitrogen atoms in the LUMO (e_g orbitals). They focused on these MO density distribution patterns and introduced electron-donating chalcogen substituents on the α-positions and high valent phosphorous(v) in the core, which destabilize the HOMO and stabilizes the LUMO, respectively (Fig. 4.17).

α-Arylchalcogen substituted Pcs were synthesized by the lithium method, and a phosphorous(v) was inserted by a reaction with excess $POBr_3$ in pyridine. After

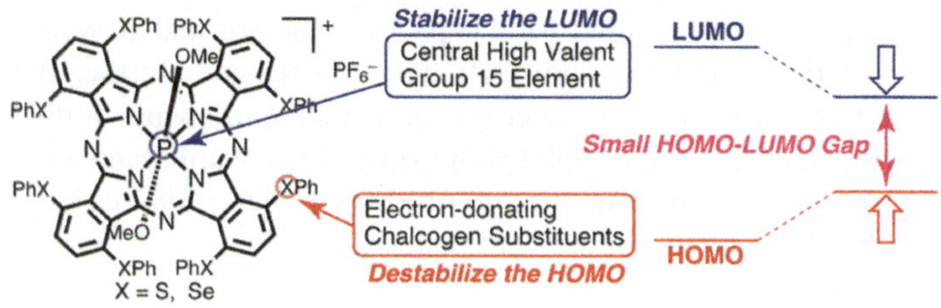

Fig. 4.17 Molecular design of NIR absorbing phosphorous(v) Pc

replacement of a counter anion with PF_6^-, stable phosphorous(v) Pc complexes (PPcs) were obtained. Both thiophenyl and selenophenyl substituted PPcs exhibited significant red-shifts of the Q band above 1000 nm (Fig. 4.18). Considering that the Q band of the α-thiophenyl substituted H_2Pc appeared at 809 nm, the high valent phosphorous(v) and electron-donating chalcogen atoms synergistically functioned to realize the observed NIR absorption.

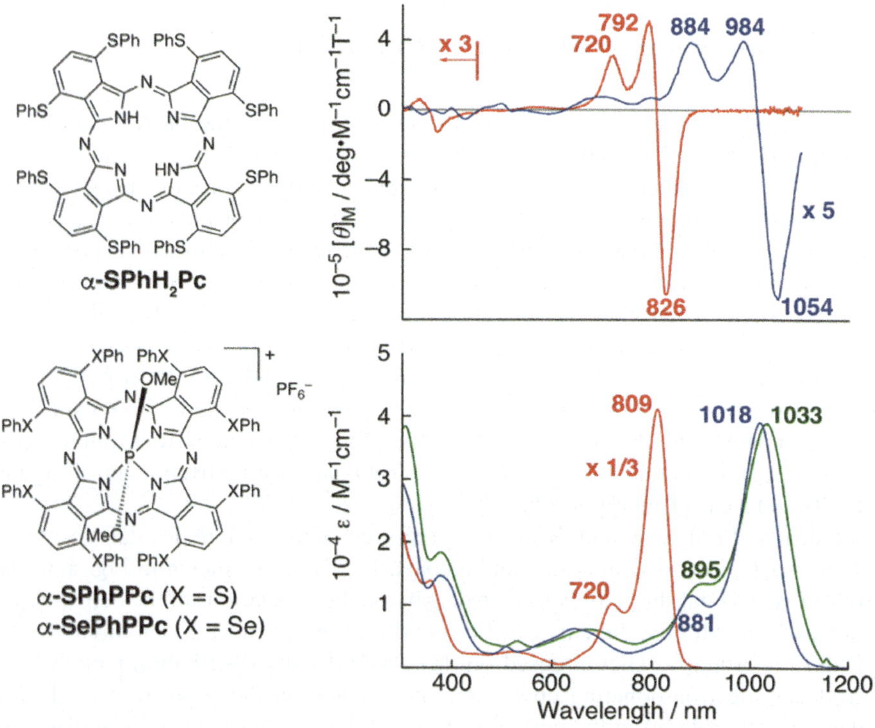

Fig. 4.18 UV/vis/NIR absorption (bottom) and MCD (top) spectra of α-SPhH_2Pc (red), α-SPhPPc (blue), and α-SePhPPc (green) in $CHCl_3$. Redrawn with permission from ref. (Furuyama and Kobayashi 2017) Copyright 2011 ACS

4.4.2 Contracted and Expanded Pc Analogues with Unique Properties

Contraction and expansion of the π-conjugated system of Pc by abstraction and addition of isoindole rings and bridging imino nitrogen atoms cause blue-shift and red-shift of the Q bands, respectively. These kinds of analogues are named *contracted Pc* and *expanded Pc* according to the nomenclature of porphyrin analogues (Sessler et al. 2017). In contrast to various expanded porphyrins (Tanaka and Osuka 2017), there are only a few expanded Pcs known to date because of their synthetic difficulties.

As described in Sect. 4.2, Pc is formed by cyclotetramerization of phthalonitrile or its derivatives. Because reactions are normally conducted at high temperatures, it is difficult to isolate oligoisoindole intermediates, which can be used for a stepwise synthesis. Considering that the availability of oligopyrrole precursors has enabled the stepwise synthesis of expanded porphyrins, this is one of the reasons why the synthesis of expanded Pcs is significantly difficult.

Superphthalocyanine (SuperPc) is a rare example of expanded Pcs. The first synthesis of SuperPc traces back more than 50 years ago. When Bloor et al. investigated the synthesis of uranyl complex of Pc, they characterized the obtained product as their target compound despite its significantly red-shifted Q band in the NIR region ranging from 800 to 1000 nm (Fig. 4.19) (Bloor et al. 1964). The reason for the false characterization of this compound is probably because H_2Pc was obtained after the removal of the central uranyl ion. Nowadays, it is known that ring contraction of SuperPc to H_2Pc occurs upon removal of the central uranyl ion. Although SuperPc is intriguing not only as an expanded Pc but also as a NIR chromophore, only the synthesis of unsubstituted and β-alkyl-substituted SuperPcs was reported (Day et al. 1975).

As Marks et al. reported, the reaction conditions are rather critical. The reaction should be conducted under dry conditions because the presence of water in reagents promotes the formation of H_2Pc as a byproduct. They also found the use of UO_2Cl_2 and *N*,*N*-dimethylformamide (DMF) is indispensable. A standard reaction scheme of SuperPc is shown in Fig. 4.19. Owing to their pioneering work, SuperPc is characterized as a Pc-like aromatic compound, but with an expanded 22π-electron conjugation (Day et al. 1975).

Recently, Furuyama and Kobayashi reported a benzo-ring-abstracted analogue of SuperPc called superporphyrazine (SuperPz) (Furuyama et al. 2012). SuperPz was synthesized from a similar uranyl template synthesis of 3,4-diarylpyrroline-2,5-diimine (Scheme 4.11). Despite the loss of the benzo-rings, SuperPz also exhibits broad absorption in the NIR region (Fig. 4.20). In an analogy to the theoretical description of the absorption properties of Pc based on Gouterman's four orbital theory, this broad NIR absorption can be assigned as a Q-like band arising from the 22π-electron aromatic conjugation. Faraday *A* terms were observed in both the Soret and Q band regions, being indicative of the degenerate excited states. In the same study, they also reported a low-symmetry SuperPz, which possesses one isoindole ring instead of a pyrrole ring. Due to the perturbation by the benzo-fusion,

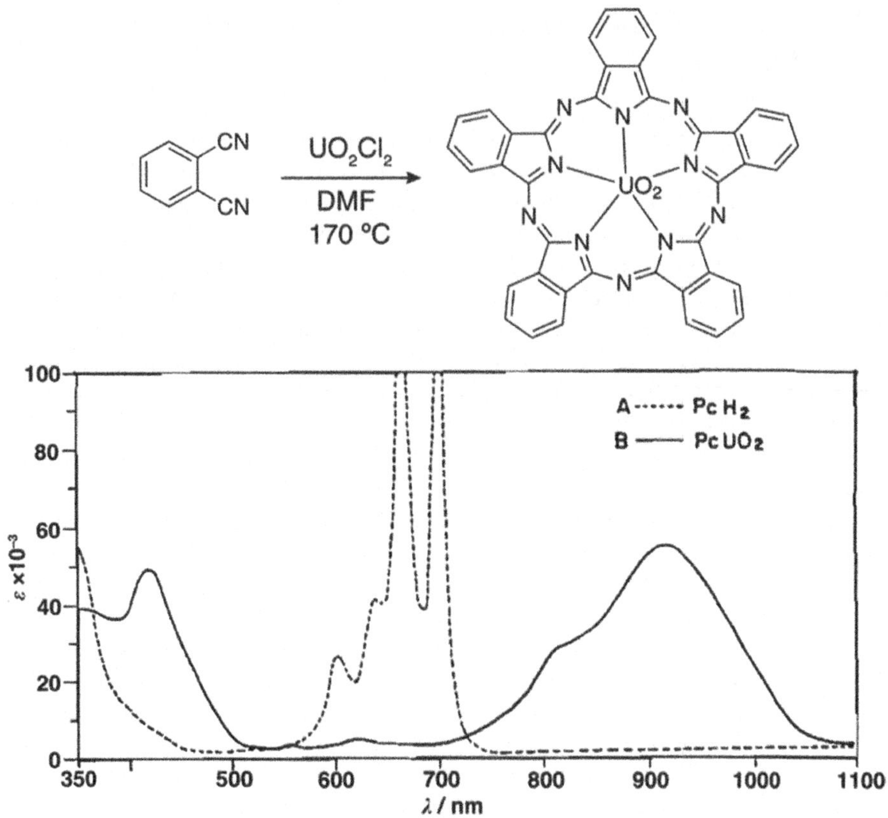

Fig. 4.19 Synthesis of SuperPc and UV/vis/NIR absorption spectra of H_2Pc and SuperPc in 1-chloronaphthalene. SuperPc is termed as $PcUO_2$ in the figure due to the false characterization. The absorption spectra are from the first paper (Bloor et al. 1964) and redrawn with permission. Copyright 1964 Canadian Science Publishing

the Q-like band was slightly red-shifted, but the splitting was rather modest so that the MCD signal corresponding to the Q-like band appeared as a pseudo-Faraday *A* term. The observed changes in the optical property by low-symmetrization of SuperPz chromophore are similar to those seen for Pcs. MO calculations based on the DFT method reproduced the degenerate LUMO for SuperPc and SuperPz and non-degenerate LUMO and LUMO+1 for low-symmetry SuperPz (Fig. 4.21).

To solve the difficulty in the synthesis of isoindole oligomers as precursors, the introduction of other heterocycles has been investigated. Two of such expanded Pc analogues have been known to date (Fig. 4.22). Torres reported the synthesis of [4 + 2] macrocycle comprising four isoindole units and two triazole units from a self-condensation of [2 + 1] linear precursor (Rodríguez-Morgade et al. 2001). This kind of structure is often referred to as hemiporphyrazine. The other example independently reported by Kobayashi and Torres is called [3 + 3] hemiporphyrazine, which

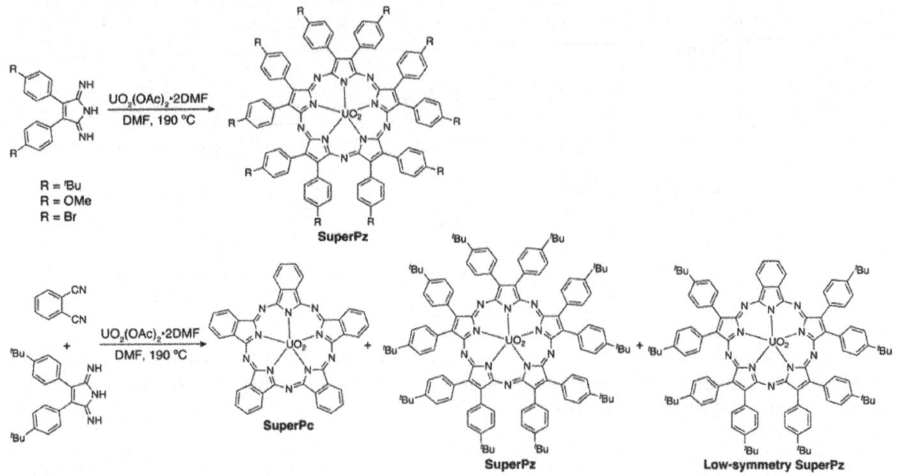

Scheme 4.11 Synthesis of SuperPz

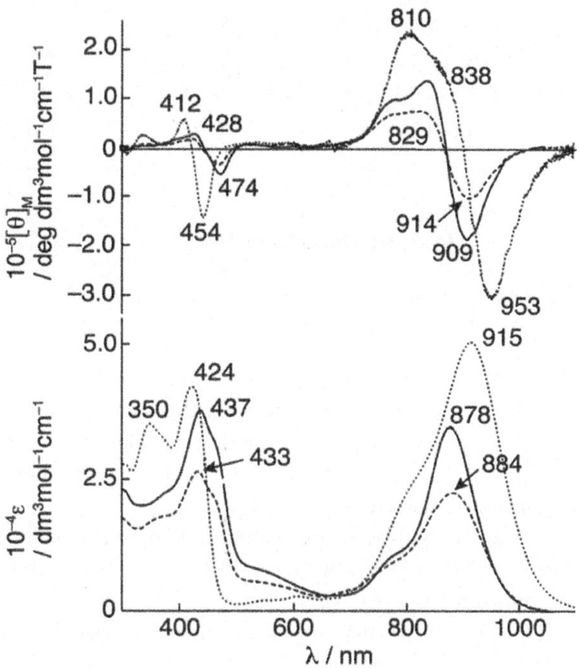

Fig. 4.20 UV/vis/NIR absorption (bottom) and MCD (top) spectra of SuperPc (dotted line), SuperPz (R = *t*Bu, solid line), and low-symmetry SuperPz (dashed line)

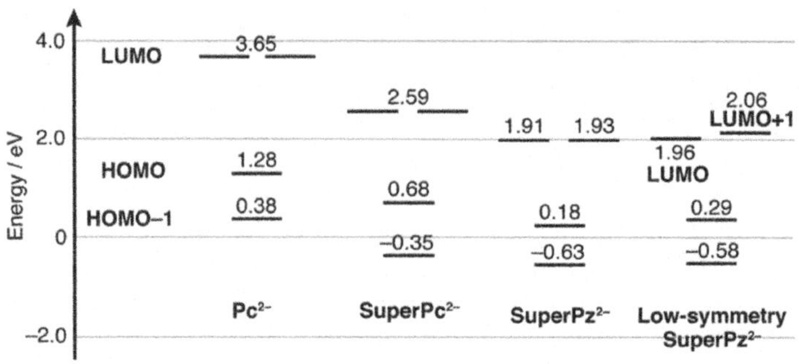

Fig. 4.21 Partial frontier MO diagram of model ligand structures (dianion forms of Pc (Pc^{2-}), SuperPc ($SuperPc^{2-}$), SuperPz ($SuperPz^{2-}$), and low-symmetry SuperPz (low-symmetry $SuperPz^{2-}$)) at B3LYP/6-31G* level

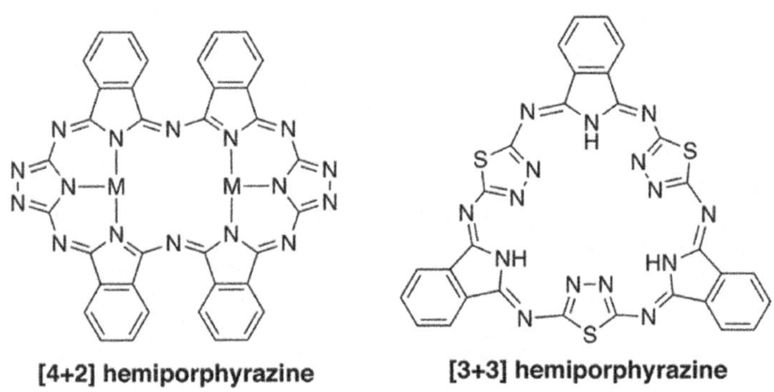

Fig. 4.22 Structures of expanded hemiporphyrazines

was synthesized from a mixed-condensation reaction of phthalonitrile or maleonitrile derivatives with 2,5-diamino-1,2,4-thiadiazole (Kobayashi et al. 2001; Islyaikin et al. 2001). Despite the expanded Pc-like structures, their optical properties are totally different from those of Pc and SuperPc due to their weak antiaromatic or nonaromatic nature.

Recently, using diamino-substituted precursors, unique antiaromatic Pc analogues were successfully synthesized. In the last part of this section, two recent examples are introduced.

During the investigation on new synthetic methods of Pc, Furuyama and Kobayashi noticed that a diamino-β-isoindigo skeleton was formed in a reaction of phthalonitrile and thiolate anions (Furuyama et al. 2015). They used this isoindigo derivative in a mixed condensation reaction with phthalonitrile to obtain a new macrocycle comprising five isoindole units and three bridging nitrogen atoms (Scheme 4.12). They named this molecule pentabenzotriazasmaragdyrin (PBTAS) after its porphyrin analogue, smaragdyrin (Pareek et al. 2012). Because of the formal 20π-electron conjugation, PBTAS exhibited antiaromatic characters, such as a strong

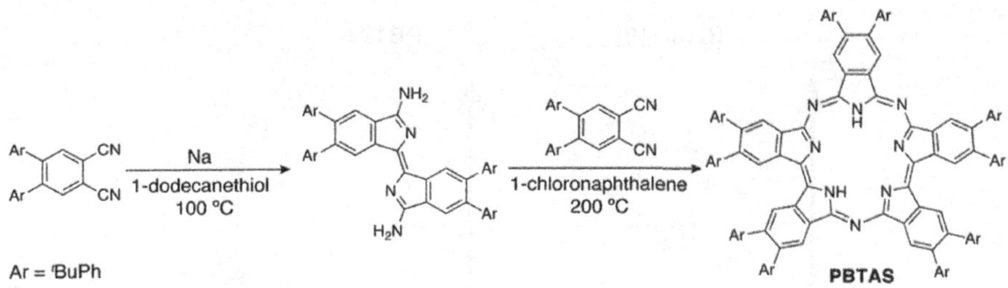

Scheme 4.12 Synthesis of PBTAS

paratropic ring current effect and broad ill-defined NIR absorption. In contrast to the sharp Q band observed for aromatic Pc and its analogues, antiaromatic compounds generally exhibit a forbidden band in the NIR region and intense bands in a much shorter wavelength region. According to Michl's perimeter model, the absorption spectrum of antiaromatic compounds mainly consists of S, N (N_1 and N_2), and P (P_1 and P_2) bands, which arise from transitions between six frontier orbitals ($h-$, $h+$, $s-$, $s+$, $l-$, $l+$). (Howeler et al. 1998). These six orbitals are derived from a symmetry perturbation of the degenerate HOMO, singly occupied molecular orbital (SOMO), and LUMO of the 4 N-electron perimeter model. The S band consisting of the $s- \rightarrow s+$ transition is forbidden because of an intrashell transition nature. The absorption in the visible region was assigned as the N and P bands, which arise from the configurational interactions of $s- \rightarrow l-$ and $h+ \rightarrow s+$ transitions and $s- \rightarrow l+$ and $h- \rightarrow s+$ transitions. The observed absorption bands of PBTAS can be assigned based on this model (Fig. 4.23).

Replacement of isoindole units with other heterocycles is also found to be a useful strategy to change the conjugation of SubPc. Recently, Ng and Kobayashi reported the synthesis of carbazosubphthalocyanine, in which one isoindole unit was replaced with a carbazole unit (Scheme 4.13) (Chan et al. 2019). Unlike regular SubPc, carbazosubphthalocyanine exhibited a planar structure. Due to the expansion with the carbazole moiety, this molecule attained a 16π-electron conjugation and exhibited antiaromatic characters, such as a paratropic ring current effect estimated based on the significant down-field shifts in the ^{11}B and ^{19}F NMR spectra, NICS (nucleus-independent chemical shift) (Chen et al. 2005), ACID (anisotropy of current density) (Geuenich et al. 2005), and UV/vis/NIR absorption spectrum. In the MCD spectrum, a negative envelope appeared corresponding to the forbidden band at 600–900 nm, whereas a peak and trough were seen for the intense bands in ascending energy (Fig. 4.24). According to Michl's prediction of the MCD sign sequence using the relative energy differences between frontier six orbitals ($h-$, $h+$, $s-$, $s+$, $l-$, $l+$), (Fleischhauer et al. 2000), the authors assigned the observed MCD spectrum. From the $|\Delta HSL| > \Sigma HL > |\Delta HL|$ relationship ($\Delta HSL = 2(\Delta HS - \Delta LS)$, $\Sigma HL = \Delta H + \Delta L$, $\Delta HL = \Delta H - \Delta L$ (Fig. 4.25)), this molecule can be classified as an orbital shift-dominated system of S-perturbed perimeters, for which positive-to-negative MCD sign sequence is expected for the N_1 and P_1 (and N_2) transitions.

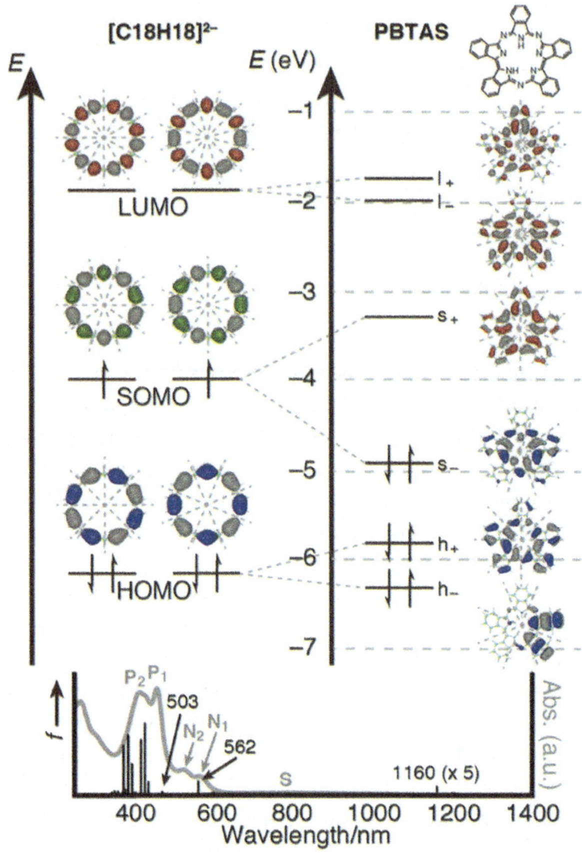

Fig. 4.23 Partial frontier MO diagram of PBTAS (top, right) and its 20π-electron annulene perimeter (top, left) at the B3LYP/6-31G(d) level. Arbitrary nodal lines are drawn on the iso-surface plots. UV/vis/NIR absorption (bottom, gray line) and calculated absorption (bottom, black sticks, TDDFT method at the B3LYP/6-31G(d) level) spectra. Reprint with permission from ref. (Furuyama et al. 2015) Copyright 2011 ACS

4.5 Summary and Outlook

In this chapter, conventional syntheses of Pc and SubPc and the structure–property relationships with a special focus on their optical properties are introduced. This chapter also emphasizes how to characterize chromophore properties both experimentally and theoretically. In the last part of this chapter, recent unique examples of Pc and related analogues to target NIR absorption and stable antiaromatic conjugated systems are covered.

Since the discovery of Pc more than a century ago, Pc has been intensively investigated as an industrial dye rather than a fundamental chromophore structure to develop new properties, so that the basic synthetic chemistry toward these goals

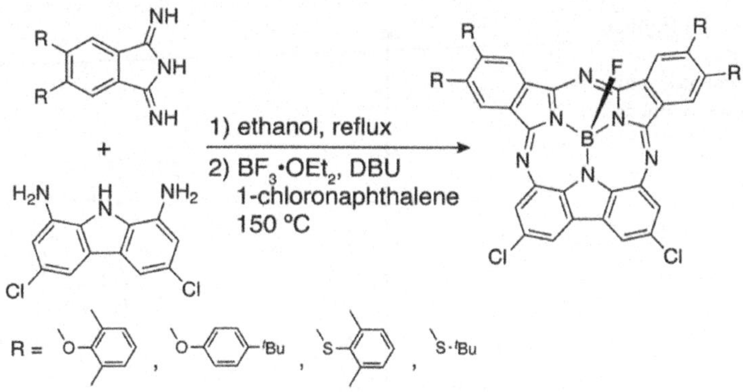

Scheme 4.13 Synthesis of carbazosubphthalocyanine

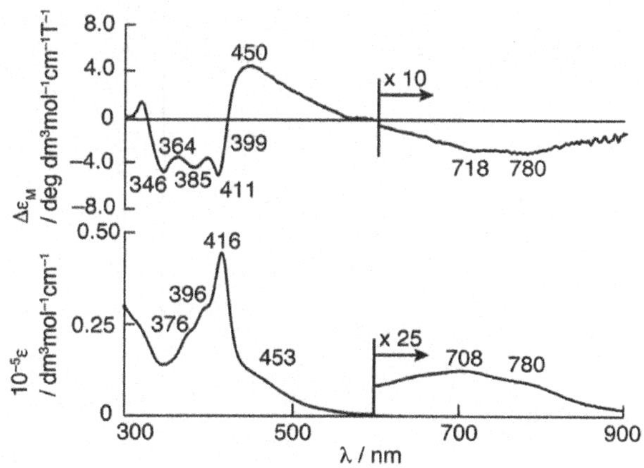

Fig. 4.24 UV/vis/NIR absorption (bottom) and MCD (top) spectra of carbazosubphthalocyanine (R = 2,6-dimethylphenoxy) in $CHCl_3$

has been less investigated compared with the rich synthetic chemistry of porphyrin and its analogues. However, considering that Pc and SubPc exhibit intense Q band in the UV/vis region, it is reasonable to use these structures to design new functional chromophores. From this viewpoint, several synthetic investigations have been performed, including the author's contribution to this research field (Shimizu and Kobayashi 2014). Although the synthetic methods have still been limited, the accumulation of knowledge will lead to a new phase of synthetic chemistry of Pc and related analogues, the glimpse of which can be seen in the recent synthetic achievements of PBTAS and carbazosubphthalocyanine. The author hopes this book chapter may become a future guide to develop new Pc-based functional chromophores.

Fig. 4.25 Partial frontier MO diagram of carbazosubphthalocyanine (R = 2,6-dimethylphenoxy) at the CAM-B3LYP/6-31G(d) level

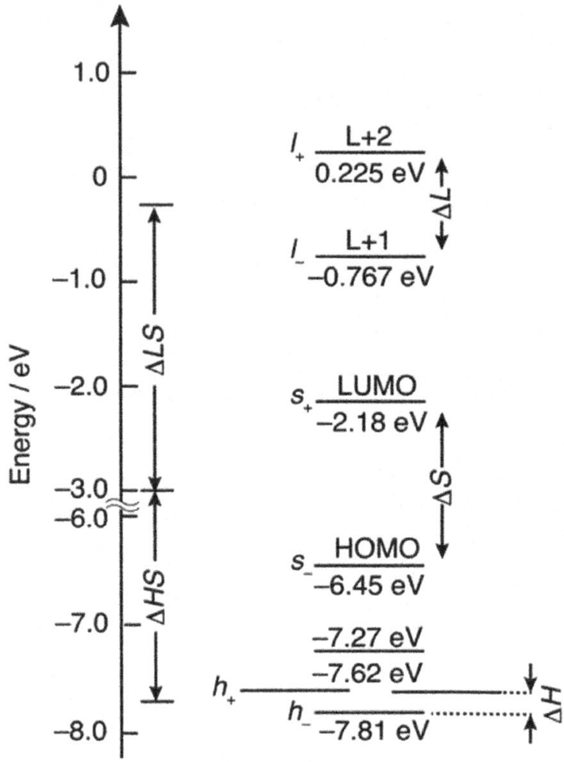

References

Braun A, Tcherniac J (1907) Über Die Produkte Der Einwirkung Von Acetanhydrid Auf Phthalamid Ber Dtsch Chem Ges 40:2709–2714

(a) Linstead RP (1934) Phthalocyanines part I A new type of synthetic colouring matters. J Chem Soc 1016–1017; (b) Byrne GT, Linstead RP, Lowe AR (1934) Phthalocyanines part II The preparation of phthalocyanine and some metallic derivatives from o-cyanobenzamide and phthalimide. J Chem Soc 1017–1022; (c) Dent CE, Linstead RP (1934) Phthalocyanines part IV Copper phthalocyanines. J Chem Soc 1027–1031; (d) Dent CE, Linstead RP, Lowe AR (1934) Phthalocyanines part VI The structure of the phthalocyanines. J Chem Soc 1033–1039; (e) Linstead RP, Rydon HN (1934) Investigation of the olefinic acids Part XV The effect of peroxides on the orientation of the addition of hydrogen bromide to vinyl- and allyl-acefic acids. J Chem Soc 2001–2003; (f) Linstead RP, Whalley M (1952) Conjugated Macrocycles .22: tetrazaporphin and its metallic derivatives. J Chem Soc 4839–4846; (g) Ficken GE, Linstead RP, Stephen E, Whalley M (1958) 789: Conjugated macrocycles. Part XXXI. Catalytic hydrogenation of tetrazaporphins, with a note on its stereochemical course. J Chem Soc 3879–3886

(a) Robertson JM (1935) 136. An X-ray study of the structure of the phthalocyanines. Part I. The metal-free, nickel, copper, and platinum compounds. J Chem Soc 615–621; (b) Robertson JM (1936) An X-ray study of the phthalocyanines - Part II Quantitative structure determination of the metal-free compound. J Chem Soc 1195–1209; (c) Robertson JM, Woodward I (1937) An X-ray study of the phthalocyanines. Part III. Quantitative structure determination of nickel phthalocyanine. J Chem Soc 219–230

Kadish KM, Smith KM, Guilard R (eds) (2003) The porphyrin handbook vol 15–20. Academic Press, San Diego
Leznoff CC, Lever ABP (eds) (1989–1996) Phthalocyanines: properties and applications vol 1–4. Wiley-VCH, New York
McKeown NB (1998) Phthalocyanine materials: synthesis. Cambridge University Press, Cambridge, Structure and Function
Kobayashi N, Fukuda T (2006) Functional Dyes. In: Kim S (ed) Elsevier, Oxford
Meller A, Ossko A (1972) Triisindolo[1,2,3-Cd-1',2',3'-Gh-1",2",3"-Kl][2,3a,5,6a,8,9a,9b]-Hena-Azaboraphenalene. Monatsh Chem 103:150–155
(a) Sessler JL, Gross Z, Furuta H (2017) Introduction: Expanded, Contracted, and Isomeric Porphyrins. Chem Rev 117:2201–2202; (b) Sessler JL, Weghorn SJ (1997) Expanded, contracted, and isomeric porphyrins. Pergamon Press, New York
(a) Claessens CG, González-Rodríguez D, Rodríguez-Morgade MS, Medina A, Torres T (2014) Subphthalocyanines, subporphyrazines, and subporphyrins: singular nonplanar aromatic systems. Chem Rev 114:2192–2277; (b) Claessens CG, González-Rodríguez D, Torres T (2002) Subphthalocyanines: singular nonplanar aromatic compounds-synthesis, reactivity, and physical properties. Chem Rev 102:835–853
(a) Shimizu S, Nakano S, Hosoya T, Kobayashi N (2011) Pyrene-fused subphthalocyanine. Chem Commun 47:316–318; (b) Nakano S, Kage Y, Furuta H, Kobayashi N, Shimizu S (2016) Pyrene-bridged boron subphthalocyanine dimers: combination of planar and bowl-shaped π-conjugated systems for creating uniquely curved π-conjugated systems. Chem –Eur J 22:7706–7710
(a) Sánchez-Molina I, Claessens CG, Grimm B, Guldi DM, Torres T (2013) Trapping fullerenes with jellyfish-like subphthalocyanines. Chem Sci 4:1338–1344; (b) Sánchez-Molina I, Grimm B, Krick Calderon RM, Claessens CG, Guldi DM, Torres T (2013) Self-assembly, host–guest chemistry, and photophysical properties of subphthalocyanine-based metallosupramolecular capsules. J Am Chem Soc 135:10503–10511
Konarev DV, Troyanov SI, Lyubovskaya RN (2015) Coordination complex of boron subphthalocyanine (BSubPc) with fluorenone pinacolate: effective π–π interaction of concave BSubPc macrocycle with fullerene C_{60}. CrystEngComm 17:3923–3926
Rhoda HM, Kayser MP, Wang Y, Nazarenko AY, Belosludov RV, Kiprof P, Blank DA, Nemykin VN (2016) Tuning up an electronic structure of the subphthalocyanine derivatives toward electron-transfer process in noncovalent complexes with C_{60} and C_{70} fullerenes: experimental and theoretical studies. Inorg Chem 55:9549–9563
(a) Sastre A, Torres T, Díaz-García MA, Agulló-López F, Dhenaut C, Brasselet S, Ledoux I, Zyss J (1996) Subphthalocyanines: novel targets for remarkable second-order optical nonlinearities. J Am Chem Soc 118:2746–2747; (b) Martín G, Rojo G, Agulló-López F, Ferro VR, García de la Vega JM, Martínez-Díaz MV, Torres T, Ledoux I, Zyss J (2002) Subphthalocyanines and subnaphthalocyanines: nonlinear quasi-planar octupolar systems with permanent polarity. J Phys Chem B 106:13139–13145
Renshaw CK, Xu X, Forrest SR (2010) A monolithically integrated organic photodetector and thin film transistor. Org Electron 11:175–178
(a) Morse GE, Castrucci JS, Helander MG, Lu Z-H, Bender TP (2011) Phthalimido-boronsubphthalocyanines: new derivatives of boronsubphthalocyanine with bipolar electrochemistry and functionality in OLEDs. ACS Appl Mater Interfaces 3:3538–3544; (b) Helander MG, Morse GE, Qiu J, Castrucci JS, Bender TP, Lu Z-H (2010) Pentafluorophenoxy boron subphthalocyanine as a fluorescent dopant emitter in organic light emitting diodes. ACS Appl Mater Interfaces 2:3147–3152
Mutolo KL, Mayo EI, Rand BP, Forrest SR, Thompson ME (2006) Enhanced open-circuit voltage in subphthalocyanine/C_{60} organic photovoltaic cells. J Am Chem Soc 128:8108–8109
(a) Duan C, Zango G, García-Iglesias M, Colberts FJM, Wienk MM, Martínez-Díaz MV, Janssen RAJ, Torres T (2016) The role of the axial substituent in subphthalocyanine acceptors for bulk-heterojunction solar cells. Angew Chem Int Ed 56:154–158; (b) Cnops K, Zango G, Genoe J,

Heremans P, Martínez-Díaz MV, Torres T, Cheyns D (2015) Energy level tuning of non-fullerene acceptors in organic solar cells. J Am Chem Soc 137:8991–8997
(a) Gouterman M (1959) Study of the effects of substitution on the absorption spectra of porphin. J Chem Phys 30:1139–1161; (b) Gouterman M (1961) Spectra of porphyrins. J Mol Spectrosc 6:138–163
Kobayashi N, Muranaka A, Mack J (2012) Circular dichroism and magnetic circular dichroism spectroscopy for organic chemists. Royal Society of Chemistry, UK
(a) Mack J, Stillman MJ, Kobayashi N (2007) Application of MCD spectroscopy to porphyrinoids. Coord Chem Rev 251:429–453; (b) Mack J, Kobayashi N, Shen Z (2012) The effect of structural modifications on the properties of porphyrinoids. In: Kadish KM, Smith KM, Guilard R (eds) Handbook of Porphyrin Science, vol 23. World Scientific, Singapore
(a) Michl J (1978) Magnetic circular dichroism of cyclic π-electron systems. 1. Algebraic solution of the perimeter model for the A and B terms of high-symmetry systems with a (4N + 2)-electron [n]annulene perimeter. J Am Chem Soc 100:6801–6811; (b) Michl J (1984) Magnetic circular dichroism of aromatic molecules. Tetrahedron 40:3845–3934
Sharman WM, van Lier JE (2003) Synthesis of phthalocyanine precursors. In: Kadish KM, Smith KM, Guilard R (eds) The Porphyrin Handbook, vol 15. Academic Press, San Diego, pp 1–60
McKeown NB (2003) The synthesis of symmetrical phthalocyanines. In: Kadish KM, Smith KM, Guilard R (eds) The Porphyrin Handbook, vol 15. Academic Press, San Diego, pp 61–160
Tomoda H, Saito S, Ogawa S, Shiraishi S (1980) Synthesis of phthalocyanines from phthalonitrile with organic strong bases. Chem Lett 9:1277–1280
Brach PJ, Grammatica SJ, Ossanna OA, Weinberger L (1970) Improved synthesis of metal-free phthalocyanines. J Heterocycl Chem 7:1403–1405
(a) Uosaki K, Ueda J (1992) Photoelectrochemical synthesis of phthalocyanine. J Chem Soc, Perkin Trans 2:1331–1333; (b) Yang CH, Lin SF, Chen HL, Chang CT (1980) Electrosynthesis of the metal phthalocyanine complexes. Inorg Chem 19:3541–3543; (c) Petit MA, Plichon V, Belkacemi H (1989). Nouv J Chim 13:459; (d) Petit MA, Thami T, Sirlin S, Lelievre D (1991). Nouv J Chim 15:71
Hall TH, Greenberg S, McArthur CR, Khouw B, Leznoff CC (1982) Nouv J Chim 6:653
Musluoglu E, Gürek A, Ahsen V, Gül A, Bekaroğlu ö, (1992) Unsymmetrical phthalocyanines with a single macrocyclic substituent. Chem Ber 125:2337–2339
Claessens CG, González-Rodríguez D, del Rey B, Torres T, Mark G, Schuchmann H-P, von Sonntag C, MacDonald JG, Nohr Ronald S (2003) Highly efficient synthesis of chloro- and phenoxy-substituted subphthalocyanines. Eur J Org Chem 2547–2551
Fukuda T, Kobayashi N (2010) UV-visible absorption sepctrscopic properties of phthalocyanines and related macrocycles. In: Kadish KM, Smith KM, Guilard R (eds) Handbook of porphyrin science, vol 9. World Scientific, Singapore
Simpson WT (1949) On the theory of the π-electron system in porphines. J Chem Phys 17:1218–1221
Longuet-Higgins HC, Rector CW, Platt JR (1950) Molecular orbital calculations on porphine and tetrahydroporphine. J Chem Phys 18:1174–1181
Mack J, Stillman MJ (2001) Assignment of the optical spectra of metal phthalocyanines through spectral band deconvolution analysis and zindo calculations. Coord Chem Rev 219–221:993–1032
Mack J, Stillman MJ (1995) Band deconvolution analysis of the absorption and magnetic circular dichroism spectral data of ZnPc(-2) recorded at cryogenic temperatures. J Phys Chem 99:7935–7945
Ledson DL, Twigg MV (1975) Acid-base behaviour of phthalocyanine. Inorg Chim Acta 13:43–46
Mack J, Kobayashi N (2011) Low symmetry phthalocyanines and their analogues. Chem Rev 111:281–321
Kobayashi N, Nakajima S-i, Ogata H, Fukuda T (2004) Synthesis, spectroscopy, and electrochemistry of tetra-tert-butylated tetraazaporphyrins, phthalocyanines, naphthalocyanines, and anthracocyanines, together with molecular orbital calculations. Chem –Eur J 10:6294–6312

(a) Kaito A, Nozawa T, Yamamoto T, Hatano M, Orii Y (1977) LCAO MO Scf π-Electron calculations on the magnetic circular dichroism of porphin, protoporphyrin, and porphyrin. Chem Phys Lett 52:154–160; (b) Tajiri A, Winkler J (1983) Magnetic circular-dichroism and molecular-orbital studies on condensed thiadiazoles. Z Naturforsch A 38:1263–1269
Miwa H, Ishii K, Kobayashi N (2004) Electronic structures of zinc and palladium tetraazaporphyrin derivatives controlled by fused benzo rings. Chem –Eur J 10:4422–4435
Fukuda T, Makarova EA, Luk'yanets EA, Kobayashi N (2004) Synthesis and spectroscopic and electrochemical studies of novel benzo- or 2,3-naphtho-fused tetraazachlorins, bacteriochlorins, and isobacteriochlorins. Chem –Eur J 10:117–133
Shimizu S, Haseba Y, Yamazaki M, Kumazawa G, Kobayashi N (2014) Control of chromophore symmetry by positional isomerism of peripheral substituents. Chem –Eur J 20:4822–4828
Liang X, Shimizu S, Kobayashi N (2014) Sizeable red-shift of absorption and fluorescence of subporphyrazine induced by peripheral push and pull substitution. Chem Commun 50:13781–13784
Shimizu S, Miura A, Khene S, Nyokong T, Kobayashi N (2011) Chiral 1,2-subnaphthalocyanines. J Am Chem Soc 133:17322–17328
(a) Furuyama T, Kobayashi N (2017) Azaporphyrin phosphorus(v) complexes: synthesis, structure, and modification of optical properties. Phys Chem Chem Phys 19:15596–15612; (b) Kobayashi N, Furuyama T, Satoh K (2011) Rationally designed phthalocyanines having their main absorption band beyond 1000 nm. J Am Chem Soc 133:19642–19645
Tanaka T, Osuka A (2017) Chemistry of meso-aryl-substituted expanded porphyrins: aromaticity and molecular twist. Chem Rev 117:2584–2640
Bloor JE, Schlabitz J, Walden CC, Demerdache A (1964) Organic complexes of uranium: Part I. The synthesis and spectrum of uranyl phthalocyanine. Can J Chem 42:2201–2208
(a) Day VW, Marks TJ, Wachter WA (1975) Large metal ion-centered template reactions: uranyl complex of cyclopentakis(2-iminoisoindoline). J Am Chem Soc 97:4519–4527; (b) Cuellar EA, Marks TJ (1981) Synthesis and characterization of metallo and metal-free octaalkylphthalocyanines and uranyl decaalkylsuperphthalocyaninates. Inorg Chem 20:3766–3770; (c) Marks TJ, Stojakovic DR (1978) Large metal ion-centered template reactions. chemical and spectral studies of the "Superphthalocyanine" Dioxocyclopentakis(1-iminoisoindolinato)uranium(VI) and its derivatives. J Am Chem Soc 100:1695–1705
Furuyama T, Ogura Y, Yoza K, Kobayashi N (2012) Superazaporphyrins: meso-pentaazapentaphyrins and one of their low-symmetry derivatives. Angew Chem Int Ed 51:11110–11114
Rodríguez-Morgade MS, Cabezón B, Esperanza S, Torres T (2001) Expanded phthalocyanine analogues: synthesis and characterization of new triazole-derived annulenes containing six heterocyclic subunits. Chem –Eur J 7:2407–2413
Kobayashi N, Inagaki S, Nemykin VN, Nonomura T (2001) A novel hemiporphyrazine comprising three isoindolediimine and three thiadiazole units. Angew Chem Int Ed 40:2710–2712
(a) Islyaikin MK, Danilova EA, Yagodarova LD, Rodríguez-Morgade MS, Torres T (2001) Thiadiazole-derived expanded heteroazaporphyrinoids. Org Lett 3:2153–2156; (b) Trukhina ON, Rodríguez-Morgade MS, Wolfrum S, Caballero E, Snejko N, Danilova EA, Gutierrez-Puebla E, Islyaikin MK, Guldi DM, Torres T (2010) Scrutinizing the chemical nature and photophysics of an expanded hemiporphyrazine: the special case of [30]Trithia-2,3,5,10,12,13,15,20,22,23,25,30-dodecaazahexaphyrin. J Am Chem Soc 132:12991–12999
Furuyama T, Sato T, Kobayashi N (2015) A bottom-up synthesis of antiaromatic expanded phthalocyanines: pentabenzotriazasmaragdyrins, i.e. norcorroles of superphthalocyanines. J Am Chem Soc 137:13788–13791
Pareek Y, Ravikanth M, Chandrashekar TK (2012) Smaragdyrins: emeralds of expanded porphyrin family. Acc Chem Res 45:1801–1816
(a) Howeler UW, Downing J, Fleischhauer J, Michl J (1998) MCD of non-aromatic cyclic π-electron systems. Part 1. The perimeter model for antiaromatic 4N-electron [n]annulene biradicals 1. J Chem Soc, Perkin Trans 2:1101–1118; (b) Fleischhauer J, Höweler U, Michl J (2000) MCD

of nonaromatic cyclic π-electron systems. 3. The perimeter model for low-symmetry "Unaromatic" and "Ambiaromatic" molecules derived from 4N-electron [n]annulenes. J Phys Chem A 104:7762–7775; (c) Fleischhauer J, Höweler U, Spanget-Larsen J, Raabe G, Michl J (2004) Magnetic circular dichroism of nonaromatic cyclic π-electron systems. 5. Biphenylene and its aza analogues. J Phys Chem A 108:3225–3234

Chan JYM, Kawata T, Kobayashi N, Ng DKP (2019) Boron(III) carbazosubphthalocyanines: core-expanded antiaromatic boron(III) subphthalocyanine analogues. Angew Chem Int Ed 58:2272–2277

Chen Z, Wannere CS, Corminboeuf C, Puchta R, PvR Schleyer (2005) Nucleus-independent chemical shifts (NICS) as an aromaticity criterion. Chem Rev 105:3842–3888

Geuenich D, Hess K, Köhler F, Herges R (2005) Anisotropy of the induced current density (ACID), a general method to quantify and visualize electronic delocalization. Chem Rev 105:3758–3772

(a) Shimizu S, Kobayashi N (2014) Structurally-modified subphthalocyanines: molecular design towards realization of expected properties from the electronic structure and structural features of subphthalocyanine. Chem Commun 50:6949–6966; (b) Shimizu S, Furuta H (2015) Core-modified phthalocyanines and subphthalocyanines: a synthetic strategy towards core-modification and novel properties arising from the inner ring-expansion. Macroheterocycles 8:332–342

Chapter 5
BODIPY Dyes and Their Analogues

Yasuhiro Kubota

Abstract This chapter discusses organoboron complexes, especially the synthesis and correlation between molecular structure and optical properties. Organoboron complexes are among the most important fluorescent dyes. Boron complexation of the ligand (dye) contributes to the enhancement of the fluorescent properties of the dye through rigidization of the chromophore, which restricts non-radiative processes. Additionally, the optical properties of organoboron complexes, such as absorption and fluorescence maxima, and fluorescence quantum yields are strongly dependent on the ligand type (dye). Therefore, boron complexation of the ligand (dye) is an efficient strategy to synthesize novel fluorescent dyes. In fact, organoboron complexes have been actively developed and applied in various fields, including optoelectronics and biomedicine. Understanding synthetic methods and the correlation between molecular structure and optical properties help us in producing materials with the desired properties. This chapter discusses BODIPY dyes, the most famous fluorescent organoboron complexes, and then outlines the four-coordinate monoboron complexes possessing anionic bidentate ligands (N^N, N^O, O^O, and other types) and multinuclear boron complexes, with a special focus on the results reported by the author.

Keywords Boron-complexation · BODIPY · Fluorescence · Pyridomethene · Near-infrared dye

5.1 Three-Coordinate Organoboron Compounds

Boron is classified as a metalloid and has a wide range of applications such as glass (borosilicate glass), ceramics, cleaning products, insecticides, drugs (bortezomib), etc. (Das et al. 2013). A boron atom is essentially three-coordinate possessing three sp^2 hybrid orbitals and a vacant p-orbital. Trivalent (three-coordinate) boron compounds generally have a trigonal-planar geometry with 120° bond angles

---

Y. Kubota (✉)
Gifu University, Gifu, Japan
e-mail: kubota@gifu-u.ac.jp

Y. Ooyama and S. Yagi (eds.), *Progress in the Science of Functional Dyes*,
https://doi.org/10.1007/978-981-33-4392-4_5

(Fig. 5.1a). Because of the vacant p-orbital, trivalent organoboron compounds are inherently electron-deficient and have strong π-electron-accepting ability through p-π* conjugation between the boron atom and a π-conjugated moiety (Wakamiya and Yamaguchi 2015). Additionally, the vacant p-orbital leads to high Lewis acidity, whereby a Lewis base or nucleophile can interact causing the boron atom to interconvert between neutral sp^2-hybridized orbitals (three-coordinate) and anionic sp^3-hybridized orbitals (four-coordinate). Furthermore, even in the excited state, a trivalent boron moiety works as a π-electron-accepting group (Wakamiya and Yamaguchi 2015). Incorporation of trivalent boron into a π-conjugated framework having strong electron-donating groups leads to an intermolecular charge transfer (ICT) transition.

On the other hand, nucleophilic attack at the vacant p-orbital results in the formation of a four-coordinate species, followed by bond cleavage. Thus, trivalent boron compounds are usually susceptible to hydrolysis by moisture. The introduction of bulky substituents such as mesityl (Mes) groups helps in increasing kinetic stability by blocking the attack of nucleophiles (Fig. 5.1b) (Entwistle and Marder 2002). In many cases, the introduction of two Mes groups is sufficient to prevent hydrolysis by moisture in the air. The Mes_2B group works as an electron-accepting moiety with π acceptor strength comparable to that of a cyano group (Entwistle and Marder 2002; Kaim and Shultz 1984).

Numerous stable trivalent boron compounds with bulky substituents have been developed (Li et al. 2017a) and applied to functional materials including electron-transporting and emissive materials for organic light-emitting diodes (OLEDs) (Turkoglu et al. 2017), anion sensors (Hudson and Wang 2009), nonlinear optical materials (Yuan et al. 2006), and two-photon absorption (TPA) and fluorescence (TPF) materials (Griesbeck et al. 2016). Trivalent boron compounds can also be stabilized by structural constraint; planarized triphenylborane with a rigidly fixed cyclic skeleton was reported as a stable compound due to destabilization of the four-coordinate intermediate and/or stabilization of B–C bonds by the chelating effect (Fig. 5.1c) (Zhou et al. 2012). Replacement of B–C bond to B–O bond or B–N bonds is also a good strategy to stabilize the trivalent boron structure (Fig. 5.1d) (Numano et al. 2016).

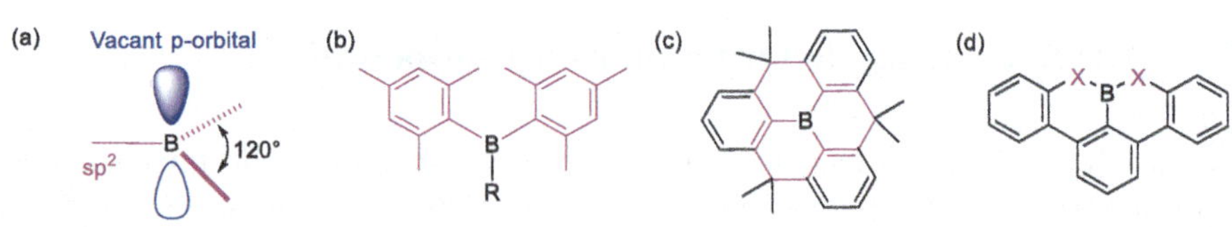

Fig. 5.1 Trivalent boron compound. **a** Trigonal-planar geometry. **b** Stabilization by introducing bulky substituents. **c** Planarized triphenylborane with rigidly fixed cyclic skeleton. **d** Triphenylborane with B–X bonds

5.2 Four-Coordinate Organoboron Complexes

5.2.1 *Boron Complexation*

Four-coordinate organoboron complexes represent one of the most important classes of fluorescent dyes (Li et al. 2013a). Boron complexation of a dye (ligand) is a simple but very effective strategy to express or increase the fluorescence quantum yield (Φ_f) of the dye (ligand). Additionally, the fluorescence properties of the boron complex are strongly dependent on the type of dye. For example, although pyridomethene does not show any fluorescence, pyridomethene-BF_2 complex which is the boron complex of pyridomethene shows a blue fluorescence in hexane (Fig. 5.2a) (Kubota et al. 2010a). Similarly, boron complexation of the bispyrrole squarylium dye leads to an increase in the Φ_f value (Fig. 5.2b) (Kubota et al. 2017).

Boron complexation is expected to have the following advantages for the enhancement of Φ_f value; (1) enhancement of the rigidity of the dye (restriction of non-radiative processes: decrease in k_{nr}), (2) prevention of hydrogen shift (inhibition of the reaction in the excited state: decrease in k_r), (3) increase in the molar extinction

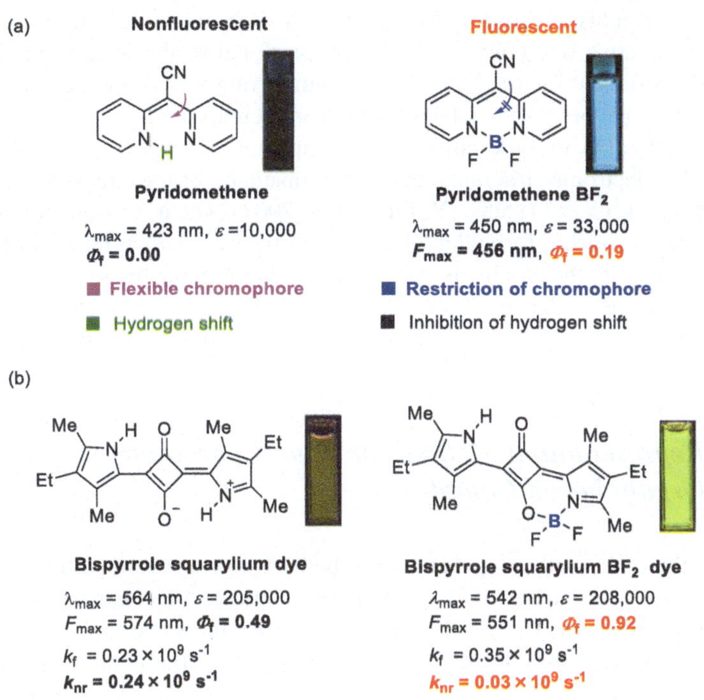

Fig. 5.2 Boron complexation. **a** Expression of fluorescence. **b** Enhancement of fluorescence quantum yield

coefficient (ε) (promotion of radiative processes: increase in k_f), and (4) change to favourable perturbation of the electronic structure.

For advantage (1), a dye having a flexible π-conjugated skeleton is prone to induce molecular rotation and vibration. Thus, a flexible dye easily promotes non-radiative processes, which increases the non-radiative rate constant (k_{nr}). Boron complexation of a flexible dye causes rigidization (restriction of vibration and rotation), which is reflected in the decrease in k_{nr}. For instance, the k_{nr} values of bispyrrole squarylium dye and the corresponding boron complex are $0.24 \times 10^9\ s^{-1}$ and $0.03 \times 10^9\ s^{-1}$, respectively, which suggest the restriction of non-radiative processes by boron complexation (Fig. 5.2b). Roughly, Φ_f is represented by the following formula: $\Phi_f = k_f/(k_f + k_{nr})$. Therefore, a decrease in k_{nr} results in an increase in Φ_f. In many cases, the main reason for the increase in Φ_f by boron complexation is considered to be due to the restriction of the flexible π-conjugated structure.

For advantage (2), in the pyridomethene skeleton as an example, hydrogen shifts such as 1,3-H shift (imine-enamine tautomerization) and 1,5-H shift are possible (Fig. 5.2a). In the excited state, a hydrogen shift results in a decrease in Φ_f. Since boron complexation of pyridomethene prevents the hydrogen shift, it contributes to the increase in Φ_f.

For advantage (3), boron complexation often leads to an increase in ε due to the increased rigidity. According to the Strickler–Berg equation (Strickler and Berg 1962), the radiative rate constant (k_f) is proportional to the integral of the molar extinction coefficient curve. Thus, boron complexation is advantageous in that it promotes radiative processes, which leads to an increase in Φ_f.

For advantage (4), in some cases, boron complexation leads to a favourable perturbation of the electronic structure, such as the inversion of the energy level between the ^{1}(n, π*) and ^{1}(π, π*) states (Yoshino et al. 2007). The n-π* transition is known to be an orbital overlap forbidden. Thus, when the $S_0 \rightarrow S_1$ transition of a dye is the n-π* transition, the ε value is very small, and consequently, Φ_f becomes small. Since in many cases, the π-π* transition is allowed, the change from $S_0 \rightarrow S_1$(n, π*) transition to $S_0 \rightarrow S_1$(π*, π*) transition is considered to enhance the Φ_f value.

5.2.2 Representation Method for Four-Coordinate Organoboron Complexes

There are three alternative representation methods to express the structure of four-coordinate organoboron complexes. Take pyridomethene-BF_2 for example. Chemical structural formulas **1a**, **1b**, and **1c** represent exactly the same molecule (Fig. 5.3). Structure **1a**, which has a negative formal charge on the boron atom and a positive formal charge on the nitrogen atom, is commonly used in organic chemistry. As shown in structure **1b**, when pyridomethene-BF_2 is represented as a coordination compound, or complex, of a metal, the bond of a neutral ligand and an anionic ligand to a metal atom can be represented as an arrowed line and a solid line, respectively.

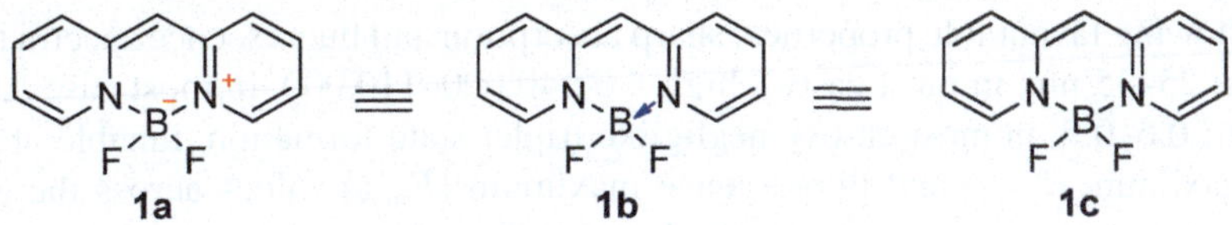

Fig. 5.3 Representation methods of organoboron complexes

In a third representation, the bond between the neutral ligand and the metal is also represented by a solid line (structure **1c**). In this chapter, the structures of four-coordinate organoboron complexes are represented as in **1c**.

In many cases, stable four-coordinate boron complexes have a bidentate ligand. Four-coordinate monoboron complexes can be roughly classified into four types depending on the type of the coordinating atom in the bidentate ligand: N^N type, O^O type, N^O type, and others. In this chapter, monoboron complexes are summarized by the four types, and then the multinuclear boron complexes are described.

5.2.3 N^N Type Organoboron Complexes

5.2.3.1 BODIPY Dye

4,4-Difluoro-4-bora-3a,4a-diaza-*s*-indacene, commonly called BODIPY, is the boron complex of dipyrrin (dipyrromethene (Wood and Thompson 2007)) (Fig. 5.4). BODIPY dye is the most famous N^N type boron complex with many reviews overviewing it (Loudet and Burgess 2007; Ulrich et al. 2008; Benstead et al. 2011; Chibani et al. 2013; Lu et al. 2014; Bañuelos 2016). BODIPY derivatives include the rigidified heptamethine cyanine dyes complexed by a nitrogen-boron-nitrogen bridge. The restriction of the flexible cyanine structure contributes to the high Φ_f of these dyes.

BODIPY dye was first reported in 1968, by Treibs and Kreuzer (1968). Although it did not attract much attention at first, it began to gain popularity in the early 1990s, due to interest in its potential applications as a fluorescent biological label (Monsma et al. 1989) and a tunable laser dye (Shah et al. 1990). The BODIPY chromophore has

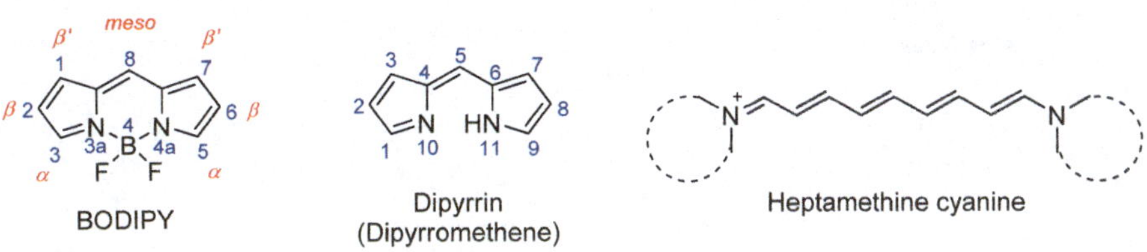

Fig. 5.4 Structure of BODIPY and dipyrrin with numbering system

the following favourable properties: sharp absorption and fluorescence spectra (half-widths: 25–35 nm, in most cases), high ε (ε: 40,000–110,000, in most cases), high Φ_f (Φ_f: 0.6–0.9, in most cases), negligible triplet-state formation, tunable absorption maximum (λ_{max}) and fluorescence maximum (F_{max}) values across the entire visible and near-infrared region, high photostability, chemical robustness, solubility in organic solvents, and insensitivity to the polarity and pH of solvent.

Nowadays, BODIPY dyes have attracted increasing attention in various fields including fluorescent probes (Boens et al. 2012; Kim et al. 2012; Culzoni et al. 2013; Yuan et al. 2013; Ni and Wu 2014; Kolemen and Akkaya 2018), fluorescent molecular rotors (FMRs) (Lee et al. 2018), circularly polarized luminescence (CPL) materials (Lu et al. 2016; Tanaka et al. 2018), dye lasers (Duran-Sampedro et al. 2014; Thorat et al. 2015), bulk heterojunction solar cells (BHJSCs) (Bessette and Hanan 2014; Bucher et al. 2017), dye-sensitized solar cells (DSSCs) (Singh and Gayathri 2014; Mao and Song 2016; Klfout et al. 2017), organic thin-film transistors (OTFTs) (Ho et al. 2019), photodynamic therapy (PDT) (Awuah and You 2012; Kamkaew et al. 2013; Zhao et al. 2015a), photodynamic inactivation (PDI) (Durantini et al. 2018), self-assembled materials (Cherumukkil et al. 2018; Solomonov et al. 2019), colorimetric sensors (Xia et al. 2018), energy transfer cassettes (Fan et al. 2013), and two-photon absorption (TPA) materials (Yang et al. 2018). Due to the excellent optical properties of BODIPY and the implementation of straightforward molecular design by regioselective post-synthetic functionalization (Boens et al. 2015; Lakshmi et al. 2016; Bodio and Goze 2019; Clarke and Hall 2019), these dyes can fit many applications.

5.2.3.2 Synthesis of BODIPY Core

Towards the synthesis of symmetrical BODIPY dyes, many routes were developed including (1) reaction of aromatic aldehydes with pyrroles (Fig. 5.5) (Cui et al. 2007), (2) reaction of acyl chlorides (Kim et al. 2015; Michel et al. 2012; Jiang et al. 2017) (aliphatic and aromatic) with pyrroles (Fig. 5.6), (3) reaction of pyrrole-2-carbaldehydes with phosphoryl chloride (Fig. 5.7) (Wu and Burgess 2008a), (4) reaction of dipyrrylketone with phosphoryl halide (Fig. 5.8) (Leen et al. 2012), (5) reaction of anhydrides with pyrroles (Wang et al. 2009), and (6) reaction of triethyl orthoformates with pyrroles (Poirel et al. 2012).

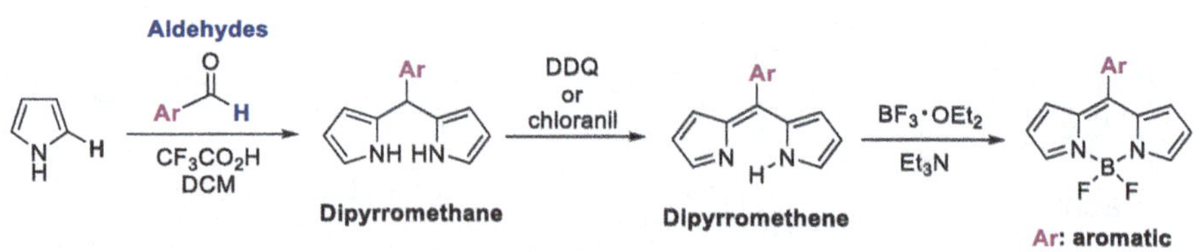

Fig. 5.5 Synthetic method of symmetrical BODIPY dyes. Method (1): reaction of aromatic aldehydes with pyrroles. Substituents on the pyrrole rings have been omitted for clarity

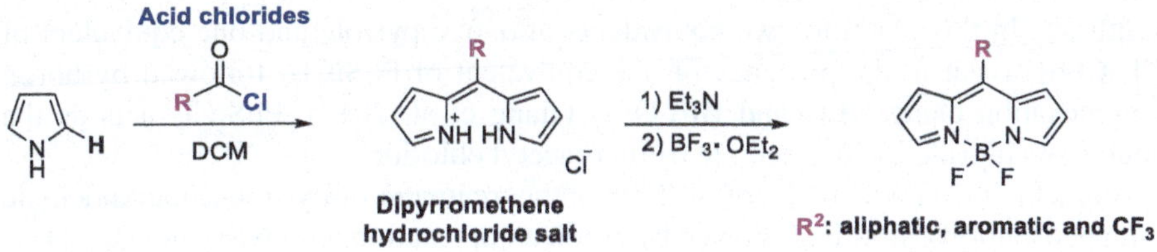

Fig. 5.6 Synthetic method of symmetrical BODIPY dyes. Method (2): reaction of acyl chlorides with pyrroles. Substituents on the pyrrole rings have been omitted for clarity

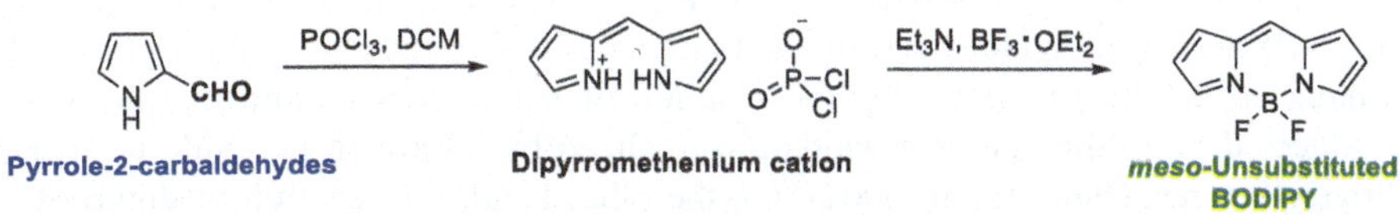

Fig. 5.7 Synthetic method of symmetrical BODIPY dyes. Method (3): reaction of pyrrole-2-carbaldehydes with phosphoryl chloride. Substituents on the pyrrole rings have been omitted for clarity

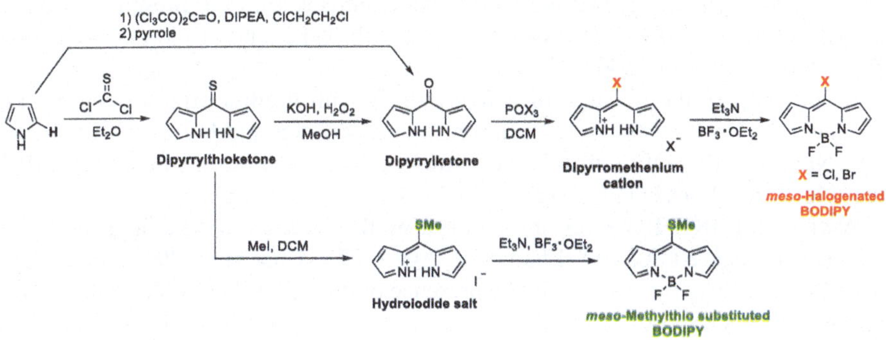

Fig. 5.8 Synthetic method of symmetrical BODIPY dyes. Method (4): reaction of dipyrrylketone with phosphoryl halide. Substituents on the pyrrole rings have been omitted for clarity

Through method (1) (Cui et al. 2007), BODIPY dyes can be synthesized in one-pot reactions without the isolation of the intermediates (Fig. 5.5). First, dipyrromethanes are prepared by the condensation of aromatic aldehydes and α-free pyrroles. Second, oxidation by DDQ gives the dipyrromethanes, and finally, boron complexation by boron trifluoride ethyl ether complex ($BF_3{\cdot}OEt_2$) forms *meso*-aryl substituted BODIPY dyes.

In method (2) (Kim et al. 2015; Michel et al. 2012; Jiang et al. 2017), condensation of acyl chlorides with α-free pyrroles gives unstable dipyrromethene hydrochloride salt intermediates. Subsequent treatment with Et_3N and $BF_3{\cdot}OEt_2$ to provides *meso*-aryl (Kim et al. 2015) or alkyl (Michel et al. 2012) substituted BODIPY dyes (Fig. 5.6). *meso*-CF_3 substituted BODIPY dyes can also be obtained by one-pot synthesis. In this reaction, two equivalents of α-free pyrrole and one equivalent of CF_3CO_2H react in the presence of one equivalent of $PhSiCl_3$, followed by boron complexation using NEt_3 and $BF_3{\cdot}OEt_2$ (Jiang et al. 2017). $PhSiCl_3$ acts as the source of chloride ion to generate trifluoroacetyl chloride.

In method (3) (Wu and Burgess 2008a), the treatment of pyrrole-2-carbaldehyde with phosphoryl chloride, followed by boron complexation with NEt_3 and $BF_3{\cdot}OEt_2$ to yield *meso*-unsubstituted BODIPY dyes (Fig. 5.7).

In method (4), halogenation of dipyrrylketone with phosphoryl halide generates the dipyrromethenium cation. Subsequent deprotonation and boron complexation gives *meso*-halogenated BODIPY dyes (Fig. 5.8) (Leen et al. 2012). Symmetric dipyrrylketones are synthesized by the reaction of α-free pyrroles with triphosgene (Zhang et al. 2015a). Similarly, the reaction of α-free pyrrole and thiophosgene produces dipyrrylthioketones, which are oxidized by hydrogen peroxide to afford dipyrrylketones (Plater et al. 2002). On the other hand, the reaction of dipyrrylthioketones with methyl iodide affords unstable hydroiodide salts, which are subsequently deprotonated and complexed with boron to form *meso*-methylthio substituted BODIPY dyes (Goud et al. 2006).

In method (5), the reaction of α-free pyrrole with cyclic carboxylic anhydrides such as glutaric anhydride and phthalic anhydride in the presence of NEt_3 and $BF_3{\cdot}OEt_2$ gives BODIPY dyes having a carboxyl group at the *meso*-position (Fig. 5.9a) (Wang et al. 2009).

In method (6), the reaction of α-free pyrrole with triethyl orthoformate in the presence of trifluoroacetic acid, followed by evaporation and boron complexation with NEt_3 and $BF_3{\cdot}OEt_2$ affords the corresponding *meso*-unsubstituted BODIPY dye (Fig. 5.9b) (Poirel et al. 2012).

Asymmetric BODIPY dyes are obtained by the reaction of a ketopyrrole (R = alkyl (Leen et al. 2009), aryl (Zhao et al. 2016), CF_3 (Choi et al. 2014), or H (Lee et al. 2011)) with α-free pyrrole in the presence of phosphoryl chloride, followed by deprotonation and boron complexation (Fig. 5.10). This method can be applied to ring-fused pyrrole derivatives such as 3-substituted indoles (Wang et al. 2014a).

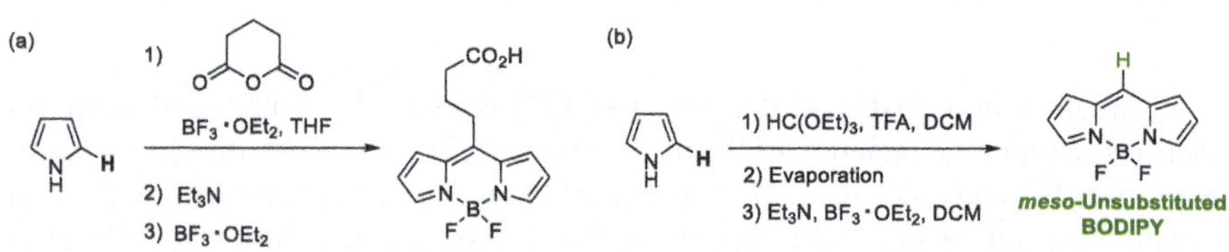

Fig. 5.9 Synthetic method of symmetrical BODIPY dyes. **a** Method (5) and **b** method (6). Substituents on the pyrrole rings have been omitted for clarity

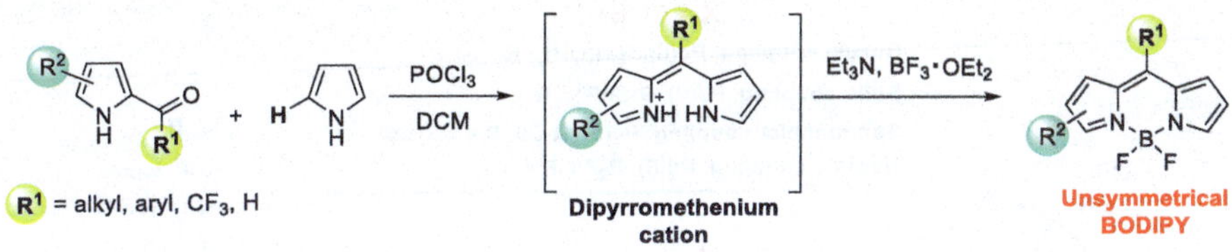

Fig. 5.10 Representative synthetic method for synthesis of asymmetric BODIPY dyes. Substituents on the pyrrole rings have been omitted for clarity

5.2.3.3 Functionalization of BODIPY Core

Selective functionalization of the BODIPY chromophore is possible through the introduction of various substituents at desired positions by regioselective post-synthetic functionalization (Boens et al. 2015; Lakshmi et al. 2016; Bodio and Goze 2019; Clarke and Hall 2019) (Fig. 5.11).

Functionalization of *meso*-position

Introduction of substituents at the *meso*-position of BODIPY dyes can be achieved by Pd-catalyzed cross-coupling (Suzuki (Leen et al. 2012), Stille (Leen et al. 2012), Sonogashira (Leen et al. 2012), and Negishi coupling (Palao et al. 2016)) of *meso*-halogenated BODIPY dyes (Fig. 5.12). The Liebeskind-Srögl (Peña-Cabrera et al. 2007) cross-coupling of *meso*-methylthio substituted BODIPY dyes is also useful to introduce substituents at the *meso*-position; the reaction characteristically proceeds under neutral conditions. Additionally, nucleophilic aromatic substitution (S_NAr) of *meso*-chlorinated BODIPY dyes achieves the introduction of N, O, and S atoms (Leen

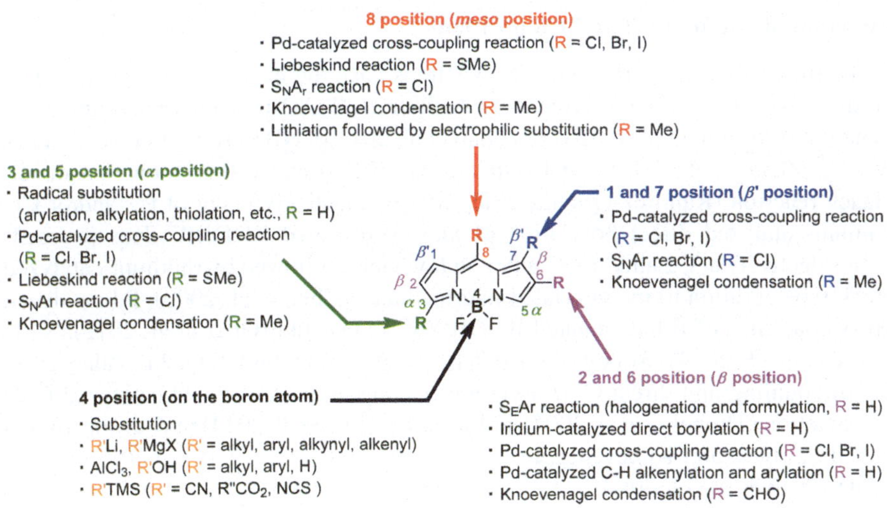

Fig. 5.11 Regioselective post-synthetic functionalization of BODIPY dyes

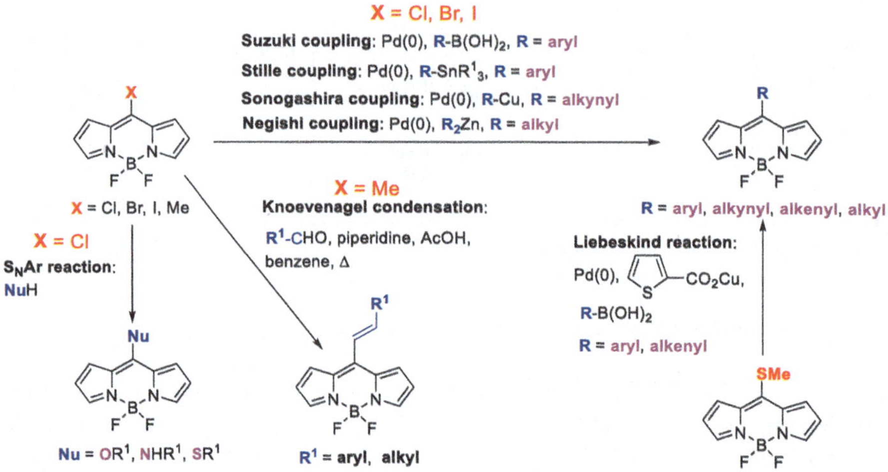

Fig. 5.12 Introduction of various substituents at the *meso*-position of BODIPY dyes. Substituents on the BODIPY core have been omitted for clarity

et al. 2012). Knoevenagel condensation of *meso*-methyl-substituted BODIPY dyes with aromatic or aliphatic aldehyde derivatives enables the introduction of alkenyl groups at the *meso*-position. 3,8-Dimethyl substituted BODIPY dyes are prone to undergo Knoevenagel condensations at C8 (*meso*-position), because more positive charge density lies on the methyl group at the C8 (Palao et al. 2013). *meso*-methyl-substituted BODIPY dyes are lithiated by LDA, followed by reaction with electrophiles such as acyl, sulfonyl, and sulfenyl chlorides, to give *meso*-functionalized BODIPY dyes (Palao et al. 2014).

Functionalization of C2 and C6 positions

In the BODIPY core, β (C2 and C6) positions have the most negative charge (Jiao et al. 2011a). Electrophilic aromatic substitution (S_EAr) reactions including halogenation (Lakshmi et al. 2015) (Br_2 (Jiao et al. 2011a), NBS (Hayashi et al. 2011a), TCCA (Zhao et al. 2015b), ICl (Ortiz et al. 2012)) and formylation (Vilsmeier-Haack reaction (Ramírez-Ornelas et al. 2016)) selectively occur at C2 and/or C6. Iridium-catalyzed direct borylation of *meso*-substituted BODIPY dyes proceeds regioselectively at C2 and/or C6, albeit in low yield, followed by rhodium-catalyzed Heck-type addition to the borylated intermediates (Chen et al. 2009). Pd-catalyzed cross-couplings of β-halogenated BODIPY dyes (Suzuki (Ortiz et al. 2012), Stille (Ahrens et al. 2014), Sonogashira (Chase et al. 2011), and Negishi (Palao et al. 2016) coupling and direct C–H arylation (Chong et al. 2015)), Pd-catalyzed C–H alkenylation (Thivierge et al. 2007) and arylation (Luo et al. 2014) of β-unsubstituted BODIPY dyes, and Knoevenagel condensation (Mao et al. 2015) of β-formylated BODIPY dyes are reported.

Functionalization of C1 and C7 positions

The introduction of substituents into β' (C1 and C7) positions (Boens et al. 2015; Leen et al. 2011) is relatively difficult due to low electron density at these positions and steric hinderance by the *meso*-substituent (Jiao et al. 2011a). S_EAr reactions at C1 and C7 hardly occur. 1,7-Halogenated BODIPY dyes can be obtained by using 4-halogenated pyrroles, in which the 5-position is unsubstituted and at least the 2-position is substituted (in many cases, such pyrroles are unstable), as the starting material of BODPY dyes or through the halogenation of 2,3,5,6,8-pentasubstituted BODIPY dyes (Leen et al. 2011). With 1,7-halogenated BODIPY dyes, Pd-catalyzed cross-coupling reactions such as Suzuki, Stille, Heck, and Sonogashira reactions are possible (Leen et al. 2011); while *meso*-substituted derivatives tend to promote reductive dehalogenation due to steric crowding of the intermediate palladium complex. Only strong nucleophilic thiolate anions are able to achieve S_NAr at C1 and C7 positions, while substitution with nitrogen, oxygen, and carbon nucleophiles fails. Knoevenagel condensation of 1,7-dimethyl-substituted BODIPY dyes is also reported (Bura et al. 2011).

Functionalization of C3 and C5 positions

The addition of radical species is an efficient strategy to introduce substituents at the α (C3 and C5) positions. For instance, regioselective introduction of substituents (radical C–H arylation) at the C3 and/or C5 positions via radical process is reported in 2015 (Verbelen et al. 2015a) (Fig. 5.13a). Reduction of aryldiazonium salts by ferrocene ($FeCp_2$) forms the corresponding radicals, which selectively react at the α position due to the increased stability of the formed radical species. The reported plausible reaction mechanism is shown in Fig. 5.13b. Similarly, radical α-regioselective alkylation (Verbelen et al. 2015b; Yu et al. 2017), benzylation (Lv et al. 2018), amination (Zhang et al. 2018a), thiolation (Lv et al. 2019), and chlorination (Zhou et al. 2015a) reactions have been reported. Functional group installation at the α-position have been reported via Pd-catalyzed cross-coupling reactions (Suzuki (Rohand et al. 2006), Stille (Rohand et al. 2006), Sonogashira (Rohand et al. 2006), Heck (Rohand et al. 2006), Negishi (Palao et al. 2016), Liebeskind-Srögl (Han et al. 2009) reaction), S_NAr reaction (Rohand et al. 2006), and Knoevenagel condensation (Deniz et al. 2008).

Functionalization on the boron atom

Typically, boron complexation of dipyrrin is performed by using excess amounts of Et_3N and $BF_3{\cdot}OEt_2$. Under these conditions, large amounts of $BF_3{\cdot}NEt_3$ are formed as by-product, which makes the purification difficult. In light of this, the use of lithium bis(trimethylsilyl)amide (LiHMDS) instead of Et_3N is a superior method for boron complexation due to a facile purification process (Lundrigan et al. 2012a). The reaction of dipyrromethenes with LiHMDS gives the lithium salts as intermediates, which are then treated with $BF_3{\cdot}OEt_2$ (1 equiv., in most cases) to obtain the corresponding BF_2-BODIPY dyes.

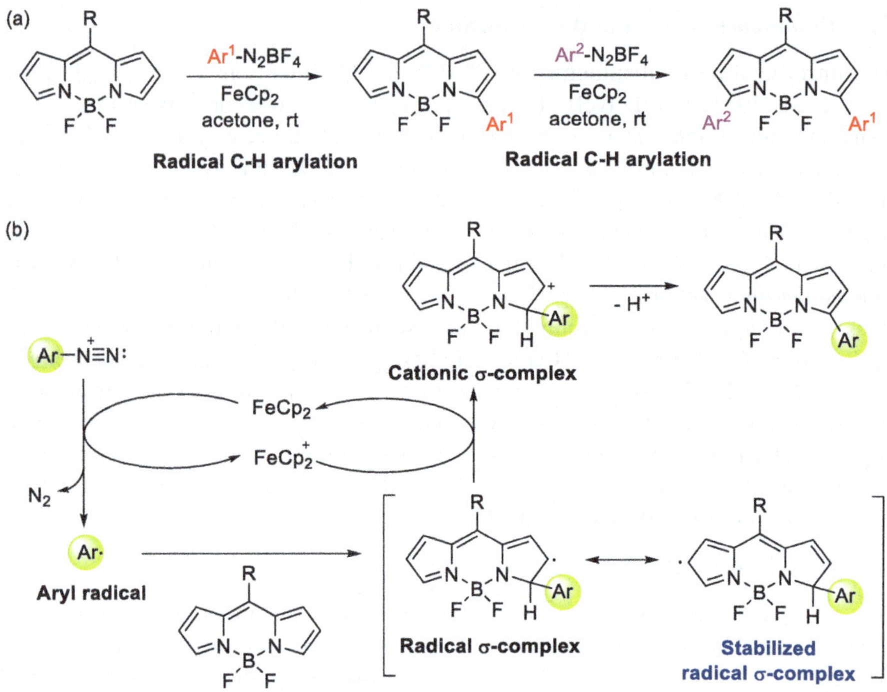

Fig. 5.13 **a** Radical C–H arylation at the α positions. **b** Plausible reaction mechanism of radical C–H arylation

Generally, the boron atom of BODIPY is substituted by two fluorine atoms (BF_2-BODIPY), although replacement of the fluorine atoms can be achieved (Fig. 5.14) (Bodio and Goze 2019). BF_2-BODIPY reacts with organic lithium reagents (RLi) or Grignard reagents (RMgX) to yield the corresponding BR_2-BODIPY dyes (R = alkyl, aryl, alkynyl) (Ulrich et al. 2006). The reaction of BF_2-BODIPY dyes with 1 equivalent of RMgX gives mono-substituted derivatives (BFR-BODIPY, R = aryl) (Goze et al. 2006). Nucleophilic substitution of BF_2-BODIPY dyes with alkoxide gives $B(OR)_2$-BODIPY dyes (R = alkyl); the reaction with bulky alkoxide (R = ^{t}Bu) produces the removal of the BF_2 unit to give the corresponding dipyrrin (Smithen et al. 2012).

The reaction of BF_2-BODIPY dyes with alcohol in the presence of $AlCl_3$ is the preferable method for the synthesis of $B(OR)_2$-BODIPY dyes (R = alkyl, aryl, H); $AlCl_3$ activates the B–F bonds and promotes the nucleophilic substitution by alcohols (Tahtaoui et al. 2007). The addition of $BF_3{\cdot}OEt$ to BF_2-BODIPY dyes also results in activation of the B–F bonds, which makes BF_2-BODIPY dyes more susceptible to nucleophiles such as RMgX (Lundrigan et al. 2014). Since Et_2AlCl works as the activator and the nucleophile, the reaction of BF_2-BODIPY dyes with Et_2AlCl yields BEt_2-BODIPY dyes; BEt_2-BODIPY dyes can be returned to the corresponding BF_2-BODIPY dyes by adding $BF_3{\cdot}OEt_2$ in moist DCM (More et al. 2014).

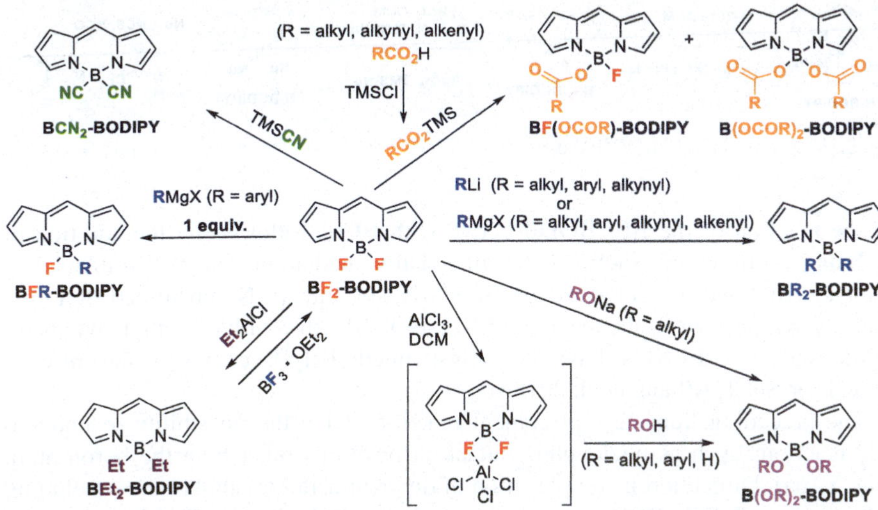

Fig. 5.14 Replacement of the fluorine atoms on the boron atom of BF_2-BODIPY

Silicon moieties such as TMS also have the potential to activate the B–F bonds. The reaction of acetic acid and TMSCl gives TMSOAc, which when reacted with BF_2-BODIPY dyes yields BF(OAc)-BODIPY and $B(OAc)_2$-BODIPY dyes (Jiang et al. 2012). The substitution reaction can be applied to other carboxylic acids including trifluoroacetic acid, acrylic acid, and propiolic acid (Durán-Sampedro et al. 2013). The reaction of BF_2-BODIPY dyes with TMSCN gives $B(CN)_2$-BODIPY dyes (Duran-Sampedro et al. 2014; Li et al. 2008a).

Boron–halogen bond strengths decrease in the order B–F >> B–Cl > B–Br > B–I (Lundrigan et al. 2012b). Thus, the stability of BODIPY dyes is expected to decrease in the following order: BF_2-BODIPY >> BCl_2-BODIPY > BBr_2-BODIPY > BI_2-BODIPY. BCl_2-BODIPY dyes are obtained by the reaction of one equivalent of BCl_3 with dipyrrin derivatives (Lundrigan et al. 2012b) or BF_2-BODIPY dyes (Lundrigan and Thompson 2013). Although BCl_2-BODIPY dyes are stable under an inert atmosphere, they decompose in air and/or moisture (Lundrigan et al. 2012b). Since BCl_2-BODIPY dyes have higher reactivity than the corresponding BF_2-BODIPY dyes, BCl_2-BODIPY dyes can be used as in situ intermediates to convert BF_2-BODIPY dyes to BR_2- and $B(OR)_2$-BODIPY dyes under mild conditions (Lundrigan and Thompson 2013). The synthesis of BBr_2-BODIPY dyes is achieved by the reaction of one equivalent of BBr_3 with BF_2-BODIPY dyes (Lundrigan et al. 2014). BBr_2-BODIPY dyes are also useful synthetic intermediates for facilitating nucleophilic substitution of BF_2-BODIPY dyes.

N-BODIPY dyes can be synthesized by the reaction of BF_2-BODIPY with BCl_3 and subsequent reaction with electron-poor amine derivatives (Fig. 5.15). In 2017, Moya and co-workers reported the formation of sp^3 N-substituted BODIPY dyes

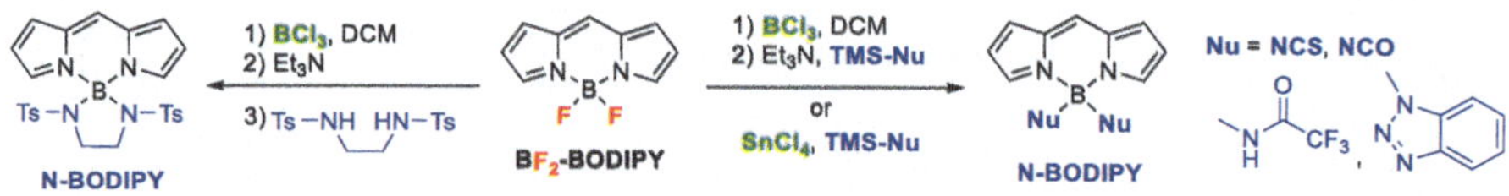

Fig. 5.15 Synthesis of N-BODIPY dyes

by the reaction of the BF_2-BODIPY dye with BCl_3, followed by the addition of Et_3N and a sulfonamide such as *N,N'*-ditosylethylenediamine and *N,N'*-ditosyl-1,2-phenylenediamine (Ray et al. 2017). Similarly, sp^3 and sp^2 N-substituted BODIPY dyes are synthesized by the reaction of BF_2-BODIPY dye with trimethylsilyl nucleophiles such as TMS-NCS, TMS-NC, or bis(trimethylsilyl)acetamide in the presence of BCl_3 or $SnCl_4$ (Zhang et al. 2018b).

The preceding fluorine replacement reactions have the possibility to undergo side reactions such as nucleophilic attack at positions other than the boron atom (Fig. 5.16a). Direct boron complexation of dipyrrin using organoboranes, including haloboranes (R_2BX, RBX_2) and boron triflates (R_2BOTf, RBFOTf), is a superior strategy to obtain BR_2-BODIPY dyes. For instance, reaction of dipyrrin with bromodimethylborane (Me_2BBr) or dibutylboron triflate (Bu_2BOTf) to give the corresponding BR_2-BODIPY dyes (R = Me or Bu) have been reported (Fig. 5.16b)

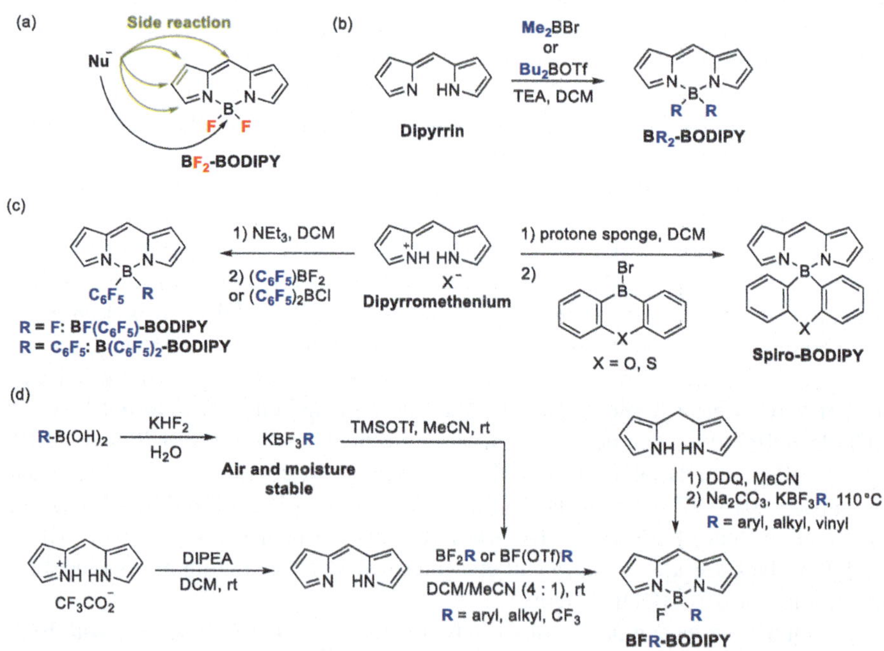

Fig. 5.16 Direct boron complexation of dipyrrin by using organoboranes. **a** Side reactions of replacement of the fluorine atoms on BF_2-BODIPY. **b** Synthesis of BR_2-BODIPY dyes from dipyrrin. **c** Synthesis of $B(C_6F_5)_2$- and Spiro-BODIPY dyes. **d** Synthesis of BFR-BODIPY dyes

(Kee et al. 2005). Additionally, the reaction of dipyrromethenium with a base such as NEt_3, and 1.8-bis(dimethylamino)naphthalene, a proton sponge, followed by reaction with $(C_6F_5)BF_2$, $(C_6F_5)_2BCl$, and bromoboraanthracene derivatives, gives $BF(C_6F_5)$-, $B(C_6F_5)_2$- (Bonnier et al. 2009) and spiro (Yuan et al. 2017)-BODIPY dyes, respectively (Fig. 5.16c). However, such direct boration has rarely been studied, probably due to the instability and unavailability of organoboranes. Recently, direct boron complexation of dipyrrin with air and moisture stable organotrifluoroborate potassium salts (KBF_3R) to yield BFR-BODIPY dyes has been reported (Fig. 5.16d) (Sawazaki et al. 2018; Wang et al. 2019).

5.2.3.4 Relationship Between Absorption and Fluorescence Wavelengths and Structure of BODIPY

The λ_{max} and F_{max} values of a BODIPY dye can be controlled by modifying its substituents. Modification of the *meso*-substituent on the BODIPY core is an effective approach to change the absorption and fluorescence wavelengths. Introduction of an electron-donating group such as methylamino (λ_{max} = 419 nm, F_{max} = 463 nm) or methoxy (λ_{max} = 452 nm, F_{max} = 487 nm) groups causes a notable hypsochromic shift when compared to a *meso*-unsubstituted derivative (λ_{max} = 504 nm, F_{max} = 511 nm) (Esnal et al. 2013) (Fig. 5.17a, b, d). Although *meso*-unsubstituted BODIPY has a cyanine-like structure, *meso*-methylanimo and *meso*-methoxy substituted BODIPY dyes receive the contribution of hemicyanine and merocyanine resonance structures, respectively (Fig. 5.17a). The hemicyanine and merocyanine structures are cross-conjugated, which interrupts the push–pull conjugation of the cyanine structure between the two pyrrole rings, consequently causing a spectral blueshift (Osorio-Martínez et al. 2012). The presence of methylthio group at the *meso*-position leads to a slight blueshift of λ_{max} (495 nm) and redshift of F_{max} (532 nm) compared to the *moso*-unsubstituted derivative (Fig. 5.17c). In the methylamino, methoxy and methylthio derivatives (λ_{max}: 419–495 nm, F_{max}: 463–532 nm, Φ_f: 0.70–0.84), methylation at both α positions led to a spectral redshift (λ_{max}: 439–531 nm, F_{max}: 485–538 nm) and enhancement of the Φ_f values (Φ_f: 0.95–1.00). This

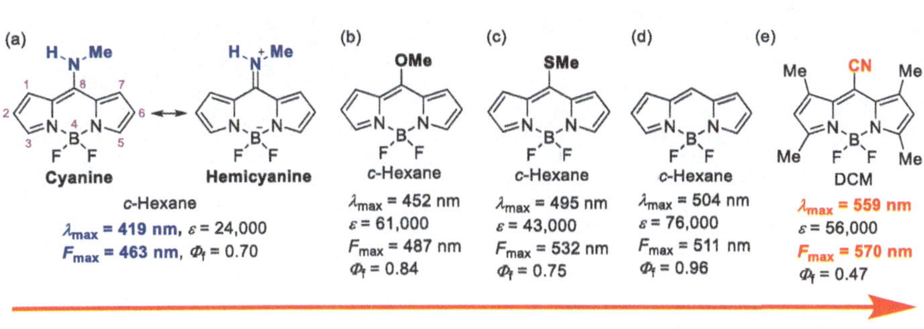

Fig. 5.17 Color tuning of BODIPY dyes by the modification of *meso*-substituent

is due to the more planar geometry of the methylated derivatives, which leads to more rigid and delocalized cyanine-like chromophores while the quenching ICT process is suppressed by the inductive electron donor character of the methyl groups (Esnal et al. 2013). On the other hand, introduction of cyano groups results in redshifted λ_{max} (559 nm) and F_{max} (570 nm) values (Yakubovskyi et al. 2016) because of the stabilization of the LUMO level (Fig. 5.17e) (Nepomnyashchii et al. 2010).

Introduction of electron-donating groups, such as phenylamino groups, at the α (3 and 5) positions leads to a redshift ($\lambda_{max} = 594$ nm, $F_{max} = 616$ nm) (Rohand et al. 2006) (Fig. 5.18a). The introduction of styryl groups causes a large bathochromic shift due to the extension of the π-conjugated system (Huang et al. 2012; Kulyk et al. 2016) (λ_{max}: 629–694 nm, F_{max}: 641–722 nm) (Fig. 5.18b). Further introduction of styryl groups at other positions ($\lambda_{max} = 802$ nm, $F_{max} = 837$ nm) (Buyukcakir et al. 2009) or a combination of strong electron-donating styryl groups and a *meso*-cyano group ($\lambda_{max} = 912$ nm) (Jiang et al. 2017) causes a greater bathochromic shift (Fig. 5.18c, d).

B,O-Chelation is also an efficient method to shift absorption to a longer wavelength. The λ_{max} of *N,N,O,O*-boron-chelated BODIPY ($\lambda_{max} = 630$ nm) is redshifted compared to that of the non *B,O*-chelated derivative ($\lambda_{max} = 550$ nm). Additionally, B–O bond formation leads to an increase in Φ_f (0.07 → 0.41) due to the restriction of the molecular rotations of two methoxyphenyl groups (Fig. 5.19a) (Kim et al. 1999).

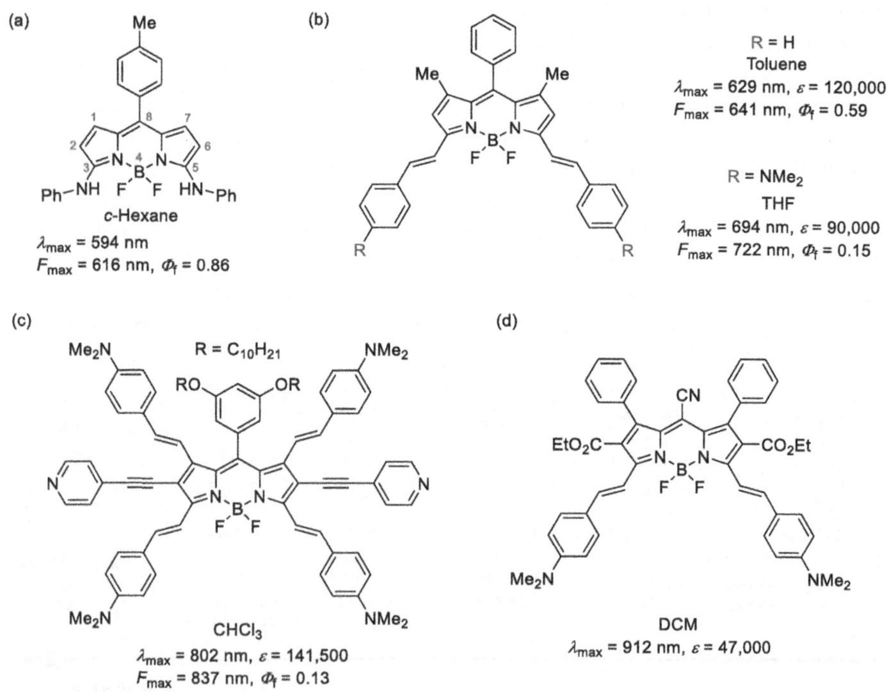

Fig. 5.18 Red shift of BODIPY dyes by the introduction of substituents at various positions

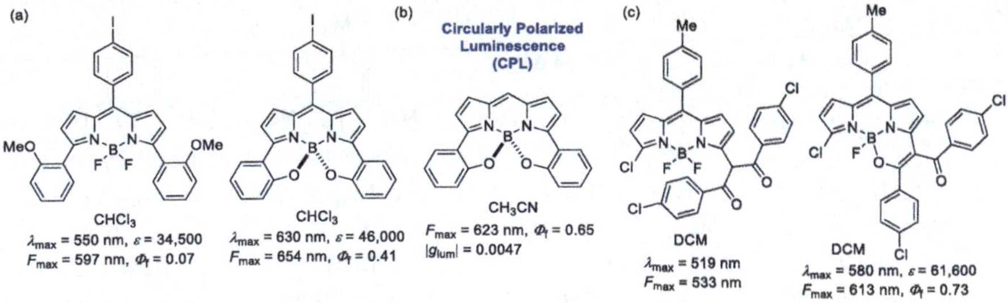

Fig. 5.19 *B,O*-Chelation of BODIPY dyes. **a** *N,N,O,O*-Boron-chelated BODIPY. **b** *N,N,O,O*-Boron-chelated BODIPY showing CPL property. **c** *N,O*-Boron-chelated BODIPY

Helically chiral *N,N,O,O*-boron-chelated BODIPY shows circularly polarized luminescence (CPL) (Fig. 5.19b) (Alnoman et al. 2016). Similarly, *N,O*-boron-chelated BODIPY ($\lambda_{max} = 580$ nm) shows redshifted absorption compared to the non-chelated derivative (λ_{max}: 519 nm) (Fig. 5.19c) (Liu et al. 2018).

5.2.3.5 Relationship Between Fluorescence Quantum Yield and Structure of BODIPY

The type and position of substituents also have an effect on the fluorescence quantum yield of BODIPY chromophores. In the case of *meso*-aryl substituted BODIPY dyes, BODIPY dye with two methyl substituents at the C1 and C7 positions ($\Phi_f = 0.60$) showed a higher Φ_f value compared to the corresponding derivative without the two methyl substituents ($\Phi_f = 0.29$) because of the prevention of the free rotation of the *meso*-tolyl group (Chen et al. 2011) (Fig. 5.20a). In the X-ray crystallographic structure, the dihedral angle of methyl derivative (81.14°) between the BODIPY core and *meso*-tolyl moiety was larger than that of nonsubstituted derivative (64.69°), which results in the prevention of free rotation and the slight blueshift in λ_{max} (**Me**: $\lambda_{max} =$ 500 nm, **H**: $\lambda_{max} = 510$ nm). Similarly, the Φ_f value of *meso*-mesityl BODIPY ($\Phi_f =$ 0.84) was higher compared to that of the corresponding phenyl derivative ($\Phi_f = 0.03$) (Yu et al. 2011) (Fig. 5.20b). On the other hand, while *meso*-methyl BODIPY showed a high Φ_f ($\Phi_f = 1.00$) value, the bulkier *meso*-*tert*-butyl BODIPY shows a low Φ_f ($\Phi_f = 0.04$) value (Jiao et al. 2015) (Fig. 5.20c). Quantum chemical calculations indicate that *meso*-*tert*-butyl BODIPY has a distorted nonplanar geometry in the S_1 excited state, which is reflected in the fast non-radiative deactivation ($k_{nr} = 8.76 \times 10^8\ s^{-1}$) and large Stokes shift (47 nm). Similar fluorescence quenching arising from the distorted nonplanar geometry was observed in the *meso*-CF_3 derivative ($\Phi_f =$ 0.008) with methyl groups at the β' positions (Fig. 5.20d) (Kim et al. 2015).

Although variance of the substituents on the boron atom of BODIPY dyes has a small effect on the absorption and fluorescence maxima, it does affect the fluorescence quantum yield, solid-state fluorescence properties, HOMO and LUMO energy levels, stability, and solubility. In BODIPY dyes, the value of k_f tends to be less

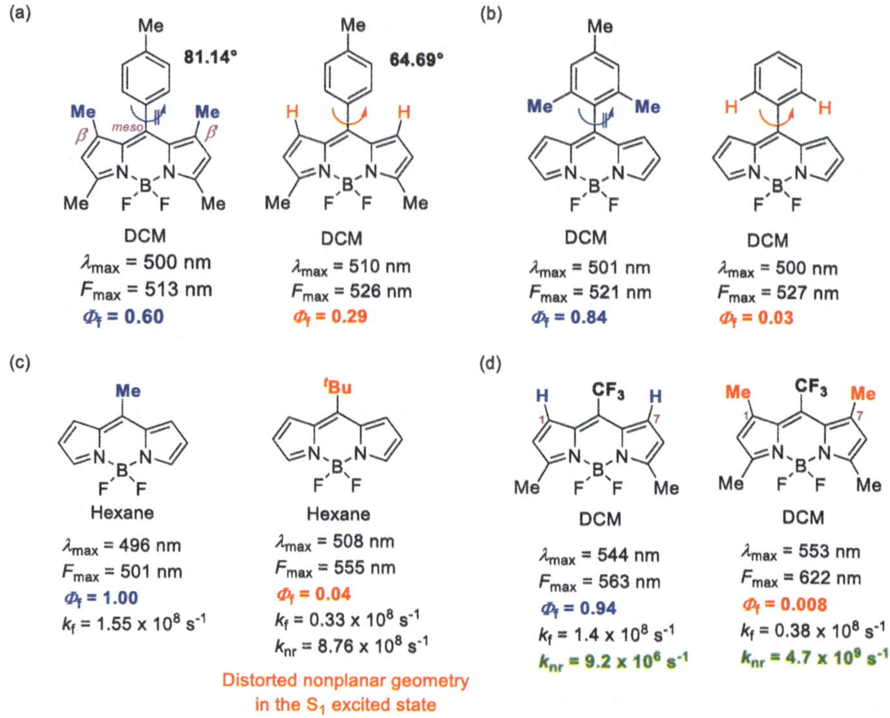

Fig. 5.20 Relationship between fluorescence quantum yield and substituents at β' and *meso*-positions. **a** Prevention of the free rotation of the *meso*-tolyl group by introduction of methyl groups at the β' positions. **b** Prevention of free rotation of the *meso*-substituent by steric repulsion. **c** Effect of the type of *meso*-substituent on fluorescence quantum yield. **d** Effect of the type of substituent at the β' positions on fluorescence quantum yield

affected by the type of substituent on the boron atom and more heavily affected by the value of k_{nr} (Goze et al. 2006; Tahtaoui et al. 2007). In many cases, the k_f value is independent of the type of substituent on the boron atom. Therefore, the decrease in Φ_f is mainly due to the increase in k_{nr}. In the case of $B(Ar)_2$-BODIPY dyes, k_{nr} tends to increase with the increasing size of the aryl substituents (Goze et al. 2006). The Φ_f value has a tendency to decrease as the solvent polarity increases because of the promotion of non-radiative processes. In the case of BF_2-BODIPY dyes, in most cases, lower Φ_f values in more polar solvents can be rationalized by the nonfluorescent excited state exhibiting ICT character in polar solvents (Qin et al. 2005). In the case of $B(Ar)_2$-BODIPY dyes, lower Φ_f values in more polar solvents are attributed to geometric changes driven by the solvophobic effect (Goze et al. 2006). The stability of BODIPY dyes towards TFA is roughly as follows: $B(CN)_2$ >> BF_2, $B(Ph)_2$ >> $B(Me)_2$, $B(OMe)_2$ (Wang et al. 2018a).

5.2.3.6 Solution Approach to the Drawback of BODIPY

Solid-state fluorescence

BODIPY dyes do have a few drawbacks, including aggregation-caused quenching (Mei et al. 2015) (ACQ) and small Stokes shifts, which limit their applications. There have been substantial efforts to improve these limitations. In terms of ACQ, dyes which show fluorescence in dilute solution generally quench or reduce the fluorescence intensity in the solid-state; this phenomenon is called ACQ or concentration quenching and is a common phenomenon in organic dyes. The main cause of ACQ is the formation of π–π stacking in the solid-state. Since BODIPY dyes have extended planar π-conjugated structures, they easily induce ACQ phenomena. In many cases, BODIPY dyes do not show fluorescence in the solid-state. Strategies to express and enhance solid-state fluorescence in BODIPY dyes include the following: (1) prevention of intermolecular interactions between neighbouring fluorophores (Kubota et al. 2010b; Ozdemir et al. 2009; Lu et al. 2012), which causes fluorescence quenching, (2) formation of emissive J-aggregates (Kim et al. 2015), and (3) solid-state entrapment of BODIPY in metal-organic frameworks (MOFs) (Glembockyte et al. 2018).

In light of strategy (1), the introduction of bulky substituents into a BODIPY core is an effective strategy to enhance solid-state fluorescence by preventing intermolecular interactions, such as π–π interactions, by steric repulsion. When considering ease of synthesis and overall effectiveness, the introduction of bulky groups is favoured at the boron atom (4-position). For instance, although the BF_2 complex did not exhibit any fluorescence in the solid-state, the BPh_2 analogue showed intense red fluorescence in the solid-state (Fig. 5.21a) (Kubota et al. 2010b); Φ_f increased as the size of the substituents at boron increased (R = F: $\Phi_f = 0.00$, R = OMe: $\Phi_f = 0.02$, R = OPh: $\Phi_f = 0.04$, R = Ph: $\Phi_f = 0.22$). In the crystal packing of the BF_2 complex, intermolecular π–π interactions (C–C: 3.57–3.64 Å) were observed. On the other hand, in the crystal structure of the BPh_2 complex, one of the phenyl rings on the boron atom was oriented almost perpendicular to the plane of the dipyrrin ring and consequently, intermolecular π–π interaction was not observed. Therefore, because the π–π interaction was avoided, the BPh_2 complex showed intense solid-state fluorescence. Introduction of a *tert*-butyl substituted phenyl group at the *meso*-position (Ozdemir et al. 2009) and triphenylsilyl phenyl groups at one or both β positions

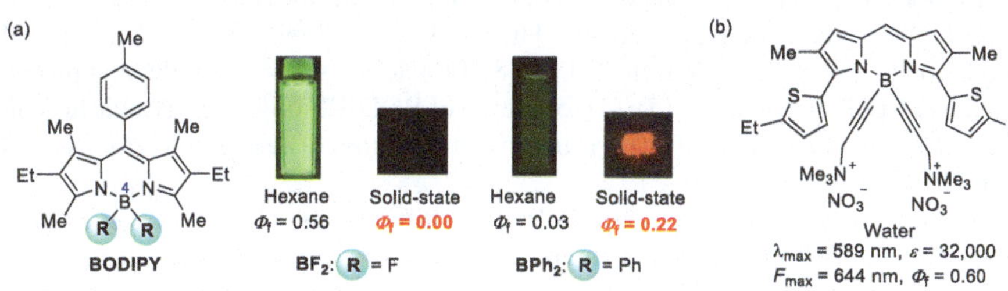

Fig. 5.21 **a** Expression of solid-state fluorescence by introducing bulky substituents on the boron atom. **b** Water-soluble BODIPY

(Lu et al. 2012) also prevented intermolecular π–π interactions and contributed the expression of fluorescence in the solid-state.

For strategy (2), careful choice and combination of suitable substituents are required to achieve the formation of emissive BODIPY J-aggregates. The combination of an electron-withdrawing *meso*-substituent, such as a trifluoromethyl or methoxycarbonyl group, and methyl groups at the β' position induced emissive *J*-aggregates (Kim et al. 2015).

For strategy (3), the incorporation of BODIPY dyes into ZIF-8 was achieved upon ion- and liquid-assisted grinding (ILAG) or accelerated ageing of 1:2 stoichiometric mixtures of ZnO and 2-methylimidazole (HMeIm) in the presence of BODIPY dye (Glembockyte et al. 2018). The resulting BODIPY@ZIF-8 showed solid-state fluorescence and a remarkable enhancement in photostability.

Enhancement of water solubility

High water solubility and suppression of the formation of nonfluorescent aggregates in water are required for biological and medical applications of BODIPY dyes. Incorporation of ionic substituents including sulfonate (Li et al. 2008b), trimethyl-(propargyl)ammonium (Poirel et al. 2014), and sulfobetaine (Sutter et al. 2018) groups into the BODIPY core is a good method to enhance water solubility. For instance, the red-emitting water-soluble ionic BODIPY dye having two trimethyl-(propargyl)ammonium groups on the boron atom has been reported (Poirel et al. 2014) (Fig. 5.21b). Neutral water-soluble BODIPY dyes, which have advantages over ionic dyes in that they avoid potential nonspecific electrostatic interactions between BODIPY dyes and biomolecules, have also been synthesized by the introduction of branched oligo(ethylene glycol)methyl ether substituents to the BODIPY core (Zhu et al. 2011).

Stokes shift

In terms of a Stokes shift (SS), a small Stokes shift leads to self-quenching and measurement error from excitation light and scattered light, which reduces detection sensitivity. BODIPY dyes generally have very small Stokes shifts (5–20 nm, in most cases) due to the rigidified fluorophore which causes limited geometric relaxation between the Franck–Condon excited state and the equilibrium excited state upon photoexcitation. In order to expand the Stokes shift, the following strategies have been reported: (1) energy transfer cassettes (Goze et al. 2007; Qu et al. 2012), (2) twisted intramolecular charge transfer (TICT) (Hu et al. 2009), (3) non-symmetrical annulation (Yang et al. 2016; Ren et al. 2018), (4) excited state intramolecular proton transfer (ESIPT) (Fei et al. 2017), (5) formation of BODIPY oligomers (Hayashi et al. 2011a; Nepomnyashchii et al. 2011), and (6) introduction of susceptible groups (Zhu et al. 2019).

Through strategy (1), large pseudo-Stokes shifts can be obtained by excitation energy transfer (EET) from donor (high energy absorbing chromophore) to acceptor (lower energy emitting fluorophore). The EET process can be classified as fluorescence resonance energy transfer (FRET) or through-bond energy transfer (TBET). While donor and acceptor are linked together by a non-conjugated spacer in the FRET system, they are linked through a conjugated spacer in the TBET system. In the case of FRET, significant spectral overlap between the donor emission and acceptor absorption spectra is required; the efficiency of FRET depends on the degree of the spectral overlap. On the other hand, in the case of TBET, such spectral overlap between donor and acceptor is not needed. Since the fluorescence spectrum of the ethynylpyrene moiety (donor) matched the absorption spectrum of the BODIPY core (acceptor) in the BODIPY dye shown in Fig. 5.22a, efficient FRET from the donor to the acceptor occurred (energy transfer efficiency: 96%) (Goze et al. 2007). Consequently, a large pseudo-Stokes shift ($SS = 164$ nm) was observed in dichloromethane with a Φ_f value of 90%. The BODIPY dye with a 5-(quinolin-2-yl)thiophen-2-yl moiety as the donor and a tetrastyryl-substituted BODIPY core as the acceptor showed a high TBET efficiency (98%), a large pseudo-Stokes shift (SS = 398 nm), NIR fluorescence (F_{max} = 732 nm), and relatively high Φ_f (0.62) in dichloromethane (Fig. 5.22b) (Qu et al. 2012).

In strategy (2), TICT-BODIPY dye exhibits fluorescence from the locally excited (LE) state in nonpolar solvents such as hexane ($F_{max} = 534$ nm, $SS = 20$ nm, $\Phi_f =$ 0.17) (Fig. 5.22c) (Hu et al. 2009). On the other hand, it shows fluorescence from the twisted intramolecular charge transfer (TICT) state in polar solvents along with a large Stokes shift, redshifted fluorescence, decreased Φ_f,, and spectral broadening of fluorescence (THF: $F_{max} = 663$ nm, $SS = 150$ nm, $\Phi_f = 0.06$).

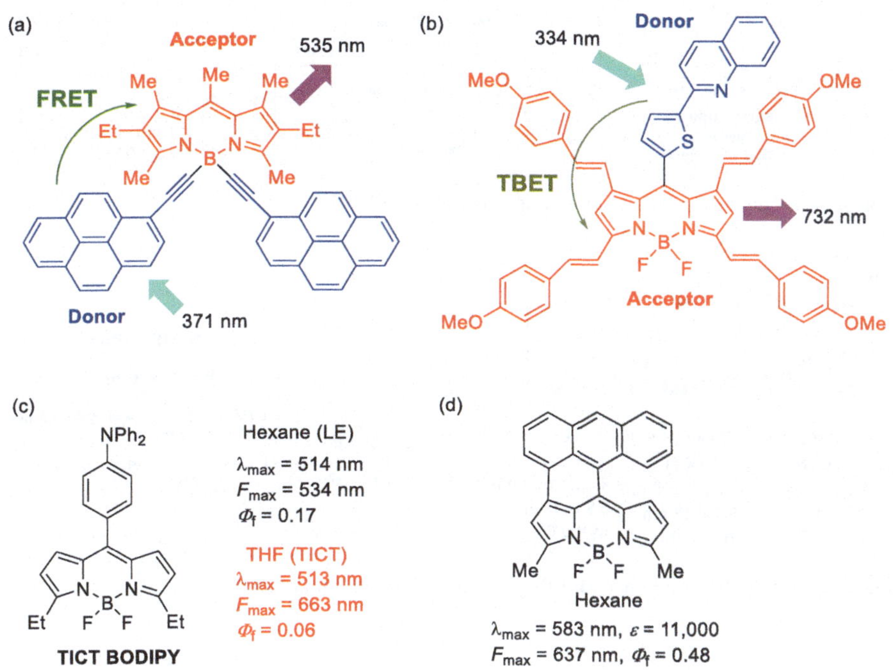

Fig. 5.22 Energy transfer cassettes for large Stokes shift by **a** FRET and **b** TBET systems. **c** TICT fluorescence. **d** Asymmetric annulation

In strategy (3), asymmetric anthracene-fused BODIPY dye shows a large Stokes shift (hexane: 54 nm, DCM: 55 nm) probably due to the improved ICT properties (Fig. 5.22d) (Yang et al. 2016). Although ICT causes broadening of the fluorescence spectrum, the asymmetric anthracene-fused BODIPY dye retained a sharp fluorescence spectrum with full width at half maximum measurements (FWHM) of 934 cm^{-1} in hexane and 1107 cm^{-1} in DCM.

For strategy (4), dramatic alternations in molecular geometry by ESIPT enable a large Stokes shift. For instance, ESIPT-BODIPY shows λ_{max} at 546 nm and exhibits dual fluorescence at 586 and 765 nm, corresponding to the LE and ESIPT states, respectively (Fig. 5.23) (Fei et al. 2017).

In strategy (5), when considering monomer ($\lambda_{max} = 497$ nm, $SS = 15$ nm), dimer ($\lambda_{max} = 526$ nm, $SS = 37$ nm), and trimer ($\lambda_{max} = 550$ nm, $SS = 37$ nm), the λ_{max} and F_{max} are redshifted and the Stokes shift tends to become larger (Fig. 5.24a) (Nepomnyashchii et al. 2011).

In terms of strategy (6), phenyl-substituted BODIPY dyes are known to show larger Stokes shifts as compared to alkyl-substituted BODIPY dyes (Fig. 5.24b) (Zhu et al. 2019). In addition, introduction of methoxy groups on the phenyl substituents further enlarge the Stokes shift. Furthermore, introduction of thienyl groups at the β-positions is even more effective at increasing the Stokes shift due to the increased geometric relaxation of the fluorophore upon photoexcitation (Chen et al. 2012).

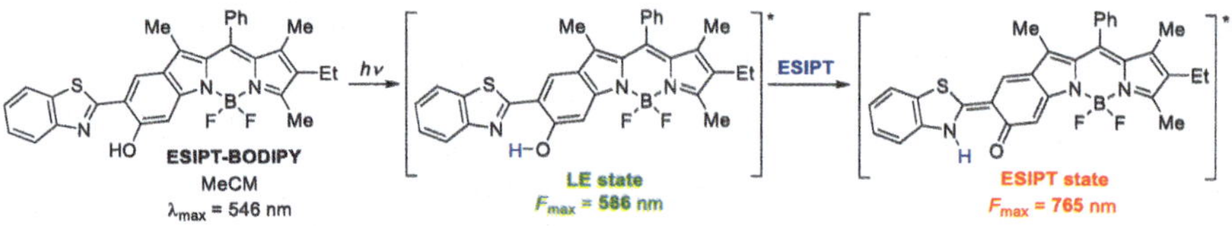

Fig. 5.23 Excited state intramolecular proton transfer (ESIPT)

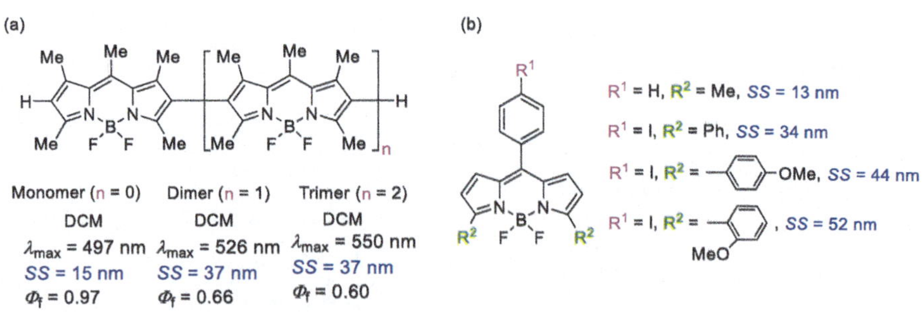

Fig. 5.24 **a** BODIPY oligomers. **b** Effect of substituents on Stokes shift

Thermal stability

When considering practical applications of BODIPY dyes, thermal stability is important. The destruction beginning temperature in an argon atmosphere (T_{Ar}) for BODIPY dyes **2–8** was in the range of 258.5–361.0 °C (Fig. 5.25a) (Bumagina et al. 2018). As alkyl-chain length increased, the T_{Ar} value was improved (**3**: R = CH_3, T_{Ar} = 260.7 °C; **6**: R = C_7H_{11}, T_{Ar} = 338.7 °C). Replacement of methyl groups by phenyl groups at positions 1, 3, 5, and 6 significantly increased thermal stability (**2**: T_{Ar} = 258.5 °C, **8**: T_{Ar} = 361.3 °C) (Fig. 5.25b). Additionally, a decrease in the symmetry of alkyl substitution tended to cause a decrease in thermal stability. The temperature of thermal decomposition in an air oxygen atmosphere (T_{O2}) for compounds **2–7** were decreased (T_{O2}: 188–254 °C) compared to T_{Ar} values of corresponding compounds (T_{Ar}: 258.5–338.0 °C), which indicates that BODIPY dyes are more thermally unstable in an air oxygen atmosphere than in an argon atmosphere. Boron complexation contributed to the enhancement of thermal stability (**3**: T_{O2} = 199 °C, **10**: T_{O2} = 190 °C) (Fig. 5.25c) (Antina et al. 2009). In contrast to BODIPY dyes, an increase in the length of alkyl chains caused a decrease in thermal stability (**11**: T_{O2} = 140 °C) in the case of uncomplexed dipyrrin ligands. The metal complex of dipyrrin exhibited a higher T_{O2} value (**12**: M = Cu, T_{O2} = 240 °C, **13**: M = Zn, T_{O2} = 280 °C) compared to the corresponding BODIPY complex (**3**: T_{O2} = 199 °C) (Fig. 5.25d) (Antina et al. 2009). The lower thermal stability of the BODIPY dye compared to the corresponding metal complex may be because of the high oxidation potential of fluorine atoms and the zwitterionic character of the BODIPY chromophore, which easily cause intramolecular oxidation by fluorine atoms.

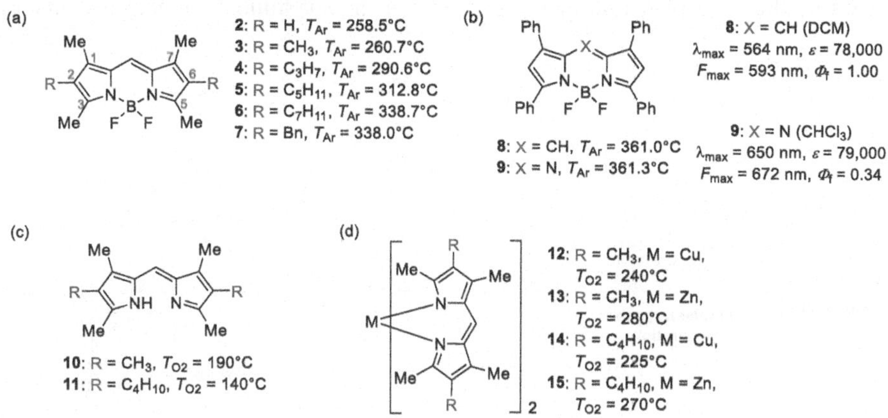

Fig. 5.25 Thermal stability of **a** BODIPYs, **b** aza-BODIPY, **c** dipyrrins and **d** dipyrrin-metal complexes

Photostability

In the photochemistry of BODIPY dyes, the grand-state triplet oxygen ($^3O_2$) induces intersystem crossing (ISC) of the excited singlet (S_1) state of BODIPY dye to generate the excited triplet (T_1) state of BODIPY dye (Jones et al. 2003). In the ISC process, the S_1 state of BODIPY dye and $^3O_2$ forms the exciplex and then it dissociates to give the T_1 state of BODIPY dye and $^3O_2$. The T_1 state of BODIPY dye further undergoes energy transfer with $^3O_2$ to give the singlet oxygen ($^1O_2$). The formed $^1O_2$ reacts primarily at the C7′–C8 double bond of the ground singlet (S_0) state of BODIPY dye to produce an unstable peroxo compound. The proposed reaction mechanism (Mula et al. 2008) of photodegradation of BODIPY dye is shown in Fig. 5.26.

According to the mechanism, the photostability of BODIPY dye is considered to depend on the ease of generation of $^1O_2$ and/or the reactivity of BODIPY dye towards $^1O_2$ (Mula et al. 2008). It is known that molecular oxygen is a ground state triplet ($^3\sum$ oxygen) possessing two low-lying excited singlet states ($^1\Delta$ oxygen and $^1\sum$ oxygen), whose excitation energies are ca. 23 and 38 kcal/mol, respectively (Turro 1991). Therefore, BODIPY dyes having excitation energies as low as 23 kcal/mol may be quenched via energy transfer to generate $^1O_2$ ($^1\Delta$). The decrease of the triplet excitation energy of BODIPY dyes may be effective to suppress the generation $^1O_2$. The reactivity is governed by steric and/or electronic factors. For the steric factor, introduction of sterically hindered substituent group R at 8-position is effective to decrease the reactivity and enhance the photostability. For the electronic factor, the lower HOMO energy level of BODIPY has an advantage in the improvement of the photostability, since the reaction between the HOMO of BODIPY and LUMO of $^1O_2$ is more suppressed. In fact, the photobleaching rate of BODIPY dyes is influenced by the electron-withdrawing capacity of the substituents; electron-deficient

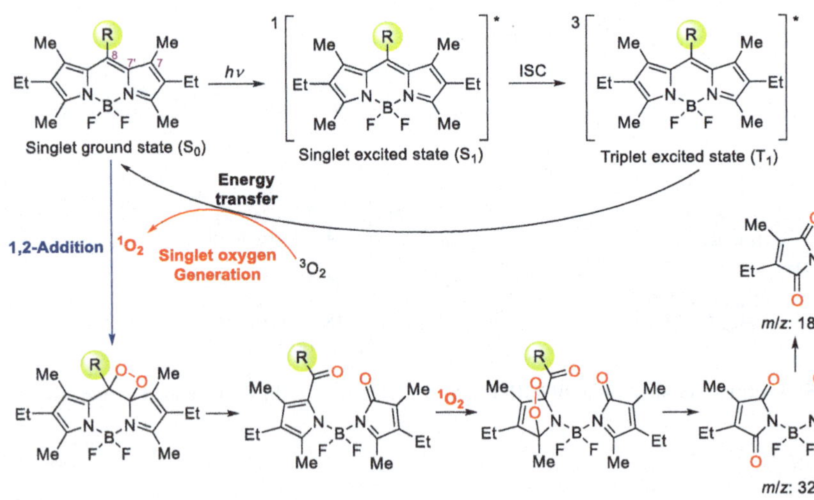

Fig. 5.26 Photodegradation of BODIPY dyes

BODIPY dyes generate less $^1O_2$, are less reactive to $^1O_2$, and are highly resistant to photobleaching (Komatsu et al. 2011). On the other hand, for the mechanism of photobleaching, the single-electron transfer (SET) from the S_1 state of BODIPY dyes to $^3\sum$ oxygen is also reported (Hinkeldey et al. 2008).

5.2.3.7 Aza-BODIPY

Aza-BODIPY (4,4′-difluoro-4-bora-3a,4a,8-triaza-*s*-indacene) dyes are BF_2 complexes of azadipyrromethenes ((*Z*)-*N*-(2*H*-pyrrol-2-ylidene)-1*H*-pyrrole-2-amines) (Ge and O'Shea 2016; Loudet et al. 2008). Although azadipyrromethene was first published in 1943 by Rogers (1943), aza-BODIPY dye was not reported until 1994 (Allik et al. 1994) and it has begun to attract much attention since the report by O'Shea et al. in 2002 (Killoran et al. 2002). The reaction of an ammonia source, such as ammonium acetate (NH_4OAc), with 1,3-diaryl-4-nitrobutan-1-one or 2,4-diaryl-4-oxobutanenitrile gives azadipyrromethenes; subsequent reaction with NEt_3 and $BF_3{\cdot}OEt_2$ yields aza-BODIPY dyes (Fig. 5.27). Almost all reported aza-BODIPY dyes are 3,3′,5,5′-tetraaryl substituted derivatives (R^1 = R^2 = aryl), probably due to the instability and/or synthetic difficulty of producing the corresponding azadipyrromethene precursors. One exception to this are 3,3′-dimethyl-,5,5′-diaryl substituted derivatives (R^1 = Me, R^2 = aryl) (Wu and O'Shea 2013). Asymmetric aza-BODIPY dyes are synthesized by the reaction of nitroso pyrroles and α-free pyrroles with different substituents (Killoran and O'Shea 2006). Generally, aza-BODIPY dyes (Killoran et al. 2002) show redshifted λ_{max} and F_{max} compared to corresponding BODIPY dyes (Wu and Burgess 2008a) due to the lower LUMO energy level (Yamane et al. 2017) of aza-BODIPY dyes (Fig. 5.25b).

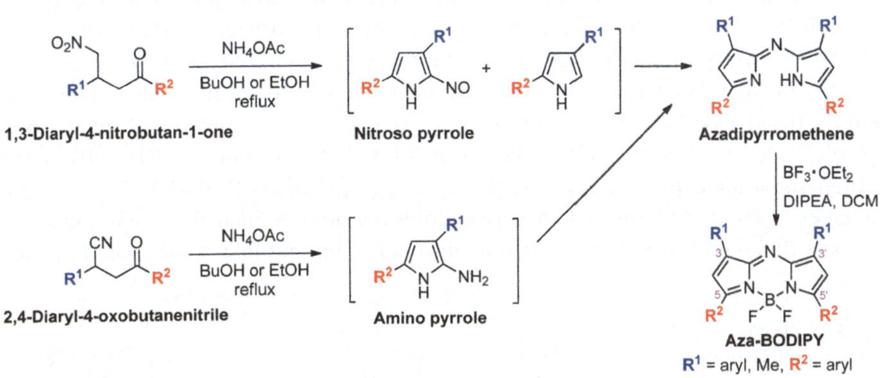

Fig. 5.27 Synthetic method of aza-BODIPY dyes

5.2.3.8 Ring-Fused BODIPY

Type of ring-fused BODIPY

Annulation of aromatic rings to the BODIPY core is an efficient strategy to achieve the extension of π-conjugation, which results in redshift of the absorption and fluorescence maxima. The π-extension of the BODIPY chromophore through the introduction of substituents sometimes leads to the promotion of non-radiative processes via molecular rotations of newly introduced substituents. On the other hand, annulation of aromatic rings to the BODIPY core forms a π-extended and rigidified BODIPY chromophore without induction of molecular rotation. Ring-fused BODIPY dyes are roughly divided into [*a*]-fused and [*b*]-fused dyes (Fig. 5.28). [*b*]-Fused BODIPY dyes tend to have higher stability compared to the corresponding [*a*]-fused dyes (Zhou et al. 2015a; Yamazawa et al. 2016; Okujima et al. 2010).

Synthesis of [*a*]-fused BODIPY

Synthetic methods (Jean-Gérard et al. 2018) for producing [*a*]-fused BODIPY dyes are roughly classified into four types based on the type of key reaction: (1) Barton–Zard reaction (Fig. 5.29a), (2) retro Diels–Alder reaction (Fig. 5.29b), (3) Paal–Knorr reaction (Fig. 5.29c), and (4) Vilsmeier–Haack reaction (Fig. 5.29d). Kang and Haugland reported the first synthesis of benzo-[*a*]-fused BODIPY dyes by using method (3) in 1995 (Kang and Haugland 1995).

For method (1), the reaction of nitroalkenes with α-isocyanoesters in the presence of a non-nucleophilic base, such as DBU, gives isoindoles (Fig. 5.29a) (Ono et al. 1996). The reactivity of nitroalkenes (nitroaromatics) towards the carbanion derived from α-isocyanoesters depends on the electrophilic character of the nitroalkene; although nitrobenzene derivatives are less reactive, polycyclic nitroaromatics such as 1-nitronaphthalene, 9-nitrophenanthrene, and 1-nitroacenaphthylene react with the carbanion to give the corresponding fused pyrrole-carboxylate derivatives. The alkoxycarbonyl group is removed by heating with potassium hydroxide in ethylene glycol to yield the unsubstituted fused pyrroles (Xu et al. 2006). The reaction of fused pyrrole-carboxylates with $LiAlH_4$ affords methyl-substituted fused pyrroles (Descalzo et al. 2008). The unsubstituted (Swavey et al. 2016, 2017) and methyl-substituted pyrroles are the precursors of [*a*]-fused BODIPY dyes. Disadvantages of method (1) include that isoindoles are not obtained from reactions with electron-deficient nitroalkenes, such as nitrobenzene, and that some isoindoles are

Fig. 5.28 Types of ring-fused BODIPY dyes

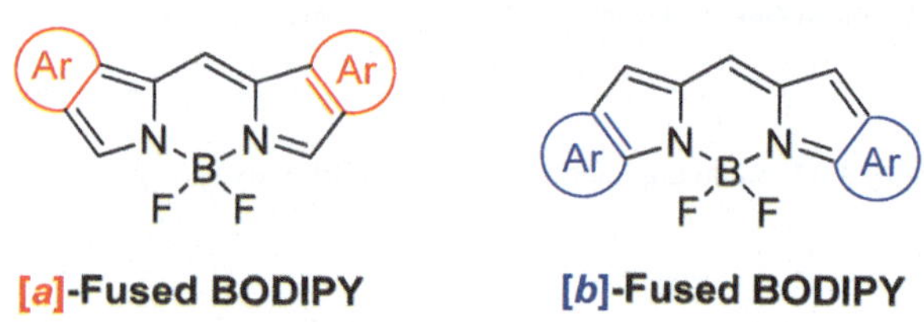

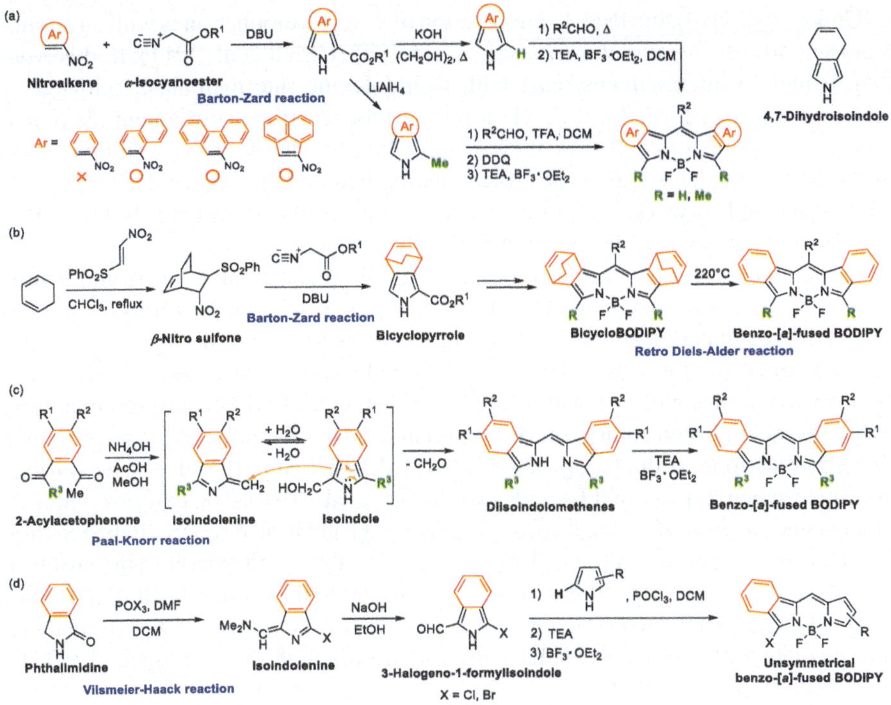

Fig. 5.29 Synthetic method of [*a*]-fused BODIPY dyes based on the type of key reaction **a** Barton–Zard reaction, **b** retro Diels–Alder reaction, **c** Paal–Knorr reaction, and **d** Vilsmeier–Haack reaction

unstable. From this perspective, 4,7-dihydroisoindoles are good equivalents of isoindoles (Filatov et al. 2007). In 2010, Cheprakov and co-workers reported the formation of benzene-fused dipyrromethanes from the reaction of 4,7-dihydroisoindoles with aldehydes, followed by oxidation of the resultant dipyrromethanes (Filatov et al. 2010). Although benzo-[*a*]-fused and naptho-[*a*]-fused metal complexes are reported, boron complexation was not mentioned in this study.

Towards method (2), Ono and co-workers (Shen et al. 2004) developed a superior method to solve the disadvantages of method (1). Diels–Alder reaction of β-sulfonylnitroethylene, which is a nitroacetylene equivalent, with cyclohexadiene gives β-nitro sulfone (Fig. 5.29b) (Ito et al. 1997). The reaction of β-nitro sulfone with α-isocyanoesters in the presence of DBU proceeds via elimination of sulfinic acid to give reactive nitroalkene intermediates, followed by a Barton–Zard reaction, to provide bicyclopyrroles. Bicyclo-BODIPY dyes, which are the precursors of benzo-[*a*]-fused BODIPY dyes, can be obtained using common synthetic methods (Shen et al. 2004) and are stable in acidic or basic solutions and on silica gel. Bicyclo-BODIPY dyes are quantitatively transformed into corresponding benzo-[*a*]-fused BODIPY dyes upon heating under reduced pressure via a retro Diels–Alder reaction without the need for further purification.

Under method (3), the Paal-Knorr reaction of 2-acylacetophenones with an ammonium salt affords diisoindolomethenes (Fig. 5.29c) (Ulrich et al. 2011). In this reaction, isoindole intermediates react with isoindolenine intermediates, followed by the elimination of formaldehyde via a retro-Aldol reaction. Subsequent air oxidation gives diisoindolomethenes (Speck and Magauer 2013). Boron complexation of diisoindolomethenes gives the corresponding benzo-[*a*]-fused BODIPY dyes. In 2016, Kubo and co-workers reported the first synthesis of naptho-[*a*]-fused BODIPY dyes by using this method (Yamazawa et al. 2016).

In method (4), a Vilsmeier–Haack reaction of phthalimidine gives isoindolenine derivatives and subsequent basic hydrolysis forms 3-halogeno-1-formylisoindoles (Fig. 5.29d) (Diana et al. 2011). Jiao and co-workers reported asymmetric halogenated benzo-[*a*]-fused BODIPY dyes via condensation reactions of 3-halogeno-1-formylisoindoles with α-unsubstituted pyrroles under $POCl_3$ catalyzed condition with subsequent boron complexation (Jiao et al. 2010). Pd-catalyzed cross-couplings or S_NAr reactions of the halogenated benzo-[*a*]-fused BODIPY dyes give various asymmetric benzo-[*a*]-fused BODIPY dyes (Jiao et al. 2010). The one-pot synthesis of an isoindole-BODIPY dye having a 2-pyrrolyl group is also reported; the reaction of 3-chloro-1-formylisoindole with excessive α-unsubstituted pyrrole in the presence of $POCl_3$ and subsequent boron complexation yields the isoindole-BODIPY dyes (Yu et al. 2012). In the reaction, unexpected S_NAr of the chlorinated dipyrromethene intermediate with the α-unsubstituted pyrrole occurs with the assistance of $POCl_3$, which acts as a Lewis acid.

As a synthetic method, except for the method (1) to (4), phenanthrene-fused BODIPY dyes are synthesized by the introduction of two biphenyl groups at C2 and C6 positions of 2,6-dichloro-BODIPY dyes via Suzuki coupling reaction, followed by bromination at the 1,7-positions and then Pd(0)-catalyzed intramolecular cyclization (Zhao et al. 2017a).

Synthesis of [*b*]-fused BODIPY

Synthetic methods to produce [*b*]-fused BODIPY dyes can be roughly classified into two types: (1) application of classical synthetic methods to ring-fused pyrroles (Figs. 5.30 and 5.31) and (2) intramolecular annulation of BODIPY dyes. Under method (1), Burgess and co-workers reported the synthesis of benzofuran-, benzothiophene- and dialin-fused BODIPY dyes by the reaction of benzofuro[3,2-*b*]pyrrole, thianaphtheno[3,2-*b*]pyrrole and 4,5-dihydro-1*H*-benz[*g*]indole with acyl chloride, followed by boron complexation, respectively (Fig. 5.30) (Chen et al. 2000). Ohe and co-workers described the synthesis of spirofluorene-indene-fused BODIPY dyes by the reaction of spiro[fluorene-9,4′(1′*H*)-indeno[1,2-*b*]pyrrole] and aromatic aldehydes, followed by oxidation and boron complexation (Kowada et al. 2010). A benzo[*e*]indole-fused BODIPY dye was synthesized from the reaction of naphthobipyrrole and triethylorthoformate in the presence of $POCl_3$, followed by boron complexation (Sarma et al. 2013). TFA-mediated condensation between thieno[2,3-*b*]pyrrole and TFA anhydride, followed by boron complexation, gives a *meso*-CF_3 substituted thieno[2,3-*b*]pyrrole type BODIPY dye (Wang et al. 2016a). Similarly, furan (Matsui et al. 2011)-, thiophene (Kubota et al. 2019)-, acenaphthylene (Jiang

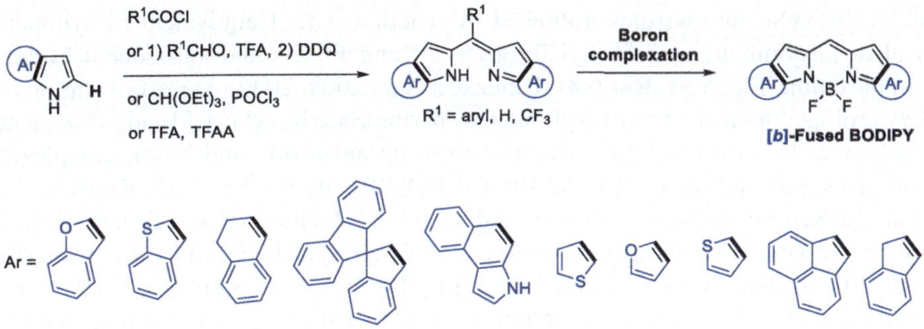

Fig. 5.30 Synthesis of [*b*]-fused BODIPY dyes by applying classical synthetic methods to ring-fused α-free pyrroles

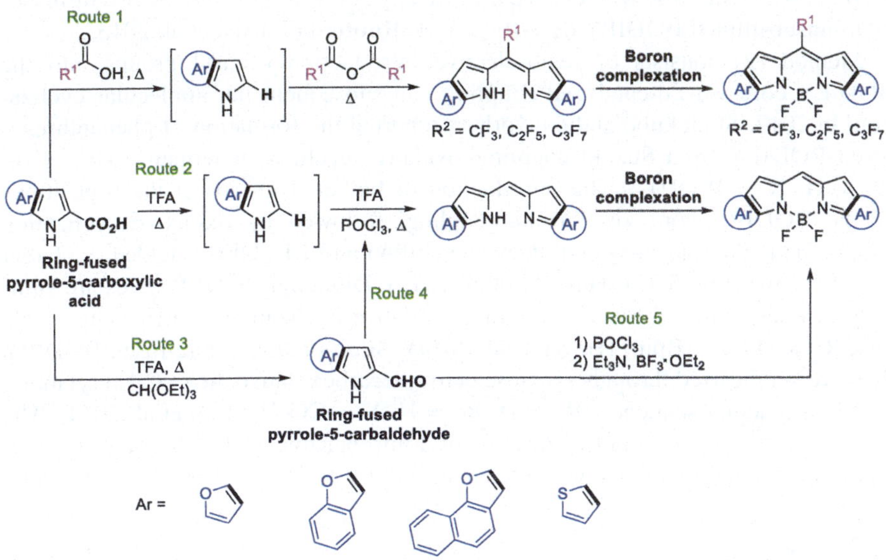

Fig. 5.31 Synthesis of [*b*]-fused BODIPY dyes by applying classical synthetic methods to ring-fused pyrrole-5-carboxylic acids and pyrrole-5-carbaldehydes

et al. 2018)- and indene (Jiang et al. 2018)-fused BODIPY dyes are synthesized from 4*H*-furo[3,2-*b*]pyrrole, 4*H*-thieno[3,2-*b*]pyrrole, 7*H*-acenaphtho[1,2-*b*]pyrrole, and 1,4-dihydroindeno[1,2-*b*]pyrrole, respectively.

Ring-fused pyrrole-5-carboxylic acid is converted to the corresponding α-free pyrrole by decarboxylation, so it is a good precursor for ring-fused BODIPY dyes (Fig. 5.31). For instance, heating a TFA solution of 4*H*-furo[3,2-*b*]pyrrole-5-carboxylic acid resulted in decarboxylation to give 4*H*-furo[3,2-*b*]pyrrole (Fig. 5.31, **Route2**) (Umezawa et al. 2008). Symmetrical and unsymmetrical *meso*-unsubstituted furan-fused BODIPY dyes are synthesized by the decarboxylation of 1 equivalent of 5-carboxylic furopyrroles, followed by reaction with 1 equivalent of symmetrical or unsymmetrical 5-formyl furopyrroles and $POCl_3$ and subsequent boron-complexation (Fig. 5.31, **Route4**) (Umezawa et al. 2008, 2009). Decarboxylation of 4*H*-furo[3,2-*b*]pyrrole-5-carboxylic acid by perfluorocarboxylic acid and subsequent reaction with corresponding perfluorocarboxylic anhydride and boron complexation gives *meso*-perfluoroalkyl substituted BODIPY dyes (Fig. 5.31, **Route1**) (Li et al. 2018a). Similarly, benzofuran- and naphtho[1,2-*b*]furan-fused BODIPY dyes are synthesized from the corresponding ring-fused pyrrole-5-carboxylic acids (Li et al. 2018b). *Meso*-CF_3 substituted furan-fused BODIPY dyes are also synthesized from 4*H*-thieno[3,2-*b*]pyrrole-5-carboxylic acid by using the same method (Awuah et al. 2011). Ring-fused pyrrole-5-carbaldehydes can be obtained by heating a TFA solution of ring-fused pyrrole-5-carboxylic acid, followed by reaction with triethyl orthoformate (Fig. 5.31, **Route3**) (Umezawa et al. 2009). The reaction of ring-fused pyrrole-5-carbaldehyde with $POCl_3$, followed by boron complexation, also furnishes *meso*-unsubstituted BODIPY dyes (Fig. 5.31, **Route5**) (Tanaka et al. 2013).

Method (2) consists of regioselective introduction of aryl groups into the BODIPY core via Pd-catalyzed coupling and subsequent intramolecular cyclization. In 2012, Shinokubo and co-workers reported the formation of phenanthrene-fused BODIPY by a Suzuki coupling–oxidative cyclization sequence (R^1 = Br, R^2 = H, X = $B(OH)_2$). The introduction of biphenyl groups at the β-positions of the BODIPY core via Suzuki coupling, followed by oxidative cyclization using [Bis(trifluoroacetoxy)iodo]benzene (PIFA) and $BF_3{\cdot}OEt_2$, yielded the target BODIPY dye (Fig. 5.32) (Hayashi et al. 2012). Dithienyl BODIPY dyes are similarly synthesized by a Suzuki coupling–oxidative cyclization sequence (R^1 = Br or I, R^2 = H, X = Bpin) (Heyer et al. 2014). Additionally, indole-fused BODIPY dyes are synthesized through a regioselective decarboxylative direct C-H arylation–Cadogan reaction sequence (R^1 = H, R^2 = H, X = CO_2H) (Luo et al. 2014). The decarboxylative cross-coupling reaction of 2-nitrobenzoic acid with a *meso*-phenyl-substituted BODIPY dye in the presence of $Pd(OAc)_2$, $PCy_3{\cdot}HBF_4$, Ag_2CO_3, and 4

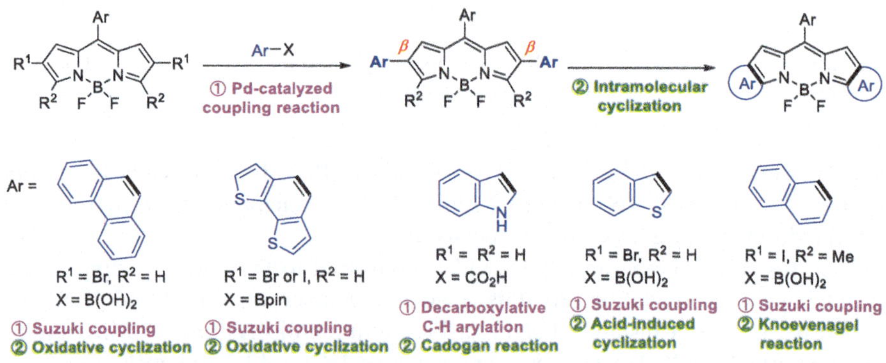

Fig. 5.32 Synthesis of [*b*]-fused BODIPY dyes by regioselective introduction of aryl groups via Pd-catalyzed coupling and subsequent intramolecular annulation

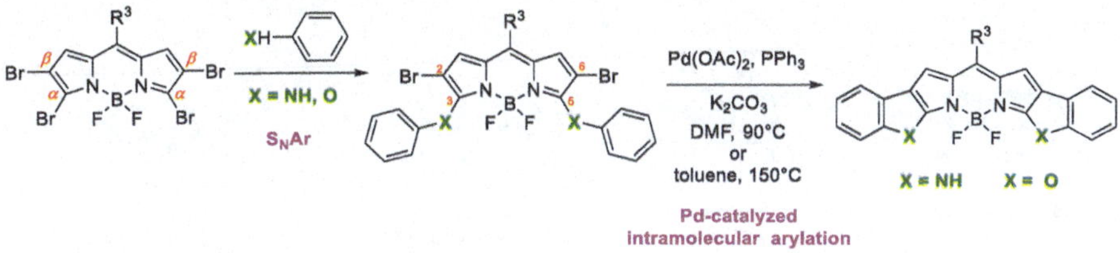

Fig. 5.33 Synthesis of [*b*]-fused BODIPY dyes by regioselective introduction of aryl groups via S_NAr reaction and subsequent pd-catalyzed intramolecular arylation

Å molecular sieves proceeds regioselectively at the β-positions to yield a BODIPY dye with two 2-nitrophenyl groups. Intramolecular reductive cyclization of the 2-nitrophenyl substituted BODIPY dye by heating in the presence of PPh_3 regioselectively forms C–N bonds. Benzothiophene (Sun et al. 2016)- and naphthalene (Zhou et al. 2017)-fused BODIPY dyes are synthesized via Suzuki coupling–acid-induced intramolecular cyclization and Suzuki coupling–Knoevenagel reaction sequences, respectively.

In 2015, Jiao and co-workers reported the synthesis of indole-fused BODIPY dyes by regioselective S_NAr reaction of 2,3,5,6-tetrabromo-BODIPY with aryl amines on the α-positions of the BODIPY core, followed by intramolecular ring fusion via Pd-catalyzed arylation at the β-positions of the BODIPY core (Fig. 5.33) (Zhou et al. 2015b). Similar procedures for phenol derivatives give benzofuran-fused BODIPY dyes (Belmonte-Vázquez et al. 2019).

Optical properties of ring-fused BODIPY

[*a*]-Fused BODIPY

The absorption and fluorescence properties of [*a*]-fused BODIPY dyes are shown in Fig. 5.34. Annulation of the benzene ring at the β and β' positions (*a* bond) of BODIPY dye leads to ca. 90-nm redshift of λ_{max} (**16** (Qin et al. 2005): λ_{max} = 512 nm, **17** (Shen et al. 2004): λ_{max} = 599 nm) (Fig. 5.34a). The redshift of λ_{max} caused by fusing the benzene ring at the *a* bond is mainly due to an increase in the HOMO energy level (Yamazawa et al. 2016; Wakamiya et al. 2013). The relatively higher Φ_f of **17** ($\Phi_f = 0.91$) compared with that of **16** ($\Phi_f = 0.29$) is probably due to the restriction of molecular rotation of the *meso*-phenyl group arising from the steric repulsion between the *meso*-phenyl group and the two hydrogen atoms on C1 and C13 of **17**. Benzo-[*a*]-fused BODIPY **17** is used in the bilayer and planar mixed heterojunction solar cells as the electron donor and with C_{60} as the electron acceptor (Chen et al. 2015); OPV devices harvest red/NIR photons with an onset of IPCE at 800 nm, and their power conversion efficiencies (PCE) reach up to 4.5%. Dibenzo-[*a*]-fused BODIPY (**19** (Yamazawa et al. 2016): λ_{max} = 761 nm) induces a remarkable redshift of more than 100 nm compared with that of the corresponding benzo-[*a*]-fused BODIPY (**18** (Ulrich et al. 2011): λ_{max} = 641 nm) owing to an increase in the HOMO energy level (Fig. 5.34b). Although **19** has an almost planar π-conjugated chromophore, the two phenyl rings of the 2*H*-benzo[*f*]isoindole are

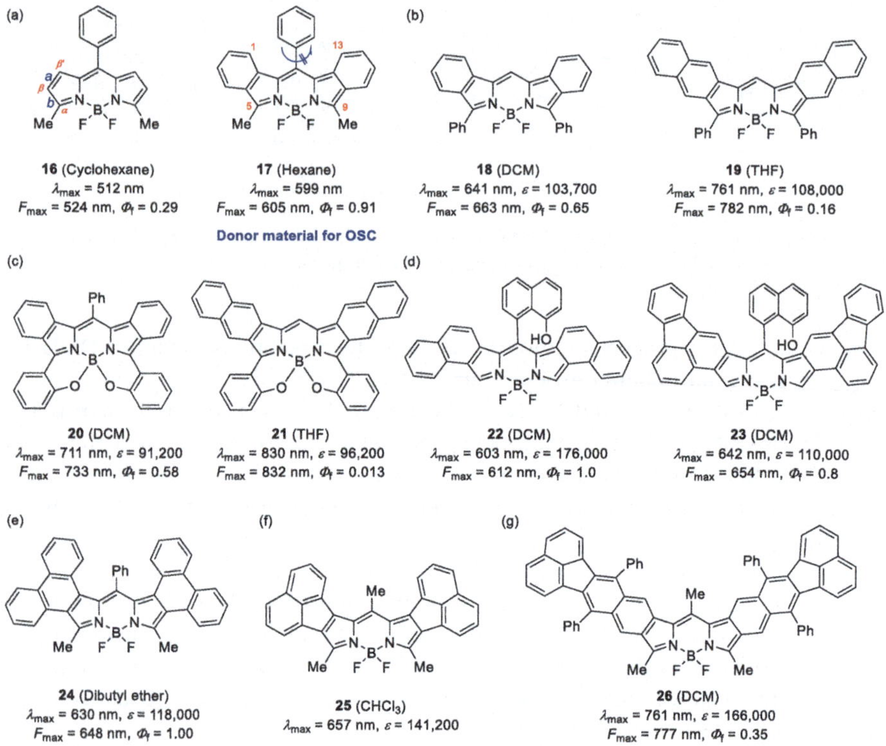

Fig. 5.34 Absorption and fluorescence properties of [*a*]-fused BODIPY dyes

tilted by 40°–50° relative to the naphtho-fused BODIPY core (Yamazawa et al. 2016). B,O-chelated derivatives (**20** (Tomimori et al. 2011): λ_{max} = 711 nm, **21** (Yamazawa et al. 2016): λ_{max} = 830 nm) show a further bathochromic shift (Fig. 5.34c).

Naphtha[1,2-*c*]pyrrole-based BODIPY **22** (Swavey et al. 2016) (λ_{max} = 603 nm) exhibits significantly blueshifted λ_{max} compared with 2*H*-benzo[*f*]isoindole-based BODIPY **19** (Fig. 5.34d). The λ_{max} of the fluoranthene-fused BODIPY **23** (Swavey et al. 2017) (λ_{max} = 642 nm) is more redshifted than that of **22** because of a decrease in the LUMO level. Although phenanthrene-fused BODIPY **24** (Descalzo et al. 2008) has a propeller-like distorted conformation in the crystalline state, it shows intense absorption (ε = 118,000) and fluorescence (Φ_f = 1.00) in solution (Fig. 5.34e). Acenaphthylene-[*a*]-fused BODIPY **25** (Ono et al. 2003) (λ_{max} = 657 nm) shows redshifted λ_{max} compared with that of **24** (λ_{max} = 630 nm) (Fig. 5.34f). Although benzo[*k*]fluoranthene-fused BODIPY **26** (Okujima et al. 2010) shows NIR absorption (λ_{max} = 761 nm) and fluorescence (F_{max} = 777 nm), it is unstable under air in room light (Fig. 5.34g).

[*b*]-Fused BODIPY

Benzene-[*b*]-fused BODIPY dye **27** (Shimogawa et al. 2013) (λ_{max} = 581 nm) shows redshifted λ_{max} compared with the non-fused BODIPY **16** (Figs. 5.34a and 5.35a). Annulation of the benzene ring at the α and β positions (*b* bond) leads to the enhancement of the nonaromatic quinoidal character of benzene rings and the electron-accepting azafulvene character in **27** (Shimogawa et al. 2013). Different from the benzene-[*a*]-fused BODIPY dyes, the main reason for the redshift of λ_{max} of benzene-[*b*]-fused BODIPY dyes is a decrease in the LUMO energy level (Wakamiya et al. 2013; Shimogawa et al. 2013). The lowest energy transition of **27** is mainly attributed to the HOMO-1 to LUMO transition, and the oscillator strength (f) is relatively low ($f = 0.055$). Because the k_f value is proportional to the f value, the nonfluorescent property of **27** is probably responsible for the low f value (Shimogawa et al. 2013). Naphthalene and phenanthrene-[*b*]-fused BODIPY dyes (**28** (Zhou et al. 2017): λ_{max} = 630 nm, **29** (Hayashi et al. 2012): λ_{max} = 673 nm) show a further spectral redshift because of the markedly stabilized LUMO energy level (Fig. 5.35b). The lower LUMO level is desirable for the molecular design of electron-transporting materials. [*b*]-Fused BODIPY **29** is used in *p–n* heterojunction solar cells as the electron acceptor with tetrabenzoporphyrin (BP) as the donor (Hayashi et al. 2012); the PCE value is 0.52%. Acenaphthylene-[*b*]-fused BODIPY **30** (Jiang et al. 2018) (λ_{max} = 648 nm) and structurally rigidified BODIPY dyes with dialin **31** (Chen et al. 2000)

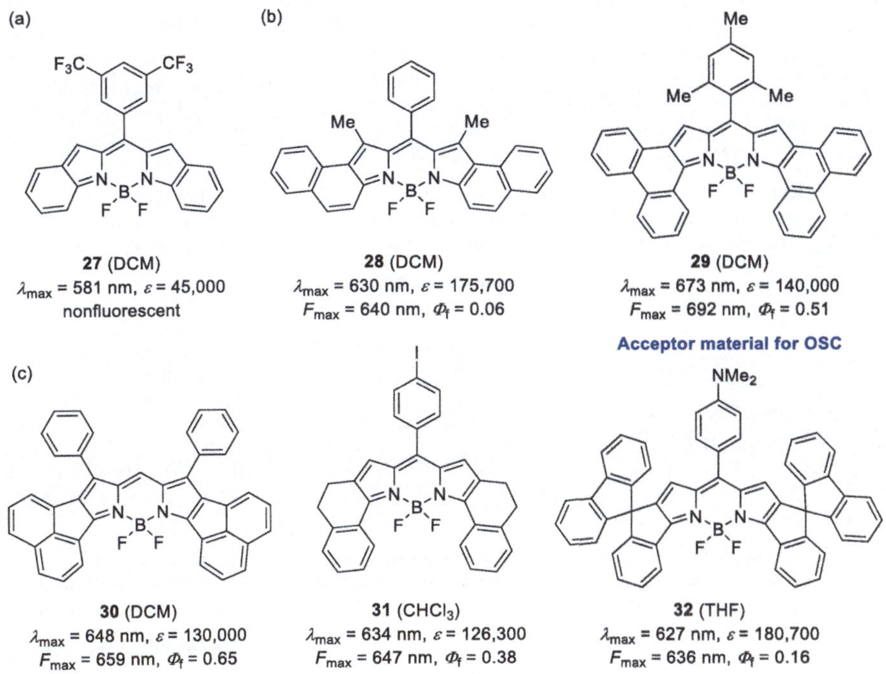

Fig. 5.35 Absorption and fluorescence properties of [*b*]-fused BODIPY dyes

($\lambda_{max} = 634$ nm) and spirofluorene **32** (Kowada et al. 2010) ($\lambda_{max} = 627$ nm) moieties have also been reported (Fig. 5.35c). In the spirofluorene derivative **32**, although the λ_{max} and F_{max} values are not changed by varying the solvent, the Φ_f value becomes lower with increasing solvent polarity (benzene: $\Phi_f = 0.78$, $CHCl_3$: $\Phi_f = 0.73$, THF: $\Phi_f = 0.16$, DMF: $\Phi_f = 0.06$); the feasibility of electron transfer from the electron donor (dimethylaminophenyl group) to the electron acceptor (BODIPY core) is estimated by the Rehm–Weller equation, which suggests that fluorescence quenching with increasing solvent polarity arises from photoinduced electron transfer (PeT) process (Kowada et al. 2010).

In 2008, Umezawa et al. first reported furan-[*b*]-fused BODIPY dyes and named it Keio Fluors (KFL) (Fig. 5.36a) (Umezawa et al. 2008). The KFL dyes have high ε and Φ_f values and cover a wide fluorescence range from yellow to the NIR region (Umezawa et al. 2009). Because of the excellent fluorescence properties, KFL dyes have been applied to NIR fluorescence probes (Matsui et al. 2011; Yang et al. 2013; Umezawa et al. 2014). For instance, **KFL-2** (Umezawa et al. 2008) shows sharp and intense absorption and fluorescence bands ($\varepsilon = 253{,}000$, $\Phi_f = 0.98$, full width at half maximum (fwhm) $= 574$ cm^{-1}). Introduction of 4-methoxyphenyl groups causes a further redshift (**KFL-4** (Umezawa et al. 2008): $\lambda_{max} = 723$ nm). X-ray single crystal analysis of **KFL-4** has revealed a planar conformation between the phenylene moiety and BODIPY core. **KFL-4** is applied as a NIR absorbing material

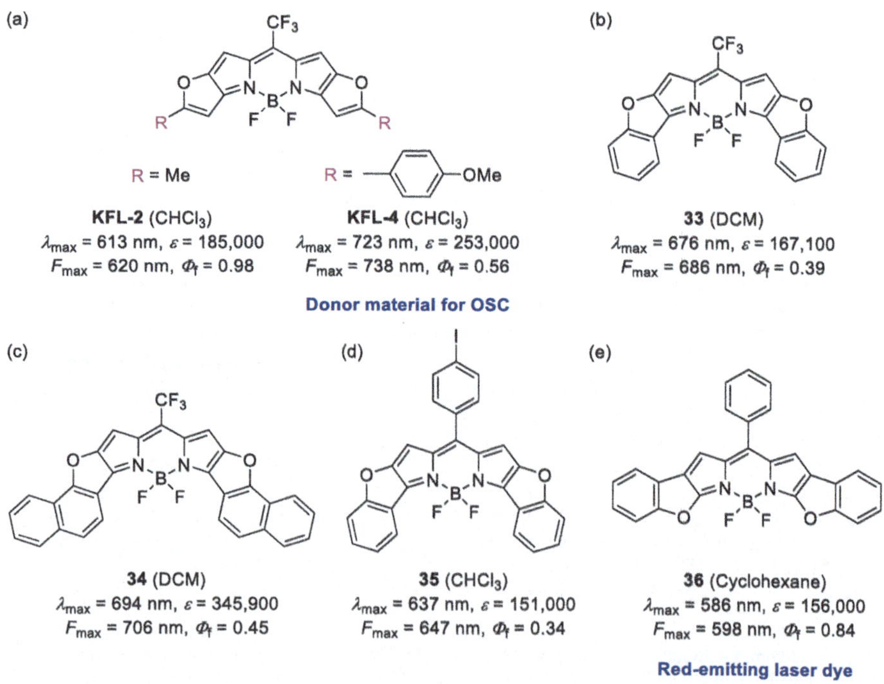

Fig. 5.36 Absorption and fluorescence properties of furan-[*b*]-fused BODIPY dyes

for vacuum processable organic solar cells; bulk heterojunction organic solar cells (BHJSCs) comprising **KFL-4** as the electron donor and C_{60} as the electron acceptor exhibit a PCE of 6.1% (Li et al. 2017b).

Benzofuran and naphtho[1,2-*b*]furan-fused BODIPY dyes exhibit a further redshifted λ_{max} (**33** (Li et al. 2018b): λ_{max} = 676 nm, **34** (Li et al. 2018b): λ_{max} = 694 nm) (Fig. 5.36b, c). Single crystal X-ray analysis suggests that the slope angle and the distance between layers are 21.08° and 21.68° and 3.5 Å and 3.9 Å for **33** and **34**, respectively, which represents typical J-aggregation. The F_{max} values of **33** (from 686 to 757 nm) and **34** (from 707 to 777 nm) are redshifted, and the Φ_f values become lower with increasing concentration in solution. The λ_{max} of benzofuro[3,2-*b*]pyrrole-based BODIPY **35** (Chen et al. 2000) (637 nm) is more redshifted than that of benzofuro[2,3-*b*]pyrrole-based BODIPY **36** (Belmonte-Vázquez et al. 2019) (586 nm) (Fig. 5.36d, e). BODIPY **36** is applied to a redemitting laser dye; the laser efficiency (Eff) and laser peak wavelength (λ_{peak}) are 44% and 647.2 nm, respectively.

The absorption and fluorescence properties of thiphene-[*b*]-fused BODIPY dyes are shown in Fig. 5.37. In 2011, You et al. reported the first synthesis of thiphene-[*b*]-fused BODIPY dyes (Awuah et al. 2011). Thiphene-fused BODIPY dye **SBDPiR731** (Awuah et al. 2011) (λ_{max} = 731 nm, F_{max} = 754 nm, Φ_f = 0.37) shows slightly redshifted λ_{max} and F_{max} values and lower Φ_f compared with those of the corresponding furan-fused derivative **KFL-4** (λ_{max} = 723 nm, F_{max} = 738 nm, Φ_f = 0.56) (Figs. 5.36a and 5.37a). Ji et al. reported that the incorporation of sulfur atoms into the π-conjugated skeleton of BODIPY is an efficient strategy to attain high quantum yields for triplet-state formation (Fig. 5.37b) (Ji et al. 2015). The triplet formation of **37** (triplet-state formation quantum yield Φ_T = 0.637) is more efficient compared with that of the non-fused derivative **38** (Φ_T = 0.061). Theoretical calculations demonstrate that the increased intersystem crossing (ISC) mechanism of **37** compared with that of **38** is caused by the participation of the sulfur atom in the lowest-lying excited states, which leads to moderate spin-orbit coupling (SOC) and the small singlet-triplet energy gap in **37**.

While thiphene-fused BODIPY **SBDPiR731** does not show singlet oxygen generation, the brominated derivatives show efficient singlet oxygen generation (Fig. 5.37a). Inclusion of heavy atoms tends to induce cytotoxicity in the absence of light. Despite the absence of heavy atom substituents, nonsubstituted (**SBDPiR690** (Awuah et al. 2013)) and trifluoromethyl-substituted (**SBDPiR688** (Watley et al. 2015)) derivatives generate singlet oxygen atoms; the singlet oxygen quantum yields (Φ_Δ) of **SBDPiR690** and **SBDPiR688** are 0.42 and 0.47, respectively. Additionally, **SBDPiR690** (Φ_f = 0.22) and **SBDPiR688** (Φ_f = 0.39) exhibit fluorescence. Such dual functional photosensitizers are rare and can be applied to not only photodynamic therapy (PDT), but also in fluorescence tumour detection (image-guided surgery). **SBDPiR690** and especially **SBDPiR688** have a remarkable dual functionality of brightness (**BT** = $\varepsilon \times \Phi_f$: **SBDPiR690**: **BT** = 26,400 M^{-1} cm^{-1}, **SBDPiR688**: **BT** = 82,290 M^{-1} cm^{-1}) and phototoxic power (**PP** = $\varepsilon \times \Phi_\Delta$: **SBDPiR690**: **PP** = 50,400 M^{-1} cm^{-1}, **SBDPiR688**: **PP** = 211,000 M^{-1} cm^{-1}) owing to the high ε and balanced decay (Φ_f and Φ_Δ). The dual functionality of **SBDPiR690** has been successfully applied to preclinical optical imaging and PDT (Watley et al. 2015).

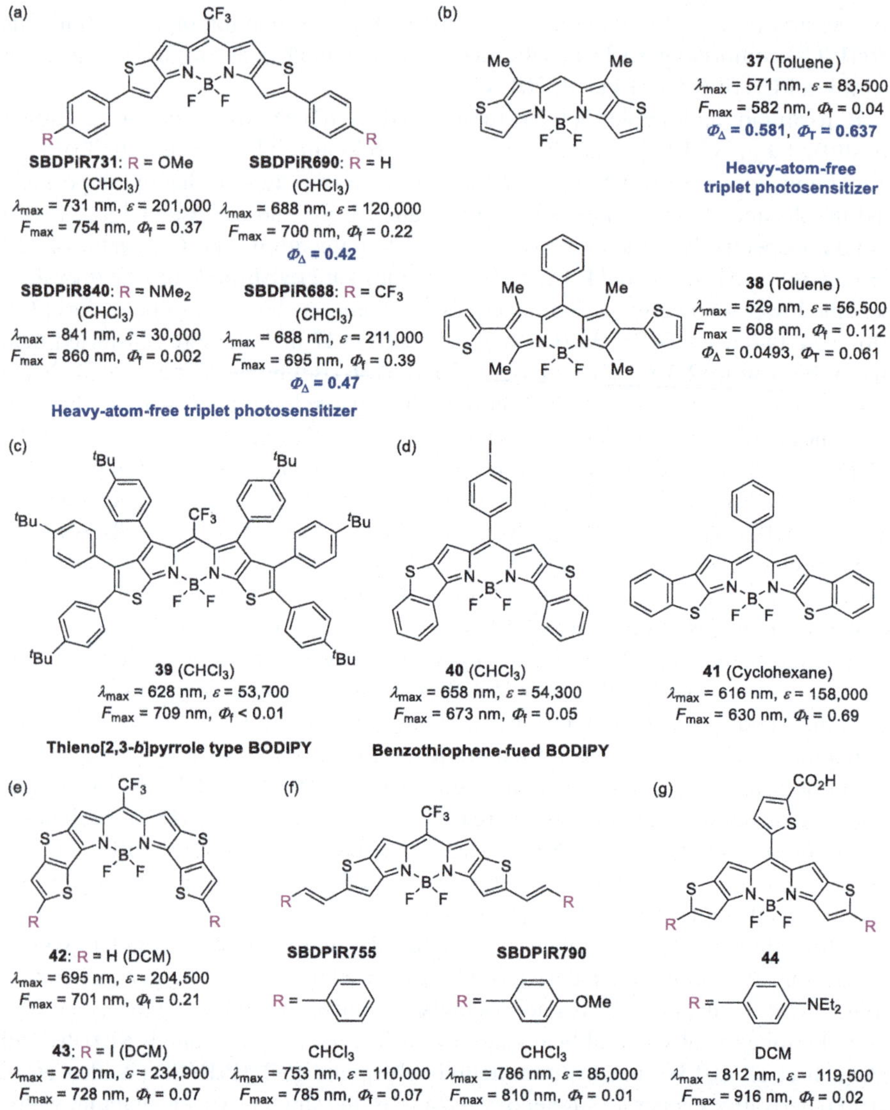

Fig. 5.37 Absorption and fluorescence properties of thiophene-[*b*]-fused BODIPY dyes

Thieno[2,3-*b*]pyrrole type BODIPY **39** (Wang et al. 2016a), which is the structural isomer of **37**, has also been reported (Fig. 5.37c). Similar to the furan-fused BODIPY dyes, thieno[3,2-*b*]pyrrole-based BODIPY **40** (Chen et al. 2000) (λ_{max} = 658 nm) shows redshifted λ_{max} compared with the isomeric thieno[2,3-*b*]pyrrole-based BODIPY **41** (Sun et al. 2016) (λ_{max} = 616 nm) (Fig. 5.37d). Thieno[3,2-*b*]thiophene-fused BODIPY **42** shows a further spectral redshift (λ_{max} = 695 nm, F_{max} = 701 nm) (Fig. 5.37e) (Sun et al. 2019). The diiodio-substituted derivative **43** exhibits efficient singlet oxygen generation with a Φ_{Δ} value of 0.32 upon irradiation with a 690-nm laser. Replacement of the aryl group with a styryl group leads to a significant redshift (**SBDPiR755**: λ_{max} = 750 nm, **SBDPiR790**: λ_{max} = 786 nm) (Fig. 5.37f) (Awuah et al. 2013). Introduction of the dimethylamino group is a more efficient approach to realize the red shift of λ_{max} (**SBDPiR840** (Awuah et al. 2013): λ_{max} = 841 nm) (Fig. 5.37a). The 2-thienyl derivative **44** (Kubota et al. 2019) (λ_{max} = 812 nm, F_{max} = 916 nm, Φ_f = 0.02) shows a blueshift of λ_{max}, redshift of F_{max}, and an increase in Φ_f compared with those of the trifluoromethyl derivative **SBDPiR840** (Fig. 5.37g). Thiphene-fused BODIPY **44** has been applied to NIR photosensitisers for dye-sensitized solar cells; the maximum IPCE and power conversion efficiency are 12.3% at 810 nm and 1.12%, respectively (Kubota et al. 2019).

Indole-fused BODIPY dyes have also been reported (Luo et al. 2014; Zhou et al. 2015b). The λ_{max} value of indole-fused BODIPY **45** (Zhou et al. 2015b) (λ_{max} = 608 nm) is more redshifted than that of the benzofuran-fused derivative **36** (λ_{max} = 586 nm) and blueshifted than that of the benzothiophene-fused derivative **41** (λ_{max} = 616 nm) (Figs. 5.36e, 5.37d and 5.38a). Installation of the methoxy group has little effect on the absorption properties, whereas the Φ_f value becomes lower (**46**: λ_{max} = 610 nm, Φ_f = 0.24). Although little variation is observed in the Φ_f value of **45** (cyclohexane: Φ_f = 0.89, acetonitrile: Φ_f = 0.96) with increasing solvent polarity, **46** (cyclohexane: Φ_f = 0.89, acetonitrile: Φ_f < 0.001) shows a considerable decrease in Φ_f with increasing solvent polarity. Unsymmetrical BODIPY **47** with the replacement of one indole moiety on **46** by a benzofuran moiety shows blueshifted

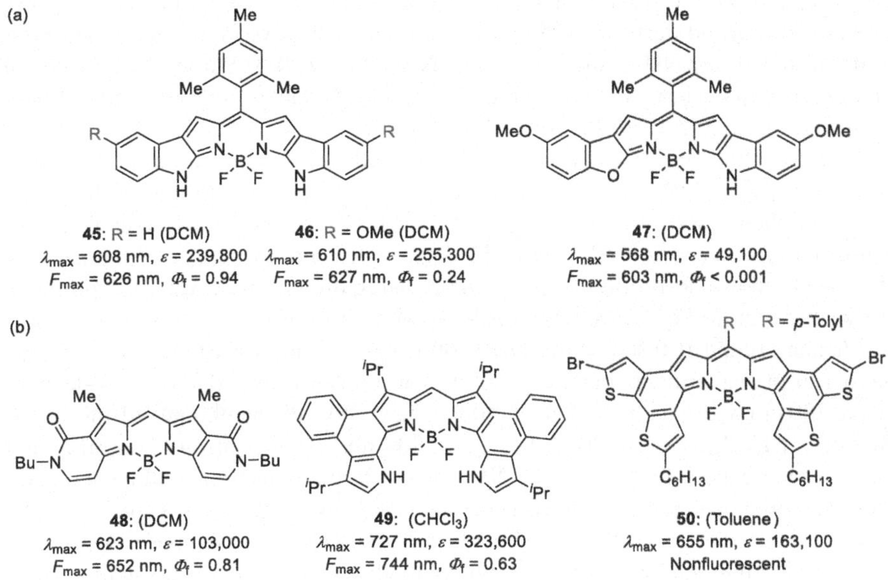

Fig. 5.38 Absorption and fluorescence properties of heterocycle-containing [*b*]-fused BODIPY dyes

λ_{max} with a considerable decrease in the ε value. The Φ_f of **47** also decreases with increasing solvent polarity (cyclohexane: $\Phi_f = 0.32$, toluene: $\Phi_f = 0.04$, DCM < 0.001). Pyridone-, benzoindole-, and benzodithiophene-fused BODIPY dyes have also been reported (**48** (Didukh et al. 2019): $\lambda_{max} = 623$ nm, **49** (Sarma et al. 2013): $\lambda_{max} = 727$ nm, **50** (Heyer et al. 2014): $\lambda_{max} = 655$ nm, respectively) (Fig. 5.38b).

Other types of ring-fused BODIPY based on naphthalene (Yang et al. 2015; Ni et al. 2016a), dihydronaphthalene (Wang et al. 2010), perylene (Jiao et al. 2011b), anthracene (Zeng et al. 2011), coumarin (Bochkov et al. 2013), porphyrin (Jiao et al. 2011c), and others (Gobo et al. 2016; Wu et al. 2017a) have been reported.

Optical properties of ring-fused aza-BODIPY

Benzo[*a*]-fused aza-BODIPY **51** (Lu et al. 2011) shows a 71-nm redshifted λ_{max} compared with the corresponding BODIPY dye **18** (Figs. 5.34b and 5.39a). Incorporation of the dimethylamino group quenches the fluorescence in polar solvents (**52** (Lu et al. 2011): $F_{max} = 830$ nm, $\Phi_f = 0.01$ in DCM, $F_{max} = 796$ nm, $\Phi_f = 0.10$ in hexane) and results in a ratiometric pH response. The addition of TFA to a DCM of **52** leads to the formation of monoprotonated species ($\lambda_{max} = 771$ nm) along with a decrease in the absorption intensity. Further sequential addition of TFA causes the quantitative formation of diprotonated species ($\lambda_{max} = 724$ nm, $F_{max} = 745$ nm) with a gradual increase in the fluorescence intensity. 1,2-Naphtho[*a*]-fused aza-BODIPY **53** (Lu et al. 2011) shows more redshifted λ_{max}.

Aza-BODIPY derivative **54** (Díaz-Moscoso et al. 2014) has a significantly blueshifted λ_{max} (439 nm) (Fig. 5.39b). Pyrazine-fused aza-BODIPY **55** (Liu et al. 2011) ($\lambda_{max} = 685$ nm) shows a hypsochromic shift compared with **51** (Fig. 5.39c). The pyrazine-fused derivative **55** can be used as a selective colorimetric and fluorometric sensor for NH_4^+; the addition of NH_4^+ to a THF solution of **55** results in fluorescence quenching and an obvious colorimetric change from green to red-pink probably because of the formation of H-aggregates.

Benzo[*c,d*]indole-based aza-BODIPY dyes with two fluorine atoms on the boron atom **56** ($\lambda_{max} = 539$ nm) show blueshifted λ_{max} compared with that of the corresponding phenyl derivative **57** ($\lambda_{max} = 602$ nm) (Fig. 5.39d) (Shimizu et al. 2015). Thiophene-fused aza-BODIPY dyes (**58**: $\lambda_{max} = 767$ nm, **59**: $\lambda_{max} = 788$ nm) show sharp and intense absorption bands in the NIR region with very low absorptions in the visible region (Fig. 5.39e) (Wu et al. 2014).

Phenanthrene-[*a*]-fused azaBODIPY **60** (Sheng et al. 2017a) ($\lambda_{max} = 745$ nm) has a near planar structure in the crystalline state and absorbs and fluoresces strongly above 700 nm with excellent photostability (Fig. 5.39f). Phenanthrene-[*b*]-fused aza-BODIPY **61** (Sheng et al. 2017b) ($\lambda_{max} = 804$ nm) shows narrow absorption and fluorescence band selectivity in the NIR region, large extinction coefficients, good thermostability, and p-type transport character (hole mobility = 0.018 $cm^2\ V^{-1}\ s^{-1}$). Annularly fused aza-BODIPY **62** (Sheng et al. 2018) has a wave conformation due to the steric hindrance among the peripheral methoxy groups in the crystalline state. Aza-BODIPY **62** ($\lambda_{max} = 878$ nm) shows excellent spectral selectivity in the NIR region (negligible absorption between 300 and 700 nm) with a narrow absorption band, large extinction coefficient, and high photostability.

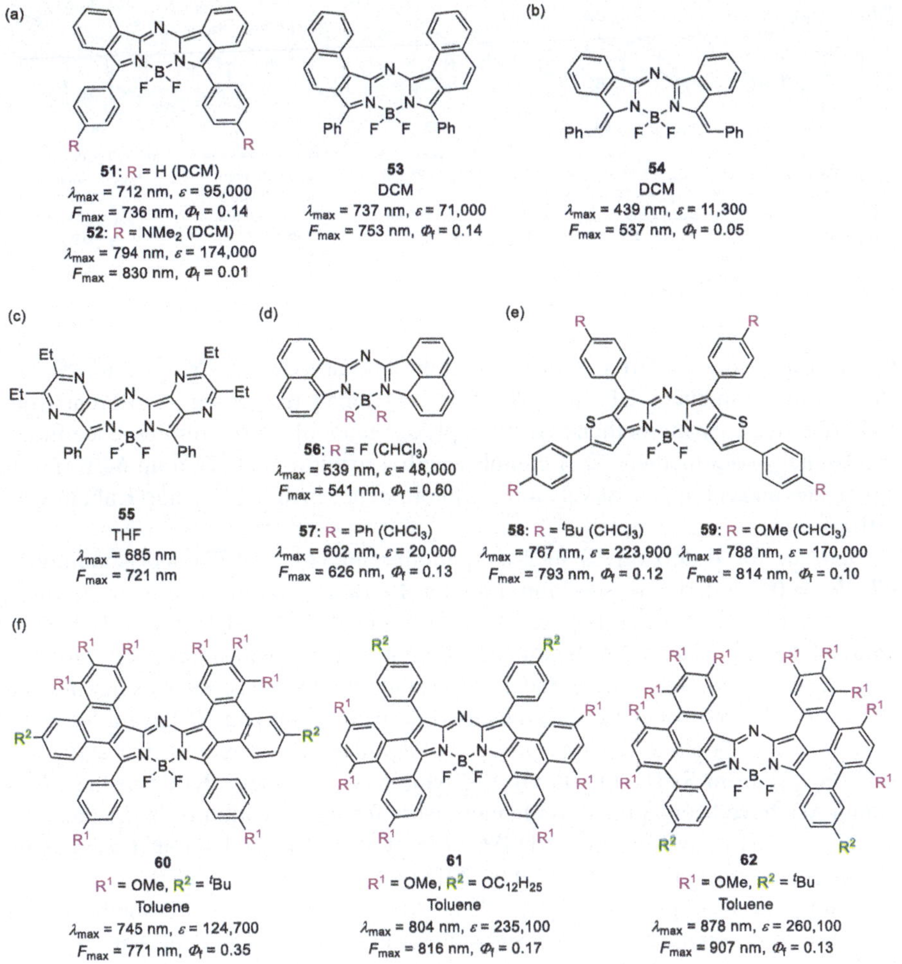

Fig. 5.39 Absorption and fluorescence properties of fused aza-BODIPY dyes

5.2.4 BODIPY Analogue

5.2.4.1 Pyridomethene Boron Complex

A pyridomethene-BF_2 complex is a six-membered ring analogue of a BODIPY dye. The pyridomethene-BF_2 complex was first reported around 1970 (Scheibe et al. 1969; Douglass et al. 1973), but it was hardly studied until the detailed optical properties were reported by the author in 2010 (Fig. 5.40a) (Kubota et al. 2010a). The pyridomethene-BF_2 complex **63** is synthesized from the reaction of 2-cyanomethylpyridine and 2-bromopyridines and subsequent boron-complexation. The pyridomethene-BF_2 complex shows blueshifted λ_{max} and F_{max} compared with

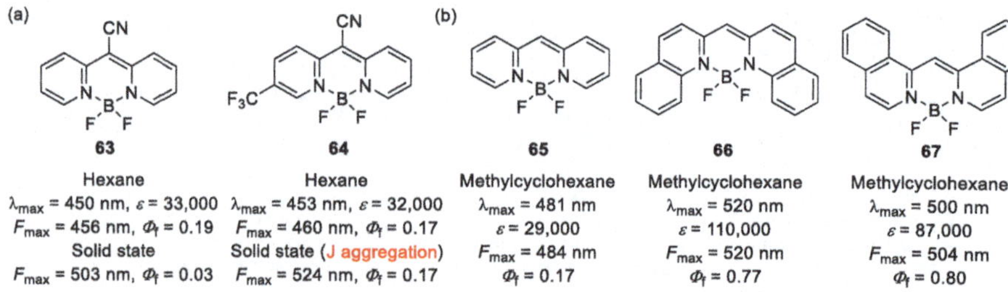

Fig. 5.40 Pyridomethene-BF_2 complexes

the corresponding BODIPY dyes in solution. Similar to BODIPY dyes, the Stokes shift of pyridomethene-BF_2 complexes is very small (*ca.* 5 nm). Different from BODIPY dyes, pyridomethene-BF_2 complexes generally show fluorescence in the solid-state; pyridomethene-BF_2 complex with a trifluoromethyl group **64** forms J-aggregates and exhibits relatively intense fluorescence in the solid-state (Kubota et al. 2010a).

Benzene-fused derivatives such as quinoline **66** ($\Phi_f = 0.77$) and isoquinoline **67** ($\Phi_f = 0.80$) derivatives exhibit a higher Φ_f value than the non-benzene fused derivative **65** ($\Phi_f = 0.17$) (Fig. 5.40b) (Golden et al. 2017). The relatively low Φ_f value of **65** is considered to be due to the relatively efficient intersystem crossing (ISC) from the S_1 (B_2) state to the T_2 (A_1) state; the ISC process is exergonic and symmetry-allowed. Because the ring-fused derivatives **66** and **67** have a lower S_1 state than the T_2 state, the ISC process is significantly decreased; the exergonic ISC process from S_1 (B_2) to T_1 (B_2) is symmetry-forbidden. Pyridomethene-BF_2 complexes have been applied as chemosensors for the detection of hydrazine (Lin and Chow 2013), sensitizers for DSSC (Lin et al. 2015), and symmetry-breaking charge transfer materials (Golden et al. 2018).

A nitrogen analogue of pyridomethene-BF_2 complex, aza-pyridomethene-BF_2 complex, was reported in 1993 (Sathyamoorthi et al. 1993). Different from the BODIPY dyes, introduction of a nitrogen atom at the *meso*-position of the pyridomethene-BF_2 complex leads to a hypsochromic shift (**65**: $\lambda_{max} = 481$ nm, **BTAA** (Bañuelos et al. 2011): $\lambda_{max} = 397$ nm) (Figs. 5.40b and 5.41a). Aza-pyridomethene-BF_2 complexes (**BTAA**: $F_{max} = 398$ nm, $\Phi_f = 0.36$) show significantly blueshifted F_{max} and relatively lower Φ_f values compared with those of the common BODIPY dyes ($F_{max} > 500$ nm, $\Phi_f > 0.6$); the decrease in Φ_f is due to an increase in both the k_f and mainly the k_{nr} values (Bañuelos et al. 2011). **BTAA** is applied to blue-emitting laser dyes; the lasing efficiency (Eff) and lasing wavelength are 18% and 427 nm, respectively, in ethanol (dye concentration: 1.5×10^{-3} M) at a pumping wavelength of 355 nm (Bañuelos et al. 2011). Aromatic ring-fused pyridomethene-BF_2 complexes have also been reported. **MQAB** (Kondakova et al. 2010) is applied as a deep-blue fluorescent dopant for OLEDs (Fig. 5.41b). Ring-fused aza-pyridomethene-BF_2 complexes tend to show fluorescence in the solid-state (Quan et al. 2012; Wang et al. 2013; Zhu et al. 2014, 2015; Gu et al. 2016). Pyridine-fused derivative **68** (Quan et al. 2012) shows strong fluorescence both in DCM (Φ_f

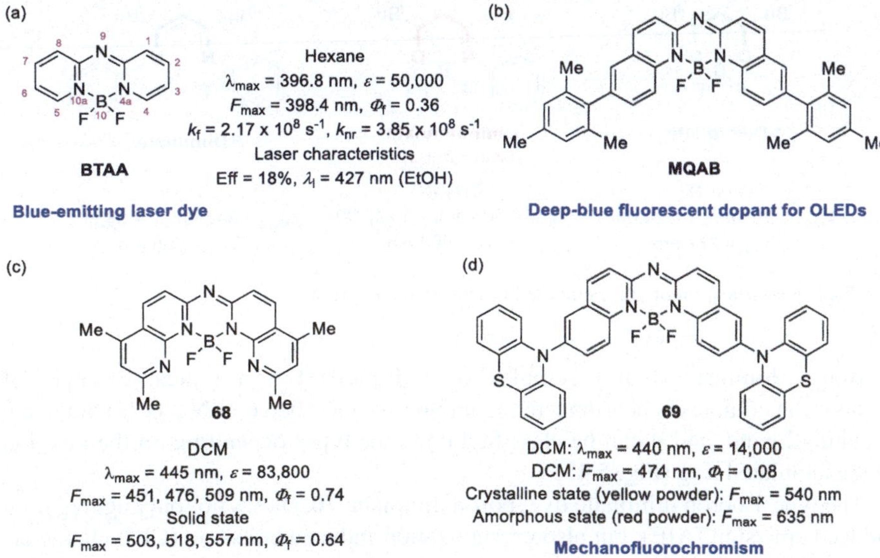

Fig. 5.41 Aza-pyridomethene-BF_2 complexes

= 0.74) and in the solid-state ($\Phi_f = 0.64$) (Fig. 5.41c). Phenothiazine-substituted derivative **69** (Zhu et al. 2014, 2015) exhibits mechanofluorochromism; its yellow powder (crystalline state) and red powder (amorphous state) show fluorescence at 540 and 635 nm, respectively, and reversible interconversion between the yellow powder and red powder is observed through grinding (crystalline state to amorphous state) and fuming with solvent vapour (amorphous state to crystalline state) (Fig. 5.41d). Fumigation of **69** with HCl/TEA vapours shows an off/on switching fluorescence through protonation-deprotonation (Zhu et al. 2014).

5.2.4.2 Boron Diiminate

Boron diketonate, boron iminoenolate (boron ketoiminate), and boron diiminate are representative examples of six-membered monoboron complexes (Fig. 5.42). When compared with the boron complexes—β-diketonate, β-iminoenolate (ketoiminate), and β-diiminate—the substitution of oxygen for more electron-donating aniline groups renders the chromophores poorer acceptors and generally results in the redshift of λ_{max} (β-diketonate: 291 nm, β-iminoenolate: 305 nm, β-diiminate: 365 nm) (Macedo et al. 2008). DFT calculations indicate that the HOMO and LUMO energies increase along the series: β-diketonate < β-iminoenolate < β-diiminate, presumably due to the greater electron-donating nature of nitrogen compared with that of oxygen. The variation in the HOMO and LUMO energies with substitution into the chromophore is the greatest for diketonates and the smallest for diiminates.

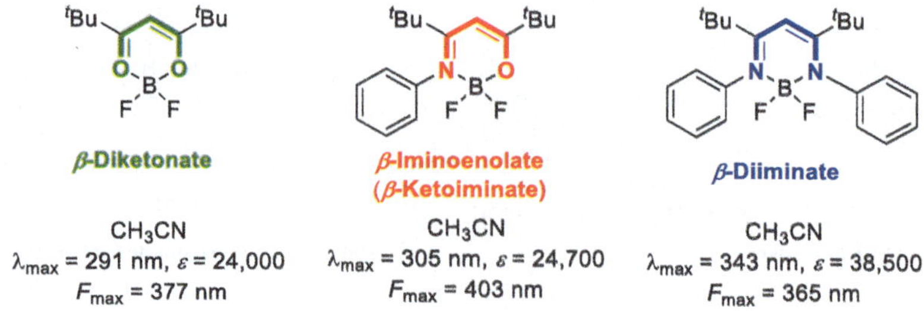

Fig. 5.42 Classification of six-membered monoboron complexes

Boron diiminate (boron complex of β-diiminate) is a typical example of monoboron complexes bearing a monoanionic six-membered N^N (⁶N^N) bidentate chelating ligand, and it can be classified into nine types depending on the position of the annulated ring (Fig. 5.43).

The type 1 boron diiminate dye, boron diiminate **70**, shows not only aggregation induced emission (AIE), but also crystallization-induced emission (CIE) characteristics (THF: $\Phi_f < 0.01$, amorphous state: $\Phi_f = 0.02$, crystalline state: $\Phi_f = 0.23$) because of the suppression of molecular motion and difference in the packing structure in the solid-states (Fig. 5.44a) (Yoshii et al. 2014a). The weaker and redshifted fluorescence properties in the amorphous state than those in the crystalline state may be due to the formation of excimers via π–π stacking of phenyl units, which often leads to a spectral redshift and fluorescence quenching (Yoshii et al. 2014b). In 2-methyl-tetrahydrofuran (2Me-THF), at a low temperature (77 K), **70** shows phosphorescence probably because of the existence of the lone-pair electrons on the two imine groups (Yoshii et al. 2014b). Introduction of the dimethylamino group (R = NMe_2) to **70** induces a large redshift of F_{max} (amorphous state: $F_{max} = 645$ nm, $\Phi_f = 0.01$, crystalline state: $F_{max} = 602$ nm, $\Phi_f = 0.08$). Iodine-substituted boron diiminates show mechanochromic luminescence (Yamaguchi et al. 2016). The mechanochromic luminescence properties depend on the position of substitution of the iodine atoms on the phenyl groups. The variability of F_{max} in the change from crystalline to amorphous states tends to be larger with increasing dihedral angles between the phenyl groups and the boron-containing six-membered ring.

For type 2 boron diiminate dye, benzothiazole-enamide-based boron difluoride complexes (**Borebt**) have been reported (Liu et al. 2015). **Borebt** dyes are synthesized by the boron-complexation of β-diiminate ligand which is obtained from the reaction of the corresponding β-iminoenolate ligand with aniline derivatives. **71** has

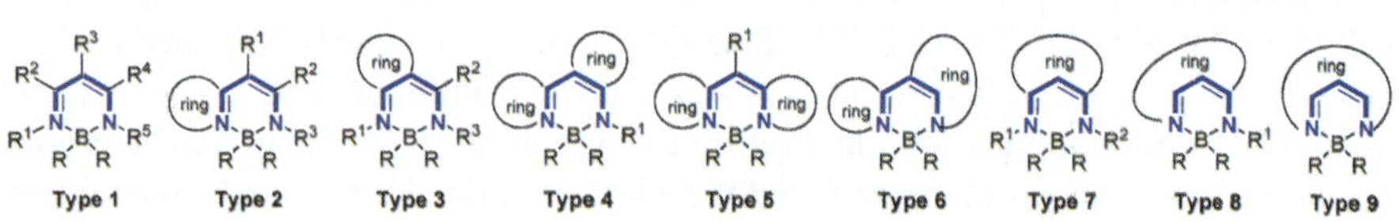

Fig. 5.43 Classification of boron complexes bearing a β-diiminate ligand

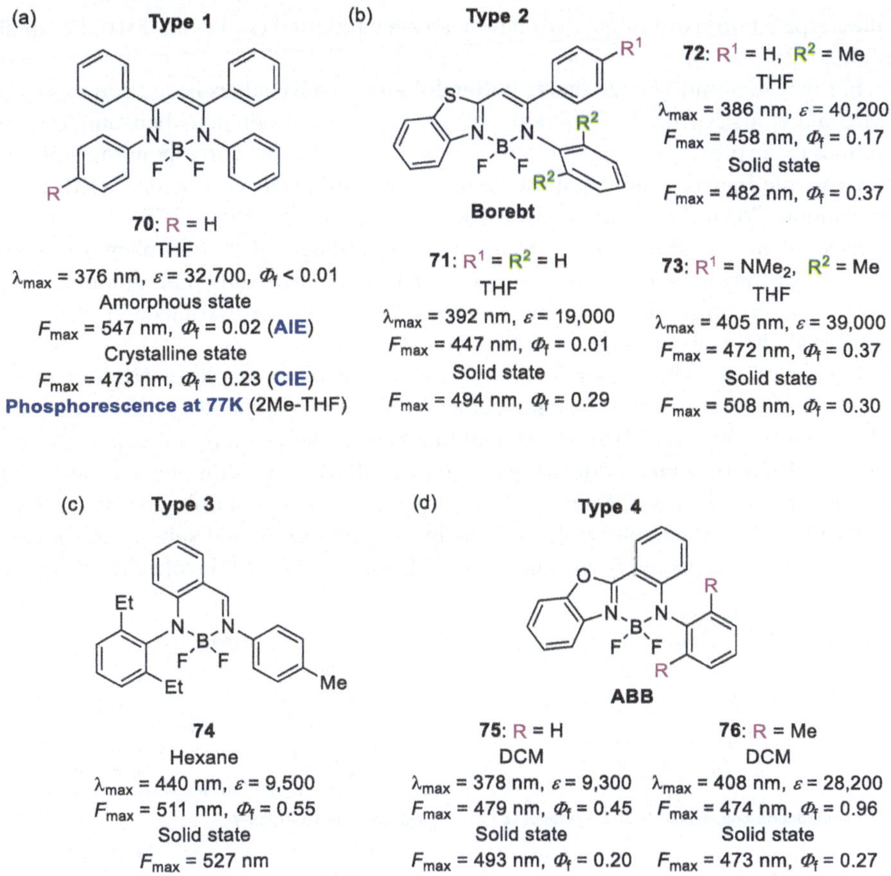

Fig. 5.44 Examples of boron diiminate **a** type 1, **b** type 2, **c** type 3, and **d** type 4

an AIE character (THF: $\Phi_f = 0.01$, solid-state: $\Phi_f = 0.29$) because of the intramolecular rotation induced non-radiative process in solution (Fig. 5.44b). Introduction of methyl groups (R^2 = Me) on the phenyl ring suppresses the intramolecular rotation of the aryl groups, subsequently enhancing the fluorescence in THF (**72**: $\Phi_f = 0.17$). Dimethyl derivative **73** shows fluorochromic properties against acidic vapours in the solid-state. Other type 2 boron diiminate dyes have also been reported (Graser et al. 2013; Wang et al. 2015a, 2015b; Yamaguchi et al. 2017; Lugovik et al. 2018).

For type 3, anilido-imine-based boron complexes have been reported (Ren et al. 2007). Condensation of 2-fluorobenzaldehyde with aniline derivatives gives imines, followed by nucleophilic aromatic displacement of the fluoride by lithiated anilines to provide anilido-imine ligands. Subsequent boron-complexation gives anilido-imine-based boron complexes. Anilido-imine-based boron complex **74** shows a relatively strong fluorescence in solution ($\Phi_f = 0.55$) in contrast to **70** and **71** (Fig. 5.44c).

Other type 3 boron diiminate dyes have also been reported (Liu et al. 2010; Perumal et al. 2012).

For type 4, anilido-benzoxazole boron difluoride (**ABB**) dyes have been reported (Meesala et al. 2015). The reaction of 2-iodo-*N*-(2-bromophenyl) benzamide with anilines in the presence of K_2CO_3 and copper iodide gives the corresponding anilido-benzoxazole ligands, and the subsequent boron complexation provides **ABB** dyes. In solution, **76** ($\Phi_f = 0.96$) shows higher Φ_f compared to that of **75** ($\Phi_f = 0.45$) because of the constraint of the rotation of the orthogonal anilido phenyl moiety caused by ortho substituents (Fig. 5.44d). **ABB** dyes show solid-state fluorescence. Other type 4 boron diiminate dyes have also been reported (Araneda et al. 2011; Frath et al. 2013; Roubinet et al. 2015; Pais et al. 2016).

For type 5, the above-mentioned pyridomethene-BF_2 complexes (Kubota et al. 2010a; Scheibe et al. 1969; Douglass et al. 1973; Golden et al. 2017, 2018; Lin and Chow 2013; Lin et al. 2015) and isoindoline-benzazole-based BF_2 complex (Wang et al. 2014b) have been reported (Figs. 5.40 and 5.45a). Isoindoline-benzazole-based BF_2 complexes **77** and **78** are synthesized from 3-iminoisoindolinone hydrochloride with 2-methylbenzothiazole or 2-methylbenzimidazole and subsequent boron-complexation, respectively (Wang et al. 2014b). The benzothiazole derivative **77**

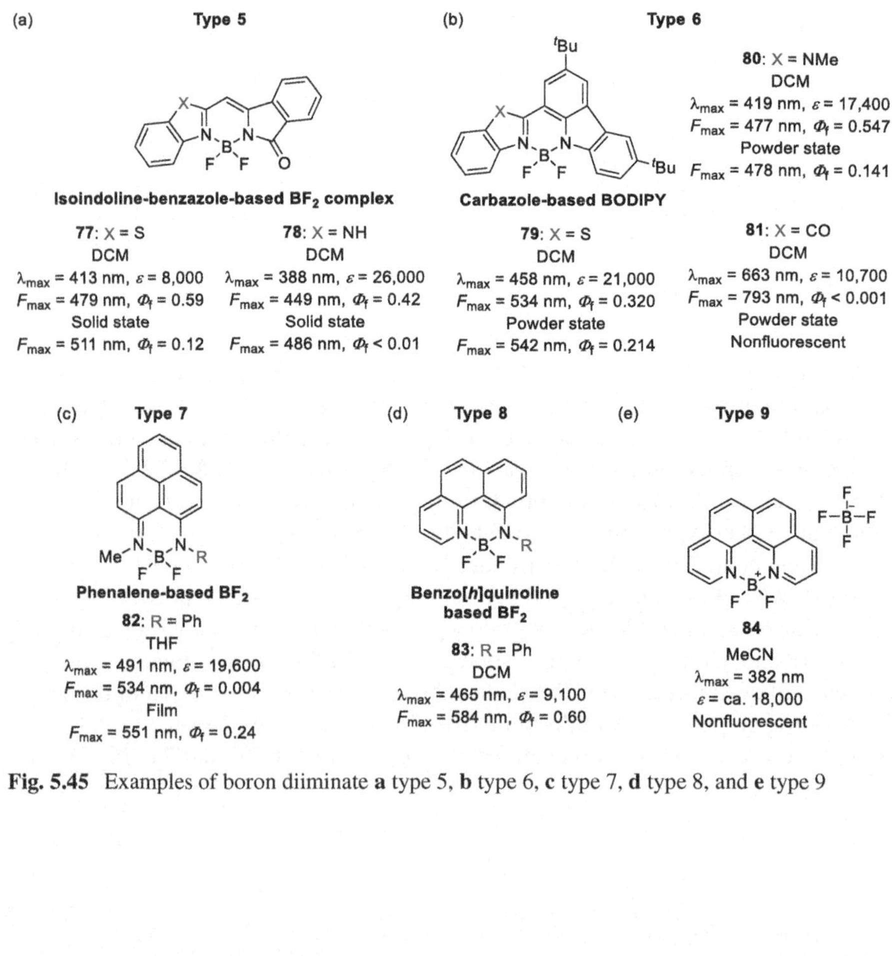

Fig. 5.45 Examples of boron diiminate **a** type 5, **b** type 6, **c** type 7, **d** type 8, and **e** type 9

shows a redshifted λ_{max} and F_{max}, lower ε, and higher Φ_f compared with those of the benzimidazole derivative **78** in solution. Benzothiazole derivative **77** exhibits a redshifted and higher Φ_f compared with benzimidazole derivative **78** in the solid-state. Other type 5 boron diiminate dyes have also been reported (Yamaguchi et al. 2017, 2018; Zhou et al. 2008; Gautam et al. 2019).

For type 6, carbazole-based BODIPY dyes are synthesized by organometallic approaches such as Suzuki coupling for benzothiazole derivative **79**, Pd-catalytic arylation for benzimidazole derivative **80**, cyclization for indolone derivative **81,** and the subsequent boron-complexation (Maeda et al. 2016). Benzothiazole and benzimidazole derivatives **79** and **80** show fluorescence both in solution and in the solid-state (Fig. 5.45b). Indolone derivative **81** shows a significantly redshifted λ_{max} because of the ICT from the carbazole moiety to the indolone moiety. Other type 6 boron diiminate dyes have also been reported (Araneda et al. 2011; Liu et al. 2005a; Ni et al. 2013; Sun et al. 2015; Más-Montoya et al. 2016; Luo et al. 2016; Maeda et al. 2019).

For type 7, phenalene-based BF_2 complex **82** has been synthesized by the reaction of 9-(phenylamino)-1*H*-phenalen-1-one with methylamine in the presence of trimethyloxonium tetrafluoroborate, followed by boron-complexation (Yan et al. 2014). Phenalene-based BF_2 complexes have AIEE properties (Fig. 5.45c). In the phenalene-based BF_2 complexes, although the type of substituent R hardly affects λ_{max} (THF: 491–492 nm), it has a strong influence on Φ_f (film: 0.24–0.89).

For type 8, benzo[*h*]quinoline-based BF_2 complexes have been synthesized by palladium-catalyzed Hartwig–Buchwald amination of 10-bromobenzo[*h*]quinoline with aniline and subsequent boron-complexation (Araneda et al. 2011). **83** has a large Stokes shift (119 nm) probably because of its unsymmetrical structure (Fig. 5.45d). Although benzo[*h*]quinoline-based BF_2 complexes have a relatively low ε (5,800–9,100), they show a relatively high Φ_f (0.60–0.66).

For type 9, quino[7,8-*h*]quinoline-based BF_2 complex with tetrafluoroborate counter ion **84** have been reported by the reaction of quino[7,8-*h*]quinoline with boron trifluoride diethyl etherate (Shaffer et al. 2012). **84** shows λ_{max} at 382 nm and does not show fluorescence (Fig. 5.45e). Other type 9 boron diiminate dyes have also been reported (Zeng et al. 2011).

5.2.4.3 Other ^{6}N^N Monoboron Complexes

A part of other examples of six-membered ring N^N (^{6}N^N) type monoboron complexes is shown in Fig. 5.46. Boron 2-picolinoylpyrrole complex (**BOPPY** (Ito et al. 2007)) shows blueshifted λ_{max} compared with those of **BODIPY** dyes (Fig. 5.46a). **BOPPY** has a large Stokes shift, whereas the Φ_f is very low. Although the λ_{max} of **BOPPY** is blueshifted with increasing solvent polarity, the F_{max} is shifted to red with increasing solvent polarity.

Amidine-based BF_2 complex **85** ($\Phi_f < 0.01$) hardly shows fluorescence in DCM, whereas it shows enhanced fluorescence in the solid-state ($\Phi_f = 0.22$) because of the

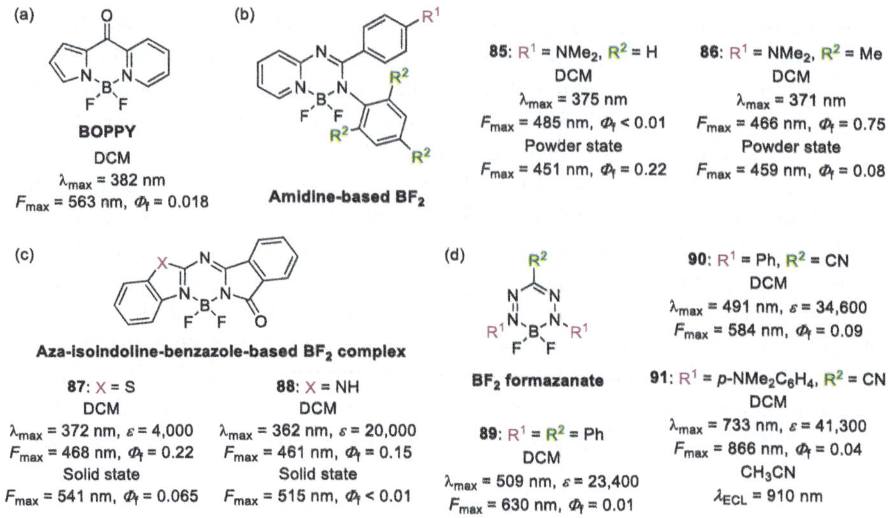

Fig. 5.46 Examples of other ^{6}N^N type monoboron complexes

restriction of intramolecular rotation (RIR) (Fig. 5.46b) (Zhao et al. 2013). Introduction of a bulky substituent group enhances the Φ_f in DCM (**86**: $\Phi_f = 0.75$) because of the suppression of intramolecular rotation, which is supported by the fact that the calculated rotation barrier (E_R) value of **86** (>8.619 a.u.) is significantly higher than that of **85** (>0.2430 a.u.). Amidine-based BF_2 complexes show mechanochromic fluorescence. Unground (crystalline) and ground (amorphous) states of **85** show fluorescence at 451 and 498 nm, respectively, and the morphological transition between crystalline and amorphous phases is observed upon every grinding/heating cycle. The mechanochromism may be due to the dissymmetric twisted propeller-like conformation, which easily results in weak intermolecular interactions and relatively loose packing.

Aza-isoindoline-benzazole-based BF_2 complexes **87** ($\lambda_{max} = 372$ nm, $\Phi_f = 0.22$) and **88** ($\lambda_{max} = 362$ nm, $\Phi_f = 0.15$) show blueshifted λ_{max} and lower Φ_f compared with those of the corresponding non-aza analogues **77** ($\lambda_{max} = 413$ nm, $\Phi_f = 0.59$) and **78** ($\lambda_{max} = 388$ nm, $\Phi_f = 0.42$) (Figs. 5.45a and 5.46c) (Wang et al. 2014b).

The optical properties of BF_2 formazanate complexes are strongly dependent on the type of substituents (R^1 and R^2) (Fig. 5.46d) (Barbon et al. 2015; Maar et al. 2019). Cyano-substituted derivative **90** ($\Phi_f = 0.09$) shows a higher Φ_f compared with that of the phenyl-substituted derivative **89** ($\Phi_f = 0.01$). In the substituent R^1 of **90**, replacement of the phenyl group with the naphthyl group leads to a spectral redshift and increase in Φ_f (R^1 = 2-naphthyl, R^2 = CN: $\lambda_{max} = 558$ nm, $F_{max} = 669$ nm, $\Phi_f = 0.32$). Regardless of the simple structure and low-molecular weight (Mw. = 383.21 g/mol), dimethylanimophenyl-substituted derivative **91** shows NIR absorption and fluorescence ($\lambda_{max} = 733$ nm, $F_{max} = 866$ nm, $\Phi_f = 0.04$) because of the enhanced electronic delocalization over the π-system that includes both the formazanate backbone and the dimethylanimophenyl group. Although **91** does not show electrochemiluminescence (ECL) in the absence of a coreactant, ECL is induced upon the addition of tripropylamine, which reaches its maximum intensity at 0.54 V relative to the Fc/Fc^{+} redox couple. The wavelength of maximum ECL intensity is centred at 910 nm with an efficiency of 17.5% relative to the benchmark $[Ru(bpy)_3][PF_6]_2$/tripropylamine system.

5.2.4.4 ^{5}N^N and ^{7}N^N Monoboron Complexes

A few examples (Li et al. 2013a; Frath et al. 2014) of five-membered ring N^N (^{5}N^N) type and seven-membered ring N^N (^{7}N^N) monoboron complexes are shown in Fig. 5.47. For the ^{5}N^N type monoboron complexes, in 1969, Hohaus and Umland reported a 8-aminoquinoline biphenylboron (BPh_2aq) dye (Fig. 5.47a) (Hohaus and Umland 1969). Although **92** has a small ε (4,100), it shows a large Stokes shift (103 nm) (Nagata and Chujo 2008). Boron-complexation of 2,5-bis(4-dimethylaminophenylazo)-1*H*-pyrrole (λ_{max} = 575 nm, ε = 59,000) gives NIR absorbing azopyrrole BF_2 complex **93** (λ_{max} = 754 nm, ε = 60,000) (Fig. 5.47b) (Li et al. 2009). Triazaborolopyridinium dyes show a relatively large Stokes shift (**94**: 89 nm) and general insensitivity to solvent or pH; triazaborolopyridinium dyes are membrane permeable and are suitable candidates of fluorescent probes for intracellular targets (Fig. 5.47c) (Hapuarachchige et al. 2011).

Thermodynamically stable *trans* Az-BF_2 complex (λ_{max} = 530 nm) isomerizes to the *cis* isomer (λ_{max} = 480 nm) upon irradiation at 570 nm; the irradiation of the *trans* Az-BF_2 complex at 570 nm yields 97% of the *cis* isomer at the photostationary state (Fig. 5.47d) (Yang et al. 2012). Irradiation at 450 nm or storing the *cis* isomer in the dark converts it back to the *trans* isomer; irradiation at 450 nm produces 80% of the *trans* isomer at the photostationary state. The well separated π–π^* transitions between the *trans* and *cis* isomers enable efficient visible light-induced isomerization. Phenanthro[9,10-*d*]imidazole-quinoline boron difluorides (**PQBDs**) show a large Stokes shift (up to 115 nm) because of the efficient charge transfer from the phenanthro[9,10-*d*]imidazole unit to the quinoline moiety and red fluorescence in the solid-state (Φ_f: up to 0.184) (Fig. 5.47e) (Li et al. 2013b). Despite **95** being planar, no π–π stacking interactions in the crystal are observed, which accounts for the relatively strong fluorescence in the solid-state.

The optical properties of **BOIMPY** are profoundly affected by the *N*-aryl substituent group (R^1 and R^2) (Fig. 5.47f) (Lee et al. 2016). Although **BOIMPY 96** has nearly co-planar arrangements of the iminopyrrolide fragment and the *N*-aryl ring with a small torsion angle (17.17° and 11.70° for two crystallographically independent molecules), **98** adopts an essentially orthogonal geometry with a large torsion angle (89.48°). **96** has a higher contribution to the quinoid-type resonance structure, whereas **98** has a higher contribution to the benzoid resonance structure. Thus, **96** shows a redshifted λ_{max} compared with that of **98**. Although **97** ($\Phi_f = 0.15$) shows relatively high Φ_f, **96** (0.02) and **98** (<0.001) hardly show fluorescence. Protonation of **96** and **98** by $HBF_4 \cdot OEt_2$ leads to a large blueshift of F_{max} (**96**: 484 nm, **98**:

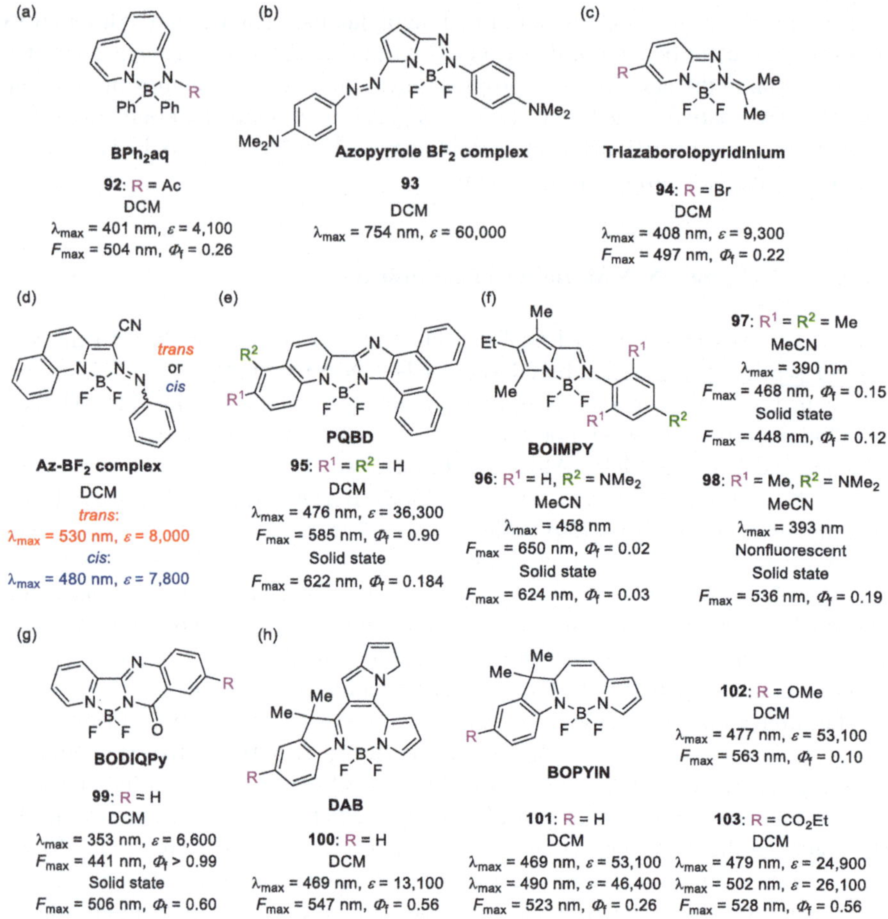

Fig. 5.47 Examples of ^{5}N^N and ^{7}N^N type monoboron complexes

465 nm) and an increase in Φ_f (**96**: 0.45, **98**: 0.09), which are similar to the LE-type emission of **97**. **BOIMPY** with hydroxy-terminated alkyl tethering groups ($R^1 = H$, $R^2 =$ *N*,*N*-bis(2-hydroxyethyl)amino) as a water solubility-enhancing group does not affect cell viability and is selectively localized at the lipid droplet with a bright fluorescence emission to allow for high-resolution imaging of subcellular organelles in living cells.

Boron difluoroquinazolinone-pyridine (**BODIQPy**) dyes show strong fluorescence both in solution (Φ_f: up to 0.99) and in the solid-state (Φ_f: up to 0.60) and a relatively large Stokes shift (Fig. 5.47g) (Zhou et al. 2018). The asymmetrical structure of **BODIQPy** dyes contributes to their efficient fluorescence properties. The HOMO and LUMO orbitals are delocalized over the quinazolinone and pyridine moieties, respectively. The effective charge transfer properties from HOMO to LUMO are responsible for the large Stokes shift. The molar extinction coefficient value of the iodine derivative (R = I: ε = 15,400) is higher than that in the nonsubstituted derivative (**99**: ε = 6,600) maybe because of the electron-donating ability of iodine. Introduction of iodine atoms does not lead to a large decrease in fluorescence quantum yield in solution (**BODIQPy**: **99**: $\Phi_f > 0.99$, R = I: $\Phi_f = 0.73$). The suppressed heavy atom effect is probably due to the strong electron-withdrawing ability of the **BODIQPy** core. In the solid-state, the interactions between the carbonyl oxygen and iodine enhance intermolecular interaction, which results in a decrease in the Φ_f value (**BODIQPy**: **99**: $\Phi_f = 0.60$, R = I: $\Phi_f = 0.02$).

In the ^{5}N^N type monoboron complexes, gelation-assisted piezochromism of boron 2-(2′-pyridyl)imidazole complexes (**BOPIMs**) (Wang et al. 2017), dual emission of isoquinolinyl–pyrazolate boron complexes (Chen et al. 2019), EL properties of pyridyl–indolate-based (Liu et al. 2005b) and pyridyl–pyrrolide-based (Chen et al. 2005) boron complexes, and solid-state fluorescence of BF_2 complexes of dipyrrolylquinoxaline (**BPQs**) (Yu et al. 2015) are also reported.

For the ^{7}N^N type monoboron complexes, 3*H*-pyrrolizine fused diazaborepin **DAB** was synthesized in 2016 (Fig. 5.47h) (Li et al. 2016). In 2019, borondifluoride-3,3-dimethyl-2-[2-(2-pyrrolyl)ethenyl] indoles (**BOPYINs**), simplified as **DAB**, were also synthesized by Knoevenagel condensation of the 3*H*-indolium salt with 2-formylpyrrole followed by coordination with $BF_3{\cdot}OEt_2$ (Zhang et al. 2019). In the crystal form, **100** and **101** possess a nonplanar saddle-shape structure. When compared with five (**BOIMPY** and **BOPIM**)-, six (**BODIPY** and **aza-BODIPY**)-, and seven (**DAB** and **BOPYIN**)-membered ring N^N boron complexes, the bond angle of the N···B–N bond increases with the number of atoms in the core ring based on Baeyer strain theory; 7-membered: ca. 110° > 6-membered: ca. 106° > 5-membered: ca. 96°. **DAB** and **BOPYINs** show a relatively larger Stokes shift. Confocal images reveal **BOPYINs** are predominantly located in the perinuclear area probably because of endocytic vesicles.

5.2.5 O^O Type Organoboron Complex

Difluoroboron β-diketonates (**BF₂bdks**) including difluoroboron dibenzoylmethane (dbm) derivatives have received considerable attention because of their high fluorescence quantum yields, tuneable fluorescent properties, two-photon-excited fluorescence (TPEF) properties, mechanofluorochromic (MFC) properties, and room temperature phosphorescence (RTP) properties (Fig. 5.48a) (Tanaka and Chujo 2015; Chen et al. 2017a). Although BF_2bdk was firstly reported in 1924 (Morgan and Tunstall 1924), it did not attract much attention until recently. BF_2bdks have high stability towards oxygen and moisture. The absorption and fluorescence spectra of BF_2bdks tend to broaden, compared with those of BODIPY dyes in solution, probably because of the flexible structure of BF_2bdks. Different from BODIPY dyes, BF_2bdks often show solid-state fluorescence.

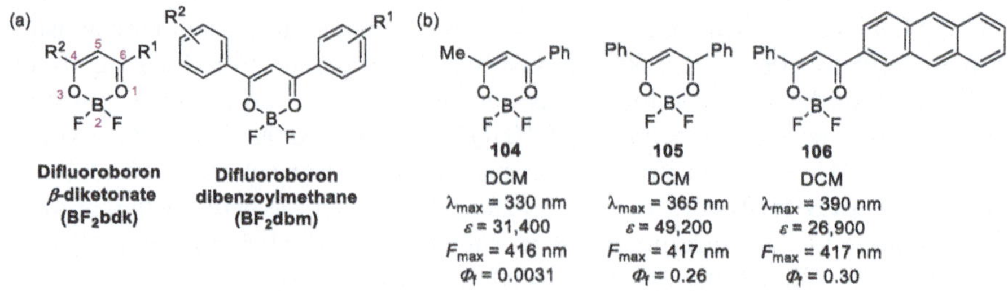

Fig. 5.48 **a** Structure and name of boron diketonate. **b** Examples of BF_2bdks

The absorption and fluorescence properties of **BF_2bdks** strongly depend on the type of substituent groups at the 4 and 6 positions. Introduction of more highly π-conjugated moieties causes a redshift of λ_{max} (**104** (Xu et al. 2013): 330 nm, **105** (Xu et al. 2013): 365 nm, **106** (Xu et al. 2013): 390 nm) because of the extension of π-conjugation (Fig. 5.48b). The introduction of electron-donating groups also induces a redshift of λ_{max} (**107** (Xu et al. 2013): 411 nm, **108** (Hu et al. 2013): 481 nm) because of the enhanced ICT character (Fig. 5.49a). The λ_{max} values of the curcumin-BF_2 complexes **109** (Bai et al. 2014) and **110** (Kamada et al. 2016) are more redshifted than that of **105** (Figs. 5.48b and 5.49b). Julolidine derivative **110** shows TPEF property; the two-photon absorption (TPA) maximum ($\lambda^2{}_{abs}$) is 990 nm with a TPA cross section (σ^{TPA}) value of *ca.* 5,000 GM.

BF_2AVB (Zhang et al. 2010) dye shows polymorphism; two different types of crystals, green-emitting prismlike crystals (F_{max} = 505 nm) and cyan-emitting needle like crystals (F_{max} = 470 nm), are obtained by slow solvent evaporation (Fig. 5.50a). Blue-emitting dendritic solids (F_{max} = 459 nm) are also formed by rapid solvent evaporation. The three **BF_2AVB** solids (prismlike: fwhm = 41 nm, needle like: fwhm = 42 nm, dendritic: fwhm = 31 nm) show narrow-band fluorescence relative to fluorescence in DCM (fwhm = 47 nm). Additionally, BF_2AVB dye shows reversible morphology-dependent fluorescence.

The boron complex of 2′-hydroxychalcone **111** (Cheng et al. 2014) shows NIR fluorescence in the crystalline state (F_{max} = 752 nm) (Fig. 5.50b). The Φ_f value

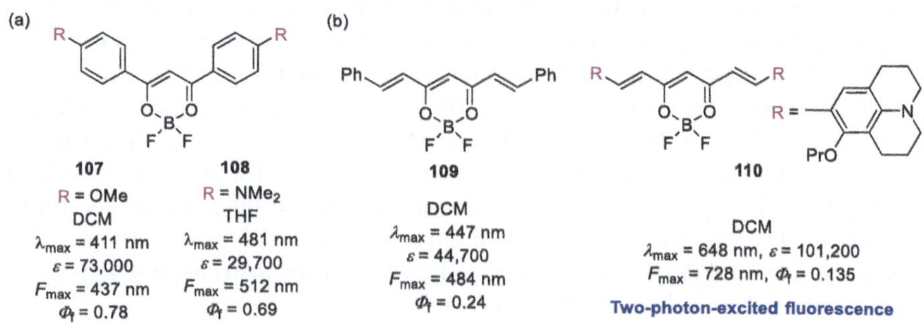

Fig. 5.49 Optical properties of boron diketonate

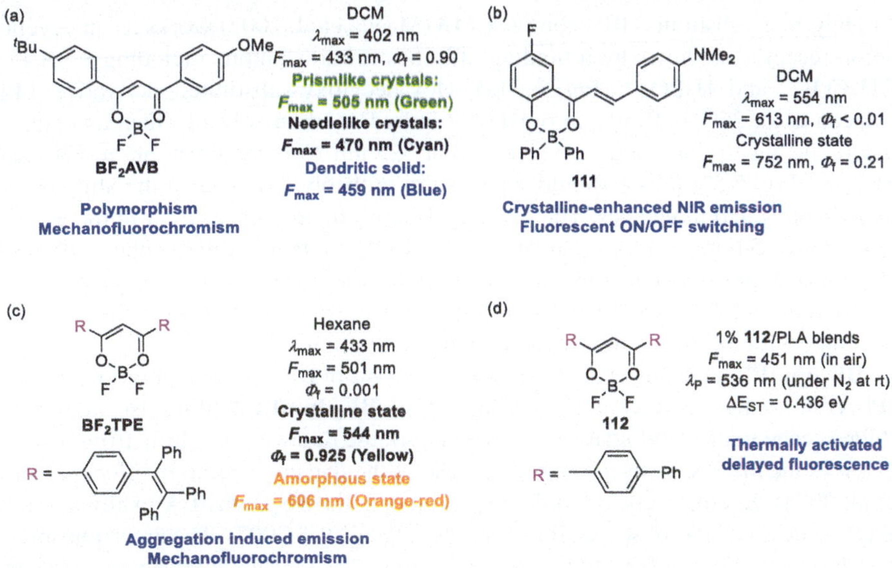

Fig. 5.50 Optical properties of boron diketonate

of **111** in the crystalline state (0.21) is higher than that in DCM (<0.01). **111** shows crystalline-enhanced emission (CEE). Additionally, reversible NIR fluorescent "ON/OFF" responses can be switched by disrupting the ordered molecular packing (OFF: mechanical grinding) and molecular repacking processes (ON: solvent annealing).

BF_2TPE dye (Gao et al. 2017) having two tetraphenylethene (TPE) moieties shows aggregation induced emission (AIE) (hexane: $\Phi_f < 0.01$, solid-state: $\Phi_f = 0.93$) because of RIR and shows fluorescent solvatochromism (hexane: F_{max} = 501 nm, DMSO: F_{max} = 658 nm) because of the pronounced ICT character in the excited state (Fig. 5.50c). Additionally, **BF_2TPE** dye exhibits reversible MFC behaviour between the crystalline (yellow: F_{max} = 544 nm) and the amorphous (orange-red: F_{max} = 606 nm) states; the emitting colour transfers between yellow and orange-red reversibly through grinding and DCM fuming treatment.

Although from the viewpoint of efficient electron-to-photon conversion under electrical excitation, phosphorescent organic light-emitting diodes (PHOLEDs) are more favourable than fluorescent organic light-emitting diodes (FOLEDs), PHOLEDs tend to be green or red. Thermally activated delayed fluorescence (TADF) is an alternative approach to harvest excited triplet energy through upconversion from the T_1 to S_1 states by thermal activation, and TADF-based OLEDs have attracted considerable attention (Uoyama et al. 2012). **112** in diluted poly(lactic acid) blends shows blue TADF because of the relatively small energy gap (ΔE_{ST}) between the S_1 and T_1 states which promotes reverse intersystem crossing (RISC) (Fig. 5.50d) (Daly et al. 2016).

Dipyrrole-substituted BF_2 complex **113** (Maeda et al. 2007) works as an acyclic anion receptor; it shows high binding affinities towards anions including F^-, Cl^-, $CH_3CO_2^-$, and $H_2PO_4^-$ (Fig. 5.51a). Hexadecyloxy-substituted derivative **114** (Maeda et al. 2007) affords an organogel in hydrocarbon solvents such as octane, where the absorption bands are observed at 525 nm with shoulders at *ca.* 470 and 555 nm (Fig. 5.51b). The shoulder peaks are probably assigned to the slipped *H*- and *J*-aggregated molecules. The organogel shows fluorescence; F_{max} in the gelatinous state (654 nm) is more redshifted compared with that in dilute octane solution (533 nm). A gel-sol transition is observed by heating the organogel above 27.5 °C. The addition of Cl^- (solid Bu_4NCl) to the octane gel at 20 °C results in a gradual decomposition of the gelatinous state to give a solution.

BF_2dpmPLA (Zhang et al. 2007) which consists of BF_2dpm and poly(lactic acid) (PLA) shows fluorescence in DCM (Fig. 5.52a). **BF_2dpmPLA** films also show blue fluorescence in the solid-state (F_{max} = 440 nm); the fluorescence shifts from blue to green as the M_n decreases probably because of the dye concentration effect (Zhang et al. 2008). Interestingly, upon deoxygenation, solid **BF_2dbmPLA** exhibits green RTP, while no RTP is observed for solutions (Zhang et al. 2007). The green phosphorescence is highly sensitive to oxygen quenching. The unusual RTP is maybe due to the restriction of triplet thermal decay pathways by the rigid polymer medium in the

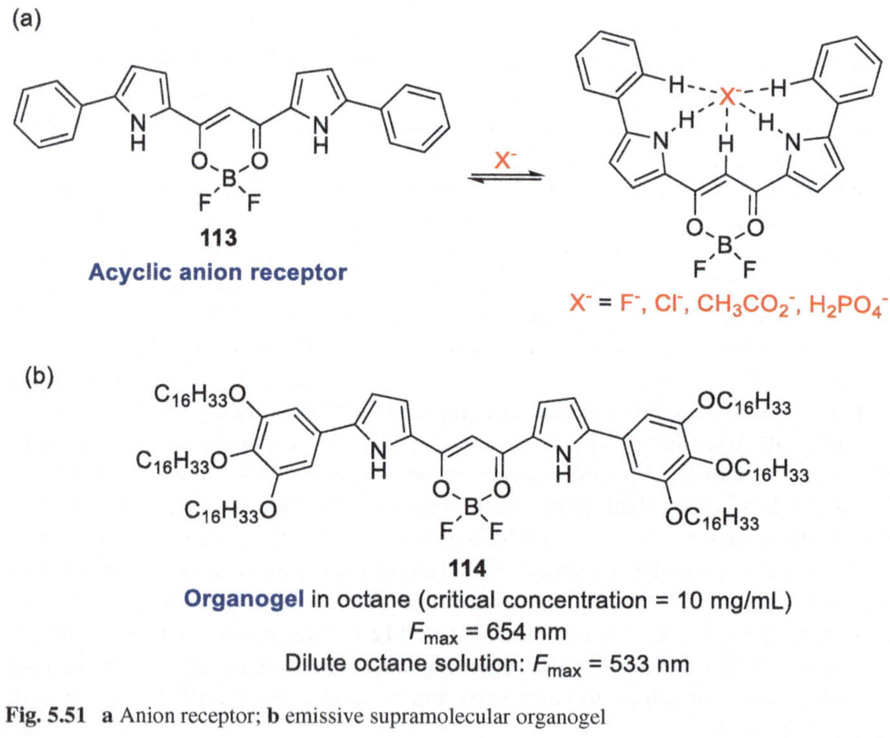

Fig. 5.51 **a** Anion receptor; **b** emissive supramolecular organogel

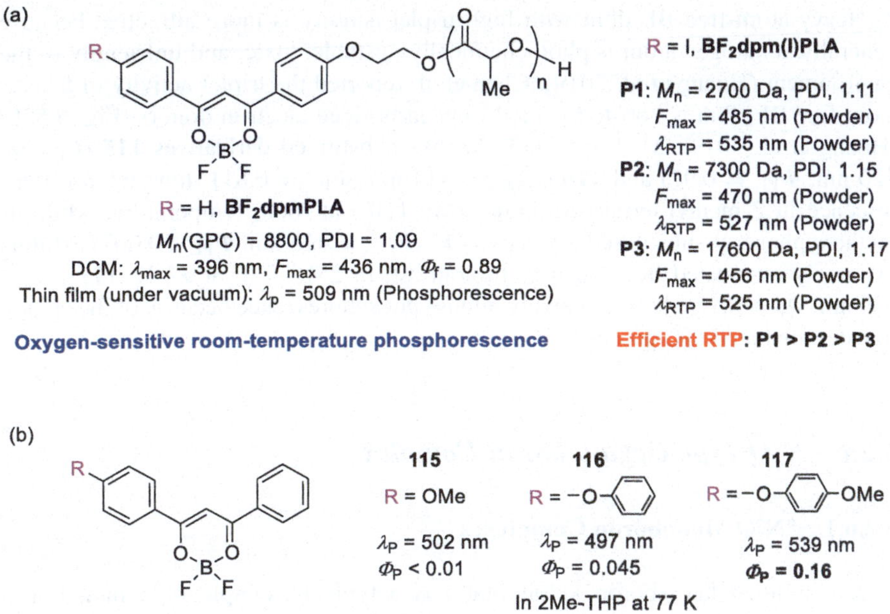

Fig. 5.52 **a** Room-temperature phosphorescence; **b** heavy-atom-free phosphorescence

solid-state. No phosphorescence is observed above the glass transition temperature ($T_g = 52$ °C) of **BF_2dbmPLA**.

Although singlet ($F_{max} = 440$ nm) and triplet emissions ($\lambda_p = 509$ nm) of BF_2dbmPLA film are clearly separated, the RTP is much weaker than the fluorescence. The weak phosphorescence intensity precludes practical applications (Zhang et al. 2009a). According to the perturbation theory, a greater spin-orbit coupling and a smaller singlet-triplet energy gap are preferable for enhancing the phosphorescence intensity relative to fluorescence (Lower and El-Sayed 1966). Fraser et al. reported the design concept for an efficient RTP (Zhang et al. 2009a); introduction of a heavy atom for enhancing spin-orbit coupling increases the rate of intersystem crossing (ISC) and control of boron dye loading (polymer-chain molecular weight) for adjusting the singlet-triplet energy gap. Iodine-substituted derivatives BF_2dpm(I)PLA with different molecular weights show fluorescence ($F_{max} = 435$ nm, $\Phi_f = 0.4$) in DCM (Fig. 5.52a); the lower Φ_f value of **BF_2dpm(I)PLA** than **BF_2dbmPLA** ($\Phi_f = 0.89$) indicates an enhanced ISC (Zhang et al. 2009a). Molecular weight dependent emission spectra are observed both in the power and film state. Although two distinct emission bands are observed in the spectra for **P2** ($F_{max} = 470$ nm, $\lambda_{RTP} = 527$ nm) and **P3** ($F_{max} = 456$ nm, $\lambda_{RTP} = 525$ nm), **P1** shows a single phosphorescence ($\lambda_{RTP} = 535$ nm) with only a small shoulder at 480 nm in the powder state. This suggests that the singlet-triplet energy gap of BF_2dpm(I)PLA decreases as the molecular weight decreases. BF_2dbm(I)PLA nanoparticles are applicable for tissue oxygen maps by using fluorescence/phosphorescence ratios (Zhang et al. 2009a).

Heavy-atom-free BF_2dbm with high triplet activity is more attractive because generally, a halogen atom is photochemically unstable, toxic, and unfriendly to the environment (Huang et al. 2019). Chen et al. reported the triplet activity of heavy-atom-free BF_2dbm is boosted via sp^3 oxygen-bridged electron donors (Fig. 5.52b) (Huang et al. 2019). Methoxy- and phenoxy-substituted derivatives **115** (F_{max} = 420 nm, Φ_{PL} = 0.97) and **116** (F_{max} = 414 nm, Φ_{PL} = 0.81) show intense fluorescence in 2-methyl-tetrahydrofuran (2Me-THF) at room temperature, while 4-methoxyphenoxy-substituted derivative **117** (F_{max} = 415 nm, $\Phi_{PL} < 0.001$) exhibits extremely attenuated fluorescence. In contrast, in 2Me-THF at 77 K, **117** (λ_P = 539 nm, Φ_P = 0.16) shows relatively intense phosphorescence because of the charge transfer mediated ISC process.

5.2.6 *N^O Type Organoboron Complex*

5.2.6.1 ^{6}N^O Monoboron Complexes

Boron iminoenolate (boron ketoiminate) is a typical example of a monoboron complex bearing a monoanionic six-membered N^O (^{6}N^O) bidentate chelating ligand, and it can be classified into seven types depending on the position of the annulated ring (Fig. 5.53).

For type 1, X-ray crystallographic results reveal that **118** (dihedral angle: 20.7°) and **119** (dihedral angle: 75.9°) have less planarity compared to **107** (dihedral angle: 2.4°) (Fig. 5.54a), and the structure of *β*-iminoenolate has a larger contribution than that of *β*-ketoiminate in the boron iminoenolates (Fig. 5.54b) (Yoshii et al. 2013). Boron iminoenolates **118** (λ_{max} = 365 nm) and **119** (λ_{max} = 360 nm) show weaker and blueshifted absorptions compared with that of the corresponding boron diketonate **107** (λ_{max} = 408 nm) because of the steric hindrance of the substituents on the nitrogen atoms and different electronegativities between the oxygen and nitrogen atoms. In contrast to **107** (Φ_f = 0.91), **118** (Φ_f = 0.01) and **119** (Φ_f < 0.01) hardly show fluorescence in THF. Considering that boron diketonate **107** has almost the same structure as boron iminoenolates **118** and **119**, the fluorescence quenching of **118** and **119** in solution would be less affected by rotational or vibrational motions derived from the *p*-methoxyphenyl groups. Although **107** shows ACQ, **118** and **119** show AIE character derived from the rotational or vibrational motions of the boron-chelating ring. **118** shows phosphorescence at around 500 nm in 2Me-THF at 77 K.

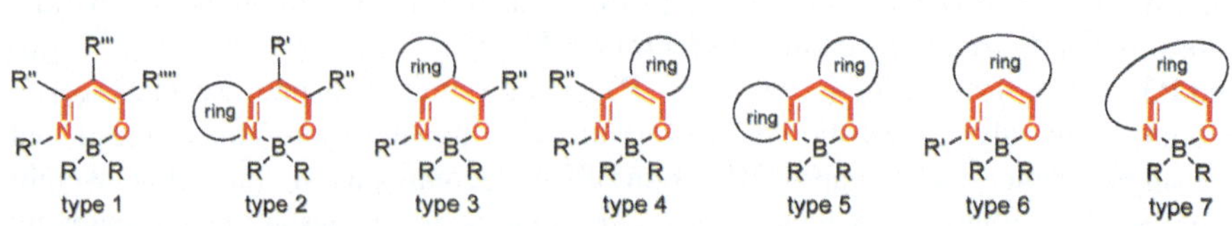

Fig. 5.53 Classification of boron complexes bearing a *β*-iminoenolate ligand

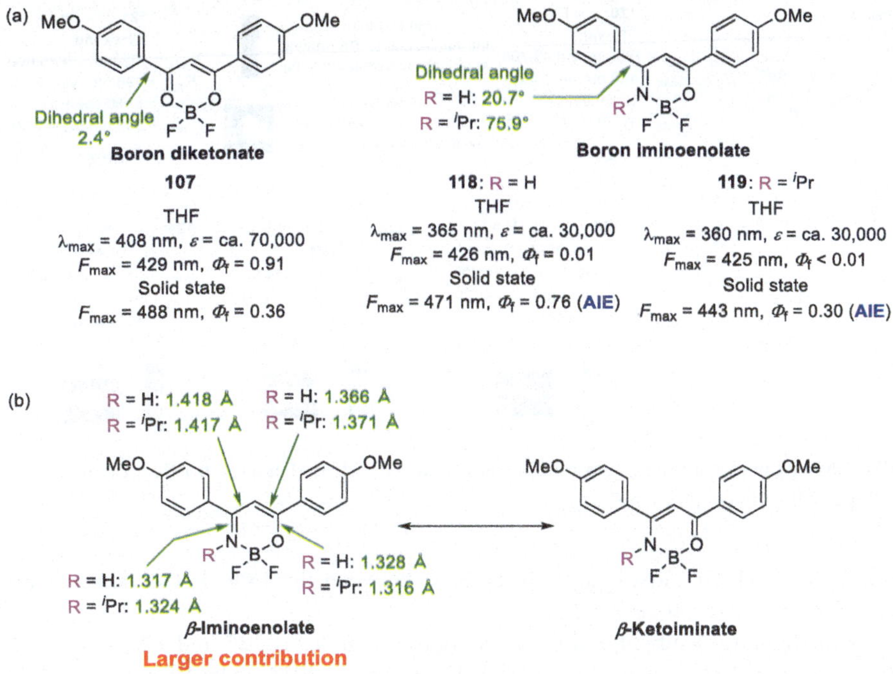

Fig. 5.54 Example of type 1 boron iminoenolate. **a** Comparison between boron diketonate and boron iminoenolate. **b** Contributing resonance structures

For type 2, thiazole–boron complexes can be easily synthesized from methylbenzothiazoles and benzoate derivatives (Kubota et al. 2012). The λ_{max} and ε values of thiazole–BPh_2 complex **121** ($\lambda_{max} = 402$ nm, $\varepsilon = 25{,}800$) are more bathochromic and lower compared with those of thiazole–BF_2 complex **120** ($\lambda_{max} = 380$ nm, $\varepsilon = 43{,}700$) due to the molecular bending of the chromophore of BPh_2 complex caused by the introduction of bulky phenyl groups at the boron atom (Fig. 5.55a). Even though the thiazole–boron complex **120** has only one phenyl ring, it shows AIE character. Regardless of the solvent polarity, **120** hardly shows fluorescence ($\Phi_f \leq 0.01$) in low-viscosity solvents (0.31–0.59 cP), while **120** exhibits fluorescence in high-viscosity solvents such as ethylene glycol ($\Phi_f = 0.05$, 23.5 cP) and glycerol ($\Phi_f = 0.12$, 1412 cP). The non-radiative process is suppressed as the solvent viscosity increases. These results indicate that the main reason for the AIE/AIEE character in thiazole–boron complexes is the restriction of intramolecular C–Ph rotation. In the solid-state, the BF_2 complex **120** ($\Phi_f = 0.26$) has a lower Φ_f value compared with that of the corresponding BPh_2 complex **121** ($\Phi_f = 0.60$) because of the formation of consecutive intermolecular CH/F interactions.

In pyrazine–boron complexes **122–124**, the F_{max} value is affected by the type of substituents on the boron atom and the phenylene moiety (F_{max}: 480–604 nm) (Fig. 5.55b) (Kubota et al. 2011). Unlike common BODIPY dyes, pyrazine–boron complexes **122–124** show a large Stokes shift (73–138 nm) owing to the flexible

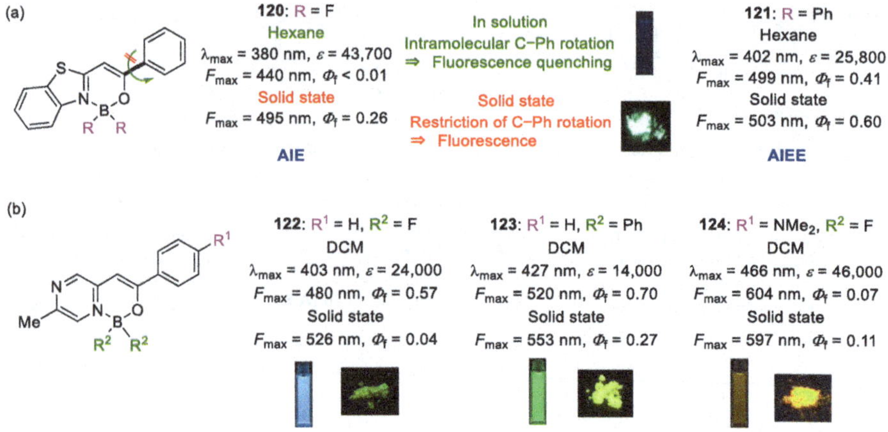

Fig. 5.55 Example of type 2 boron iminoenolate. **a** AIE/AIEE character. **b** Effect of substituent groups on the optical properties

fluorophore which causes large differences between the Franck–Condon and the equilibrium excited state structures.

Dimethylamino-substituted pyrazine–boron complexes **125** and **126** show fluorescence solvatochromism (Fig. 5.56a) (Kubota et al. 2014). The BF_2 complex **125** shows only one fluorescence arising from the LE state in the nonpolar solvent (hexane) and relatively polar solvents (THF, DCM, acetone, and acetonitrile), while dual fluorescence, corresponding to LE and TICT states, is observed in less polar solvents (toluene, 1,4-dioxane, and $CHCl_3$). Although the BPh_2 complex **126** does not show dual fluorescence, its F_{max} value is redshifted with increasing solvent polarity. The linear relationship between the ln k_{nr} and ν_f values indicates that the observed fluorescence arises from one emitting excited state, and the linear correlation between the ν_f and $E_T(30)$ indicates that the emitting excited state is a zwitterionic molecule. The Lippert–Mataga plot reveals that the excited state has a considerably large dipole moment (ca. 19.4 D).

In pyrimidine boron complexes, although non-, trifluoromethyl-, and cyano-substituted derivatives show AIEE due to the lower Φ_f in solution (DCM: $\Phi_f \leq 0.05$), methoxy- and dimethylamino-substituted derivatives show ACQ because of the higher Φ_f in solution (DCM: Φ_f: 0.52–0.78) (Fig. 5.56b) (Kubota et al. 2013, 2015a). The low Φ_f of non-, trifluoromethyl-, and cyano-substituted derivatives is thought to be due to the rotation of the C–Ar bond. In contrast, the introduction of electron-donating groups such as methoxy and dimethylamino groups is expected to restrict the C–Ar rotation in solution given that the unpaired electron of the methoxy or dimethylamino group contributes to the formation of the quinoid-type resonance structures with C–C double bond character in the C–Ar bond, as shown in **128**. The k_{nr} values of the methoxy and dimethylamino-derivatives are significantly smaller than those of the non- and trifluoromethyl-derivatives, which substantiates the restriction of the C–Ar rotation by introducing methoxy or dimethylamino groups. The F_{max} value

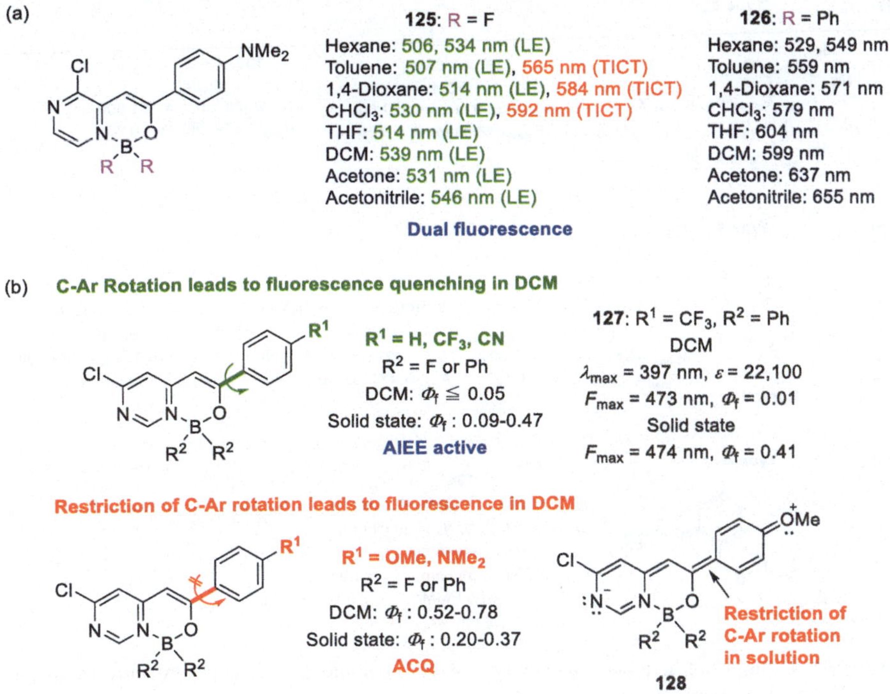

Fig. 5.56 Example of type 2 boron iminoenolate. **a** Fluorescence solvatochromism. **b** Effect of substituent groups on the AIEE and ACQ properties

of the BPh_2 complexes is blueshifted (474–618 nm), and the Φ_f values are higher (0.16–0.47) than those of the corresponding BF_2 complexes (F_{max}: 486–629 nm, Φ_f: 0.09–0.30) in the solid-state, plausibly due to inhibition of the intermolecular interactions.

For other type 2 boron iminoenolates, reversible off–on solid-state luminescence switching by acid/base fuming processes (Liao et al. 2015), mechanochromic fluorescence (Suenaga et al. 2017), and fluorescent sensory properties towards acid in xerogel-based films (Wu et al. 2017b) have also been reported.

For type 3, indolenine derivative **129** ($\lambda_{max} = 536$ nm) shows blueshifted λ_{max} compared with that of the benzothiazole derivative **130** ($\lambda_{max} = 558$ nm) (Fig. 5.57a) (Zyabrev et al. 2012). N^O-chelated merocyanine boron complexes **129** ($F_{max} = 569$ nm, $\Phi_f = 0.65$) and **130** ($F_{max} = 579$ nm, $\Phi_f = 0.65$) show a relatively strong fluorescence compared with that of the corresponding O^O-chelated boron complexes (indolenine derivative: $F_{max} = 556$ nm, $\Phi_f = 0.03$, benzothiazole derivative: $F_{max} = 556$ nm, $\Phi_f = 0.03$) maybe due to the prevention of the S_0–S_1 conical intersection associated with bond twisting which promotes the photoisomerisations or non-radiative processes.

For type 4, boranil complexes, which are boron complexes of anils (aniline-imines), are reported. The reaction of aldehydes and anilines in the presence

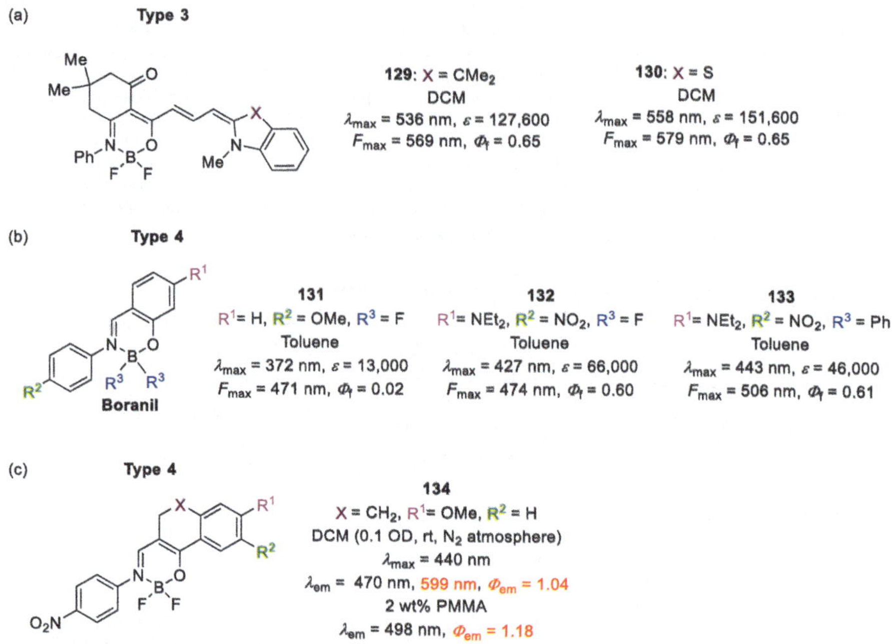

Fig. 5.57 Example of **a** type 3 and **b** type 4 boron iminoenolate. **c** Heavy-atom-free room temperature phosphorescence

of *p*-TsOH gives anils, and the subsequent boron-complexation provides boranil complexes (Frath et al. 2011). The fluorescence stems from an intraligand charge transfer (ILCT) state (Fig. 5.57b). Metal- and heavy atom-free β-hydroxyvinylimine boron complexes show efficient and highly tuneable phosphorescent emission at room temperature both in solution and in doped thin PMMA films (Koch et al. 2014). In DCM, **134** shows fluorescence at 470 nm ($\tau_0^F = 2.94$ ns) and phosphorescence at 599 nm ($\tau_0^P = 1.38$ μs) (Fig. 5.57c). Additionally, the emission of **134** in DCM is bathochromically shifted over 150 nm with increasing concentration (509 nm at 4.33 $\times$ 10^{-9} M and 659 nm at 4.33 $\times$ 10^{-5} M) probably due to the formation of excimer. The absolute quantum yield value of **134** in DCM and PMMA exceeds 100% because of singlet fission favoured by amalgamation of factors such as permanent molecular dipole moment, strong π–π stacking interactions, and the presence of a functional group that aids to promote the radical character in the excited state. For other type 4 boron iminoenolates, fluorescent labelling for bovine serum albumin (BSA) (Frath et al. 2012), bioimaging in HeLa cells (Zhang et al. 2017), and fluorescent probes for specific imaging of lipid droplets in living cells (Zhao et al. 2019) have been reported.

For type 5, benzoxazole- (**135**), benzothiazole- (**136**), and benzimidazole (**137**)-based BPh_2 complexes have been synthesized from the corresponding ligands with BPh_3 (Fig. 5.58a) (Li et al. 2011; Zhang et al. 2015b). These boron complexes show strong fluorescence not only in solution, but also in the solid-state. The λ_{max} and

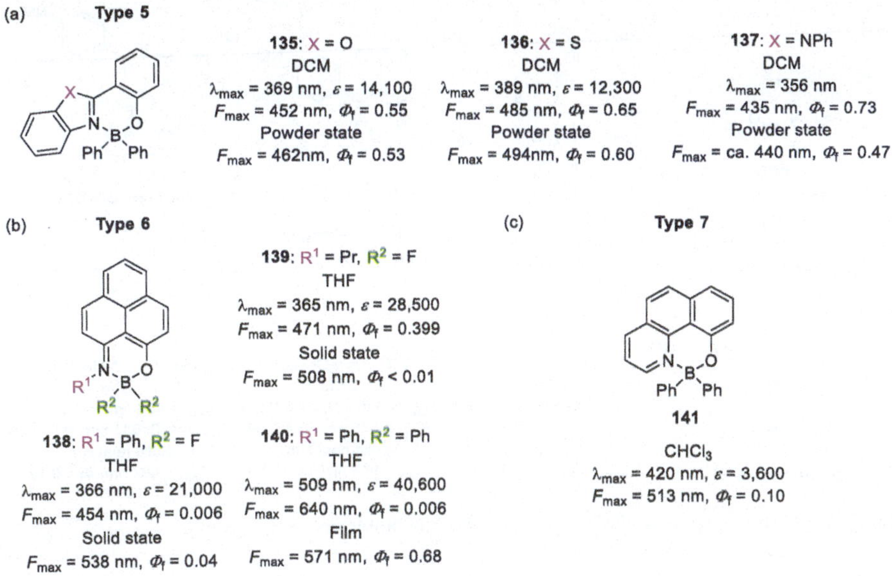

Fig. 5.58 Examples of boron iminoenolate **a** type 5, **b** type 6, and **c** type 7

F_{max} values are redshifted in the order of **137** < **135** < **136**. OLED, which employs **137** as the emitting material, shows deep-blue EL with Commission Internationale de L'Eclairage (CIE) of ($X = 0.16$, $Y = 0.09$) (Zhang et al. 2015b). In the benzimidazole-based boron complex, the substitution of dibenzoborole moiety into BPh_2 moiety enhances the molecular rigidity, which is reflected in the higher melting point and decomposition temperature. Other type 5 boron iminoenolates have been reported to be applied in OLEDs (Li et al. 2012; Zhang et al. 2018c; Salla et al. 2019).

For type 6, the reaction of 9-alkoxy-1*H*-phenalen-1-one with amines gives 9-(substitutedamino)-1*H*-phenalen-1-one, and the subsequent boron-complexation provides **DFBPLY** dyes (Fig. 5.58b). **DFBPLY** dye **138** (Yan et al. 2010) (λ_{max} = 366 nm) shows significantly blueshifted λ_{max} compared with that of the corresponding N^N derivative **82** ($\lambda_{max} = 491$ nm) (Fig. 5.45c). In **DFBPLY** dyes, nonsubstituted (R = H, $\Phi_f = 0.344$) and propyl (**139**: R = Pr, $\Phi_f = 0.399$) derivatives show relatively strong fluorescence in THF, and the phenyl (**138**: R = Ph, $\Phi_f = 0.006$) derivative hardly shows fluorescence. The reaction of **DFBPLY** dye **138** with PhMgBr gives the corresponding BPh_2 complex **140** (Yan et al. 2014).

For type 7, benzo[*h*]quinoline-based BPh_2 complex **141** is synthesized by the reaction of 10-hydroxybenzo[*h*]quinoline with BPh_3 (Fig. 5.58c) (Tokoro et al. 2010). **141** has very small ε. Polymers containing benzo[*h*]quinoline-based BPh_2 complex show enhanced Φ_f (0.16–0.18).

Few other examples of ^{6}N^O type monoboron complexes are shown in Fig. 5.59. Difluoro[amidopyrazinato-*O*,*N*]borons (**APBs**) are synthesized by the boron-complexation of amidopyrazines (Fig. 5.59a) (Hachiya et al. 2010). *tert*-Butyl

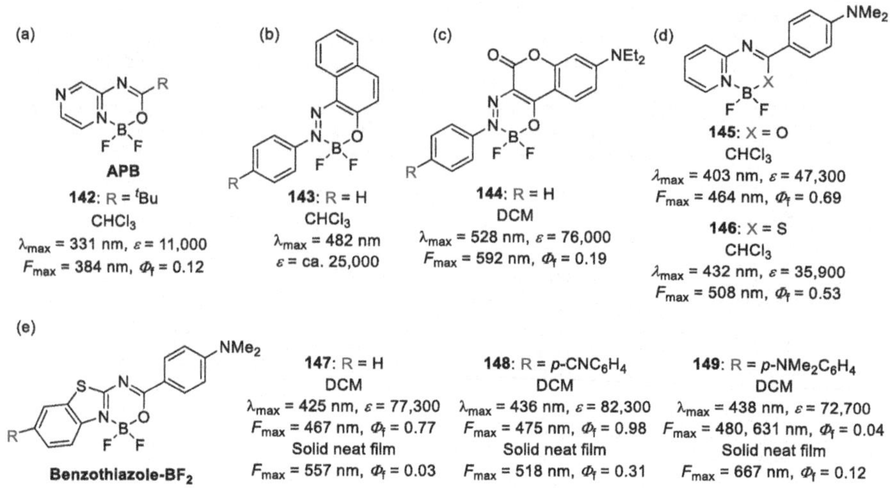

Fig. 5.59 Examples of other ^{6}N^O type monoboron complexes

derivative **142** shows a λ_{max} at 331 nm with an ε value of 11,000. Introduction of the phenyl group into the APB core causes a spectral redshift and increases ε (**APB**: R = Ph: λ_{max} = 353 nm, ε = 27,000, F_{max} = 402 nm, Φ_f = 0.15). In the APBs, the 4-cyanophenyl derivative (**APB**: R = *p*-CNC$_6$H$_4$: λ_{max} = 354 nm, ε = 21,000, F_{max} = 408 nm, Φ_f = 0.004) hardly shows fluorescence, and the 4-methoxyanophenyl derivative (**APB**: R = *p*-OMeC$_6$H$_4$: λ_{max} = 373 nm, ε = 29,000, F_{max} = 429 nm, Φ_f = 0.16) shows a spectral redshift due to the ICT property.

Azo-naphthyl-BF$_2$ complex **143** is synthesized by the reaction of 1-phenylazo-2-naphthol with BF$_3$·OEt in glacial acetic acid (Fig. 5.59b) (Jiménez et al. 2015). X-ray single crystal analysis of **143** reveals that the boron-containing six-membered ring is a nearly planar conformation. Azo-coumarin-BF$_2$ complex **144** shows fluorescence (R = H: λ_{max} = 528 nm, F_{max} = 592 nm, Φ_f = 0.19) (Fig. 5.59c) (Tathe and Sekar 2016). Introduction of the nitro group into **144** leads to a spectral redshift (R = NO$_2$: λ_{max} = 570 nm, F_{max} = 616 nm, Φ_f = 0.12). Benzamide-BF$_2$ complex **145** shows a relatively bright fluorescence (Fig. 5.59d) (Jędrzejewska et al. 2016). The corresponding thiobenzamide derivative **146** shows the redshift of both λ_{max} and F_{max} with a moderate drop of Φ_f. In the benzothiazole BF$_2$ complexes, introduction of an aryl group leads to a slight redshift of λ_{max} (**147**: 425 nm, **148**: 436 nm, **149**: 438 nm) (Fig. 5.59e) (Potopnyk et al. 2019). Dimethylamino derivative **149** shows dual fluorescence in DCM and THF.

For other ^{6}N^O type monoboron complexes, amide-type BF$_2$ complexes based on pyridine (Yamaji et al. 2017), pyridazine (Yamaji et al. 2017), pyrazine (Yamaji et al. 2017), 1,3-thiazole (Potopnyk et al. 2018), 1,3,4-thiadiazole (Zhang et al. 2018d), and 1,8-naphthyridine (Wu et al. 2012) are reported.

5.2.6.2 ^{5}N^O and ^{7}N^O Monoboron Complexes

Few examples of ^{5}N^O and ^{7}N^O type monoboron complexes are shown in Fig. 5.60. For ^{5}N^O monoboron complexes, 8-hydroxyquinoline-based boron complex **BPh$_2$q**, a boron analogue of tris(8-hydroxyquinolinato)aluminium (Alq$_3$), is synthesized by the reaction of 8-hydroxyquinoline with triphenylborane (Fig. 5.60a) (Hohaus and Umland 1969; Anderson et al. 2000; Cui et al. 2005). **BPh$_2$q 150** ($\lambda_{max} = 395$ nm, $\varepsilon = 3{,}200$ (Anderson et al. 2000), $F_{max} = 504$ nm, $\Phi_f = 0.30$ (Cui et al. 2005)) shows slightly redshifted λ_{max} and blueshifted F_{max} and a higher Φ_f compared with those of Alq$_3$ ($\lambda_{max} = 387$ nm, $\varepsilon = 7{,}000$, $F_{max} = 526$ nm, $\Phi_f = 0.17$) (Pohl and Anzenbacher 2003) in DCM. As a result of evaluating **BPh$_2$q 150** as the emitting or electron-transporting layers, **BPh$_2$q 150** is found to possess good electron-transporting properties (Wu et al. 2000). Although the introduction of an electron-withdrawing group into **BPh$_2$q** fluorophore leads to the blueshift of F_{max} and increases Φ_f (**152**: $F_{max} = 498$ nm, $\Phi_f = 0.30$), the introduction of an electron-donating group into **BPh$_2$q** fluorophore leads to a redshift of F_{max} and decreases Φ_f (**153**: $F_{max} = 625$ nm, $\Phi_f = 0.001$) (Qin et al. 2006). **BPh$_2$q** derivatives show fluorescence from quinoline-based intraligand charge transfer (ILCT) excited state.

For ^{7}N^O monoboron complexes, in 2008, **154** was synthesized from the condensation of imidazolinone with 2-formylpyrrole and the subsequent boron-complexation as an undesired compound, along with corresponding the ^{6}N^N monoboron complex **155** (Fig. 5.60b) (Wu and Burgess 2008b). Although **155** shows a strong fluorescence, **154** hardly shows fluorescence.

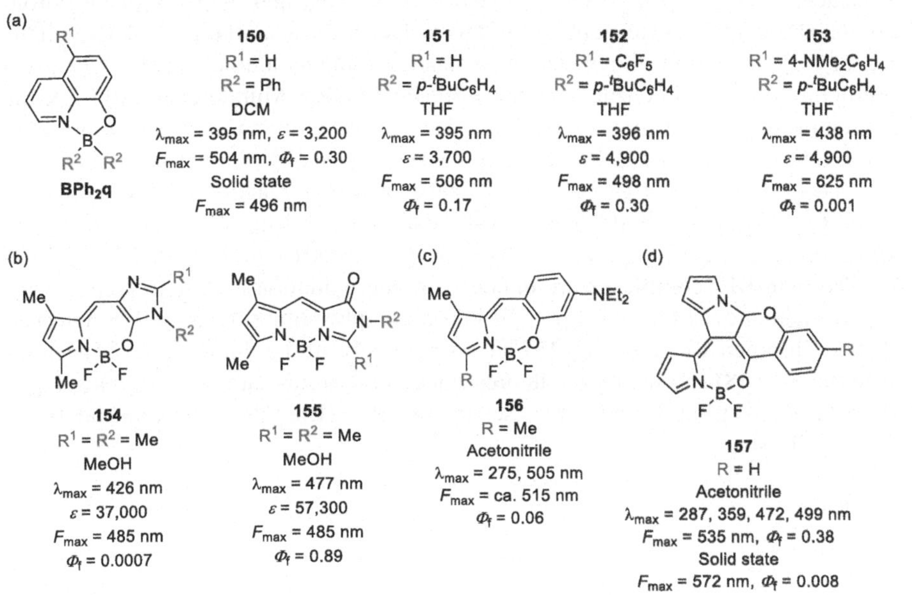

Fig. 5.60 Examples of ^{5}N^O and ^{7}N^O type monoboron complexes

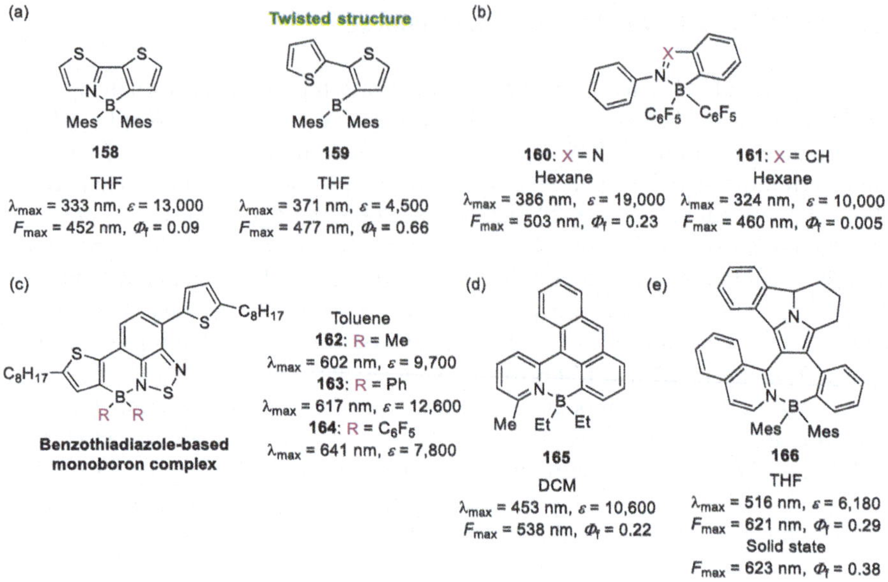

Fig. 5.61 Examples of ^{5}N^C, ^{6}N^C, and ^{7}N^C type monoboron complexes

The reaction of 2-methylaminobenzaldehyde with 2,4-dimethylpyrrole in the presence of TFA, followed by the reaction with DDQ and the subsequent boron-complexation gives corresponding BODIPY dye, while a similar procedure with the use of 4-diethylaminosalicylaldehyde instead of 2-methylaminobenzaldehyde gives a difluoroboroazaoxobenzazulene dye **156** (Fig. 5.60c) (Murale et al. 2011). X-ray crystallographic results reveal **156** has a distorted seven-membered ring with an N–B–O–C torsion angle of 72.8°. **156** shows absorption at 275 (S_0–S_2) and 505 (S_0–S_1) nm and fluorescence at ca. 515 nm. Knoevenagel-type reaction of **156** with aldehydes (e.g. 2-thienylcarbaldehyde) gives redshifted styryl-like compounds (e.g. R = (*E*)-2-(thiophen-2-yl)vinyl: $\lambda_{max} = 577$ nm, $F_{max} = 601$ nm, $\Phi_f = 0.18$).

Chroman-BF_2 complexes are obtained by the oxidation of 5-(chromen-4-one)-dipyrromethanes by DDQ followed by boron-complexation (Fig. 5.60d) (Singh et al. 2013). Chroman-BF_2 complex **157** shows relatively strong fluorescence in acetonitrile ($\Phi_f = 0.38$) and negligible fluorescence in the solid-state ($\Phi_f = 0.008$). Introduction of a *tert*-butyl group increases the Φ_f (R = ^{t}Bu: 0.50 in acetonitrile, 0.084 in the solid-state).

5.2.7 N^C Monoboron Complex

Four-coordinate organoboron complexes with a N^C bidentate ligand have been reported (Fig. 5.61). For ^{5}N^C type, in 2006, Yamaguchi et al. reported boryl-substituted thienylthiazoles (Fig. 5.61a) (Wakamiya et al. 2006). Dimesitylboryl-substituted thienylthiazole **158** shows slightly redshifted λ_{max} (333 nm) and lower Φ_f (0.09) compared with those of 2-thienyl-2-thiazle ($\lambda_{max} = 313$ nm, $\Phi_f = 0.11$), which is the N^C bidentate ligand of **158**. The boron-complexation of 2-thienyl-2-thiazle not only decreases the reduction potential, but also stabilizes the produced radical anion. 3-Boryl-substituted bithiophene **159** has a significantly twisted structure with a dihedral angle between the two thiophenes of 56.0 (Wakamiya et al. 2007). Although **159** has a relatively lower k_f value ($k_f = 5.5 \times 10^7$ s^{-1}) which is consistent with the less-allowed S_0 to S_1 transition ($\varepsilon = 4{,}500$), the k_{nr} value is also low ($k_{nr} = 2.8 \times 10^7$ s^{-1}). Due to the lower k_{nr} value, **159** shows a relatively high Φ_f value ($\Phi_f = 0.66$). Additionally, **159** shows a relatively large Stokes shift (106 nm) presumably because of the change from the twisted structure to the planar structure in the excited state. Because the S_0 to S_1 transition of **159** is essentially assigned to the charge transfer transition from the bithiophene moiety to the boron moiety, the extension of π-conjugation and the introduction of an electron-donating group in the bithiophene moiety leads to the redshift of F_{max} (THF: $F_{max} = 510$–660 nm, $\Phi_f = 0.38$–0.93). The 3-boryl-substituted bithiophene derivatives also show fluorescence in the solid-state (film: $F_{max} = 486$–657 nm, $\Phi_f = 0.30$–0.87).

In 2007, Kawashima et al. reported 2-borylazobenzenes (Fig. 5.61b) (Yoshino et al. 2007). Although azobenzene derivatives do not show fluorescence, the boron complexes, which are 2-borylazobenzenes, show fluorescence (e.g.: **160**, $F_{max} = 503$ nm, $\Phi_f = 0.23$) because of the prevention of photoisomerization of the azobenzene moiety, enhancement of the molecular rigidity, and the change in nature of the transition between the lowest excited state and the ground state from the optically forbidden n–π* transition to the allowed π–π* transition (Yoshino et al. 2013). In the 2-borylazobenzenes, the introduction of a strong electron-withdrawing pentafluorophenyl group into the boron atom is important for intense fluorescence. Replacement of the pentafluorophenyl groups to less electron-withdrawing *p*-fluorophenyl groups leads to fluorescence quenching because of the change in the S_0 to S_1 transition to a forbidden π–π* transition caused by spatial separation. Boron-substituted aromatic aldimine **161** shows blueshifted absorption ($\lambda_{max} = 324$ nm) and fluorescence ($F_{max} = 460$ nm) compared with those of **160** (Yoshino et al. 2013). Boron-substituted aromatic aldimines can be applied to cyanide ion sensing. For instance, the reaction of **161** with cyanide ions gives cyanide adduct, in which a cyanide ion binds to the imine carbon atom, subsequently leading to fluorescence quenching.

For ^{6}N^C type, benzothiadiazole-based monoboron complexes have been reported (Fig. 5.61c) (Crossley et al. 2015). BCl_2 derivatives (R = Cl) are synthesized by electrophilic C–H borylation which is the reaction of the corresponding non-borylated ligand with BCl_3. Because initial borylation reduces the nucleophilicity of the other nitrogen atom in benzothiadiazole, the diboron complex is not obtained. Although the BCl_2 derivative is stable to non-protic Lewis bases, it undergoes slow hydrolysis. The reaction of the BCl_2 derivative with $AlMe_3$, $Zn(Ph)_2$, and $Zn(C_6F_5)_2$ gives the corresponding BMe_2 **162**, BPh_2 **163,** and $B(C_6F_5)_2$ **164** derivatives, respectively; these boron complexes are stable for months in wet solvent. **162**, **163**, and **164** show significantly redshifted λ_{max} compared with that of the corresponding non-borylated ligand (λ_{max} = 471 nm, ε = 15,700). Benzothiadiazole-based monoboron complexes are non-emissive in solvent. In contrast, ^{6}N^C type boron complex **165** shows fluorescence (Fig. 5.61d) (Liu et al. 2017).

^{7}N^C type boron complex **166** shows fluorescence both in solution and in the solid-state (Fig. 5.61e) (Zeng et al. 2019). For **166**, upon increasing temperature, the fluorescence peak slightly blueshifts with the appearance of a new peak at around 515 nm because of the dissociation of the B ← N bond and the formation of the open isomer. The temperature-dependent emission spectral change of **166** is fully reversible. The addition of 60 equivalents of TBAF to the THF solution of **166** converts **166** to its F^--bound open form, accompanied by the λ_{max} and F_{max} change from 516 nm to ca. 480 nm and 621 nm to ca. 510 nm, respectively. Other N^C boron complexes have also been reported (Rao et al. 2012).

5.3 Tridentate Boron Complex

Subphthalocyanines (SubPcs) are ring-contracted phthalocyanines (Claessens et al. 2002, 2014; Shimizu and Kobayashi 2014). In 1972, SubPcs were serendipitously discovered (Fig. 5.62a) (Meller and Ossko 1972). SubPcs have a cone-shaped 14π-electron structure geometry along with an axial substituent group (Claessens et al. 2002, 2014). The cone-shaped structure makes SubPcs intrinsically noncentrosymmetric, which may induce chirality. SubPcs can be synthesized by the cyclotrimerization reaction of phthalonitriles in the presence of a boron trihalide (Claessens et al.

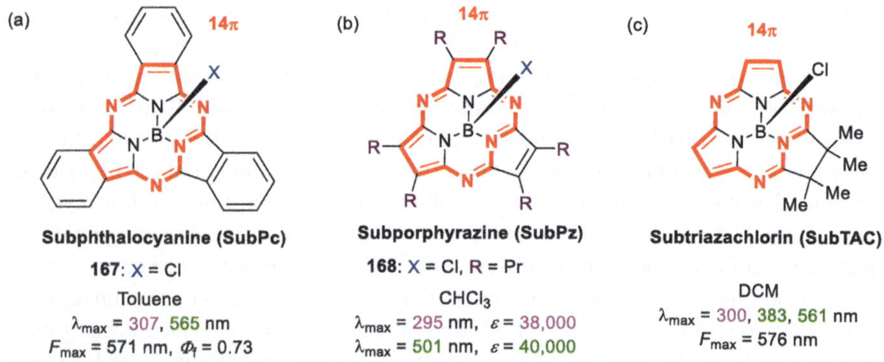

Fig. 5.62 **a** Subphthalocyanine. **b** Subporphyrazine. **c** Subtriazachlorin. Absorption maxima corresponding to Soret-like and Q-like bands are shown in pink and green, respectively

2002). When the starting phthalonitriles do not belong to the C_{2v} symmetry group, SubPcs are formed as a mixture of C_1 and C_3 isomers; each of these isomers is a racemic mixture of enantiomers (Claessens et al. 2002). The halogen atom in the axial position can be easily displaced by oxygen, sulfur, nitrogen, and carbon nucleophiles (Guilleme et al. 2011). Kobayashi et al. reported the synthesis of unsymmetrical phthalocyanines by the reaction of the structurally distorted SubPc with 1,3-diiminoisoindoline derivatives (Kobayashi et al. 1990). **167** shows a blueshifted Soret band (B transition) ($\lambda_{max} = 307$ nm, $\varepsilon =$ ca. 40,000) and Q band ($\lambda_{max} = 565$ nm, $\varepsilon =$ 88,600) compared with those of phthalocyanines (Fulford et al. 2012). The blueshift is due to the reduction in the number of π-electrons in the aromatic circuit from 18π (phthalocyanines) to 14π (SubPcs) (Claessens et al. 2014). The relatively lower ε value of SubPcs compared with that of phthalocyanines may be due to the nonplanar structure. Although axial substituents have little effect on λ_{max}, peripheral donor and acceptor substituents tend to shift the Q band towards longer wavelengths (Claessens et al. 2014). C_3-Symmetric SubPc having peripheral nitro and *tert*-butyl groups is resolved into two enantiomeric isomers by using an optically active HPLC column, and their circular dichroism (CD) spectra are measured (Kobayashi and Nonomura 2002). Generally, SubPcs show a very small Stokes shift (typically around 10–15 nm), and the Φ_f values are typically around 0.2–0.5 (Claessens et al. 2014).

Subporphyrazine (SubPz), which is also called subtriazaporphyrin (SubTAP), is less studied compared to SubPc (Fig. 5.62b) (Claessens et al. 2002, 2014; Shimizu and Kobayashi 2014). In 2005, Torres et al. reported the synthesis of SubPzs by a boron template cyclotrimerization of maleonitriles, which is a synthetic method similar to that of SubPcs (Rodríuez-Morgade et al. 2005). **168** shows two intense absorption bands at $\lambda_{max} = 295$ ($\varepsilon = 38{,}000$) and 501 nm ($\varepsilon = 40{,}000$) corresponding to the Soret and Q transitions, respectively (Rodríuez-Morgade et al. 2005). SubPzs tend to show low Φ_f values (typically less than 0.05) (Higashino et al. 2013). Ring-fused SubPzs such as tri(benzo[*b*]thiopheno)subporphyrazine (BTSubPz) (Shang et al. 2014), subazaphenalenephthalocyanine (SubAPPc) (Zhu et al. 2010), and subnaphthalocyanine (SubNc) (Rauschnabel and Hanack 1995) have also been reported.

Subtriazachlorin (SubTAC), which is sp^3-hybridized SubPz, shows markedly split Q bands which consist of a sharp band at 561 nm ($\varepsilon =$ ca. 50,000) and a weak band at 383 nm ($\varepsilon =$ ca. 8,000) (Fig. 5.62c) (Shimizu et al. 2012). The low intensity of the higher-energy Q band ($\lambda_{max} = 383$ nm) can be ascribed to the configuration interaction between the energetically close Soret band transition ($\lambda_{max} = 300$ nm) and the higher-energy Q band transition. Because the Q band absorptions mainly comprise transitions from the HOMO to the LUMO and LUMO + 1, the observed split nature of the Q band absorptions of SubTAC is considered to be due to the non-degeneration of the LUMO and LUMO + 1 (Shimizu and Kobayashi 2014).

In 2006, Osuka et al. reported the first synthesis of benzosubporphyrin (BzSubP) by the thermal self-condensation of 2-(3-oxo-2,3-dihydro-1*H*-isoindol-1-yl)acetic acid in the presence of boric acid as a template (Inokuma et al. 2006). **169** has a bowl-shaped geometry, and the depth of the bowl, which is defined as the distance from the lowest edge of the macrocycle to the boron atom, is 2.33 Å. **169** shows a

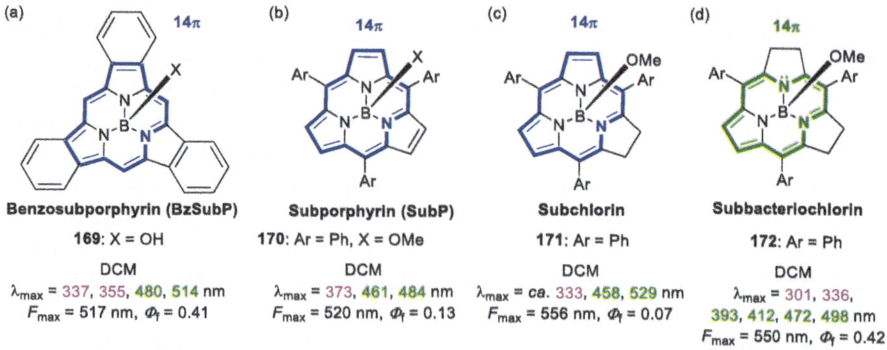

Fig. 5.63 **a** Benzosubporphyrin. **b** *meso*-Aryl-substituted subporphyrins. **c** Subchlorin. **d** Subbacteriochlorin. Absorption maxima corresponding to Soret-like and Q-like bands are shown in pink and green, respectively

sharp Soret band at 355 nm (ε = ca. 170,000) with a shoulder at 337 nm and Q bands at 480 and 514 nm (ε = ca. 90,000) (Fig. 5.63a). The strong and sharp Soret bands of **169** are similar to those of porphyrins and differ from the rather broad Soret bands of SubPcs. **169** shows green fluorescence (F_{max} = 517 nm).

In 2007, *meso*-aryl-substituted subporphyrins (SubPs) were firstly synthesized by the reaction of preformed tripyrrolylborane with aryl aldehyde in refluxing propionic acid (Kobayashi et al. 2007) or the reaction of pyridine-tri-*N*-pyrrolylborane with aryl aldehyde (Inokuma et al. 2007). The reduction of **170** with *p*-tosylhydrazide and Raney nickel gives Subchlorin **171** (Tsurumaki et al. 2008) and subbacteriochlorin **172** (Hayashi et al. 2011b), respectively. Similar to BzSubps, SubPs, subchlorins, and subbacteriochlorins have a bowl-like structure. SubP **170** shows an intense Soret-like band at 373 nm (ε = ca. 160,000) and two relatively weak Q-like bands at 461 and 484 nm (ε = ca. 10,000) (Fig. 5.63b). Compared with **170**, subchlorin **171** shows a less intense and blueshifted Soret-like band at ca. 333 nm (ε = ca. 60,000) and intensified and redshifted Q-like bands at ca. 458 (ε = ca. 15,000) and ca. 529 nm (ε = ca. 30,000) (Fig. 5.63c). Subbacteriochlorin **172** shows Soret-like bands at 301 and 336 nm (ε = ca. 30,000) and Q-like bands at 393, 412, 472, and 498 nm (ε = ca. 10,000) (Fig. 5.63d). Subbacteriochlorin **172** (Φ_f = 0.42) exhibits a higher Φ_f compared to that of **170** (Φ_f = 0.13) and subchlorin **171** (Φ_f = 0.07) probably because of the suppression of the intersystem crossing (**170**: $k_{nr} = 4.4 \times 10^8$ s^{-1}, **171**: $k_{nr} = 3.2 \times 10^8$ s^{-1}, **172**: $k_{nr} = 9.8 \times 10^7$ s^{-1}) (Hayashi et al. 2011b). The slower intersystem crossing of **172** could probably be attributed to the lowered energy level of its triplet state which leads to an increased energy gap between the singlet and triplet states.

Few other examples of tridentate boron complexes are shown in Fig. 5.64. 1,6-Bis(2-hydroxyphenyl)pyridine boron complex **(dppy)BF** shows strong blue emission at *ca.* 445 nm, and it has been applied to EL devices (Fig. 5.64a) (Li et al. 2000). Pyridine-based boron complexes with ONO- or NNN-tridentate ligands **173** or **174** show higher Φ_f in the solid-state than in solution (Fig. 5.64b, c) (Glotzbach et al.

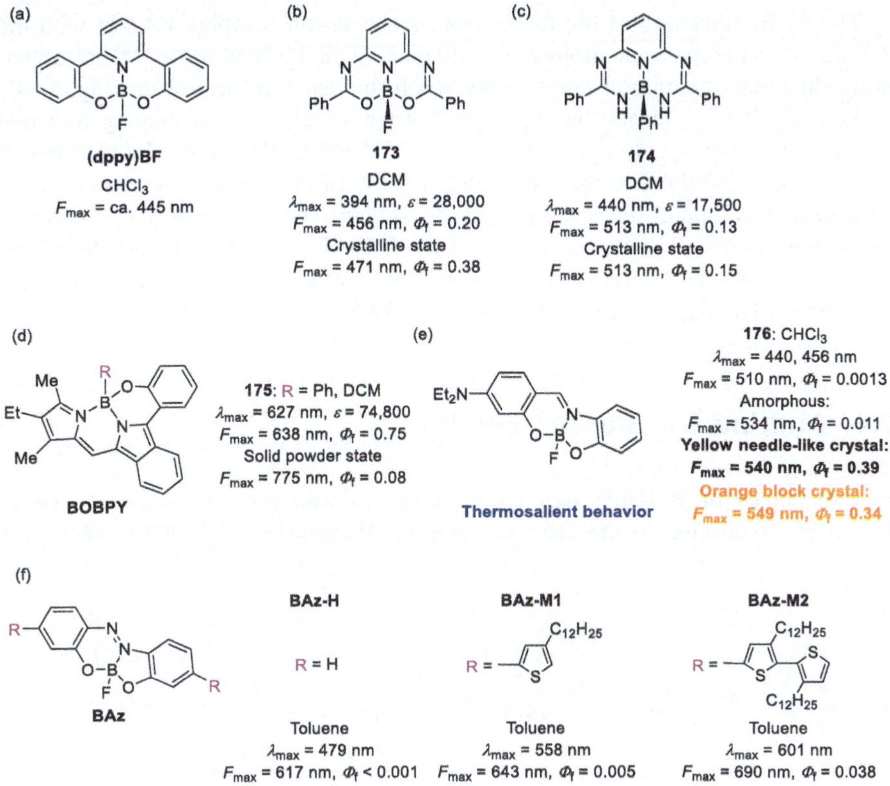

Fig. 5.64 Examples of tridentate boron complexes

2013). N_2O-Type benzopyrromethene boron complexes, named **BOBPY**, have an axial substituted group which provides steric protection and prevents the formation of aggregates (Fig. 5.64d) (Chen et al. 2017b). **BOBPY** dyes show fluorescence both in solution and in the solid-state.

Azomethine–boron complex **176** hardly shows fluorescence in $CHCl_3$ (Φ_f = 0.0013) because of the large structural relaxation (conformational change to more bent structure) in the excited state and molecular motions (Fig. 5.64e) (Ohtani et al. 2017). In the crystallization of **176**, two types of crystal polymorphs (**176Y**: yellow needle like crystal and **176O**: orange block crystal) are obtained. **176** shows not only AIEE (amorphous: $\Phi_f = 0.011$) but also CIEE (**176Y**: $\Phi_f = 0.39$, **176O**: $\Phi_f = 0.34$). Crystal-to-crystal transformation between **176Y** and **176O** occurs; **176Y** changes to **176O** at a high temperature, and **176O** changes to **176Y** at a low temperature. In the **176Y** and **176O**, thermosalient phenomena, which are defined as thermally induced hopping and fragmentation in molecular crystals, are observed maybe due to the loosely fused azomethine structure which leads to very fast crystal-to-crystal phase transition.

The S_0–S_1 transition of the fused azobenzene–boron complex **BAz-H** with the 2,2′-dihydroxyazobenzene moiety as a tridentate ligand is a permitted π–π* transition, while that of common azobenzenes is a forbidden n–π* transition (Fig. 5.64f) (Gon et al. 2018). Despite having a rigid structure and without having movable substituents which lead to molecular rotations, **BAz-H** hardly shows fluorescence in solution ($\Phi_f < 0.001$). The nonfluorescent property of **BAz-H** may be caused by the structural relaxation of the π-conjugated fluorophore in the excited state. Owing to the strong electron-accepting property of the **BAz-H** moiety, the mono (**BAz-M1**: $\lambda_{max} = 558$ nm) and bithiophene-substituted (**BAz-M2**: $\lambda_{max} = 601$ nm) derivatives show redshifted λ_{max} compared with that of **BAz-H** ($\lambda_{max} = 479$ nm).

5.4 Multinuclear Boron Complex

Few examples of BODIPY-based multinuclear boron complexes are shown in Fig. 5.65. According to the DFT calculation (Wakamiya et al. 2013), fusion of

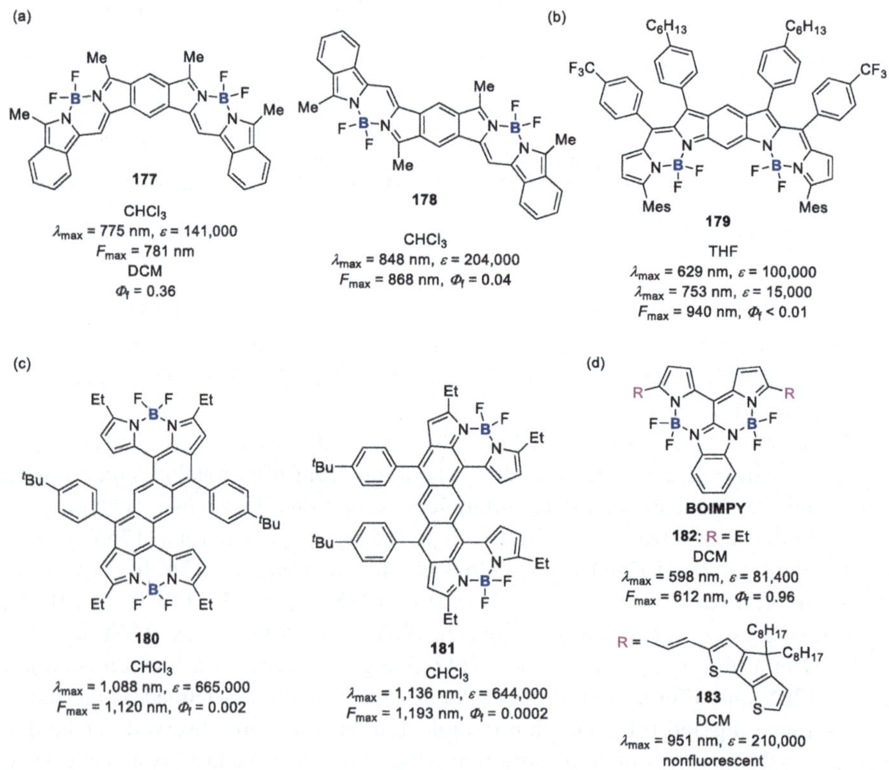

Fig. 5.65 BODIPY-based multinuclear boron complexes

the benzene ring to the *a* bond in the BODIPY skeleton increases the HOMO level, while the benzene-fusing at the *b* bond decreases the LUMO level. *Syn*-bis-benzoBODIPY **177** has a strong absorption and fluorescence at 775 and 781 nm, and *anti*-bis-benzoBODIPY **178** shows more redshifted λ_{max} (848 nm) and F_{max} (868 nm) (Fig. 5.65a) (Nakamura et al. 2012). Benzo[*b*]-fused BODIPY dimer **179** shows an intense absorption at 629 nm along with a broad shoulder band at 753 nm; the central benzene ring significantly deviates from the aromatic benzene geometry, and the quinoid character is enhanced in the benzodipyrrole moiety (Fig. 5.65b) (Wakamiya et al. 2013). The *para*- and *meta*-quinodimethane-bridged BODIPY dimers **180** and **181** show good stability towards oxidation in air due to the electron-withdrawing BF_2 unit (Fig. 5.65c) (Ni et al. 2016b). BODIPY dimers **180** and **181** have a small amount of diradical character and exhibit an intense NIR absorption. Bis(borondifluoride)-8-imidazodipyrromethene **BOIMPY 182** shows intense and sharp absorption and fluorescence spectra (Fig. 5.65d) (Patalag et al. 2016). The introduction of arylvinyl groups into the **BOIMPY** core leads to a further spectral redshift (**182**: λ_{max} = 598 nm, **183**: λ_{max} = 951 nm). Other multinuclear type boron complexes containing BODIPY cores have also been reported (Köhler et al. 2004; Sakamoto et al. 2010; Yokoi et al. 2014; Albrett et al. 2014; Chua et al. 2018; Uno et al. 2018; Tay et al. 2019; Szyszko et al. 2019).

Daltrozzo and Zumbusch et al. firstly reported pyrrolopyrrole–cyanine (PPCy) dyes (Fig. 5.66) (Fischer et al. 2007). PPCy dyes are excellent NIR chromophores and fluorophores which are synthesized by the condensation reaction of diketopyrrolopy-rrole with heteroarylacetonitrile compounds (Fischer et al. 2007, 2009). **BF_2-PPCy** and **BPh_2-PPCy** show strong and narrow NIR absorption and fluorescence and have only negligible absorption in the visible range; the type of heterocyclic rings (e.g.: **BF_2-PPCy 1**: λ_{max} = 754 nm, **BF_2-PPCy 2**: λ_{max} = 684 nm) and the substituent R on the boron atom (e.g.: **BF_2-PPCy 3**: λ_{max} = 789 nm, **BPh_2-PPCy 3**: λ_{max} = 864 nm) strongly affect the λ_{max} (684–864 nm) and F_{max} (708–881 nm) while maintaining the high ε, Φ_f, and photostability (Fig. 5.66a–c) (Fischer et al. 2009). Introduction of substituents on the PCCy core also affects the optical properties (Fischer et al. 2011a). PCCy dye is applied to NIR fluorescence labelling reagents; the PPCy dye with a carboxy group is linked to the *N*-terminus of an Arg_9 peptide which is well known as a cell penetrating peptide, and internalization of the peptide bound PPCy is demonstrated using live cell microscopy (Fischer et al. 2010). Introduction of a sulfonate group to the hydrophobic PPCy core leads to good solubility in pure water (mM range) (Wiktorowski et al. 2014). Water-soluble PPCy dyes with two aminophenyl donors have been applied to NIR fluorescent pH indicators for strong acidity (Wiktorowski et al. 2015).

Wang et al. reported some PPCy dyes to form NIR fluorescent J-aggregates when they co-precipitate with amphiphilic diblock copolymers to form colloidal nanoparticles in water (Yang et al. 2017); the packing configuration of PPCy in NPs is determined by factors including the structure, the concentration of the dye, and the ultrasonication treatment. The nanoparticles composed of spirofluorene-flanked PPCy dye exhibit effective tumour-targeting and high fluorescence contrast for in vivo bioimaging of xenografted tumour-model mice (Huang et al. 2018a).

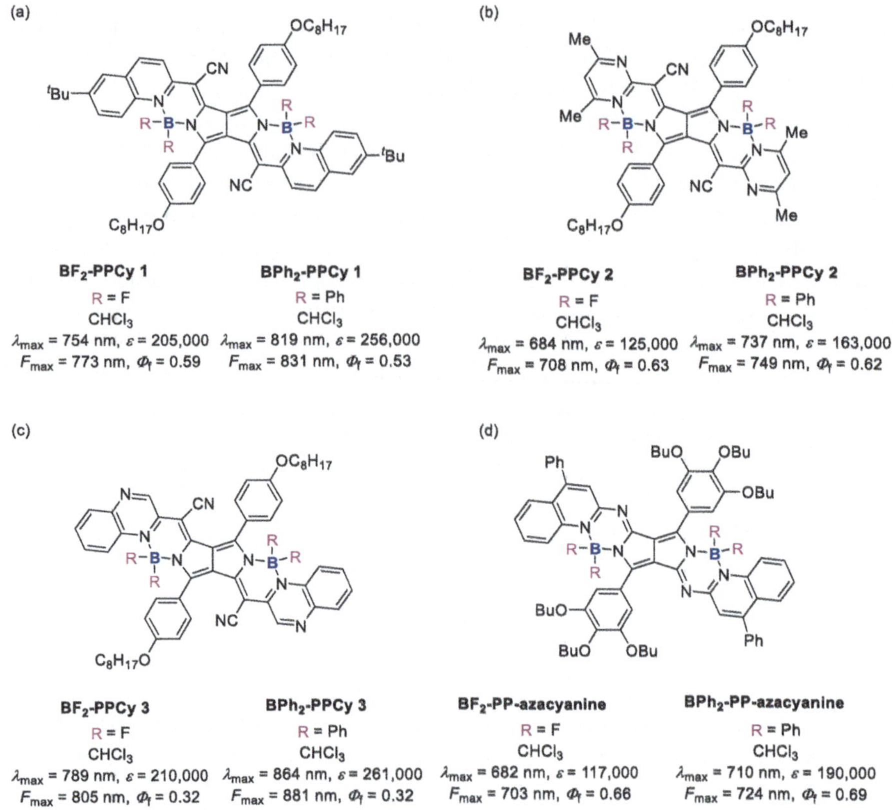

Fig. 5.66 Pyrrolopyrrole–cyanine (PPCy) dyes

Colloidal nanoparticles of PPCy dyes were also applied as theranostic agents capable of achieving both diagnosis and therapy at the same time in clinical cancer treatment (Huang et al. 2018b). Nanoparticles (30–200 nm) tend to selectively accumulate in tumours during blood circulation. The nanoparticles are preferably accumulated in a tumour site probably because of passive targeting via the enhanced permeability and retention (EPR) effect in tumour xenograft models of nude mice. The nanoparticles can perform NIR fluorescence imaging under NIR light irradiation at an excitation wavelength of 745 nm and photothermal therapy (PTT) under continuous 808 nm laser irradiation.

The aza analogues of PPCy dye, which are called pyrrolopyrrole aza-BODIPY (PPAB) or pyrrolopyrrole–azacyanine (PP–azacyanine), are also synthesized from diketopyrrolopyrrole with heteroaromatic amines in the presence of $TiCl_4$ (Shimizu et al. 2013) or $POCl_3$ (Fig. 5.66d) (Marks et al. 2014). AIE-active PPAB with the triphenylethylene group (Li et al. 2017c), application of the PPAB dimer to BHJ-OPV (Kage et al. 2018), and application of PPAB nanoparticles to PTT, photoacoustic imaging (PAI), and NIR fluorescence imaging (Wu et al. 2019) have been reported.

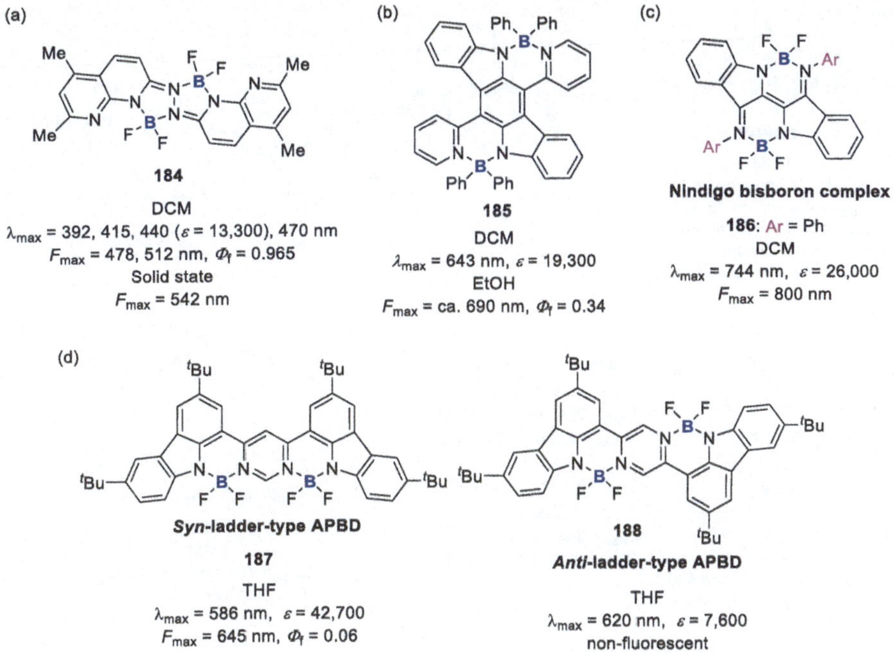

Fig. 5.67 Examples of N^N type bisboron complexes

Examples of N^N type bisboron complexes are shown in Figs. 5.67, 5.68 and 5.69. 1,8-Naphthyridine-based bisboron complex **184** shows well-resolved vibronic peaks at 392, 415, 440, and 470 nm with the vibration progressions ranging from 1370 to 1450 cm^{-1}, which are close to C=C and C=N stretching (Fig. 5.67a) (Li et al. 2010). Indolo[3,2-*b*]carbazole-based bisboron complex **185** shows a wide absorption spectrum with three λ_{max} at ca. 340 nm, ca. 480 nm, and 643 nm (ε = 19,300) (Fig. 5.67b) (Curiel et al. 2012). Although nindigo (indigo *N*,*N'*-diaryldiimine)-based bisboron complexes are isolable, they gradually decompose in solution to the corresponding monoboron complexes (Fig. 5.67c). The decomposition is accelerated with increasing steric hindrance of the aryl substituents Ar) (Nawn et al. 2013). *Anti*-ladder-type APBD **188** shows redshifted and broader absorption compared with those of the *syn*-ladder type APBD **187** (Fig. 5.67d) (Hao et al. 2014).

Bis(difluoroboron)1,2-bis((1*H*-pyrrol-2-yl)methylene)hydrazine (**BOPHY**) dye was first reported in 2014 (Fig. 5.68a) (Tamgho et al. 2014; Yu et al. 2014). BOPHY dye can be synthesized from 2-formylpyrrole via hydrazine-Schiff base linked bispyrrole and shows excellent fluorescence properties in solution. Compared to the parent BODIPY dye (*c*-hexane: λ_{max} = 504 nm, F_{max} = 511 nm) (Esnal et al. 2013), fully unsubstituted BOPHY dyes show blueshifted λ_{max} (DCM: λ_{max} = 424, 442 nm, F_{max} = 465, 493 nm) (Tamgho et al. 2014), a relatively large Stokes shift, and solid-state fluorescence (F_{max} = 543 nm) (Yu et al. 2014). α-Benzene (Wang et al. 2016b), β-benzene (Yu et al. 2014), β-thiophene (Wang et al. 2016b), and β-furan (Zhou et al.

(a)

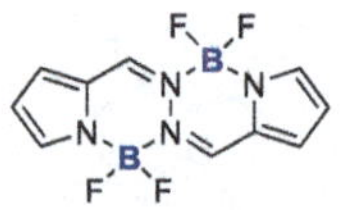

BOPHY

DCM

λ_{max} = 424 nm, ε = 40,900

λ_{max} = 442nm, ε = 38,600

F_{max} = 465, 493 nm, Φ_f = 0.95

Solid powder

F_{max} = 543 nm, Φ_f = 0.12

(b)

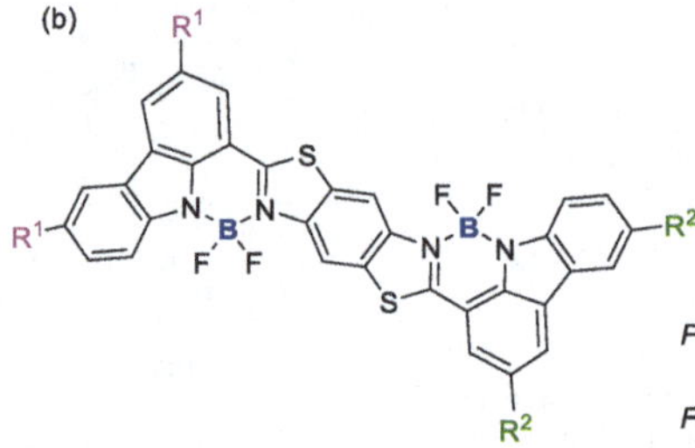

189

$R^1 = R^2 = {}^tBu$

DCM

λ_{max} = 517 nm

F_{max} = 548 nm, Φ_f = 0.445

Solid state

F_{max} = 588 nm, Φ_f = 0.064

(c)

BNBP

190: $R^1 = R^2$ = F

$CHCl_3$

λ_{max} = 501 nm, ε = 18,600

F_{max} = 528 nm, Φ_f = 0.98

191: R^1 = F, R^2 = Ph

$CHCl_3$

λ_{max} = 524 nm, ε = 11,700

F_{max} = 582 nm, Φ_f = 0.22

192: $R^1 = R^2$ = Ph

$CHCl_3$

λ_{max} = 545 nm, ε = 11,600

F_{max} = 603 nm, Φ_f = 0.75

(d)

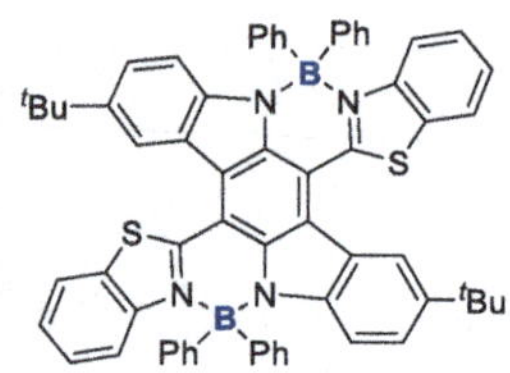

193

DCM

λ_{max} = 768 nm, ε = 18,000

F_{max} = 812 nm, Φ_f = 0.012

Fig. 5.68 Examples of N^N type bisboron complexes

(a)

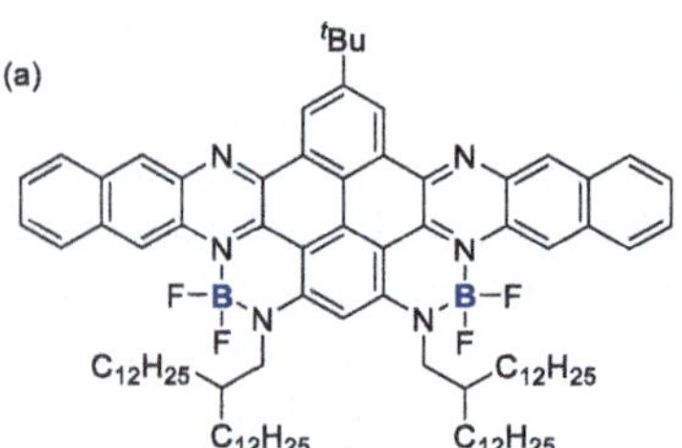

194

Toluene

λ_{max} = 686 nm, ε = 113,000

F_{max} = 705 nm, Φ_f = 0.02

(b) **BOPPY**

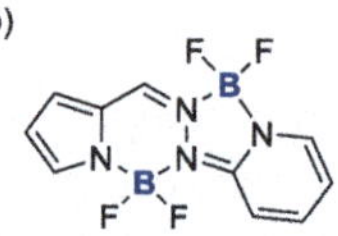

195

DCM

λ_{max} = 383 nm, ε = 29,500

λ_{max} = 413 nm, ε = 28,200

F_{max} = 432, 462 nm, Φ_f = 0.79

Solid powder

F_{max} = 484, 557 nm, Φ_f = 0.21

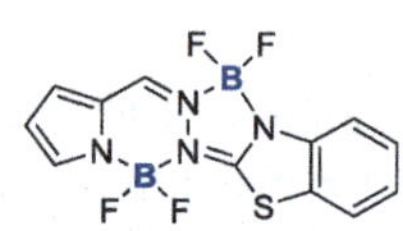

196

DCM

λ_{max} = 378 nm, ε = 29,500

λ_{max} = 394 nm, ε = 28,200

F_{max} = 409, 432, 463 nm, Φ_f = 0.62

Solid powder

F_{max} = 480, 507 nm, Φ_f = 0.27

(c)

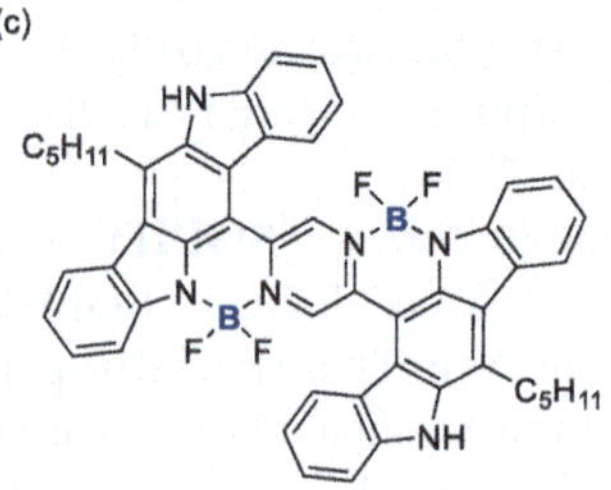

197

DCM

λ_{max} = 845 nm

ε = 15,700

film

λ_{max} = 910 nm

(d)

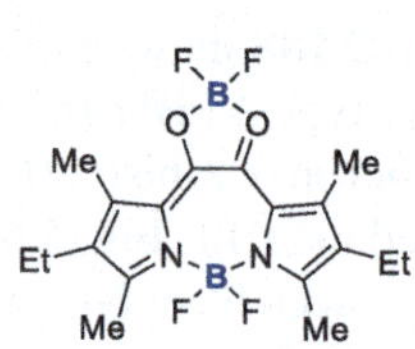

198

$CHCl_3$

λ_{max} = ca. 465 nm

ε = ca. 15,000

Fig. 5.69 Examples of N^N type bisboron complexes

2016)-fused BOPHY dyes have been reported, and these fused derivatives show more redshifted absorption and fluorescence maxima. BOPHY dyes have been applied to electron donor material for bulk heterojunction (BHJ) solar cells (Mirloup et al. 2015), pH sensors (Jiang et al. 2015), and triplet–triplet annihilation upconversion (Zhang and Zhao 2016).

In carbazole-based BODIPY dimer, the unsymmetrical derivative ($R^1 = {}^t$Bu, R^2 = Mes: $F_{max} = 609$ nm, $\Phi_f = 0.110$) tends to show higher Φ_f than the symmetrical derivative (**189**: $F_{max} = 588$ nm, $\Phi_f = 0.064$) in the solid-state (Fig. 5.68b) (Maeda et al. 2017).

A double B ← N bridged bipyridyl (BNBP) shows intense fluorescence (Fig. 5.68c) (Dou et al. 2016). BNBP is an electron-deficient building block for polymer electron acceptors. The B ← N bridging units endow BNBP with a fixed planar configuration and low-lying LUMO/HOMO energy levels. The polymer based on BNBP and thiophene units (P-BNBP-T) exhibits high electron mobility, low-lying LUMO/HOMO energy levels, and strong absorbance in the visible region, which is desirable for polymer electron acceptors. Preliminarily, all-polymer solar cell (all-PSC) devices with P-BNBP-T as the electron acceptor and PTB7 as the electron donor exhibit a PCE of 3.38%. All-PSC devices with polymers based on BNBP and 4,4-difluoro-4*H*-cyclopenta[2,1-b:3,4-b′]dithiophene (fCDT) units (P-BNBP- fCDT) as electron acceptors show a PCE of 3.83% (Zhao et al. 2017a). By changing the substituents from fluorine atoms to phenyl groups, the λ_{max} and F_{max} are redshifted and the electron affinities are degraded due to the electron-withdrawing ability of the fluorine atoms ($E_{ox, onset}$ vs. Fc/Fc$^+$: **190**: 0.38 V, **191**: 0.60 V, **192**: 0.83 V, $E_{red, onset}$ vs. Fc/Fc$^+$: **190**: −1.87 V, **191**: −1.72 V, **192**: −1.61 V) (Wang et al. 2018b).

Bisbron complex **193** shows well separated two reductive and two oxidative electron transfer processes (Fig. 5.68d) (Zhu et al. 2018). Spectroelectrochemical measurements reveal multicolour electrochromism with excellent recyclability. The radical cation and dication of **193** show absorption at 1180 nm corresponding to HOMO-β to LUMO-β transition of the delocalized radical cation and multiple absorption bands covering a broad visible light region, respectively. The radical anion of **193** shows a highly intensive absorption at 677 nm and a weak absorption at ca. 940 nm, corresponding to the HOMO-1-α to LUMO-α and HOMO-β to LUMO-β transitions of the radical anion, respectively. The dianion of **193** shows absorption at 544 nm, corresponding to the HOMO to LOMO transition.

B ← N-containing azaacene **194** shows λ_{max} at 686 nm with an ε value of 113,000 (Fig. 5.69a) (Min et al. 2018). The introduction of B ← N units into azaacene leads to significant perturbations of the electronic structure, that is, unique LUMO delocalized over the entire acene skeleton, decreased aromaticity of the B ← N-adjacent rings, low-lying LUMO energy levels, and high electron affinities. Solution-processed OFET devices based on **194** show unipolar n-type characteristics with an electron mobility of up to 0.21 cm^2 V^{-1} s^{-1}.

BOPPY dyes which are unsymmetrical bis(BF_2) dyes containing both pyrrole and *N*-heteroarene derivatives are prepared via a simple one-pot procedure from the reaction of pyrrole-2-carboxaldehyde with 2-hydrazinylpyridine or 2-hydrazinylbenzo[*d*]thiazole and the subsequent boron-complexation (Fig. 5.69b) (Yu et al. 2018). **BOPPY** dyes are highly fluorescent both in solution and solid powder states. **BOPPY** dyes show good two-photon absorption cross sections of more than 900 GM in the NIR region and are applied to one-photon microscopy (OPM) and two-photon microscopy (TPM) imaging of living cells. Indolo[3,2-*b*]carbazole-based bisboron complex **197** shows NIR absorption (Fig. 5.69c) (Mula et al. 2018).

An expanded ring size (seven-membered ring) distorts the perfect tetrahedral structure of the boron bridge, making it more labile to a fluoride attack (Fig. 5.69d). Because of the relatively easy B ← N bond dissociation character, **198** is applied to chromogenic sensors for fluoride and cyanide ions (Guliyev et al. 2012). The gradual addition of tetrabutylammonium fluorides into the chloroform solution of **198** decreases the broad absorption at 465 nm with a concomitant increase in new absorptions at 565 and 600 nm. NMR and MS spectrometric analyses suggest that at lower concentrations of fluoride ions, initially, the BF_2 unit bridging two pyrroles is detached. Furthermore, addition of excess amounts of fluoride ion results in the appearance of absorption at around 520 nm with a concomitant disappearance of absorptions at 565 and 600 nm. Higher concentrations result in the removal of the BF_2 unit bridging two oxygen atoms in the form of BF_4^-.

Examples of N^O type bisboron complexes are shown in Figs. 5.70, 5.71 and 5.72. Pyrimidine-based diboron complexes bearing β-iminoenolate ligands show redshifted λ_{max} (**199**: 489 nm) compared with that of the corresponding monoboron complex (397 nm) (Fig. 5.70a) (Kubota et al. 2016). The diboron complexes with D–π–A structures, that is, the dimethylamino derivatives ($R^1 = NMe_2$), show redshifted λ_{max} values ($\lambda_{max} = 531$–541 nm) compared with that of the parent compound **199** due to the ICT character. In the dimethylamino derivatives, the ε values decrease with increasing ICT character ($R^2 = OMe$: $\varepsilon = 119{,}600$, $R^2 = CN$: $\varepsilon = 73{,}600$). The F_{max} and Φ_f values are redshifted and decrease with increasing ICT character ($R^2 = OMe$: $F_{max} = 583$ nm, $\Phi_f = 0.84$, $R^2 = CN$: $F_{max} = 639$ nm, $\Phi_f = 0.06$) in DCM. The dimethylamino derivatives exhibit positive fluorescence solvatochromism; the cyano derivative ($R^2 = CN$) exhibits the most dramatic redshift of F_{max} with increasing solvent polarity (from 551 nm in hexane to 710 nm in acetonitrile). Pyrimidine-based diboron complexes show fluorescence in the crystal state ($F_{max} = 578$–706 nm, $\Phi_f = 0.06$–0.27). In the case of trifluoromethyl derivative **200**, toluene inclusion crystal and ethyl acetate-inclusion crystals are obtained via recrystallization. The colours of the crystals are obviously different. The toluene inclusion crystal ($F_{max} = 668$ nm, $\Phi_f = 0.16$) exhibits a blueshifted F_{max} and higher Φ_f compared with those of the original trifluoromethyl derivative ($F_{max} = 694$ nm, $\Phi_f = 0.08$) in the crystal state. In contrast, the F_{max} (709 nm) and Φ_f (0.04) values of the ethyl acetate-inclusion crystal are redshifted and lower, respectively. The X-ray crystallographic results indicate that although the lower Φ_f of the ethyl acetate-inclusion crystal is due to the formation of consecutive π–π interactions between fluorophores, consecutive π–π interactions between the fluorophores and toluene molecules in the toluene inclusion crystal have

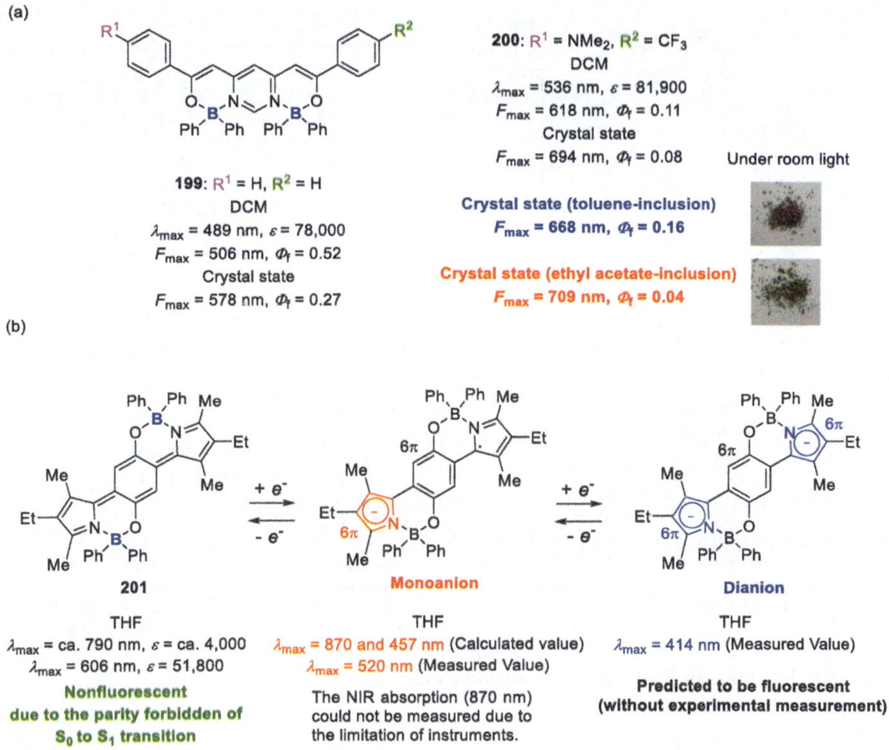

Fig. 5.70 Examples of N^O type bisboron complexes; **a** pyrimidine-based diboron complexes; and **b** quinoid-type bisboron complexes

an insignificant effect on the decrease in Φ_f. The redshifted F_{max} of ethyl acetate-inclusion crystal may be due to the formation of the CH···O interactions between the fluorophores and the ethyl acetate molecules.

Quinoid-type bisboron complex **201** shows weak absorption at 606 nm (ε = 51,800) and ca. 790 nm (ε = ca. 4,000), corresponding to the allowed $S_0 \rightarrow S_2$ and forbidden $S_0 \rightarrow S_1$ transitions, respectively (Fig. 5.70b) (Kubota et al. 2015b). The $S_0 \rightarrow S_1$ transition of **201** is mostly attributed to the HOMO–LUMO transitions. Despite the fact that the HOMO and LOMO spatially overlap, the transition is not allowed ($f = 0.00$). According to Laporte's parity selection rule, gerade–gerade and ungerade–ungerade optical transitions are forbidden. Because the HOMO and LUMO of **201** are ungerade, the forbidden transition between the HOMO and LUMO can be attributed to the forbidden parity caused by the highly symmetrical structure.

In contrast to benzene-1,4-diol analogues **202**, **204**, and **205**, quinoid-type bisboron complexes do not exhibit fluorescence (Figs. 5.70b and 5.71a, c). According to Kasha's rule, only the lowest excited state is a candidate for the initiation of emission. Therefore, the fluorescence of **201** can occur from the S_1 state, even when the allowed S_0 to S_2 absorption occurs. Additionally, according to the Strickler–Berg

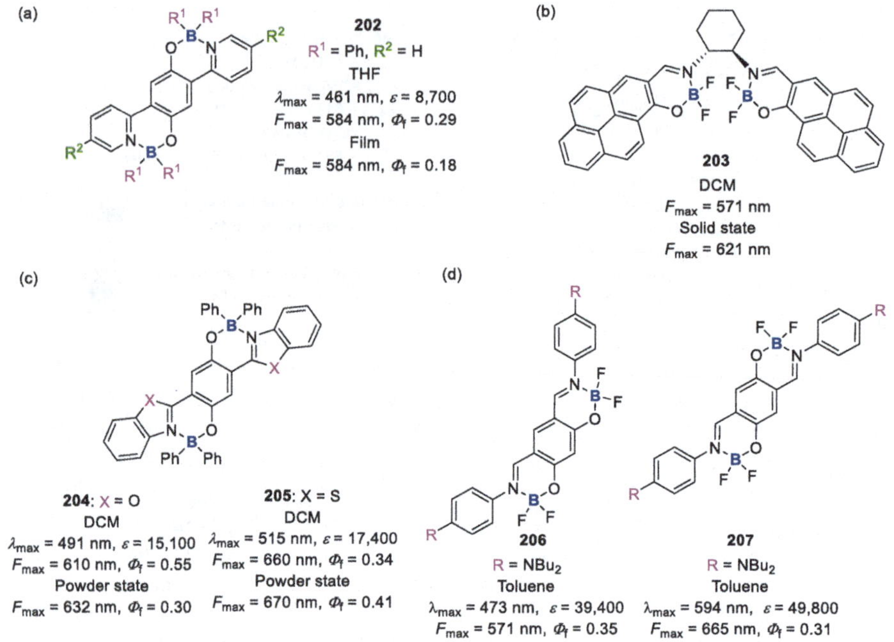

Fig. 5.71 Examples of N^O type bisboron complexes

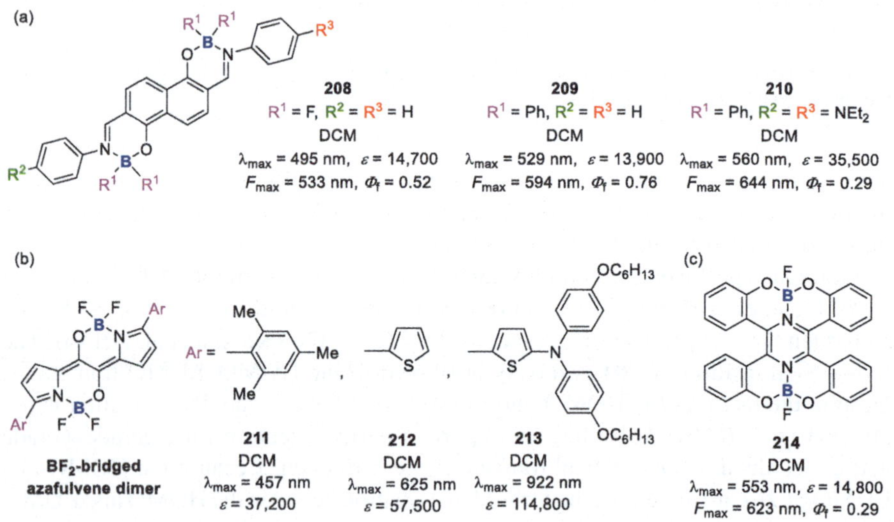

Fig. 5.72 Examples of N^O type bisboron complexes

equation, k_f is proportional to the integral of the molar extinction coefficient curve. Thus, low ε values decrease the k_f value, and consequently, the Φ_f value decreases. Because the S_0 to S_1 transition of **201** is predicted to be forbidden by the TDDFT calculation ($f = 0.00$), the nonfluorescent property of the quinoid-type bisboron complexes is probably because of the forbidden S_0 to S_1 transition that significantly decreases the k_f values. The cyclic voltammograms indicate that bisboron complex **201** is reduced to the corresponding aromatic dianion. The dianion and monoanion of **201** show λ_{max} at 414 and 520 nm, respectively. Although TDDFT results indicate that the monoanion has λ_{max} at 870 nm, NIR absorption is not observed because of the limitations of the measuring instruments. The dianion of **201** has a benzene-1,4-diol-based bisboron structure with two aromatic pyrrole moieties. The fact that the S_0 to S_1 transition of the dianion is allowed predicts that it may show fluorescence because benzene-1,4-diol-based bisboron complexes **202**, **204**, and **205** show fluorescence.

Benzene-1,4-diol-based bisboron complex **202** shows broad absorption at 461 nm and fluorescence both in solution and the thin film state (Fig. 5.71a) (Zhang et al. 2009b). Bisboron complex **202** (T_{d5} = 378 °C) has a higher decomposition temperature with a 5% weight loss (T_{d5}) compared to the corresponding monoboron complex (T_{d5} = 247 °C). Simple double layer EL devices fabricated using **202** as both the emitter and electron-transporting layers display good performance. Pyrene-based bisboron complex **203** shows red fluorescence in the solid-state and behaves as both an emitting and electron-transporting material in OLEDs (Fig. 5.71b) (Zhou et al. 2010). In the benzene-1,4-diol-based bisboron complexes **204** and **205**, the two phenyl groups coordinated to each boron atom effectively keep the fluorophores apart in the solid-state (Fig. 5.71c) (Li et al. 2011). Thus, bisboron complexes **204** and **205** exhibit solid-state fluorescence. OLEDs employing **204** or **205** as non-doped emitters exhibit red and NIR electroluminescence. Salicylaldimine-based bisboron complex **207** shows redshifted λ_{max} compared with that of the structural isomer **206** (Fig. 5.71d) (Frath et al. 2016).

The photophysical properties of bis(boranil) complexes **208–210** can be easily tuned in a wide range by variation of substituents (Fig. 5.72a) (Urban et al. 2017). BF_2–bridged azafulvene dimers are reported to be strong electron-accepting units (Fig. 5.72b). Introduction of electron-donating groups enables the formation of D–A–D chromophores to provide the intense NIR absorption (Shimogawa et al. 2018). Especially, **213** with strong electron-donating groups shows an absorption maximum at 922 nm in DCM, 914 nm in 1 wt% PMMA films, and 1066 and 1174 nm in neat films. Because of the low-lying LUMO and moderate HOMO levels, BF_2-bridged azafulvene dimer **213** has a high photostability and strong resistance to oxidation. Boron-fused double (Li et al. 2017a) helicene **214** shows two reversible reductions and a low LUMO energy (Fig. 5.72c). It has been employed as a cathode active material in lithium ion batteries and has shown moderate performance, capacity, and cycle stability (Oda et al. 2019).

For O^O type bisboron complexes, Ono et al. reported **DHND-BF$_2$** (Ono et al. 2009) and **QA-BF$_2$** (Moriya et al. 2019). **DHND-BF$_2$** is the BF_2 complex of 6,11-dihydroxy-5,12-naphthacenedione (DHND). The T_{d5} of **DHND-BF$_2$** (325 °C) is

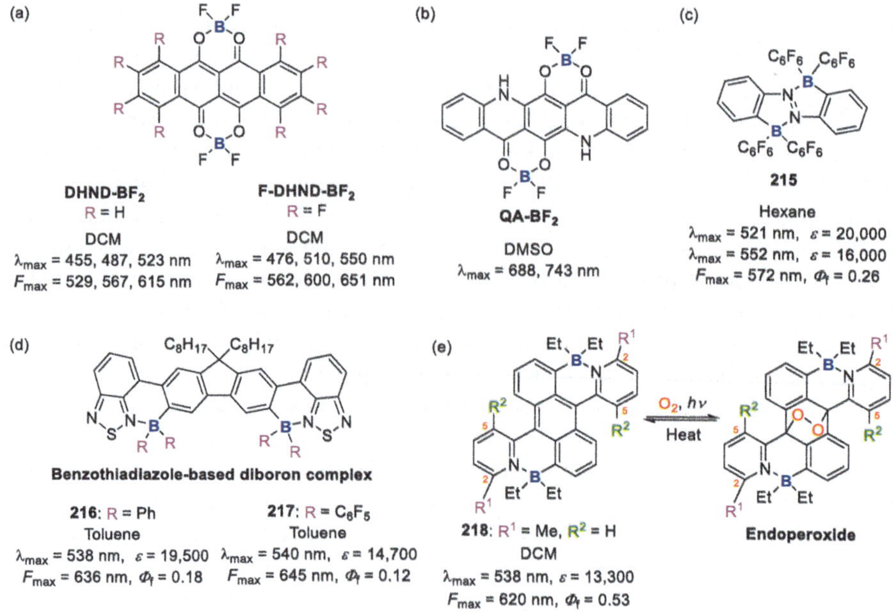

Fig. 5.73 Examples of O^O and N^C type bisboron complexes

higher than that of the corresponding fluorine derivative **F-DHND-BF$_2$** (R = F: 277 °C) (Fig. 5.73a) (Ono et al. 2009). **DHND-BF$_2$** shows a sharp absorption peak at 523 nm along with a vibrational peak. The λ_{max} values of **DHND-BF$_2$** are blueshifted compared with those of **F-DHND-BF$_2$** (λ_{max} = 550 nm). In the crystal of **F-DHND-BF$_2$**, the molecules are arranged to form a tapelike network with short F···π and F···F contacts, leading to a dense crystal packing. **F-DHND-BF$_2$** shows n-type semiconducting behaviour in devices with a bottom contact configuration.

The BF_2 chelation of quinacridonequinone (QQ) generates the BF_2 complex of QQ (**QQ-BF$_2$**) (Moriya et al. 2019). Because the generated **QQ-BF$_2$** possesses a strong electron affinity, it is easily reduced to produce **QA-BF$_2$** which is the BF_2 complex of 6,13-dihydroxyquinacridone (**QA-OH**) (Fig. 5.73b). **QA-BF$_2$** is stable in the solid-state, whereas it undergoes hydrolysis in solution to give **QA-OH** in argon. **QA-BF$_2$** shows a weak and broadened absorption in the NIR region.

For N^C type bisboron complexes, 2,2′-diborylazobenzene **215** shows a spectral redshift (F_{max} = 572 nm) compared with that of 2-borylazobenzene **161** (Figs. 5.61b and 5.73c) (Yoshino et al. 2013). Benzothiadiazole-based diboron complexes **216** and **217** show a relatively large Stokes shift (Fig. 5.73d) (Crossley et al. 2015). Anthracene-based boron complex **218** has a highly twisted structure, low-lying LUMO, and relatively high Φ_f (0.53) (Fig. 5.73e) (Liu et al. 2017). In the presence of light, **218** reacts with O_2 without an external photosensitizer, resulting in selective and reversible formation of the corresponding endoperoxide probably because of the release of steric strain upon peroxide formation. When compared with **218**, the

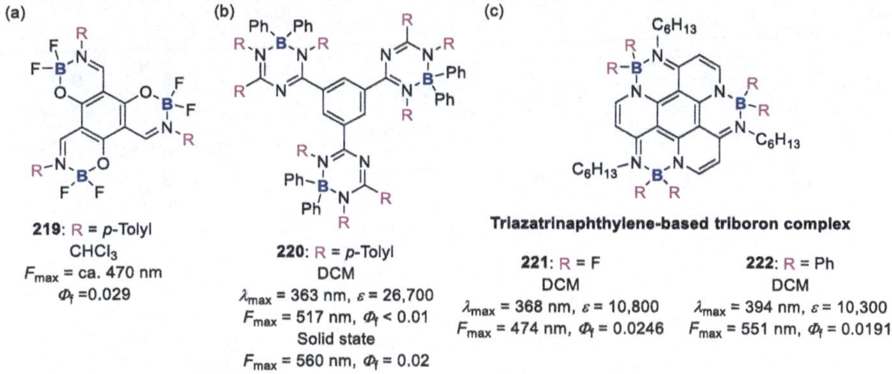

Fig. 5.74 Triboron complexes

5-methyl derivative ($R^1 = H$, $R^2 = Me$) slowly reacts with O_2 to give its endoperoxide because of the prevention of the effective planarization and limited carbocation stabilization of the generally accepted zwitterionic intermediate in the oxygenation reaction (Liu et al. 2019). The rate of thermal release of singlet oxygen from the endoperoxide of **218** is also faster than that of the 5-methyl derivative maybe because of the weaker B $\leftarrow$ N bond.

Examples of trisboron complexes are shown in Fig. 5.74. Triboron complex **219** is weakly fluorescent in $CHCl_3$ (Fig. 5.74a) (Riddle et al. 2006). The fluorescence intensity of **219** increases with the increasing volume fraction of hexane in $CHCl_3$-hexane mixed solvents because of the formation of aggregates. The λ_{max} of tris-triazapentadiene boron complex **220** (363 nm) is almost similar to that of the corresponding monoboron complex (353 nm) because of the specific steric properties which cause the limited electron delocalization involving the central benzene ring and the neighbouring triazapentadiene boron systems (Fig. 5.74b) (Glotzbach et al. 2015). In the crystal state, triazatrinaphthylene-based triboron complexes have a slightly distorted disc-shaped skeleton (Fig. 5.74c) (Qiu et al. 2016). The λ_{max} of BPh_2 complex **222** (λ_{max}: 394 nm, HOMO: −5.37 eV, LUMO: −2.93 eV) is redshifted compared with that of BF_2 complex **221** (λ_{max}: 368 nm, HOMO: −5.97 eV, LUMO: −3.17 eV) probably because of the relatively stronger electron-donating effect of phenyl groups as compared with that of fluorine atoms, which is reflected in the elevation of the HOMO energy level. The ligands of **221** and **222** are nonfluorescent because of the strong PeT effect, and **221** and **222** show weak fluorescence probably because of the confinement of the nitrogen lone-pair electrons through the formation of the B–N coordinated bonds to avoid the PeT effect.

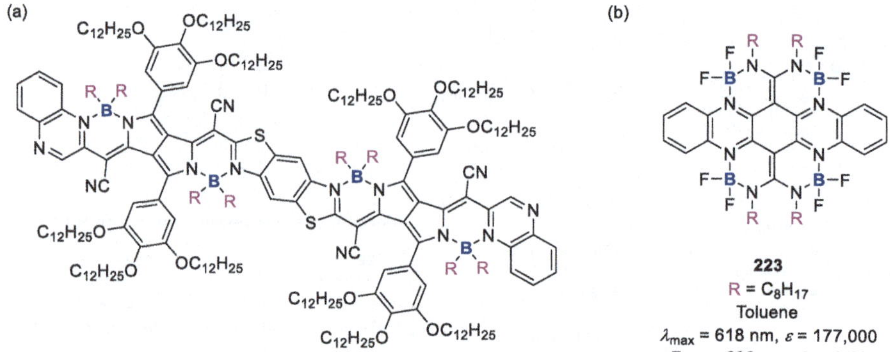

Bis-BF_2-PPCy
R = F
$CHCl_3$
λ_{max} = 849 nm, ε = 574,000
F_{max} = 867 nm

Bis-BPh_2-PPCy
R = Ph
$CHCl_3$
λ_{max} = 941 nm, ε = 531,000
F_{max} = 966 nm

223
R = C_8H_{17}
Toluene
λ_{max} = 618 nm, ε = 177,000
F_{max} = 630 nm, Φ_f = 0.50

Fig. 5.75 Tetraboron complexes

Bis(pyrrolopyrrole) cyanines **Bis(PPCy)s** are selective NIR absorbers with very narrow absorption bands, extremely high absorption coefficients, high photochemical stability, and the lack of any strong absorption in the visible spectral range (Fig. 5.75a) (Fischer et al. 2011b). Tetraboron complex **223** shows strong red fluorescence and unipolar electron-transporting semiconductor characteristics (Fig. 5.75b) (Min et al. 2019).

References

Ahrens J, Haberlag B, Scheja A, Tamm M, Bröring M (2014) Conjugated BODIPY DYEmers by metathesis reactions. Chem Eur J 20:2901–2912

Albrett AM, Thomas KE, Maslek S, Młodzianowska A, Conradie J, Beavers CM, Ghosh A, Brothers PJ (2014) Mono- and diboron corroles: factors controlling stoichiometry and hydrolytic reactivity. Inorg Chem 53:5486–5493

Allik TH, Hermes RE, Sathyamoorthi G, Boyer JH (1994) Spectroscopy and laser performance of new BF2-complex dyes in solution. Proc SPIE Int Soc Opt Eng 2115:240–248

Alnoman RB, Rihn S, O'Connor DC, Black FA, Costello B, Waddell PG, Clegg W, Peacock RD, Herrebout W, Knight JG, Hall MJ (2016) Circularly polarized luminescence from helically chiral N,N,O,O-boron-chelated dipyrromethenes. Chem Eur J 22:93–96

Anderson S, Weaver MS, Hudson AJ (2000) Materials for organic electroluminescence: aluminium vs. boron. Synth Met 111–112:459–463

Antina EV, Gusev GB, Rumyantsev EV, Dudina NA (2009) Thermal properties of ligands, salts and metal complexes of linear oligopyrroles. Russ J Gen Chem 79:1900–1909

Araneda JF, Piers WE, Heyne B, Parvez M, McDonald R (2011) High Stokes shift anilido-pyridine boron difluoride dyes. Angew Chem Int Ed 50:12214–12217

Awuah SG, You Y (2012) Boron dipyrromethene (BODIPY)-based photosensitizers for photodynamic therapy. RSC Adv 2:11169–11183
Awuah SG, Polreis J, Biradar V, You Y (2011) Singlet oxygen generation by novel NIR BODIPY dyes. Org Lett 13:3884–3887
Awuah SG, Das SK, D'Souza F, You Y (2013) Thieno–pyrrole–fused BODIPY intermediate as a platform to multifunctional NIR agents. Chem Asian J 8:3123–3132
Bai G, Yu C, Cheng C, Hao E, Wei Y, Mu X, Jiao L (2014) Syntheses and photophysical properties of BF_2 complexes of curcumin analogues. Org Biomol Chem 12:1618–1626
Bañuelos J (2016) BODIPY dye, the most versatile fluorophore ever? Chem Rec 16:335–348
Bañuelos J, Arbeloa FL, Martinez V, Liras M, Costela A, Morenoc IG, Arbeloa IL (2011) Difluoroboron-triaza-anthracene: a laser dye in the blue region. Theoretical simulation of alternative difluoro-boron-diaza-aromatic systems. Phys Chem Chem Phys 13:3437–3445
Barbon SM, Staroverov VN, Gilroy JB (2015) Effect of extended π conjugation on the spectroscopic and electrochemical properties of boron difluoride formazanate complexes. J Org Chem 80:5226–5235
Belmonte-Vázquez JL, Avellanal-Zaballa E, Enríquez-Palacios E, Cerdán L, Esnal I, Bañuelos J, Villegas-Gómez C, Arbeloa IL, Peña-Cabrera E (2019) Synthetic approach to readily accessible benzofuran-fused borondipyrromethenes as red-emitting laser dyes. J Org Chem 84:2523–2541
Benstead M, Mehl GH, Boyle RW (2011) 4,4′-Difluoro-4-bora-3a,4a-diaza-s-indacenes (BODIPYs) as components of novel light active materials. Tetrahedron 67:3573–3601
Bessette A, Hanan GS (2014) Design, synthesis and photophysical studies of dipyrromethene-based materials: insights into their applications in organic photovoltaic devices. Chem Soc Rev 43:3342–3405
Bochkov AY, Akchurin IO, Dyachenko OA, Traven VF (2013) NIR-fluorescent coumarin-fused BODIPY dyes with large Stokes shifts. Chem Commun 49:11653–11655
Bodio E, Goze C (2019) Investigation of B-F substitution on BODIPY and aza-BODIPY dyes: development of B-O and B-C BODIPYs. Dyes Pigm 160:700–710
Boens N, Leen V, Dehaen W (2012) Fluorescent indicators based on BODIPY. Chem Soc Rev 41:1130–1172
Boens N, Verbelen B, Dehaen W (2015) Postfunctionalization of the BODIPY core: synthesis and spectroscopy. Eur J Org Chem 6577–6595
Bonnier C, Piers WE, Ali AAS, Thompson A, Parvez M (2009) Perfluoroaryl-substituted boron dipyrrinato complexes. Organometallics 28:4845–4851
Bucher L, Desbois N, Harvey PD, Sharma GD, Gros CP (2017) Porphyrins and BODIPY as building blocks for efficient donor materials in bulk heterojunction solar cells. Solar RRL 1:1700127
Bumagina NA, Kritskaya AY, Antina EV, Berezin MB, V'yugin AI (2018) Effect of alkyl, aryl, and meso-aza substitution on the thermal stability of BODIPY. Russ J Inorg Chem 63:1326–1332
Bura T, Retailleau P, Ulrich G, Ziessel R (2011) Highly substituted bodipy dyes with spectroscopic features sensitive to the environment. J Org Chem 76:1109–1117
Buyukcakir O, Bozdemir OA, Kolemen S, Erbas S, Akkaya EU (2009) Tetrastyryl-Bodipy dyes: convenient synthesis and characterization of elusive near IR fluorophores. Org Lett 11:4644–4647
Chase DT, Young BS, Haley MM (2011) Incorporating BODIPY fluorophores into tetrakis(arylethynyl)benzenes. J Org Chem 76:4043–4051
Chen J, Burghart A, Derecskei-Kovacs A, Burgess K (2000) 4,4-Difluoro-4-bora-3a,4a-diaza-*s*-indacene (BODIPY) dyes modified for extended conjugation and restricted bond rotations. J Org Chem 65:2900–2906
Chen H-Y, Chi Y, Liu C-S, Yu J-K, Cheng Y-M, Chen K-S, Chou P-T, Peng S-M, Lee G-H, Carty AJ, Ye S-J, Chen C-T (2005) Rational color tuning and luminescent properties of functionalized boron-containing 2-pyridyl pyrrolide complexes. Adv Funct Mater 15:567–574
Chen J, Mizumura M, Shinokubo H, Osuka A (2009) Functionalization of boron dipyrrin (BODIPY) dyes through iridium and rhodium catalysis: a complementary approach to α- and β-substituted BODIPYs. Chem Eur J 15:5942–5949

Chen Y, Wan L, Zhang D, Bian Y, Jiang J (2011) Modulation of the spectroscopic property of Bodipy derivates through tuning the molecular configuration. Photochem Photobiol Sci 10:1030–1038
Chen Y, Zhao J, Guo H, Xie L (2012) Geometry relaxation-induced large Stokes shift in red-emitting borondipyrromethenes (BODIPY) and applications in fluorescent thiol probes. J Org Chem 77:2192–2206
Chen JJ, Conron SM, Erwin P, Dimitriou M, McAlahney K, Thompson ME (2015) High-efficiency BODIPY-based organic photovoltaics. ACS Appl Mater Interfaces 7:662–669
Chen P-Z, Niu L-Y, Chen Y-Z, Yang Q-Z (2017a) Difluoroboron β-diketonate dyes: spectroscopic properties and applications. Coord Chem Rev 350:196–216
Chen N, Zhang W, Chen S, Wu Q, Yu C, Wei Y, Xu Y, Hao E, Jiao L (2017b) Sterically protected N_2O-type benzopyrromethene boron complexes from boronic acids with intense red/near-infrared fluorescence. Org Lett 19:2026–2029
Chen D-G, Ranganathan R, Lin J-A, Huang C-Y, Ho M-L, Chi Y, Chou P-T (2019) Ratiometric tuning of luminescence: interplay between the locally excited and interligand charge-transfer states in pyrazolate-based boron compounds. J Phys Chem C 123:4022–4028
Cheng X, Li D, Zhang Z, Zhang H, Wang Y (2014) Organoboron compounds with morphology-dependent NIR emissions and dual-channel fluorescent ON/OFF switching. Org Lett 16:880–883
Cherumukkil S, Vedhanarayanan B, Das G, Praveen VK, Ajayaghosh A (2018) Self-assembly of bodipy-derived extended π-systems. Bull Chem Soc Jpn 91:100–120
Chibani S, Le Guennic B, Charaf-Eddin A, Laurenta AD, Jacquemin D (2013) Revisiting the optical signatures of BODIPY with ab initio tools. Chem Sci 4:1950–1963
Choi S, Bouffard J, Kim Y (2014) Aggregation-induced emission enhancement of a meso-trifluoromethyl BODIPY via J-aggregation. Chem Sci 5:751–755
Chong H, Lin H-A, Shen M-Y, Liu C-Y, Zhao H, Yu H-H (2015) Step-economical syntheses of functional BODIPY-EDOT π-conjugated materials through direct C–H arylation. Org Lett 17:3198–3201
Chua MH, Kim T, Lim ZL, Gopalakrishna TY, Ni Y, Xu J, Kim D, Wu J (2018) BODIPY-based antiaromatic macrocycles: facile synthesis by knoevenagel condensation and unusual aggregation-enhanced two-photon absorption properties. Chem Eur J 24:2232–2241
Claessens CG, González-Rodríguez D, Torres T (2002) Subphthalocyanines: singular nonplanar aromatic compoundssynthesis, reactivity, and physical properties. Chem Rev 102:835–853
Claessens CG, González-Rodríguez D, Rodríguez-Morgade MS, Medina A, Torres T (2014) Subphthalocyanines, subporphyrazines, and subporphyrins: singular nonplanar aromatic systems. Chem Rev 114:2192–2277
Clarke RG, Hall MJ (2019) Chapter Three—Recent developments in the synthesis of the BODIPY dyes. Adv Heterocycl Chem 128:181–261
Crossley DL, Cade IA, Clark ER, Escande A, Humphries MJ, King SM, Vitorica-Yrezabal I, Ingleson MJ, Turner ML (2015) Enhancing electron affinity and tuning band gap in donor–acceptor organic semiconductors by benzothiadiazole directed C-H borylation. Chem Sci 6:5144–5151
Cui Y, Liu Q-D, Bai D-R, Jia W-L, Tao Y, Wang S (2005) Organoboron compounds with an 8-hydroxyquinolato chelate and its derivatives: substituent effects on structures and luminescence. Inorg Chem 44:601–609
Cui A, Peng X, Fan J, Chen X, Wu Y, Guo B (2007) Synthesis, spectral properties and photostability of novel boron–dipyrromethene dyes. J Photochem Photobiol A 186:85–92
Culzoni MJ, de la Peña AM, Machuca A, Goicoechea HC, Babiano R (2013) Rhodamine and BODIPY chemodosimeters and chemosensors for the detection of Hg^{2+}, based on fluorescence enhancement effects. Anal Methods 5:30–49
Curiel D, Más-Montoya M, Usea L, Espinosa A, Orenes RA, Molina P (2012) Indolocarbazole-based ligands for ladder-type four-coordinate boron complexes. Org Lett 14:3360–3363
Daly ML, DeRosa CA, Kerr C, Morris WA, Fraser CL (2016) Blue thermally activated delayed fluorescence from a biphenyl difluoroboron β-diketonate. RSC Adv 6:81631–81635

Das BC, Thapa P, Karki R, Schinke C, Das S, Kambhampati S, Banerjee SK, Van Veldhuizen P, Verma A, Weiss LM, Evans T (2013) Boron chemicals in diagnosis and therapeutics. Future Med Chem 5:653–676

Deniz E, Isbasar GC, Bozdemir OA, Yildirim LT, Siemiarczuk A, Akkaya EU (2008) Bidirectional switching of near IR emitting boradiazaindacene fluorophores. Org Lett 10:3401–3403

Descalzo AB, Xu H-J, Xue Z-L, Hoffmann K, Shen Z, Weller MG, You XZ, Rurack K (2008) Phenanthrene-fused boron–dipyrromethenes as bright long-wavelength fluorophores. Org Lett 10:1581–1584

Diana P, Martorana A, Barraja P, Montalbano A, Carbone A, Cirrincione G (2011) Nucleophilic substitutions in the isoindole series as a valuable tool to synthesize derivatives with antitumor activity. Tetrahedron Lett 67:2072–2080

Díaz-Moscoso A, Emond E, Hughes DL, Tizzard GJ, Coles SJ, Cammidge AN (2014) Synthesis of a class of core-modified aza-BODIPY derivatives. J Org Chem 79:8932–8936

Didukh NO, Yakubovskyi VP, Zatsikha YV, Rohde GT, Nemykin VN, Kovtun YP (2019) Flexible BODIPY platform that offers an unexpected regioselective heterocyclization reaction toward preparation of 2-pyridone[*a*]-fused BODIPYs. J Org Chem 84:2133–2147

Dou C, Long X, Ding Z, Xie Z, Liu J, Wang L (2016) An electron-deficient building block based on the B $\leftarrow$ N unit: an electron acceptor for all-polymer solar cells. Angew Chem Int Ed 55:1436–1440

Douglass JE, Barelski PM, Blankenship RM (1973) Diazaboracyclic cations. III. A homomorph of 9,10-dihydroanthracene. J Heterocycl Chem 10:255–257

Duran-Sampedro G, Esnal I, Agarrabeitia AR, Prieto JB, Cerdán L, García-Moreno I, Costela A, Lopez-Arbeloa I, Ortiz M (2014) First highly efficient and photostable E and C derivatives of 4,4-difluoro-4-bora-3a,4a-diaza-s-indacene (BODIPY) as dye lasers in the liquid phase, thin films, and solid-state rods. Chem Eur J 20:2646–2653

Durán-Sampedro G, Agarrabeitia AR, Cerdán L, Pérez-Ojeda ME, Costela A, García-Moreno I, Esnal I, Bañuelos J, Arbeloa IL, Ortiz MJ (2013) Carboxylates versus fluorines: boosting the emission properties of commercial BODIPYs in liquid and solid media. Adv Funct Mater 23:4195–4205

Durantini AM, Heredia DA, Durantini JE, Durantini EN (2018) BODIPYs to the rescue: potential applications in photodynamic inactivation. Eur J Med Chem 144:651–661

Entwistle CD, Marder TB (2002) Boron chemistry lights the way: optical properties of molecular and polymeric systems. Angew Chem Int Ed 41:2927–2931

Esnal I, Valois-Escamilla I, Gómez-Durán CFA, Urías-Benavides A, Betancourt-Mendiola ML, López-Arbeloa I, Bañuelos J, García-Moreno I, Costela A, Peña-Cabrera E (2013) Blue-to-orange color-tunable laser emission from tailored boron-dipyrromethene dyes. ChemPhysChem 14:4134–4142

Fan J, Hu M, Zhan P, Peng X (2013) Energy transfer cassettes based on organic fluorophores: construction and applications in ratiometric sensing. Chem Soc Rev 42:29–43

Fei Q, Gu X, Liu Y, Shi B, Liu H, Xu G, Li C, Shi P, Zhao C (2017) Near-infrared fluorescent dyes with large Stokes shifts: light generation in BODIPYs undergoing excited state intramolecular proton transfer. Org Biomol Chem 15:4072–4076

Filatov MA, Cheprakov AV, Beletskaya IP (2007) A facile and reliable method for the synthesis of tetrabenzoporphyrin from 4,7-dihydroisoindole. Eur J Org Chem 3468–3475

Filatov MA, Lebedev AY, Mukhin SN, Vinogradov SA, Cheprakov AV (2010) π-extended dipyrrins capable of highly fluorogenic complexation with metal ions. J Am Chem Soc 132:9552–9554

Fischer GM, Ehlers AP, Zumbusch A, Daltrozzo E (2007) Near-infrared dyes and fluorophores based on diketopyrrolopyrroles. Angew Chem Int Ed 46:3750–3753

Fischer GM, Isomäki-Krondahl M, Göttker-Schnetmann I, Daltrozzo E, Zumbusch A (2009) Pyrrolopyrrole cyanine dyes: a new class of near-infrared dyes and fluorophores. Chem Eur J 15:4857–4864

Fischer GM, Jüngst C, Isomäki-Krondahl M, Gauss D, Möller HM, Daltrozzo E, Zumbusch A (2010) Asymmetric PPCys: strongly fluorescing NIR labels. Chem Commun 5289–5291

Fischer GM, Klein MK, Daltrozzo E, Zumbusch A (2011a) Pyrrolopyrrole cyanines: effect of substituents on optical properties. Eur J Org Chem 3421–3429
Fischer GM, Daltrozzo E, Zumbusch A (2011b) Selective NIR chromophores: bis(pyrrolopyrrole) cyanines. Angew Chem Int Ed 50:1406–1409
Frath D, Azizi S, Ulrich G, Retailleau P, Ziessel R (2011) Facile synthesis of highly fluorescent *boranil* complexes. Org Lett 13:3414–3417
Frath D, Azizi S, Ulrich G, Ziessel R (2012) Chemistry on boranils: an entry to functionalized fluorescent dyes. Org Lett 14:4774–4777
Frath D, Poirel A, Ulrich G, De Nicola A, Ziessel R (2013) Fluorescent boron(iii) iminocoumarins (Boricos). Chem Commun 49:4908–4910
Frath D, Massue J, Ulrich G, Ziessel R (2014) Luminescent materials: locking π-conjugated and heterocyclic ligands with boron(III). Angew Chem Int Ed 53:2290–2310
Frath D, Benelhadj K, Munch M, Massue J, Ulrich G (2016) Polyanils and polyboranils: synthesis, optical properties, and aggregation-induced emission. J Org Chem 81:9658–9668
Fulford MV, Jaidka D, Paton AS, Morse GE, Brisson ERL, Lough AJ, Bender TP (2012) Crystal Structures, reaction rates, and selected physical properties of halo-boronsubphthalocyanines (halo = fluoride, chloride, and bromide). J Chem Eng Data 57:2756–2765
Gao H, Xu D, Liu X, Han A, Zhou L, Zhang C, Li Z, Dang J (2017) Tetraphenylethene-based β-diketonate boron complex: efficient aggregation-induced emission and high contrast mechanofluorochromism. Dyes Pigm 139:157–165
Gautam R, Petritis SJ, Tomat E (2019) Redox-switchable cyan fluorescence of a BODIPY analog inspired by propentdyopent pigments. Eur J Inorg Chem 68–72
Ge Y, O'Shea DF (2016) Azadipyrromethenes: from traditional dye chemistry to leading edge applications. Chem Soc Rev 45:3846–3864
Glembockyte V, Frenette M, Mottillo C, Durantini AM, Gostick J, Štrukil V, Friščić T, Cosa G (2018) Highly photostable and fluorescent microporous solids prepared via solid-state entrapment of boron dipyrromethene dyes in a nascent metal–organic framework. J Am Chem Soc 140:16882–16887
Glotzbach C, Kauscher U, Voskuhl J, Kehr NS, Stuart MCA, Fröhlich R, Galla HJ, Ravoo BJ, Nagura K, Saito S, Yamaguchi S, Würthwein E-U (2013) Fluorescent modular boron systems based on NNN- and ONO-tridentate ligands: self-assembly and cell imaging. J Org Chem 78:4410–4418
Glotzbach C, Gödeke N, Fröhlich R, Daniliuc C-G, Saito S, Yamaguchi S, Würthwein E-U (2015) Fluorescent benzene-centered mono-, bis- and tris-triazapentadiene–boron complexes. Dalton Trans 44:9659–9671
Gobo Y, Yamamura M, Nakamura T, Nabeshima T (2016) Synthesis and chiroptical properties of a ring-fused BODIPY with a skewed chiral π skeleton. Org Lett 18:2719–2721
Golden JH, Facendola JW, Sylvinson MRD, Baez CQ, Djurovich PI, Thompson ME (2017) Boron dipyridylmethene (DIPYR) dyes: shedding light on pyridine-based chromophores. J Org Chem 82:7215–7222
Golden JH, Estergreen L, Porter T, Tadle AC, Sylvinson MRD, Facendola JW, Kubiak CP, Bradforth SE, Thompson ME (2018) Symmetry-breaking charge transfer in boron dipyridylmethene (DIPYR) dimmers. ACS Appl Energy Mater 1:1083–1095
Gon M, Tanaka K, Chujo Y (2018) A highly efficient near-infrared-emissive copolymer with a N = N double-bond π-conjugated system based on a fused azobenzene-boron complex. Angew Chem Int Ed 57:6546–6551
Goud TV, Tutary A, Biellmann J-F (2006) Synthesis of 8-heteroatom-substituted 4,4-difluoro-4-bora- 3a,4a-diaza-s-indacene dyes (BODIPY). Tetrahedron 62:5084–5091
Goze C, Ulrich G, Mallon LJ, Allen BD, Harriman A, Ziessel R (2006) Synthesis and photophysical properties of borondipyrromethene dyes bearing aryl substituents at the boron center. J Am Chem Soc 128:10231–10239
Goze C, Ulrich G, Ziessel R (2007) Tetrahedral boron chemistry for the preparation of highly efficient "cascatelle" devices. J Org Chem 72:313–322

Graser M, Kopacka H, Wurst K, Ruetz M, Kreutz CR, Müller T, Hirtenlehner C, Monkowius U, Knör G, Bildstein B (2013) Efficient fluorophores based on pyridyl-enolato and enamido difluoroboron complexes: simple alternatives to boron-dipyrromethene (bodipy) dyes. Inorg Chim Acta 405:116–120

Griesbeck S, Zhang Z, Gutmann M, Lühmann T, Edkins RM, Clermont G, Lazar AN, Haehnel M, Edkins K, Eichhorn A, Blanchard-Desce M, Meinel L, Marder TB (2016) Water-Soluble triarylborane chromophores for one- and two-photon excited fluorescence imaging of mitochondria in cells. Chem Eur J 22:14701–14706

Gu L, Liu R, Shi H, Wang Q, Song G, Zhu X, Yuan S, Zhu H (2016) Synthesis, luminescent properties of aza-boron-diquinomethene difluoride complexes and their application for fluorescent security inks. J Fluoresc 26:407–412

Guilleme J, González-Rodríguez D, Torres T (2011) Triflate-subphthalocyanines: versatile, reactive intermediates for axial functionalization at the boron atom. Angew Chem Int Ed 50:3506–3509

Guliyev R, Ozturk S, Sahin E, Akkaya EU (2012) Expanded bodipy dyes: anion sensing using a bodipy analog with an additional difluoroboron bridge. Org Lett 14:1528–1531

Hachiya S, Inagaki T, Hashizume D, Maki S, Niwa H, Hirano T (2010) Synthesis and fluorescence properties of difluoro[amidopyrazinato-O, N]boron derivatives: a new boron-containing fluorophore. Tetrahedron Lett 51:1613–1615

Han J, Gonzalez O, Aguilar-Aguilar A, Peña-Cabrera E, Burgess K (2009) 3- and 5-functionalized BODIPYs via the Liebeskind-Srogl reaction. Org Biomol Chem 7:34–36

Hao Q, Yu S, Li S, Chen J, Zeng Y, Yu T, Yang G, Li Y (2014) Locked planarity: a strategy for tailoring ladder-type π-conjugated anilido-pyridine boron difluorides. J Org Chem 79:459–464

Hapuarachchige S, Montaño G, Ramesh C, Rodriguez D, Henson LH, Williams CC, Kadavakkollu S, Johnson DL, Shuster CB, Arterburn JB (2011) Design and synthesis of a new class of membrane-permeable triazaborolopyridinium fluorescent probes. J Am Chem Soc 133:6780–6790

Hayashi Y, Yamaguchi S, Cha WY, Kim D, Shinokubo H (2011a) Synthesis of directly connected BODIPY oligomers through Suzuki-Miyaura coupling. Org Lett 13:2992–2995

Hayashi S, Tsurumaki E, Inokuma Y, Kim P, Sung YM, Kim D, Osuka A (2011b) Synthesis and properties of boron(III)-coordinated subbacteriochlorins. J Am Chem Soc 133:4254–4256

Hayashi Y, Obata N, Tamaru M, Yamaguchi S, Matsuo Y, Saeki A, Seki S, Kureishi Y, Saito S, Yamaguchi S, Shinokubo H (2012) Facile synthesis of biphenyl-fused BODIPY and its property. Org Lett 14:866–869

Heyer E, Retailleau P, Ziessel R (2014) α-fused dithienyl BODIPYs synthesized by oxidative ring closure. Org Lett 16:2330–2333

Higashino T, Rodríuez-Morgade MS, Osuka A, Torres T (2013) Peripheral arylation of subporphyrazines. Chem Eur J 19:10353–10359

Hinkeldey B, Schmitt A, Jung G (2008) Comparative photostability studies of BODIPY and fluorescein dyes by using fluorescence correlation spectroscopy. ChemPhysChem 9:2019–2027

Ho D, Ozdemir R, Kim H, Earmme T, Usta H, Kim C (2019) BODIPY-based semiconducting materials for organic bulk heterojunction photovoltaics and thin-film transistors. ChemPlusChem 84:18–37

Hohaus H, Umland F (1969) Borchelate und Bormetallchelate, I. Borchelate mit Chelatbildnern der Pyridin- und Chinolin-Reihe und ihren *N*-Oxiden. Chem Ber 102:4025–4031

Hu R, Lager E, Aguilar-Aguilar A, Liu J, Lam JWY, Sung HHY, Williams ID, Zhong Y, Wong KS, Peña-Cabrera E, Tang BZ (2009) Twisted intramolecular charge transfer and aggregation-induced emission of BODIPY derivatives. J Phys Chem C 113:15845–15853

Hu J, He Z, Wang Z, Li X, You J, Gao G (2013) A simple approach to aggregation-induced emission in difluoroboron dibenzoylmethane derivatives. Tetrahedron Lett 54:4167–4170

Huang L, Yu X, Wu W, Zhao J (2012) Styryl Bodipy-C_{60} dyads as efficient heavy-atom-free organic triplet photosensitizers. Org Lett 14:2594–2597

Huang S, Yang C, Huang J, Wang X, Wang M (2018a) Near-infrared fluorescent pyrrolopyrrole cyanine derivatives and colloidal nanoparticles with tunable optical properties for in vivo bioimaging. Dyes Pigm 154:269–274

Huang S, Liu W, Huang J, Wang X, Yang C, Bohra H, Liu Q, Wang M (2018b) Theranostic colloidal nanoparticles of pyrrolopyrrole cyanine derivatives for simultaneous near-infrared fluorescence cancer imaging and photothermal therapy. ACS Appl Bio Mater 1:1109–1117

Huang W, Zhang X, Chen B, Miao H, Trindle CO, Wang Y, Luo Y, Zhang G (2019) Boosting the triplet activity of heavy-atom-free difluoroboron dibenzoylmethane via sp^3 oxygen-bridged electron donors. Chem Commun 55:67–70

Hudson ZM, Wang S (2009) Impact of donor–acceptor geometry and metal chelation on photophysical properties and applications of triarylboranes. Acc Chem Res 42:1584–1596

Inokuma Y, Kwon JH, Ahn TK, Yoo M-C, Kim D, Osuka A (2006) Tribenzosubporphines: synthesis and characterization. Angew Chem Int Ed 45:961–964

Inokuma Y, Yoon ZS, Kim D, Osuka A (2007) meso-Aryl-substituted subporphyrins: synthesis, structures, and large substituent effects on their electronic properties. J Am Chem Soc 129:4747–4761

Ito S, Murashima T, Ono N (1997) A new synthesis of pyrroles fused with polycyclic skeletons. J Chem Soc Perkin Trans 1:3161–3165

Ito F, Nagai T, Ono Y, Yamaguchi K, Furuta H, Nagamura T (2007) Photophysical properties of 2-picolinoylpyrrole boron complex in solutions. Chem Phys Lett 435:283–288

Jean-Gérard L, Vasseur W, Scherninski F, Andrioletti B (2018) Recent advances in the synthesis of [*a*]-benzo-fused BODIPY fluorophores. Chem Commun 54:12914–12929

Jędrzejewska B, Zakrzewska A, Mlostoń G, Budzák Š, Mroczyńska K, Grabarz AM, Kaczorowska MA, Jacquemin D, Ośiałowski B (2016) Synthesis and photophysical properties of novel donor-acceptor N-(pyridin-2-yl)-substituted benzo(thio)amides and their difluoroboranyl derivatives. J Phys Chem A 120:4116–4123

Ji S, Ge J, Escudero D, Wang Z, Zhao J, Jacquemin D (2015) Molecular structure-intersystem crossing relationship of heavy-atom-free BODIPY triplet photosensitizers. J Org Chem 80:5958–5963

Jiang X-D, Zhang J, Furuyama T, Zhao W (2012) Development of mono- and di-AcO substituted BODIPYs on the boron center. Org Lett 14:248–251

Jiang X-D, Su Y, Yue S, Li C, Yu H, Zhang H, Sun C-L, Xiao L-J (2015) Synthesis of mono-(*p*-dimethylamino)styryl-containing BOPHY dye for a turn-on pH sensor. RSC Adv 5:16735–16739

Jiang X-D, Fang T, Liu X, Xi D (2017) Synthesis of meso-CF-substituted BODIPY Compounds with redshifted absorption. Eur J Org Chem 5074–5079

Jiang X-D, Liu X, Fang T, Sun C, Xiao L (2018) Synthesis and photophysical properties of long wavelength absorbing BODIPY/aza-BODIPY bearing a five-membered ring. Tetrahedron Lett 59:546–549

Jiao L, Yu C, Liu M, Wu Y, Cong K, Meng T, Wang Y, Hao E (2010) Synthesis and functionalization of asymmetrical benzo-fused BODIPY dyes. J Org Chem 75:6035–6038

Jiao L, Pang W, Zhou J, Wei Y, Mu X, Bai G, Hao E (2011a) Regioselective stepwise bromination of boron dipyrromethene (BODIPY) dyes. J Org Chem 76:9988–9996

Jiao C, Huang K-W, Wu J (2011b) Perylene-fused BODIPY dye with near-IR absorption/emission and high photostability. Org Lett 13:632–635

Jiao C, Zhu L, Wu J (2011c) BODIPY–fused porphyrins as soluble and stable near–IR dyes. Chem Eur J 17:6610–6614

Jiao L, Yu C, Wang J, Briggs EA, Besley NA, Robinson D, Ruedas-Rama MJ, Orte A, Crovetto L, Talavera EM, Alvarez-Pez JM, Van der Auweraer M, Boens N (2015) Unusual spectroscopic and photophysical properties of meso-tert-butylBODIPY in comparison to related alkylated BODIPY dyes. RSC Adv 5:89375–89388

Jiménez C, Farfan N, Romero-Avila M, Santillan R, Malfant I, Lacroix PG (2015) Light induced nonlinear optical switch in boronated chromophores: a theoretical search towards high contrast switches in the azobenzene series. J Organomet Chem 799–800:215–222

Jones G II, Kumar S, Klueva O, Pacheco D (2003) Photoinduced electron transfer for pyrromethene dyes. J Phys Chem A 107:8429–8434

Kage Y, Mori S, Ide M, Saeki A, Furuta H, Shimizu S (2018) Blackening of aza-BODIPY analogues by simple dimerization: panchromatic absorption of a pyrrolopyrrole aza-BODIPY dimer. Mater Chem Front 2:112–120

Kaim W, Shultz A (1984) p-Phenylenediboranes: mirror images of p-phenylenediamines? Angew Chem Int Ed 23:615–616

Kamada K, Namikawa T, Senatore S, Matthews C, Lenne P-F, Maury O, Andraud C, Ponce-Vargas M, Le Guennic B, Jacquemin D, Agbo P, An DD, Gauny SS, Liu X, Abergel RJ, Fages F, D′Aléo A (2016) Boron difluoride curcuminoid fluorophores with enhanced two-photon excited fluorescence emission and versatile living-cell imaging properties. Chem Eur J 22:5219–5232

Kamkaew A, Lim SH, Lee HB, Kiew LV, Chung LY, Burgess K (2013) BODIPY dyes in photodynamic therapy. Chem Soc Rev 42:77–88

Kang HC, Haugland RP (1995) US 5433896

Kee HL, Kirmaier C, Yu L, Thamyongkit P, Youngblood WJ, Calder ME, Ramos L, Noll BC, Bocian DF, Scheidt WR, Birge RR, Lindsey JS, Holten D (2005) Structural control of the photodynamics of boron–dipyrrin complexes. J Phys Chem B 109:20433–20443

Killoran J, O'Shea DF (2006) Impact of a conformationally restricted receptor on the BF_2 chelated azadipyrromethene fluorosensing platform. Chem Commun 1503–1505

Killoran J, Allen L, Gallagher JF, Gallagher WM, O'Shea DF (2002) Synthesis of BF_2 chelates of tetraarylazadipyrromethenes and evidence for their photodynamic therapeutic behavior. Chem Commun 1862–1863

Kim H, Burghart A, Welch MB, Reibenspies J, Burgess K (1999) Synthesis and spectroscopic properties of a new 4-bora-3a,4a-diaza-s-indacene (BODIPY) dye. Chem Commun 1889–1890

Kim HN, Ren WX, Kim JS, Yoon J (2012) Fluorescent and colorimetric sensors for detection of lead, cadmium, and mercury ions. Chem Soc Rev 41:3210–3244

Kim S, Bouffard J, Kim Y (2015) Tailoring the solid-state fluorescence emission of BODIPY dyes by meso substitution. Chem Eur J 21:17459–17465

Klfout H, Stewart A, Elkhalifa M, He H (2017) BODIPYs for dye-sensitized solar cells. ACS Appl Mater Interfaces 9:39873–39889

Kobayashi N, Nonomura T (2002) First observation of the circular dichroism spectra of chiral subphthalocyanines with C_3 symmetry. Tetrahedron Lett 43:4253–4255

Kobayashi N, Kondo R, Nakajima S, Osa T (1990) New route to unsymmetrical phthalocyanine analogs by the use of structurally distorted subphthalocyanines. J Am Chem Soc 112:9640–9641

Kobayashi N, Takeuchi Y, Matsuda A (2007) meso-Aryl subporphyrins. Angew Chem Int Ed 46:758–760

Koch M, Perumal K, Blacque O, Garg JA, Saiganesh R, Kabilan S, Balasubramanian KK, Venkatesan K (2014) Metal-free triplet phosphors with high emission efficiency and high tenability. Angew Chem Int Ed 53:6378–6382

Köhler T, Hodgson MC, Seidel D, Veauthier JM, Meyer S, Lynch V, Boyd PDW, Brothers PJ, Sessler JL (2004) Octaethylporphyrin and expanded porphyrin complexes containing coordinated BF_2 groups. Chem Commun 1060–1061

Kolemen S, Akkaya EU (2018) Reaction-based BODIPY probes for selective bio-imaging. Coord Chem Rev 354:121–134

Komatsu T, Oushiki D, Takeda A, Miyamura M, Ueno T, Terai T, Hanaoka K, Urano Y, Mineno T, Nagano T (2011) Rational design of boron dipyrromethene (BODIPY)-based photobleaching-resistant fluorophores applicable to a protein dynamics study. Chem Commun 47:10055–10057

Kondakova ME, Deaton JC, Pawlik TD, Giesen DJ, Kondakov DY, Young RH, Royster TL, Comfort DL, Shore JD (2010) Highly efficient fluorescent-phosphorescent triplet-harvesting hybrid organic light-emitting diodes. J Appl Phys 107:014515

Kowada T, Yamaguchi S, Ohe K (2010) Highly fluorescent BODIPY dyes modulated with spirofluorene moieties. Org Lett 12:296–299

Kubota Y, Tsuzuki T, Funabiki K, Ebihara M, Matsui M (2010a) Synthesis and fluorescence properties of a pyridomethene–BF2 complex. Org Lett 12:4010–4013

Kubota Y, Uehara J, Funabiki K, Ebihara M, Matsui M (2010b) Strategy for the increasing the solid-state fluorescence intensity of pyrromethene–BF_2 complexes. Tetrahedron Lett 51:6195–6198

Kubota Y, Hara H, Tanaka S, Funabiki K, Matsui M (2011) Synthesis and fluorescence properties of novel pyrazine-boron complexes bearing a β-iminoketone ligand. Org Lett 13:6544–6547

Kubota Y, Tanaka S, Funabiki K, Matsui M (2012) Synthesis and fluorescence properties of thiazole-boron complexes bearing a β-ketoiminate ligand. Org Lett 14:4682–4685

Kubota Y, Ozaki Y, Funabiki K, Matsui M (2013) Synthesis and fluorescence properties of pyrimidine mono- and bisboron complexes. J Org Chem 78:7058–7067

Kubota Y, Sakuma Y, Funabiki K, Matsui M (2014) Solvatochromic fluorescence properties of pyrazine-boron complex bearing a β-iminoenolate ligand. J Phys Chem A 118:8717–8729

Kubota Y, Kasatani H, Takai K, Funabiki K, Matsui M (2015a) Strategy to enhance solid-state fluorescence and aggregation-induced emission enhancement effect in pyrimidine boron complexes. Dalton Trans 44:3326–3341

Kubota Y, Niwa T, Jin J, Funabiki K, Matsui M (2015b) Synthesis, absorption, and electrochemical properties of quinoid-type bisboron complexes with highly symmetrical structures. Org Lett 17:3174–3177

Kubota Y, Kasatani K, Niwa T, Sato H, Funabiki K, Matsui M (2016) Chem Eur J 22:1816–1824

Kubota Y, Tsukamoto M, Ohnishi K, Jin J, Funabiki K, Matsui M (2017) Synthesis and fluorescence properties of novel squarylium–boron complexes. Org Chem Front 4:1522–1527

Kubota Y, Kimura K, Jin J, Manseki K, Funabiki K, Matsui M (2019) Synthesis of near-infrared absorbing and fluorescing thiophene-fused BODIPY dyes with strong electron-donating groups and their application in dye-sensitised solar cells. New J Chem 43:1156–1165

Kulyk B, Taboukhat S, Akdas-Kilig H, Fillaut J-L, Boughaleb Y, Sahraoui B (2016) Nonlinear refraction and absorption activity of dimethylaminostyryl substituted BODIPY dyes. RSC Adv 6:84854–84859

Lakshmi V, Rao MR, Ravikanth M (2015) Halogenated boron-dipyrromethenes: synthesis, properties and applications. Biomol Chem 13:2501–2517

Lakshmi V, Sharma R, Ravikanth M (2016) Functionalized boron-dipyrromethenes and their applications. Rep Org Chem 6:1–24

Lee J-S, Kim HK, Feng S, Vendrell M, Chang Y-T (2011) Accelerating fluorescent sensor discovery: unbiased screening of a diversity-oriented BODIPY library. Chem Commun 47:2339–2341

Lee B, Park BG, Cho W, Lee HY, Olasz A, Chen C-H, Park SB, Lee D (2016) BOIMPY: fluorescent boron complexes with tunable and environment-responsive light-emitting properties. Chem Eur J 22:17321–17328

Lee S-C, Heo J, Woo HC, Lee J-A, Seo YH, Lee C-L, Kim S, Kwon O-P (2018) Fluorescent molecular rotors for viscosity sensors. Chem Eur J 24:13706–13718

Leen V, Braeken E, Luckermans K, Jackers C, Van der Auweraer M, Boens N, Dehaen W (2009) A versatile, modular synthesis of monofunctionalized BODIPY dyes. Chem Commun 4515–4517

Leen V, Miscoria D, Yin S, Filarowski A, Ngongo JM, Van der Auweraer M, Boens N, Dehaen W (2011) 1,7-disubstituted boron dipyrromethene (BODIPY) dyes: synthesis and spectroscopic properties. J Org Chem 76:8168–8176

Leen V, Yuan P, Wang L, Boens N, Dehaen W (2012) Synthesis of meso-halogenated BODIPYs and access to meso-substituted analogues. Org Lett 14:6150–6153

Li Y, Liu Y, Bu W, Guo J, Wang Y (2000) A mixed pyridine–phenol boron complex as an organic electroluminescent material. Chem Commun 1551–1552

Li L, Nguyen B, Burgess K (2008a) Functionalization of the 4,4-difluoro-4-bora-3a,4a-diaza-s-indacene (BODIPY) core. Bioorg Med Chem Lett 18:3112–3116

Li L, Han J, Nguyen B, Burgess K (2008b) Syntheses and spectral properties of functionalized, water-soluble BODIPY derivatives. J Org Chem 73:1963–1970

Li Y, Patrick BO, Dolphin D (2009) Near-infrared absorbing azo dyes: synthesis and X-ray crystallographic and spectral characterization of monoazopyrroles, bisazopyrroles, and a boron–azopyrrole complex. J Org Chem 74:5237–5243

Li H-J, Fu W-F, Li L, Gan X, Mu W-H, Chen W-Q, Duan X-M, Song H-B (2010) Intense one- and two-photon excited fluorescent bis(BF_2) core complex containing a 1,8-naphthyridine derivative. Org Lett 12:2924–2927

Li D, Wang K, Huang S, Qu S, Liu X, Zhu Q, Zhang H, Wang Y (2011) Brightly fluorescent red organic solids bearing boron-bridged π–conjugated skeletons. J Mater Chem 21:15298–15304

Li D, Zhang H, Wang C, Huang S, Guo J, Wang Y (2012) Construction of full-color-tunable and strongly emissive materials by functionalizing a boron-chelate four-ring-fused π-conjugated core. J Mater Chem 22:4319–4328

Li D, Zhang H, Wang Y (2013a) Four-coordinate organoboron compounds for organic light-emitting diodes (OLEDs). Chem Soc Rev 42:8416–8433

Li W, Lin W, Wang J, Guan X (2013b) Phenanthro[9,10-*d*]imidazole-quinoline boron difluoride dyes with solid-state red fluorescence. Org Lett 15:1768–1771

Li Y, Yan J, Xiao S, Zhang B, Lu F, Cheng K, Li D, Feng Y, Zhang N (2016) A novel synthesis of stable 3*H*-pyrrolizine fused diazaborepin. Tetrahedron Lett 57:3226–3230

Li S-Y, Sun Z-B, Zhao C-H (2017a) Charge-transfer emitting triarylborane π-electron systems. Inorg Chem 56:8705–8717

Li T, Meyer T, Ma Z, Benduhn J, Körner C, Zeika O, Vandewal K, Leo K (2017b) Small molecule near-infrared boron dipyrromethene donors for organic tandem solar cells. J Am Chem Soc 139:13636–13639

Li L, Wang L, Tang H, Cao D (2017c) A facile synthesis of novel near-infrared pyrrolopyrrole aza-BODIPY luminogens with aggregation-enhanced emission characteristics. Chem Commun 53:8352–8355

Li T, Benduhn J, Li Y, Jaiser F, Spoltore D, Zeika O, Ma Z, Neher D, Vandewal K, Leo K (2018a) Boron dipyrromethene (BODIPY) with meso-perfluorinated alkyl substituents as near infrared donors in organic solar cells. J Mater Chem A 6:18583–18591

Li Y, Qiao Z, Li T, Zeika O, Leo K (2018b) Highly efficient deep-red-to near-infrared-absorbing and emissive benzo/naphtho[b]furan-fused boron dipyrromethene (BODIPY). ChemPhotoChem 2:1017–1021

Liao C-W, Rao MR, Sun S-S (2015) Structural diversity of new solid-state luminophores based on quinoxaline-β-ketoiminate boron difluoride complexes with remarkable fluorescence switching properties. Chem Commun 51:2656–2659

Lin Y-D, Chow TJ (2013) A pyridomethene–BF_2 complex-based chemosensor for detection of hydrazine. RSC Adv 3:17924–17929

Lin Y-D, Ke B-Y, Chang YJ, Chou P-T, Liau K-L, Liu C-Y, Chow TJ (2015) Pyridomethene–BF_2 complex/phenothiazine hybrid sensitizer with high molar extinction coefficient for efficient, sensitized solar cells. J Mater Chem A 3:16831–16842

Liu D, Mudadu MS, Thummel R, Tao Y, Wang S (2005) From blue to red: syntheses, structures, electronic and electroluminescent properties of tunable luminescent N, N chelate boron complexes. Adv Funct Mater 15:143–154

Liu X, Ren Y, Xia H, Fan X, Mu Y (2010) Synthesis, structures, photoluminescent and electroluminescent properties of boron complexes with anilido-imine ligands. Inorg Chim Acta 363:1441–1447

Liu H, Mack J, Guo Q, Lu H, Kobayashi N, Shen Z (2011) A selective colorimetric and fluorometric ammonium ion sensor based on the H-aggregation of an aza-BODIPY with fused pyrazine rings. Chem Commun 47:12092–12094

Liu Q, Wang X, Yan H, Wu Y, Li Z, Gong S, Liu P, Liu Z (2015) Benzothiazole-enamide-based BF_2 complexes: luminophores exhibiting aggregation-induced emission, tunable emission and highly efficient solid-state emission. J Mater Chem C 3:2953–2959

Liu K, Lalancette RA, Jäkle F (2017) B-N Lewis pair functionalization of anthracene: structural dynamics, optoelectronic properties, and O_2 sensitization. J Am Chem Soc 139:18170–18173

Liu Y, Niu L-Y, Liu X-L, Chen P-Z, Yao Y-S, Chen Y-Z, Yang Q-Z (2018) Synthesis of N, O, B-chelated dipyrromethenes through an unexpected intramolecular cyclisation: enhanced near-infrared emission in the aggregate/solid state. Chem Eur J 24:13549–13555

Liu K, Lalancette RA, Jäkle F (2019) Tuning the structure and electronic properties of B-N fused dipyridylanthracene and implications on the self-sensitized reactivity with singlet oxygen. J Am Chem Soc 141:7453–7462

Loudet A, Burgess K (2007) BODIPY dyes and their derivatives: syntheses and spectroscopic properties. Chem Rev 107:4891–4932

Loudet A, Bandichhor R, Wu L, Burgess K (2008) Functionalized BF_2 chelated azadipyrromethene dyes. Tetrahedron 64:3642–3654

Lower SK, El-Sayed MA (1966) The triplet state and molecular electronic processes in organic molecules. Chem Rev 66:199–241

Lu H, Shimizu S, Mack J, Shen Z, Kobayashi N (2011) Synthesis and spectroscopic properties of fused-ring-expanded aza-boradiazaindacenes. Chem Asian J 6:1026–1037

Lu H, Wang Q, Gai L, Li Z, Deng Y, Xiao X, Lai G, Shen Z (2012) Tuning the solid-state luminescence of BODIPY derivatives with bulky arylsilyl groups: synthesis and spectroscopic properties. Chem Eur J 18:7852–7861

Lu H, Mack J, Yanga Y, Shen Z (2014) Structural modification strategies for the rational design of red/NIR region BODIPYs. Chem Soc Rev 43:4778–4823

Lu H, Mack J, Nyokong T, Kobayashi N, Shen Z (2016) Optically active BODIPYs. Coord Chem Rev 318:1–15

Lugovik KI, Eltyshev AK, Suntsova PO, Slepukhin PA, Benassi E, Belskaya NP (2018) Highlights on the road towards highly emitting solid-state luminophores: two classes of thiazole-based organoboron fluorophores with the AIEE/AIE effect. Chem Asian J 13:311–324

Lundrigan T, Thompson A (2013) Conversion of *F*-BODIPYs to *Cl*-BODIPYs: enhancing the reactivity of *F*-BODIPYs. J Org Chem 78:757–761

Lundrigan T, Baker AEG, Longobardi LE, Wood TE, Smithen DA, Crawford SM, Cameron TS, Thompson A (2012a) An improved method for the synthesis of F-BODIPYs from dipyrrins and bis(dipyrrin)s. Org Lett 14:2158–2161

Lundrigan T, Crawford SM, Cameron TS, Thompson A (2012b) *Cl*-BODIPYs: a BODIPY class enabling facile *B*-substitution. Chem Commun 48:1003–1005

Lundrigan T, Cameron TS, Thompson A (2014) Activation and deprotection of F-BODIPYs using boron trihalides. Chem Commun 50:7028–7031

Luo L, Wu D, Li W, Zhang S, Ma Y, Yan S, You J (2014) Regioselective decarboxylative direct C-H arylation of boron dipyrromethenes (BODIPYs) at 2,6-positions: a facile access to a diversity-oriented BODIPY library. Org Lett 16:6080–6083

Luo H-X, Niu Y, Jin X, Cao X-P, Yao X, Ye X-S (2016) Indolo-quinoline boron difluoride dyes: synthesis and spectroscopic properties. Org Biomol Chem 14:4185–4188

Lv F, Yu Y, Hao E, Yu C, Wang H, Jiao L, Boens N (2018) Copper-catalyzed α-benzylation of BODIPYs via radical-triggered oxidative cross-coupling of two C-H bonds. Chem Commun 54:9059–9062

Lv F, Tang B, Hao E, Liu Q, Wang H, Jiao L (2019) Transition-metal-free regioselective cross-coupling of BODIPYs with thiols. Chem Commun 55:1639–1642

Maar RR, Zhang R, Stephens DG, Ding Z, Gilroy JB (2019) Near-infrared photoluminescence and electrochemiluminescence from a remarkably simple boron difluoride formazanate dye. Angew Chem Int Ed 58:1052–1056

Macedo FP, Gwengo C, Lindeman SV, Smith MD, Gardinier JR (2008) β-diketonate, β-ketoiminate, and β-diiminate complexes of difluoroboron. Eur J Inorg Chem 3200–3211

Maeda H, Haketa Y, Nakanishi T (2007) Aryl-substituted C_3-bridged oligopyrroles as anion receptors for formation of supramolecular organogels. J Am Chem Soc 129:13661–13674

Maeda C, Todaka T, Ueda T, Ema T (2016) Color-tunable solid-state fluorescence emission from carbazole-based BODIPYs. Chem Eur J 22:7508–7513

Maeda C, Todaka T, Ueda T, Ema T (2017) Synthesis of carbazole-based BODIPY dimers showing red fluorescence in the solid state. Org Biomol Chem 15:9283–9287

Maeda C, Nagahata K, Takaishi K, Ema T (2019) Synthesis of chiral carbazole-based BODIPYs showing circularly polarized luminescence. Chem Commun 55:3136–3139

Mao M, Song Q-H (2016) ^{18}F-labeling of arenes and heteroarenes for applications in positron emission tomography. Chem Rec 16:719–733

Mao M, Zhang X, Cao L, Tong Y, Wu G (2015) Design of Bodipy based organic dyes for high-efficient dye-sensitized solar cells employing double electron acceptors. Dyes Pigm 117:28–36

Marks T, Daltrozzo E, Zumbusch A (2014) Azacyanines of the pyrrolopyrrole series. Chem Eur J 20:6494–6504

Más-Montoya M, Usea L, Ferao AE, Montenegro MF, de Arellano CR, Tárraga A, Rodríguez-López JN, Curiel D (2016) Single heteroatom fine-tuning of the emissive properties in organoboron complexes with 7-(azaheteroaryl)indole systems. J Org Chem 81:3296–3302

Matsui A, Umezawa K, Shindo Y, Fujii T, Citterio D, Oka K, Suzuki K (2011) A near-infrared fluorescent calcium probe: a new tool for intracellular multicolour Ca2+ imaging. Chem Commun 47:10407–10409

Meesala Y, Kavala V, Chang H-C, Kuo T-S, Yao C-F, Lee W-Z (2015) Synthesis, structures and electrochemical and photophysical properties of anilido-benzoxazole boron difluoride (ABB) complexes. Dalton Trans 44:1120–1129

Mei J, Leung NLC, Kwok RTK, Lam JWY, Tang BZ (2015) Aggregation-induced emission: together we shine, united we soar! Chem Rev 115:11718–11940

Meller A, Ossko A (1972) Phthalocyaninartige bor-komplexe. Monatsh Chem 103:150–155

Michel BW, Lippert AR, Chang CJ (2012) A reaction-based fluorescent probe for selective imaging of carbon monoxide in living cells using a palladium-mediated carbonylation. J Am Chem Soc 134:15668–15671

Min Y, Dou C, Tian H, Geng Y, Liu J, Wang L (2018) n-type azaacenes containing B $\leftarrow$ N units. Angew Chem Int Ed 57:2000–2004

Min Y, Dou C, Tian H, Liu J, Wang L (2019) A disk-type polyarene containing four B $\leftarrow$ N units. Chem Commun 55:3638–3641

Mirloup A, Huaulmé Q, Leclerc N, Lévêque P, Heiser T, Retailleau P, Ziessel R (2015) Thienyl-BOPHY dyes as promising templates for bulk heterojunction solar cells. Chem Commun 51:14742–14745

Monsma FJ, Barton AC, Kang HC, Brassard DL, Haughland RP, Sibley DR (1989) Characterization of novel fluorescent ligands with high affinity for D_1 and D_2 dopaminergic receptors. J Neurochem 52:1641–1644

More AB, Mula S, Thakare S, Sekar N, Ray AK, Chattopadhyay S (2014) Masking and demasking strategies for the BF_2–BODIPYs as a tool for BODIPY fluorophores. J Org Chem 79:10981–10987

Morgan GT, Tunstall RB (1924) CCLIII.—Researches on residual affinity and coordination. Part XXI. Boron β-diketone difluorides. J Chem Soc Trans 125:1963–1967

Moriya K, Shimada R, Ono K (2019) Difluoroboron chelation to quinacridonequinone: a synthetic method for air-sensitive 6,13-dihydroxyquinacridone via boron complexes. Chem Asian J 14:1452–1456

Mula S, Ray AK, Banerjee M, Chaudhuri T, Dasgupta K, Chattopadhyay S (2008) Design and development of a new pyrromethene dye with improved photostability and lasing efficiency: theoretical rationalization of photophysical and photochemical properties. J Org Chem 73:2146–2154

Mula S, Leclerc N, Lévêque P, Retailleau P, Ulrich G (2018) Synthesis of indolo[3,2-*b*]carbazole-based boron complexes with tunable photophysical and electrochemical properties. J Org Chem 83:14406–14418

Murale DP, Lee KM, Kim K, Churchill DG (2011) Facile "one pot" route to the novel benzazulene-type dye class: asymmetric, derivatizable, 5-7-6 fused ring puckered half BODIPY design. Chem Commun 47:12512–12514

Nagata Y, Chujo Y (2008) Main-chain-type N, N'-chelate organoboron aminoquinolate polymers: synthesis, luminescence, and energy transfer behavior. Macromolecules 41:3488–3492

Nakamura M, Tahara H, Takahashi K, Nagata T, Uoyama H, Kuzuhara D, Mori S, Okujima T, Yamada H, Uno H (2012) π-fused bis-BODIPY as a candidate for NIR dyes. Org Biomol Chem 10:6840–6849

Nawn G, Oakley SR, Majewski MB, McDonald R, Patrick BO, Hicks RG (2013) Redox-active, near-infrared dyes based on 'Nindigo' (indigo-N, N'-diarylimine) boron chelate complexes. Chem Sci 4:612–621

Nepomnyashchii AB, Cho S, Rossky PJ, Bard AJ (2010) Dependence of electrochemical and electrogenerated chemiluminescence properties on the structure of BODIPY dyes. Unusually large separation between sequential electron transfers. J Am Chem Soc 132:17550–17559

Nepomnyashchii AB, Bröring M, Ahrens J, Bard AJ (2011) Synthesis, photophysical, electrochemical, and electrogenerated chemiluminescence studies. Multiple sequential electron transfers in BODIPY monomers, dimers, trimers, and polymer. J Am Chem Soc 133:8633–8645

Ni Y, Wu J (2014) Far-red and near infrared BODIPY dyes: synthesis and applications for fluorescent pH probes and bio-imaging. Org Biomol Chem 12:3774–3791

Ni Y, Zeng W, Huang K-W, Wu J (2013) Benzene-fused BODIPYs: synthesis and the impact of fusion mode. Chem Commun 49:1217–1219

Ni Y, Kannadorai RK, Peng J, Yu SW-K, Chang Y-T, Wu J (2016a) Naphthalene-fused BODIPY near-infrared dye as a stable contrast agent for *in vivo* photoacoustic imaging. Chem Commun 52:11504–11507

Ni Y, Lee S, Son M, Aratani N, Ishida M, Samanta A, Yamada H, Chang Y-T, Furuta H, Kim D, Wu J (2016b) A diradical approach towards BODIPY-based dyes with intense near-infrared absorption around $\lambda = 1100$ nm. Angew Chem Int Ed 55:2815–2819

Numano M, Nagami N, Nakatsuka S, Katayama T, Nakajima K, Tatsumi S, Yasuda N, Hatakeyama T (2016) Synthesis of boronate-based benzo[fg]tetracene and benzo[hi]hexacene via demethylative direct borylation. Chem Eur J 22:11574–11577

Oda S, Shimizu T, Katayama T, Yoshikawa H, Hatakeyama T (2019) Tetracoordinate boron-fused double [5]helicenes as cathode active materials for lithium batteries. Org Lett 21:1770–1773

Ohtani S, Gon M, Tanaka K, Chujo Y (2017) A flexible, fused, azomethine-boron complex: thermochromic luminescence and thermosalient behavior in structural transitions between crystalline polymorphs. Chem Eur J 23:11827–11833

Okujima T, Tomimori Y, Nakamura J, Yamada H, Uno H, Ono N (2010) Synthesis of π-expanded BODIPYs and their fluorescent properties in the visible–near–infrared region. Tetrahedron 66:6895–6900

Ono N, Hironaga H, Ono K, Kaneko S, Murashima T, Ueda T, Tsukamura C, Ogawa T (1996) A new synthesis of pyrroles and porphyrins fused with aromatic rings. J Chem Soc Perkin Trans 1:417–423

Ono N, Yamamoto T, Shimada N, Kuroki K, Wada M, Utsunomiya R, Yano T, Uno H, Murashima T (2003) A new synthesis of functional dyes from 2-Acenaphtho[1,2-c]pyrrole. Heterocycles 61:433–447

Ono K, Hashizume J, Yamaguchi H, Tomura M, Nishida J, Yamashita Y (2009) Synthesis, crystal structure, and electron-accepting property of the BF_2 complex of a dihydroxydione with a perfluorotetracene skeleton. Org Lett 11:4326–4329

Ortiz MJ, Agarrabeitia AR, Duran-Sampedro G, Prieto JB, Lopez TA, Massad WA, Montejano HA, García NA, Arbeloa IL (2012) Synthesis and functionalization of new polyhalogenated BODIPY dyes. Study of their photophysical properties and singlet oxygen generation. Tetrahedron 68:1153–1162

Osorio-Martínez CA, Urías-Benavides A, Gómez-Durán CFA, Bañuelos J, Esnal I, Arbeloa IL, Peña-Cabrera E (2012) 8-AminoBODIPYs: cyanines or hemicyanines? The effect of the coplanarity of the amino group on their optical properties. J Org Chem 77:5434–5438

Ozdemir T, Atilgan S, Kutuk I, Yildirim LT, Tulek A, Bayindir M, Akkaya EU (2009) Solid-state emissive BODIPY dyes with bulky substituents as spacers. Org Lett 11:2105–2107

Pais VF, Ramírez-López P, Romero-Arenas A, Collado D, Nájera F, Pérez-Inestrosa E, Fernández R, Lassaletta JM, Ros A, Pischel U (2016) Red-emitting tetracoordinate organoboron chelates: synthesis, photophysical properties, and fluorescence microscopy. J Org Chem 81:9605–9611

Palao E, Agarrabeitia AR, Bañuelos-Prieto J, Lopez TA, López-Arbeloa I, Armesto D, Ortiz MJ (2013) 8-functionalization of alkyl-substituted-3,8-dimethyl BODIPYs by knoevenagel condensation. Org Lett 15:4454–4457

Palao E, de la Moya S, Agarrabeitia AR, Esnal I, Bañuelos J, López-Arbeloa I, Ortiz MJ (2014) Selective lateral lithiation of methyl BODIPYs: synthesis, photophysics, and electrochemistry of new meso derivatives. Org Lett 16:4364–4367

Palao E, Duran-Sampedro G, de la Moya S, Madrid M, García-López C, Agarrabeitia AR, Verbelen B, Dehaen W, Boens N, Ortiz MJ (2016) Exploring the application of the Negishi reaction of HaloBODIPYs: generality, regioselectivity, and synthetic utility in the development of BODIPY laser dyes. J Org Chem 81:3700–3710

Patalag LJ, Jones PG, Werz DB (2016) BOIMPYs: rapid access to a family of red-emissive fluorophores and NIR dyes. Angew Chem Int Ed 55:13340–13344

Peña-Cabrera E, Aguilar-Aguilar A, González-Domínguez M, Lager E, Zamudio-Vázquez R, Godoy-Vargas J, Villanueva-García F (2007) Simple, general, and efficient synthesis of meso-substituted borondipyrromethenes from a single platform. Org Lett 9:3985–3988

Perumal K, Garg JA, Blacque O, Saiganesh R, Kabilan S, Balasubramanian KK, Venkatesan K (2012) β-iminoenamine-BF_2 complexes: aggregation-induced emission and pronounced effects of aliphatic rings on radiationless deactivation. Chem Asian J 7:2670–2677

Plater MJ, Aiken S, Bourhill G (2002) A new synthetic route to donor–acceptor porphyrins. Tetrahedron 58:2405–2413

Pohl R, Anzenbacher P (2003) Emission color tuning in AlQ_3 complexes with extended conjugated chromophores. Org Lett 5:2769–2772

Poirel A, De Nicola A, Retailleau P, Ziessel R (2012) Oxidative coupling of 1,7,8-unsubstituted BODIPYs: synthesis and electrochemical and spectroscopic properties. J Org Chem 77:7512–7525

Poirel A, Retailleau P, De Nicola A, Ziessel R (2014) Synthesis of water-soluble red-emitting thienyl-BODIPYs and bovine serum albumin labeling. Chem Eur J 20:1252–1257

Potopnyk MA, Lytvyn R, Danyliv Y, Ceborska M, Bezvikonnyi O, Volyniuk D, Gražulevičius JV (2018) Correction to N, O π-conjugated 4-substituted 1,3-thiazole BF2 complexes: synthesis and photophysical properties. J Org Chem 83:1095–1105

Potopnyk MA, Volyniuk D, Luboradzki R, Ceborska M, Hladka I, Danyliv Y, Gražulevičius JV (2019) Application of the Suzuki-Miyaura reaction for the postfunctionalization of the benzo[4,5] thiazolo[3,2-c][1,3,5,2]oxadiazaborinine core: an approach toward fluorescent dyes. J Org Chem 84:5614–5626

Qin W, Baruah M, Van der Auweraer M, De Schryver FC, Boens N (2005) Photophysical properties of borondipyrromethene analogues in solution. J Phys Chem A 109:7371–7384

Qin Y, Kiburu I, Shah S, Jäkle F (2006) Luminescence tuning of organoboron quinolates through substituent variation at the 5-position of the quinolato moiety. Org Lett 8:5227–5230

Qiu F, Zhang F, Tang R, Fu Y, Wang X, Han S, Zhuang X, Feng X (2016) Triple boron-cored chromophores bearing discotic 5,11,17-triazatrinaphthylene-based ligands. Org Lett 18:1398–1401

Qu X, Liu Q, Ji X, Chen H, Zhou Z, Shen Z (2012) Enhancing the Stokes' shift of BODIPY dyes via through-bond energy transfer and its application for Fe^{3+}-detection in live cell imaging. Chem Commun 48:4600–4602

Quan L, Chen Y, Lv X-J, Fu W-F (2012) Aggregation-induced photoluminescent changes of naphthyridine-BF_2 complexes. Chem Eur J 18:14599–14604

Ramírez-Ornelas DE, Alvarado-Martínez E, Bañuelos J, López-Arbeloa I, Arbeloa T, Mora-Montes HM, Pérez-García LA, Peña-Cabrera E (2016) FormylBODIPYs: privileged building blocks for multicomponent reactions. The case of the Passerini reaction. J Org Chem 81:2888–2898

Rao Y-L, Amarne H, Wang S (2012) Photochromic four-coordinate N, C-chelate boron compounds. Coord Chem Rev 256:759–770
Rauschnabel J, Hanack M (1995) New derivatives and homologues of subphthalocyanine. Tetrahedron Lett 36:1629–1632
Ray C, Díaz-Casado L, Avellanal-Zaballa E, Bañuelos J, Cerdán L, García-Moreno I, Moreno F, Maroto BL, López-Arbeloa Í, de la Moya S (2017) *N*-BODIPYs come into play: smart dyes for photonic materials. Chem Eur J 23:9383–9390
Ren Y, Liu X, Gao W, Xia H, Ye L, Mu Y (2007) Boron complexes with chelating anilido-imine ligands: synthesis, structures and luminescent properties. Eur J Inorg Chem 1808–1814
Ren T-B, Xu W, Zhang W, Zhang X-X, Wang Z-Y, Xiang Z, Yuan L, Zhang X-B (2018) A general method to increase Stokes shift by introducing alternating vibronic structures. J Am Chem Soc 140:7716–7722
Riddle JA, Lathrop SP, Bollinger JC, Lee D (2006) Schiff base route to stackable pseudo-triphenylenes: stereoelectronic control of assembly and luminescence. J Am Chem Soc 128:10986–10987
Rodríuez-Morgade MS, Esperanza S, Torres T, Barberá J (2005) Synthesis, characterization, and properties of subporphyrazines: a new class of nonplanar, aromatic macrocycles with absorption in the green region. Chem Eur J 11:354–360
Rogers MAT (1943) 156. 2:4-diarylpyrroles. Part I. Synthesis of 2:4-diarylpyrroles and 2:2′:4:4′-tetra-arylazadipyrromethines. J Chem Soc 590–596
Rohand T, Qin W, Boens N, Dehaen W (2006) Palladium-catalyzed coupling reactions for the functionalization of BODIPY dyes with fluorescence spanning the visible spectrum. Eur J Org Chem 4658–4663
Rohand T, Baruah M, Qin W, Boens N, Dehaen W (2006) Functionalisation of fluorescent BODIPY dyes by nucleophilic substitution. Chem Commun 266–268
Roubinet B, Massif C, Moreau M, Boschetti F, Ulrich G, Ziessel R, Renard P-Y, Romieu A (2015) New 3-(heteroaryl)-2-iminocoumarin-based borate complexes: synthesis, photophysical properties, and rational functionalization for biosensing/biolabeling applications. Chem Eur J 21:14589–14601
Sakamoto N, Ikeda C, Nabeshima T (2010) Cation recognition and pseudorotaxane formation of tris-dipyrrin BF_2macrocycles. Chem Commun 46:6732–6734
Salla CAM, dos Santos JM, Farias G, Bortoluzi AJ, Curcio SF, Cazati T, Izsák R, Neese F, de Souza B, Bechtold IH (2019) New boron(III) blue emitters for all-solution processed OLEDs: molecular design assisted by theoretical modeling. Eur J Inorg Chem 2247–2257
Sarma T, Panda PK, Setsune J (2013) Bis-naphthobipyrrolylmethene derived BODIPY complex: an intense near-infrared fluorescent dye. Chem Commun 49:9806–9808
Sathyamoorthi G, Soong M-L, Ross TW, Boyer JH (1993) Fluorescent tricyclic β-azavinamidine–BF_2 complexes. Heteroat Chem 4:603–608
Sawazaki T, Shimizu Y, Oisaki K, Sohma Y, Kanai M (2018) Convergent and functional-group-tolerant synthesis of B-organo BODIPYs. Org Lett 20:7767–7770
Scheibe G, Daltrozzo E, Wörz O, Heiss J (1969) Das Franck-Condon-Prinzip und die Lichtabsorption von Merocyaninen. Z Phys Chem N F 64:97–114
Shaffer KJ, McLean TM, Waterland MR, Wenzel M, Plieger PG (2012) Structural characterisation of difluoro-boron chelates of quino[7,8-*h*]quinoline. Inorg Chim Acta 380:278–283
Shah M, Thangaraj K, Soong ML, Wolford L, Boyer JH, Politzer I, Pavlopoulos TG (1990) Pyrromethene–BF_2 complexes as laser dyes:1. Heteroat Chem 1:389–399
Shang H, Zhao L, Qi D, Chen C, Jiang J (2014) The first five-membered-heterocycle-fused subphthalocyanine analogues: chiral tri(benzo[b]thiopheno)subporphyrazines. Chem Eur J 20:16266–16272
Shen Z, Röhr H, Rurack K, Uno H, Spieles M, Schulz B, Reck G, Ono N (2004) Boron-diindomethene (BDI) dyes and their tetrahydrobicyclo precursors—en route to a new class of highly emissive fluorophores for the red spectral range. Chem Eur J 10:4853–4871

Sheng W, Cui J, Ruan Z, Yan L, Wu Q, Yu C, Wei Y, Hao E, Jiao L (2017a) [*a*]-phenanthrene-fused BF_2 azadipyrromethene (AzaBODIPY) dyes as bright near-infrared fluorophores. J Org Chem 82:10341–10349

Sheng W, Zheng Y-Q, Wu Q, Wu Y, Yu C, Jiao L, Hao E, Wang J-Y, Pei J (2017b) Synthesis, properties, and semiconducting characteristics of BF2 complexes of β, β-bisphenanthrene-fused azadipyrromethenes. Org Lett 19:2893–2896

Sheng W, Wu Y, Yu C, Bobadova-Parvanova P, Hao E, Jiao L (2018) Synthesis, crystal structure, and the deep near-infrared absorption/emission of bright AzaBODIPY-based organic fluorophores. Org Lett 20:2620–2623

Shimizu S, Kobayashi N (2014) Structurally-modified subphthalocyanines: molecular design towards realization of expected properties from the electronic structure and structural features of subphthalocyanine. Chem Commun 50:6949–6966

Shimizu S, Otaki T, Yamazaki Y, Kobayashi N (2012) Synthesis and properties of β, β-sp3-hybridized subphthalocyanine analogues. Chem Commun 48:4100–4102

Shimizu S, Iino T, Araki Y, Kobayashi N (2013) Pyrrolopyrrole aza-BODIPY analogues: a facile synthesis and intense fluorescence. Chem Commun 49:1621–1623

Shimizu S, Murayama A, Haruyama T, Iino T, Mori S, Furuta H, Kobayashi N (2015) Benzo[c, d]indole-containing Aza-BODIPY dyes: asymmetrization-induced solid-state emission and aggregation-induced emission enhancement as new properties of a well-known chromophore. Chem Eur J 21:12996–13003

Shimogawa H, Mori H, Wakamiya A, Murata Y (2013) Impacts of dibenzo- and dithieno-fused structures at the *b, g* bonds in the BODIPY skeleton. Chem Lett 42:986–988

Shimogawa H, Murata Y, Wakamiya A (2018) NIR-absorbing dye based on BF_2-bridged azafulvene dimer as a strong electron-accepting unit. Org Lett 20:5135–5138

Singh SP, Gayathri T (2014) Evolution of BODIPY dyes as potential sensitizers for dye-sensitized solar cells. Eur J Org Chem 4689–4707

Singh RS, Yadav M, Gupta RK, Pandey R, Pandey DS (2013) Luminescent N, O-chelated chroman-BF2 complexes: structural variants of BODIPY. Dalton Trans 42:1696–1707

Smithen DA, Baker AEG, Offman M, Crawford SM, Cameron TS, Thompson A (2012) Use of F-BODIPYs as a protection strategy for dipyrrins: optimization of BF_2 removal. J Org Chem 77:3439–3453

Solomonov AV, Marfin YS, Rumyantsev EV (2019) Design and applications of dipyrrin-based fluorescent dyes and related organic luminophores: from individual compounds to supramolecular self-assembled systems. Dyes Pigm 162:517–542

Speck K, Magauer T (2013) The chemistry of isoindole natural products. Beilstein J Org Chem 9:2048–2078

Strickler SJ, Berg RA (1962) Relationship between absorption intensity and fluorescence lifetime of molecules. J Chem Phys 37:814–822

Suenaga K, Tanaka K, Chujo Y (2017) Design and luminescence chromism of fused boron complexes having constant emission efficiencies in solution and in the amorphous and crystalline states. Eur J Org Chem 5191–5196

Sun L, Zhang F, Wang X, Qiu F, Xue M, Tregnago G, Cacialli F, Osella S, Beljonne D, Feng X (2015) Geometric and electronic structures of boron(III)-cored dyes tailored by incorporation of heteroatoms into ligands. Chem Asian J 10:709–714

Sun Z-B, Guo M, Zhao C-H (2016) Synthesis and properties of benzothieno[*b*]-fused BODIPY dyes. J Org Chem 81:229–237

Sun Y, Qu Z, Zhou Z, Gai L, Lu H (2019) Thieno[3,2-*b*]thiophene fused BODIPYs: synthesis, near-infrared luminescence and photosensitive properties. Org Biomol Chem 17:3617–3622

Sutter A, Elhabiri M, Ulrich G (2018) Fluorescent pH-responsive probes based on water-soluble boron-dipyrromethene (BODIPY) derivatives, featuring long-wavelength emission. Chem Eur J 24:11119–11130

Swavey S, Coladipietro M, Burbayea A, Krause JA (2016) Two-step synthetic route toward asymmetric and symmetric boron dipyrromethenes: synthesis, optical properties, and electrochemistry. Eur J Org Chem 4429–4435

Swavey S, Quinn J, Coladipietro M, Cox KG, Brennaman MK (2017) Tuning the photophysical properties of BODIPY dyes through extended aromatic pyrroles. RSC Adv 7:173–179

Szyszko B, Chmielewski PJ, Przewoźnik M, Białek MJ, Kupietz K, Białońska A, Latos-Grażyński L (2019) Diphenanthrioctaphyrin(1.1.1.0.1.1.1.0): conformational switching controls the stereochemical dynamics of the topologically chiral system. J Am Chem Soc 141:6060–6072

Tahtaoui C, Thomas C, Rohmer F, Klotz P, Duportail G, Mély Y, Bonnet D, Hibert M (2007) Convenient method to access new 4,4-dialkoxy- and 4,4-diaryloxy-diaza-s-indacene dyes: synthesis and spectroscopic evaluation. J Org Chem 72:269–272

Tamgho I-S, Hasheminasab A, Engle JT, Nemykin VN, Ziegler CJ (2014) A new highly fluorescent and symmetric pyrrole–BF_2 chromophore: BOPHY. J Am Chem Soc 136:5623–5626

Tanaka T, Chujo Y (2015) Recent progress of optical functional nanomaterials based on organoboron complexes with β-diketonate, ketoiminate and diiminate. NPG Asia Mater 7:e223

Tanaka K, Yamane H, Yoshii R, Chujo Y (2013) Efficient light absorbers based on thiophene-fused boron dipyrromethene (BODIPY) dyes. Bioorg Med Chem 21:2715–2719

Tanaka H, Inoue Y, Mori T (2018) Circularly polarized luminescence and circular dichroisms in small organic molecules: correlation between excitation and emission dissymmetry factors. ChemPhoto-Chem 2:386–402

Tathe AB, Sekar N (2016) Red emitting coumarin—azo dyes: synthesis, characterization, linear and non-linear optical properties-experimental and computational approach. J Fluoresc 26:1279–1293

Tay ACY, Frogley BJ, Ware DC, Conradie J, Ghosh A, Brothers PJ (2019) Tetrahedral pegs in square holes: stereochemistry of diboron porphyrazines and phthalocyanines. Angew Chem Int Ed 58:3057–3061

Thivierge C, Bandichhor R, Burgess K (2007) Spectral dispersion and water solubilization of BODIPY dyes via palladium-catalyzed C–H functionalization. Org Lett 9:2135–2138

Thorat KG, Kamble P, Mallah R, Ray AK, Sekar N (2015) Congeners of pyrromethene-567 dye: perspectives from synthesis, photophysics, photostability, laser, and TD-DFT theory. J Org Chem 80:6152–6164

Tokoro Y, Nagai A, Chujo Y (2010) Synthesis of π-conjugated polymers containing organoboron benzo[*h*]quinolate in the main chain. Macromolecules 43:6229–6233

Tomimori Y, Okujima T, Yano T, Mori S, Ono N, Yamada H, Uno H (2011) Synthesis of π-expanded O-chelated boron–dipyrromethene as an NIR dye. Tetrahedron 67:3187–3193

Treibs A, Kreuzer F-H (1968) Difluorboryl-komplexe von di- und tripyrrylmethenen. Justus Liebigs Ann Chem 718:208–223

Tsurumaki E, Saito S, Kim KS, Lim JM, Inokuma Y, Kim D, Osuka A (2008) Synthesis and characterization of meso-Aryl-substituted subchlorins. J Am Chem Soc 130:438–439

Turkoglu G, Cinar ME, Ozturk T (2017) Triarylborane-based materials for OLED applications. Molecules 22:1522

Turro NJ (1991) Modern molecular photochemistry. University Science Books, CA

Ulrich G, Goze C, Goeb S, Retailleau P, Ziessel R (2006) New fluorescent aryl- or ethynylaryl-boron-substituted indacenes as promising dyes. New J Chem 30:982–986

Ulrich G, Ziessel R, Harriman A (2008) The chemistry of fluorescent bodipy dyes: versatility unsurpassed. Angew Chem Int Ed 47:1184–1201

Ulrich G, Goeb S, De Nicola A, Retailleau P, Ziessel R (2011) Chemistry at boron: synthesis and properties of red to near-IR fluorescent dyes based on boron-substituted diisoindolomethene frameworks. J Org Chem 76:4489–4505

Umezawa K, Nakamura Y, Makino H, Citterio D, Suzuki K (2008) Bright, color-tunable fluorescent dyes in the visible–near-infrared region. J Am Chem Soc 130:1550–1551

Umezawa K, Matsui A, Nakamura Y, Citterio D, Suzuki K (2009) Bright, color-tunable fluorescent dyes in the Vis/NIR region: establishment of new "tailor-made" multicolor fluorophores based on borondipyrromethene. Chem Eur J 15:1096–1106

Umezawa K, Citterio D, Suzuki K (2014) New trends in near-infrared fluorophores for bioimaging. Anal Sci 30:327–349

Uno H, Honda T, Kitatsuka M, Hiraoka S, Mori S, Takase M, Okujima T, Nakae T (2018) Benzene-fused bis(acenaphthoBODIPY)s, stable near-infrared-selective dyes. RSC Adv 8:14072–14083

Uoyama H, Goushi K, Shizu K, Nomura H, Adachi C (2012) Highly efficient organic light-emitting diodes from delayed fluorescence. Nature 492:234–238

Urban M, Durka K, Jankowski P, Serwatowski J, Luliński S (2017) Highly fluorescent red-light emitting bis(boranils) based on naphthalene backbone. J Org Chem 82:8234–8241

Verbelen B, Boodts S, Hofkens J, Boens N, Dehaen W (2015a) Radical C-H arylation of the BODIPY core with aryldiazonium salts: synthesis of highly fluorescent red-shifted dyes. Angew Chem Int Ed 54:4612–4616

Verbelen B, Rezende LCD, Boodts S, Jacobs J, Van Meervelt L, Hofkens J, Dehaen W (2015b) Radical C-H alkylation of BODIPY dyes using potassium trifluoroborates or boronic acids. Chem Eur J 21:12667–12675

Wakamiya A, Yamaguchi S (2015) Designs of functional π-electron materials based on the characteristic features of boron. Bull Chem Soc Jpn 88:1357–1377

Wakamiya A, Taniguchi T, Yamaguchi S (2006) Intramolecular B-N coordination as a Scaffold for electron-transporting materials: synthesis and properties of boryl-substituted thienylthiazoles. Angew Chem Int Ed 45:3170–3173

Wakamiya A, Mori K, Yamaguchi S (2007) 3-boryl-2,2′-bithiophene as a versatile core skeleton for full-color highly emissive organic solids. Angew Chem Int Ed 46:4273–4276

Wakamiya A, Murakami T, Yamaguchi S (2013) Benzene-fused BODIPY and fully-fused BODIPY dimer: impacts of the ring-fusing at the b bond in the BODIPY skeleton†. Chem Sci 4:1002–1007

Wang D, Fan J, Gao X, Wang B, Sun S, Peng X (2009) Carboxyl BODIPY dyes from bicarboxylic anhydrides: one-pot preparation, spectral properties, photostability, and biolabeling. J Org Chem 74:7675–7683

Wang Y-W, Descalzo AB, Shen Z, You X-Z, Rurack K (2010) Dihydronaphthalene-fused boron-dipyrromethene (BODIPY) dyes: insight into the electronic and conformational tuning modes of BODIPY fluorophores. Chem Eur J 16:2887–2903

Wang D, Liu R, Chen C, Wang S, Chang J, Wu C, Zhu H, Waclawik ER (2013) Synthesis, photophysical and electrochemical properties of aza-boron-diquinomethene complexes. Dyes Pigm 99:240–249

Wang F, Guo Z, Li X, Li X, Zhao C (2014a) Development of a small molecule probe capable of discriminating cysteine, homocysteine, and glutathione with three distinct turn-on fluorescent outputs. Chem Eur J 20:11471–11478

Wang X, Liu H, Cui J, Wu Y, Lu H, Lu J, Liu Z, He W (2014b) New J Chem 38:1277–1283

Wang X, Wu Y, Liu Q, Li Z, Yan H, Ji C, Duan J, Liu Z (2015a) Aggregation-induced emission (AIE) of pyridyl-enamido-based organoboron luminophores. Chem Commun 51:784–787

Wang X, Liu Q, Yan H, Liu Z, Yao M, Zhang Q, Gong S, He W (2015b) Piezochromic luminescence behaviors of two new benzothiazole–enamido boron difluoride complexes: intra- and inter-molecular effects induced by hydrostatic compression. Chem Commun 51:7497–7500

Wang J, Li J, Chen N, Wu Y, Hao E, Wei Y, Mu X, Jiao L (2016a) Synthesis, structure and properties of thiophene-fused BODIPYs and azaBODIPYs as near-infrared agents. New J Chem 40:5966–5975

Wang J, Wu Q, Yu C, Wei Y, Mu X, Hao E, Jiao L (2016b) Aromatic ring fused BOPHYs as stable red fluorescent dyes. J Org Chem 81:11316–11323

Wang S, Lan H, Xiao S, Tan R, Lu Y (2017) Highly fluorescent non-conventional boron-difluoride-based π organogel with gelation-assisted piezochromism. Chem Asian J 12:198–202

Wang M, Vicente MCH, Mason D, Bobadova-Parvanova P (2018a) Stability of a series of BODIPYs in acidic conditions: an experimental and computational study into the role of the substituents at boron. ACS Omega 3:5502–5510

Wang T, Dou C, Liu J, Wang L (2018b) Effects of the substituents of boron atoms on conjugated polymers containing B ← N units. Chem Eur J 24:13043–13048

Wang Z, Cheng C, Kang Z, Miao W, Liu Q, Wang H, Hao E (2019) Organotrifluoroborate salts as complexation reagents for synthesizing BODIPY dyes containing both fluoride and an organo substituent at the boron center. J Org Chem 84:2732–2740

Watley RL, Awuah SG, Bio M, Cantu R, Gobeze HB, Nesterov VN, Das SK, D'Souza F, You Y (2015) Dual functioning thieno-pyrrole fused BODIPY dyes for NIR optical imaging and photodynamic therapy: singlet oxygen generation without heavy halogen atom assistance. Chem Asian J 10:1335–1343

Wiktorowski S, Rosazza C, Winterhalder MJ, Daltrozzo E, Zumbusch A (2014) Water-soluble pyrrolopyrrole cyanine (PPCy) NIR fluorophores. Chem Commun 50:4755–4758

Wiktorowski S, Daltrozzo E, Zumbusch A (2015) Water-soluble pyrrolopyrrole cyanine (PPCy) near-infrared fluorescent pH indicators for strong acidity. RSC Adv 5:29420–29423

Wood TE, Thompson A (2007) Advances in the chemistry of dipyrrins and their complexes. Chem Rev 107:1831–1861

Wu L, Burgess K (2008a) A new synthesis of symmetric boraindacene (BODIPY) dyes. Chem Commun 4933–4935

Wu L, Burgess K (2008b) Syntheses of highly fluorescent GFP-chromophore analogues. J Am Chem Soc 130:4089–4096

Wu D, O'Shea DF (2013) Synthesis and properties of BF_2-3,3′-dimethyldiarylazadipyrromethene near-infrared fluorophores. Org Lett 15:3392–3395

Wu Q, Esteghamatian M, Hu N-X, Popovic Z, Enright G, Tao Y, D'Iorio M, Wang S (2000) Synthesis, structure, and electroluminescence of BR_2q (R = Et, Ph, 2-naphthyl and q = 8-hydroxyquinolato). Chem Mater 12:79–83

Wu Y-Y, Chen Y, Gou G-Z, Mu W-H, Lv X-J, Du M-L, Fu W-F (2012) Large stokes shift induced by intramolcular charge transfer in N, O-chelated naphthyridine–BF2 complexes. Org Lett 14:5226–5229

Wu Y, Cheng C, Jiao L, Yu C, Wang S, Wei Y, Mu X, Hao E (2014) β-thiophene-fused BF_2-azadipyrromethenes as near-infrared dyes. Org Lett 16:748–751

Wu Y, Mack J, Xiao X, Li Z, Shen Z, Lu H (2017a) N − bridged annulated BODIPYs: synthesis of highly fluorescent blueshifted dyes. Chem Asian J 12:2216–2220

Wu Z, Sun J, Zhang Z, Yang H, Xue P, Lu R (2017b) Nontraditional π gelators based on β-iminoenolate and their difluoroboron complexes: effect of halogens on gelation and their fluorescent sensory properties towards acids. Chem Eur J 23:1901–1909

Wu C, Huang X, Tang Y, Xiao W, Sun L, Shao J, Dong X (2019) Pyrrolopyrrole aza-BODIPY near-infrared photosensitizer for dual-mode imaging-guided photothermal cancer therapy. Chem Commun 55:790–793

Xia S, Shen J, Wang J, Wang H, Fang M, Zhou H, Tanasova M (2018) Ratiometric fluorescent and colorimetric BODIPY-based sensor for zinc ions in solution and living cells. Sens Act B 258:1279–1286

Xu H-J, Shen Z, Okujima T, Ono N, You X-Z (2006) Synthesis and spectroscopic characterization of *meso*-tetraarylporphyrins with fused phenanthrene rings. Tetrahedron Lett 47:931–934

Xu S, Evans RE, Liu T, Zhang G, Demas JN, Trindle CO, Fraser CL (2013) Aromatic difluoroboron β-diketonate complexes: effects of π-conjugation and media on optical properties. Inorg Chem 52:3597–3610

Yakubovskyi VP, Didukh NO, Zatsikha YV, Kovtun YP (2016) A new approach to the synthesis of meso-CN-substituted BODIPYs. ChemistrySelect 1:1462–1466

Yamaguchi M, Ito S, Hirose A, Tanaka K, Chujo Y (2016) Modulation of sensitivity to mechanical stimulus in mechanofluorochromic properties by altering substituent positions in solid-state emissive diiodo boron diiminates. J Mater Chem C 4:5314–5319

Yamaguchi M, Ito S, Hirose A, Tanaka K, Chujo Y (2017) Control of aggregation-induced emission *versus* fluorescence aggregation-caused quenching by bond existence at a single site in boron pyridinoiminate complexes. Mater Chem Front 1:1573–1579

Yamaguchi M, Tanaka K, Chujo Y (2018) Design of conjugated molecules presenting short-wavelength luminescence by utilizing heavier atoms of the same element group. Chem Asian J 13:1342–1347

Yamaji M, Kato S, Tomonari K, Mamiya M, Goto K, Okamoto H, Nakamura Y, Tani F (2017) Blue fluorescence from BF2 complexes of N, O-benzamide ligands: synthesis, structure, and photophysical properties. Inorg Chem 56:12514–12519

Yamane H, Ohtani S, Tanaka K, Chujo Y (2017) Synthesis of furan-substituted aza-BODIPYs having near-infrared emission. Tetrahedron Lett 58:2989–2992

Yamazawa S, Nakashima M, Suda Y, Nishiyabu R, Kubo Y (2016) 2,3-naphtho-fused BODIPYs as near-infrared absorbing dyes. J Org Chem 81:1310–1315

Yan W, Wan X, Chen Y (2010) Phenalenyl-based boron–fluorine complexes: synthesis, crystal structures and solid-state fluorescence properties. J Mol Struct 968:85–88

Yan W, Hong C, Long G, Yang Y, Liu Z, Bian Z, Chen Y, Huang C (2014) Synthesis, crystal structures and photophysical properties of novel boron-containing derivatives of phenalene with bright solid-state luminescence. Dyes Pigm 106:197–204

Yang Y, Hughes RP, Aprahamian I (2012) Visible light switching of a BF_2-coordinated azo compound. J Am Chem Soc 134:15221–15224

Yang SK, Shi X, Park S, Ha T, Zimmerman SC (2013) A dendritic single-molecule fluorescent probe that is monovalent, photostable and minimally blinking. Nat Chem 5:692–697

Yang L, Liu Y, Ma C, Liu W, Li Y, Li L (2015) Naphthalene-fused BODIPY with large Stokes shift as saturated-red fluorescent dye for living cell imaging. Dyes Pigm 122:1–5

Yang L, Liu Y, Zhou X, Wu Y, Ma C, Liu W, Zhang C (2016) Asymmetric anthracene-fused BODIPY dye with large Stokes shift: synthesis, photophysical properties and bioimaging. Dyes Pigm 126:232–238

Yang C, Wang X, Wang M, Xu K, Xu C (2017) Robust colloidal nanoparticles of pyrrolopyrrole cyanine J-aggregates with bright near-infrared fluorescence in aqueous media: from spectral tailoring to bioimaging applications. Chem Eur J 23:4310–4319

Yang J, Rousselin Y, Bucher L, Desbois N, Bolze F, Xu H-J, Gros CP (2018) Two-photon absorption properties and structures of BODIPY and its dyad. Triad Tetrad ChemPlusChem 83:838–844

Yokoi H, Wachi N, Hiroto S, Shinokubo H (2014) Oxidation of 2-amino-substituted BODIPYs providing pyrazine-fused BODIPY trimers. Chem Commun 50:2715–2717

Yoshii R, Nagai A, Tanaka K, Chujo Y (2013) Highly emissive boron ketoiminate derivatives as a new class of aggregation-induced emission fluorophores. Chem Eur J 19:4506–4512

Yoshii R, Hirose A, Tanaka K, Chujo Y (2014a) Boron diiminate with aggregation-induced emission and crystallization-induced emission-enhancement characteristics. Chem Eur J 20:8320–8324

Yoshii R, Hirose A, Tanaka K, Chujo Y (2014b) Functionalization of boron diiminates with unique optical properties: multicolor tuning of crystallization-induced emission and introduction into the main chain of conjugated polymers. J Am Chem Soc 136:18131–18139

Yoshino J, Kano N, Kawashima T (2007) Synthesis of the most intensely fluorescent azobenzene by utilizing the B–N interaction. Chem Commun 559–561

Yoshino J, Kano N, Kawashima T (2013) Fluorescent azobenzenes and aromatic aldimines featuring an N-B interaction. Dalton Trans 42:15826–15834

Yu C, Jiao L, Yin H, Zhou J, Pang W, Wu Y, Wang Z, Yang G, Hao E (2011) α-/β-formylated boron–dipyrrin (BODIPY) dyes: regioselective syntheses and photophysical properties. Eur J Org Chem 5460–5468

Yu C, Xu Y, Jiao L, Zhou J, Wang Z, Hao E (2012) Isoindole-BODIPY dyes as red to near-infrared fluorophores. Chem Eur J 18:6437–6442

Yu C, Jiao L, Zhang P, Feng Z, Cheng C, Wei Y, Mu X, Hao E (2014) Highly fluorescent BF2 complexes of hydrazine-schiff base linked bispyrrole. Org Lett 16:3048–3051

Yu C, Hao E, Li T, Wang J, Sheng W, Wei Y, Mu X, Jiao L (2015) Dipyrrolylquinoxaline difluoroborates with intense red solid-state fluorescence. Dalton Trans 44:13897–13905

Yu Y, Jiao L, Wang J, Wang H, Yu C, Hao E, Boens N (2017) Bu4NI/tBuOOH catalyzed, α-regioselective cross-dehydrogenative coupling of BODIPY with allylic alkenes and ethers. Chem Commun 53:581–584

Yu C, Huang Z, Wang X, Miao W, Wu Q, Wong W-Y, Hao E, Xiao Y, Jiao L (2018) A family of highly fluorescent and unsymmetric bis(BF_2) chromophore containing both pyrrole and *N*-heteroarene derivatives: BOPPY. Org Lett 20:4462–4466

Yuan Z, Entwistle CD, Collings JC, Albesa-Jové D, Batsanov AS, Howard JAK, Taylor NJ, Kaiser HM, Kaufmann DE, Poon S-Y, Wong W-Y, Jardin C, Fathallah S, Boucekkine A, Halet J-F, Marder TB (2006) Synthesis, crystal structures, linear and nonlinear optical properties, and theoretical studies of (p-R-phenyl)-, (p-R-phenylethynyl)-, and (E)-[2-(p-R-phenyl)ethenyl]dimesityl boranes and related compounds. Chem Eur J 12:2758–2771

Yuan L, Lin W, Zheng K, He L, Huang W (2013) Far-red to near infrared analyte-responsive fluorescent probes based on organic fluorophore platforms for fluorescence imaging. Chem Soc Rev 42:622–661

Yuan K, Wang X, Mellerup SK, Kozin I, Wang S (2017) Spiro-BODIPYs with a diaryl chelate: impact on aggregation and luminescence. J Org Chem 82:13481–13487

Zeng L, Jiao C, Huang X, Huang K-W, Chin W-S, Wu J (2011) Anthracene-fused BODIPYs as near-infrared dyes with high photostability. Org Lett 13:6026–6029

Zeng C, Yuan K, Wang N, Peng T, Wu G, Wang S (2019) The opposite and amplifying effect of B ← N coordination on photophysical properties of regioisomers with an unsymmetrical backbone. Chem Sci 10:1724–1734

Zhang C, Zhao J (2016) Triplet excited state of diiodoBOPHY derivatives: preparation, study of photophysical properties and application in triplet–triplet annihilation upconversion. J Mater Chem C 4:1623–1632

Zhang G, Chen J, Payne SJ, Kooi SE, Demas JN, Fraser CL (2007) Multi-emissive difluoroboron dibenzoylmethane polylactide exhibiting intense fluorescence and oxygen-sensitive room-temperature phosphorescence. J Am Chem Soc 129:8942–8943

Zhang G, Kooi SE, Demas JN, Fraser CL (2008) Emission color tuning with polymer molecular weight for difluoroboron dibenzoylmethane-polylactide. Adv Mater 20:2099–2104

Zhang G, Palmer GM, Dewhirst MW, Fraser CL (2009a) A dual-emissive-materials design concept enables tumour hypoxia imaging. Nat Mater 8:747–751

Zhang Z, Bi H, Zhang Y, Yao D, Gao H, Fan Y, Zhang H, Wang Y, Wang Y, Chen Z, Ma D (2009b) Luminescent boron-contained ladder-type π-conjugated compounds. Inorg Chem 48:7230–7236

Zhang G, Lu J, Sabat M, Fraser CL (2010) Polymorphism and reversible mechanochromic luminescence for solid-state difluoroboron avobenzone. J Am Chem Soc 132:2160–2162

Zhang Y, Shao X, Wang Y, Pan F, Kang R, Peng F, Huang Z, Zhang W, Zhao W (2015a) Dual emission channels for sensitive discrimination of Cys/Hcy and GSH in plasma and cells. Chem Commun 51:4245–4248

Zhang Z, Zhang H, Jiao C, Ye K, Zhang H, Zhang J, Wang Y (2015b) 2-(2-hydroxyphenyl)benzimidazole-based four-coordinate boron-containing materials with highly efficient deep-blue photoluminescence and electroluminescence. Inorg Chem 54:2652–2659

Zhang P, Liu W, Niu G, Xiao H, Wang M, Ge J, Wu J, Zhang H, Li Y, Wang P (2017) Coumarin-based boron complexes with aggregation-induced emission. J Org Chem 82:3456–3462

Zhang H, Chen X, Lan J, Liu Y, Zhou F, Wu D, You J (2018a) Silver-mediated direct C-H amination of BODIPYs for screening endoplasmic reticulum-targeting reagents. Chem Commun 54:3219–3222

Zhang G, Wang M, Fronczek FR, Smith KM, Graça M, Vicente H (2018b) Lewis-acid-catalyzed BODIPY boron functionalization using trimethylsilyl nucleophiles. Inorg Chem 57:14493–14496

Zhang Z, Zhang Z, Zhang H, Wang Y (2018c) 2-(2-hydroxyphenyl)imidazole-based four-coordinate organoboron compounds with efficient deep blue photoluminescence and electroluminescence. Dalton Trans 47:127–134

Zhang K, Zheng H, Hua C, Xin M, Gao J, Li Y (2018d) Novel fluorescent N, O-chelated fluorine-boron benzamide complexes containing thiadiazoles: synthesis and fluorescence characteristics. Tetrahedron 74:4161–4167

Zhang T, Yan J, Hu Y, Liu X, Wen L, Zheng K, Zhang N (2019) A simple central seven-membered BOPYIN: synthesis, structural, spectroscopic properties, and cellular imaging application. Chem Eur J 25:9266–9271

Zhao D, Li G, Wu D, Qin X, Neuhaus P, Cheng Y, Yang S, Lu Z, Pu X, Long C, You J (2013) Regiospecific N-heteroarylation of amidines for full-color-tunable boron difluoride dyes with mechanochromic luminescence. Angew Chem Int Ed 52:13676–13680

Zhao J, Xu K, Yang W, Wang Z, Zhong F (2015a) The triplet excited state of bodipy: formation, modulation and application. Chem Soc Rev 44:8904–8939

Zhao N, Xuan S, Fronczek FR, Smith KM, Vicente MGH (2015b) Stepwise polychlorination of 8-chloro-BODIPY and regioselective functionalization of 2,3,5,6,8-pentachloro-BODIPY. J Org Chem 80:8377–8383

Zhao C, An J, Zhou L, Fei Q, Wang F, Tan J, Shi B, Wang R, Guo Z, Zhu W-H (2016) Transforming the recognition site of 4-hydroxyaniline into 4-methoxyaniline grafted onto a BODIPY core switches the selective detection of peroxynitrite to hypochlorous acid. Chem Commun 52:2075–2078

Zhao N, Xuan S, Zhou Z, Fronczek FR, Smith KM, Vicente MGH (2017a) Synthesis and spectroscopic and cellular properties of near-IR [*a*]phenanthrene-fused 4,4-difluoro-4-bora-3a,4a-diaza-*s*-indacenes. J Org Chem 82:9744–9750

Zhao R, Min Y, Dou C, Liu J, Wang L (2017b) A new electron-rich unit for polymer electron acceptors: 4,4-difluoro-4H-cyclopenta[2,1-b:3,4-b']dithiophene. Chem Eur J 23:9486–9490

Zhao N, Ma C, Yang W, Yin W, Wei J, Li N (2019) Facile construction of boranil complexes with aggregation-induced emission characteristics and their specific lipid droplet imaging applications. Chem Commun 55:8494–8497

Zhou Y, Xiao Y, Li D, Fu M, Qian X (2008) Novel fluorescent fluorine–boron complexes: synthesis, crystal structure, photoluminescence, and electrochemistry properties. J Org Chem 73:1571–1574

Zhou Y, Kim JW, Kim MJ, Son W-J, Han SJ, Kim HN, Han S, Kim Y, Lee C, Kim S-J, Kim DH, Kim J-J, Yoon J (2010) Novel bi-nuclear boron complex with pyrene ligand: red-light emitting as well as electron transporting material in organic light-emitting diodes. Org Lett 12:1272–1275

Zhou Z, Wakamiya A, Kushida T, Yamaguchi S (2012) Planarized triarylboranes: stabilization by structural constraint and their plane-to-bowl conversion. J Am Chem Soc 134:4529–4532

Zhou X, Yu C, Feng Z, Yu Y, Wang J, Hao E, Wei Y, Mu X, Jiao L (2015a) Highly regioselective α-chlorination of the BODIPY chromophore with copper(II) chloride. Org Lett 17:4632–4635

Zhou X, Wu Q, Feng Y, Yu Y, Yu C, Hao E, Wei Y, Mu X, Jiao L (2015b) Tandem regioselective substitution and palladium-catalyzed ring fusion reaction for core-expanded boron dipyrromethenes with red-shifted absorption and intense fluorescence. Chem Asian J 10:1979–1986

Zhou L, Xu D, Gao H, Zhang C, Ni F, Zhao W, Cheng D, Liu X, Han A (2016) β-furan-fused bis(difluoroboron)-1,2-bis((1H-pyrrol-2-yl)methylene)hydrazine fluorescent dyes in the visible deep-red region. J Org Chem 81:7439–7447

Zhou Z, Zhou J, Gai L, Yuan A, Shen Z (2017) Naphtho[*b*]-fused BODIPYs: one pot Suzuki–Miyaura–Knoevenagel synthesis and photophysical properties. Chem Commun 53:6621–6624

Zhou J, Liu L, Pan Y, Zhu Q, Lu Y, Wei J, Luo K, Fu Y, Zhong C, Peng Y, Song Z (2018) Asymmetric difluoroboron quinazolinone-pyridine dyes with large Stokes shift: high emission efficiencies both in solution and in the solid state. Chem Eur J 24:17897–17901

Zhu H, Shimizu S, Kobayashi N (2010) Subazaphenalenephthalocyanine: a subphthalocyanine analogue bearing a six-membered ring unit. Angew Chem Int Ed 49:8000–8003

Zhu S, Zhang J, Vegesna G, Luo F-T, Green SA, Liu H (2011) Highly water-soluble neutral BODIPY dyes with controllable fluorescence quantum yields. Org Lett 13:438–441

Zhu X, Liu R, Li Y, Huang H, Wang Q, Wang D, Zhu X, Liu S, Zhu H (2014) An AIE-active boron-difluoride complex: multi-stimuli-responsive fluorescence and application in data security protection. Chem Commun 50:12951–12954

Zhu X, Huang H, Liu R, Jin X, Li Y, Wang D, Wang Q, Zhu H (2015) Aza-boron-diquinomethene complexes bearing N-aryl chromophores: synthesis, crystal structures, tunable photophysics, the protonation effect and their application as pH sensors. J Mater Chem C 3:3774–3782

Zhu C, Ji X, You D, Chen TL, Mu AU, Barker KP, Klivansky LM, Liu Y, Fang L (2018) Extraordinary redox activities in ladder-type conjugated molecules enabled by B ← N coordination-promoted delocalization and hyperconjugation. J Am Chem Soc 140:18173–18182

Zhu X-Y, Yao H-W, Fu Y-J, Guo X-F, Wang H (2019) Effect of substituents on Stokes shift of BODIPY and its application in designing bioimaging probes. Anal Chim Acta 1048:194–203

Zyabrev K, Dekhtyar M, Vlasenko Y, Chernega A, Slominskii Y, Tolmachev A (2012) New 2,2-difluoro-1,3,2(2*H*)oxazaborines and merocyanines derived from them. Dyes Pigm 92:749–757

Part II
Properties of Functional Dyes

Chapter 6
Molecular and Crystal Structures of Polymorphic Organic Dyes and Coloured Organic Compounds

Shinya Matsumoto and Jiyong Hwang

Abstract Molecular and crystal structures were reviewed on several dyes and coloured organic compounds with four or more polymorphic forms whose structural data are available from the online database. The difference in the molecular and crystal structures between polymorphs of the target compounds was briefly described. The result indicated that rigid dye chromophores, structural flexibility and weak intermolecular interactions such as hydrogen bonding and halogen interactions are important in terms of the occurrence of the polymorphs of organic dyes.

Keywords Polymorphs · Molecular structure · Crystal structure

6.1 Introduction

Polymorphism, a phenomenon in which the same molecule or chemical composite exhibits two or more different crystal structures (Bernstein 2002; Hilfiker 2006; Threlfal 1995; Dunitz and Bernstein 1995; Bernstein et al. 1999; Nangia 2008; Price 2013; Cruz-Cabeza and Bernstein 2014), is essential in several industries, such as pharmaceutics (Hilfiker 2006) and dyes, particularly for pigments and functional dye applications (Zollinger 2003; Faulkner and Schwartz 2009; Hunger and Schmidt 2018; Whitaker 1977, 1995; Erk and Hengelsberg 2003; Engel 2003; Law 1993). This is reflected by numerous patents involving polymorphs and their crystallisation methods. Many research endeavours have been devoted to understanding polymorphism in terms of crystal growth, polymorphic occurrence and structure–property relationships. In particular, the prediction of polymorphic crystal structures has been considered a challenging research subject (Price 2014; Reilly et al. 2016; Gavezzotti 2007). Polymorphs can generally be categorised into two types, namely conformational polymorphs and packing polymorphs (Nangia 2008; Cruz-Cabeza and Bernstein 2014), both of which are commonly observed in dye polymorphs.

S. Matsumoto (✉) · J. Hwang
Graduate School of Environment and Information Sciences, Yokohama National University, Yokohama, Japan
e-mail: matsumoto-shinya-py@ynu.ac.jp

Y. Ooyama and S. Yagi (eds.), *Progress in the Science of Functional Dyes*,
https://doi.org/10.1007/978-981-33-4392-4_6

A typical example of polymorphism in the dye industry is the application of organic dyes as pigments (Zollinger 2003; Faulkner and Schwartz 2009; Hunger and Schmidt 2018; Whitaker 1977, 1995). Many examples of polymorphs are applied as organic pigments, where several polymorphs of the same dye molecule are available on the market as different products with different colours and other physicochemical properties. Another important example is functional dye applications (Zollinger 2003), such as the use of organic dyes as photo-conducting materials (Law 1993). Many patents and publications have been reported on the application of near-infrared (NIR)-absorbing dyes to the charge generation layer of the Carlson process in electrophotography. Much of the research and development of potential dyes has been performed on powder and thin-film states. These efforts have made it possible to replace inorganic compounds with organic dyes as charge generation materials, resulting in the availability of several phthalocyanine (Pc) compounds on the market (Law 1993).

The importance and examples of dye polymorphs are described in books on dyestuffs (Zollinger 2003), organic pigments (Faulkner and Schwartz 2009; Hunger and Schmidt 2018; Whitaker 1977, 1995) and functional dyes (Freeman and Peters 2000; Würthner 2005; Kim 2006). Some reviews on dye polymorphs and their molecular and crystal structures have also been published (Engel 1996; Erk et al. 2004; Lincke 2000, 2002; Paulus et al. 2007). Along with the development of single-crystal X-ray analysis, the number of structural reports on dye polymorphs has markedly increased. Similarly, the development of structural analysis should also be highlighted based on carefully measured powder diffraction data (Hunger and Schmidt 2018). This inspired us to analyse the crystal structure of target polymorphs whose single-crystal growth is difficult under any conditions.

This chapter reviews the molecular and crystal structures of several dyes and coloured organic compounds with four or more polymorphs whose structural data are available from the Cambridge Crystallographic Data Centre (CCDC). The differences in the molecular and crystal structures between polymorphs are briefly described. The number of polymorphs was checked through the Cambridge Structural Database (CSD) (The Cambridge Structural Database. The Cambridge Crystallographic Data Centre. https://www.ccdc.cam.ac.uk/solutions/csd–system/components/csd/), and a keyword search was performed through SciFinder (SciFinder. The American Chemical Society. https://www.cas.org/products/scifinder) using a combination of several related keywords. The results were verified from the viewpoint of coloured compounds and atomic coordinate availability. Some candidates with phase transitions induced by ambient conditions, such as a temperature change, were excluded, as well as the polymorphs observed in molecular complexes, including solvates. Further, some compounds whose polymorphs were reported to be prepared at room temperature were selected. Figure 6.1 shows the chemical structures of the compounds reviewed in this chapter.

The authors re-produced the molecular and crystal structure images for the target compounds to obtain uniform viewing directions. For the molecular geometries, two images are depicted, one viewed parallel and the other perpendicular to the chromophore or related conjugated system. The molecular arrangements are basically

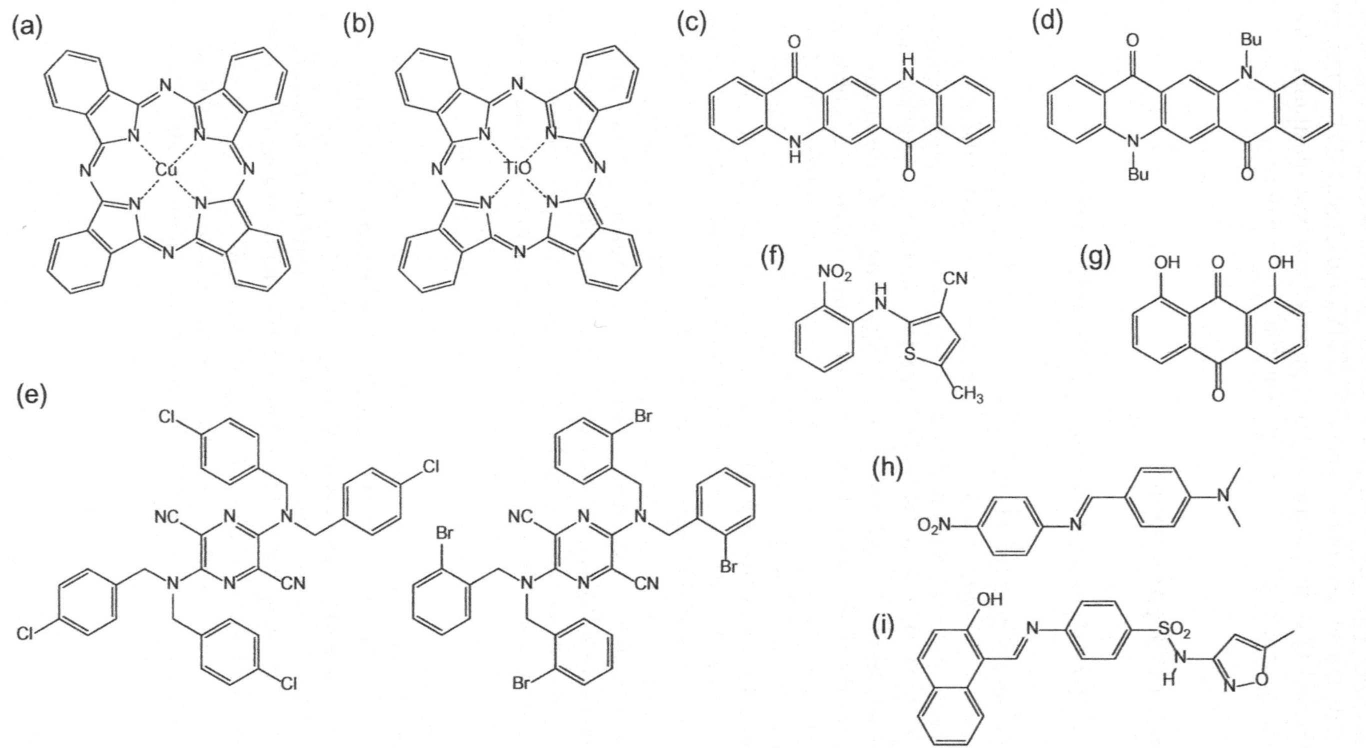

Fig. 6.1 Chemical structures of dyes and related compounds reviewed in this chapter with their CSD codes: **a** CuPc (CUPOCY), **b** TiOPc (BITSAY), **c** QA (QNACRD), **d** *N*,*N*'-butylated QA (WAMFAS), **e** two 2,5-diamino-3,6-dicyano pyrazine dyes (KELFOX and KELGEO), **f** 5-methyl-2-[(2-nitrophenyl)amino]-3-thiophenecarbonitrile: ROY (QAXMEH), **g** 1,8-dihydroxyanthraquinone (DHANQU), **h** *N*-(*p*-dimethylaminobenzylidene)-*p*-nitroaniline (MABZNA) and **i** 1-[(*E*)-({4-[(5-methyl-1,2-oxazol-3-yl)sulphamoyl]phenyl}imino)methyl]naphthalen-2-olate (ZULGUE)

illustrated with a focus on the characteristic low-dimensional molecular arrangement and their further alignment, except for copper Pc (CuPc) and quinacridone (QA). In the case of CuPc and QA, the reported figures provided the best illustrations for outlining the molecular arrangement of the polymorphs of these dyes.

6.2 Examples of Organic Dyes

6.2.1 CuPc (CUPOCY) and Oxotitanyl Pc (BITSAY)

Pcs are one of the most significant classes of organic dyes from both industrial and scientific perspectives. This compound was accidentally discovered in the beginning of twentieth century, after which many derivatives were studied for pigment applications (Zollinger 2003; Hunger and Schmidt 2018; Erk and Hengelsberg 2003; Erk et al. 2004). Among the numerous Pc compounds, organic pigments based on CuPc have significant market importance. In addition to their applications as colouring materials, Pcs have also been investigated for their applicability as functional dyes. Much research has been devoted to fundamental and application-based investigations on Pcs, particularly for optoelectronics. The most successful example of Pcs for functional dye applications is their use as organic photoconductors (OPCs). Pcs have been investigated as potential charge generation materials for electrophotography (Zollinger 2003; Law 1993). In parallel with research on OPCs, their application to solar cells (Kim et al. 2009; Ameri et al. 2009) and semiconductors (Gsänger et al. 2016; Chung and Diao 2016) has also been pursued. An important characteristic of Pc compounds in these applications is that they can be prepared in polymorphic solid phases such as powders and films with NIR absorption or semiconducting properties. Oxotitanyl Pc (TiOPc), a representative OPC compound, has been reported to exhibit several polymorphs (Engel 2003; Xerox 1994). Among them, the Y-form has excellent OPC properties.

Numerous structural reports have been published on Pc compounds. Engle summarised them in both a study report (Engel 1996) and in part of the handbook of Pcs (Engel 2003). The polymorph occurrence of Pc compounds was also reviewed in these articles. This section was written by referencing several books (Hunger and Schmidt 2018; Erk and Hengelsberg 2003; Engel 2003) and from a review by Erk et al. (2004). The polymorphic crystal structures of Pcs, particularly Pcs related to pigment applications, are summarised in these references. The analysed crystal structures of four polymorphs of CuPc and five polymorphs of TiOPc are described in this section.

6.2.1.1 CuPc

CuPc is the most significant type of organic pigment covering blue and green hues. The CuPc pigment is known to have 11 polymorphs (Hunger and Schmidt 2018), although structural analyses have been completed for its α-, β-, γ- and ε-forms (Erk and Hengelsberg 2003; Erk et al. 2004), of which the α- and β-forms are commercially important. The industrial preparation methods of these four polymorphs are described in detail in an excellent pigment book (Hunger and Schmidt 2018). For pigment applications, the polymorphs of CuPc are prepared using acid pasting, milling and solvent treatment. Vacuum sublimation is generally used to grow single crystals of Pcs of analytical quality for X-ray single-crystal structural analysis.

The crystal structures of Pcs, including CuPc, were first reported by Robertson et al. in the early stage of X-ray single-crystal structural analysis (Robertson 1935). The structure of the β-form with good quality was determined by Brown (1968), after which other analytical reports with atomic coordinates on the β-form were available (Jiang et al. 2018). Metal-free Pc and other metal Pc compounds have been considered isostructural (Fábián and Kálmán 2004) in their α- and β-forms, but the α-form of CuPc varies (Erk and Hengelsberg 2003; Erk et al. 2004) from the known α-form Pc structure (Ashida et al. 1966). The accurate crystal structure of the α-form of CuPc was determined by single-crystal X-ray data (Erk 2004) and also confirmed by different groups using the electron diffraction of small single crystals formed on a KCl plate (Hoshino et al. 2003). In the latter study, the lattice constants were directly determined by electron diffraction analysis, the crystal structure was estimated according to the molecular packing energies and the results were confirmed by several experimental considerations, including Rietveld analysis of the powder sample. The crystal structures of the γ- and ε-forms have also been analysed by the combination of Rietveld and energy minimisation methods (Erk et al. 2004).

Figures 6.2 and 6.3 present the molecular conformation and molecular arrangements of the four CuPc polymorphs, respectively, and Table 6.1 lists the corresponding crystallographic parameters. Figure 6.3 was prepared by reference to Fig. 3.2 in Hunger and Schmidt (2018). In all forms, the Pc ring is almost planar. The triclinic α-form (CUPOCY14 (Erk 2004)) creates simple dye-stacking columns along the *a*-axis, which are arranged into a three-dimensional (3D) structure. The molecular planes are tilted towards the stacking axis by approximately 25°. The β-form (CUPOCY10 (Brown 1968)), which is isomorphous to the β-forms of other Pc structures, shows a well-known herringbone arrangement. The γ-form (CUPOCY16 (Erk et al. 2004)) is isomorphous to the known α-form Pc structure, where the molecules align in a herringbone fashion with a tilt angle of approximately 25°. In the ε-form (CUPOCY15 (Erk et al. 2004)), the molecules are also arranged in a herringbone fashion. In the β- and ε-forms, the tilt angle of the molecular plane towards the stacking direction is approximately 46°. In these four polymorphs, the molecules are stacked at a small interplanar distance of approximately 3.4 Å with different degrees of overlap between the two stacking molecules.

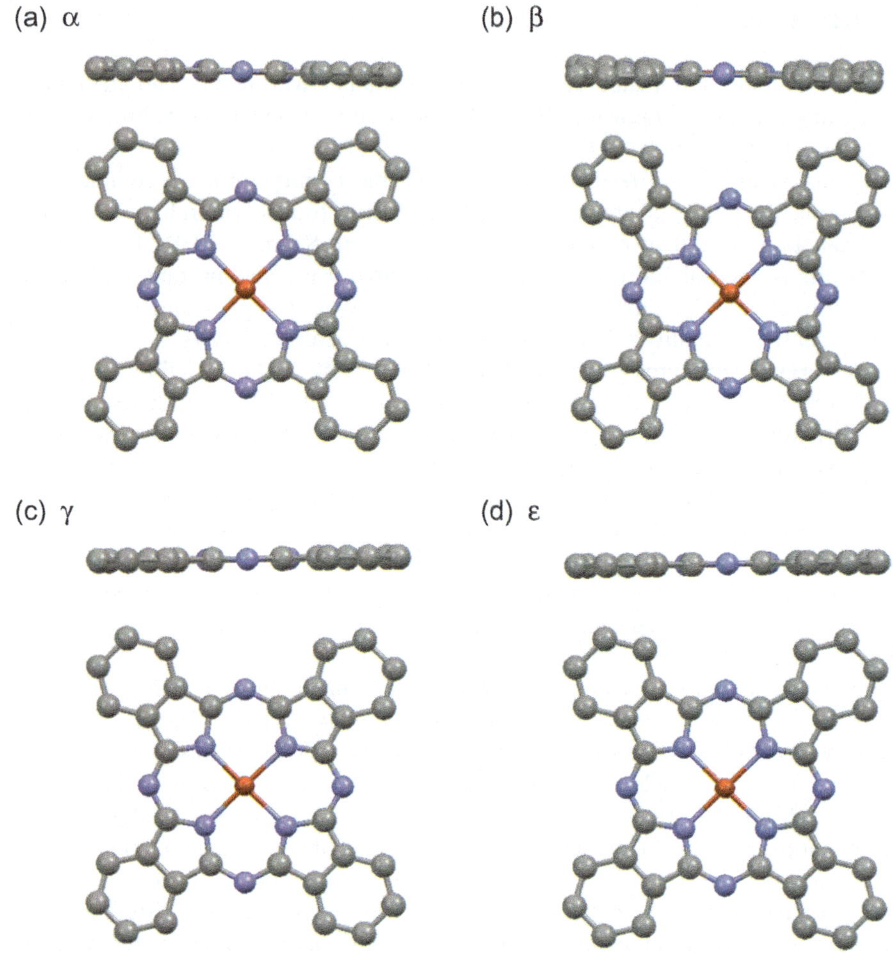

Fig. 6.2 Molecular geometries of the four polymorphs of CuPc viewed parallel (upper) and perpendicular (lower) to the Pc ring: **a** α-form (CUPOCY14), **b** β-form (CUPOCY10), **c** γ-form (CUPOCY15) and **d** ε-form (CUPOCY16)

6.2.1.2 TiOPc

TiOPc is a recognised representative of functional dyes considering its practical application as an OPC and its potential for other optoelectronic applications such as solar cells. This compound is also known to form several polymorphs. In the 1990s, a new polymorph of TiOPc, called the Y-form, was reported, along with its excellent photo-conducting properties for electrophotography (Watanabe et al. 1990). Three other polymorphs—the I-, II- (Hiller et al. 1982) and C-forms (Okada et al. 1993)—and the IV-form have also been reported (Bluhm et al. 1992). These polymorphs have distinct physicochemical properties for use as OPCs. For example, the absorption

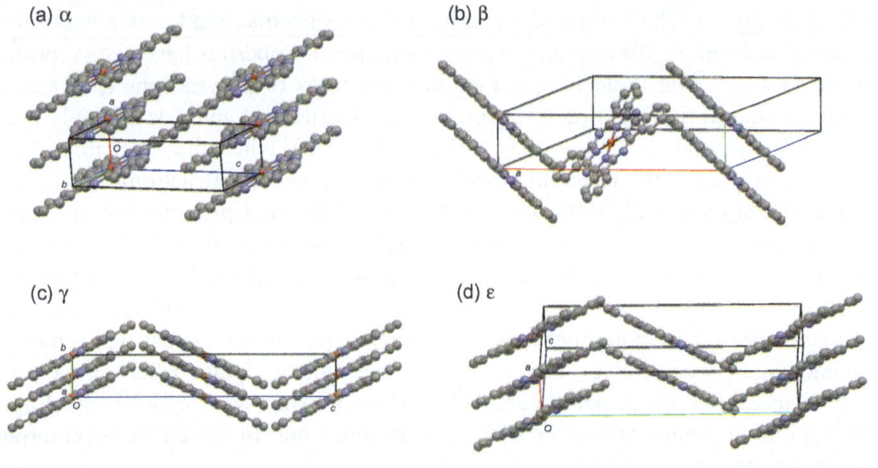

Fig. 6.3 Packing motifs of the four polymorphs of CuPc: **a** α-form (CUPOCY14), **b** β-form (CUPOCY10), **c** γ-form (CUPOCY15) and **d** ε-form (CUPOCY16). This figure was prepared by reference to Fig. 3.2 in Hunger and Schmidt (2018)

Table 6.1 Crystallographic parameters of the four polymorphs of CuPc

Polymorph	α	β	γ	ε
Space group	*P*-1	*P*2$_1$/*a*	*C*2/*n*	*P*2$_1$/*c*
Z	1	2	4	2
a/Å	3.805	19.407	26.33 (3)	5.000 (3)
b/Å	12.959	4.790	3.813 (2)	23.071 (3)
c/Å	12.043	14.628	23.71 (2)	10.571 (5)
α/°	90.64	90	90	90
β/°	95.26	120.56	94.28 (10)	96.02 (6)
γ/°	90.72	90	90	90
V/Å^3	591.27	1166	2374.23	1212.73
d/g/cm^3	1.618	1.639	1.612	1.578
Refcode	CUPOCY14	CUPOCY10	CUPOCY15	CUPOCY16
References	Erk (2004)	Brown (1968)	Erk et al. (2004)	Erk et al. (2004)

spectra of the I-, II-, C- and Y-forms in pigment-dispersed or dye-deposited films are known to exhibit quite different shapes (Watanabe et al. 1990; Mizuguchi et al. 1995). The band of the I-form is slightly broadened with a small bathochromic shift from that observed in a solvent such as chloronaphthalene. The II-form shows a doubly split band shape, where the longer wavelength peak is largely shifted towards the bathochromic NIR region. The C-form exhibits a hypsochromic absorption band, which implies H-aggregate formation in its crystalline state. The photo-conducting properties of the C-form have not been examined because this form does not absorb in the NIR region. The Y-form shows a broad bathochromic band with a maximum in the bathochromic NIR region. No physicochemical properties have been reported for the IV-form. The molecular and crystal structures of polymorphs I, II, C and Y were reviewed by Engel (2003). Single-crystal structural analysis has only been performed on polymorphs I and II (Hiller et al. 1982), while the other three have been analysed using the Rietveld method based on powder diffraction data (Okada et al. 1993; Bluhm et al. 1992; Oka et al. 1992). Table 6.2 provides the crystallographic parameters for the five TiOPc polymorphs. The molecule of TiOPc has a bent macrocyclic shape akin to a shuttlecock, along with a Ti–O bond. The Pc ring in all forms adopts a slightly distorted conformation, as shown in Fig. 6.4. The molecular arrangements of these polymorphs are, therefore, analysed by considering the molecular packing between convex–convex and concave–concave pairs, where convex and concave mean the metal-axial ligand side and the Pc-ring side, respectively (Engel 2003). Figure 6.5 summarises the molecular arrangements of the TiOPc polymorphs considering this structural feature.

In the I-form (BITSAY (Hiller et al. 1982)), one molecule overlaps with four molecules in the neighbouring layer at one terminal phenyl ring to form convex–convex overlapping. On the concave side of the molecule, two molecules overlap significantly. The II-form (BITSAY01 (Hiller et al. 1982)) exhibits convex–convex overlapping between one molecule and two stacked molecules, which occurs at the two phenyl rings. Two molecules also significantly overlap to form concave–concave overlapping. This polymorph is considered isostructural with polymorph II of oxyvanadium Pc (Engel 2003). In the Y-form (BITSAY02 (Oka et al. 1992)), one molecule forms a convex–convex overlap with four molecules in the neighbouring layer, similar to that observed in the I-form, but with a different molecular arrangement. The molecules also form concave–concave overlapping with a large

Table 6.2 Crystallographic parameters of the five polymorphs of TiOPc

Polymorph	I	II	Y	IV	C
Space group	$P2_1/c$	P-1	$P2_1/c$	P-1	Cc
Z	4	2	4	2	4
a/Å	13.411 (6)	12.166 (4)	13.85	10.83	25.19
b/Å	13.230 (3)	12.584 (5)	13.92	13.12	3.85
c/Å	13.810 (4)	8.641 (3)	15.14	9.96	25.46
α/°	90	96.28 (3)	90	72.28	90
β/°	103.72 (3)	95.03 (4)	120.22	77.25	90.3
γ/°	90	67.86 (4)	90	104.48	90
V/Å^3	2381.2	1216.4	2522.19	1235.81	2469.12
d/g/cm^3	1.608	1.574	1.52	1.549	1.54
Refcode	BITSAY	BITSAY01	BITSAY02	BITSAY03	BITSAY04
References	Hiller et al. (1982)	Hiller et al. (1982)	Oka et al. (1992)	Bluhm et al. (1992)	Okada et al. (1993)

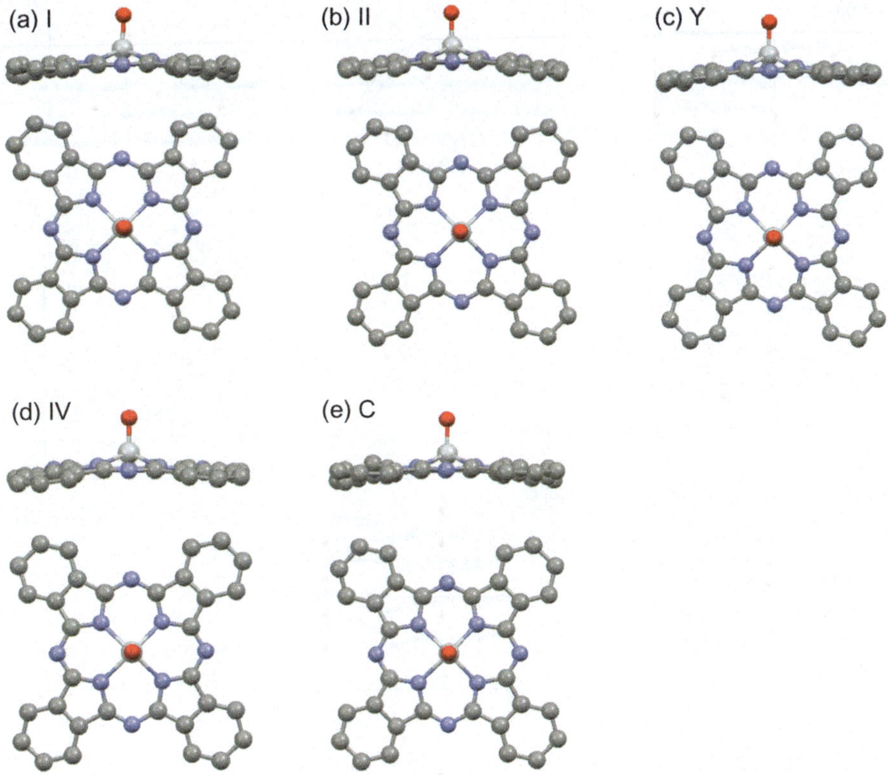

Fig. 6.4 Molecular geometries of the five polymorphs of TiOPc viewed parallel (upper) and perpendicular (lower) to the Pc ring: **a** I-form (BITSAY), **b** II-form (BITSAY01), **c** Y-form (BITSAY02), **d** IV-form (BITSAY03) and **e** C-form (BITSAY04)

molecular overlap. Several reports have stated that this polymorph includes water molecules (Oka et al. 1992; Fujimaki 1991). Further, the structure of the IV-form (BITSAY03 (Bluhm et al. 1992)) is characterised by the two overlapping structures. In the convex–convex overlapping, one molecule overlaps with a combination of one and two neighbouring molecules by two and one phenyl rings, respectively. The molecule also forms concave–concave overlapping with a molecule in the neighbouring layer. The structure of the C-form (BITSAY04 (Okada et al. 1993)) is quite different from those of the other four reported polymorphs. In general, shuttlecock-type Pc compounds exhibit interactions between similar sides of the molecule, such as the convex–convex overlap, in their molecular arrangement (Engel 2003). In this polymorph, however, the opposite sides of different molecules interact with each other. The molecules are connected through the axial oxygen atom and Ti atom to form a one-dimensional (1D) chain via convex–concave overlapping. This 1D structure has also been observed in the monoclinic polymorph of PbPc (Engel 2003). Recently, single-crystal structural analysis of the C-form grown on a substrate by

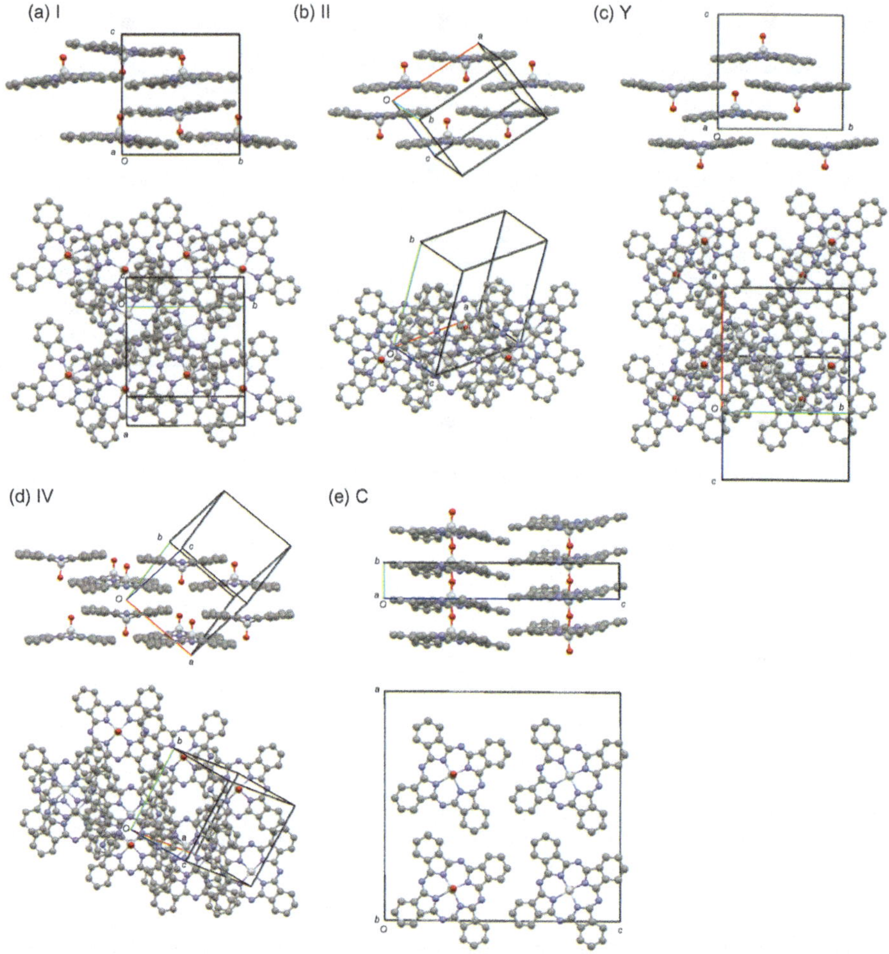

Fig. 6.5 Packing motifs of the five polymorphs of TiOPc: **a** I-form (BITSAY), **b** II-form (BITSAY01), **c** Y-form (BITSAY02), **d** IV-form (BITSAY03) and **e** C-form (BITSAY04). For all polymorphs except for the C-form, the molecular arrangements are illustrated based on their convex–convex and concave–concave overlapping. A columnar arrangement of molecules is depicted for the C-form

vacuum deposition was reported, although the space groups was different (C2/c) (Yoon et al. 2020).

CuPc and TiOPc as two examples, and polymorph reports on other Pcs clearly indicate that a simple, rigid dye chromophore has significant potential to exhibit polymorphism with a variety of molecular arrangements with or without structural deformation. Polymorphism in Pcs also shows that polymorphism in functional dyes has a notable impact on the development of practical organic functional materials.

6.2.2 *QA (QNACRD) and Its* N,N'*-Dibutylated Derivative (WAMFAS)*

6.2.2.1 QA

QA (QA, 5,12-dihydroquinolino[2,3-b]acridine-7,14-dione) is another important dye with applications to organic pigments intensively studied in the middle of the twentieth century (Zollinger 2003; Faulkner and Schwartz 2009; Hunger and Schmidt 2018; Lincke 2000, 2002; Labana and Labana 1967). Pc, QA and diketopyrrolopyrrole were regarded as the three major organic dye chromophores in the twentieth century because of their impact on the realms of both industrial applications and chemistry (Zollinger 2003). Many polymorphs of QA in the form of powders have been reported in the literature and patents, but, currently, many can be considered as groupable into one of the four polymorphic forms with successfully analysed structures (Paulus et al. 2007). The three polymorphic forms α, β and γ were reported in the early stage of the development of QA pigment applications. Several structural reports based on single crystals are available for the β- and γ-forms (Paulus et al. 1989; Nishimura et al. 2006; Potts et al. 1994; Mizuguchi et al. 2002), and many attempts have been made towards the structural analysis of the α-form (Leusen 1994, 1996; Lincke and Finzel 1996). In 2007, Schmidt et al. showed that the reported α-form includes two polymorphs, α^{I} and α^{II}, and analysed the structure of the α^{I}-form using the Rietveld method based on powder diffraction data (Paulus et al. 2007). The structure of the α^{II}-form was reportedly difficult to analyse because of its small crystal size and poor crystallinity. Finally, the α^{II}-form structure was investigated using 3D electron diffraction combined with X-ray diffraction measurements and DFT calculations (Gorelik et al. 2016), although the structure is considered to require further experimental consideration. Table 6.3 lists the crystallographic parameters of these four polymorphs of QA. Numerous methods have been reported for the preparation of the reported QA crystal forms, particularly for pigment preparation. For crystallographic analysis, vapour sublimation has generally been used. The α-form is metastable and has no direct commercial importance, although it is used as an intermediate to prepare the reddish-violet β- and red γ-forms, which are important pigments. Some substituted QAs have also been applied as notable. For QA, the formation of solid solutions (mixed crystals) is also an essential technical and scientific characteristic. Some solid solutions are also available on the market as organic pigments (Hunger and Schmidt 2018). These QA pigments cover a wide range of colours from orange to violet with good to excellent fastness.

Figure 6.6 shows the molecular structures of the four polymorphic forms of QA. Among them, the structure of the α^{II}-form should be cautiously interpreted because of the limited data accuracy arising from the measurement method and sample crystallinity, although the determined structure has been confirmed theoretically. It is also noted that the quality of the reported analyses is diverse and depends largely on the quality of the measured crystals, which may render it difficult to quantitatively discuss their molecular structures. Nevertheless, Fig. 6.6 shows no significant

Table 6.3 Crystallographic parameters of the four polymorphs of QA

Polymorph	α^{I}	β	γ	α^{II}
Space group	*P*-1	$P2_1/c$	$P2_1/c$	$P2_1/c$
Z	1	2	2	2
a/Å	3.802 (1)	5.692 (1)	13.697 (9)	7.1
b/Å	6.612 (3)	3.975 (1)	3.881 (3)	28.4
c/Å	14.485 (6)	30.02 (4)	13.4020 (10)	3.9
α/°	100.68 (8)	90	90	90
β/°	94.40 (6)	96.76 (6)	100.44 (1)	110
γ/°	102.11 (5)	90	90	90
$V/\text{Å}^3$	346.7 (1)	674.5 (9)	700.6 (7)	734
$d/\text{g/cm}^3$	1.493	1.538	1.480	–
Refcode	QNACRD06	QNACRD07	QNACRD08	CCDC 1425922
References	Paulus et al. (2007)	Paulus et al. (2007)	Paulus et al. (2007)	Gorelik et al. (2016)

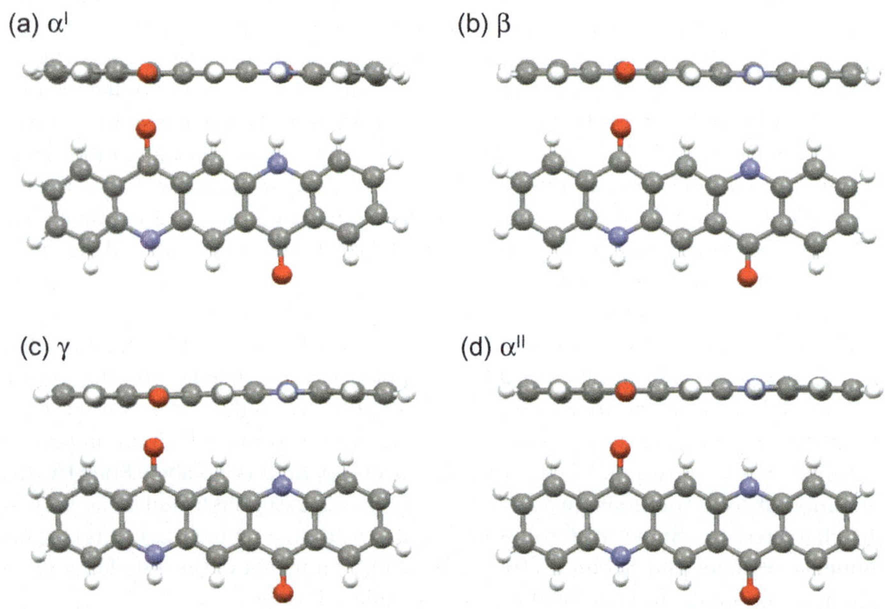

Fig. 6.6 Molecular geometries of the four polymorphs of QA viewed parallel (upper) and perpendicular (lower) to the QA ring: **a** α^{I}-form (QNACRD06), **b** β-form (QNACRD07), **c** γ-form (QNACRD08) and **d** α^{II}-form (CCDC 1425922)

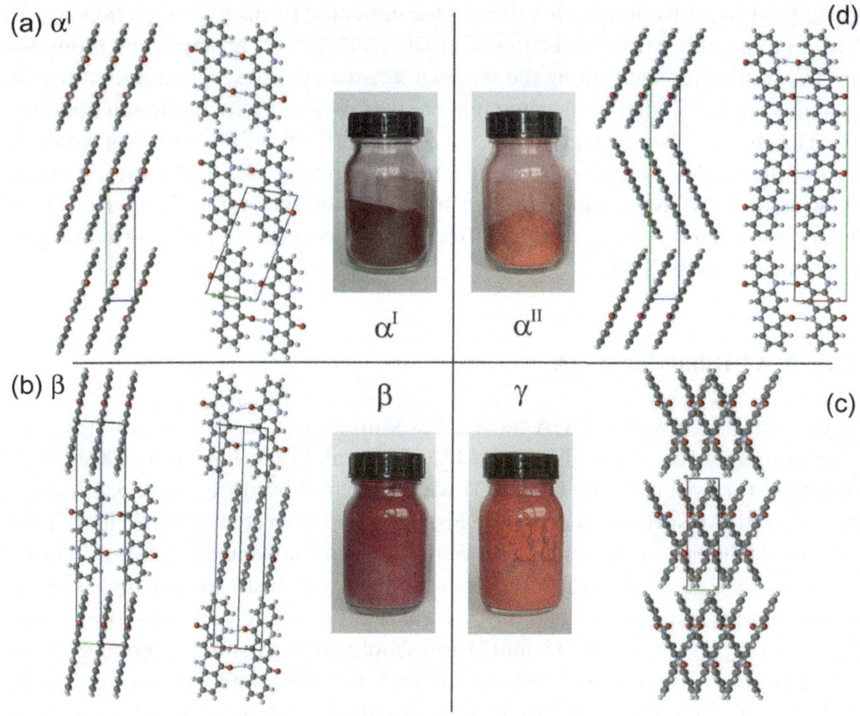

Fig. 6.7 Packing motifs of the four polymorphs of QA: **a** α^{I}-form (QNACRD06), **b** β-form (QNACRD07), **c** γ-form (QNACRD08) and **d** α^{II}-form (CCDC 1425922). This figure was reproduced from Fig. 6.5 in Gorelik et al. (2016), published by the Royal Society of Chemistry

difference in the planarity of the chromophores in these four forms. This implies that the colour difference between these polymorphs is dependent on their crystal structure, particularly on hydrogen bonding and the molecular arrangement (Hunger and Schmidt 2018; Paulus et al. 2007). Figure 6.7 illustrates their characteristic molecular arrangements with the colours of the powder samples from Gorelik et al. (2016).

The molecular arrangements of the α^{I}-, α^{II}- and β-forms consist of chains formed by a hydrogen-bonding network with adjacent molecules. In this structure, one molecule is bonded to two neighbouring molecules by two complementary hydrogen bonds each. These chains are packed to form a two-dimensional (2D) molecular layer. The difference in the molecular arrangements of these three forms arises from the mutual arrangement of these molecular layers. In the α^{I}-form, the 2D layers are simply aligned along the *c*-axis, whereas they align in a herringbone fashion in the α^{II}-form, although the latter has been reported to exhibit the severe structural disorder in this layer arrangement (Gorelik et al. 2016). In the β-form, the molecules in neighbouring layers are skewed by approximately 70°. This structure has been reported as unique to the β-form because no similar structure has been observed in any other QA pigment (Hunger and Schmidt 2018). The molecular arrangement of the γ-form is completely different from the other three. One molecule forms hydrogen bonds with four different neighbouring molecules to create a criss-cross arrangement along the *c*-axis. The molecules stack along the *b*-axis to form a 1D stacking column; thus, the 1D columns are considered to arrange in a criss-cross pattern by hydrogen bonding. These 1D columns also arrange along the *a*-axis to form a 2D molecular layer, in which the molecules also tend to form an almost planar chain-like arrangement. These structural differences in hydrogen bonding patterns and molecular arrangements are considered reasons for the different colours of these polymorphic forms (Hunger and Schmidt 2018).

6.2.2.2 *N,N'*-Dibutylated QA

The polymorph occurrence of QA derivatives with more than four polymorphs has also been reported for *N,N'*-dibutylated QA (Ye et al. 2005; Fan et al. 2009). Four polymorphic forms (i.e. A, B, C and D) were prepared by different crystallisation conditions, and two solvated forms were also obtained (Fan et al. 2009). The B-form was obtained from its dichloromethane solution by slow solvent evaporation. The C- and D-forms were grown from a chloroform solution by diffusing petroleum ether and methanol vapour, respectively, and the A-form was grown by vacuum sublimation.

Figure 6.8 depicts the A-, B-, C- and D-form molecular structures reproduced from the related figures in Fan et al. (2009) using their CIF files, and Table 6.4 lists their crystallographic parameters. The molecular structural differences mainly arise from the conformation of the butyl groups. The QA core has been reported to be slightly distorted in the A-, B- and C-forms. The average distances from the mean plane consisting of the QA chromophore have been estimated as 0.074 Å for the A-form, 0.080 Å for the B-form, 0.098 Å for the C-form, and 0.059 and 0.044 Å for the two independent molecules of the D-form. These values indicate that the plane of the QA core of the two independent D-form units was also slightly distorted. No significant differences were observed in the bond lengths of these four polymorphic forms. Among them, only in the A-form are both butyl groups projected onto the same side against the QA core. The butyl groups also adopted a simple zig-zag conformation in this form. In the other forms, both butyl groups were projected onto the opposite side of the QA core. In the D-form, there are two independent half-molecular units, one with a simple zig-zag conformation of the butyl groups, as observed in the A-form, and the other with butyl groups projected out of the QA core, almost perpendicular, with a slightly elongated zig-zag conformation. The molecules in the B- and C-forms show similar butyl group zig-zag conformations with bent terminal methyl groups.

The molecular arrangements of the four polymorphic forms of *N,N'*-dibutylated QA are different, as shown in Fig. 6.9. The B- and C-forms, which have similar conformations, also exhibit 2D structural similarity in terms of their molecular arrangement. In the A-form, two molecules stack to form a dimer-like structure with a distance of 3.492 Å between QA rings. The crystal structure of this form can be interpreted according to the arrangement of this dimer-like unit, such as in anthracene (Kitaigorodskii 1973). The molecules also form six weak hydrogen bonds with four

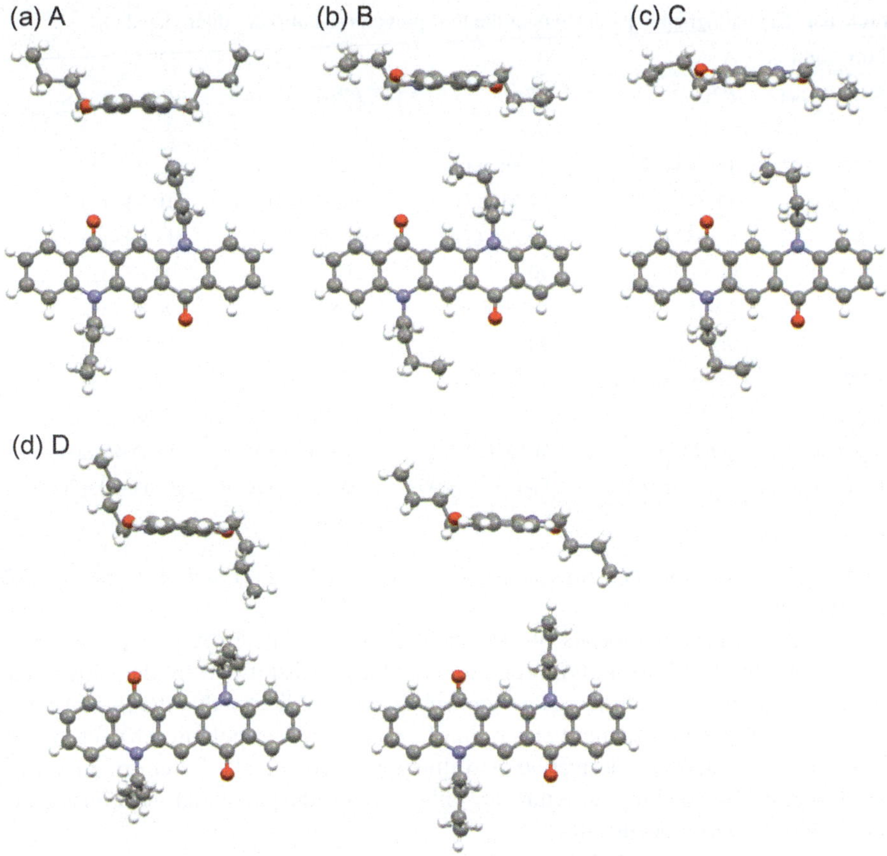

Fig. 6.8 Molecular geometries of the four polymorphs of *N*,*N'*-butylated QA viewed parallel (upper) and perpendicular (lower) to the QA ring: **a** A-form (WAMFAS01), **b** B-form (WAMFAS02), **c** C-form (WAMFAS03) and **d** D-form (WAMFAS04)

adjacent molecules. These two structural features determined the 3D structure of the A-form. 2D structural similarity occurs in the crystal structures of the B- and C-forms, where their molecules form a 2D molecular layer as shown in Fig. 6.9b, c. Their 3D structures are characterised by the stacking of these 2D layers. In the B-form, the molecules stack in a staircase fashion with a distance of 3.463 Å between QA rings, where the stacked molecules are slipped along the long molecular axis. In this form, hydrogen bonding has no considerable contribution to the layered stacking structure. In the C-form, the molecules stack with alternate slipping along the long molecular axis to form a 'sawtooth' arrangement with two stacking distances of 3.494 and 3.527 Å. Weak hydrogen bonding was observed in these structural features. The two independent molecules in the D-form arrange in separate stacking columns, as shown in Fig. 6.9d. In these two columns, the molecules are stacked with slipping along the long molecular axis at distances between QA cores of 3.433 and 3.396

Table 6.4 Crystallographic parameters of the four polymorphs of *N*,*N*'-dibutylated QA

Polymorph	A	B	C	D
Space group	$P2_1/n$	$P2_1/c$	$P2_1/c$	P-1
Z	4	2	4	2
a/Å	13.443 (3)	7.2362 (14)	7.5410 (15)	7.1938 (14)
b/Å	11.745 (2)	14.737 (3)	14.766 (3)	10.337 (2)
c/Å	15.632 (3)	10.360 (2)	20.209 (5)	15.022 (3)
α/°	90	90	90	94.67 (3)
β/°	114.52 (3)	106.60 (3)	108.08 (3)	90.58 (3)
γ/°	90	90	90	101.87 (3)
V/Å^3	2245.6 (8)	1058.7 (4)	2139.2 (8)	1089.1 (4)
d/g/cm^3	1.256	1.332	1.318	1.295
Refcode	WAMFAS01	WAMFAS02	WAMFAS03	WAMFAS04
References	Fan et al. (2009)	Fan et al. (2009)	Fan et al. (2009)	Fan et al. (2009)

Å. These two separate columns interact by weak hydrogen bonding to form a 3D arrangement.

The reports on polymorphs of QA and *N*,*N*'-dibutylated QA indicate that QA has the potential to form polymorphs by the QA ring itself and by the substituted butyl groups. Analysis of the QA polymorphs clearly indicates that the formation of different hydrogen bonds might play a significant role in their polymorph occurrence. The *N*,*N*'-butylated QA derivative also illustrates that the alkyl conformation and variations in the stacking structure are other important parameters to consider in terms of polymorph occurrence.

6.2.3 2,5-Diamino-3,6-Dicyano Pyrazine Dyes (KELFOX and KELGEO)

2,5-Diamino-3,6-dicyanopyrazine, an organic dye based on a pyrazine skeleton, was largely developed by Matsuoka's group (Matsuoka 2000). Many pyrazine dyes have been reported, some of which are known to exhibit strong fluorescence in solution (Matsuoka 2000). Some pyrazine dyes have been reported to exhibit polymorphs with different colours and/or emissions (Matsuoka 2000). In particular, two 2,5-diamino-3,6-dicyanopyrazine dyes with benzyl substituents were found to have more than four differently coloured polymorphs. The bulky benzyl substituents were introduced to this dye chromophore to impede the negative effects of stacking interactions between chromophores on solid-state fluorescence towards the development of a novel material for organic electroluminescent devices (Shirai et al. 1998; Kim et al. 1998). The colour difference between the polymorphs is considered to correspond to their different molecular conformations (Matsumoto et al. 2006). A recent report on the

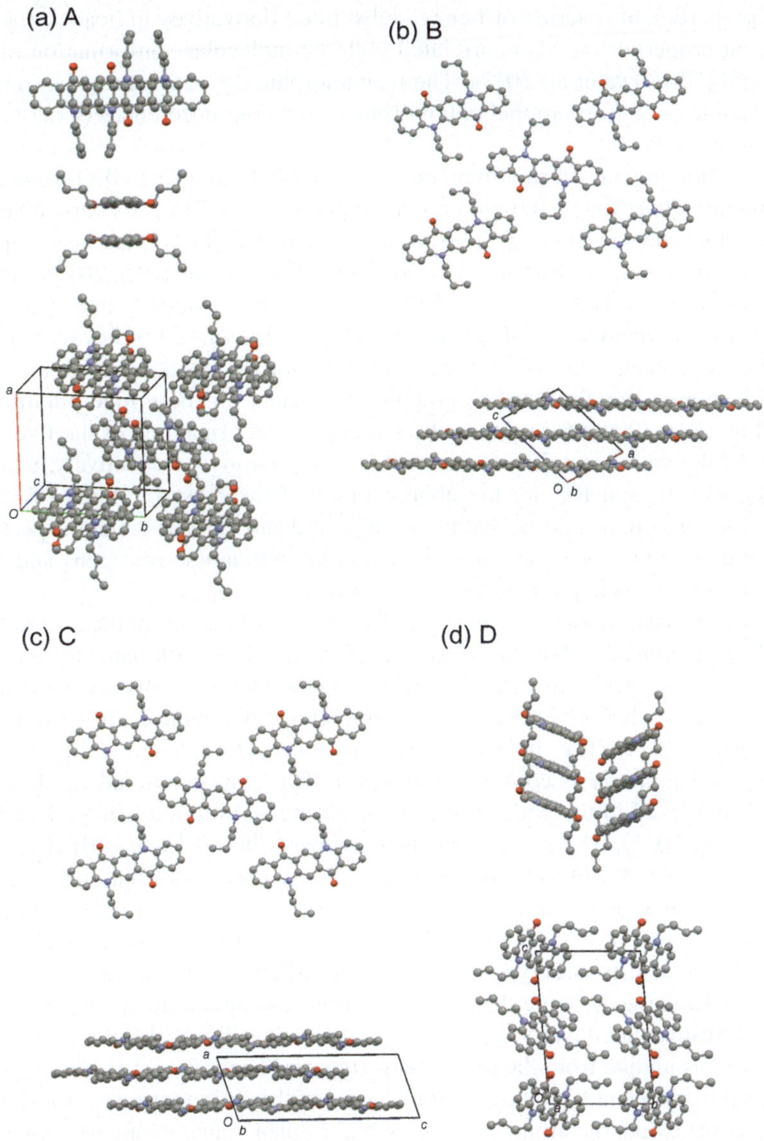

Fig. 6.9 Packing motifs of the four polymorphs of *N*,*N'*-butylated QA: **a** A-form (WAMFAS01), **b** B-form (WAMFAS02), **c** C-form (WAMFAS03) and **d** D-form (WAMFAS04). For the A-form, the stacking dimers and their arrangement are illustrated. The arrangements of the B- and C-forms are viewed perpendicular (upper) and parallel (lower) to the QA ring. For the D-form, the two columns formed by the two independent molecules are viewed from two directions

optical properties of a series of benzyl-substituted derivatives indicated that their fluorescent properties are also correlated with the molecular conformation of each polymorph (Hirosawa et al. 2017). The pentamorphic derivative has been reported as an example of conformational polymorphs with colour differences (Cruz-Cabeza and Bernstein 2014).

Two 2,5-diamino3,6-dicyanopyrazine dyes (KELFOX and KELGEO) were found to exhibit more than four differently coloured polymorphs. The *p*-Cl derivative (2,5-bis(bis(4-chlorobenzyl)amino)-3,6-dicyanopyrazine, KELFOX) has been reported to have five polymorphs (Matsumoto et al. 2006; Akune et al. 2015, 2017a), as well as an amorphous solid (Akune et al. 2017a) and benzene solvate (Akune et al. 2016), and the *o*-Br derivative (2,5-bis(bis(2-bromobenzyl)amino)-3,6-dicyanopyrazine, KELGEO) has been reported to have four polymorphs (Matsumoto et al. 2006). Table 6.5 summarizes the crystallographic parameters for these nine polymorphic forms. Figures 6.10 and 6.11 depict the molecular conformations of the five forms of the *p*-Cl derivative and four forms of the *o*-Br derivative, respectively, where Y, YO, DO, RO, R, and RV are the abbreviations of the crystal colours of yellow, yellowish-orange, dark-orange, reddish-orange, red and reddish-violet, respectively. The chromophore is composed of a pyrazine ring with amino nitrogens and cyano groups and adopts a planar geometry in all forms.

The structural similarity in terms of molecular conformation for the crystal forms of the 12 analogous 2,5-diamino3,6-dicyanopyrazine dyes with benzyl groups was examined (Akune 2017) using the RMSD tool in the Mercury software according to the methods described in a review on conformational polymorphism (Cruz-Cabeza and Bernstein 2014). The molecular structures of the *p*-Cl Y and *p*-Cl DO forms were grouped similarly as conformation type 1. This conformational similarity has also been observed in the yellow crystals of other *p*-substituted benzyl derivatives (Akune et al. 2015). The red and reddish forms of the *p*-Cl and *o*-Br derivatives (*p*-Cl R, *p*-Cl RV, *o*-B RO and *o*-Br R) were categorised as conformation type 2. This structural type has also been confirmed for the molecular structures of orange- and red-coloured forms of other *p*- and *o*-derivatives. The YO forms of both derivatives are categorised as structural type 3, in which four benzyl groups spread upwards and downwards from the pyrazine skeleton. This group also applies to the orange form of the *p*-F derivative and the red form of the *o*-I derivative. The molecular shape of the *o*-Br Y form is unique from the other forms (type 4). These four structural types are considered to correlate with the electronic states of the amino nitrogen atoms. In the type 1 conformation, the amino groups form a trigonal planar geometry, suggesting sp^2-like hybridisation of the amino nitrogen atom. Meanwhile, a tetragonal geometry occurs in the type 2 conformation, indicating that the amino nitrogen atoms adopt sp^3-like hybridisation. In type 3, the *p*- and *o*-derivatives have different geometries at the amino position, where the former adopts a trigonal geometry and the latter adopts a tetragonal geometry. The yellow phase of the *o*-Br derivative also has trigonal planar amino groups, but the phenyl ring conformations are different from those of types 2 and 3. These structural differences in the electronic states of the amino nitrogen atom have also been reported to be correlated with the optical properties of these crystal forms (Matsumoto et al. 2006; Hirosawa et al. 2017).

Table 6.5 Crystallographic parameters of (a) the five polymorphs of the *p*-Cl pyrazine dye and (b) four polymorphs of the *o*-Br pyrazine dye

(a)

Polymorph	Y	YO	DO	R	RV
Space group	$C2/c$	$P2_1/n$	$P2_1/n$	$P2_1/n$	$P2_1/c$
Z	4	2	2	2	2
a/Å	23.485 (5)	7.680 (2)	14.6554 (5)	9.441 (2)	4.70709 (16)
b/Å	5.974 (1)	18.549 (4)	6.00711 (18)	11.084 (2)	26.1212 (12)
c/Å	22.593 (7)	11.229 (3)	17.9996 (5)	15.303 (3)	12.1528 (5)
α/°	90	90	90	90	90
β/°	102.69 (1)	92.71 (2)	90.1994 (14)	104.380 (9)	93.267 (3)
γ/°	90	90	90	90	90
V/Å^3	3092.3 (1)	1597.9 (6)	1584.2 (8)	1551.1 (5)	1491.82 (11)
d/g/cm^3	1.414	1.368	1.380	1.410	1.466
Refcode	KELFOX01	KELFOX02	KELFOX03	KELFOX	KELFOX04
References	Matsumoto et al. (2006)	Akune et al. (2015)	Akune et al. (2017)	Matsumoto et al. (2006)	Akune et al. (2017)

(b)

Polymorph	Y	YO	RO	R
Space group	P-1	P-1	$P2_1/n$	P-1
Z	1	1	2	1
a/Å	8.55 (4)	7.6207 (9)	12.831 (2)	8.013 (3)
b/Å	10.05 (3)	10.209 (2)	7.241 (1)	9.204 (4)
c/Å	10.93 (7)	11.121 (1)	17.990 (1)	11.328 (5)
α/°	103.5 (2)	79.85 (1)	90	77.00 (2)
β/°	107.3 (3)	84.07 (2)	108.516 (5)	87.33 (2)
γ/°	112.8 (5)	71.51 (1)	90	80.09 (3)
V/Å^3	759 (8)	806.6 (2)	1584.9 (4)	801.8 (6)
d/g/cm^3	1.830	1.721	1.752	1.732
Refcode	KELGEO02	KELGEO03	KELGEO01	KELGEO
References	Matsumoto et al. (2006)	Matsumoto et al. (2006)	Matsumoto et al. (2006)	Matsumoto et al. (2006)

Figures 6.12 and 6.13 show the 1D column structures and molecular arrangement of the five forms of the *p*-Cl derivative and four forms of the *o*-Br derivative, respectively. The Y and DO forms of the *p*-Cl derivative form the same packing arrangement in their 1D columns. These two forms are regarded as 3D isostructures except for a small difference in the packing of the phenyl rings of the adjacent benzyl groups. In the Y-form, the phenyl rings stack in a face-to-face manner, while the DO-form adopts herringbone packing. This confined structural difference is considered an important factor in their significantly different fluorescence quantum yields (Y: 64%, DO: 14%)

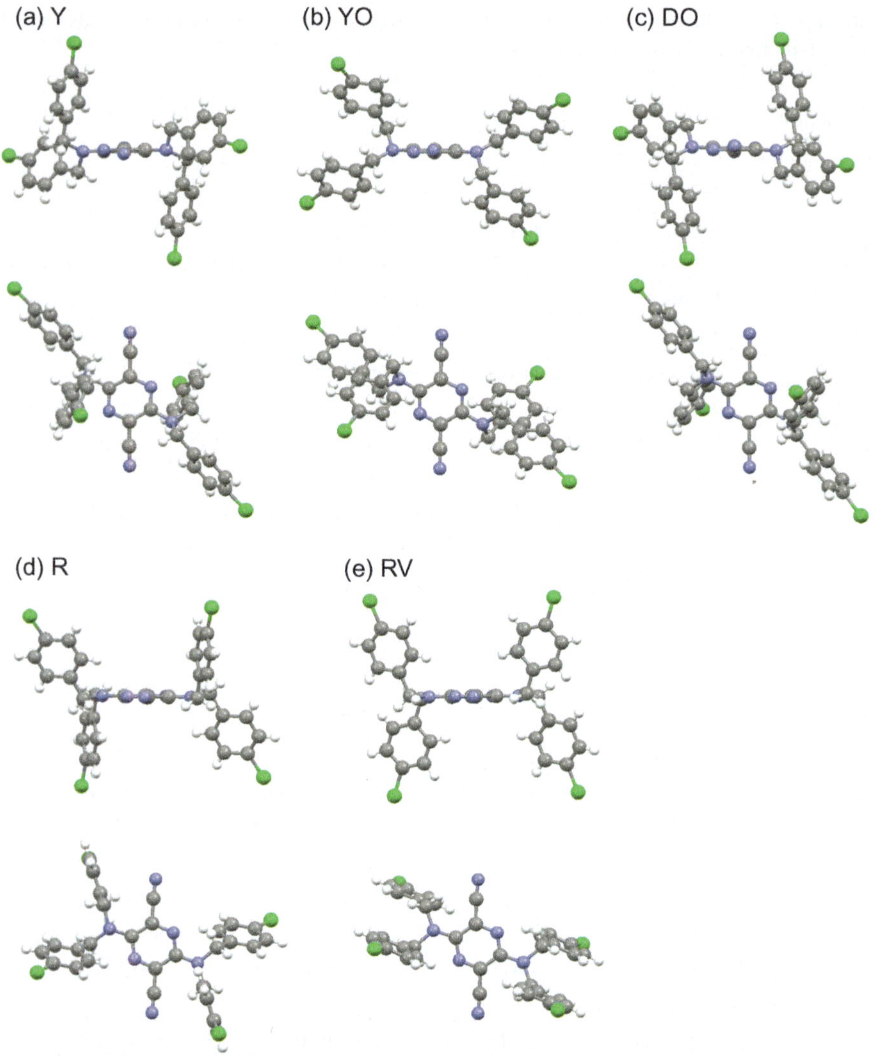

Fig. 6.10 Molecular geometries of the five polymorphs of the *p*-Cl pyrazine dye viewed parallel (upper) and perpendicular (lower) to the pyrazine ring: **a** Y-form (KELFOX01), **b** YO-form (KELFOX02), **c** DO-form (KELFOX03), **d** R-form (KELFOX) and **e** RV-form (KELFOX04)

(Akune et al. 2017a). The isostructural feature of these two forms was also confirmed by Hirshfeld surface analysis, which found that the relative contributions of the Cl atom interactions to the Hirshfeld surface area were almost the same. The four red and reddish-orange forms (R and RV of *p*-Cl, RO and R of *o*-Br) grouped as type 2 also form 1D columns, although they have no structural similarity, as reflected by their completely different crystal structures. Various halogen interactions have been

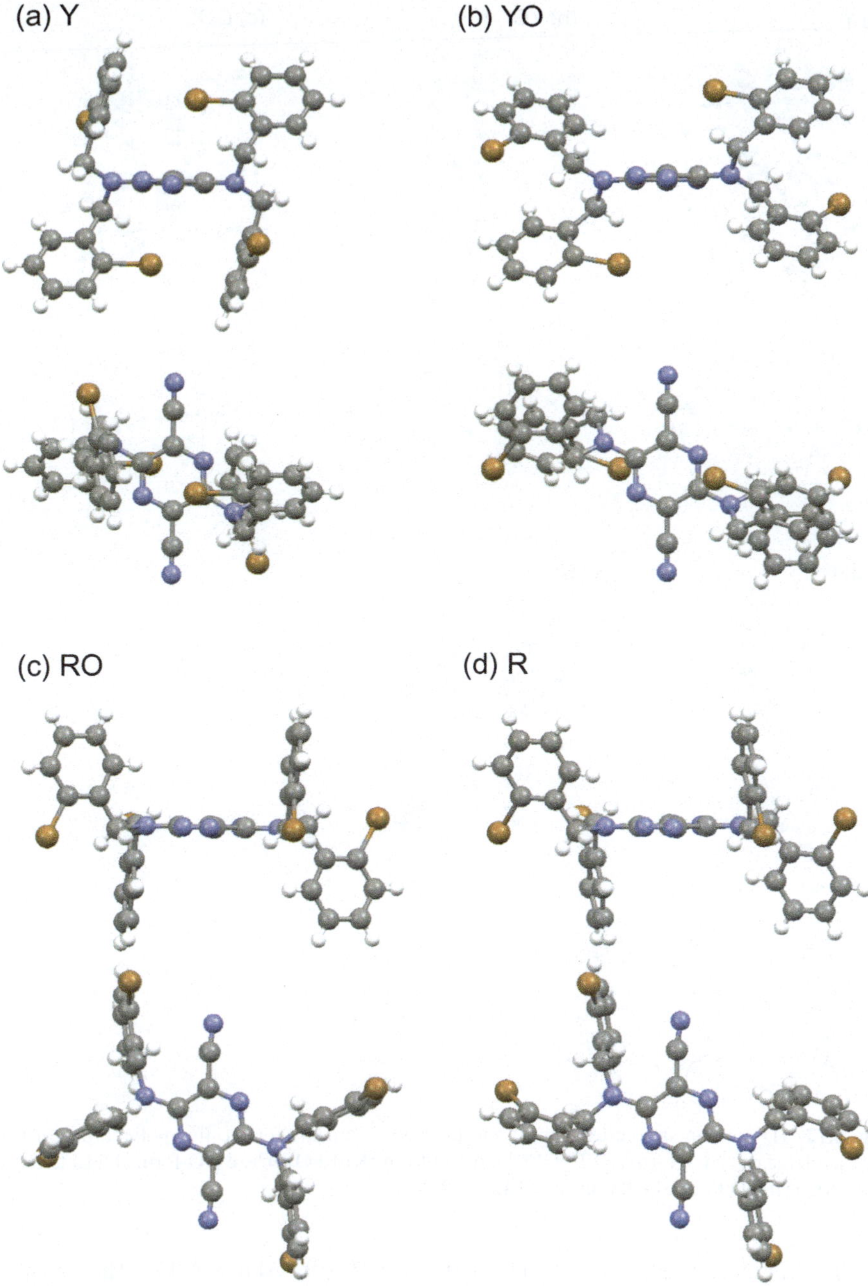

Fig. 6.11 Molecular geometries of the four polymorphs of the *o*-Br pyrazine dye viewed parallel (upper) and perpendicular (lower) to the pyrazine ring: **a** Y-form (KELGEO02), **b** YO-form (KELGEO03), **c** RO-form (KELGEO01) and **d** R-form (KELGEO)

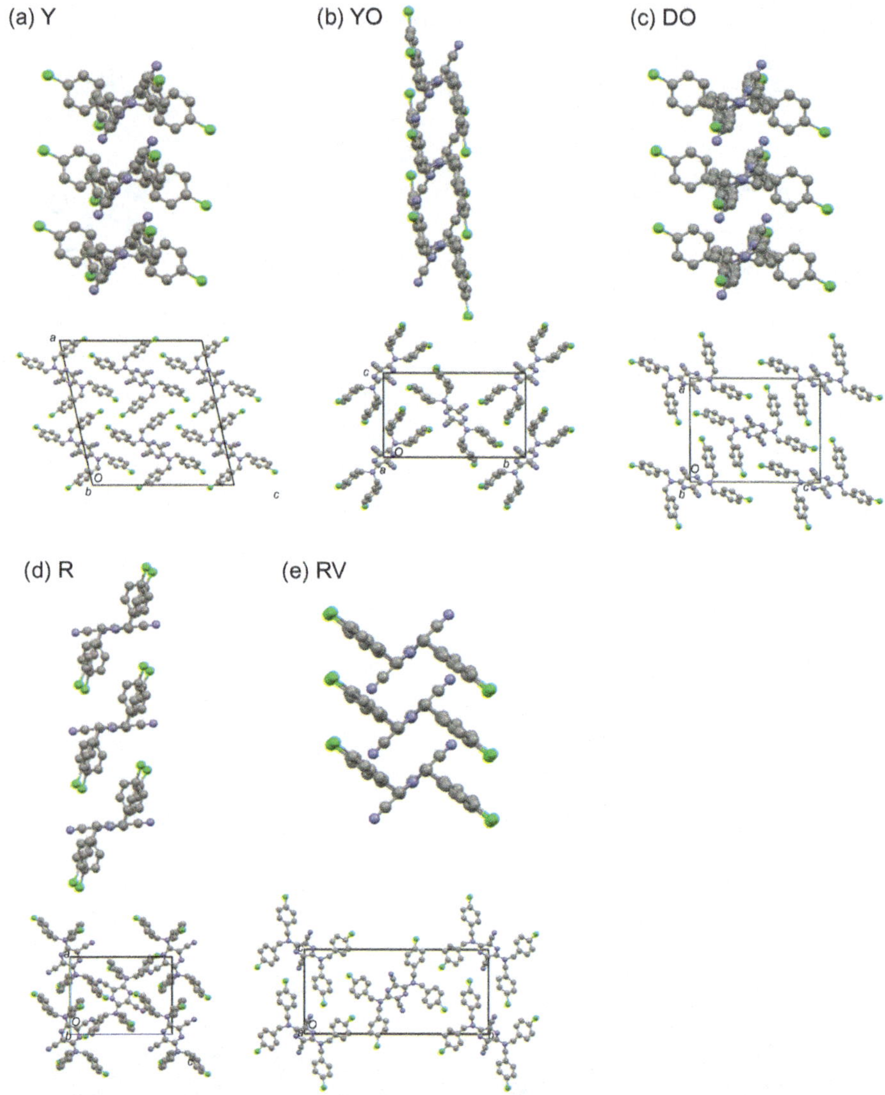

Fig. 6.12 1D column structures (upper) and their arrangements for the five polymorphs of the *p*-Cl pyrazine dye: **a** Y-form (KELFOX01), **b** YO-form (KELFOX02), **c** DO-form (KELFOX03), **d** R-form (KELFOX) and **e** RV-form (KELFOX04)

observed in these four forms (Akune et al. 2015, 2017b; Akune 2017). The two YO forms grouped in type 3 appear to form similar 1D columns, but the column arrangements are completely different. This difference is determined by the different motifs of the intermolecular halogen–N interaction. In the YO form of the *p*-Cl derivative, one dye forms four Cl–N interactions through its two cyano nitrogen atoms and two

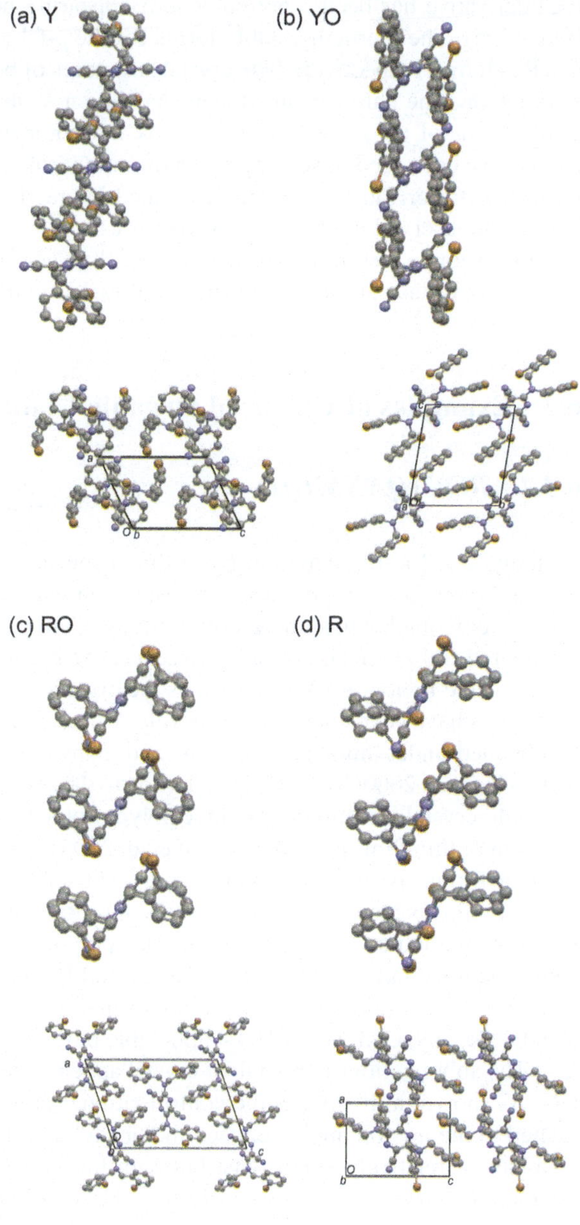

Fig. 6.13 1D column structures (upper) and their arrangements for the four polymorphs of the *o*-Br pyrazine dye: **a** Y-form (KELGEO02), **b** YO-form (KELGEO03), **c** RO-form (KELGEO01) and **d** R-form (KELGEO)

terminal Cl atoms to form a 3D column arrangement. Conversely, in the YO form of the *o*-Br derivative, one molecule forms two complementary Br–N interactions between two neighbouring molecules to form the column arrangement along the *b*-axis. This structural difference is considered to arise from the substituted position of the halogen atom. Unlike the YO form of the *o*-Br derivative, the YO form of the *p*-Cl derivative has been reported as a disappearing polymorph (Dunitz and Bernstein 1995). The thermally stable forms for the *p*-Cl and *o*-Br derivative are the Y- and RO-forms, respectively. The crystal structures of both are characterised by close packing, and the halogen substituents have no notable energy impact. In the meta-stable forms of the *p*-Cl derivative, weak intermolecular interactions involving Cl atoms have been recognised to play an important role in the molecular arrangement. For the *o*-Br derivatives, various Br-related interactions, such as Br–N and Br–Br, have been observed in their crystal structures.

This example clearly indicates that weak halogen interactions with respect to Cl and Br atoms have an impact on structural variation in a crystalline state.

6.3 Examples of Coloured Organic Compounds

6.3.1 *ROY (QAXMEH)*

5-Methyl-2-[(2-nitrophenyl)amino]-3-thiophenecarbonitrile (QAXMEH), known as ROY, is an organic compound synthesised as an intermediate for the development of a novel compound with antipsychotic activity (Calligaro et al. 1997). This compound is coloured, although it is not categorised as a dye because of its weak light absorption in the visible region. ROY is known as the organic small molecule whose polymorphic forms have been the most extensively structurally analysed, where all forms can be obtained under ambient conditions and many can also be crystallised concomitantly (Yu et al. 2000; Yu 2010). In addition to the seven polymorphic forms analysed by single-crystal measurements, three polymorphic forms have been characterised in the form of thin films (Yu 2010; Chen et al. 2005). Among the seven analysed polymorphic forms, five are stable and two are meta-stable under ambient conditions. The three film forms have been reported to be unstable, instead of transforming to other forms within hours to days (Yu 2010). This polymorphic system is representative of conformational polymorphs (Cruz-Cabeza and Bernstein 2014).

Figure 6.14 shows the molecular conformations of the seven forms of ROY—Y, ON, OP, R, YN, ORP and YT04—and Table 6.6 lists their crystallographic parameters. The abbreviations denote their crystal appearance, such as 'orange needles' for ON. As shown in the figure, the conformational differences in the seven forms are related to the torsion angle between the phenyl and thiophene rings. Several forms also have a relatively small twist between the phenyl ring and nitro moiety. The torsion angle seems to be related to the crystal colour, indicating that the colours of the polymorphs are dependent on the electronic state of the molecules in the crystal (Yu 2002). Meanwhile, no significant differences have been observed in the molecular bond lengths or angles of the different polymorphs (Yu 2010).

ROY has been reported to form intramolecular hydrogen bonds between the amino and nitro groups in its crystal forms. The molecular arrangements are thus considered to form mainly by van der Waals interactions and weak hydrogen bonding (Yu 2010;

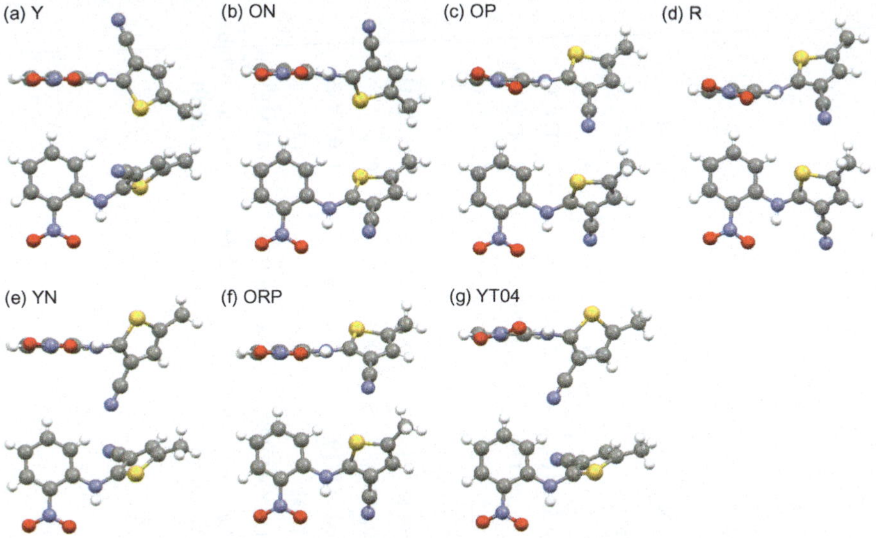

Fig. 6.14 Molecular geometries of the seven polymorphs of ROY viewed parallel (upper) and perpendicular (lower) to the phenyl ring: **a** Y-form (QAXMEH25), **b** ON-form (QAXMEH24), **c** OP-form (QAXMEH27), **d** R-form (QAXMEH26), **e** YN-form (QAXMEH28), **f** ORP-form (QAXMEH29) and **g** YT04-form (QAXMEH30)

Dunitz and Gavezzotti 2005). Figure 6.15 depicts the ROY polymorph molecular arrangements. Their crystal structures were interpreted in detail by the energy of the intermolecular interactions in a molecular pair (Dunitz and Gavezzotti 2005) and Hirshfeld surface analysis (McKinnon et al. 2007), both of which indicated that the seven forms exhibit different molecular arrangements with no formation of similar molecular pairs. A previous study on all forms except for YT04 revealed that the major interactions are dispersion contributions, and notable structural similarities are not observed (Dunitz and Gavezzotti 2005). The latter paper (McKinnon et al. 2007) reported that the R- and ORP-forms create similar molecular dimers by the formation of weak cyclic hydrogen bonds. A similarity in their crystal structures with respect to the stacking motif is apparent, where the phenyl and thiophene rings are independently stacked. The arrangements of the ON- and YN-forms are both characterised by 1D chain formation, but their stacking motifs are different. The arrangement of the OP-form is regarded as a brick-wall 2D structure (Smith 1974). In the YT04-form, the molecules are aligned in a structure resembling a molecular tube along the *c*-axis, where the tubes are two-dimensionally aligned to form the crystal structure. The stacking between nitrophenyl rings and between nitrophenyl and thiophene rings is important for interpreting the molecular arrangement of the Y-form according to an investigation of molecular pairwise interactions (Dunitz and Gavezzotti 2005). Based on an analysis of the radial distribution of the molecular centres of mass in the seven forms (Yu 2010), the ON-, YN- and ORP-forms have

Table 6.6 Crystallographic parameters of the seven polymorphs of ROY

Polymorph	Y	ON	OP	R	YN	ORP	YT04
Space group	$P2_1/n$	$P2_1/c$	$P2_1/n$	*P*-1	*P*-1	*Pbca*	$P2_1/n$
Z	4	4	4	2	2	8	4
a/Å	8.5001(8)	3.9453 (7)	7.9760 (9)	7.4918 (5)	4.5918 (8)	13.177 (3)	8.2324 (4)
b/Å	16.413 (2)	18.685 (1)	13.319 (2)	7.7902 (5)	11.249 (2)	8.0209 (18)	11.8173 (5)
c/Å	8.5371 (5)	16.3948 (4)	11.676 (1)	11.9110 (8)	12.315 (2)	22.801 (5)	12.3121 (6)
α/°	90	90	90	75.494 (6)	71.194 (3)	90	90
β/°	91.767 (7)	93.830 (5)	104.683 (8)	77.806 (6)	89.852 (4)	90	102.505 (1)
γ/°	90	90	90	63.617 (6)	88.174 (3)	90	90
V/Å^3	1190.5 (4)	1205.9 (3)	1199.9 (4)	598.88 (7)	601.85 (19)	2409.8 (9)	1169.36 (9)
d/g/cm^3	1.447	1.428	1.435	1.438	1.431	1.429	1.473
Refcode	QAXMEH25	QAXMEH24	QAXMEH27	QAXMEH26	QAXMEH28	QAXMEH29	QAXMEH30
References	Yu et al. (2000), Yu (2010)	Yu et al. (2000), Yu (2010)	Yu et al. (2000), Yu (2010)	Yu et al. (2000), Yu (2010)	Yu et al. (2000), Yu (2010)	Yu et al. (2000), Yu (2010)	Yu et al. (2000), Yu (2010)

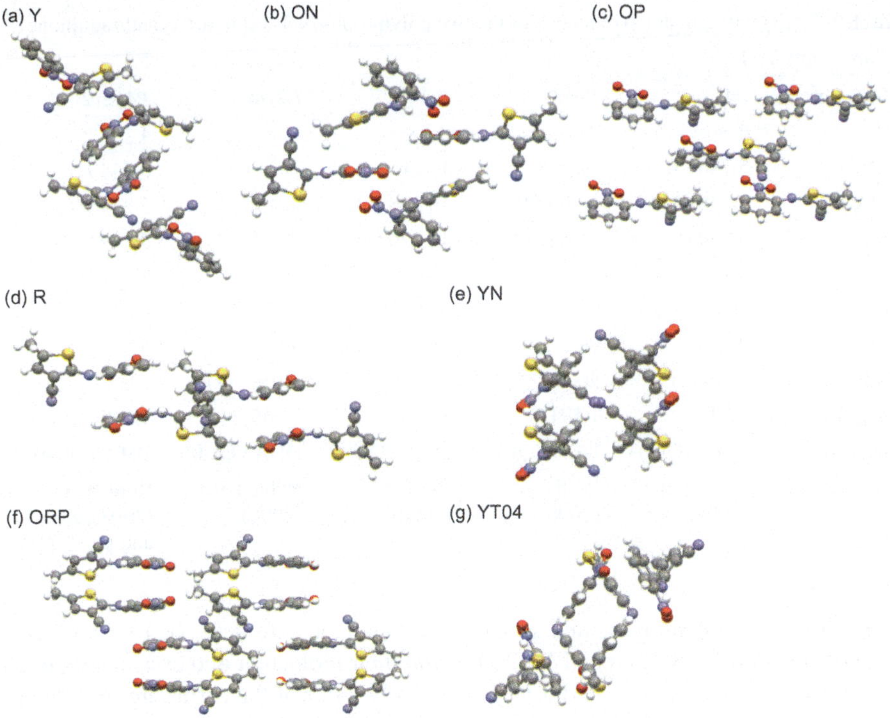

Fig. 6.15 Characteristic molecular arrangements of the seven polymorphs of ROY: **a** Y-form (QAXMEH25), **b** ON-form (QAXMEH24), **c** OP-form (QAXMEH27), **d** R-form (QAXMEH26), **e** YN-form (QAXMEH28), **f** ORP-form (QAXMEH29) and **g** YT04-form (QAXMEH30)

close stacking pairs. However, no structural similarities between their stacking pairs were found.

This example stimulated us to think that torsional flexibility is a piece of important parameters with respect to polymorph occurrence.

6.3.2 1,8-Dihydroxyanthraquinone (DHANQU)

Many anthraquinone dyes have been developed and used for dyeing, printing and other colouration purposes (Zollinger 2003; Faulkner and Schwartz 2009; Hunger and Schmidt 2018; Whitaker 1977, 1995). We searched the CSD for the anthraquinone skeleton with respect to the number of reported polymorphs and found five polymorphs of 1,8-dihydroxyanthraquinone with structural data. 1,8-Dihydroxyanthraquinone is not used as a colorant but instead as a potential pharmaceutical. Its analogous compounds, alizarin and quinizarin, are important in dye chemistry. Table 6.7 lists the crystallographic parameters for its five polymorphs;

Table 6.7 Crystallographic parameters of the five polymorphs of 1,8-dihydroxyanthraquinone

Polymorph	1	2	3	4	5
Space group	$P4_1$	$Pca2_1$	P-1	$P2_1/n$	$P4_12_12$
Z	4	8	8	4	4
a/Å	5.746 (3)	21.578 (2)	10.211 (2)	7.2930 (5)	5.7440 (6)
b/Å	5.746 (3)	3.766 (2)	10.308 (2)	9.5001 (7)	5.7440 (6)
c/Å	31.39 (5)	24.683 (2)	19.776 (3)	14.721 (1)	31.393 (3)
α/°	90	90	78.058 (8)	90	90
β/°	90	90	83.905 (9)	91.634 (2)	90
γ/°	90	90	88.995 (11)	90	90
V/Å^3	1036 (2)	2005.8 (9)	2024.9 (5)	1019.5 (1)	1035.8 (2)
d/g/cm^3	1.539	1.591	1.576	1.565	1.540
Refcode	DHANQU03	DHANQU04	DHANQU05	DHANQU06	DHANQU07
References	Rohl et al. (2008)	Rohl et al. (2008)	Rohl et al. (2008)	Rohl et al. (2008)	Rohl et al. (2008), Zain and Ng (2005)

their preparation conditions and structural comparisons are described in the literature (Rohl et al. 2008; Zain and Ng 2005), and their molecular and crystal structures are illustrated in Figs. 6.16 and 6.17, respectively. Four of the forms are reported to have an orange colour, and one exhibits a red colour. The anthraquinone moiety was found to adopt a planar geometry in the four forms except for the 3-form.

The thermodynamically stable 1-form was first reported in 1957 (Jagannadham 1957), and its structural analyses were reported in 1962 (Prakash 1962) and 1965 (Prakash 1965). In the tetragonal 1-form, intermolecular C = O - - - O = C interactions along the molecular short axis parallel to the C = O groups form linear 1D molecular chains, which stack to form a characteristic brick wall-type molecular arrangement (Smith 1974) with a distance between neighbouring layers of approximately 3.23 Å. The stacked molecules also form weak hydrogen bonds (C–H - - - O) with neighbouring layers to create a 3D structure, where one molecule interacts with four neighbouring molecules. The overall molecular arrangement is a typical herringbone type. In the orthorhombic 2-form, two independent molecules are present in the unit cell, which are separately π–π stacked along the b-axis to form a columnar structure. The two columns are hydrogen bonded to each other between hydroxyl and carbonyl groups. Weak C–H - - - O hydrogen bonds also form between neighbouring columns to create a 3D molecular arrangement. The 3-form belongs to the triclinic P-1 space group and has been reported as a disappearing polymorph (Dunitz and Bernstein 1995). Four independent molecular units exist, three of which have a distorted anthraquinone ring, but all four are composed of two pairs of π-stacked dimers in an approximately anti-parallel arrangement. Both dimers are almost perpendicular to each other, as shown in Fig. 6.16c; these dimers further stack to form a 2D molecular arrangement, which then aligns to form a 3D structure. Moderate and weak

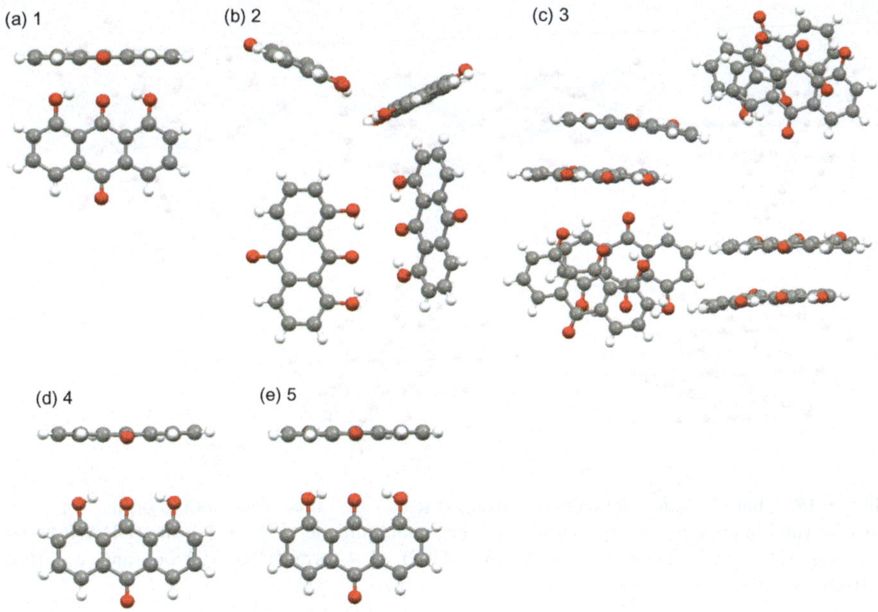

Fig. 6.16 Molecular geometries of the five polymorphs of 1,8-dihydroxyanthraquinone: **a** 1-form (DHANQU03), **b** 2-form (DHANQU04), **c** 3-form (DHANQU05), **d** 4-form (DHANQU06) and **e** 5-form (DHANQU07). For 1-, 2-, 4- and 5-forms, the molecules are viewed parallel (upper) and perpendicular (lower) to the anthraquinone ring. For the 3-form, the figure shows two pairs of dimers composed of four independent molecular units. The left and right dimers are independent, and this figure was drawn in such a way that the left dimer is viewed parallel (upper) and perpendicular (lower) to the anthraquinone ring

hydrogen bonds contribute to these structural characteristics. In the monoclinic 4-form, the molecules are parallelly stacked with slight slippage along the molecular long axis to form a 1D molecular column. These molecules are correlated with an inversion centre. The nearest four 1D columns align with one column symmetry related by a twofold screw axis to form a 3D structure, the arrangement of which resembles a herringbone. Many CH - - - O = C interactions occur between the columns. The arrangement of molecules in the tetragonal 5-form is similar to that in the 1-form; Hirshfeld surface analysis obtained almost the same results in terms of intermolecular interactions between these phases. In the 5-form, the molecules are located on the two-fold axis. The trial made for the analysis of the 1-form by imposing the structure of the 5-form was reported to be failed. The 5-form was also found to transform to the 1-form at low temperatures.

The polymorph structures of 1,8-dihydroxyanthlaquinone are reminiscent of the polymorphs of QA, where the structure of polymorphs is interpreted by hydrogen-bonding and stacking patterns.

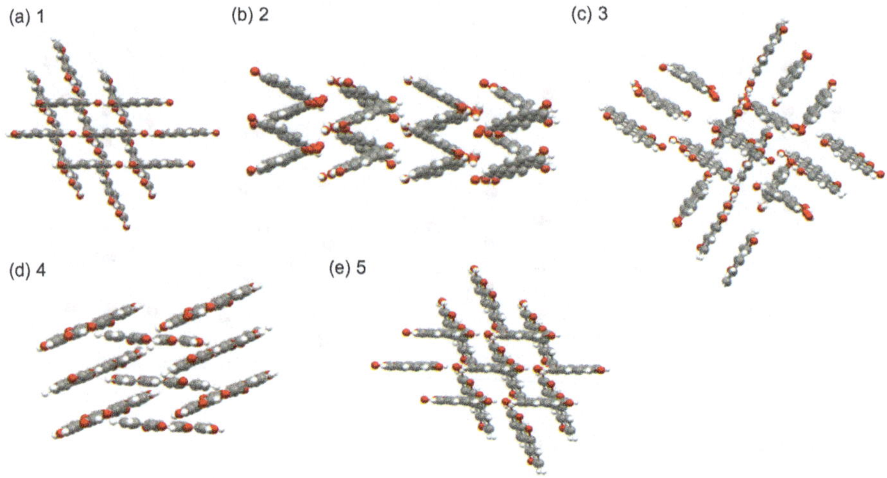

Fig. 6.17 Characteristic molecular arrangements of the five polymorphs of 1,8-dihydroxyanthraquinone viewed parallel to the anthraquinone ring: **a** 1-form (DHANQU03), **b** 2-form (DHANQU04), **c** 3-form (DHANQU05), **d** 4-form (DHANQU06) and **e** 5-form (DHANQU07)

6.3.3 Two Azomethine Compounds

Some azomethine compounds are known to have polymorphs. This section describes the molecular and crystal structures of two coloured azomethine compounds, which have been reported to have more than four polymorphs.

6.3.3.1 *N*-(*p*-Dimethylaminobenzylidene)-*p*-Nitroaniline (MABZNA)

The simple azomethine compound *N*-(*p*-dimethylaminobenzylidene)-*p*-nitroaniline was studied as an example to consider the effect of intramolecular charge transfer on the planarity of a chromophoric system (Nakai et al. 1976). During the structural analysis, four differently coloured polymorphs were found and structurally characterised (Nakai et al. 1976, 1981). Table 6.8 lists their crystallographic parameters. The orange I- and II-forms are obtained from their acetone and benzene–acetone solutions, respectively, via slow evaporation. The reddish-orange III-form can be obtained concomitantly from experimental batches for the II- or red IV-form, and the IV-form is prepared from its acetone solution at 273 K. These four forms have been reported to transform into the V-form at 403 K, although its structure was discussed based on a comparison of powder diffraction results (Nakai et al. 1981). New analytical results on this compound obtained at 100 K were recently reported (Valdivia-Berroeta et al. 2017).

The molecular conformations of the *N*-(*p*-dimethylaminobenzylidene)-*p*-nitroaniline polymorphs are shown in Fig. 6.18. Two independent molecules exist in

Table 6.8 Crystallographic parameters of the four polymorphs of *N*-(*p*-dimethylaminobenzylidene)-*p*-nitroaniline

Polymorph	I	II	III	IV
Space group	*P*-1	*P*-1	$P2_1/c$	$P2_12_12_1$
Z	4	4	4	16
a/Å	9.509 (1)	7.507 (1)	3.989 (1)	64.297 (13)
b/Å	16.200 (1)	16.704 (1)	10.621 (1)	10.789 (1)
c/Å	9.505 (1)	11.051 (1)	31.932 (2)	7.731 (1)
α/°	91.69 (1)	92.79 (1)	90	90
β/°	107.52 (1)	90.30 (1)	95.10 (1)	90
γ/°	101.04 (1)	96.58 (1)	90	90
V/Å^3	1364.5	1374.8	1347.7	5363.1
d/g/cm^3	1.311	1.301	1.327	1.334
Refcode	MABZNA	MABZNA01	MABZNA02	MABZNA03
References	Nakai et al. (1976)	Nakai et al. (1981)	Nakai et al. (1981)	Nakai et al. (1981)

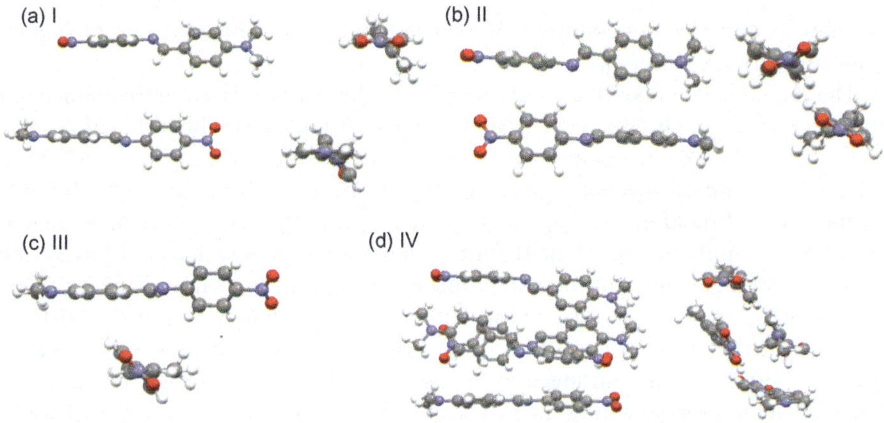

Fig. 6.18 Molecular geometries of the four polymorphs of *N*-(*p*-dimethylaminobenzylidene)-*p*-nitroaniline viewed from the short (left for I-, II- and IV-forms; upper for III-form) and long molecular directions (right for I-, II- and IV-forms; lower for III-form): **a** I-form (MABZNA), **b** II-form (MABZNA01), **c** III-form (MABZNA02) and **d** IV-form (MABZNA03)

the I- and II-forms, which have no significant differences in bond lengths or angles. The twisted angles of the two phenyl rings to the plane composed of the azomethine unit and two bonded carbon atoms are similar in both forms, where the aniline and nitrophenyl rings are twisted out of the plane by approximately 45° and 10°, respectively. The molecular conformation of the III-form is similar to those of the I- and II-forms. The IV-form unit cell contains four independent molecules, which occur

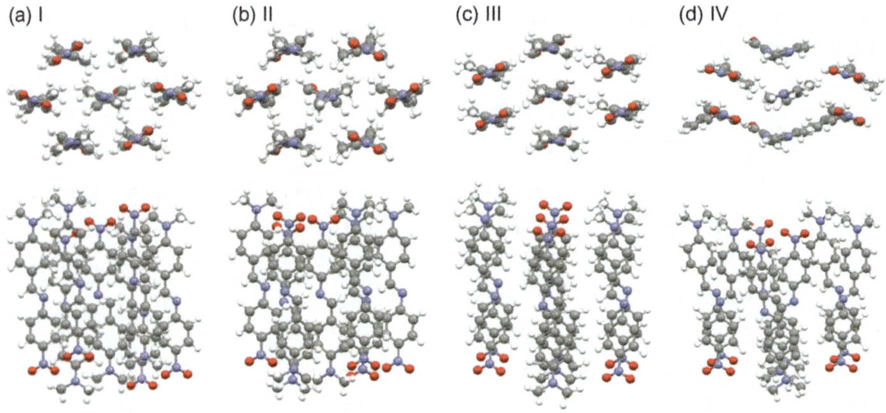

Fig. 6.19 Packing motifs of the four polymorphs of *N*-(*p*-dimethylaminobenzylidene)-*p*-nitroaniline viewed from the direction of the long molecular axis (upper) and perpendicular to the long axis (lower): **a** I-form (MABZNA), **b** II-form (MABZNA01), **c** III-form (MABZNA02) and **d** IV-form (MABZNA03)

as two pairs of molecules connected by an inversion centre, although their molecular conformations are different.

The crystal structures of these polymorphs are characterised by a similar molecular arrangement. Figure 6.19 depicts the arrangements for seven neighbouring molecules viewed from the long molecular axis as well as perpendicular direction to this axis. Clearly, they form a brick wall-type assembly (Smith 1974) from this view, although the molecules do not form a complete π–π stacking arrangement. The arrangement of the I-form is similar to that of the II-form, where the molecules stack to form a brick wall-type arrangement with neighbouring molecules on the same plane arranged in opposite directions, as indicated in the figure. The arrangement of the III-form is slightly different from that of the I- and II-forms; the molecules form stacking columns with molecules arranged in opposite directions in neighbouring columns. The molecular arrangement of the IV-form is roughly similar to that of the III-form. An additional structural report was found for this compound with respect to low-temperature measurements at 100 K (Valdivia-Berroeta et al. 2017), which indicated a molecular and crystal structure similar to that of the II-form.

6.3.3.2 1-[(*E*)-({4-[(5-Methyl-1,2-Oxazol-3-yl)Sulphamoyl]Phenyl}Imino)Methyl]Naphthalen-2-Olate (ZULGUE)

Another polymorphic azomethine compound, 1-[(*E*)-({4-[(5-Methyl-1,2-oxazol-3-yl)sulphamoyl]phenyl}imino)methyl]naphthalen-2-olate, was derived from sulphamethoxazole, which is a known antibiotic. Sulphamethoxazole is also known to exhibit four polymorphs and a hemihydrate (Tahir et al. 2015). This Schiff base synthesised from sulphamethoxazole by 2-hydroxy-1-naphtalaldehyde was found to crystallise in five polymorphic forms in the same methanol solution. The five forms (I–V) are orange, although the colour of one form has been reported as dark. Table 6.9 lists their crystallographic parameters. All forms were analysed at room temperature. The IV- and V-forms consist of two independent molecules.

Figure 6.20 shows the molecular geometries of the five forms. All the molecules form π-conjugation between the central phenyl and naphthalene rings, which was confirmed by the nearly planar conformation between these two units and the azomethine unit. The conjugation between these units can also be characterised by the bond lengths, indicating a significant contribution from the quinoidal structure. A major structural difference in the molecular structures of the five forms was found in the conformation of the sulphonamide unit, which is largely bent and almost perpendicular to the azomethine group in all forms.

Figure 6.21 depicts the molecular arrangements of the five polymorphs. The basic structural element in the crystal structures of these polymorphic forms is a π–π stacked molecular pair. In addition, hydrogen bonding plays an important role in forming the 3D structures. In the I-form, the stacked pairs are aligned nearly parallel to form a 1D arrangement with slight slipping along the long molecular axis. This 1D arrangement aligns in a herringbone-like manner through weak intermolecular interactions to form a 3D structure. In the II-form, the stacked units assemble with neighbouring stacked units similar to puzzle pieces to form 2D molecular sheets, which align along the *a*-axis to form a 3D crystal structure. The molecules in the III-form also form a stacked pair similar to that in the II-form. This stacked unit further arranges in a translational operation to form 2D-like molecular sheets, which align along the *c*-axis to form a 3D crystal structure. In both the IV- and V-forms, two

Table 6.9 Crystallographic parameters of the five polymorphs of 1-[(*E*)-({4-[(5-methyl-1,2-oxazol-3-yl)sulphamoyl]phenyl}imino)methyl]naphthalen-2-olate

Polymorph	I	II	III	IV	V
Space group	$P2_1/c$	$C2/c$	P-1	P-1	P-1
Z	4	8	2	4	4
a/Å	12.1116 (18)	27.545 (2)	8.3554 (7)	7.4491 (16)	10.9669 (5)
b/Å	11.2257 (15)	12.0217 (10)	10.0022 (8)	14.2402 (11)	11.6578 (6)
c/Å	14.8557 (18)	13.7703 (10)	12.7955 (10)	18.3131 (16)	15.8105 (8)
α/°	90	90	67.259 (5)	87.191 (6)	90.949 (2)
β/°	94.573 (5)	118.646 (4)	80.749 (4)	84.895 (5)	106.494 (1)
γ/°	90	90	86.833 (4)	77.080 (5)	94.680 (2)
V/Å^3	2013.4 (5)	4001.7 (6)	973.37 (14)	1885.1 (13)	1930.06 (17)
d/g/cm^3	1.344	1.353	1.39	1.436	1.402
Refcode	ZULGEU01	ZULGEU02	ZULGEU03	ZULGEU04	ZULGEU05
References	Tahir et al. (2015)	Tahir et al. (2015)	Tahir et al. (2015)	Tahir et al. (2015)	Tahir et al. (2015)

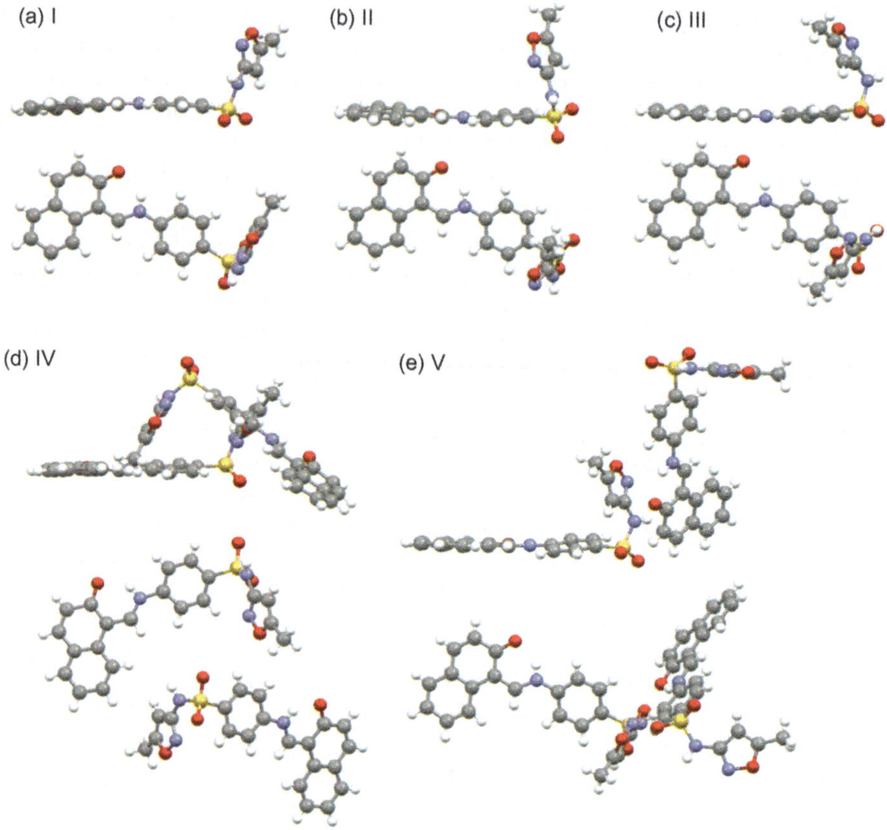

Fig. 6.20 Molecular geometries of the five polymorphs of 1-[(*E*)-({4-[(5-methyl-1,2-oxazol-3-yl)sulphamoyl]phenyl}imino)methyl]naphthalen-2-olate viewed parallel (upper) and perpendicular (lower) to the azomethine moiety: **a** I-form (ZULGUE01), **b** II-form (ZULGUE02), **c** III-form (ZULGUE03), **d** IV-form (ZULGUE04) and **e** V-form (ZULGUE05). The V-form is drawn with a focus on one independent molecule (left)

independent molecules exist in the unit cell. The two independent molecules of the IV-form stack to form separate 1D columns, which connect to form a 2D layer. These 2D layers alternatively align along the *b*-axis to form the 3D structure. In the V-form, two independent V-shaped molecules stack to form a 1D chain. These 1D chains align to form 2D molecular layers nearly parallel to the *bc*-plane, which arrange along the *a*-axis with an inversion operation to form a 3D structure by hydrogen bonding interactions. This arrangement resembles a herringbone pattern when viewed, as shown in Fig. 6.21e.

These examples of polymorphs seemed to suggest that it is an important issue to take the number of crystallographically independent molecules into account in consideration of polymorph occurrence.

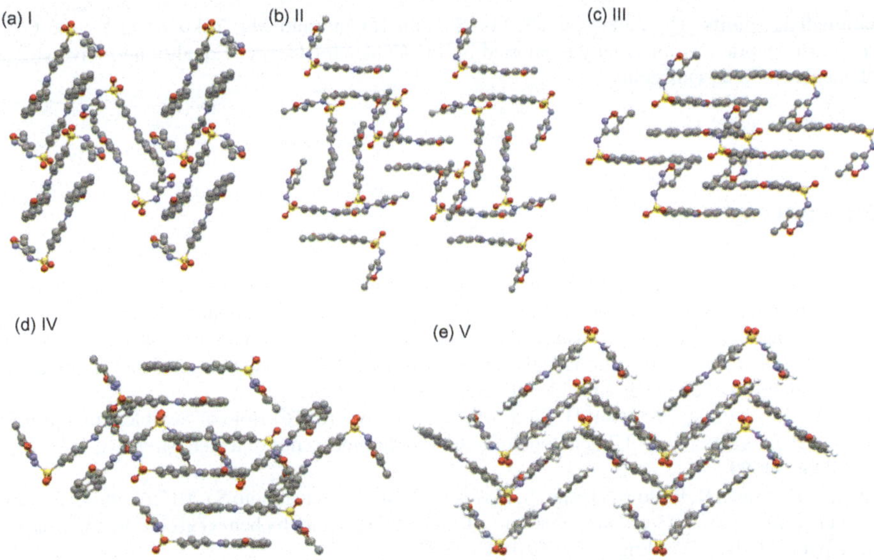

Fig. 6.21 Packing motifs of the five polymorphs of 1-[(*E*)-({4-[(5-methyl-1,2-oxazol-3-yl)sulphamoyl]phenyl}imino)methyl]naphthalen-2-olate viewed parallel to the azomethine moiety: **a** I-form (ZULGUE01), **b** II-form (ZULGUE02), **c** III-form (ZULGUE03), **d** IV-form (ZULGUE04) and **e** V-form (ZULGUE05)

6.4 Summary

This chapter reviewed the polymorph structures of organic dyes and coloured organic compounds with more than four analysed polymorphs. The structures of the polymorphs of CuPc and TiOPc clearly indicate that rigid dye chromophores can exhibit structural variations in terms of molecular arrangement without a flexible structural moiety or hydrogen bonding. The two reviewed QA derivatives show that hydrogen bonding and a flexible substituent play important roles in their polymorphic occurrence. Other examples also suggest that structural flexibility and weak intermolecular interactions such as hydrogen bonding and halogen interactions are important in terms of the occurrence of polymorphism.

Polymorphism in dye chemistry has significant importance from the perspective of materials development. Many polymorph reports on pigments and functional dyes have clearly shown that polymorphism results in different solid properties from one chemical component. They have also demonstrated that the discovery of novel polymorphs has the potential to lead to dramatic improvements in certain physicochemical solid properties or the creation of novel materials. Thus, in the research and development of functional dyes, molecular design should consider the potential crystal structure and the possibility of polymorphic occurrence in addition to the electronic states of the dye molecule itself.

Acknowledgements The authors thank Ms. Sayumi Hatano and Ms. Yuko Kishi for the CSD search. This work was financially supported by the Center for Gender Equality and Diversity at Yokohama National University.

References

Akune Y (2017) Study on the occurrence of polymorphs and the optical properties in a series of benzylated diaminodicyanopyrazine dyes. PhD thesis, Yokohama National University

Akune Y, Gontani H, Hirosawa R, Koseki A, Matsumoto S (2015) The effects of molecular flexibility and substituents on conformational polymorphism in a series of 2,5-diamino-3,6-dicyanopyrazine dyes with highly flexible groups. CrystEngComm 17:5789–5800

Akune Y, Hirosawa R, Takahashi H, Shiro M, Matsumoto S (2016) Role of flexible bulky groups and weak interactions involving halogens in the vapoluminescence of a metal-free dye. RSC Adv 6:74506–74509

Akune Y, Hirosawa R, Endo N, Hatano S, Hosokai T, Sato H, Matsumoto S (2017a) Tuning of fluorescence efficiency via local modification of the crystal structure by benzyl groups in polymorphs of a pyrazine dye. CrystEngComm 19:1947–1952

Akune Y, Hirosawa R, Koseki A, Matsumoto S (2017b) Role of halogen substituents in a series of polymorphic 2,5-diamino-3,6-dicyanopyrazine derivatives with highly flexible groups. Z Kristallogr Cryst Mater 232:395–405

Ameri T, Dennler G, Lungenschmied C, Brabec CJ (2009) Organic tandem solar cells: a review. Energy Environ Sci 2:347–363

Ashida M, Uyeda N, Suito E (1966) Unit cell metastable-form constants of various phthalocyanines. Bull Chem Soc Jpn 39:2616–2624

Bernstein J (2002) Polymorphism in molecular crystals. Oxford University Press, Oxford

Bernstein J, Davey RJ, Henck JO (1999) Concomitant polymorphs. Angew Chem Int Ed 38:3440–3461

Bluhm T, Mayo J, Hamer G, Martin T (1992) The application of Rietveld analysis to crystal structures of titanyl phthalocyanine. Proc SPIE Int Soc Opt Eng 1670:160–171

Brown CJ (1968) Crystal structure of β-copper phthalocyanine. J Chem Soc A 2488–2493

Calligaro DO, Fairhurst J, Hotten TM, Moore NA, Tupper DE (1997) The synthesis and biological activity of some known and putative metabolites of the atypical antipsychotic agent olanzapine (LY170053). Bioorg Med Chem Let 7:25–30

Chen S, Guzei IA, Yu L (2005) New polymorphs of ROY and new record for coexisting polymorphs of solved structures. J Am Chem Soc 127:9881–9885

Chung H, Diao Y (2016) Polymorphism as an emerging design strategy for high performance organic electronics. J Mater Chem C 4:3915–3933

Cruz-Cabeza AJ, Bernstein J (2014) Conformational polymorphs. Chem Rev 114:2170–2191

Dunitz JD, Bernstein J (1995) Disappearing polymorphs. Acc Chem Res 28:193–200

Dunitz JD, Gavezzotti A (2005) Toward a quantitative description of crystal packing in terms of molecular pairs: application to the hexamorphic crystal system, 5-Methyl-2-[(2-nitrophenyl)amino]-3-thiophenecarbonitrile. Cryst Growth Des 5:2180–2189

Engel MK (1996) Single-crystal and solid-state molecular structures of phthalocyanine complexes. Rep Kawamura Inst Chem Res 11–54

Engel MK (2003) Single crystal structures of phthalocianine complexes and related macrocycles. In: Kadish K, Guillard R, Smith KM (eds) Handbook of porphyrins and phthalocyanines, vol 20. Elsevier Science

Erk P (2004) Experimental crystal structure determination. CSD Communication

Erk P, Hengelsberg H (2003) Phthalocyanine dyes and pigments. In: Kadish K, Guillard R, Smith KM (eds) Handbook of porphyrins and phthalocyanines, vol 19. Elsevier Science, pp 105–149

Erk P, Hengelsberg H, Haddow MF, Gelder RV (2004) The innovative momentum of crystal engineering. CrystEngComm 6:474–483
Fábián L, Kálmán A (2004) Isostructurality in one and two dimensions: isostructurality of polymorphs. Acta Cryst B 60:547–558
Fan Y, Zhao Y, Ye L, Li B, Yang G, Wang Y (2009) Polymorphs and pseudopolymorphs of N, N-di(n-butyl)quinacridone: Structures and solid-state luminescence properties. Cryst Growth Des 9:1421–1430
Faulkner EB, Schwartz RHJ (eds) (2009) High performance pigment, second revised and expanded edition. Wiley-VCH
Freeman HS, Peters AT (eds) (2000) Colorants for non-textile applications. Elsevier
Fujimaki Y (1991) High photosensitivity of an organic photoreceptor. In: IS&T's 7th international congress on non-impact printing technologies, pp 269–275
Gavezzotti A (2007) Molecular aggregation. Oxford University Press, Oxford
Gorelik TE, Czech C, Hammer SM, Schmidt MU (2016) Crystal structure of disordered nanocrystalline αII quinacridone determined by electron diffraction. CrystEngComm 18:529–535
Gsänger M, Bialas D, Huang L, Stolte M, Würthner F (2016) Organic semiconductors based on dyes and color pigments. Adv Mater 28:3615–3645
Hilfiker R (ed) (2006) Polymorphism in the pharmaceutical industries. Wiley-VCH
Hiller W, Strahle J, Kobel W, Hanack M (1982) Polymorphie, Leitfähigkeit und Kristallstrukturen von Oxo-phthalocyaninato-titan(IV). Z Kristallogr Cryst Mater 159:173–183
Hirosawa R, Akune Y, Endo N, Hatano S, Hosokai T, Sato H, Matsumoto S (2017) A variety of solid-state fluorescence properties of pyrazine dyes depending on terminal substituent. Dyes Pigm 146:576–581
Hoshino A, Takenaka Y, Miyaji H (2003) Redetermination of the crystal structure of [alpha]-copper phthalocyanine grown on KCl. Acta Cryst B 59:393–403
Hunger K, Schmidt MU (2018) Industrial organic pigments, forth completely revised edition. Wiley-VCH
Jagannadham AV (1957) Kuerzere Originalmitteilungen und Notizen Space-Group of 1,8-Dihydroxyanthraquinone, C14H8O4. Z Kristallogr 108:457–458
Jiang H, Hu P, Ye J, Ganguly R, Li Y, Long Y, Fichou D, Hu W, Kloc C (2018) Hole mobility modulation in single-crystal metal phthalocyanines by changing the metal-π/π-π interactions. Angew Chem Int Ed 57:10112–10117
Kim SH (ed) (2006) Functional dyes. Elsevier
Kim JH, Shin SR, Matsuoka M, Fukunishi K (1998) Self-assembling of fluorescent dyes and their solid state spectra. Dyes Pigm 39:341–357
Kim I, Haverinen HM, Wang ZG, Madakuni S, Kim Y, Li J, Jabbour GE (2009) Efficient organic solar cells based on planar metallophthalocyanines. Chem Mater 21:4256–4260
Kitaigorodskii AI (1973) Molecular crystals and molecules. Academic Press
Labana SS, Labana LL (1967) Quinacridones. Chem Rev 67:1–18
Law KY (1993) Organic photoconductive materials: recent trends and developments. Chem Rev 93:449–486
Leusen FJJ (1994) In: 15th European crystallographic meeting ECM15, Dresden, 18 August–2 September 1994. Abstracts. Z Kristallogr 8:161
Leusen FJJ (1996) Ab initio prediction of polymorphs. J Cryst Growth 166:900–903
Lincke G (2000) A review of thirty years of research on quinacridones. X-ray crystallography and crystal engineering. Dyes Pigm 44:101–122
Lincke G (2002) On quinacridones and their supramolecular mesomerism within the crystal lattice. Dyes Pigm 52:169–181
Lincke G, Finzel HU (1996) Studies on the structure of alpha-Quinacridone. Cryst Res Technol 31:441–452
Matsumoto S, Uchida Y, Yanagita M (2006) A series of polymorphs with different colors in fluorescent 2,5-diamino-3,6-dicyanopyrazine dyes. Chem Lett 35:654–655

Matsuoka M (2000) Multifunctional dye materials from new dicyanopyrazine chromophores. In: Freeman HS, Peters AT (eds) Colorants for non-textile applications. Elsevier Science, pp 339–381
McKinnon JJ, Fabbiani FPA, Spackman MA (2007) Comparison of polymorphic molecular crystal structures through hirshfeld surface analysis. Cryst Growth Des 7:755–769
Mizuguchi J, Rihs G, Karfunkel HR (1995) Solid-state spectra of titanylphthalocyanine as viewed from molecular distorsion. J Phys Chem 99:16217–16227
Mizuguchi J, Sasaki T, Tojo K (2002) Refinement of the crystal structure of 5,7,12,14-tetrahydro[2,3-b]-quinolinoacridine (γ-form), C20H12N2O2, at 223K. Z. Kristallogr NCS 217:249–250
Nakai H, Shiro M, Ezumi K, Sakata S, Kubota T (1976) The crystal and molecular structures of p-nitrobenzylidene-p-dimethylaminoaniline and p-Dimethylaminobenzylidene-p-nitroaniline. Acta Crys B 32:1827–1833
Nakai H, Ezumi K, Shiro M (1981) The structures of polymorphs of N-(p-dimethylaminobenzylidene)-p-nitroaniline. Acta Cryst B 37:193–197
Nangia A (2008) Conformational polymorphism in organic crystals. Acc Chem Res 41:595–604
Nishimura N, Senju T, Mizuguchi J (2006) 5,7,12,14-Tetra-hydro[2,3-b]quinolinoacridine (β form). Acta Cryst E 62:4683–4685
Oka K, Okada O, Nukada K (1992) Study of the crystal structure of titanylphthalocyanine by Rietveld analysis and intermolecular energy minimization method. Jpn J Appl Phys 31:2181–2184
Okada O, Oka K, Iijima M (1993) Study of the crystal structure of titanylphthalocyanine by Rietveld analysis II. Jpn J Appl Phys 32:3556–3660
Paulus EF, Dietz E, Kroh A, Prokschy F (1989) In: 12th European crystallographic meeting collected abstracts, vol 2, pp 23–24
Paulus EF, Leusen FJJ, Schmidt MU (2007) Crystal structures of quinacridones. CrystEngComm 9:131–143
Potts GD, Jones W, Bullock JF, Andrews SJ, Maginn SJ (1994) The crystal structure of quinacridone: an archetypal pigment. J Chem Soc Chem Commun 2565–2566
Prakash A (1962) The crystal structure of 1,8-dihydroxy anthraquinone. Indian J Phys 36:654–656
Prakash A (1965) The crystal and molecular structure of chrisazin. Z Kristallogr 122:272–282
Price SL (2013) Why don't we find more polymorphs? Acta Cryst B 69:313–328
Price SL (2014) Predicting crystal structures of organic compounds. Chem Soc Rev 43:2098–2111
Reilly AM et al (2016) Report on the sixth blind test of organic crystal structure prediction methods. Acta Cryst B 72:439–459
Robertson JM (1935) An X-ray study of the structure of the phthalocyanines. Part I. The metal-free, nickel, copper, and platinum compounds. J Chem Soc 615–621
Rohl AL, Moret M, Kaminsky W, Claborn K, McKinnon JJ, Kahr B (2008) Hirshfeld surfaces identify inadequacies in computations of intermolecular interactions in crystals: pentamorphic 1,8-dihydroxyanthraquinone. Cryst Growth Des 8:4517–4525
SciFinder. The American Chemical Society. https://www.cas.org/products/scifinder
Shirai K, Yanagisawa A, Takahashi H, Fukunishi K, Matsuoka M (1998) Syntheses and fluorescent properties of 2,5-diamino-3,6-dicyanopyrazine dyes. Dyes Pigm 39:49–68
Smith DL (1974) Structure of dyes and dye aggregates-evidence from crystal structure analysis. Photo Sci Eng 18:309–322
Tahir MN, Shafiq Z, Shad HA, Rehman ZU, Karim A, Naseer MM (2015) Polymorphism in a sulfamethoxazole derivative: coexistence of five polymorphs in methanol at room temperature. Cryst Growth Des 15:4750–4755
Threlfal TL (1995) Analysis of organic polymorphs, a review. Analyst 120:2435–2460
The Cambridge Structural Database. The Cambridge Crystallographic Data Centre. https://www.ccdc.cam.ac.uk/solutions/csd–system/components/csd/
Valdivia-Berroeta GA, Smith SJ, Michaelis DJ, Johnson JA (2017) (E)-N,N-dimethyl-4-(((4-nitrophenyl)imino)methyl)aniline. CSD Communication
Watanabe K, Kinoshita A, Hirose N, Itami A (1990) A new charge generation material for organic photoconductors: Y-form Titanylphthalocyanine. Konica Tech Rep 3:108–114

Whitaker A (1977) X-ray powder diffraction. In: Venkataraman K (ed) The analytical chemistry of synthetic dyes, pp 269–298
Whitaker A (1995) X-ray powder diffraction of synthetic organic colorants. In: Peters AT, Freeman HS (eds) Analytical chemistry of synthetic colorants, pp 1–48
Würthner F (ed) (2005) Supramolecular dye chemistry. Springer
Xerox (1994) Titanium phthalocyanine imaging member. US Patent 5,371,213
Ye K, Wang J, Sun H, Liu Y, Mu Z, Li F, Jiang S, Zhang J, Zhang H, Wang Y, Che CM (2005) Supramolecular structures and assembly and luminescent properties of quinacridone derivatives. J Phys Chem B 109:8008–8016
Yoon Y, Koo JY, Oh J, Kim S, Choi HC, Yoon SM (2020) Surface-guided polymorphism control of titanyl phthalocyanine single crystals. Inorg Chem Front 7:2178–2187
Yu L (2002) Color changes caused by conformational polymorphism: optical-crystallography, single-crystal spectroscopy, and computational chemistry. J Phys Chem A 106:544–550
Yu L (2010) Polymorphism in molecular solids: an extraordinary system of red, orange, and yellow crystals. Acc Chem Res 43:1257–1266
Yu L, Stephenson GA, Mitchell CA, Bunnell CA, Snorek SV, Bowyer JJ, Borchardt TB, Stowell JG, Byrn SR (2000) Thermochemistry and conformational polymorphism of a hexamorphic crystal system. J Am Chem Soc 122:585–591
Zain SM, Ng SW (2005) 1,8-Dihydroxy-9,10-anthraquinone. Acta Cryst E 61:2921–2923
Zollinger H (2003) Color chemistry, 3rd revised edition. Wiley-VCH

Chapter 7
Photochromism

Seiya Kobatake

Abstract Photochromic compounds can be divided into two types, T- and P-types. T-type photochromic compounds include azobenzene, spiropyran, hexaarylbiimidazole, spirooxazine, naphthopyran, and the donor-acceptor Stenhouse adduct, as introduced here. In P-type photochromic compounds, there are furylfulgide and diarylethene. Diarylethene derivatives have the most excellent photochromic compounds, such as thermal stability of both isomers, high fatigue-resistance, high response, high sensitivity, high coloration quantum yield, and high reactivity even in the crystalline phase. In this chapter, the photochromic properties of these compounds have been focused.

Keywords Photochromism · Photoswitching · T-type · P-type · Diarylethene · Photoactuator

7.1 Introduction

Photochromism is a phenomenon in which the color changes reversibly by light, and a compound exhibiting such a phenomenon is called a photochromic compound. The history of photochromism is old and the first report of academic research is reaction with oxygen by irradiation to tetracene solution in air, reported by Fritzsche (1867). The produced endoperoxide desorbs oxygen at room temperature to return to the original compound. Regarding the history of photochromism, it is written in detail in some literature (Crano and Guglielmetti 1999; Dürr and Bouas-Laurent 2003; Tian and Zhang 2016). The photochromic compounds exhibit changes in various chemical and physical properties such as absorption spectra, refractive indices, dielectric constants, oxidation-reduction potentials, and geometrical structures by photoirradiation. The photoinduced rapid properties changes lead to their use in various optoelectronic devices, such as optical memories, photoswitching, image recording, photomechanical devices, and so on.

S. Kobatake (✉)
Graduate School of Engineering, Osaka City University, Sugimoto, Sumiyoshi-ku, Osaka 558-8585, Japan
e-mail: kobatake@a-chem.eng.osaka-cu.ac.jp

Y. Ooyama and S. Yagi (eds.), *Progress in the Science of Functional Dyes*,
https://doi.org/10.1007/978-981-33-4392-4_7

Photochromic compounds can be roughly divided into two types, T- and P-types (Fig. 7.1a). T-type photochromic compounds change to colored isomers by photoirradiation, and the colored isomers are thermally unstable to return to the initial colorless isomers. In contrast, P-type photochromic compounds change to thermally stable colored isomers by photoirradiation. Thus, the P-type photochromic compounds can be transformed between two isomers by only photoirradiation. Among the T-type photochromic compounds, there are some compounds that the colored states are thermally stable and the colorless states produced by photoirradiation are thermally unstable to return to the initial colored isomers. This phenomenon is called negative photochromism (Fig. 7.1b). Recently, the donor–acceptor Stenhouse adducts (DASA) have been reported to exhibit photochromism upon visible light, not ultraviolet (UV) light. The photoswitching is that the colored state is thermally stable and reacts to colorless state upon irradiation with visible light. The photogenerated colorless isomers are thermally unstable to result in returning the initial colored isomers, as shown in Fig. 7.2 (Helmy et al. 2014; Hemmer et al. 2016). They may be used for application to bioimaging using visible light.

Typical photochromic compounds are shown in Fig. 7.3. The molecular structure change accompanying the photochromic reaction includes *trans-cis* isomerization reaction, bond dissociation/formation reaction, cyclization/ring-opening reaction, hydrogen transfer reaction, and so on. Azobenzene, hexaarylbiimidazole, spiropyran, spirooxazine, and naphthopyran are classified as T-type photochromic compounds. On the other hand, furylfulgide and diarylethene are classified as P-type

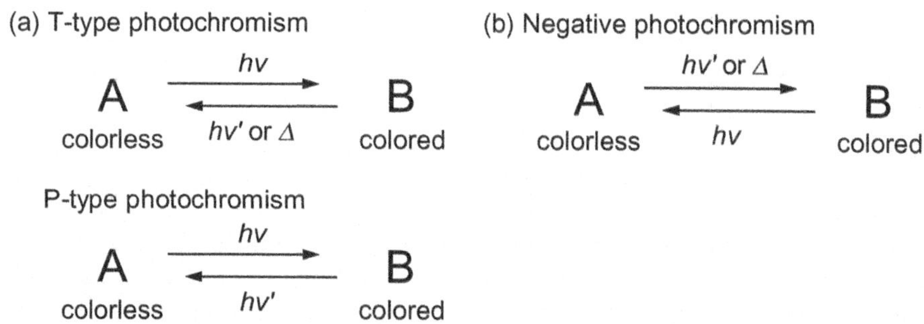

Fig. 7.1 Photochromic compounds: **a** T- and P-type photochromism, **b** negative photochromism

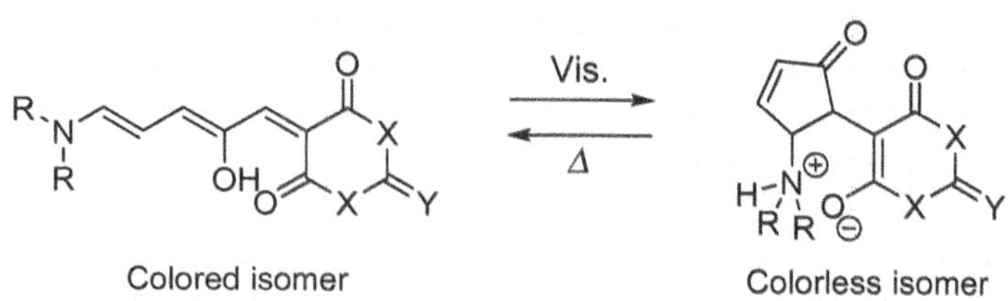

Fig. 7.2 Negative photochromism of the donor-acceptor Stenhouse adducts

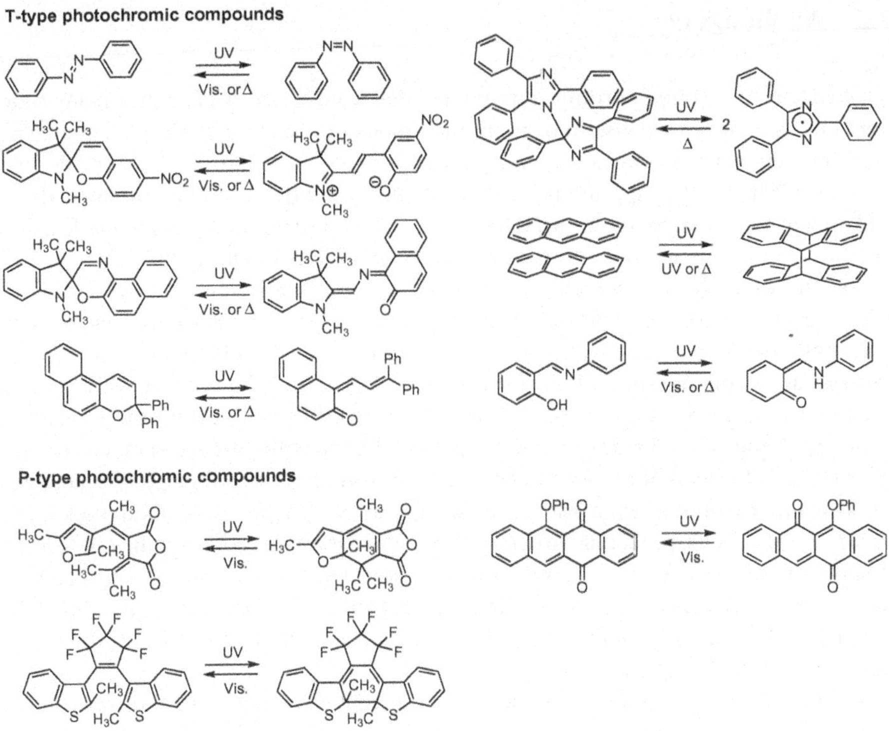

Fig. 7.3 Typical photochromic compounds

photochromic compounds. Due to the difference between T-type and P-type, various applications have been studied according to its characteristics. In the T-type, since the colored state produced by photoirradiation returns to its original colorless state in a few minutes at room temperature, such photoresponsive materials can be used for ophthalmic lenses and color-changing ornaments such as T-shirts, straps, and beads that were colored with UV light in outdoor. Since the photochromic reaction is caused by a change in molecular structure, not only color change but also various physical properties of the molecules themselves change reversibly, such as dielectric constant, refractive index, and redox potential. Therefore, it is expected to be applied to electronics and photonics devices such as optical switching elements, memory materials, display materials, and so on, which utilize changes in physical properties. In particular, as a memory material, it is necessary for the recorded state to be storage stable, and thermally stable P-type photochromic compounds have been studied. Furthermore, it is possible to create actuators that can be moved in response to light and that are expected as a next generation material called a photoactuator.

7.2 Azobenzene

Azobenzene is a photochromic compound that causes *trans-cis* photoisomerization, but its history began with dyes. In the nineteenth century, *p*-aminoazobenzene (Aniline Yellow) was synthesized as a yellow dye among various azobenzene derivatives synthesized. It is applied to medicines and synthetic colorants, dichroic dyes, and optical recording materials used for CD, DVD, and so on. These are not utilization of *trans-cis* photoisomerization but have been utilized as functional dyes.

On the other hand, molecular functional materials utilizing the properties of molecules themselves by photoisomerization are well known. Upon irradiation with UV light, azobenzene causes photoisomerization from a pale yellowish colored *trans*-form to an orange-colored *cis*-form. The *cis*-form is unstable and quickly returns to the *trans*-form by heating. As an example of skillfully exploiting the molecular structure change of azobenzene, switching of inclusion ability of alkali metal ion was investigated (Shinkai et al. 1983). Azobenzene having crown ether greatly changes the structure by photoisomerization and greatly affects inclusion of alkali metal ion. Figure 7.4 shows its molecular structure. Only when it is isomerized to the *cis*-isomer, it shows a complex formation. When it returns to the *trans*-form, the capability of the complex formation decreases. Inclusion capabilities of Na^+, K^+, Rb^+, and Cs^+ are different depending on the ring size of the crown ether, but the *trans*-form has no capability of the complex formation.

On the other hand, photo-alignment, photomechanical function, and photo-mass transfer of liquid crystal are well known as macroscopic physical property changes using *trans-cis* isomerization. When polarized light is irradiated to a polymer film having azobenzene, molecules whose polarization direction and molecular long axis coincide are isomerized to form a *cis*-form. On the other hand, molecules whose polarization direction is orthogonal to the molecular long axis cannot absorb light to result in no isomerization. The photoisomerized *cis*-isomer returns to the *trans*-form but it is isomerized repeatedly when oriented in the original orientation. However, the isomerization does not occur if it is oriented in such a direction where the molecule cannot absorb polarized light. Thus, by repeated isomerization, the major axis of the molecule converges in the direction orthogonal to the direction of polarization. In

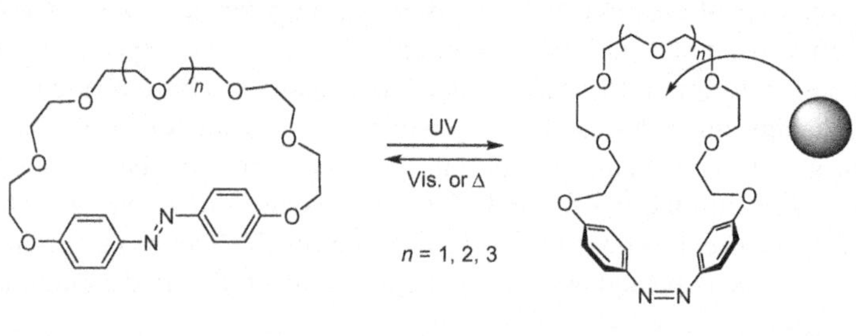

Fig. 7.4 Photochromism of azobenzene bearing a crown ether. The *cis*-form produced upon irradiation with UV light has a capability of the complex formation of metal ions

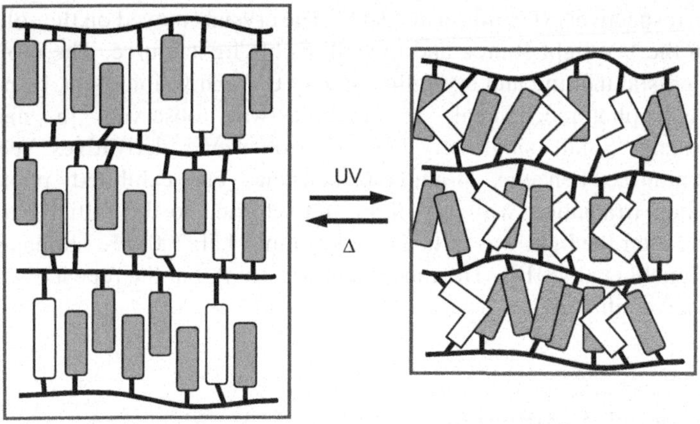

Fig. 7.5 Shrinkage of poly(oxymethylsilylene) crosslinked with azobenzene by photoirradiation

this way, molecules can be aligned by light, dichroism and birefringence appear in the film. This phenomenon is called optical alignment of liquid crystal and is utilized as a technique for producing the oriented film of liquid crystal in the electronics field (Ichimura 2000).

Attempts are also made to develop mechanical functions leading to macroscopic deformation associated with photoisomerization of azobenzene. It has been found that shrinkage of the polymer form occurs due to light-induced phase transition between order and disorder of liquid crystal polymer having poly(oxymethylsilylene) as the crosslinking site of azobenzene (Fig. 7.5) (Finkelmann et al. 2001). Thus, with the photoisomerization, the nematic phase changes to an isotropic phase, and shrinkage of about 20% occurs in the orientation direction of the mesogen. If shrinkage occurs near the surface of the film, the film can bend toward the incident light. When the film containing azobenzene is irradiated with polarized light, the bending direction can be changed according to the direction of polarization (Yu et al. 2003).

In photoisomerization of azobenzene, a mass transfer phenomenon has been found where molecules are moved more greatly by repeatedly reversible isomerization. This example is known as a photoinduced surface relief formation (Rochon et al. 1995; Kim et al. 1995). Interferential exposure from two directions using polarized light produces periodic intensity, and irregularities of micrometer size are periodically formed accordingly. The irregularity can be eliminated by heating at a temperature higher than the glass transition temperature of the polymer or irradiating uniform light. The phenomenon occurs due to the large movement of molecules by photoirradiation. It has been observed not only in polymer systems but also in low molecular amorphous systems. These azobenzene type materials are expected to be applied to photo-alignment of liquid crystals, holographic memory, and so on.

The crystals of 3,3'-dimethylazobenzene exhibited the crawling motion on a glass surface upon irradiation with UV and visible light simultaneously from the left and right sides, respectively (Uchida et al. 2015). The crystals moved on the solid surface away from the UV light source and toward visible light source. The movement is due to the crystal melting and crystallization by UV and visible light, respectively.

The real photomechanical deformation was observed in azobenzene single crystals. Small single crystals of azobenzene derivatives such as 4-(dimethylamino)azobenzene, 4-aminoazobenzene, etc exhibited photoinduced bending upon irradiation with UV light and returned to the initial flat at room temperature after the light was turned off (Koshima et al. 2009; Koshima and Ojima 2012; Taniguchi et al. 2019). The movement was short, but the repeat cycles of 100 or more were confirmed.

7.3 Hexaarylbiimidazole

Hexaarylbiimidazole is cleaved to two molecules by UV light irradiation to form radicals. As a result, two molecules are dissipated in the solution, so it is not always bound to the original cleaved molecule. Therefore, the reaction rate is complicated and highly dependent on the concentration of the solution. Hexaarylbiimidazole shows photochromism even in crystals. X-ray structural analysis of hexaarylbiimidazole having chloride after UV light irradiation confirmed the presence of two radical pairs generated by cleavage of C–N bond (Fig. 7.6a) (Kawano et al. 1999). These radical pairs return to the original dimer. However, in unsubstituted hexaphenylbiimidazole crystals, two radicals generated by cleavage form a bond between molecules

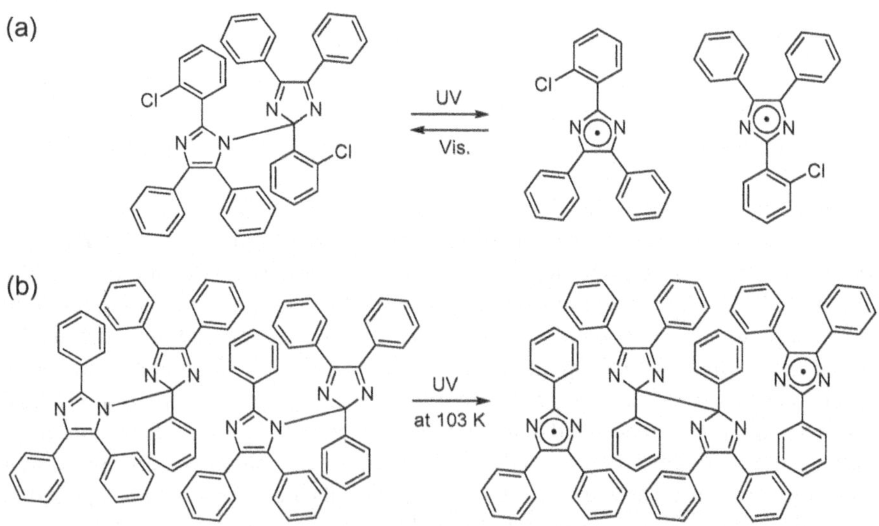

Fig. 7.6 Photochromism of **a** hexaphenylbiimidazole having *o*-chloride and **b** unsubstituted hexaphenylbiimidazole in the crystals

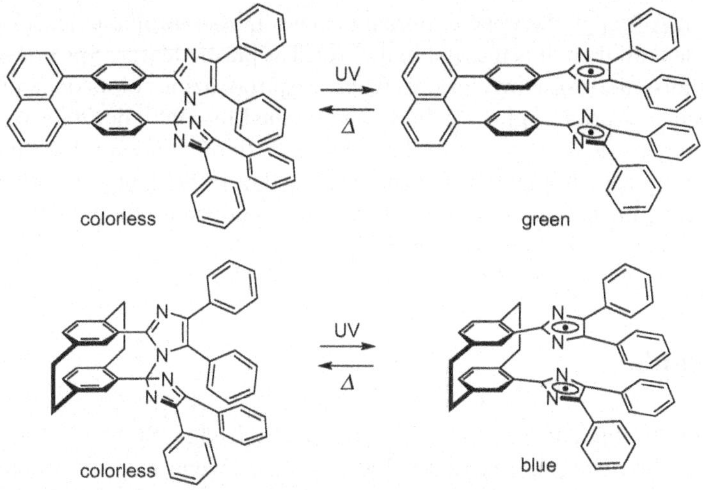

Fig. 7.7 Photochromism of crosslinked triphenylbiimidazoles

(Fig. 7.6b) (Kawano et al. 2000). The bond formed here is a C–C bond, resulting in the formation of isomers having different bonding positions.

In recent years, crosslinked biimidazoles have been synthesized in order to suppress dissipation of two molecules accompanying photoisomerization (Fig. 7.7) (Iwahori et al. 2007; Kishimoto and Abe 2009). These have excellent characteristics that radicals generated by irradiation with UV light do not dissipate, the back reaction at room temperature is extremely fast within 1 s, and the repeating durability is much higher. By realizing a high-speed photochromic compound, it can be dispersed in a polymer film and partially change the refractive index at a high speed. Therefore, it is expected to be applied to a holographic material capable of displaying three-dimensional information of an object in real time.

7.4 Spiropyran, Spirooxazine, Naphthopyran

Spiropyran has been well known as typical photochromic compound for a long time, and colored isomers photogenerated by irradiation with UV light show T-type photochromism returning thermally to colorless. Depending on the type of solvent used, it shows solvatochromism which color of the colored isomers differs largely (Klajn 2014; Kortekaas and Browne 2019). Dissociation to zwitterion by UV light irradiation is a major feature. An application of the T-type photochromic compound is automatic light-controllable ophthalmic lenses. It is colorless indoors, but when it goes outdoors it is colored with sunlight UV light to result in sunglasses. When returning indoors, it fades naturally and returns to transparent lenses. In 1982, plastic light-controllable sunglasses lenses using spirooxazine were brought on the market in the USA. Although there were inorganic glass light-controllable lenses that used silver halide until then, it is the first in the world as plastic lenses. Spiropyran differs in that photoisomers form zwitterions, whereas spirooxazine colors on neutral open-ring molecules. Various spirooxazines were studied by 1990, but they were not as good as inorganic glass light-controllable lenses. After that, researches on dimming materials using naphthopyran were conducted, and in 1999, light-controllable lenses that developed naphthopyran to gray or brown developed on the market from each company, which is now mainstream.

7.5 Fulgide

The history of fulgide is old, and it was first synthesized by Stobbe in 1905 (Stobbe 1911). The photochromic reaction belongs to a pericyclic reaction which isomerizes between hexatriene-cyclohexadiene, which is known as the 6π-electron cyclic reaction. Fulgide developed by Stobbe is a thermally unstable T-type photochromic compound, and side reactions from the colored state proceed (Fig. 7.8a). On the other hand, P-type fulgide was synthesized by Heller in 1981 (Heller and Oliver 1981; Darcy et al. 1981). In particular, the P-type fulgide is called furylfulgide because of the presence of the furyl group as the aromatic ring (Fig. 7.8b). As a P-type photochromic compound, it is historically a very important compound, which was noticed in the 1980s. Many researches and developments have been made because it can be applied to optical memories (Yokoyama 2000), but research on P-type photochromic compounds gradually shifted to diarylethene, which was discovered in the late 1980s, because fulgide has drawback of showing *cis-trans* isomerization and low repeated durability.

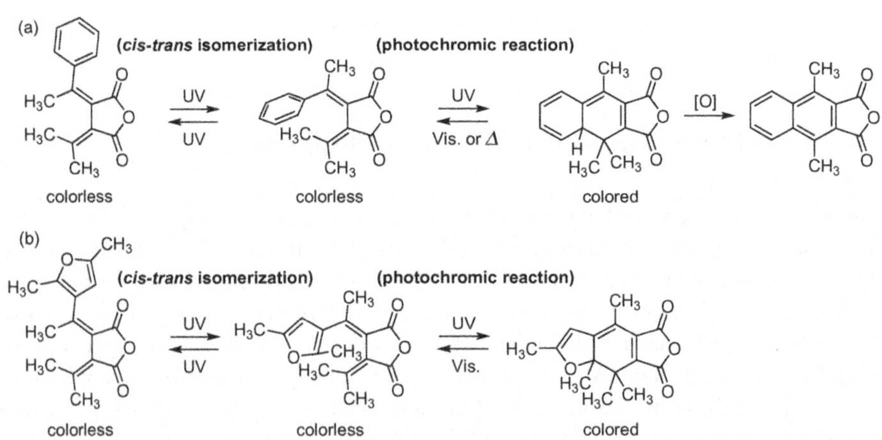

Fig. 7.8 Photochemical reactions of **a** T-type fulgide and **b** P-type furylfulgide

7.6 Diarylethene

Diarylethene is a P-type photochromic compound that was found by Irie in the late 1980s and has developed rapidly to date. The history of discovery for diarylethene is described in detail in the literature (Irie 2015). It is ring-closing and ring-opening reactions which isomerizes between a hexatriene and a cyclohexadiene by alternating irradiation with UV and visible light and shows the conrotatory reaction of a 6π-electron system according to the Woodward-Hoffmann rule. In the colorless open-ring isomer, there are two conformations of antiparallel and parallel types. Because the lifetime is short in the excited state, these conformational changes do not occur, and the reaction proceeds only from the photoreactive antiparallel conformation. The ratio of antiparallel to parallel conformations in the ground state of diarylethene shown in Fig. 7.9 is approximately 1:1, and the photocyclization quantum yield is approximately 0.5. That is, if it can be fixed to the antiparallel conformation, a high photocyclization quantum yield can be obtained and a highly efficient photoreaction system is accomplished. The compound shown in Fig. 7.10 has an interaction between S–N and H–N. It is almost fixed to the antiparallel conformation to result in extraordinarily high photocyclization quantum yield of approximately unity (Fukumoto et al. 2011).

Since the photocycloreversion reaction is a bleaching process, diarylethenes having a high photocycloreversion quantum yield mean fast fading, and diarylethenes having a small photocycloreversion quantum yield mean slow in discoloration and stable to light. These requirements differ depending on the application to be applied. The effect of the substituent on the photocycloreversion quantum yield is shown in Fig. 7.11. The photocycloreversion quantum yield decreases with increase in length of π-conjugation, largely changes with polarity of substituent at reaction positions, and greatly decreases due to oxidation of thiophene ring. In this way, the reactivity

Fig. 7.9 Antiparallel and parallel conformation of a diarylethene and its photochromic reactions

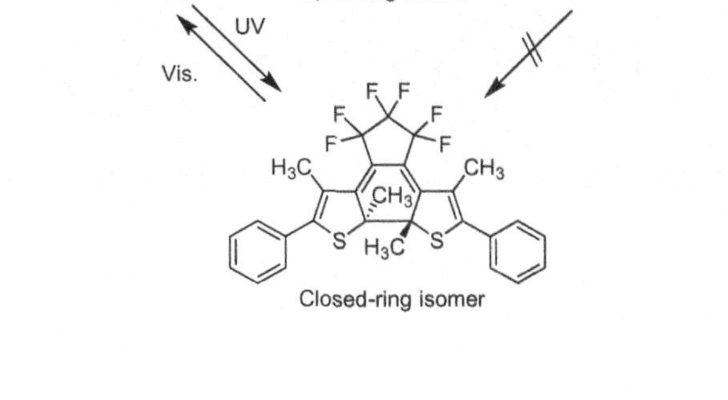

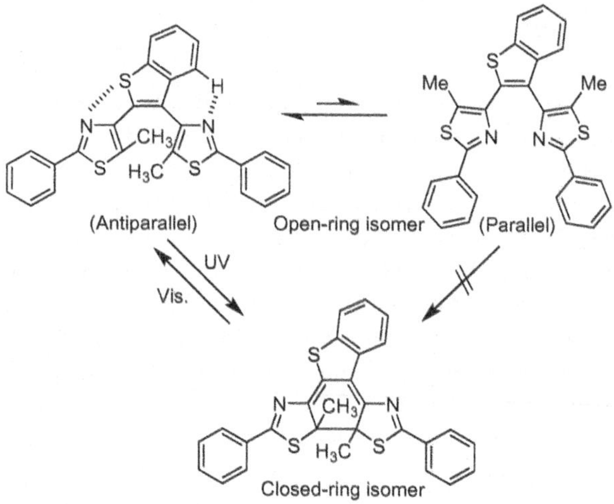

Fig. 7.10 Diarylethene showing photocyclization quantum yield of approximately unity

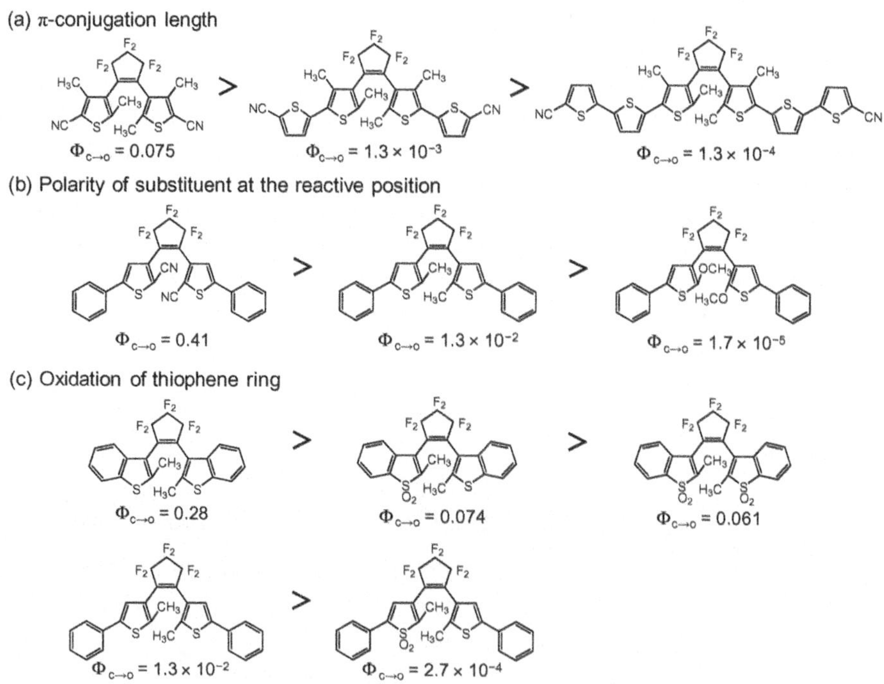

Fig. 7.11 Factors affecting photocycloreversion quantum yield of diarylethenes

can be controlled by changing the substituent group, and it can be used properly according to the application.

Diarylethene is a P-type photochromic compound but exhibits a thermal cycloreversion reaction when a substituent is introduced. Figure 7.12 shows the degree of thermal cycloreversion reactivity. The thermal cycloreversion reactivity increases with increasing aromatic stability of the aryl group, increases with the electron-withdrawing property of the substituent bonded to the aryl group, and increases with increasing bulkiness of substituent at the reactive position. These thermal reactivities are determined by the energy difference (ΔH) between the closed-ring form and the open-ring form in the ground state. When the ΔH value is less than 60 kJ mol^{-1}, the closed-ring form is thermally stable.

Diarylethene is capable of switching π-conjugation by UV and visible light irradiation. It can switch the interaction of two stable radicals. In other words, it means that the magnetic properties can be switched by light. In the open-ring form shown in Fig. 7.13, an interaction between two radicals is not observed by electron spin resonance (ESR), and only isolated radicals of both are observed. On the other hand, in the closed-ring form, the interaction between the radicals is observed by the ESR spectrum. It is in the OFF state in the open-ring form and the ON state in the closed-ring form. The difference in the interaction between the open-ring form and the closed-ring form can be changed by 150 times or more (Matsuda et al. 2001).

Fluorescence photoswitching can be accomplished by photocyclization and cycloreversion reactions. A system in which the open-ring form is fluorescent (Fig. 7.14a) (Fukaminato et al. 2003), a system in which the closed-ring form is

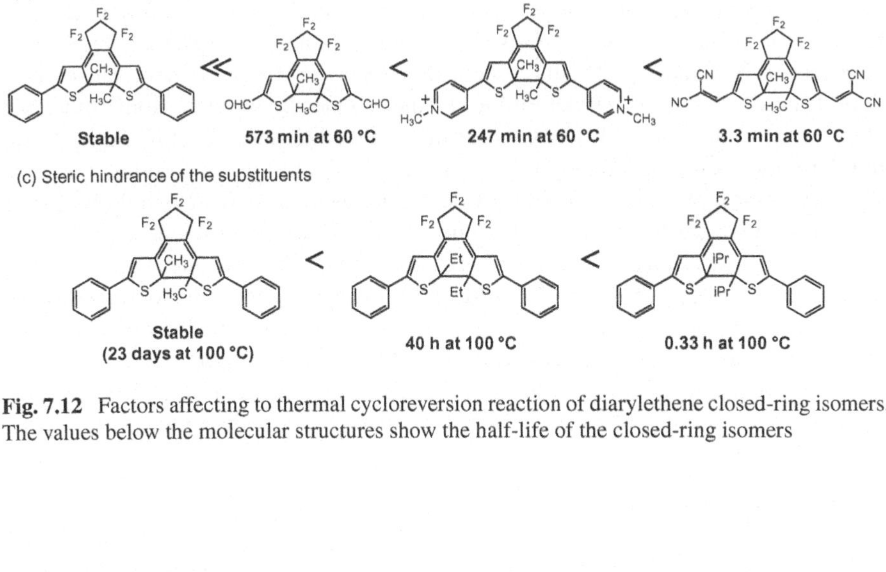

Fig. 7.12 Factors affecting to thermal cycloreversion reaction of diarylethene closed-ring isomers. The values below the molecular structures show the half-life of the closed-ring isomers

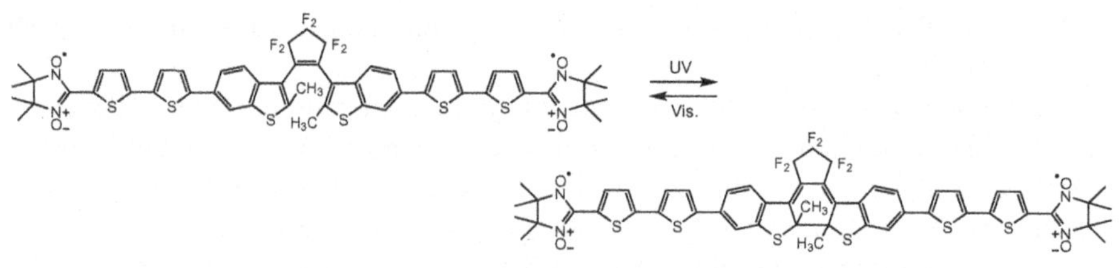

Fig. 7.13 Photochromism of diarylethene having two stable radicals

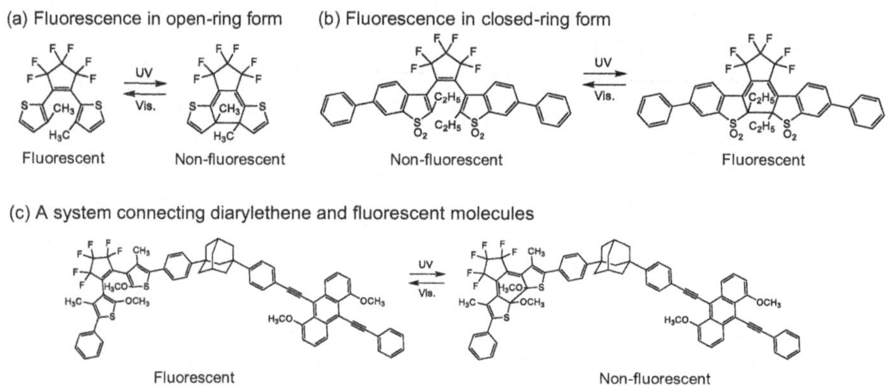

Fig. 7.14 Fluorescence photoswitching molecules using diarylethene: **a** a system exhibiting fluorescence in the open-ring form, **b** a system exhibiting fluorescence in the closed-ring form, **c** a system connecting diarylethene and fluorescent molecules

fluorescent (Fig. 7.14b) (Uno et al. 2011), and a system in which diarylethene and fluorescent molecules are bonded (Fig. 7.14c) (Irie et al. 2002) have been reported. In the system in which diarylethene and fluorescent molecules are bonded, when diarylethene is the open-ring form, fluorescence is emitted from a fluorescent molecule, but when isomerized to the closed-ring form, fluorescence is quenched by fluorescence resonance energy transfer (Förster resonance energy transfer; FRET) from the excited state of the fluorescent molecule to the diarylethene closed-ring form. The FRET efficiency is determined by the distance between the donor molecule (fluorescent molecule) and the acceptor molecule (diarylethene closed-ring form) and the fluorescence spectrum of the donor molecule and the absorption spectrum of the acceptor molecule. Therefore, by designing the diarylethene and fluorescence molecules, it is possible to control the fluorescence ON/OFF efficiency.

Unique surface physical properties using thin films of diarylethenes have been reported. Metallic deposition properties change with photoisomerization of diarylethene (Tsujioka et al. 2008). In the amorphous film of the diarylethene open-ring form shown in Fig. 7.15a, deposition of Mg does not occur, but Mg is vapor-deposited when it is photoisomerized in the closed-ring form. That is, Mg is vapor-deposited only at a specific location by photoisomerization only at a specific location

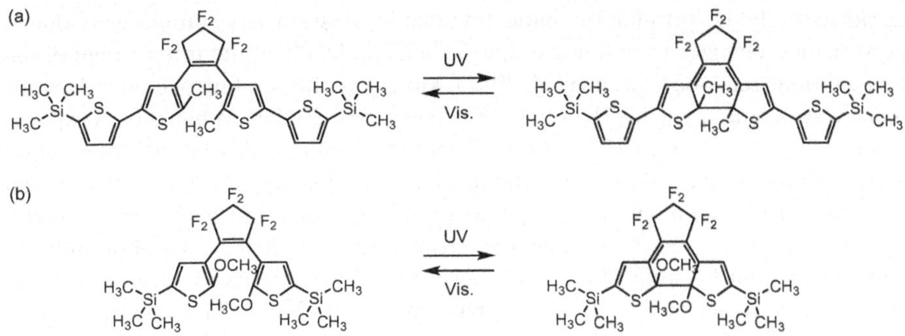

Fig. 7.15 **a** Diarylethene showing a different metal deposition character in the open- and closed-ring forms and **b** diarylethene showing photoinduced crystal surface change

using a photomask. These vapor deposition characteristics are due to the fact that the glass transition temperature (T_g) of the amorphous film changes largely between the open-ring and closed-ring forms, because it is deposited at temperature lower than T_g or its vicinity and is not deposited at temperature higher than T_g. This is due to the mobility of the molecules near the surface, and on the surface where molecules move dynamically, the deposited metal atoms reattach and hinder metal deposition.

The characteristic of interest as a peculiar phenomenon of the diarylethene crystal surface is to control water repellent on the surface by photoirradiation. On the surface of the single crystal consisting of diarylethene shown in Fig. 7.15b, a flat surface is formed in the open-ring form, but when it is photoisomerized, needle-like fine crystals consisting of the closed-ring form appeared on the surface of the diarylethene single crystal. The specific surface shows lotus effect like the lotus leaf surface and petal effect like the rose petal surface by changing the photoirradiation condition and it has super water repellency and water drop pinchable surface. These surfaces can be controlled by light, and it is an example that skillfully controls crystal growth of minute crystals (Uchida et al. 2010).

To apply photochromic diarylethene single crystals to functional materials in photonics, electronics, mechanics, and medical fields, the materials are required to change large physical property by photoirradiation. Among them, photoinduced crystal shape deformation of diarylethene crystals was observed reversibly (Kobatake et al. 2007). It means that macroscopic mechanical movement of materials based on molecular-scale structure changes of individual molecules was realized in molecular crystals of diarylethene derivatives. In the photoinduced crystal shape changes, there are contraction/expansion, bending, twisting, and so on, which depend on the crystal structure and the size of the crystal. The rod-like crystals consisting of diarylethene derivatives show the crystal bending away from the incident UV light or toward the incident UV light (Kitagawa et al. 2015a). The bending velocity depends on the crystal thickness, and the curvature change against the crystal thickness was well fitted to Timoshenko's bimetal model (Kitagawa and Kobatake 2013, 2014). When the irradiation power was changed, the bending velocity is proportional to the power of the UV irradiation, indicating that local strain acts cumulatively due to the structural changes in individual diarylethene molecule at the initial stage of the photochromic reaction (Hirano et al. 2017). Moreover, the bending behavior depends on the wavelength of the incident light (Kitagawa et al. 2015b). When the crystal was irradiated under polarized UV light, the bending speed depends on the polarization angles (Hirano et al. 2019). A unique mechanical behavior of a molecular crystal was induced by combination of a photochromic reaction and a reversible single-crystal-to-single-crystal phase transition (Kitagawa et al. 2017). The photoinduced reversible crystal twisting of a diarylethene crystal was observed upon alternating irradiation with UV and visible light (Kitagawa et al. 2013). The crystal twisting takes place in both a left-handed helix and a right-handed helix. The direction of the twisting depends on the face irradiated with UV light. The control of the photomechanical twisting of a diarylethene crystal was studied from the viewpoint of illumination direction (Kitagawa et al. 2018). Changing the UV illumination direction with respect to the crystal resulted in different twisting modes, ranging from helicoid to cylindrical, as shown in Fig. 7.16. The control of photomechanical crystal deformation by illumination direction provides a convenient and useful way to generate a variety of photomechanical motions from a single crystal.

It is an important task of how to apply the photomechanical behavior to practical use. Several demonstrations using photomechanical behavior of diarylethene crystals are described here. Cocrystals composed of a diarylethene and perfluoronaphthalene

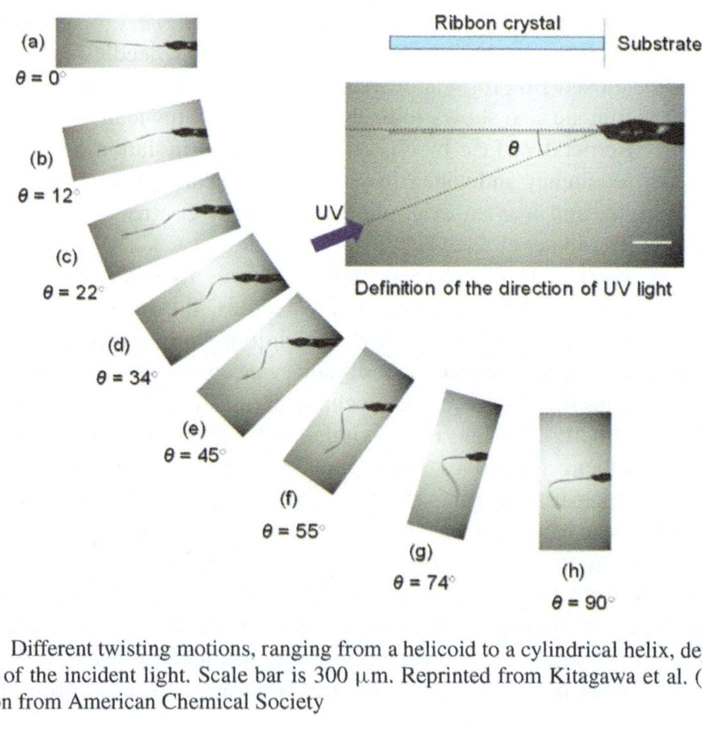

Fig. 7.16 Different twisting motions, ranging from a helicoid to a cylindrical helix, depending on the angle of the incident light. Scale bar is 300 μm. Reprinted from Kitagawa et al. (2018) with permission from American Chemical Society

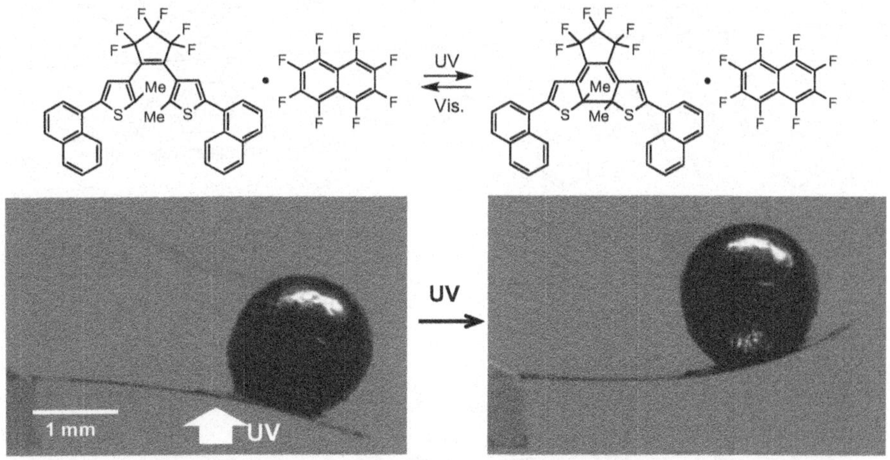

Fig. 7.17 Photomechanical work of molecular crystal cranes composed on a diarylethene and perfluoronaphthalene. UV light was irradiated from the lower side of the crystal: 2-mm lead ball, 46.77 mg; crystal, 0.17 mg. Adapted from Morimoto and Irie (2010) with permission from American Chemical Society

with a length of 1−5 mm exhibited photoreversible bending behavior over 250 times (Morimoto and Irie 2010). The crystals were able to lift a heavy metal that was 200 times heavier than the crystals, like "molecular crystal cranes", as shown in Fig. 7.17. The photogenerated maximum stress in the crystals was estimated to be 44 MPa, which is 100 times larger than that of muscle (ca. 0.3 MPa). Young's modulus of the crystal was measured to be 11 GPa. The relatively large Young's modulus enabled the crystals to perform large mechanical work.

Mixed crystals composed of two diarylethenes also exhibited more than 1000 reversible bending cycles upon alternating irradiation with UV and visible light without any crystal destruction (Terao et al. 2012). The crystal exhibited a bending motion across a wide temperature range from 4.6–370 K and even in water. The rod-like crystal bent toward the incident light source irrespective of the irradiation direction. The edge of the rod-like crystal could be rotated when the light intensity of the UV and visible lights was controlled, which resulted in actual mechanical work to rotate a gearwheel.

The bending behavior of a gold-coated diarylethene crystal was investigated (Kitagawa and Kobatake 2015). The photoreversible current ON/OFF switching of an electric circuit by the gold-coated diarylethene crystal was demonstrated as shown in Fig. 7.18. The current was able to flow in a stable manner even when the current value was over 0.5 mA. The current switching could be repeated over 10 cycles. Thus, gold-coated diarylethene crystals could be used as real photoactuator in the mesoscopic region.

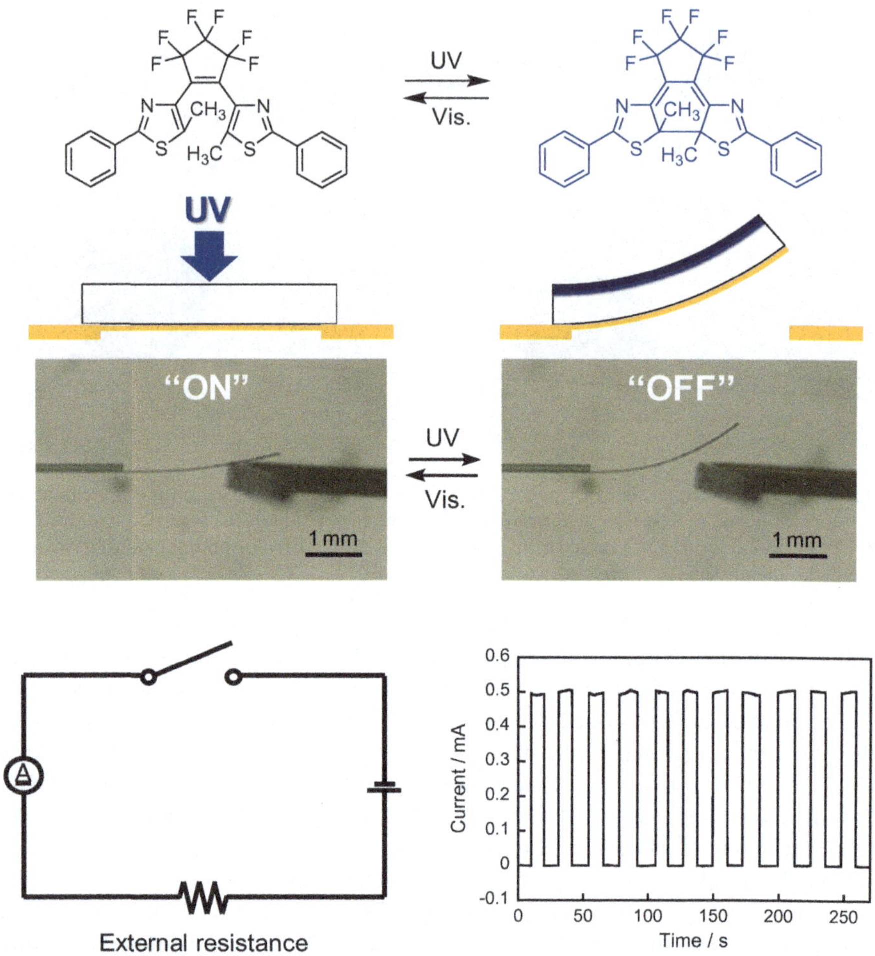

Fig. 7.18 Real current switching by photoinduced bending. The gold was coated under the crystal. The electric circuit containing external resistance was prepared. The crystal thickness is 6.2 μm. The gold thickness is 21 nm. The applied voltage is 1.0 V. The external resistance is 2 kΩ. The current ON/OFF repetition is shown in the right bottom. Adapted from Kitagawa and Kobatake (2015) with permission from Royal Society of Chemistry

Diarylethene is the most studied worldwide as a P-type photochromic compound and is likely to have new possibilities in the future. Diarylethene has high repetitive durability, various functions by introducing a substituent to the molecule, ease to synthesize, and high photochromic reactivity not only in solution but also in a polymer medium and in a crystalline state. The basic properties of diarylethene found in the latter half of the 1980s were reported in a review paper in 2000 (Irie 2000) and the functional properties including subsequent applications are summarized in a review paper in 2014 (Irie et al. 2014). The field of electronics and optics is about to proceed to molecular-scale optoelectronics in the future. Not only diarylethene but also other photochromic compounds are important for optical switching of physical properties using molecular functions.

References

Crano JC, Guglielmetti RJ (1999) Organic photochromic and thermochromic compounds, vols 1, 2. Prenum Press, New York

Darcy PJ, Heller HG, Strydom PJ, Whittall I (1981) Photochromic heterocyclic fulgides. Part 2. Electrocyclic reactions of (*E*)-α-2,5-dimethyl-3-furylethylidene(alkyl-substituted methylene)succinic anhydrides. J Chem Soc Perkin Trans 1:202–205

Dürr H, Bouas-Laurent H (2003) Photochromism: molecules and systems. Elsevier

Finkelmann H, Nishikawa E, Pereira GG, Warner M (2001) A new opto-mechanical effect in solids. Phys Rev Lett 87:015501

Fritzsche M (1867) Comptes Rendus Acad Sci Paris 69:1035–1037

Fukaminato T, Kawai T, Kobatake S, Irie M (2003) Fluorescence of photochromic 1,2-bis(3-methyl-2-thienyl)ethene. J Phys Chem B 107:8372–8377

Fukumoto S, Nakashima T, Kawai T (2011) Photon-quantitative reaction of a dithiazolylarylene in solution. Angew Chem Int Ed 50:1565–1568

Heller HG, Oliver S (1981) Photochromic heterocyclic fulgides. Part 1. Rear rangement reactions of (*E*)-α-3-furylethylidene(isopropylidene)succinic anhydride. J Chem Soc Perkin Trans 1:197–201

Helmy S, Leibfarth FA, Oh S, Poelma JE, Hawker CJ, de Alaniz JR (2014) Photoswitching using visible light: a new class of organic photochromic molecules. J Am Chem Soc 136:8169–8172

Hemmer JR, Poelma SO, Treat N, Page ZA, Dolinski ND, Diaz YJ, Tomlinson W, Clark KD, Hooper JP, Hawker C, de Alaniz JR (2016) Tunable visible and near infrared photoswitches. J Am Chem Soc 138:13960–13966

Hirano A, Hashimoto T, Kitagawa D, Kono K, Kobatake S (2017) Dependence of photoinduced bending behavior of diarylethene crystals on ultraviolet irradiation power. Cryst Growth Des 17:4819–4825

Hirano A, Kitagawa D, Kobatake S (2019) Photomechanical bending behavior of photochromic diarylethene crystals induced under polarized light. CrystEngComm 21:2495–2501

Ichimura K (2000) Photoalignment of liquid-crystal systems. Chem Rev 100:1847–1874

Irie M (2000) Diarylethenes for memories and switches. Chem Rev 100:1685–1716

Irie M (2015) Discovery and development of photochromic diarylethenes. Pure Appl Chem 87:617–626

Irie M, Fukaminato T, Sasaki T, Tamai N, Kawai T (2002) A digital fluorescent molecular photoswitch. Nature 420:759–760

Irie M, Fukaminato T, Matsuda K, Kobatake S (2014) Photochromism of diarylethene molecules and crystals: memories, switches, and actuators. Chem Rev 114:12174–12277

Iwahori F, Hatano S, Abe J (2007) Rational design of a new class of diffusion-inhited HABI with fast back-reaction. J Phys Org Chem 20:857–863

Kawano M, Sano T, Abe J, Ohashi Y (1999) The first in situ direct observation of the light-induced radical pair from a hexaarylbiimidazolyl derivative by X-ray crystallography. J Am Chem Soc 121:8106–8107

Kawano M, Sano T, Abe J, Ohashi Y (2000) In situ observation of molecular swapping in a crystal by X-ray analysis. Chem Lett 29:1372–1373

Kim DY, Tripathy SK, Li L, Kumar J (1995) Laser-induced holographic surface relief gratings on nonlinear optical polymer films. Appl Phys Lett 66:1166–1168

Kishimoto Y, Abe J (2009) A Fast photochromic molecule that colors only under UV light. J Am Chem Soc 131:4227–4229

Kitagawa D, Kobatake S (2013) Crystal thickness dependence of photoinduced crystal bending of 1,2-bis(2-methyl-5-(4-(1-naphthoyloxymethyl)phenyl)-3-thienyl)perfluorocyclopentene. J Phys Chem C 117:20887–20892
Kitagawa D, Kobatake S (2014) Crystal thickness dependence of the photoinduced crystal bending of 1-(5-methyl-2-(4-(p-vinylbenzoyloxymethyl)phenyl)-4-thiazolyl)-2-(5-methyl-2-phenyl-4-thiazolyl)perfluorocyclopentene. Photochem Photobiol Sci 13:764–769
Kitagawa D, Kobatake S (2015) Photoreversible current ON/OFF switching by the photoinduced bending of gold-coated diarylethene crystals. Chem Commun 51:4421–4424
Kitagawa D, Nishi H, Kobatake S (2013) Photoinduced twisting of a photochromic diarylethene crystal. Angew Chem Int Ed 52:9320–9322
Kitagawa D, Iwaihara C, Nishi H, Kobatake S (2015a) Quantitative evaluation of photoinduced bending speed of diarylethene crystals. Crystals 5:551–561
Kitagawa D, Tanaka R, Kobatake S (2015b) Dependence of photoinduced bending behavior of diarylethene crystals on irradiation wavelength of ultraviolet light. Phys Chem Chem Phys 17:27300–27305
Kitagawa D, Kawasaki K, Tanaka R, Kobatake S (2017) Mechanical behavior of molecular crystals induced by combination of photochromic reaction and reversible single-crystal-to-single-crystal phase transition. Chem Mater 29:7524–7532
Kitagawa D, Tsujioka H, Tong F, Dong X, Bardeen CJ, Kobatake S (2018) Control of photomechanical crystal twisting by illumination direction. J Am Chem Soc 140:4208–4212
Klajn R (2014) Spiropyran-based dynamic materials. Chem Soc Rev 43:148–184
Kobatake S, Takami S, Muto H, Ishikawa T, Irie M (2007) Rapid and reversible shape changes of molecular crystals on photoirradiation. Nature 446:778–781
Kortekaas L, Browne WR (2019) The evolution of spiropyran: fundamentals and progress of an extraordinarily versatile photochrome. Chem Soc Rev 48:3406–3424
Koshima H, Ojima N (2012) Photomechanical bending of 4-aminoazobenzene crystals. Dyes Pigm 92:798–801
Koshima H, Ojima N, Uchimoto H (2009) Mechanical motion of azobenzene crystals upon photoirradiation. J Am Chem Soc 131:6890–6891
Matsuda K, Matsuo M, Irie M (2001) Photoswitching of intramolecular magnetic interaction using diarylethene with oligothiophene π-conjugated chain. J Org Chem 66:8799–8803
Morimoto M, Irie M (2010) A diarylethene cocrystal that converts light into mechanical work. J Am Chem Soc 132:14172–14178
Rochon P, Batalla E, Natansohn A (1995) Optically induced surface gratings on azoaromatic polymer films. Appl Phys Lett 66:136–138
Shinkai S, Minami T, Kusano Y, Manabe O (1983) Photoresponsive crown ethers. 8. Azobenzenophane-type switched-on crown ethers which exhibit an all-or-nothing change in ion-binding ability. J Am Chem Soc 105:1851–1856
Stobbe H (1911) Die fulgide. Justus Liebig's Ann Chem 380:1–2
Taniguchi T, Asahi T, Koshima H (2019) Electrically tunable-focusing liquid crystal microlens array with simple electrode. Crystals 9:431
Terao F, Morimoto M, Irie M (2012) Light-driven molecular-crystal actuators: rapid and reversible bending of rodlike mixed crystals of diarylethene derivatives. Angew Chem Int Ed 51:901–904
Tian H, Zhang J (2016) Photochromic materials. Wiley-VCH
Tsujioka T, Sesumi Y, Takagi R, Masui K, Yokojima S, Uchida K, Nakamura S (2008) Selective metal deposition on photoswitchable molecular surfaces. J Am Chem Soc 130:10740–10747
Uchida K, Nishikawa N, Izumi N, Yamazoe S, Mayama H, Kojima Y, Yokojima S, Nakamura S, Tsujii K, Irie M (2010) Phototunable diarylethene microcrystalline surfaces: lotus and petal effects upon wetting. Angew Chem Int Ed 49:5942–5944
Uchida E, Azumi R, Norikane Y (2015) Light-induced crawling of crystals on a glass surface. Nature Commun 6:7310

Uno K, Niikura H, Morimoto M, Ishibashi Y, Miyasaka H, Irie M (2011) In situ preparation of highly fluorescent dyes upon photoirradiation. J Am Chem Soc 133:13558–13564
Yokoyama Y (2000) Fulgides for memories and switches. Chem Rev 100:1717–1740
Yu Y, Nakano M, Ikeda T (2003) Directed bending of a polymer film by light. Nature 425:145

Chapter 8
Red and Near-IR Fluorescent Two-Photon Absorption Dyes

Tsutomu Ishi-i

Abstract The two-photon excitation technique has several advantages such as reduced autofluorescence and deeper penetration in tissues, less photodamage and photobleaching, and higher three-dimensional resolution, which cannot be achieved by linear one-photon excitation. Because of these advantages, two-photon absorption (TPA) of organic molecules has been developed in biological and materials science fields. In particular, fluorescent TPA dyes have attracted the attention of chemists as well as biologists because of their application in biological imaging. At present, various types of fluorescent TPA dyes that have large TPA cross sections as well as good fluorescence quantum yields are accessible. However, the fluorescent color has mostly been restricted to the shorter wavelength region of blue and green light, which competes with the autofluorescence from intrinsic biomolecules. To achieve efficient two-photon imaging, the longer wavelength emission of red and near-IR light is required for two-photon excitation because this light lies in the biological optical window. In this chapter, we focus first on fluorescent TPA dyes that have red and near-IR light-emitting properties. Then, we summarize the recent advances made in the development of these dyes for applications in biological systems. In the last section, the unique features of two-photon absorption in aggregate systems are summarized.

Keywords Two-photon absorption · Red fluorescence · Near-IR fluorescence · Biological imaging · Aggregation

8.1 Introduction

Two-photon absorption (TPA) is a nonlinear optical phenomenon (Marder 2006). In contrast to linear one-photon absorption, two photons that are approximately twice the wavelength (or approximately half the energy) are simultaneously absorbed upon photoexcitation (Fig. 8.1). Thus, the red and near-infrared (IR) light regions can be

T. Ishi-i (✉)
Department of Biochemistry and Applied Chemistry, National Institute of Technology, Kurume College, 1-1-1 Komorino, Kurume 830-8555, Japan
e-mail: ishi-i@kurume-nct.ac.jp

Y. Ooyama and S. Yagi (eds.), *Progress in the Science of Functional Dyes*,
https://doi.org/10.1007/978-981-33-4392-4_8

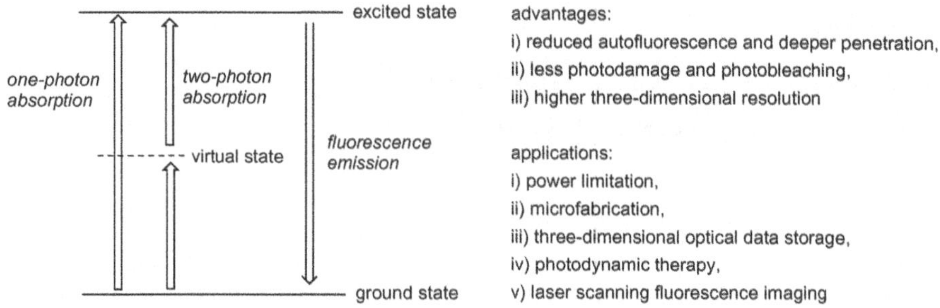

Fig. 8.1 Three-state model of two-photon absorption, and its advantages and applications

used as two-photon excitation sources compared with the ultraviolet and visible (blue and green light) light regions used in one-photon excitation. Longer wavelength red and near-IR light are particularly useful for biological applications because it falls within the biological optical window where the maximal penetration of light in biological tissues occurs (Weissleder 2001). The photoexcitation in the TPA process is confined to the focal point to show a high spatial resolution in three dimensions because the probability of TPA is proportional to the square of the light intensity. Thus, two-photon excitation has several advantages: (i) a reduced autofluorescence and deeper penetration in tissues, (ii) less photodamage and photobleaching, and (iii) a higher three-dimensional resolution by a focused laser beam; these cannot be achieved by linear one-photon excitation (Fig. 8.1). Because of these advantages, TPA organic molecules have been developed for application in biological and material sciences fields, such as for optical power limitation (He et al. 1995; Ehrlich et al. 1997), microfabrication (Cumpston et al. 1999; Kawata et al. 2001), three-dimensional optical data storage (Parthenopoulos and Rentzepis 1989; Kawata and Kawata 2000; Dy et al. 2007), photodynamic therapy (Frederiksen et al. 2001; Gu et al. 2017; Shen et al. 2016), and two-photon laser scanning fluorescence imaging (Denk et al. 1990; Zipfel et al. 2003a; Kim and Cho 2015).

In scientific history, TPA was predicted theoretically by Göpper–Mayer in the 1930s (Göppert-Mayer 1931). TPA efficiency has been expressed as the TPA cross section in which the unit of GM (1 GM = 10^{-50} cm^4 s photon^{-1}) originated from Göpper–Mayer. Later, an experimental demonstration was succeeded through two-photon excitation and subsequent upconverted fluorescence for an inorganic compound (Kaiser and Garrett 1961). Although TPA organic molecules have been used in biological and material sciences fields since the 1980s and 1990s (He et al. 1995; Ehrlich et al. 1997; Cumpston et al. 1999; Kawata et al. 2001; Parthenopoulos and Rentzepis 1989; Kawata and Kawata 2000; Dy et al. 2007; Frederiksen et al. 2001; Gu et al. 2017; Shen et al. 2016; Denk et al. 1990; Zipfel et al. 2003a; Kim and Cho 2015), highly efficient TPA was desired to achieve large TPA cross sections. In 1998, an important finding in organic chemistry was reported by Bredas, Marder, Perry, and Webb's group (Albota et al. 1998). A large TPA cross section can be achieved based on the concept that the TPA nature is ascribed to intramolecular charge transfer by using model compounds of bis(styryl)benzene derivatives that have electron donor, electron acceptor, and π-spacer moieties. After that breakthrough, a number of organic compounds including donor–acceptor, donor–acceptor–donor, acceptor–donor–acceptor, donor–π–acceptor, and donor–π–donor molecules were synthesized, and the relationship between the molecular structure and the TPA property was investigated to provide guidelines for the development of materials with large TPA cross sections (He et al. 2008; Pawlicki et al. 2009; Kim and Cho 2009; Terenziani et al. 2008). In addition to the structure–property relationship, TPA applications for different fields have been comprehensively reviewed. Here, we mainly focus on fluorescent TPA dyes emitting longer wavelength red and near-IR light. Additionally, we summarize the recent advances made in their development for applications in biological systems. In the last section, the unique features of TPA in aggregate systems are summarized.

8.2 Fluorescent Two-Photon Absorption Dyes

8.2.1 History and Demands

Fluorescent TPA dyes have attracted the attention of chemists as well as biologists because of their applications in biological imaging. Microscopic observation combined with near-IR two-photon excitation provides versatile advantages such as increased imaging depth and reduced photodamage and photobleaching. Since two-photon microscopy was introduced by Denk and Webb (Denk et al. 1990; Zipfel et al. 2003a), many light-emitting TPA dyes have been designed and prepared. Currently, various types of TPA dyes that exhibit large TPA cross sections and good fluorescence quantum yields are accessible (Kim and Cho 2015; Ventelon et al. 2001; Werts et al. 2004; Kim et al. 2008; Wang et al. 2010; Zhang et al. 2011; Heo et al. 2016; Tang et al. 2016; Maeda et al. 2016). The fluorescent color emitted from TPA dyes has mostly been restricted to the shorter wavelength blue and green regions (Ventelon et al. 2001; Werts et al. 2004; Kim et al. 2008; Wang et al. 2010; Zhang et al. 2011; Heo et al. 2016; Tang et al. 2016; Maeda et al. 2016), which competes with the autofluorescence from intrinsic biomolecules, such as nicotinamide adenine, riboflavin, and flavoproteins, aromatic amino acids even at the low level of the TPA cross section (Zipfel et al. 2003b). To achieve efficient two-photon microscopic imaging, the emission of red and near-IR light is required upon two-photon excitation because it is within the biological optical window (Weissleder 2001). However, the emission of longer wavelength red and near-IR light with two-photon excitation is rare compared to blue and green wavelengths because an elongated π-system and/or an increased intramolecular charge transfer in fluorescent TPA dyes are required. In this section, we focus on fluorescent TPA dyes that have a red or near-IR light-emitting nature.

8.2.2 *Red- and Near-IR-Light-Emitting Two-Photon Absorption Dyes*

Red light emission was first reported by Fréchet and Prasad's group. Bichromophoric molecule **1** was combined with a triphenylamine-oxadiazole-based TPA chromophore, and the red light-emitting Nile Red chromophore was designed, which prevented the difficulty of adding a red emission property to the TPA dye and achieved intense red color emission indirectly by efficient intramolecular fluorescence resonance energy transfer (Brousmiche et al. 2003, 2004). The red emission intensity increased eightfold compared to the direct excitation of the Nile Red moiety (Fig. 8.2).

In 2004, Mataka and Ishi-i's group reported the first example of red fluorescent TPA dyes by the direct two-photon excitation of donor–acceptor type triphenylamine-benzothiadiazole fluorescent dyes **2a-f** (Fig. 8.3) (Kato et al. 2004). The red light emission and TPA nature (130–330 GM) can be controlled by an additional π-spacer between the donor and acceptor moieties (Kato et al. 2006). Up to 800 MG, the TPA cross section was improved by developing star-burst structure **3** (Fig. 8.3). Then, Ishi-i's group developed the triphenylamine-benzothiadiazole skeleton for functional TPA materials in the fields of photodynamic therapy (Ishi-i et al. 2007) and three-dimensional data storage (Ishi-i et al. 2009). In dyes **4a** and **4b**, the combination of the TPA nature of the peripheral triphenylamine-benzothiadiazole dye and the singlet oxygen sensitizing ability of the porphyrin core facilitated efficient singlet oxygen

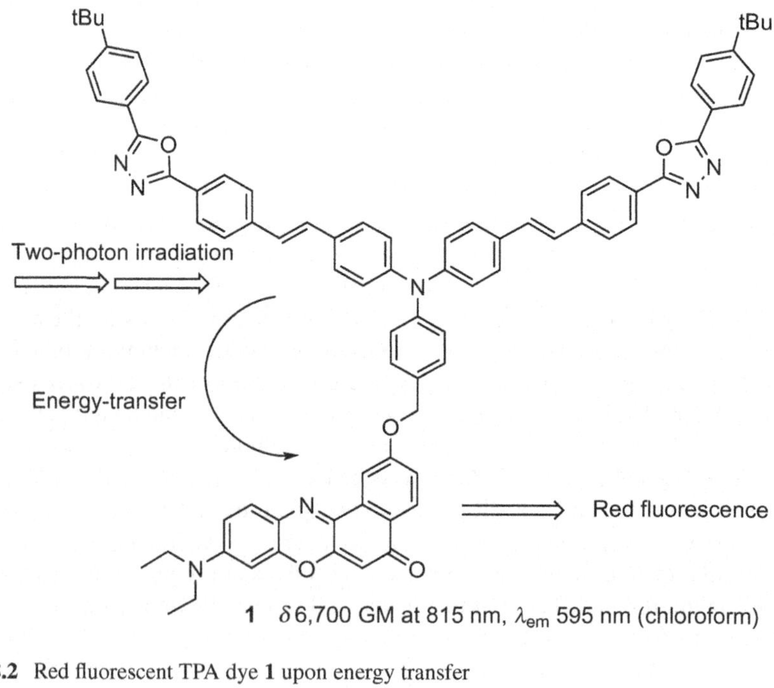

1 δ 6,700 GM at 815 nm, λ_{em} 595 nm (chloroform)

Fig. 8.2 Red fluorescent TPA dye **1** upon energy transfer

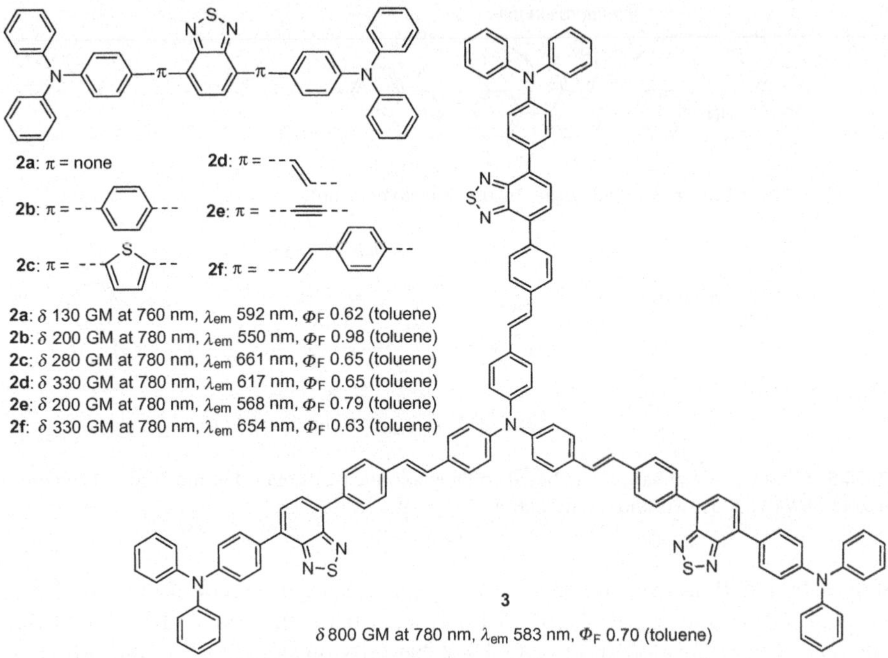

Fig. 8.3 Benzothiadiazole-based red fluorescent TPA dyes **2a-f** and **3**

generation upon near-IR two-photon excitation (Fig. 8.4) (Ishi-i et al. 2007). The efficiency of two-photon-induced singlet oxygen generation was higher by two orders of magnitude than that of tetraphenylporphyrin, indicating the excellent performance

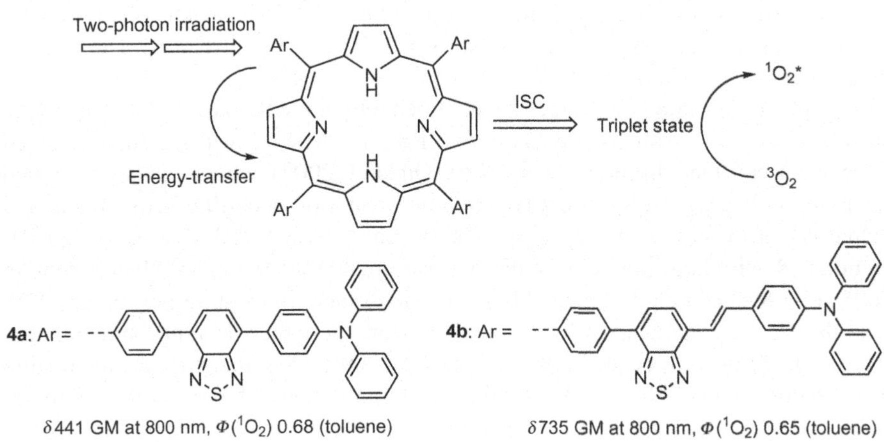

Fig. 8.4 Singlet oxygen generation by two-photon excitation of the benzothiadiazole-porphyrin TPA dyes **4a** and **4b**

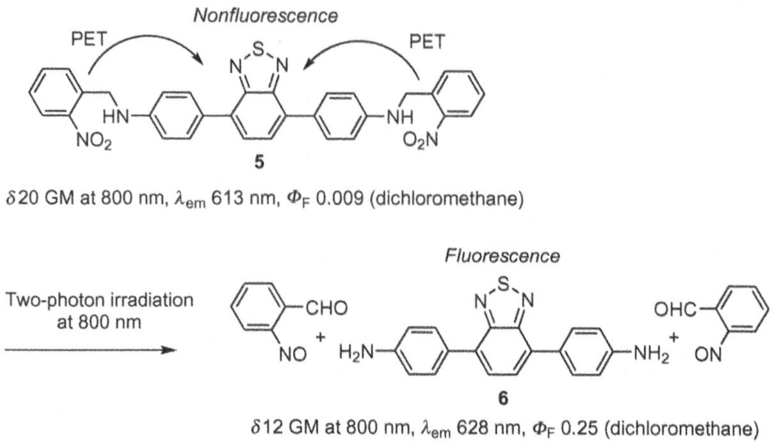

Fig. 8.5 Fluorescence enhancement based on the photorelease of quencher moieties of benzothiadiazole TPA dye **5** by two-photon irradiation

of dyes **4a** and **4b** as two-photon singlet oxygen sensitizers (Frederiksen et al. 2001; Gu et al. 2017; Shen et al. 2016). A fluorescent benzothiadiazole-based TPA dye was successfully developed into an OFF–ON emission switching system (Fig. 8.5) (Ishi-i et al. 2009). The benzothiadiazole-based dye, **5**, provides significant fluorescence quenching by introducing photoreleasing nitrobenzyl quencher moieties through a photo-induced electron transfer (PET) mechanism. The fluorescence intensity is recovered by photoreleasing the quencher moieties upon irradiation, indicating possible development for three-dimensional data storage systems using two-photon excitation (Parthenopoulos and Rentzepis 1989; Kawata and Kawata 2000; Dy et al. 2007). Recently, the attractive TPA-active red light emission observed in the triphenylamine-benzothiadiazole system was achieved even in an aqueous system by using the aggregate emission concept (Ishi-i et al. 2012, 2014, 2015a, b). The detailed findings are described in Sect. 8.3.

At almost the same time as the report of benzothiadiazole-based TPA dyes (Kato et al. 2004, 2006), Müllen and his co-workers synthesized water-soluble and red fluorescent perylene diimides **7** (Fig. 8.6) (Qu et al. 2004), which can be developed for living cell imaging by direct two-photon excitation according to Hofkens and Schryver's group (Margineanu et al. 2004). An excellent TPA dye of the Eu(III) complex, **8**, with high-purity red emission was created by Zhang and Wang group in 2005 (Fig. 8.6) (Fu et al. 2005). After this early stage, interesting fluorescent TPA dyes that had red and near-IR light-emitting nature were then designed and prepared (He et al. 2008; Lee et al. 2005; Guo et al. 2009). Very recently, an interesting near-IR fluorescent TPA dye, **9**, was designed and prepared based on the combination of a donor–acceptor motif with an excited state intramolecular proton transfer (ESIPT) concept (Fig. 8.6). Yokogawa, Kamada, and Yamaguchi's group reported an acceptor–π–donor–π–acceptor-type dye in which electron-accepting boryl groups

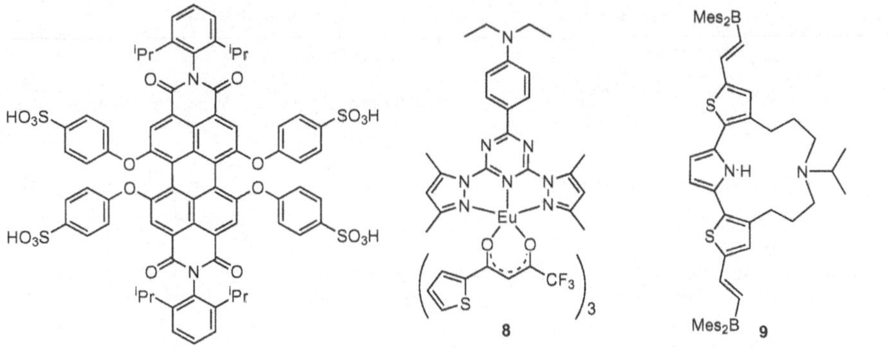

7	8	9
δ 50 GM at 840 nm, λ_{em} 619 nm, Φ_F 0.58 (water)	δ 157 GM at 808 nm, λ_{em} 614 nm, Φ_F 0.52 (toluene)	δ 698 GM at 881 nm (THF), λ_{em} 708 nm, Φ_F 0.55 (acetone)

Fig. 8.6 Red and near-IR fluorescent TPA dyes **7**, **8**, and **9**

were introduced into an ESIPT-active dithienylpyrrole moiety containing a dialkylamino strap (Suzuki et al. 2018). The electron-accepting units induce red-shifted emission from the ESIPT state and enhance the TPA character. The dye provides a bright near-IR fluorescence emission (Φ_F, 0.55) at approximately 700 nm upon near-IR two-photon excitation (698 GM at 881 nm). The large Stokes shift arising from ESIPT enables a close energy between the near-IR two-photon excitation and the near-IR emission.

As described above, the TPA dyes were constructed by connecting donor and/or acceptor moieties in a linear structure and a branched star-shaped structure. In addition, the typically used fluorescent dyes, such as coumarin, acedan, and Nile Red, were developed for red and near-IR fluorescent applications based on a strategy of extended π-conjugation and/or enhanced intramolecular charge transfer characteristics. Recent advances made in their development for applications in biological systems are presented in Sect. 8.2.3.

8.2.3 Biological Application of Red- and Near-IR-Light-Emitting Two-Photon Absorption Dyes

As described above, the emission of red and near-IR light has versatile advantages for biological applications because of the increased imaging depth and reduced photodamage and photobleaching. In this section, recent examples of red and near-IR light-emissive TPA dyes for applications in biological systems are summarized. As described in the introduction part of Sect. 8.1, TPA nature has been expressed as TPA cross sections (δ). In fluorescent TPA dyes, the two-photon action cross section ($\Phi_F\delta$), which is determined by the combination of the fluorescence quantum yield (Φ_F) and the TPA cross section, has been used to indicate the two-photon-induced

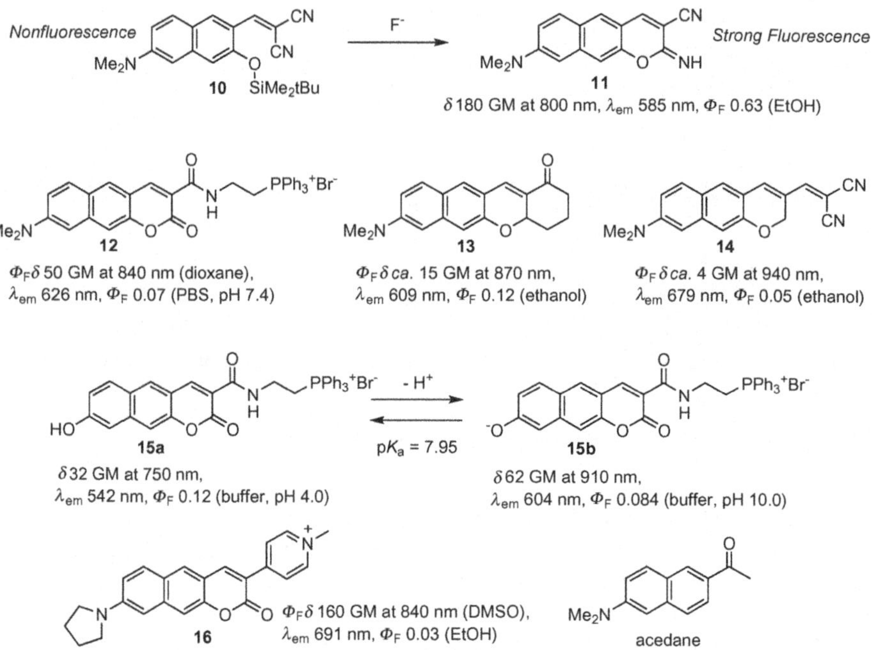

Fig. 8.7 Red fluorescent TPA dyes **10–16** bearing the π-extended coumarin chromophore for biological imaging

light-emitting nature. In this section, the properties of fluorescent TPA dyes are indicated by the two parameters, $\Phi_F\delta$ and δ.

A π-extended imino-coumarin dye, **11**, that has a donor–acceptor structure was reported by Kim and Ahn's group (Kim et al. 2012a). The coumarin dye emitted reddish-yellow light with a higher quantum yield of 0.63 and has a good TPA cross section of 180 GM. This π-extended dye **11** was transformed from the precursor coumarin-based dye **10** by treatment with fluoride ions (Fig. 8.7). This system can be developed for fluorescent imaging of fluoride ions in live zebrafish using two-photon fluorescence microscopy (Kim et al. 2012b). Kim's group reported imaging of mitochondria trafficking by π-extended coumarin dye **12**, which emitted red light upon two-photon excitation (Fig. 8.7) (Sarkar et al. 2014). π-Extended acedan derivatives **13** and **14** were designed and prepared as red-fluorescent TPA dyes by Kim, Mook-Jung, and Ahn's group (Fig. 8.7) (Kim et al. 2015). This new dye system has several advantages regarding the two-photon excited emission behavior, including minimal autofluorescence during tissue imaging compared to the parent acedan dye. This newly designed dye structure was combined with a dicyanoethylene functionality to develop fluorescent probe **14** for amyloid-β plaques, a key biomarker of Alzheimer's disease. This probe enabled in vivo imaging of amyloid-β plaques in a disease mouse model with negligible background signal. Kim group developed ratiometric TPA probe **15** based on the π-extended coumarin dye for analysis of pH in mitochondria (Sarkar et al. 2016). This dye has a pKa value of 7.95 and exhibits a yellow to red emission color change in response to pH alterations from 6.0 to 9.0 (Fig. 8.7). Ratiometric two-photon fluorescence microscopy imaging revealed the heterogeneity of pH values with respect to the specific location of mitochondria within the cells.

A systematic study of deep-tissue imaging (brain, kidney, liver, lung, and spleen mouse tissues) by two-photon microscopy using far-red-emitting dyes **16** was reported by Ahn's group (Jun et al. 2017). The TPA dye was designed and prepared based on a π-extended coumarin chromophore that has donor and acceptor moieties (Fig. 8.7). The intramolecular charge transfer property arising from the donor–acceptor structure provided the attractive far-red emission and TPA. The two-photon action cross section was 160 GM, which is comparable to those of the precedent TPA dyes (e.g., acedan) widely used in two-photon fluorescence microscopy that emit short-wavelength fluorescence. For deep depth (~150 μm) imaging of kidney tissue, only the far-red dye **16** can maintain a high signal to background ratio. In contrast, the ratio decreased in acedan with short-wavelength green emission.

Nile Red-based red-emissive TPA dye **17** was reported by Lin's group (Kong et al. 2017). The biotin moiety connected to the Nile Red chromophore acted as the tumor-specific ligand (Fig. 8.8). A moderate two-photon action cross section of 15.4 GM was obtained upon two-photo excitation at 800 nm and subsequent monitoring of the far-red light emission at 655 nm. The dye can be used for imaging biotin receptor-positive cancer cells over biotin-negative cells. Three-dimensional imaging of living tumor tissues at a penetration depth of approximately 90 μm was successfully performed under red-emitting two-photon manners. Xu, Yuan, and Liang' group reported 22 phthalazinone derivatives, **18**, which were designed for two-photon fluorescence microscopy imaging (Yang et al. 2016). By coupling electron-donating aromatic groups, such as dialkylaminophenyl, thienyl, naphthyl, and quinolyl, some derivatives have efficient emission in the red region together with a moderate TPA nature ($\delta =$ ~200 MG) (Fig. 8.8). Selected derivatives were applied for two-photon fluorescence microscopic imaging of mouse brain slices.

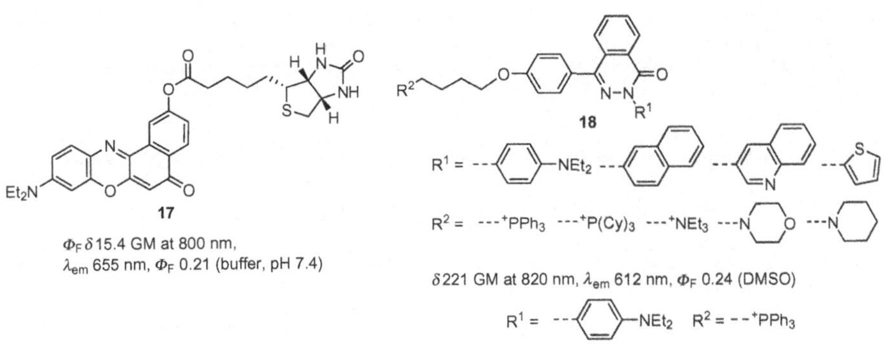

Fig. 8.8 Red fluorescent TPA dyes **17** and **18** for biological imaging

The detection of biological chemical species, such as sulfur dioxide (SO_2), hydrogen sulfide (H_2S), carbon monoxide (CO), and nitric oxide (NO), is important to elucidate physiological and pathological processes (Xu et al. 2016; Chen et al. 2016). Red-emissive TPA dyes have been developed for the detection of chemical species. The fluorescence detection of NO was reported by Liu's group (Mao et al. 2016). The far-red fluorescent TPA dye **19** was composed of the Nile Red chromophore as the fluorescent TPA-active unit and the *o*-phenylenediamine moiety as the receptor of NO. The dye is weakly fluorescent due to the PET effect of the *o*-phenylenediamine unit. When the dye reacts with NO, *o*-phenylenediamine transforms into triazole, and the product, **20**, indicated enhanced far-red light emission at 650 nm because of the inhibition of the PET effect (Fig. 8.9). The two-photon action cross section increased to 38 GM (at 820 nm) due to the emission enhancement. Thus, this dye can be developed for visualizing NO generation in a lipopolysaccharide-mediated inflammation process.

In addition to the NO species, the *o*-phenylenediamine moiety is also responsive to reactive carbonyl species, such as dehydroascorbic acid and methylglyoxal, and to reactive oxygen/nitrogen species, such as hydrogen peroxide and peroxynitrite, thereby leading to a low selectivity for NO monitoring. This problem was addressed on the basis of the *N*-nitrosation of an aromatic secondary amine by the same group

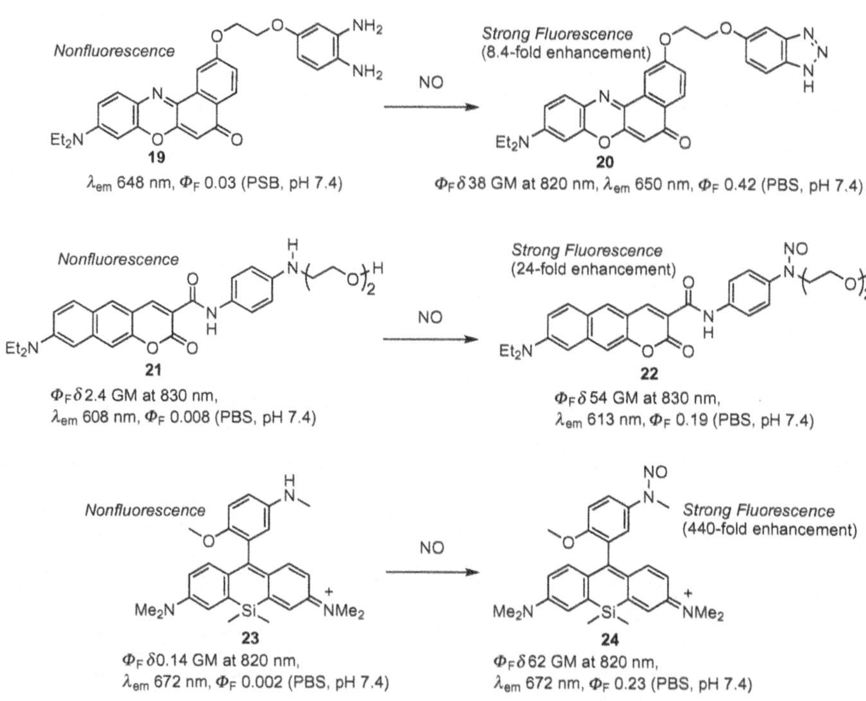

Fig. 8.9 Fluorescence detection of NO species based on red fluorescent TPA dyes **19**, **21**, and **23**

(Mao et al. 2017a). The newly designed TPA dye, **21**, which is composed of the amine-substituted benzocoumarin chromophore as the red-emissive TPA unit, reacts with NO, resulting in the corresponding *N*-nitroso derivative, **22**, together with enhanced red light emission at 613 nm and a two-photon action cross section of 54 GM (at 830 nm) (Fig. 8.9). This system can be applied to monitoring NO in ischemia–reperfusion injuries in mice kidneys by two-photon fluorescence microscopy. A similar strategy of NO detection is used in silicon–rhodamine-based TPA dye **23** (Mao et al. 2017b). This dye provides very weak fluorescence emission at 672 nm with a two-photon action cross section of 0.14 GM. After the reaction with NO, a significant emission enhancement of 440-fold was achieved to produce product **24**, which has a large two-photon action cross section of 62 GM (Fig. 8.9). This dye can be used to monitor NO in the development of tumors in a xenograft mouse model.

A TPA fluorescence probe for H_2S detection was reported by Fan and Peng group (Sun et al. 2013). A donor–acceptor dicyanomethylene-benzopyran dye **25**, which had a terminal azide group, is nonfluorescent due to PET quenching. When the dye was treated with sodium hydrogen sulfide as the H_2S precursor, the resulting dye, **26**, which had a terminal amino group, provided a 65-fold turn-on response of fluorescence because of the inhibition of the PET effect (Fig. 8.10). The far-red fluorescence can be detected upon two-photon excitation. This probe can be developed for tracking H_2S in living mice using two-photon fluorescence microscopy. The dicyanomethylene-benzopyran chromophore was also developed for cysteine detection as reported by Dong's group (Wang et al. 2016). The donor–acceptor dye **27** displays fluorescence enhancement of near-IR light (702 nm) with an up to 35-fold increase after the addition of cysteine (Fig. 8.10). This cysteine recognition can be developed for live-cell imaging. The fluorescence detection of SO_2 in brain tissues and zebrafishes was reported by Lin's group (Ma et al. 2017). Imaging was based on a ratiometric fluorescence change from red light emission at 645 nm to blue emission at 450 nm, which is based on the Michael addition reaction of coumarin–benzopyrylium-conjugated TPA dye **29** with SO_2 (Fig. 8.10). Three-dimensional fluorescence imaging of SO_2 in living brain tissue of mice can be performed at a depth of 80 μm. Endogenous CO imaging by a two-photon technique was reported by Lin's group (Liu et al. 2017). Nile Red-based Pd complex **31** has a significantly weak fluorescent at 660 nm with a fluorescence quantum yield of 0.0047 because of the strong quenching effect of Pd. When dye **31** reacts with CO molecule, the released free dye, **32**, becomes fluorescent. In the presence of 100 equivalents of CO, the far-red fluorescence is increased by 60-fold (Fig. 8.10). This CO response system enables the monitoring of CO in both zebrafish embryos and living mice by two-photon near-IR light excitation at 760 nm.

An efficient TPA in the second near-IR region was reported by Kawamata and Konishi's group (Niko et al. 2015). A pyrene-based acceptor–π–acceptor dye, **33**, exhibited bright red light emission at 650 nm and had an efficient two-photon action cross section of 304 GM at 1050 nm (Fig. 8.11). In this TPA, a femtosecond fiber laser was developed as an inexpensive excitation source in the second near-IR region. This attractive TPA system can be applied for mitochondria imaging in Hek293 cells. The styrylpyridine-salt-based TPA dye, **34**, was reported by Yu and He's group (Guo et al.

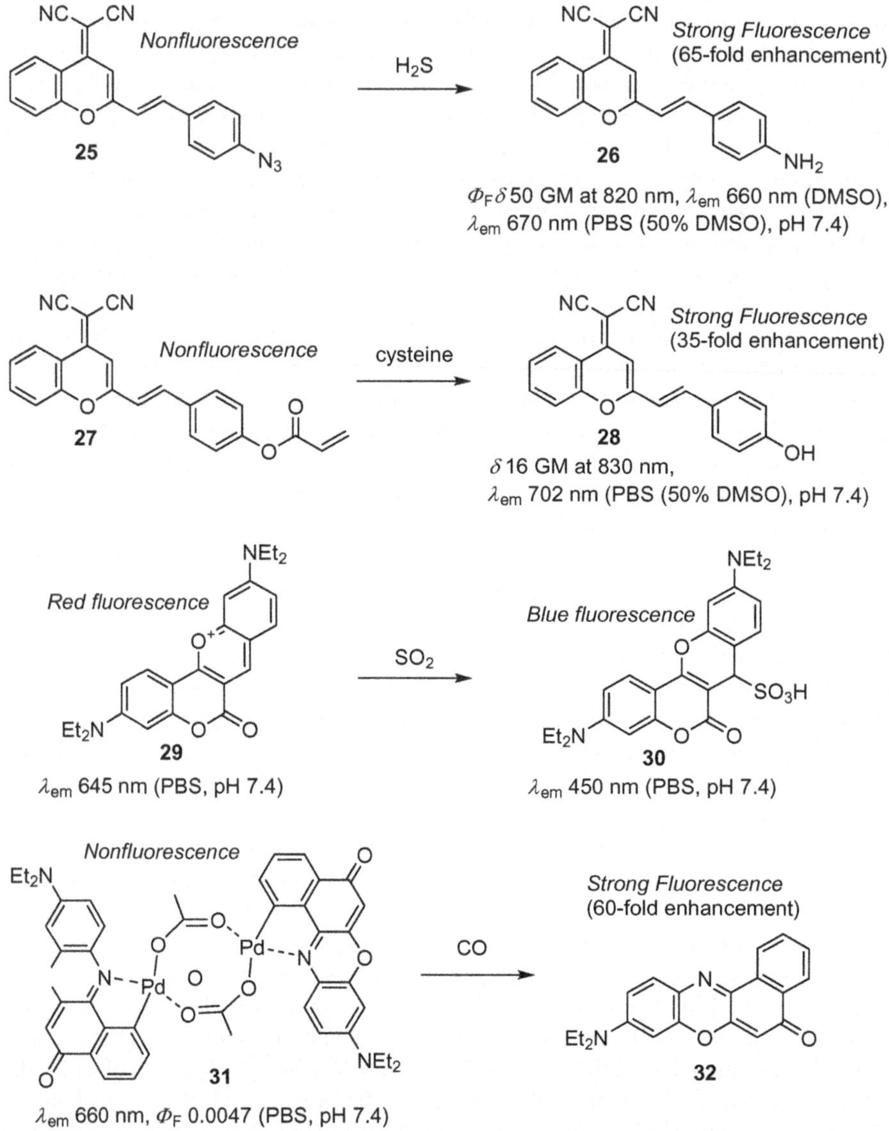

Fig. 8.10 Fluorescence detection of chemical species based on red fluorescent TPA dyes **25**, **27**, **29**, and **31**

2016). The acceptor–π–donor–π–acceptor conjugation exhibits red light emission at approximately 600 nm in polar solvents together with high TPA cross sections of 814–1454 GM (Fig. 8.11). The amphipathic property, which is provided by the hydrophilic cation pyridinium moieties and hydrophobic alky chains, provides a higher affinity

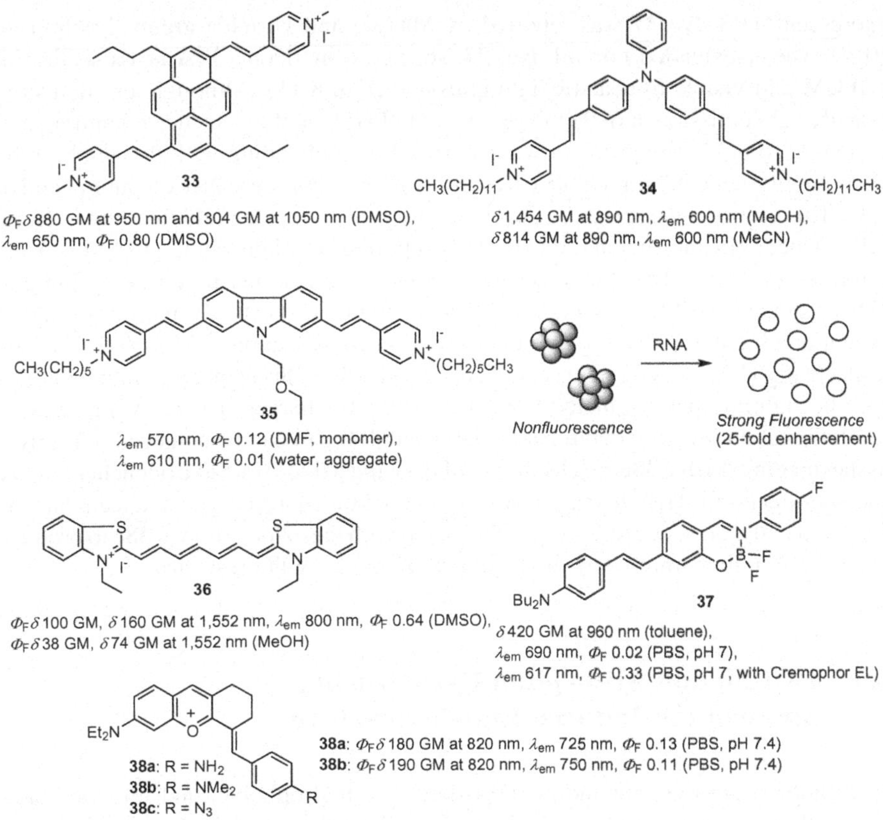

Fig. 8.11 Red and near-IR fluorescent TPA dyes **33–38** for biological imaging

to membranes. This red-emissive TPA dye was used for the visualization of plasma membranes in HeLa cells and in muscle and hepatic tissues.

Yu and Lin's group reported interesting RNA detection based on the aggregation/disaggregation equilibrium of red fluorescent TPA dye (Liu et al. 2016). The carbazole-based dye, **35**, which has two pyridinium ion moieties, formed a weakly red fluorescent aggregate. In contrast, in the presence of RNA, a strong 25-fold emission enhancement was provided due to the disaggregation (Fig. 8.11). The dye can image RNA in living cells of mouse liver tissue upon 800 nm two-photon excitation. Significant fluorescence was observed up to a 100-μm penetration depth.

Berezin and Achilefu's group reported a series of cyanine-based TPA dyes with a near-IR light-emitting nature (Berezin et al. 2011). One cyanine dye, **36**, indicated a good two-photon action cross section of 100 GM in DMSO (Fig. 8.11). The parameter is highly sensitive to the solvent polarity. Polar environments resulted in a decrease in the two-photon action cross section, which is ascribed to the weak fluorescence intensity arising from the intramolecular charge transfer character. This near-IR fluorescent TPA dye enables ex vivo mouse kidney imaging. Near-IR light fluorescent TPA dye **37** was reported by Massue and Ulrich's group (Frath et al. 2017). The π-extended boranil dye, **37**, showed a moderate TPA cross section of 420 GM and visible to near-IR light emission (Fig. 8.11). Although an interesting near-IR light emission at 690 nm was shown in PBS, the fluorescence quantum yield decreased to 0.02. The problem was improved by embedding with the amphiphilic block copolymer of Cremophor EL, resulting in a good fluorescence quantum yield of 0.33. This embedded dye can be developed for real-time widefield imaging in HeLa cells. A near-IR fluorescent TPA dye was reported by Zhou, Sun, and Liu's group (Zhou et al. 2017). The donor–acceptor–donor type cationic pyran dyes, **38a** and **38b**, emitted near-IR fluorescence at 725 and 750 nm, respectively, upon two-photon excitation together with large two-photon action cross sections of 180 and 190 GM, respectively (Fig. 8.11). In HeLa and A375 cells, dye **38a** displayed mitochondria-specific staining that originated from the positively charged nature. A high spatial resolution at an imaging depth between 90 and 400 μm can be achieved in rat liver tissue imaging. In dye **38c**, an electron-withdrawing azide group as a quencher moiety was introduced for H_2S detection. A turn-on type of fluorescence enhancement was observed in the presence of H_2S, which reacted with the azide group in **38c** to produce the corresponding amino derivative, **38a**, with intensive fluorescence.

8.3 Two-Photon Absorption Dyes Bearing Aggregation-Induced Emission Nature

Most fluorescent dyes, including TPA dyes, are hydrophobic, and this facilitates aggregation in aqueous media. The intermolecular interactions in the aggregate state usually favor the formation of a nonradiative deactivation channel, resulting in significant quenching of emission (Langhals et al. 1989). Further, the efficiency of the longer wavelength emission of donor–acceptor-type dyes decreases in polar aqueous media because a highly polarized excited state that originates from the donor–acceptor characteristic through intramolecular charge transfer or twisted intramolecular charge transfer increases the formation of a nonradiative deactivation channel (Reichardt 1994). However, recently, unusual light-emitting systems have been discovered that derive their properties from aggregation-induced emission (AIE) in which the aggregation of nonfluorescent or weakly fluorescent compounds results in an enhancement of emission efficiency (Fig. 8.12) (Luo et al. 2001; Mei et al. 2015; For details of AIE, see Chap. 9 in this book). The quenched state in aqueous media changes to an emissive state owing to the restriction of the intramolecular rotation/vibration and the polarization that arises from aggregate formation. Thus, enhanced emission systems have been created by a combination of two-photon-induced emission with AIE to obtain a larger two-photon action cross section. In this section, two-photon absorbing AIE dyes with red and near-IR light are summarized as well as their biological applications.

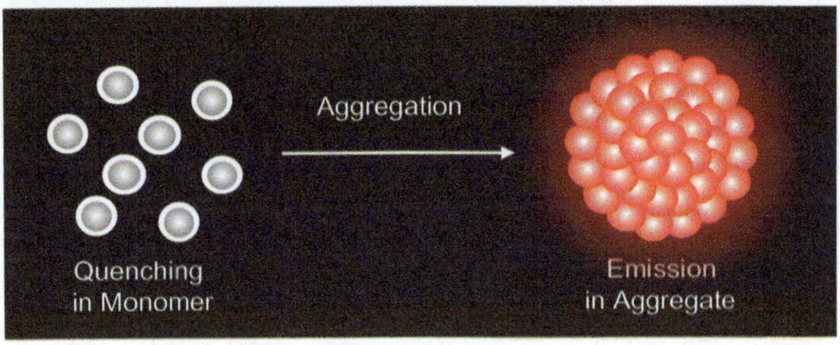

Fig. 8.12 Schematic explanation for the emission enhancement based on the aggregation of TPA dyes

As described in Sect. 8.2.2, triphenylamine-benzothiadiazole-based donor–acceptor-type TPA dyes indicate AIE in polar aqueous media (Ishi-i et al. 2012, 2014, 2015a, b). In a THF/water medium, the emission of dyes **39** (Ishi-i et al. 2012), **40** (Ishi-i et al. 2014), **41** (Ishi-i et al. 2015a), and **42** (Ishi-i et al. 2015b) was quenched in a low volume of water, whereas the emission was recovered and increased in a high volume of water. In a low volume of water, dye molecules exist in monomeric form. In contrast, the dye molecules aggregated in a high volume of water (Fig. 8.13). This aggregation efficiently prevents the fluorescence quenching problem caused by the excited state polarization. The aggregates provide a less polar hydrophobic space inside the aggregate structure, thus avoiding excited-state polarization and subsequent fluorescence quenching. Furthermore, the restriction of intramolecular rotation/vibration contributes to the emission enhancement, as found in typical AIE dyes (Mei et al. 2015).

Red-light-emitting TPA dye that has an enhanced emission nature was reported by Prasad's group. Bis[(aminostyryl)styryl]anthracene donor–π–donor dye, **43**, exhibits significant increases in the TPA cross section and fluorescence quantum yield in the aggregate state (Fig. 8.13). The two-photon action cross section increased from 15.5 GM (monomer) to 101 MG (aggregate) (Kim et al. 2006). This enhancement is due to the planarization of π-conjugation, locking of torsional motion by aggregation, and less quenching interaction by loose packing of partially distorted molecules. This novel TPA dye, which was encapsulated into organically modified silica nanoparticles, was developed for HeLa cell imaging (Kim et al. 2007).

The structure–property relationship of TPA and the AIE nature were systematically studied by Hua and Tian's group. Multibranched triarylamine end-capped triazine TPA dyes, **44a** and **44b** (Jiang et al. 2010), star-burst triarylamine donor–acceptor–donor TPA dyes, **45a** and **45b** (Wang et al. 2011), and triphenylamine-distyrylanthracene-based dendrimer TPA dye, **46** (Xu et al. 2014), exhibited aggregation-induced orange–red emission and large TPA cross sections (Fig. 8.14). A significantly large two-photon action cross section of 4,400 GM, arising from the enhanced fluorescence quantum yield (from 0.07 to 0.85) based on AIE was obtained in the dendrimer system of **46** (Xu et al. 2014).

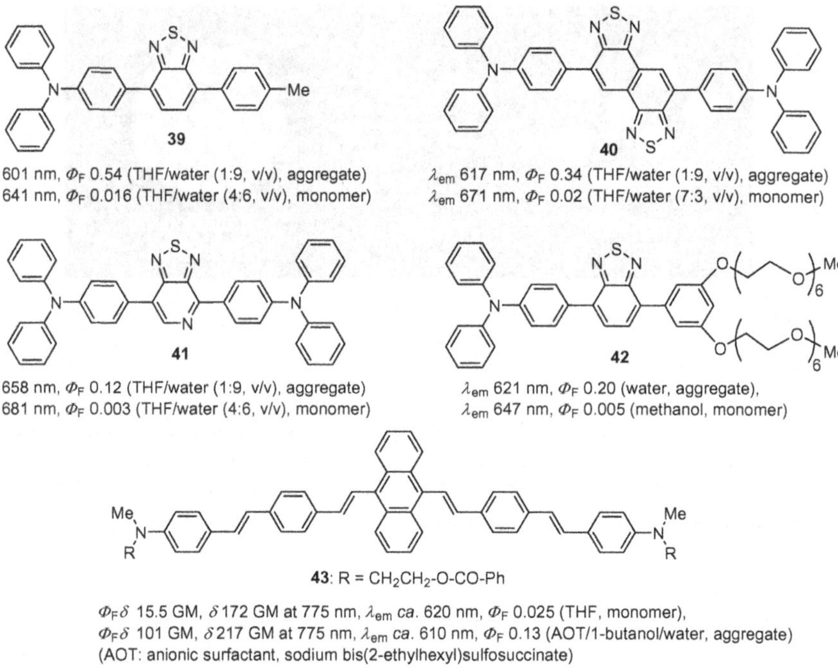

Fig. 8.13 Red fluorescent TPA dyes **39–43** with aggregation-induced emission nature

Tang's group designed and prepared a donor–acceptor type of TPA dye, **47**, composed of an electron-donating triphenylamine moiety and an electron-accepting benzylidene imidazolone moiety (Jiang et al. 2017). The dye exhibited a strong emission enhancement upon aggregate formation in an aqueous medium together with a good TPA nature (Fig. 8.14). The dye can be developed for two-photon imaging of lipid droplets in live cells. Compared to one-photon absorption dyes, TPA dye **47** has the advantages of improved three-dimensional resolution, less photobleaching and autofluorescence, and a deeper penetration depth.

Liu's group reported the encapsulation of two-photon-absorbing AIE dye **44a** into nanoparticle dots (AIE dots) by simple nanoprecipitation with a phospholipid matrix (1,2-distearoyl-*sn*-glycero-3-phosphoethanolamine-*N*-[methoxy(polyethylene glycol)-2000, DSPE-PEG$_{2000}$) (Fig. 8.15) (Li et al. 2011). The resulting AIE dots preserved large TPA and AIE properties with a TPA cross section of 2,015 GM and a fluorescence quantum yield of 0.24 (Fig. 8.14). This encapsulation enhanced the stability and biocompatibility of the aggregate, leading to biological application. Imaging of a MCF-7 breast cancer cell is succeeded by using folic acid-functionalized AIE dots. Recently, these phospholipid-based AIE dots have been widely used as a versatile matrix in many biological applications, as described below.

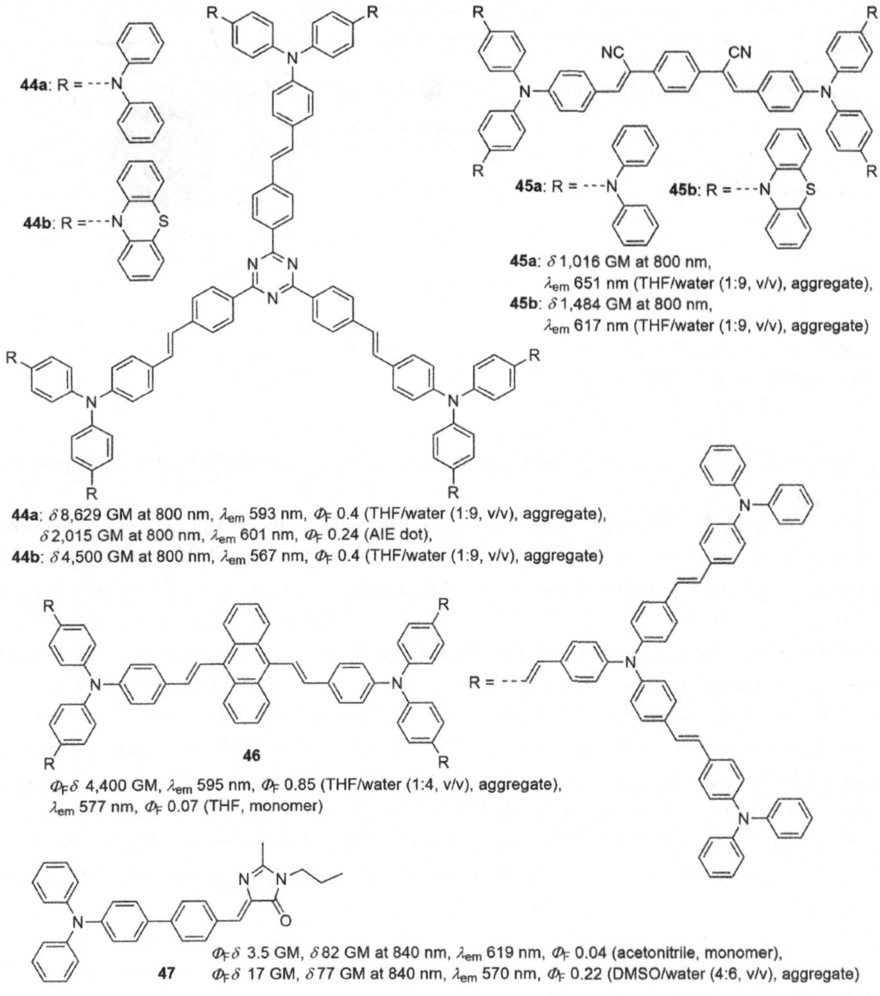

Fig. 8.14 Red fluorescent TPA dyes **44–47** with aggregation-induced emission nature

Liu and Tang's group reported a two-photon-absorbing AIE dye, **48**, which is composed of a dicyanomethylene-*4H*-chromene acceptor unit, triphenylamine donor units, and tetraphenylethene AIE units (Fig. 8.16). This dye can efficiently emit far-red/near-IR light fluorescence by two-photon excitation in the aggregate state (Geng et al. 2012). In the lipid-based AIE dots that encapsulate dye **48**, the TPA cross section was calculated based on the dot concentration to be significantly large at 2.3×10^6 GM. This AIE dot was developed for imaging of MCF-7 breast cancer cells. Then, a similar TPA dye, **49**, which has four tetraphenylethene moieties, was reported by Tang's group (Nicol et al. 2017). The biotinylated AIE dots incorporating dye

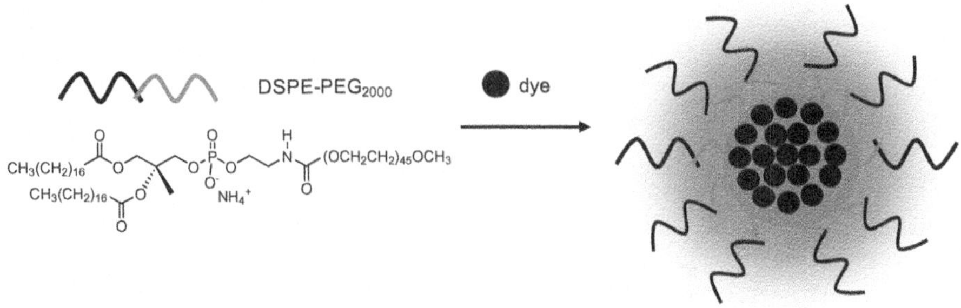

Fig. 8.15 Schematic representation of encapsulation of TPA dye molecules into nanoparticle dots with a phospholipid matrix

49 indicated selective mitochondria staining in live cancer cells by the two-photon technique.

Far-red/near-IR light-emitting AIE dye **50**, which is composed of a donor–acceptor type of triphenylamine-fumaronitrile moiety as the TPA-active unit and tetraphenylethene moieties as the active AIE units, was reported by Liu and Tang's group (Fig. 8.16) (Li et al. 2013). Amine-functionalized AIE dots that encapsulate dye **50** enabled long-term tracing of C6 glioma cells in vivo for up to 21 days and deep tumor imaging at a depth up to 550 μm ex vivo, which were ascribed to the good TPA nature and bright fluorescence of AIE. These AIE dots were developed for three-dimensional dynamic imaging with a high resolution of blood vessels of mouse brain under two-photon excitation (Wang et al. 2014). Then, some blood vasculature images were studied by using some AIE dots. The blood vasculature inside the mouse ear skin was visualized by AIE dots that have diketopyrrolopyrrole-based TPA dye, **51**, incorporated into them (Fig. 8.16) (Gao et al. 2015). The aggregate of **51** has a large TPA cross section of 8,100 GM, which is larger than those (several dozen and hundreds of GM) of precedent two-photon-absorbing AIE dyes (Kim et al. 2006; Wang et al. 2012) and is comparable to that of triazine-based dye **44a** (Jiang et al. 2010). This high TPA nature is suitable for blood vasculature imaging.

An accurate measurement of the capillary diameters in mouse ears was successful for the first time in AIE dots that incorporate donor–acceptor-type triphenylamine-benzothiadiazole TPA dye, **52**, bearing tetraphenylethene active AIE units (Fig. 8.16) (Qin et al. 2018). A brilliant red fluorescence emitted from the aggregate of dye **52** with a quantum yield of 0.385 enabled the measurement of capillary diameters. These AIE dots were also developed for in vivo deep-tissue imaging of mouse livers and for real-time imaging of blood vessels of mouse brains with deep penetration and high contrast by the two-photon technique. Two-photon imaging of mouse brain vessels was performed using AIE dots that incorporated red-emissive azabenzanthrone dye **53** (Fig. 8.16) (Zang et al. 2018). The vessels were clearly imaged at vertical depths of up to 280 μm.

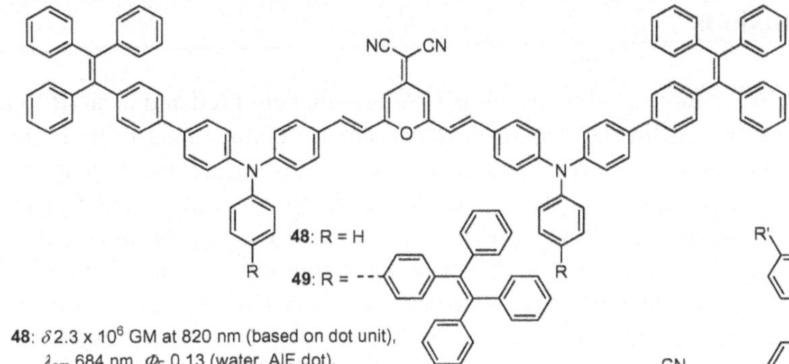

48: δ 2.3 x 10^6 GM at 820 nm (based on dot unit), λ_{em} 684 nm, Φ_F 0.13 (water, AIE dot),
49: δ 313 GM at 830 nm, λ_{em} 680 nm, Φ_F 0.125 (water, AIE dot)

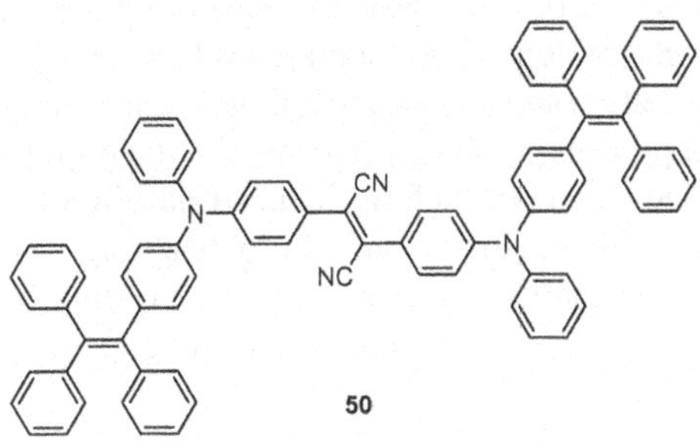

50

δ 6.7 x10^5 GM at 810 nm (based on dot unit), λ_{em} 671 nm, Φ_F 0.24 (water, AIE dot)

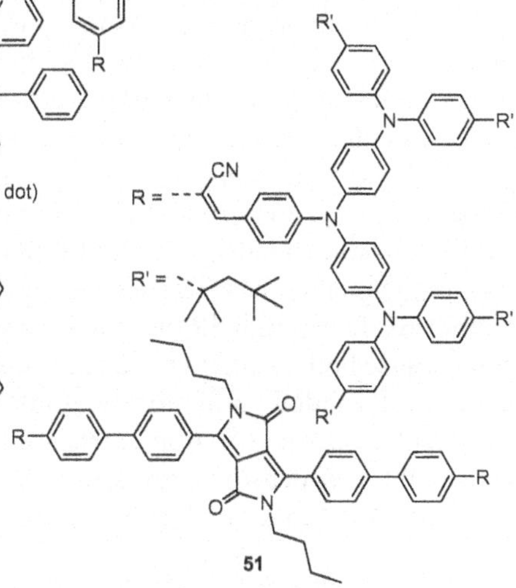

51

δ 8,100 GM at 800 nm, Φ_F 0.001 (chloroform, monomer), δ 5.34 x 10^5 GM at 810 nm (based on dot unit), λ_{em} 627 nm, Φ_F 0.12 (water, AIE dot)

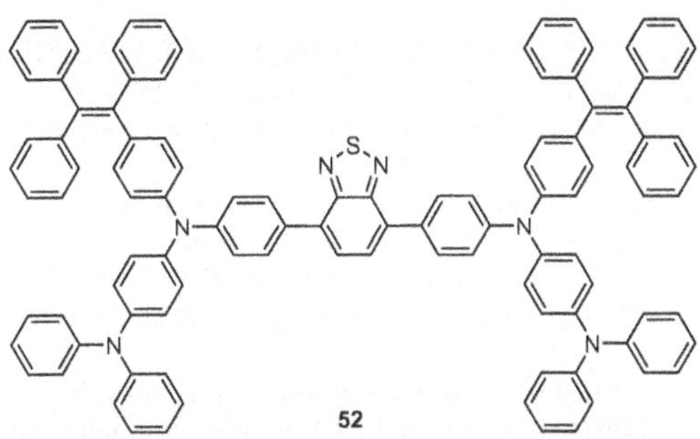

52

δ 310 GM at 900 nm (THF/water (1:99, v/v), aggregate), λ_{em} 636 nm, Φ_F 0.385 (THF/water (1:9, v/v), aggregate), λ_{em} 630 nm (water, AIE dot)

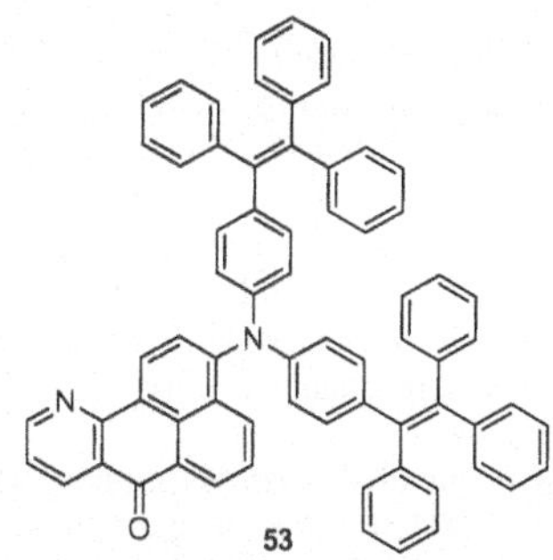

53

δ 10 GM at 1,040 nm, λ_{em} 649 nm, Φ_F 0.082 (THF/water (1:99, v/v), aggregate)

Fig. 8.16 Red fluorescent TPA dyes **48–53** incorporated into nanoparticle dots

8.4 Conclusions

In this chapter, we summarized fluorescent TPA dyes that emit red and near-IR light and the recent advances made in their development for applications in biological systems. The TPA nature is ascribed to the intramolecular charge transfer characteristic arising from the conjugation of electron–donor, electron–acceptor, and π-spacer moieties. The fluorescence efficiency can be improved by using the aggregation-induced emission concept, leading to enhanced light-emitting systems. The red and near-IR light emission of longer wavelength can be achieved by elongated π-systems and/or an enhanced intramolecular charge transfer nature. The created fluorescent TPA dyes can be developed for biological monitoring of fluoride ions, sulfur dioxide, nitrogen oxide, carbon monoxide, and hydrogen sulfide as well as for biological imaging of mitochondria, amyloid-β plaques, biotin receptors, mouse brain slices, and plasma membranes. The success of these biological applications is based on the combined effect of near-IR two-photon excitation and red/near-IR fluorescence emission because both excitation and emission processes can be performed in the biological optical window, providing versatile advantages, such as reduced autofluorescence and deeper penetration in tissues, less photodamage and photobleaching, and a higher three-dimensional resolution. The two-photon excited fluorescent technique will contribute further to the elucidation of living systems and disease diagnosis.

References

Albota M, Beljonne D, Brédas J-L, Ehrlich JE, Fu J-Y, Heikal AA, Hess SE, Kogej T, Levin MD, Marder SR, McCord-Maughon D, Perry JW, Röckel H, Rumi M, Subramaniam G, Webb WW, Wu X-L, Xu C (1998) Design of organic molecules with large two-photon absorption cross sections. Science 281:1653

Berezin MY, Zhan C, Lee H, Joo C, Akers WJ, Yazdanfar S, Achilefu S (2011) Two-photon optical properties of near-infrared dyes at 1.55 μm excitation. J Phys Chem B 115:11530

Brousmiche DW, Serin JM, Fréchet JMJ, He GS, Lin T-C, Chung S-J, Prasad PN (2003) Fluorescence resonance energy transfer in a novel two-photon absorbing system. J Am Chem Soc 125:1448

Brousmiche DW, Serin JM, Fréchet JMJ, He GS, Lin T-C, Chung S-J, Prasad PN, Kannan R, Tan L-S (2004) Fluorescence resonance energy transfer in novel multiphoton absorbing dendritic structures. J Phys Chem B 108:8592

Chen X, Wang F, Hyun JY, Wei T, Qiang J, Ren X, Shin I, Yoon J (2016) Recent progress in the development of fluorescent, luminescent and colorimetric probes for detection of reactive oxygen and nitrogen species. Chem Soc Rev 45:2976

Cumpston BH, Ananthavel SP, Barlow S, Dyer DL, Ehrlich JE, Erskine LL, Heikal AA, Kuebler SM, Lee I-YS, McCord-Maughon D, Qin J, Röckel H, Rumi M, Wu XL, Marder SR, Perry JW (1999) Two-photon polymerization initiators for three-dimensional optical data storage and microfabrication. Nature 398:51

Denk W, Strickers JH, Webb WW (1990) Two-photon laser scanning fluorescence microscopy. Science 248:73

Dy JT, Maeda R, Nagatsuka Y,Ogawa K, Kamada K, Ohta K, Kobuke Y (2007) A photochromic porphyrin–perinaphthothioindigo conjugate and its two-photon absorption properties. Chem Commun 5170

Ehrlich JE, Wu XL, Lee I-YS, Hu Z-Y, Röckel H, Marder SR, Perry JW (1997) Two-photon absorption and broadband optical limiting with bis-donor stilbenes. Opt Lett 22:1843

For details of AIE, see Chap. 10 in this book

Frath D, Didier P, Mély Y, Massue J, Ulrich G (2017) Vectorization and intracellular distribution of a two-photon-absorbing, near-infrared-emitting π-extended boranil dye. Chem Photo Chem 1:109

Frederiksen PK, Jørgensen M, Ogilby PM (2001) Two-photon photosensitized production of singlet oxygen. J Am Chem Soc 123:1215

Fu L-M, Wen X-F, Ai X-C, Sun Y, Wu Y-S, Zhang J-P, Wang Y (2005) Efficient two-photon-sensitized luminescence of a europium(III) complex. Angew Chem Int Ed 44:747

Gao Y, Feng G, Jiang T, Goh C, Ng L, Liu B, Li B, Yang L, Hua J, Tian H (2015) Biocompatible nanoparticles based on diketo-pyrrolo-pyrrole (DPP) with aggregation-induced red/NIR emission for in vivo two-photon fluorescence imaging. Adv Funct Mater 25:2857

Geng J, Li K, Ding D, Zhang X, Qin W, Liu J, Tang BZ, Liu B (2012) Lipid-PEG-folate encapsulated nanoparticles with aggregation induced emission characteristics: cellular uptake mechanism and two-photon fluorescence imaging. Small 8:3655

Göppert-Mayer M (1931) Über Elementarakte mit zwei Quantensprüngen. Ann Phys (Berlin) 401:273

Gu B, Wu W, Xu G, Feng G, Yin F, Han P, Chong J, Qu J, Yong K-T, Liu B (2017) Precise two-photon photodynamic therapy using an efficient photosensitizer with aggregation-induced emission characteristics. Adv Mater 29:1701076

Guo EQ, Ren PH, Zhang YL, Zhang HC, Yang WJ (2009) Diphenylamine end-capped 1,4-diketo-3,6-diphenylpyrrolo[3,4-*c*]pyrrole (DPP) derivatives with large two-photon absorption cross-sections and strong two-photon excitation red fluorescence. Chem Commun 5859

Guo L, Zhang R, Sun Y, Tian M, Zhang G, Feng R, Li X, Yu X, He X (2016) Styrylpyridine salts-based red emissive two-photon turn-on probe for imaging the plasma membrane in living cells and tissues. Analyst 141:3228

He GS, Bhawalkar JD, Zhao CF, Prasad PN (1995) Optical limiting effect in a two-photon absorption dye doped solid matrix. Appl Phys Lett 67:2433

He GS, Tan L-S, Zheng Q, Prasad PN (2008) Multiphoton absorbing materials: molecular designs, characterizations, and applications. Chem Rev 108:1245

Heo CH, Sarkar AR, Baik SH, Jung TS, Kim JJ, Kang H, Mook-Jung I, Kim HM (2016) A quadrupolar two-photon fluorescent probe for *in vivo* imaging of amyloid-β plaques. Chem Sci 7:4600

Ishi-i T, Taguri Y, Kato S, Shigeiwa M, Gorohmaru H, Maeda S, Mataka S (2007) Singlet oxygen generation by two-photon excitation of porphyrin derivatives having two-photon-absorbing benzothiadiazole chromophores. J Mater Chem 17:3341

Ishi-i T, Nakamura N, Mine T, Imamura S, Shigeiwa M, Gorohmaru H, Maeda S (2009) Fluorescent two-photon absorption benzothiadiazole dyes having photoreleasing quenchers. Chem Lett 38:1042

Ishi-i T, Ikeda K, Kichise Y, Ogawa M (2012) Red-light-emitting system based on aggregation of donor-acceptor derivatives in polar aqueous media. Chem Asian J 7:1553

Ishi-i T, Hashimoto R, Ogawa M (2014) Aggregation of naphthobisthiadiazole-based donor-acceptor-donor dyes that restrict quenching in solution and emit red light in polar aqueous media. Asian J Org Chem 3:1074

Ishi-i T, Ikeda K, Ogawa M, Kusakaki Y (2015a) Light-emitting properties of donor–acceptor and donor–acceptor–donor dyes in solution, solid, and aggregated states: structure–property relationship of emission behavior. RSC Adv 5:89171

Ishi-i T, Kitahara I, Yamada S, Sanada Y, Sakurai K, Tanaka A, Hasebe N, Yoshihara T, Tobita S (2015b) Amphiphilic benzothiadiazole–triphenylamine-based aggregates that emit red light in water. Org Biomol Chem 13:1818

Jiang Y, Wang Y, Hua J, Tang J, Li B, Qian S, Tian H (2010) Multibranched triarylamine end-capped triazines with aggregation-induced emission and large two-photon absorption cross-sections. Chem Commun 46:4689

Jiang M, Gu X, Lam JWY, Zhang Y, Kwok RTK, Wong KS, Tang BZ (2017) Two-photon AIE bio-probe with large Stokes shift for specific imaging of lipid droplets. Chem Sci 8:5440

Jun YW, Kim HR, Reo YJ, Dai M, Ahn KH (2017) Addressing the autofluorescence issue in deep tissue imaging by two-photon microscopy: the significance of far-red emitting dyes. Chem Sci 8:7696

Kaiser W, Garrett CGB (1961) Two-photon excitation in CaF_2: Eu^{2+}. Phys Rev Lett 7:229

Kato S, Matsumoto T, Ishi-i T, Thiemann T, Shigeiwa M, Gorohmaru H, Maeda S, Yamashita Y, Mataka S (2004) Strongly red fluorescent novel donor–π-bridge–acceptor–π-bridge–donor (D–π–A–π–D) type 2,1,3-benzothiadiazoles with enhanced two-photon absorption cross-sections. Chem Commun 2342

Kato S, Matsumoto T, Shigeiwa M, Gorohmaru H, Maeda S, Ishi-i T, Mataka S (2006) Novel 2,1,3-benzothiadiazole-based red-fluorescent dyes with enhanced two-photon absorption cross-sections. Chem Eur J 12:2303

Kawata S, Kawata Y (2000) Three-dimensional optical data storage using photochromic materials. Chem Rev 100:1777

Kawata S, Sun H-B, Tanaka T, Takada K (2001) Finer features for functional microdevices. Nature 412:697

Kim M, Cho BR (2009) Two-photon materials with large two-photon cross sections. Structure–property relationship. Chem Commun 153

Kim HM, Cho BR (2015) Small-molecule two-photon probes for bioimaging applications. Chem Rev 115:5014

Kim S, Zheng Q, He GS, Bharadi DJ, Pudavar HE, Baev A, Prasad PN (2006) Aggregation-enhanced fluorescence and two-photon absorption in nanoaggregates of a 9,10-bis[4′-(4″-aminostyryl)styryl] anthracene derivative. Adv Funct Mater 16:2317

Kim S, Pudavar HE, Bonoiu A, Prasad PN (2007) Aggregation-enhanced fluorescence in organically modified silica nanoparticles: a novel approach toward high-signal-output nanoprobes for two-photon fluorescence bioimaging. Adv Mater 19:3791

Kim HM, Jeong BH, Hyon J-Y, An MJ, Seo MS, Hong JH, Lee KJ, Kim CH, Joo T, Hong S-C, Cho BR (2008) Two-photon fluorescent turn-on probe for lipid rafts in live cell and tissue. J Am Chem Soc 130:4246

Kim D, Sambasivan S, Nam H, Kim KH, Kim JY, Joo T, Lee K, Kim K, Ahn KH (2012a) Reaction-based two-photon probes for *in vitro* analysis and cellular imaging of monoamine oxidase activity. Chem Commun 48:6833

Kim D, Singha S, Wang T, Seo E, Lee JH, Lee S-J, Kim KH, Ahn KH (2012b) *In vivo* two-photon fluorescent imaging of fluoride with a desilylation-based reactive probe. Chem Commun 48:10243

Kim D, Moon H, Baik SH, Singha S, Jun YW, Wang T, Kim KH, Park BS, Jung J, Mook-Jung I, Ahn KH (2015) Two-photon absorbing dyes with minimal autofluorescence in tissue imaging: application to *in vivo* imaging of amyloid-β plaques with a negligible background signal. J Am Chem Soc 137:6781

Kong X, Dong B, Zhang N, Wang C, Song X, Lin W (2017) A unique red-emitting two-photon fluorescent probe with tumor-specificity for imaging in living cells and tissues. Talanta 174:357

Langhals H, Potrawa T, Noth H, Linti G (1989) The influence of packing effects on the solid-state fluorescence of diketopyrrolopyrroles. Angew Chem Int Ed Engl 28:478

Lee SK, Yang WJ, Choi JJ, Kim CH, Jeon S-J, Cho BR (2005) 2,6-Bis[4-(*p*-dihexylaminostyryl) styryl]anthracene derivatives with large two-photon cross sections. Org Lett 7:323

Li K, Jiang Y, Ding D, Zhang X, Liu Y, Hua J, Feng S-S, Liu B (2011) Folic acid-functionalized two-photon absorbing nanoparticles for targeted MCF-7 cancer cell imaging. Chem Commun 47:7323

Li K, Qin W, Ding D, Tomczak N, Geng J, Liu R, Liu J, Zhang X, Liu H, Liu B, Tang BZ (2013) Photostable fluorescent organic dots with aggregation-induced emission (AIE Dots) for noninvasive long-term cell tracing. Sci Rep 3:1150

Liu Y, Meng F, He L, Yu X, Lin W (2016) Fluorescence behavior of a unique two-photon fluorescent probe in aggregate and solution states and highly sensitive detection of RNA in water solution and living systems. Chem Commun 52:8838

Liu K, Kong X, Ma Y, Lin W (2017) Rational design of a robust fluorescent probe for the detection of endogenous carbon monoxide in living zebrafish embryos and mouse tissue. Angew Chem Int Ed 56:13489

Luo J, Xie Z, Lam JWY, Cheng L, Chen H, Qiu C, Kwok HS, Zhan X, Liu Y, Zhu D, Tang BZ (2001)Aggregation-induced emission of 1-methyl-1,2,3,4,5-pentaphenylsilole. Chem Commun 1740

Ma Y, Tang Y, Zhao Y, Gao S, Lin W (2017) Two-photon and deep-red emission ratiometric fluorescent probe with a large emission shift and signal ratios for sulfur dioxide: ultrafast response and applications in living cells, brain tissues, and zebrafishes. Anal Chem 89:9388

Maeda H, Kowada T, Kikuta J, Furuya M, Shirazaki M, Mizukami S, Ishii M, Kikuchi K (2016) Real-time intravital imaging of pH variation associated with osteoclast activity. Nat Chem Biol 12:579

Mao Z, Feng W, Li Z, Zeng L, Lv W, Liu Z (2016) NIR in, far-red out: developing a two-photon fluorescent probe for tracking nitric oxide in deep tissue. Chem Sci 7:5230

Mao Z, Jiang H, Li Z, Zhong C, Zhang W, Liu Z (2017a) An *N*-nitrosation reactivity-based two-photon fluorescent probe for the specific *in situ* detection of nitric oxide. Chem Sci 8:4533

Mao Z, Jiang H, Song X, Hu W, Liu Z (2017b) Development of a silicon-rhodamine based near-infrared emissive two-photon fluorescent probe for nitric oxide. Anal Chem 89:9620

Marder SR (2006) Organic nonlinear optical materials: where we have been and where we are going. Chem Commun 131

Margineanu A, Hofkens J, Cotlet M, Habuchi S, Stefan A, Qu J, Kohl C, Müllen K, Vercammen J, Engelborghs Y, Gensch T, De Schryver FC (2004) Photophysics of a water−soluble rylene dye: comparison with other fluorescent molecules for biological applications. J Phys Chem B 108:12242

Mei J, Leung NLC, Kwok RTK, Lam JWY, Tang BZ (2015) Aggregation-induced emission: together we shine, united we soar! Chem Rev 115:11718

Nicol A, Qin W, Kwok RTK, Burkhartsmeyer JM, Zhu Z, Su H, Luo W, Lam JWY, Qian J, Wong KS, Tang BZ (2017) Functionalized AIE nanoparticles with efficient deep-red emission, mitochondrial specificity, cancer cell selectivity and multiphoton susceptibility. Chem Sci 8:4634

Niko Y, Moritomo H, Sugihara H, Suzuki Y, Kawamata J, Konishi G (2015) A novel pyrene-based two-photon active fluorescent dye efficiently excited and emitting in the 'tissue optical window (650–1100 nm).' J Mater Chem B 3:184

Parthenopoulos DA, Rentzepis PM (1989) Three-dimensional optical storage memory. Science 245:843

Pawlicki M, Collins HA, Denning RG, Anderson HL (2009) Two-photon absorption and the design of two-photon dyes. Angew Chem Int Ed 48:3244

Qin W, Zhang P, Li H, Lam JWY, Cai Y, Kwok RTK, Qian J, Zheng W, Tang BZ (2018) Ultrabright red AIEgens for two-photon vascular imaging with high resolution and deep penetration. Chem Sci 9:2705

Qu J, Kohl C, Pottek M, Müllen K (2004) Ionic perylenetetracarboxdiimides: highly fluorescent and water-soluble dyes for biolabeling. Angew Chem Int Ed 43:1528

Reichardt C (1994) Solvatochromic dyes as solvent polarity indicators. Chem Rev 94:2319

Sarkar AR, Heo CH, Lee HW, Park KH, Suh YH, Kim HM (2014) Red emissive two-photon probe for real-time imaging of mitochondria trafficking. Anal Chem 86:5638

Sarkar AR, Heo ChH, Xu L, Lee HW, Si HY, Byun JW, Kim HM (2016) A ratiometric two-photon probe for quantitative imaging of mitochondrial pH values. Chem Sci 7:766

Shen Y, Shuhendler AJ, Ye D, Xua J-J, Chen H-Y (2016) Two-photon excitation nanoparticles for photodynamic therapy. Chem Soc Rev 45:6725

Sun W, Fan J, Hu C, Cao J, Zhang H, Xiong X, Wang J, Cui S, Sun S, Peng X (2013) A two-photonfluorescent probe with near-infrared emission for hydrogen sulfide imaging in biosystems. Chem Commun 49:3890

Suzuki N, Suda K, Yokogawa D, Kitoh-Nishioka H, Irle S, Ando A, Abegão LMG, Kamada K, Fukazawa A, Yamaguchi S (2018) Near infrared two-photon-excited and -emissive dyes based on a strapped excited-state intramolecular proton-transfer (ESIPT) scaffold. Chem Sci 9:2666

Tang Y, Kong X, Xu A, Dong B, Lin W (2016) Development of a two-photon fluorescent probe for imaging of endogenous formaldehyde in living tissues. Angew Chem Int Ed 55:3356

Terenziani F, Katan C, Badaeva E, Tretiak S, Blanchard-Desce M (2008) Enhanced two-photon absorption of organic chromophores: theoretical and experimental assessments. Adv Mater 20:4641

Ventelon L, Charier S, Moreaux L, Mertz J, Blanchard-Desce M (2001) Nanoscale push-push dihydrophenanthrene derivatives as novel fluorophores for two-photon-excited fluorescence. Angew Chem Int Ed 40:2098

Wang X, Nguyen DM, Yanez CO, Rodriguez L, Ahn H-Y, Bondar MV, Belfield KD (2010) High-fidelity hydrophilic probe for two-photon fluorescence lysosomal imaging. J Am Chem Soc 132:12237

Wang B, Wang Y, Hua J, Jiang Y, Huang J, Qian S, Tian H (2011) Starburst triarylamine donor–acceptor–donor quadrupolar derivatives based on cyano-substituted diphenylaminestyrylbenzene: tunable aggregation-induced emission colors and large two-photon absorption cross sections. Chem Eur J 17:2647

Wang Y, Liu T, Bu L, Li J, Yang C, Li X, Tao Y, Yang W (2012) Aqueous nanoaggregation-enhanced one- and two-photon fluorescence, crystalline J-aggregation-induced red shift, and amplified spontaneous emission of 9,10-bis(*p*-dimethylaminostyryl)anthracene. J Phys Chem C 116:15576

Wang D, Qian J, Qin W, Qin A, Tang BZ, He S (2014) Biocompatible and photostable AIE dots with red emission for *In Vivo* two-photon bioimaging. Sci Rep 4:4279

Wang J, Li B, Zhao W, Zhang X, Luo X, Corkins ME, Cole SL, Wang C, Xiao Y, Bi X, Pang Y, McElroy CA, Bird AJ, Dong Y (2016) Two-photon near infrared fluorescent turn-on probe toward cysteine and its imaging applications. ACS Sens 1:882

Weissleder R (2001) A clearer vision for in *vivo* imaging. Nat Biotechnol 19:316

Werts MVH, Gmouth S, Mongin O, Pons T, Blanchard-Desce M (2004) Strong modulation of two-photon excited fluorescence of quadripolar dyes by (de)protonation. J Am Chem Soc 126:16294

Xu B, Zhang J, Fang H, Ma S, Chen Q, Sun H, Imc C, Tian W (2014) Aggregation induced enhanced emission of conjugated dendrimers with a large intrinsic two-photon absorption cross-section. Polym Chem 5:479

Xu W, Zeng Z, Jiang J-H, Chang Y-T, Yuan L (2016) Discerning the chemistry in individual organelles with small-molecule fluorescent probes. Angew Chem Int Ed 55:13658

Yang L, Zhu Y, Shui M, Zhou T, Cai Y, Wang W, Xu F, Niu Y, Wang C, Zhang J-L, Xu P, Yuan L, Liang L (2016) Rational design of fluorescent phthalazinone derivatives for one- and two-photon imaging. Chem Eur J 22:12363

Zang Q, Yu J, Yu W, Qian J, Hu R, Tang BZ (2018) Red-emissive azabenzanthrone derivatives for photodynamic therapy irradiated with ultralow light power density and two-photon imaging. Chem Sci 9:5165

Zhang JF, Lim CS, Bhuniya S, Cho BR, Kim JS (2011) A highly selective colorimetric and ratiometric two-photon fluorescent probe for fluoride ion detection. Org Lett 13:1190

Zhou L, Wang Q, Tan Y, Lang MJ, Sun H, Liu X (2017) Rational development of near-infrared fluorophores with large stokes shifts, bright one-photon, and two-photon emissions for bioimaging and biosensing applications. Chem Eur J 23:8736

Zipfel WR, Williams RM, Webb WW (2003a) Nonlinear magic: multiphoton microscopy in the biosciences. Nat Biotechnol 21:1369

Zipfel WR, Williams RM, Christie R, Nikitin AY, Hyman BT, Webb WW (2003b) Live tissue intrinsic emission microscopy using multiphoton-excited native fluorescence and second harmonic generation. Proc Natl Acad Sci USA 100:7075

Chapter 9
Molecular Designs for Solid-State Luminescent Properties and Recent Progresses on the Development of Functional Luminescent Solid Materials

Kazuo Tanaka, Masayuki Gon, and Yoshiki Chujo

Abstract Luminescent organic dyes are known to be an essential building block for developing modern organic optoelectronic devices. In these devices, organic dyes are normally used as a film. However, most of the organic dyes suffer from the quenching effect induced in the condensed state, named as concentration quenching or aggregation-caused quenching (ACQ). Therefore, so far, various strategies have been proposed for overcoming ACQ. In this chapter, we survey conventional strategies for suppressing ACQ and obtaining intense solid-state emission mainly regarding boron-containing materials which are known to be a platform for designing luminescent dyes. A series of typical examples to exhibit solid-state emission are presented in each part. Initially, luminescent organic– inorganic hybrids are illustrated. ACQ is often induced through non-specific intermolecular interactions in the condensed state. In order to realize the solution-like situation where each dye molecule is isolated from each other, the transparent hybrid matrices were useful as a scaffold. Recent progresses on the development of conjugated polymer-based hybrids by employing hybrid molecules as a building block is also demonstrated. The similar strategy for isolating dye molecules is accomplished by introducing steric substituents and structures near the chromophore moiety. Particularly, this strategy is conventionally applied for obtaining luminescent polymer films. Moreover, by assembling heterogeneous types of dyes at the cardo structure in the polymer main chains, the multiple emission bands originating from each dye were able to be simultaneously observed from the film samples. These results on simultaneous multiple emission bands are explained. Next topics are regarding aggregation-induced emission (AIE)-active molecules. By suppressing molecular motions at the steric moiety, intense

K. Tanaka (✉) · M. Gon · Y. Chujo
Department of Polymer Chemistry, Graduate School of Engineering, Kyoto University, Nishikyo-ku, Katsura 615-8510, Japan
e-mail: tanaka@poly.synchem.kyoto-u.ac.jp

M. Gon
e-mail: gon@poly.synchem.kyoto-u.ac.jp

Y. Chujo
e-mail: chujo@poly.synchem.kyoto-u.ac.jp

Y. Ooyama and S. Yagi (eds.), *Progress in the Science of Functional Dyes*,
https://doi.org/10.1007/978-981-33-4392-4_9

emission can be observed in the aggregation state. Furthermore, it was shown that emission color from some of the AIE-active molecules was sensitive to molecular distributions in the solid state. Based on these characteristics, stimuli-responsive luminochromism was obtained. Basic design strategies and recent applications are described.

Keywords Boron · Carborane · Solid-state luminescence · Luminochromism · Element-block · Aggregation-induced emission

9.1 Introduction

Organic luminescent dyes are versatile for developing advanced optoelectronic devices because of their property tunability according to preprogrammed designs and intrinsic superior material properties to inorganics, such as lightness, solubility, compatibility in various media, and processability. Therefore, not only electroluminescent panels but also flexible displays have been developed based on organic materials, and some of them have been already commercialized. Meanwhile, in common devices, luminescent dyes are used as a film. Therefore, efficient solid-state luminescence is the essential requirement to obtain practical products. However, most of the organic dyes show critical decreases in their emission characteristics in the condensed state by non-specific intermolecular interactions in the ground and/or excited states even though intense emission can be observed in the diluted solution. This phenomenon is called as concentration quenching or aggregation-caused quenching (ACQ).

In this chapter, the design strategies for obtaining solid-state luminescence from on organic optical materials and particularly boron-containing dyes, as an example, are described. So far, there are several strategies for obtaining efficient solid-state emission as shown in reviews (Anthony 2012; Shaikh et al. 2015; Carayon and Fery-Forgues 2017; Varughese 2014; Xue et al. 2016; Shimizu and Hiyama 2010). By introducing twisting structures (Zhu et al. 2018; Wang et al. 2018; Srujana and Radhakrishnan 2018; Takayanagi et al. 2018; Xu et al. 2018; Zhu et al. 2018; Yu et al. 2017) or heteroatom units (Namba et al. 2018; Shimizu et al. 2017; Munch et al. 2019; Shao et al. 2019; Sun et al. 2018; Jia and Wu 2017) in the conjugation system, ACQ is able to be suppressed. Intermolecular interaction can be disturbed by introducing bulky substituents directly into molecular structures to some extent (Sun et al. 2019; Yan et al. 2017; Iida and Yamaguchi 2009; Hariharan et al. 2017; Huang et al. 2018; Peng et al. 2019; Gu et al. 2017; Hirosawa et al. 2017). This strategy is also applicable for obtaining luminescent polymers having extended π-conjugation through main chains. In particular, the boron-containing units have been utilized as a versatile platform not only for obtaining intense luminescence but also for suppressing ACQ (Zhao et al. 2006; Wang et al. 2015; Ohsedo et al. 2015; Yan et al. 2014; Wakamiya et al. 2007; Shimizu et al. 2015; Kalluvettukuzhy and Thilagar 2017; Zhang et al. 2018; Zhao et al. 2007); Nie et al. 2019; Weber et al. 2012). In this chapter, initially, polymer-based examples to offer intense emission as a solid material including a film are illustrated. By loading luminescent materials in the transparent polymer or organic–inorganic hybrid matrices, intermolecular interaction and subsequently ACQ can be suppressed. As a result, solid-state emission was able to be observed. Additionally, it was found that some of the luminophores are capable of simultaneously exhibiting each emission band in diverse wavelength regions from the single materials. As a consequence, multiple color emission is detectable. Several studies on these mixture films are introduced mainly from our researches. Next, the recent topics on the development of solid-state luminescent molecules are explained. In particular, since the first report on the aggregation-induced emission (AIE)-active molecule (Luo et al. 2001), which shows enhanced emission only in the condensed state, a wide variety of solid-state luminescent materials have been produced mainly with tetraphenylethene (Tang et al. 2015). Moreover, stimuli-responsive luminochromic materials, which show emission color changes in response to external stimuli or environment alteration, have been also obtained based on the AIE-active molecules. Their useful characters as a functional optical material and prediction of the new AIE-active dyes are described.

9.2 Mixture Materials

As mentioned in the introduction, ACQ is mainly induced through intermolecular interactions in the condensed state. Therefore, the simple and facile strategy for suppressing ACQ is to mix luminescent dyes with transparent matrices. From this standpoint, conventional polymers and organic–inorganic hybrids are a suitable scaffold for obtaining luminescent materials. Additionally, material properties, such as rigidity, elasticity, and thermal stability, are originated from the type of matrices. Therefore, it is relatively easy to obtain desired multi-functional luminescent materials based on dye-containing mixtures.

Organic–inorganic hybrids have attracted attention because multiple characteristics originating from both organic and inorganic components (Gon et al. 2017). In particular, hybridization with inorganic components is recently recognized as one of effective strategies for obtaining robust products without crucial changes of intrinsic optical properties of the loaded molecules. Therefore, applications of the dye-doped hybrids have been accomplished for creating advanced optical devices, such as light-emitting diodes and dye-lasers. By maintaining the isolation state of organic dyes in the products, ACQ can be effectively suppressed. Therefore, during the material preparation, it is critically important to maintain homogeneous dispersion of organic dyes. However, due to the polarity mismatch between organic dyes and inorganic components (and sometimes organic matrices), unexpected aggregation, followed by ACQ, is often induced. Therefore, the suppression of aggregation between the dyes at the molecular level should be essential to obtain desired materials.

White-light emissive hybrids have been manufactured by loading each RGB light-emitting organoboron dye (1,3-diketonate BF_2 complex, boron dipyrromethene

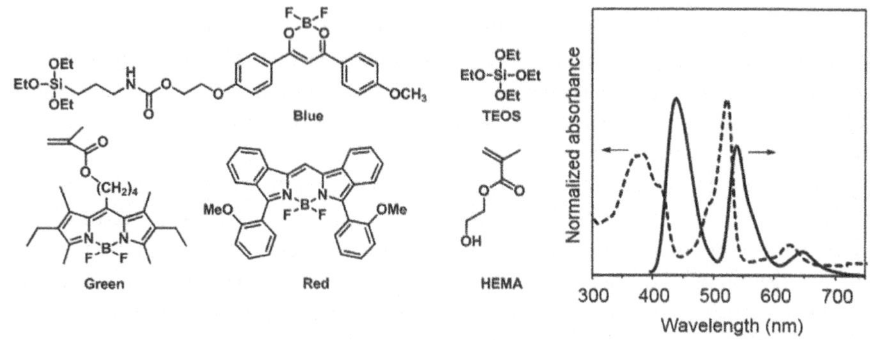

Fig. 9.1 Chemical structures of the reagents for fabricating the white-light-emitting hybrid and optical spectra of the product. Reproduced from Ref. Kajiwara et al. (2013) with permission from The Royal Society of Chemistry

(BODIPY), and boron di(iso)iodomethane as blue, green, and red light-emitting dyes, respectively) into poly(2-hydroxyethyl methacylate) (PHEMA)–silica hybrids (Fig. 9.1) (Kajiwara et al. 2013). In the preliminary study, simple mixing with these luminescent dyes in the hybrid as well as polymer matrices, aggregation occurred and multiple emission colors were not obtained. Therefore, chemical modification was added to each dye for controlling the localization in the product. In order to anchor the blue luminescent boron complex at the inorganic domain, the ethoxyl silane group was introduced. The green luminescent BODIPY dye was intended to be dispersed in the polymer component by incorporating into the polymer main chains. It was presumed that the small amount red luminescent dye might be located largely in the critical regions between organic and inorganic domains. In addition, to inhibit aggregation during sol–gel reactions for hybrid formation, the microwave-heating method was employed (Okada et al. 2014a,b).

The product showed significant white-light emission consists of each luminescent property of three kinds of the doped dyes according to the comparison of the photoluminescence (PL) spectra. The emission efficiency was 47%. Furthermore, the color tuning was capable of changing the feed ratios of the dyes. PL spectra of two dyes in various concentrations and Stern–Volmer plots, in which the energy transfer efficiency can be estimated from the slope of the approximate line, indicated that the energy transfer efficiencies between the dyes in the hybrids were lower than that in the solution state. By fixing well-dispersion state in the material, interactions should be prevented. It should be mentioned that the product presented high stability against photo-degradation owing to the inorganic component. From these data, it can be said that pure white-light-emitting hybrids were obtained by employing the hybrid matrices.

The sol–gel reactions are conventionally used for the fabrication of hybrid materials. However, due to high hydrophobicity of commodity organic dyes, severe optimization for reaction conditions is inevitable to obtain homogenous products. To solve this problem, new idea has been proposed on the material design including various elements and inorganic components by employing element-blocks, which are defined as a minimum functional unit containing heteroatoms (Chujo and Tanaka 2015; Gon et al. 2018a). By the combination, connection, and assembly of element-blocks, it is presumable that various types of functional materials possessing unique functions derived from each element can be obtained. On the basis of this idea regarding element-block materials, functional hybrids have been developed.

Polyhedral oligomeric silsesquioxane (POSS, Fig. 9.2) has eight organic side chains at each vertex in the rigid silica cube (Tanaka and Chujo 2012, 2013a, b). From this chemical structure, POSS is called as "a molecular hybrid". In the commodity hybrids, organic, and inorganic components are mixed at the molecular or nano levels. From this fact, it was presumed that homogeneous materials with POSS might show similar properties to commodity hybrids with silicate. Owing to good solubility of POSS derivatives in various conventional solvents, it is easy to realize homogenous mixture materials with polymers (Tanaka et al. 2009). Especially, it has been clearly indicated that POSS derivatives contributed to enhance thermal and mechanical properties of the matrices in the conventional polymer films (Tanaka et al. 2010, 2017; Ueda et al. 2016, 2017, 2018, 2019). From these data, POSS is recognized as a facile element-block for manufacturing hybrids not with the sol–gel method but with simple mixing.

In general, molecular arrangement critically affects emission performances in a solid state. Unique cubic structure of POSS can arrange organic substituents at radial positions and avoid intermolecular π-π interaction which causes ACQ of luminescent dyes. To confirm the effect of POSS on the emission properties, the luminophore-integrated POSS named **iPrPh3POSS** was synthesized, and its optical properties were measured (Fig. 9.3a) (Gon et al. 2016). Accordingly, **iPrPh3POSS** prevented ACQ and intense blue emission was observed even in the solid state. Additionally, **iPrPh3POSS** exhibited high thermal stability owing to a thermally stable inorganic POSS core and large molecular weight. The decomposition temperature of **iPrPh3POSS** with a 15 wt% weight loss is over 500 °C under N_2 atmosphere, and the POSS showed bright emission even at 200 °C in air. If the luminescent dye was not attached on the POSS (**iPrPh3TMS**), the dye molecule was melted at under 200 °C (melting point = 159 °C), and the emission was largely quenched. The radially integrated structure on the POSS has an externally sparse dendrimer-like conformation. Therefore, intermolecular interaction should be suppressed and the

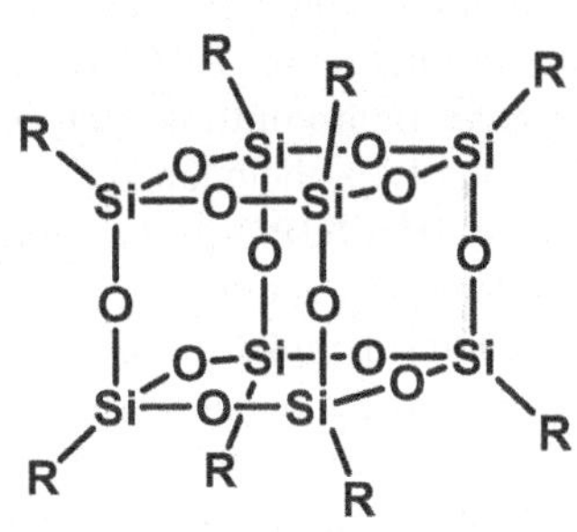

Fig. 9.2 Chemical structure of POSS

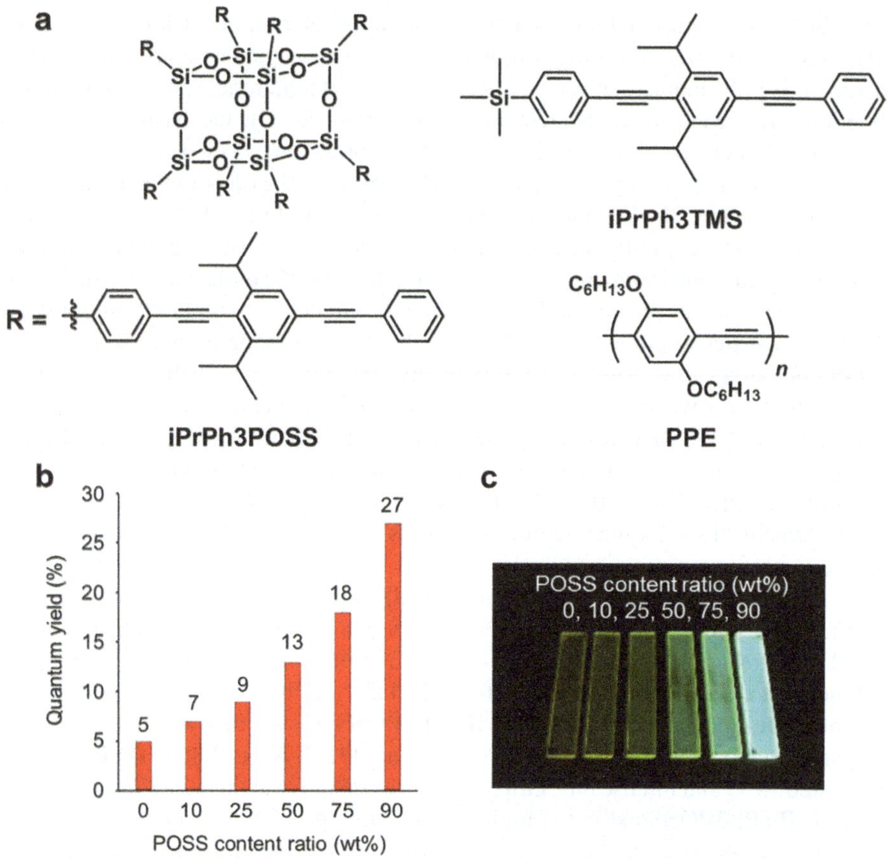

Fig. 9.3 **a** Chemical structures of the modified POSS and **b** luminescent properties and **c** appearances of the conjugated polymer hybrids with variable concentrations of POSS. Reproduced from Ref. Gon et al. (2019) with permission from The Royal Society of Chemistry

crystallinity is lowered. Focusing on the low crystallinity and sterically bulky structure, it was considered that the radially integrated structure on the POSS is suitable for dispersing π-conjugated polymer in the film state to overcome the ACQ problem. Due to strong intermolecular π-π interaction, ACQ is a serious matter for application of π-conjugated polymer to organic light-emitting devices. The hybrid materials were prepared with **iPrPh3POSS** and poly(*p*-phenylene-ethylene) (PPE) (Fig. 9.3b) (Gon et al. 2019a). As a consequence, **iPrPh3POSS** was able to release aggregation of PPE and homogeneous hybrid film was obtained at an arbitrary ratio. As increasing the POSS content ratio in the hybrid films, the emission intensity was gradually enhanced and the absolute emission efficiencies were improved to be from 5% (0 wt%: pristine PPE film) to 27% (10 wt% PPE and 90 wt% **iPrPh3POSS** included hybrid film) when PPE was excited (excitation wavelength = 450 nm) (Fig. 9.3b). This clearly indicates that the ACQ of PPE was suppressed by hybridization with

iPrPh3POSS, and luminescent films were obtained (Fig. 9.3c). Moreover, the homogeneous hybridization brought additional functions to the luminescent hybrid films. In the hybrid films, the emission from **iPrPh3POSS** was completely quenched even when **iPrPh3POSS** were excited (excitation wavelength = 328 nm), meaning excited energy of **iPrPh3POSS** is effectively transferred to PPE. Even in the 10 wt% PPE and 90 wt% **iPrPh3POSS** hybrid film, the emission from **iPrPh3POSS** almost disappeared. In contrast, the emission from PPE was dramatically enhanced. Such an effective energy transfer is usually realized by direct chemical modification of light harvesting units to the luminophores (Sato et al. 1999). In summary, by simply mixing the designed POSS and conjugated polymers, the functional hybrid, where effective energy transfer can proceed inside materials, was obtained. Therefore, it is assumed that this strategy has high generality, and indeed bright emissive hybrid materials composed of **iPrPh3POSS** and commodity π-conjugated polymers, such as polyfluorene or regiorandom poly(3-hexylthiophene), were obtained. The mechanism should be potentially useful for manufacturing solid-state emissive materials showing wavelength conversion properties.

As mentioned above, it was revealed that POSS has high affinity to π-conjugated polymer. It is assumed that low polarity and the sterically bulky cubic structure could be favorable for exhibiting high compatibility. To demonstrate applicability of POSS for combining different types of polymer functionalities mediated by POSS, elastic polyurethane and the luminescent π-conjugated polymer were focused. Polyurethane has polar repeating units composed of urethane bonds and phase separation potentially occurs in the composition of polar polyurethane and non-polar π-conjugated polymer. To improve the compatibility between these different kinds of polymers, POSS-capped polyurethane (**PUPOSS**) was prepared (Fig. 9.4a) (Gon et al. 2019b; Chen 2016). POSS was modified only at the ends of polyurethane chains to keep the elasticity. As a luminescent π-conjugated polymer, polyfluorene having two 2-ethylhexyl groups at 9,9-substitution position (PF) was used.

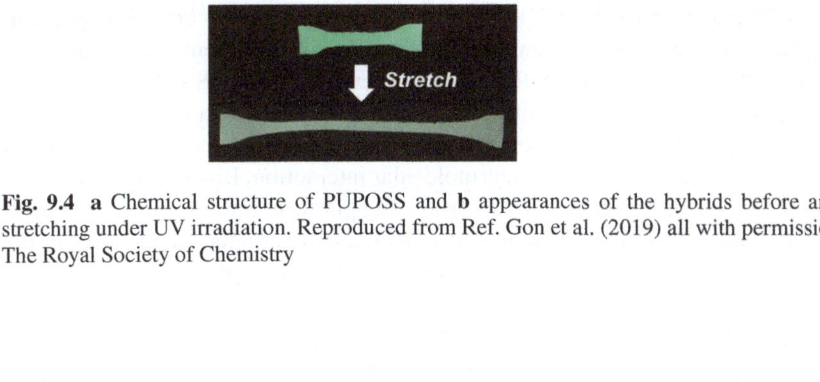

Fig. 9.4 **a** Chemical structure of PUPOSS and **b** appearances of the hybrids before and after stretching under UV irradiation. Reproduced from Ref. Gon et al. (2019) all with permission from The Royal Society of Chemistry

From simply mixing in the solution and casting on the substrate, the homogeneous hybrid film was obtained from **PUPOSS** and 0.5 wt% PF. The 0.5 wt% loading ratio was enough to observe luminescence from the product. When the pristine polyurethane was used, phase separation occurred in 0.5 wt% PF loading ratio, indicating that the POSS contributes to improve compatibility with polyurethane. The hybrid film from **PUPOSS** and 0.5 wt% PF had good elasticity which was almost the same with the pristine polyurethane. Moreover, it is disclosed that the hybrid film enhanced the mechanical property and absolute emission efficiencies. This means that both components should be mixed at the nano level and the aggregates of PF are partially suppressed by POSS. It is known that PF shows not only blue emission from a locally excited state (LE) but also yellow-greenish emission from an excimer state in the solid and the film state (Scott et al. 1999). Since the excimer emission was not observed in the solution state, the yellow-greenish emission from our materials should be originated from intermolecular or intramolecular interaction of the PF chains. Hence, it was presumed that elongation of the elastic hybrid film released *micro*-order aggregates and subsequently induced emission color change. Indeed, after elongation of the hybrid film, the emission color was altered from yellow green to white green and it was observable by the naked eye (Fig. 9.4b). From the PL spectra, it was shown that the blue emission from the LE state increased and the yellow-greenish emission in the excimer state decrease after stretching. The phenomenon is called mechanochromic luminescence and is a key character for constructing mechanical force sensors. These results represent that different functions of the polymers are successfully hybridized by utilizing POSS as a mediator. In this section, it is proved that the unique properties of POSS can combine the distinctly different polymer functions, such as luminescent properties and stimuli-responsiveness, and offer novel polymer materials.

9.3 Steric Effect

Another strategy for avoiding ACQ in the condensed state is the modification with the steric groups around chromophores. As a representative example, POSS has been used for this purpose. The modified polyphenylene ethynylenes, which often show critical ACQ in film similarly to other conventional conjugated polymers due to long main-chain conjugation, were synthesized with the POSS side chains (Fig. 9.5) (Miyake and Chujo 2008). From the evaluation of emission efficiencies, significant emission was detected in film. In addition, emission color was able to be altered by changing the introduction ratio of POSS. The degree of inter-chain interaction might be dominant for emission color variation.

In the dye molecules, the steric substituents can play an effective role in suppression of ACQ by disturbing intermolecular interaction. Boron complexes with π-conjugated ligands are known to be a promising platform for designing a luminescent dye because of rigid and planar structures because structural relaxation in the excited state can be effectively suppressed by the boron complexation (Tanaka and

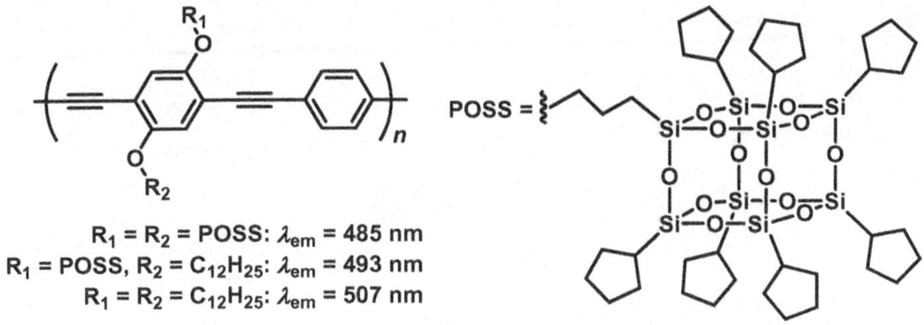

Fig. 9.5 Chemical structures and optical properties of polyphenylene ethynylenes with or without POSS side chains

Chujo 2012b). Therefore, as mentioned above, a huge variety of boron-containing luminescent dyes have been developed. However, these structural features are also favorable for stacking formation and aggregation in various media. Thus, most of the luminescent boron complexes suffer from ACQ in the solid state. By introducing the steric substituents near the boron complex moiety, intermolecular interaction can be prohibited (Zhao et al. 2006, 2007; Wakamiya et al. 2007; Yamane et al. 2015). Thus, highly efficient solid-state luminescence was detected from several reports.

The introduction of the steric structure is also applicable for the design of solid-state luminescent materials. Polyfluorene has the cardo structure at the 9 positions, and alkyl substituents are introduced at this position for improving solubility (Fig. 9.6) (Yeo et al. 2013). The cardo carbon is regarded as a key unit for dye assembly without any interactions including electronic ones. Based on this idea, multiple kinds of BODIPYs were tethered to the cardo carbon in fluorene-containing polymers, and their optical properties were evaluated (Yeo et al. 2015a,b). By the comparison of UV-vis absorption and PL spectra in the solution state between each BODIPY derivative and the polymers, similar shapes of spectra were obtained, indicating that electronic interactions between the main chain and the side chains and the side chains themselves should be negligible. Interestingly, the significant emission bands originating from the BODIPY moieties at the side chains as well as the main-chain conjugation were detected although emission efficiencies were lowered. Owing to the steric distributions of the BODIPY moieties, emission properties of the luminescent units including both main and side chains could be preserved with some extent even in the condensed state. Finally, dual emission properties were expressed.

9.4 Aggregation-Induced Emission

Based on the above strategies to isolate dye molecules for avoiding ACQ, environment resistance can be enhanced. From the different standpoint, it can be said that it is difficult to obtain stimuli-responsiveness from the above materials. In other words,

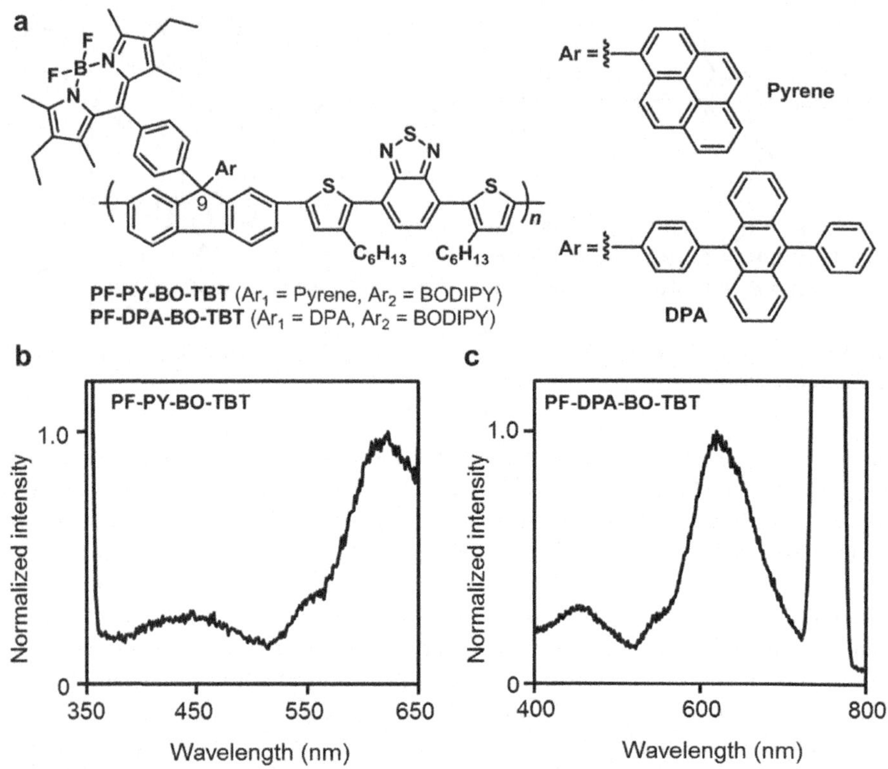

Fig. 9.6 **a** Chemical structures of the cardo fluorene-containing polymers and their solid-state luminescent properties containing **b** pyrene and **c** DPA substituents. Reprinted with permission from Ref. Yeo et al. (2013). Copyright 2015 Wiley–VCH Verlag GmbH & Co. KGaA

in order to simultaneously obtain solid-state emission and stimuli-responsiveness, establishment of the new strategy is essential. To meet this demand, AIE-active molecules, which can show different luminescent properties depending on environment, have attracted attention as a platform for designing solid-state luminescent materials with stimuli-responsivity (Tang et al. 2015). In particular, molecular interactions would be drastically changed by altering molecular morphology in the solid state. In the crystalline state, each molecule tends to make an interaction with the neighboring molecules at the specified spots, while molecular interactions would be randomly formed in amorphous. If crystalline polymorphs are able to be obtained, different motifs of molecular interaction would be expected. Diverse optical properties are expected to be expressed under each different condition, leading to turn ON/OFF sensing and luminochromism. In this part, AIE-active boron element-blocks, as an example, and their applications for stimuli-responsive luminochromic materials are explained.

9.4.1 o-Carborane Materials

o-Carborane ($B_{10}C_2H_{12}$) is an icosahedral cluster consists of three-center two-electron bonds. Because of intrinsic electron-deficiency, *o*-carborane can work as a strong electron-accepting unit when bonded with the π-conjugated moiety at the carbon atom in the cluster. Subsequently, significant emission bands from the intramolecular charge transfer (ICT) state can be observed from the aryl-modified *o*-carborane derivatives. Particularly, it should be emphasized that *o*-carborane can play a critical role in suppressing ACQ. It is likely that the steric structure could be favorable for disturbing intermolecular interaction in crystals and aggregates, leading to preservation of intense luminescence. In 2009, it was discovered that *o*-carborane can work as a fluorescence quencher only in the solution state (Fig. 9.7a) (Kokado and Chujo 2009). The alternating copolymers containing *o*-carborane and several kinds of aromatic compounds as a comonomer were prepared, and interestingly, intense emission was observed not from solutions but from film samples (Fig. 9.7b), indicating that these polymers should have AIE properties. From the series of mechanistic studies, it was concluded that emission annihilation should occur in the solution

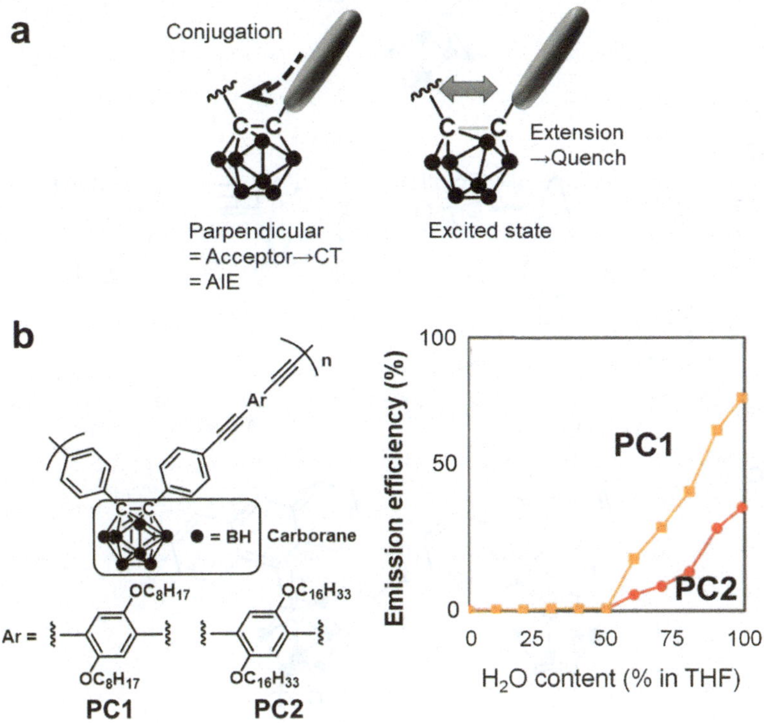

Fig. 9.7 **a** Emission quenching by *o*-carborane and **b** AIE behaviors of the *o*-carborane-containing main-chain-type conjugation polymers. Reprinted with permission from Ref. Tanaka et al. (2017). Copyright 2009 American Chemical Society

through molecular vibration at the C–C bond in *o*-carborane, while the emission band attributable to the transition from the ICT state should be exhibited in the aggregation state by restricting energy-consumable molecular motions (Tanaka and Chujo 2017). Owing to steric hindrances around the *o*-carborane unit, ACQ could be also suppressed.

After this finding, another unique mechanical motion in the excited state was found from the crystalline sample of the anthracene-tethered *o*-carborane dyad (Fig. 9.8a). According to the AIE-active *o*-carborane-containing copolymers, it was presumed that any mechanical motions including the rotation at the *o*-carborane unit should be suppressed in the crystal because of structural restriction. On the other hand, surprisingly, significant change was observed in the PL spectra by altering measurement temperature (Fig. 9.8b) (Naito et al. 2017a). At room temperature, the clear ICT emission band was observed from the crystalline sample, whereas both the ICT and LE emission bands were obtained at 77 K where any molecular motions

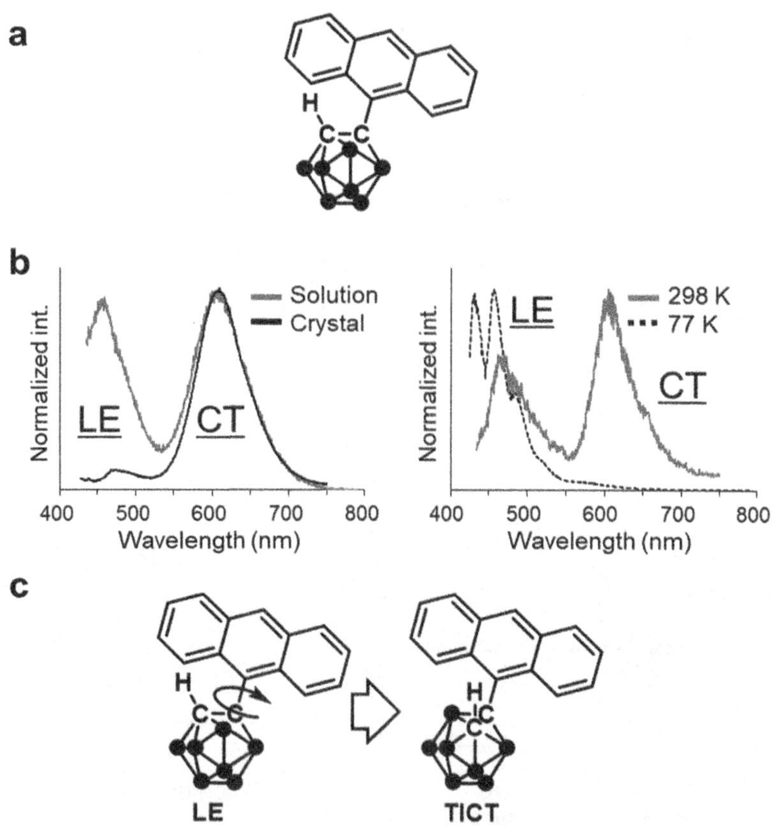

Fig. 9.8 **a** Chemical structure of the anthracene—o-carborane dyad and **b** PL spectra under various conditions. **c** Plausible molecular rotation in crystal after photo-excitation. Reprinted with permission from Ref. Naito et al. (2017). Copyright 2017 Wiley–VCH Verlag GmbH & Co. KGaA

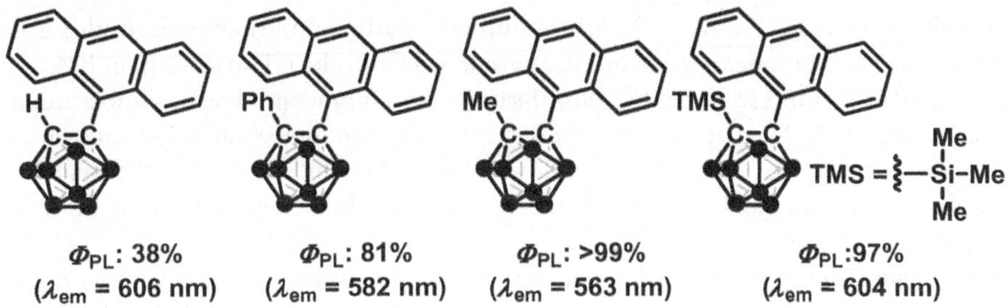

Fig. 9.9 Chemical structures and optical properties of the anthracene—*o*-carborane dyads

should be suppressed. From the quantum calculations, it was shown that the LE and ICT emission bands should be induced from the parallel and perpendicular conformations between the π-plane of the anthracene moiety and the C–C bond in *o*-carborane, respectively. These data mean that both conformations should exist in the ground state. After photo-excitation, the perpendicular conformation, in which the ICT emission can be presented, should be dominant through the molecular rotation at the *o*-carborane unit even in the crystal packing (Fig. 9.8c). Because of the sphere structure, molecular rotation could be allowed in the condensed state. From these results, it was assumed that solid-state emission could be feasible by completely suppressing molecular motions including vibration and rotation. Based on this idea, the series of the modified *o*-carboranes having various types of substituents at the adjacent carbon and the aryl moiety were synthesized.

The series of anthracene—*o*-carborane dyads with various substituents, such as methyl and trimethylsilyl (TMS) at the adjacent carbon were prepared and their optical properties were compared with the hydrogen-substituted dyad (Fig. 9.9) (Naito et al. 2017b). All dyads showed the AIE properties because larger emission efficiencies were observed in the aggregation than those in the solutions. Moreover, much larger emission bands in the crystalline state were obtained, indicating that the dyads had crystalline-induced emission (CIE) properties. It should be noted that almost quantitative values of emission efficiency were obtained from the methyl and TMS-substituted dyads. The bulky substituents at the adjacent carbon in *o*-carborane should effectively prevent the *o*-carborane unit from molecular vibration in the crystal packing. Thus, non-radiative decay processes should be closed. Finally, intense emission was presented in the crystalline state. Diverse emission color was detected depending on the structures of the substituents (H: orange, methyl: yellow, TMS: orange). As mentioned above, the electronic interaction between *o*-carborane and the aryl moiety critically depends on the angle between the direction of the C–C bond in *o*-carborane and the hypothetical plane involving anthracene (0°: minimum, 90°: maximum). In the larger substituents, the molecular conformation should be fixed at the perpendicular conformation. Then, the emission band appeared in relatively longer wavelength region. In the absence of the substituent, free rotation is capable. Therefore, the TICT process proceeded, resulting in the longer wavelength emission.

To investigate generality of the aryl-modified *o*-carborane structure for highly efficient solid-state luminescent materials, pyrene was introduced instead of anthracene (Fig. 9.10) (Nishino et al. 2016). Similarly to the anthracene dyads, almost quantitative emission efficiencies were obtained. In addition, emission color was varied (H: orange, methyl: green, TMS: yellow). These data indicated that the aryl—*o*-carborane unit should be the luminescent element-block with intense solid-state emission properties.

By controlling the width of π-conjugated system in the aryl moiety, it is likely that emission color tuning is capable. However, it was presumed that ACQ could appear due to the formation of π-stacking at the extended conjugated planes. Therefore, the dual *o*-carborane-substituted acenes were designed. For obtaining various color emissions, a wide variety of the aryl moieties were introduced instead of anthracene and pyrene (Fig. 9.11) (Naito et al. 2017c). All carboranes showed clear AIE behaviors in diverse wavelength regions. Blue color emissions with almost quantitative efficiencies were obtained from the naphthalene derivatives, indicating that

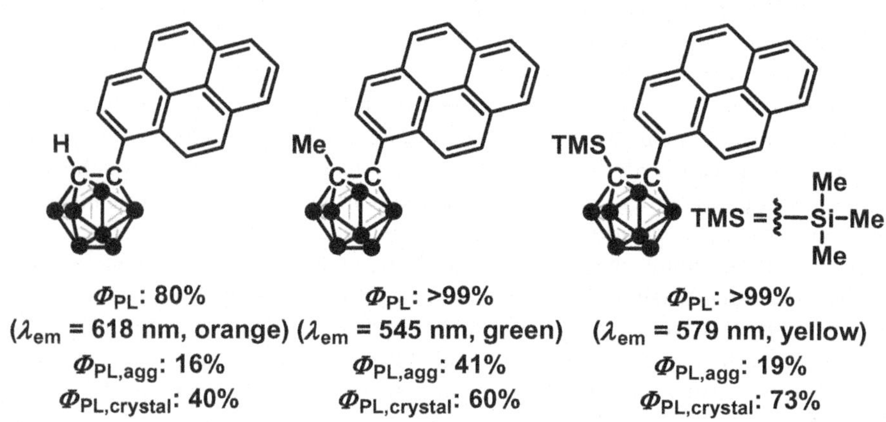

Fig. 9.10 Chemical structures and optical properties of the pyrene—*o*-carborane dyads

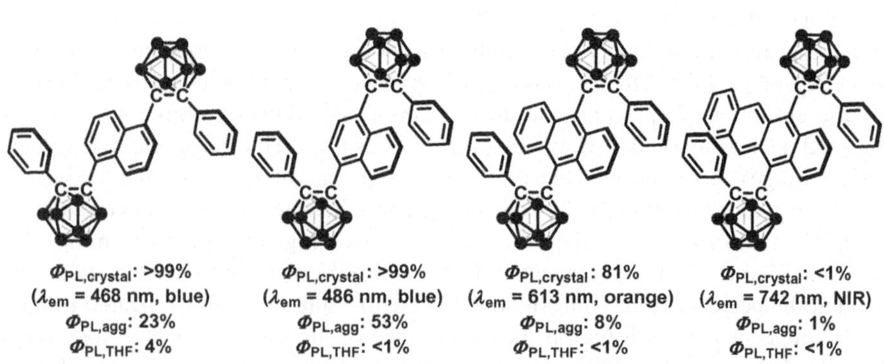

Fig. 9.11 Chemical structures and optical properties of the triads

o-carborane should be responsible for preventing ACQ by inhibiting intermolecular interaction and be the solid-state emission-inducible element-block. Corresponding to increase of the number of benzene ring in the acene moiety, emission color was shifted to the longer wavelength region. Finally, the near-infrared (NIR) emission was detected from the tetracene derivative. One of the problems for practical usages of the NIR-emissive dyes is low photostability, while the tetracene derivative showed high resistance under UV light irradiation. It was clearly indicated that the *o*-carborane substituents can also play a significant role in the improvement of stability. These molecules are promised to be applied as a robust luminescent material in optical devices.

The above researches indicated that introduction of multiple *o*-carboranes should be effective for expressing AIE properties. The triphenylamine derivatives with variable numbers of the *o*-carborane unit (1–3) were prepared (Fig. 9.12) (Nishino et al. 2017a). All triphenylamines exhibited the typical AIE assigned to the ICT emission. In the solution state, emission efficiencies were very low (less than 1%), while emission enhancements were obtained in solid. With the single *o*-carborane substituent, slight enhancement was observed by the aggregation formation (4%). In contrast, it was clearly shown that the multiple *o*-carborane substituents were able to enhance solid-state emission properties (bis-substitution: 30%, tri-substitution: 25%). ACQ was increasingly suppressed by adding the number of the *o*-carborane substituent in triphenylamine. As a result, clear AIE behaviors were induced.

Pyrene is well known to form the excimer, which presents the broad emission band in the longer wavelength region than that of the LE emission, under higher concentrated condition. Similarly to the LE emission, critical ACQ appeared in the case of the excimer emission in the condensed state. When we investigated the temperature dependency of solid-state luminescence of the pyrene—*o*-carborane dyad with the ethynyl spacer, the new broad emission band was detected only at 77 K (Fig. 9.13) (Nishino et al. 2017b). At room temperature, orange emission

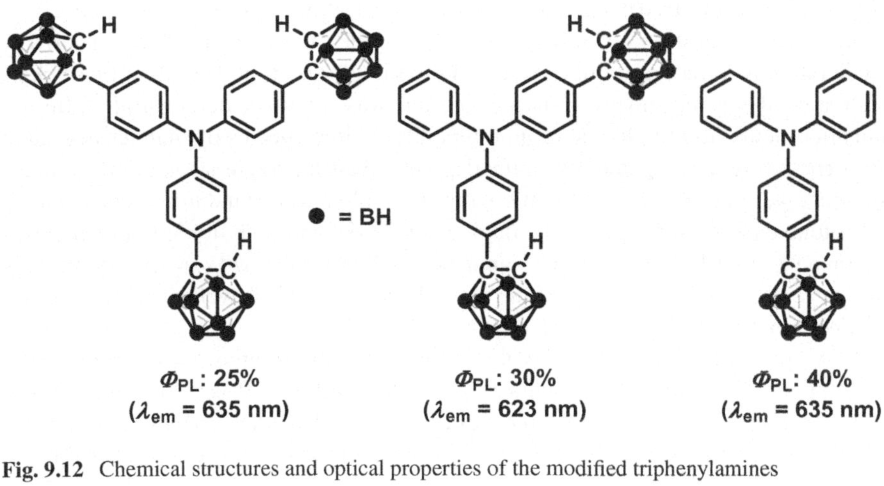

Fig. 9.12 Chemical structures and optical properties of the modified triphenylamines

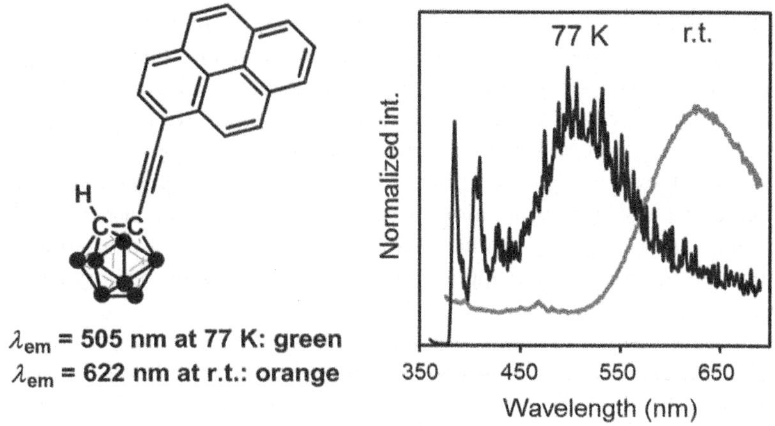

Fig. 9.13 Chemical structure of the pyrene dyad with the ethynyl linker and PL spectra at room temperature and 77 K. Reprinted with permission from Ref. Nishino et al. (2017). Copyright 2017 Wiley-VCH Verlag GmbH & Co. KGaA

with the ICT character was observed from the crystalline sample, meanwhile drastic color change to green was observed because of the appearance of the new emission band, indicating that this molecule has the unexpected thermochromic luminescent property. In the solution state, the LE emission was obtained. Moreover, the new emission band possessed relatively longer lifetime. From these data, it was assumed that the new emission band should be attributable to the excimer emission. Other unique properties were found in this pyrene dyad with the ethynyl spacer.

The pyrene dyad with the ethynyl spacer shows the time-dependent emission enhancement (TDEE) phenomena (Nishino et al. 2019). Emission intensity from the dyad increased in THF, acetone, and dichloromethane by increasing incubation time (Fig. 9.14). From the mechanistic studies, it was revealed that agglomeration of the dyad gradually occurred even in the solution state, followed by expression of excimer luminescence. Additionally, it was observed that the rates of TDEE were sensitively accelerated by a trace amount of water in the sample solution. Based on these characteristics, the water sensing system in acetone was able to be demonstrated. Before and after incubation for 96 h at room temperature, time courses of changes in optical properties were investigated. From the degree of TDEE and emission color changes, the plots were prepared. Finally, by using these plots as a standard, water contents in acetone can be estimated by the degrees of TDEE and emission color changes in the range from 0.1 wt% to 2.0 wt% and from 2.0 wt% to 20 wt%, respectively. This study is on the basis of a very unique assembling property as well as ACQ-resistant excimer emission.

It was presumed that the dyad structure involving the ethynyl spacer might be the excimer emission-inducible element-block in the solid state. To evaluate the validity of this speculation, the next molecule was designed and synthesized (Fig. 9.15) (Ochi and Tanaka 2019). Efficient solid-state excimer emission was observed from the

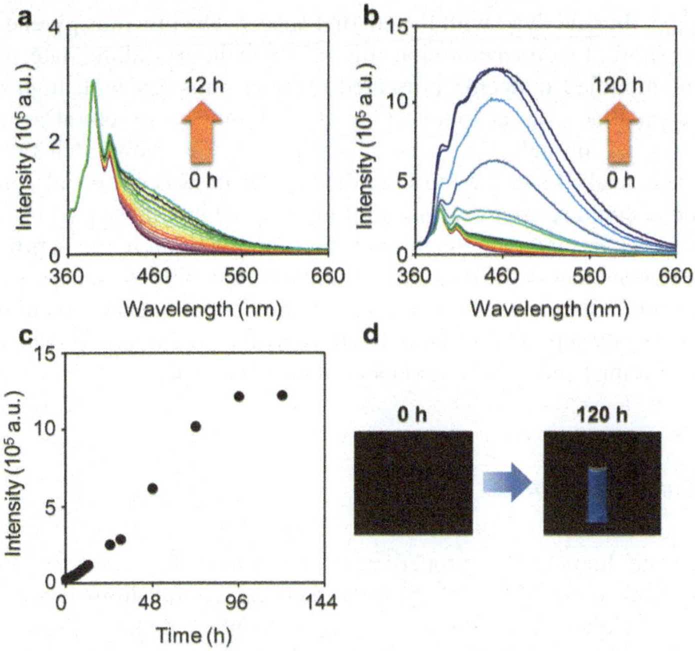

Fig. 9.14 Time courses of the changes in PL spectra of the pyrene dyad with the ethynyl spacer in THF at 298 K **a** for 12 h and **b** 120 h. **c** Intensity change from the solution. **d** Pictures under UV irradiation (365 nm). Reprinted with permission from Ref. Nishino et al. (2019). Copyright 2019 Wiley–VCH Verlag GmbH & Co. KGaA

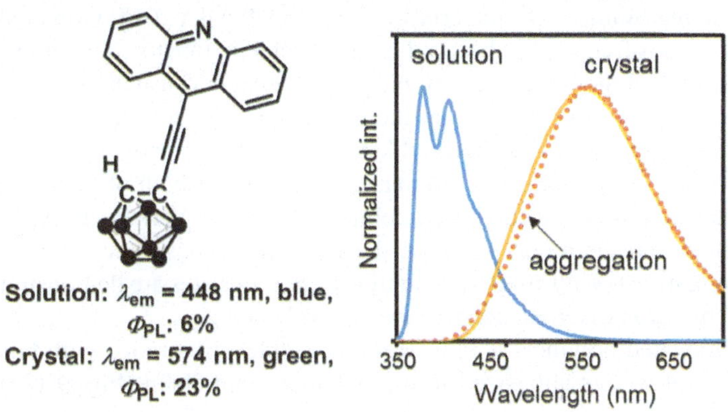

Fig. 9.15 Chemical structure of the acridine dyad with the ethynyl linker and PL spectra under various conditions. Reprinted with permission from Ref. Ochi et al. (2019). Copyright 2019 Wiley–VCH Verlag GmbH & Co. KGaA

acridine—*o*-carborane dyad with the ethynyl spacer. The previous pyrene-modified *o*-carborane showed excimer emission only at 77 K in the crystalline state, meanwhile the acridine-modified molecule presented excimer emission with high efficiency (23%) in crystalline state at room temperature. From the structural analysis with the single crystal through X-ray crystallography, it was shown that two acridine moieties were stacked and the third acridine molecule was out of alignment. This packing mode could restrict exciton splitting over the columnar packing structure. Moreover, molecular interactions through the nitrogen atom in the acridine moiety and the hydrogen atom in the *o*-carborane unit has possibility to suppress molecular motions. From these results, improvement of emission efficiency was implied. The results in this study support the claim that the ethynyl—*o*-carborane skeleton should work as the excimer-inducible component in the solid state.

9.4.2 Boron Complex

The solid-state luminescent properties of the class of boron complexes with *β*-diketonate analogue involving diketone, ketoimine, and diimine are explained (Fig. 9.16). The diketone compound is one of simple and stable ligand involving π-conjugation for the complexation with various types of metal cations (Tanaka and Chujo 2015). In particular, by boron complexation, molecular rigidity would be improved. As a result, intense emission can be obtained. Moreover, by applying organic synthetic methods, a wide variety of derivatives including polymers have been developed. Thus, boron diketonates are a typical conjugated element-block for constructing luminescent materials (Chujo and Tanaka 2015; Gon et al. 2018a). However, ACQ is frequently observed in the boron diketonate-containing materials. Thereby, chemical modification with bulky substituents is usually needed to obtain solid-state emission. In this section, the transformation from the ACQ dye to the AIE-active molecule based on boron diketonate is initially explained. We will illustrate that boron ketoiminate and diiminate, where one or both of the oxygen atoms are replaced to nitrogen at the boron complexation, can work as an AIE-inducible element-block. Furthermore, based on these element-blocks, the series of stimuli-responsive materials with solid-state luminescent properties are demonstrated. Wide versatility of boron diketonate derivatives is explained.

The conjugated polymers having various complexation ratios with boron in the diketonate unit were synthesized in the polymer main chain (Fig. 9.17a) (Tanaka

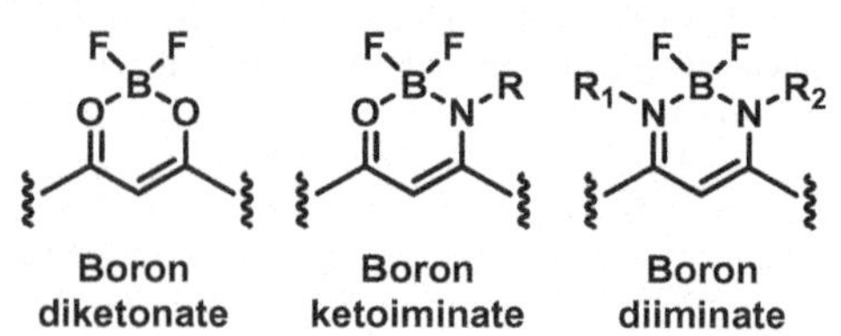

Fig. 9.16 Chemical structures of boron diketonate, ketoiminate, and diiminate

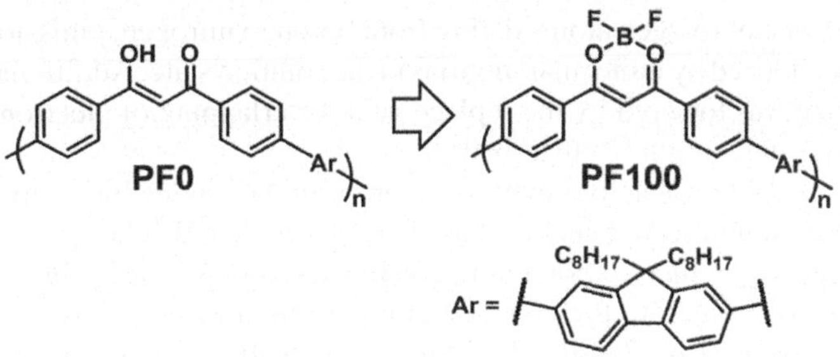

	$l_{abs.}$ (nm)	$l_{PL,film}$ (nm)	$F_{PL,THF}$ (%)	$F_{PL,film}$ (%)
PF0	404	450	10	0.4
PF21	407	554	33	1.9
PF41	413	552	31	1.3
PF58	461	561	24	1.2
PF71	464	563	22	1.0
PF100	470	540	21	1.9

Fig. 9.17 Chemical structures and optical properties of boron diketonate polymers with variable complexation ratios

et al. 2013). Similarly to conventional conjugated polymers, intense emission was only observed in the diluted solution, whereas ACQ occurred in the film samples. For constructing robust main-chain conjugation involving boron, the ketoiminate structure-containing polymers were designed (Fig. 9.17b) (Yoshii et al. 2014a). From the optical measurements, reverse behaviors, that is AIE, to conventional conjugated polymers were found. The small molecules were synthesized with or without the substituent on the nitrogen atom (Fig. 9.18) (Yoshii et al. 2013). In summary, the resulting boron ketoiminates were an AIE-active molecule. By fixing molecular motions under high viscosity or frozen conditions, significant emission was observed, meaning that intramolecular motions would induce emission annihilation in the solution state. Comparing to the B–O bond, the B–N one has lower energy. Thereby, by

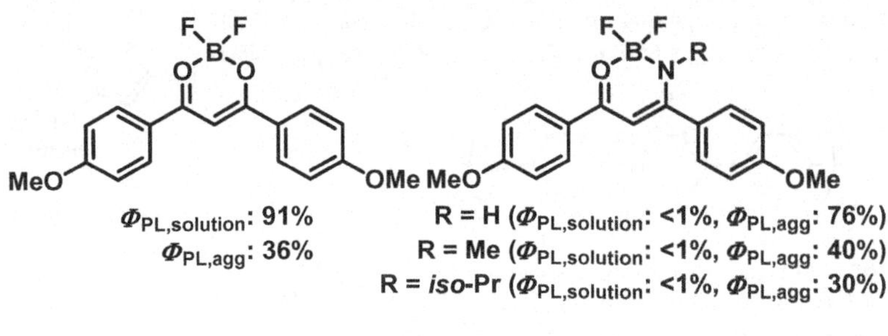

Fig. 9.18 Chemical structures and optical properties of boron diketonate and ketoiminates

replacing the one of oxygen atoms dating from boron to nitrogen, emission annihilation could be induced by molecular motions in the solution state. Additionally, molecular symmetry was lowered by the replacement. Overlapping of electron orbitals in the condensed state might be disturbed.

By utilizing the boron ketoiminate skeleton as an AIE-inducible element-block, a wide variety of luminescent materials have been developed. By changing comonomer units in the above polymers, color tuning of AIE was accomplished (Fig. 9.17) (Yoshii et al. 2014a). By changing the connection point, luminescent films were obtained (Fig. 9.19) (Yoshii et al. 2014b). Electronic conjugation was often limited at the conventional AIE structures, such as tetraphenylethene and silole derivatives, due to molecular distortion at the phenyl rings (Luo and Tang 2001; Dong 2007). In contrast, molecular planarity is relatively maintained in boron ketoiminates having AIE properties. Thus, this structure could be favorable as an element-block for constructing AIE-active conjugated polymers and their property tuning.

Based on the strategy for fixing solid-state emission of the AIE behavior of boron ketoiminate, multi-state emissive molecules, which can provide continuous emission intensity in any state with high environment resistance (Suenaga and Tanaka 2017). In particular, as is often the case with stimuli-responsive luminochromic materials, emission intensity often decreases after chromism. Fused boron ketoiminates (FBKIs) were designed and synthesized (Fig. 9.20). From optical measurements in solution, crystal, and amorphous, similar emission efficiencies were observed. Moreover, typical mechanochromic luminescence was observed by grinding the crystalline sample. It should be mentioned that emission efficiencies were not significantly influenced during the physical treatments. By grinding, the highly ordered structure would be destroyed, and amorphous states appear. Emission efficiencies of

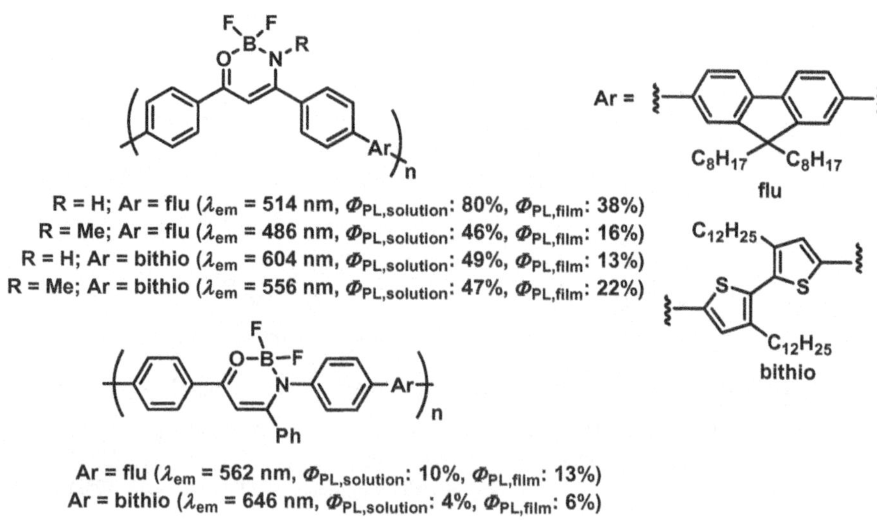

Fig. 9.19 Chemical structures and optical properties of boron ketoiminate polymers

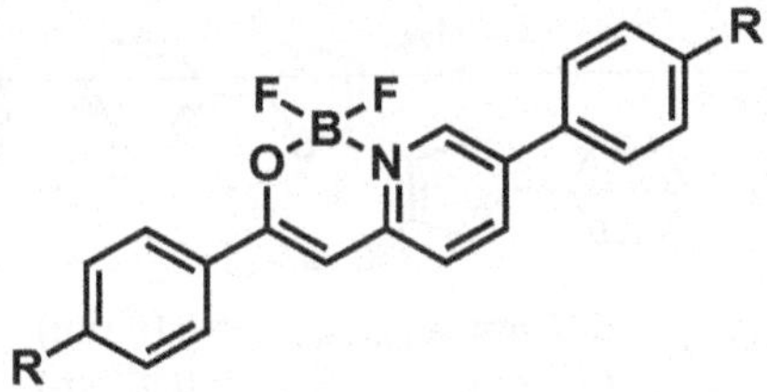

FBKI1: R = H ($\Phi_{PL,solution}$: 17%, $\Phi_{PL,crystal}$: 25%, $\Phi_{PL,ground}$: 20%)
FBKI2: R = OMe ($\Phi_{PL,solution}$: 58%, $\Phi_{PL,crystal}$: 68%, $\Phi_{PL,ground}$: 64%)

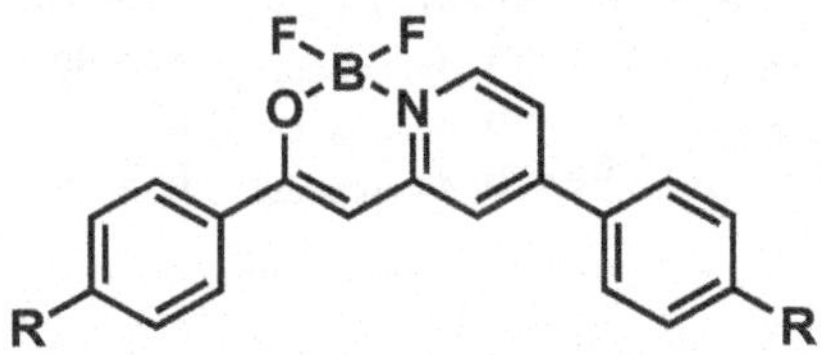

FBKI3: R = H ($\Phi_{PL,solution}$: 35%, $\Phi_{PL,crystal}$: 38%, $\Phi_{PL,ground}$: 24%)
FBKI4: R = OMe ($\Phi_{PL,solution}$: 80%, $\Phi_{PL,crystal}$: 81%, $\Phi_{PL,ground}$: 57%)

Fig. 9.20 Chemical structures and optical properties of FBKIs

commodity dyes in amorphous are usually lower than those in crystal due to non-specific intermolecular interaction and molecular distortions. In contrast, FBKIs can present environment-sensitive luminescent chromism as well as solid-state emission because of rigid and planar skeletons.

Precise control of the direction of color changes in mechanochromic luminescence by the substituents was achieved (Yoshii et al. 2015). The series of the triads composed of bis boron ketoiminates and the bithiophene moiety were synthesized, and solid-state emission was observed (Fig. 9.21). The pristine molecule showed red emission at the initial state, meanwhile blue-shift of the emission band was induced by grinding in the spectra, and finally yellow emission was obtained from the ground sample. On the other hand, the iodine-substituted triad showed yellow emission at the initial state, and the red-shifted emission band was obtained after grinding. From the structural and optical analyses with the series of triads having different types of the substituents at the both ends, it was revealed that the size of the substituents should be dominant to determine the direction of luminescent color changes (blue- or red-shifts) in the mechanochromic luminescent behaviors of the triads. With smaller substituents including hydrogen, relatively condensed packing can be formed in the initial state. By the stabilization through π-stacking, red emission tends to be observed. By grinding the crystalline sample, regular structures would be collapsed. At this situation, stabilization by molecular interaction disappears, and then blue-shift should be induced. On the other hand, in the presence of bulky substituents at both ends, it is likely that stacking interaction could be disturbed because of steric hindrances. Then, yellow emission is observed at the initial state. After transition to amorphous triggered by the mechanical treatment, intermolecular interaction

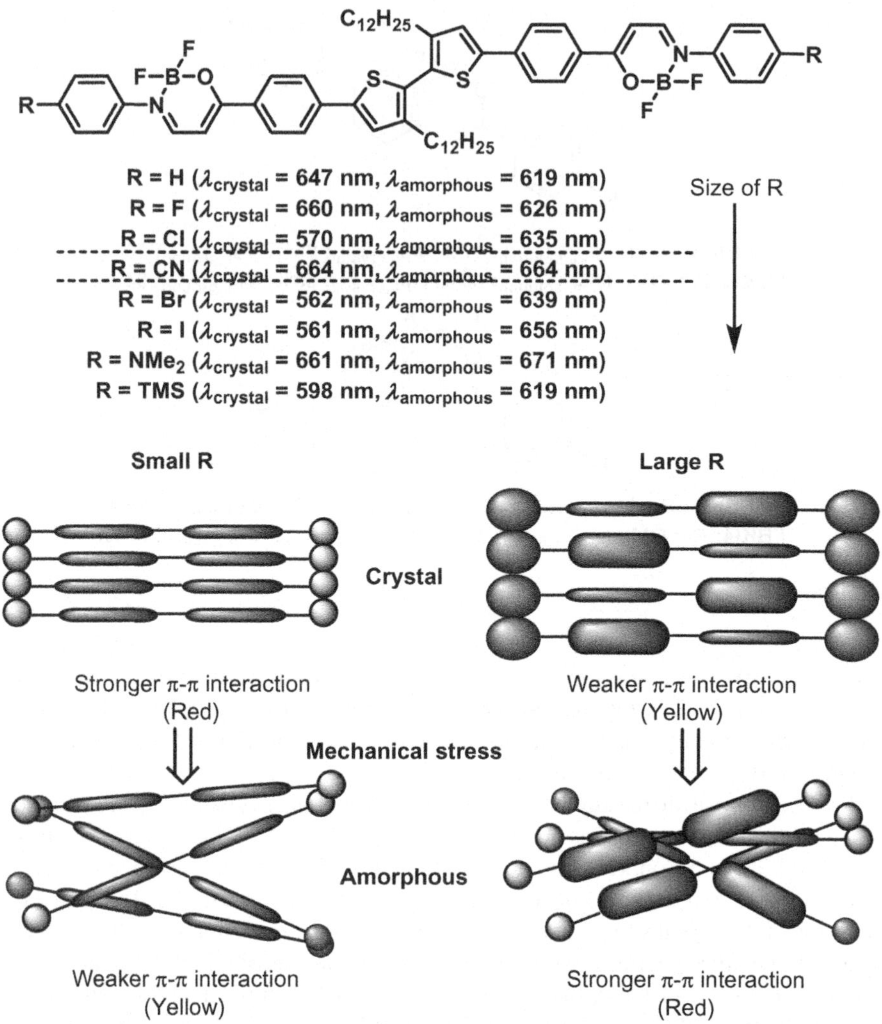

Fig. 9.21 Optical properties of boron ketoiminates and plausible mechanism of their mechanochromic luminescence

could be recovered, resulting in the red-shifted emission. After grinding, the initial luminescent color was obtained by heating. Reversible changes were observed.

In the practical usages of solid-state luminescent materials with mechanochromic properties as a sensor for mechanical forces, thermal resistance is required. By adding mechanical stresses to the solid sample, generation of fraction heat is inevitable. In the above materials, reversible changes were capable triggered by heating. In other words, there is a possibility that luminescent color changes induced by mechanical

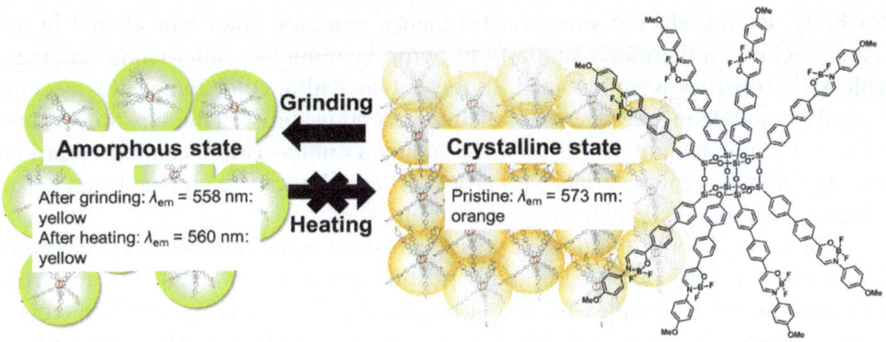

Fig. 9.22 Schematic illustration and optical properties of the thermal-resistant mechanochromic luminescent hybrid

stresses are spoiled by fraction heating. In order to obtain thermal-resistant mechano-luminescent chromic materials which can show luminochromism only by mechanical forces, the hybrid-type material was designed (Fig. 9.22) (Suenaga et al. 2017). By employing the hybrid element-block, we aimed to reinforce thermal stability. Boron ketoiminates were tethered to POSS, and it was expected that phase transition from amorphous to the initial crystalline state can be inhibited. The modified luminescent POSS showed mechanochromic luminescence, and emission color was preserved from heating. By the combination with different types of element-blocks, the problem in the conventional material can be overcome.

By increasing molecular flexibility, influence on luminescent behaviors was examined. The boron diiminate complexes were designed, where another oxygen in boron ketoiminate was replaced with nitrogen (Fig. 9.23) (Yoshii et al. 2014c). The resulting complexes exhibited very slight and weak emission in solution and aggregation, respectively. The molecular motions were not efficiently suppressed even in amorphous. Interestingly, in the crystalline state, drastic emission enhancement was observed. This fact means that boron diiminates have CIE properties and it is proposed that incorporation into crystal packing should be necessary for suppressing molecular motions and subsequently emission annihilation due to high molecular

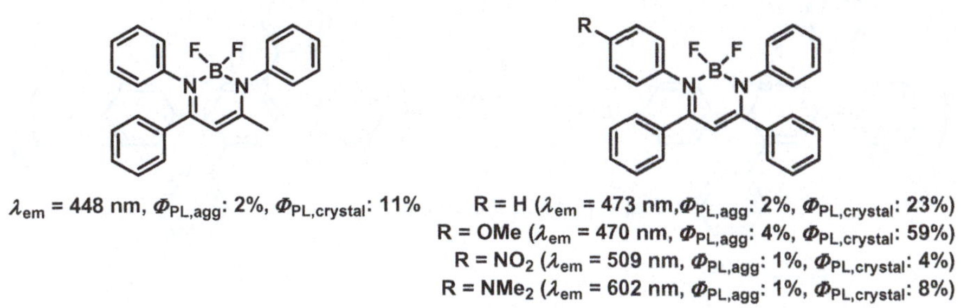

Fig. 9.23 Chemical structures and optical properties of boron diiminates

flexibility. By introducing various substituents, emission color was altered in the visible region. Furthermore, similarly to boron ketoiminates, alternating polymers with AIE properties were obtained. In the polymer film, color tuning was accomplished by modulating the electron-donating/accepting ability of the substituents into the boron diiminate unit as well as the comonomer units. These data represent that film-type luminescent materials with property tunability can be obtained, and boron diiminate is also an AIE-inducible element-block.

In the case of boron diiminates, drastic emission color variation was observed. Yellow emission was observed from the pristine complex, while red emission from the amine-modified complex (Yoshii et al. 2014d). These data mean that the electronic properties of boron diiminate can be readily tuned by the substitution effect. Indeed, the alternating copolymers composed of boron diiminates were prepared, and diverse emission color was observed similarly to the monomers. Based on these data, emission properties were monitored by fuming acid gas to the polymer film involving the amine-modified complex. Before the treatment, yellow emission was observed from the film, meanwhile in the presence of trifluoroacetic acid vapor, emission color of the film turned to red (Fig. 9.24a) (Yoshii et al. 2014d). Moreover, by treating with triethylamine vapor, emission color was recovered to the original yellow emission. These data mean that the film-type acid–base sensor can be obtained.

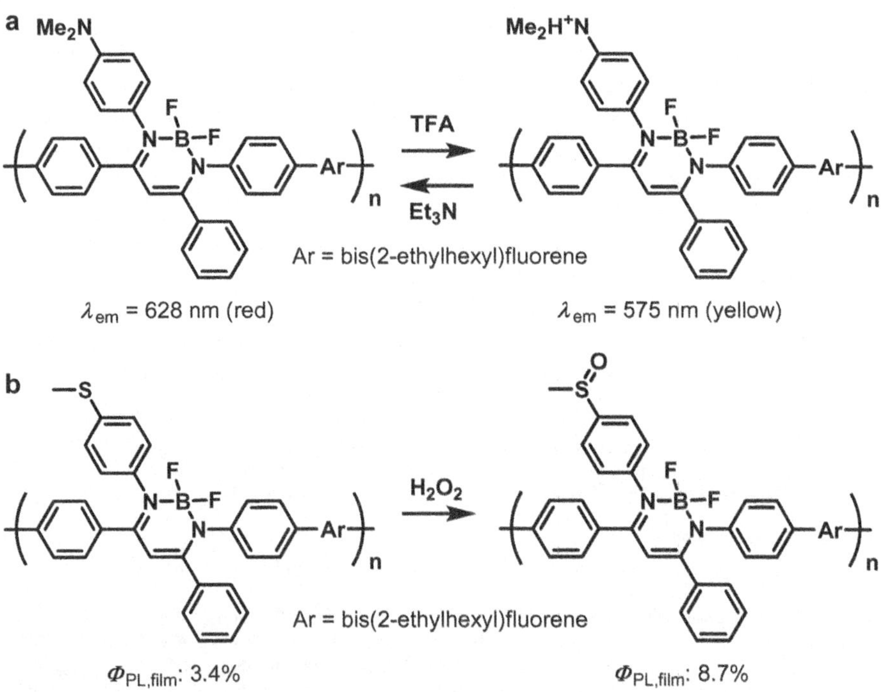

Fig. 9.24 Chemical structures and plausible reaction schemes of film-type sensors for **a** acid vapor and **b** hydrogen peroxide

To induce the changes in luminescent properties of the film, the polymers having the methyl sulfide group as the side group were reported (Fig. 9.24b) (Hirose et al. 2015). The methyl sulfide group can be smoothly oxidized by various oxidizers, and the electron-donating property is oppositely converted to the electron-accepting nature. In order to utilize this reaction for modulating electronic properties, followed by emission characters, the methyl sulfide-modified polymer at the boron complex was synthesized, and their photophysical properties were monitored during oxidation. The film showed weak emission before the treatment with hydrogen peroxide which is the biomolecules concerning metabolism and carcinogenesis. Emission intensity gradually increased during soaking the film into the hydrogen peroxide solution. By soaking, the methyl sulfide group should be oxidized. Finally, enhanced emission was observed. The electronic property of the substituent was contrary changed from electron donating to accepting one. Therefore, emission intensity was enhanced. The film-type sensor for the bio-significant molecule was accomplished based on oxidation-induced AIE.

By replacing boron with gallium which is a heavier atom in the group 13 element than boron, influence on solid-state luminescent properties was evaluated (Fig. 9.25) (Ito et al. 2016). Since gallium atom is larger than that of boron, lower degree of intermolecular interaction was expected in crystal, leading to the improvement of emission efficiency. Indeed, gallium diiminate presented larger emission efficiency

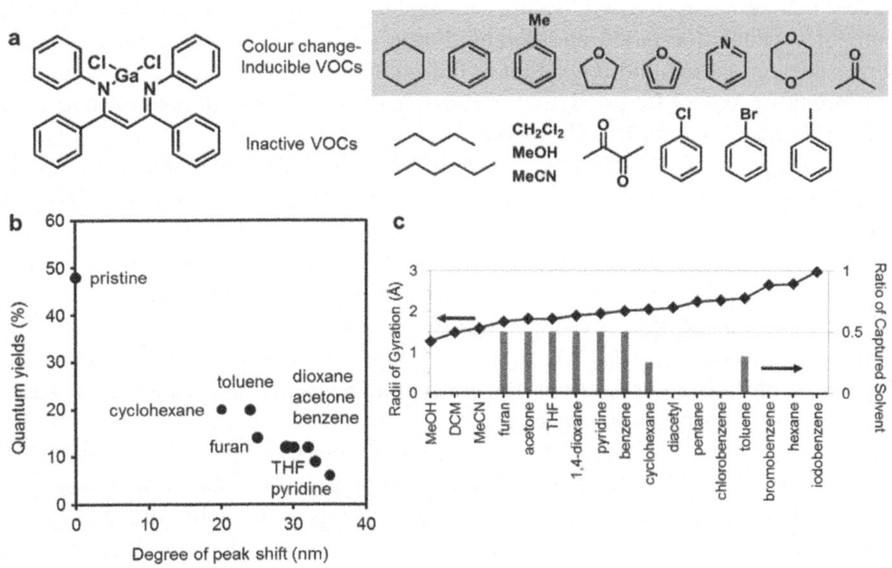

Fig. 9.25 **a** Chemical structure of gallium diiminate and various types of VOCs used in the study. **b** The emission quantum yields of the crystalline samples including VOCs. **c** The number of captured VOCs into the crystalline samples and their radius of gyration. Reproduced from Ref. Ito et al. (2016) with permission from The Royal Society of Chemistry

in the crystal than the boron complex. In addition, environment-sensitive luminescent character regarding vapochromic properties was discovered. By exposing the crystal powder of gallium diiminate to vapor of relatively smaller sizes of volatile organic compounds (VOCs), emission color was changed from blue to green, blue color emission was maintained in the presence of larger sizes of VOC vapors. From the mechanistic studies, the crystal–crystal transition was induced by smaller sized VOCs, resulting in luminescent chromism. Because of larger atomic size of gallium than that of boron, pore structures should be created in the crystalline surfaces. As a consequence, structural alterations are able to be induced when VOCs are inserted. This material is expected to be a sensor for detecting air pollutants.

9.5 Rational Design for AIE-Active Molecules

Except for the above examples, it is shown that AIE-active molecules are versatile for stimuli-responsive luminescent materials for the applications to organic optoelectronic devices and bioprobes (Tang 2015). Therefore, development of new AIE-active molecules is of great significance for improving these properties in the products. So far, most of the AIE-active molecules seem to have multiple phenyl rings to induce emission annihilation in solution as well as to suppress ACQ. Recently, it was demonstrated that new AIE-active molecules are able to be designed from scratch through computer calculation based on "flexible boron complexes" which show relatively larger degree of structural relaxation in the excited state (Gon et al. 2019c). In this section, this protocol for predicting AIE-active dyes and the brilliant new dye with the highly planar structure is illustrated.

According to the mechanism on AIE behaviors, flexible conjugated boron complexes can have possibility to show AIE properties. In the solution state, vigorous molecular motions occur, resulting in emission annihilation, while emission can be observed in aggregation (Fig. 9.26) (Yamaguchi et al. 2017). On the basis of this assumption, new boron complexes were explored by computer calculations. As a result, the optimized structures in the ground and excited states were estimated with density functional theory (DFT) and time-dependent DFT calculations, respectively. By comparing both conformations, the boron complex which showed relatively large difference was selected and synthesized.

BPI was proposed to be a potential candidate from the comparisons (Fig. 9.27) (Yamaguchi et al. 2017). In the ground state, the planar conformation was obtained, meanwhile the bent conformation was suggested as the most stable structure in the excited state. To fix the conformation, FBPI was also designed and synthesized as a control compound. From the optical measurements, the clear AIE and CIE characters were observed only from BPI. From the width of the Stokes shift, it was supported that large structural relaxation should proceed in the excited state of BPI. In the case of FBPI, intense emission with small Stokes shift was obtained only from the solution, and critical ACQ was observed similarly to other conventional organic dyes. This result indicates that the AIE-active molecule can be theoretically predicted.

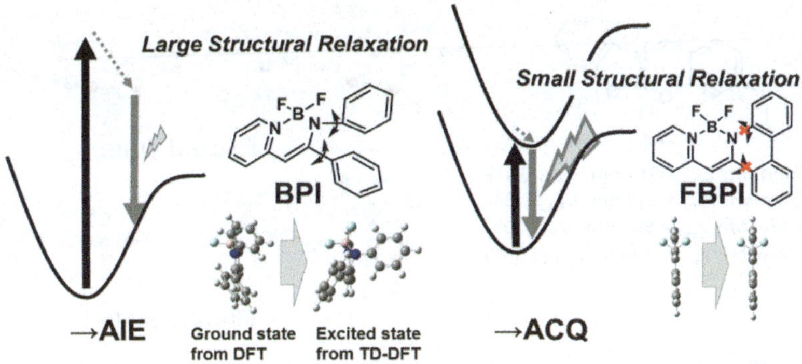

Fig. 9.26 Chemical structures of **BPI** and **FBPI** and plausible models of the difference in the degree of structural relaxation in the excited states. Reproduced from Ref. Yamaguchi et al. (2017) with permission from The Royal Society of Chemistry

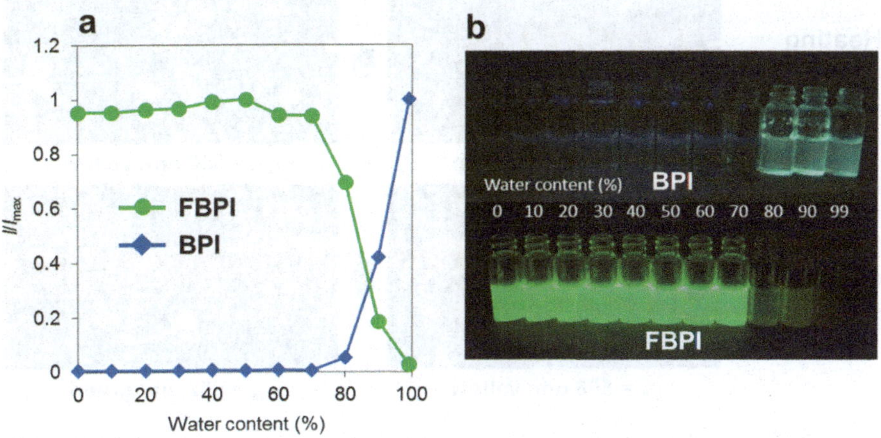

Fig. 9.27 **a** Changes in intensity ratios by the aggregation formation in the THF solutions with variable water concentrations and **b** their appearances under UV irradiation (365 nm). Reproduced from Ref. Yamaguchi et al. (2017) with permission from The Royal Society of Chemistry

Finally, based on this design protocol, the new AIE-active dye without exocyclic phenyl groups was invented. Through the comparison of the optimized structures in the ground and excited states, the azomethine compound **Az** was obtained (Fig. 9.28) (Ohtani et al. 2017). **Az** showed AIE and CIE, similarly to BPI. Furthermore, reversible crystal–crystal transitions in the cooling and heating cycle between crystal polymorphs with different luminescent color were observed. Correspondingly, thermochromic luminescence and thermosalient effect were simultaneously observed during the transition (Fig. 9.29). Owing to the planar structure of **Az**, not only sensitive luminescent chromism toward environment changes but also mechanical motions could be observed. This molecule could be a building block for obtaining conjugated

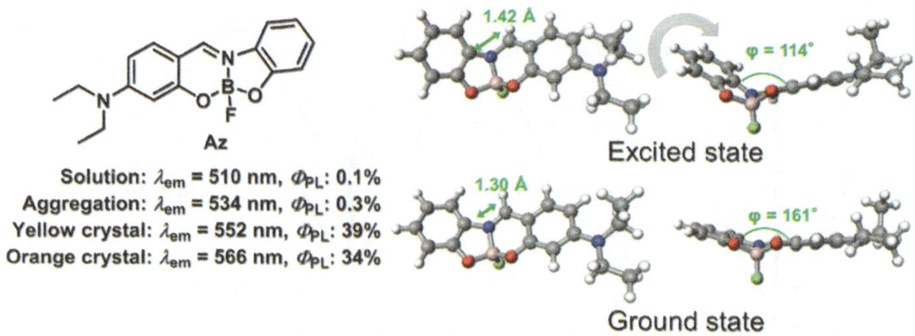

Fig. 9.28 Chemical structure, optical properties, and optimized structures of **Az**. Reprinted with permission from Ref. Ohtani el al. (2017). Copyright 2017 Wiley–VCH Verlag GmbH & Co. KGaA

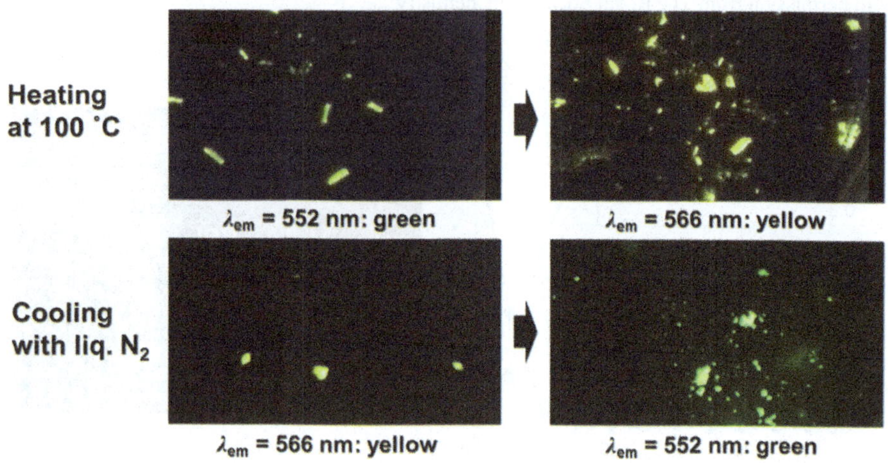

Fig. 9.29 Thermosalient effects of **Az** with thermochromic luminescence. Reprinted with permission from Ref. Ohtani el al. (2017). Copyright 2017 Wiley–VCH Verlag GmbH & Co. KGaA

molecules and polymers having extended main-chain conjugation because of the planar structure (Gon et al. 2018b).

9.6 Conclusion

In this chapter, the strategies for obtaining solid-state emission are reviewed and representative examples are introduced mainly from boron complexes. As mentioned in the introduction of this chapter, solid-state luminescent materials play a key role in modern optoelectronic devices. Moreover, in biotechnology and biomedical fields, development of the luminescent probes is increasingly needed for detecting aggregation substances or solid organisms. The design strategies presented here would be useful as a guideline not only for obtaining intense luminescent solid and films but also for fabricating optically functional solid materials.

References

Anthony SP (2012) Organic solid-state fluorescence: strategies for generating switchable and tunable fluorescent materials. ChemPlusChem 77(7):518–531

Carayon C, Fery-Forgues S (2017) 2-Phenylbenzoxazole derivatives: a family of robust emitters of solid-state fluorescence. Photochem Photobiol Sci 16(7):1020–1035

Chen S et al (2016) Waterborne POSS-silane-urethane hybrid polymer and the fluorinated films. Polymer 103:27–35

Chujo Y, Tanaka K (2015) New polymeric materials based on element-blocks. Bull Chem Soc Jpn 88(5):633–643

Dong Y et al (2007) Aggregation-induced emissions of tetraphenylethene derivatives and their utilities as chemical vapor sensors and in organic light-emitting diodes. Appl Phys Lett 91(1):011111–011113

Gon M, Sato K, Tanaka K, Chujo Y (2016) Controllable intramolecular interaction of 3D arranged π-conjugated luminophores based on a POSS scaffold, leading to highly thermally-stable and emissive materials. RSC Adv 6(82):78652–78660

Gon M, Tanaka K, Chujo Y (2017) Creative synthesis of organic-inorganic molecular hybrid materials. Bull Chem Soc Jpn 90(5):463–474

Gon M, Tanaka K, Chujo Y (2018a) Recent progress in the development of advanced element-block materials. Polym J 50(1):109–126

Gon M, Tanaka K, Chujo Y (2018b) A highly efficient near-infrared-emissive copolymer with a N=N double-bond π-conjugated system based on a fused azobenzene-boron complex. Angew Chem Int Ed 57(22):6546–6551

Gon M, Tanaka K, Chujo Y (2019c) Concept of Excitation-Driven Boron Complexes and Their Applications for Functional Luminescent Materials. Bull Chem Soc Jpn 92(1):7–18

Gon M, Sato K, Kato K, Tanaka K, Chujo Y (2019a) Preparation of bright-emissive hybrid materials based on light-harvesting poss having radially integrated luminophores and commodity . Mater Chem Front 3(2):314–320

Gon M, Kato K, Tanaka K, Chujo Y (2019b) Elastic and mechanofluorochromic hybrid films with poss-capped polyurethane and polyfluorene. Mater Chem Front 3(6):1174–1180

Gu P-Y et al (2017) Understanding the structure-determining solid fluorescence of an azaacene derivative. J Mater Chem C 5(34):8869–8874

Hariharan PS et al (2017) Polymorphism and benzene solvent controlled stimuli responsive reversible fluorescence switching in triphenylphosphoniumfluorenylide crystals. New J Chem 41(20):4592–4598

Hirosawa R et al (2017) A variety of solid-state fluorescence properties of pyrazine dyes depending on terminal substituents. Dyes Pigm 146:576–581

Hirose A, Tanaka K, Yoshii R, Chujo Y (2015) Film-type chemosensors based on boron diminate polymers having oxidation-induced emission properties. Polym Chem 6(31):5590–5595

Huang R et al (2018) Constructing full-color highly emissive organic solids based on an X-shaped tetrasubstituted benzene skeleton. J Phys Chem C 122(19):10510–10518

Iida A, Yamaguchi S (2009) Intense solid-state blue emission with a small Stokes' shift: π-stacking protection of the diphenylanthracene skeleton. Chem Commun 21:3002–3004

Ito S, Hirose A, Yamaguchi M, Tanaka K, Chujo Y (2016) Size-discrimination for volatile organic compounds utilizing gallium diiminate by luminescent chromism of crystallization-induced emission via encapsulation-triggered crystal-crystal transition. J Mater Chem C 3(24):5564–5571

Jia J, Wu Y (2017) Alkyl length dependent reversible mechanofluorochromism of phenothiazine derivatives functionalized with formyl group. Dyes Pigm 147:537–543

Kajiwara Y, Nagai A, Tanaka K, Chujo Y (2013) Efficient simultaneous emission from RGB-emitting organoboron dyes incorporated into organic-inorganic hybrids and preparation of white light-emitting materials. J Mater Chem C 1(29):4437–4444

Kalluvettukuzhy NK, Thilagar P (2017) Bistable polyaromatic aminoboranes: bright solid state emission and mechanochromism. Organometallics 36(14):2692–2701

Kokado K, Chujo Y (2009) Emission via aggregation of alternating polymers with *o*-carborane and *p*-phenylene–ethynylene sequences. Macromolecules 42(5):1418–1420

Luo J, Tang BZ et al (2001) Aggregation-induced emission of 1-methyl-1,2,3,4,5-pentaphenylsilole. Chem Commun 18:1740–1741

Miyake J, Chujo Y (2008) Persistent and emission color tunable poly(phenylene–ethynylene)s covered with polyhedral oligomeric silsesquioxanes. J Polym Sci Part a: Polym Chem 46(24):8112–8116

Munch M et al (2019) Ethynyl-tolyl extended 2-(2'-hydroxyphenyl)benzoxazole dyes: solution and solid-state Excited-State Intramolecular Proton Transfer (ESIPT) emitters. Eur J Org Chem 2019:1134–1144

Naito H, Nishino K, Morisaki Y, Tanaka K, Chujo Y (2017a) Solid-state emission of the anthracene–*o*-carborane dyad from the twisted-intramolecular charge transfer in the crystalline state. Angew Chem Int Ed 56(1):254–259

Naito H, Nishino K, Morisaki Y, Tanaka K, Chujo Y (2017b) Highly-efficient solid-state emissions of anthracene-*o*-carborane dyads with various substituents and their thermochromic luminescence properties. J Mater Chem C 4(38):10047–10054

Naito H, Nishino K, Morisaki Y, Tanaka K, Chujo Y (2017c) Luminescence color tuning from blue to near infrared of stable luminescent solid materials based on bis-*o*-carborane-substituted oligoacenes. Chem Asian J 12(16):2134–2138

Namba T et al (2018) Rhodium-catalyzed cascade synthesis of benzofuranylmethylidene-benzoxasiloles: elucidating reaction mechanism and efficient solid-state fluorescence. Chem Eur J 24(28):7161–7171

Nie Y et al (2019) Highly efficient aggregation-induced emission and stimuli-responsive fluorochromism triggered by carborane-induced charge transfer state. Inorg Chem Commun 106:1–5

Nishino K, Yamamoto H, Ochi J, Tanaka K, Chujo Y (2019) Time-dependent emission enhancement of the etynylpyrene–o-carborane dyad and its application as a luminescent color sensor for evaluating water contents in organic solvents. Chem Asian J 14(9):1577–1581

Nishino K, Yamamoto H, Tanaka K, Chujo Y (2016) Development of solid-state emissive materials based on multi-functional *o*-carborane-pyrene dyads. Org Lett 18(16):4064–4067

Nishino K, Uemura K, Tanaka K, Chujo Y (2017a) Enhancement of aggregation-induced emission by introducing multiple *o*-carborane substitutions into triphenylamine. Molecules 22(11):2009–2018

Nishino K, Yamamoto H, Tanaka K, Chujo Y (2017b) Solid-state themochromic luminescence through twisted intramolecular charge transfer and excimer formation of a carborane–pyrene dyad with an ethynyl spacer. Asian J Org Chem 6(12):1818–1822

Ochi J, Tanaka CY (2019) Improvement of solid-state excimer emission of the aryl–ethynyl–o-carborane skeleton by acridine introduction. Eur J Org Chem 19:2984–2988

Ohsedo Y, Saruhashi K, Watanabe H (2015) A new family of light-emissive symmetric squarylium dyes in the solid state. Dyes Pigm 122:134–138

Ohtani S, Gon M, Tanaka K, Chujo Y (2017) A flexible fused azomethine-boron complex: thermochromic luminescence and thermosalient behavior in structural transitions between crystalline polymorphs. Chem Eur J 23(49):11827–11833

Okada H, Tanaka K, Chujo Y (2014a) Preparation of environmentally resistant conductive silica-based polymer hybrids containing tetrathiafulvalen-tetracyanoquinodimethane charge-transfer complexes. Polym J 46(11):800–805

Okada H, Tanaka K, Chujo Y (2014b) Regulation of responsiveness of phosphorescence toward dissolved oxygen concentration by modulating polymer contents in organic–inorganic hybrid materials. Bioorg Med Chem 22(12):3141–3145

Peng L et al (2019) Tuning emission wavelength of polymorphous crystal via controllable alkyl chain stacking and its vapor- and thermo-responsive fluorescence. Chem Eur J 25(34):8043–8052

Sato T, Jiang D-L, Aida T (1999) A blue-luminescent dendritic rod: poly(phenyleneethynylene) within a light-harvesting dendritic envelope. J Am Chem Soc 121(45):10658–10659

Scott JC, Klalrner G, Miller RD, Miller DC (1999) Electrical and Photoinduced Degradation of Polyfluorene Based Films and Light-Emitting Devices. Macromolecules 32(2):361–369

Shaikh AC et al (2015) Highly emissive organic solids with remarkably broad color tunability based on *N*, C-chelate, four-coordinate organoborons. Chem Commun 51(89):16115–16118

Shao B, Qian H, Li Q, Aprahamian I (2019) Structure property analysis of the solution and solid-state properties of bistable photochromic hydrazones. J Am Chem Soc 141(20):8364–8371

Shimizu S et al (2015) Benzo [*c*, d] indole-containing Aza-BODIPY dyes: asymmetrization-induced solid-state emission and aggregation-induced emission enhancement as new properties of a well-known chromophore. Chem Eur J 21(37):12996–13003

Shimizu M, Hiyama T (2010) Organic fluorophores exhibiting highly efficient photoluminescence in the solid state. Chem Asian J 5(7):1516–1531

Shimizu M, Ryuse D, Kinoshita T (2017) Germanium-bridged 2-phenylbenzoheteroles as luminophores exhibiting highly efficient solid-state fluorescence. Chem Eur J 23(58):14623–14630

Srujana P, Radhakrishnan TP (2018) Establishing the critical role of oriented aggregation in molecular solid state fluorescence enhancement. Chem Eur J 24(8):1784–1788

Suenaga K, Tanaka K, Chujo Y (2017) Heat-resistant mechanoluminescent chromism of the hybrid molecule based on boron ketoiminate modified octa-substituted polyhedral oligomeric silsesquioxane. Chem Eur J 23(6):1409–1414

Suenaga K, Tanaka K (2017) Chujo Y (2017) Design and luminescent chromism of fused boron complexes having constant emission efficiencies in solution and in the amorphous and crystalline states. Eur J Org Chem 35:5191–5196

Sun Y et al (2018) Efficient solution- and solid-state fluorescence for a series of 7-diethylaminocoumarin amide compounds. Asian J Org Chem 7(1):197–202

Sun Q et al (2019) Subtly manipulating the end group structures of DPP-centered dyes for the diverse aggregate fluorescence and stimuli-responsive behaviors. Dyes Pigm 165:193–199

Takayanagi N et al (2018) Interpretation of absorption spectra of some bisazomethine dyes in a crystalline state in terms of conformational change and exciton interaction. Bull Chem Soc Jpn 91(10):1498–1505

Tanaka K, Chujo Y (2013b) Unique properties of amphiphilic POSS and their applications. Polym J 45(3):247–254

Tanaka K, Chujo Y (2015) Recent progress of optical functional nanomaterials based on organoboron complexes with β-diketonate. Ketoiminate and Diiminate. NPG Asia Mater 7:e223

Tanaka K, Chujo Y et al (2017) Development of solid-state emissive *o*-carborane and theoretical investigation of the mechanism of the aggregation-induced emission behaviors of organoboron "element-blocks." Faraday Discuss 196:31–42

Tanaka K, Adachi S, Chujo Y (2009) Structure-property relationship of octa-substituted POSS in thermal and mechanical reinforcements of conventional polymers. J Polym Sci Part a: Polym Chem 47(21):5690–5697

Tanaka K, Adachi S, Chujo Y (2010) Side-chain effect of octa-substituted POSS fillers on refraction in polymer composites. J Polym Sci Part a: Polym Chem 48(24):5712–5717

Tanaka K, Chujo Y (2012a) Advanced functional materials based on polyhedral oligomeric silsesquioxane (POSS). J Mater Chem 22(5):1733–1746

Tanaka K, Chujo Y (2012b) Advanced Luminescent Materials Based on Organoboron Polymers. Macromol Rapid Commun 33(15):1235–1255

Tanaka K, Tamashima K, Nagai A, Okawa T, Chujo Y (2013) Facile modulation of optical properties of diketonate-containing polymers by regulating complexation ratios with boron. Macromolecules 46(8):2969–2975

Tanaka K, Chujo Y (2013a) Chemicals-inspired biomaterials; developing biomaterials inspired by material science based on POSS. Bull Chem Soc Jpn 86(11):1231–1239

Tanaka K, Kozuka H, Ueda K, Jeon J-H, Chujo Y (2017) POSS-based molecular fillers for simultaneously enhancing thermal and viscoelasticity of poly(methyl methacrylate) films. Mater Lett 203:62–67

Tang BZ et al (2015) Aggregation-induced emission: together we shine, united we soar! Chem Rev 115(21):11718–11940

Ueda K, Tanaka K, Chujo Y (2016) Remarkably high miscibility of octa-substituted poss with commodity conjugated polymers and molecular fillers for the improvement of homogeneities of polymer matrices. Polym J 48(12):1133–1139

Ueda K, Tanaka K, Chujo Y (2017) Synthesis of POSS derivatives having dual types of alkyl substituents and their application as a molecular filler for low-refractive and highly durable materials. Bull Chem Soc Jpn 90(2):205–209

Ueda K, Tanaka K, Chujo Y (2018) Fluoroalkyl POSS with dual functional groups as a molecular filler for lowering refractive indices and improving thermomechanical properties of PMMA. Polymers 10(12):1332

Ueda K, Tanaka K, Chujo Y (2019) Optical, electrical and thermal properties of organic−inorganic hybrids with conjugated polymers based on poss having heterogeneous substituents. Polymers 11(1):44

Varughese S (2014) Non-covalent routes to tune the optical properties of molecular materials. J Mater Chem C 2(18):3499–3516

Wakamiya A, Mori K, Yamaguchi S (2007) 3-Boryl-2,2'-bithiophene as a versatile core skeleton for full-color highly emissive organic solids. Angew Chem Int Ed 46(23):4273–4276

Wang C et al (2015) Charge-transfer emission in organoboron-based biphenyls: effect of substitution position and conformation. J Org Chem 80(21):10914–10924

Wang Y et al (2018) Twisted donor−acceptor cruciform luminophores possessing substituent-dependent properties of aggregation-induced emission and mechanofluorochromism. J Phys Chem C 122(4):2297–2306

Weber L et al (2012) Luminescence properties of *c*-diazaborolyl-ortho-carboranes as donor-acceptor systems. Chem Eur J 18(27):8347–8357

Xu D et al (2018) Twisted donor–acceptor cruciform fluorophores exhibiting strong solid emission, efficient aggregation-induced emission and high contrast mechanofluorochromism. Dyes Pigm 150:293–300

Xue S, Qiu X, Sun Q, Yang W (2016) Alkyl length effects on solid-state fluorescence and mechanochromic behavior of small organic luminophores. J Mater Chem C 4(8):1568–1578

Yamaguchi M, Ito S, Hirose A, Tanaka K, Chujo Y (2017) Control of aggregation-induced emission versus fluorescence aggregation-caused quenching by bond existence at a single site in boron pyridinoiminate complexes. Mater Chem Front 1(8):1573–1579

Yamane H, Tanaka K, Chujo Y (2015) Simple and valid strategy for the enhancement of the solid-emissive property based on boron dipyrromethene. Tetrahedron Lett 56(48):6786–6790

Yan W et al (2014) Synthesis, crystal structures and photophysical properties of novel boron-containing derivatives of phenalene with bright solid-state luminescence. Dyes Pigm 106:197–204

Yan H et al (2017) Design, synthesis and aggregation induced emission properties of two bichromophores with a triphenylamine-coumarin dyad structure. Dyes Pigm 146:479–490

Yeo H, Tanaka K, Chujo Y (2013) Effective light-harvesting antennae based on bodipy-tethered cardo polyfluorenes via rapid energy transferring and low concentration quenching. Macromolecules 46(7):2599–2605

Yeo H, Tanaka K, Chujo Y (2015b) Synthesis of dual-emissive polymers based on ineffective energy transfer through cardo fluorene-containing conjugated polymers. Polymer 60:228–233

Yeo H, Tanaka K, Chujo Y (2015a) Energy transfer through heterogeneous dyes-substituted fluorene-containing alternating copolymers and their dual-emission properties in the films. J Polym Sci Part a: Polym Chem 53(17):2026–2035

Yoshii R, Nagai A, Tanaka K, Chujo Y (2014b) Boron ketoiminate-based polymers: fine-tuning of the emission color and expression of strong emission both in the solution and film state. Macromol Rapid Commun 35(15):1315–1319

Yoshii R, Suenaga K, Tanaka K, Chujo Y (2015) Mechanofluorochromic Materials Based on Aggregation-Induced Emission-Active Boron Ketoiminates: Regulation of the Direction of the Emission Color Changes. Chem Eur J 21(19):7231–7237

Yoshii R, Nagai A, Tanaka K, Chujo Y (2013) Highly emissive boron ketoiminate derivatives as new class of aggregation-induced emission fluorophores. Chem Eur J 19(14):4506–4512

Yoshii R, Tanaka K, Chujo Y (2014a) Conjugated polymers based on tautomeric units: regulation of main-chain conjugation and expression of aggregation induced emission property via boron-complexation. Macromolecules 47(7):2268–2278

Yoshii R, Hirose A, Tanaka K, Chujo Y (2014c) Boron diiminate with aggregation-induced emission and crystallization-induced emission enhancement characteristics. Chem. Eur. J. 20(27):8320–8324

Yoshii R, Hirose A, Tanaka K, Chujo Y (2014d) Functionalization of boron diiminates with unique optical properties: multicolor tuning of crystallization-induced emission and introduction into the main-chain of conjugated polymers. J Am Chem Soc 136(52):18131–18139

Yu H-W, Kim B-S, Matsumoto S (2017) Effect of alkoxy side chain length on the solid-state fluorescence behaviour of bisazomethine dyes possessing dipropylamino terminal group. Dyes Pigm 136:131–139

Zhang M-Y et al (2018) Solid-state emissive triarylborane-based [2.2] paracyclophanes displaying circularly polarized luminescence and thermally activated delayed fluorescence. Org Lett 20(15):6868−6871

Zhao C-H et al (2006) Highly emissive organic solids containing 2,5-diboryl-1,4-phenylene unit. J Am Chem Soc 128(50):15934–15935

Zhao C-H, Wakamiya A, Yamaguchi S (2007) Highly emissive poly(aryleneethynylene)s containing 2,5-diboryl-1,4-phenylene as a building unit. Macromolecules 40(11):3898–3900

Zhu H et al (2018b) Twisted donor−π−acceptor carbazole luminophores with substituent-dependent properties of aggregated behavior (aggregation-caused quenching to aggregation-enhanced emission) and mechanoresponsive luminescence. J Phys Chem C 122(34):19793–19800

Zhu H et al (2018a) Branched triphenylamine luminophores: aggregation-induced fluorescence emission, and tunable near-infrared solid-state fluorescence characteristics via external mechanical stimuli. Dyes Pigm 151:140–148

Chapter 10
Circularly Polarized Luminescence (CPL) Based on Planar Chiral [2.2]Paracyclophane

Yasuhiro Morisaki

Abstract In this chapter, chiroptical properties, especially circularly polarized luminescence (CPL) properties of optically active molecules based on planar chiral [2.2]paracyclophane are mainly introduced. In addition, practical optical resolution methods of disubstituted and tetrasubstituted [2.2]paracyclophane molecules are also focused on. The enantiopure [2.2]paracyclophane compounds have been used as chiral building blocks to synthesize the optically active molecules by means of optical resolution. The [2.2]paracyclophane-based molecules are π-stacked molecules, which construct optically active second-ordered structures, such as V-, X-, triangle-shaped, and one-handed double helical structures, due to the orientation of stacked π-electron systems. Photoexcitation allows them to emit bright CPL with good photoluminescence (PL) quantum efficiencies and large dissymmetry factors (g_{lum} values). Thus, planar chiral [2.2]paracyclophane is the ideal scaffold to achieve excellent CPL properties.

Keywords Circularly polarized luminescence · [2.2]Paracyclophane · Planar chirality

10.1 Introduction: Circularly Polarized Luminescence (CPL)

Circularly polarized luminescence (CPL) is a chiroptical (chiral + optical) property that is responsible for the difference in luminescence intensity between left-handed and right-handed emissions. Circular dichroism (CD) is another chiroptical property that is responsible for the difference in molar extinction coefficients between left- and right-polarized light. CPL and CD spectroscopy provide important information about the orientation of luminophores and chromophores, respectively (Berova et al. 2000). Many optically active molecules exhibit CPL and CD in the excited and

Y. Morisaki (✉)
School of Biological and Environmental Sciences, Kwansei Gakuin University, Sanda, Japan
e-mail: ymo@kwansei.ac.jp

Y. Ooyama and S. Yagi (eds.), *Progress in the Science of Functional Dyes*,
https://doi.org/10.1007/978-981-33-4392-4_10

ground states, respectively, but these phenomena cannot be always observed even if the molecules possess a chiral source.

CPL is evaluated by the anisotropic factor (dissymmetric factor), referred to as g_{lum} value (by Berova et al. 2000; Riehl and Richardson 1986; Riehl and Muller 2012), which can be calculated as the following Eq. (10.1):

$$g_{lum} = \Delta I/I \tag{10.1}$$

ΔI = (emission intensity of left-handed CPL) − (emission intensity of right-handed CPL) and I = emission intensity

The g_{lum} value can also be expressed as

$$g_{lum} = 4|\boldsymbol{\mu}||\mathbf{m}|\cos\theta/\left(|\boldsymbol{\mu}|^2|\mathbf{m}|^2\right) \tag{10.2}$$

Here, $\boldsymbol{\mu}$ and **m** represent electric and magnetic transition dipole moments, respectively, and the θ represents the angle between them (Berova et al. 2000; Riehl and Richardson 1986; Riehl and Muller 2012). Thus, the maximum absolute g_{lum} value is calculated to be $|g_{lum}| = +2$.

Ideally, forbidden electric transitions and allowed magnetic transitions are preferred. Generally, the g_{lum} values of chiral lanthanide complexes are much larger than those of chiral organic molecules. The emission of lanthanides is derived from the Laporte forbidden f-f transition (Muller 2014); therefore, the $\boldsymbol{\mu}$ is small and **m** is large. It is reported that the chiral Eu(III) complex, tetrakis(3-heptafluoro-butylryl-(+)-camphorato) Eu(III), has a g_{lum} value of +1.38 (Lunkley et al. 2008).

Many of g_{lum} values of organic molecules reported are within the range 10^{-3}–10^{-5} order due to the much smaller **m**. Therefore, Eq. (10.2) can be simply converted to the approximated Eq. (10.3).

$$g_{lum} = 4|\mathbf{m}|\cos\theta/|\boldsymbol{\mu}| \tag{10.3}$$

Considering the photoluminescence quantum efficiency (Φ_{PL}) and emission intensity, an allowed electronic transition is preferable. It is a challenging task to achieve a large **m** and small $\boldsymbol{\mu}$, and their linear orientation (their angle $\theta = 0°$ or $180°$) in the organic molecules. Organic CPL dyes are attractive luminescent materials due to their light-weight, as well as the ease of fabrication, functional group modification, and emission color control. These dyes are expected to be promising candidates for next-generation emissive materials such as light-emitting diodes for three-dimensional displays, security inks, light for plant growth, bioimaging materials, etc.

A wide variety of organic CPL emitters have been reported thus far, and the last five years have seen a marked increase in the number of related manuscripts because of the widespread use of CPL spectrometers (Maeda and Bando 2013; Sánchez-Carnerero et al. 2015; Tanaka et al. 2018; Chem and Yan 2018). The CPL-emitting small organic molecules in the dispersed state are briefly introduced below. In 1967, CPL from organic molecules was reported for cyclic ketones with a central chirality (Emeris and Oosterhoff 1967); a representative example is shown in Fig. 10.1. Because of the forbidden n-π* transition, the molecule emitted CPL with a very large g_{lum} value of $+3.5 \times 10^{-2}$. This value was long considered the champion value for a long time in the small organic molecules in a diluted solution, although the Φ_{PL} was quite low ($\Phi_{\text{PL}} = 1 \times 10^{-5}$). Various CPL emitters with the central chirality have been reported to show large g_{lum} values. As the representative examples, naphthyl (Amako et al. 2014) and naphthalimide (Sheng et al. 2016) emissive units were connected on chiral cyclic scaffolds, which showed CPL emission from the intramolecular aggregate (excimer) of the aromatic units with $|g_{\text{lum}}|$ values of 9.4×10^{-3} ($\Phi_{\text{PL}} = 0.02$) and 1.4×10^{-2}, respectively. In 2017, the champion g_{lum} value was updated for the cylindrical molecule **4** with cylinder helicity (Fig. 10.2), and it was on the order of 10^{-1} ($|g_{\text{lum}}| = 1.5 \times 10^{-1}$ and $\Phi_{\text{PL}} = 0.80$) (Sato et al. 2017). Theoretical studies revealed a large **m**, as well as linearly oriented **m** and $\boldsymbol{\mu}$ ($\theta = 180°$).

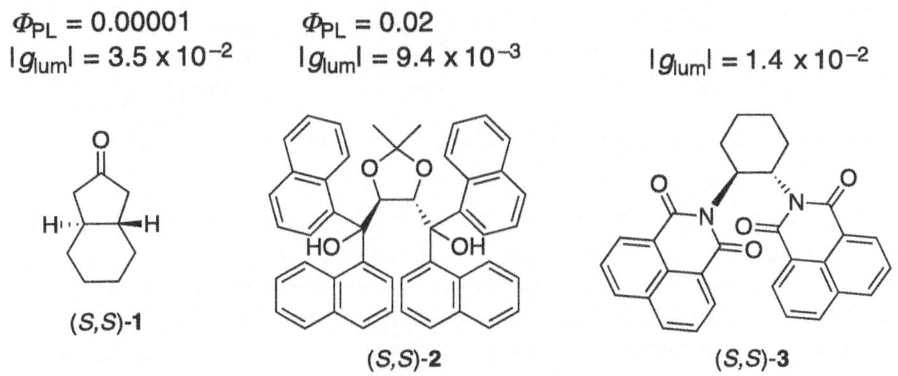

Fig. 10.1 Representative examples of CPL emitters with central chirality

Fig. 10.2 Cylindrical molecules with an excellent CPL profile

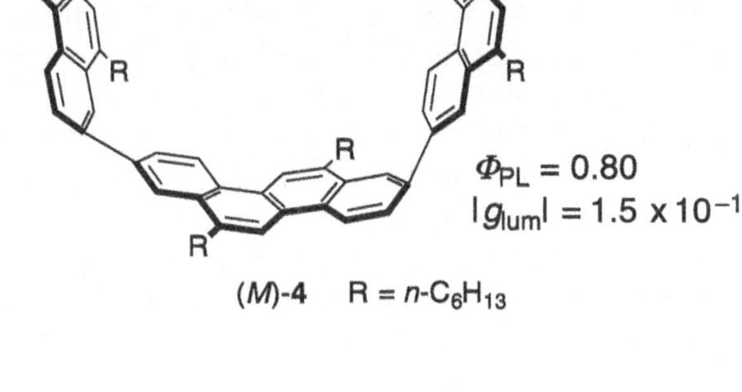

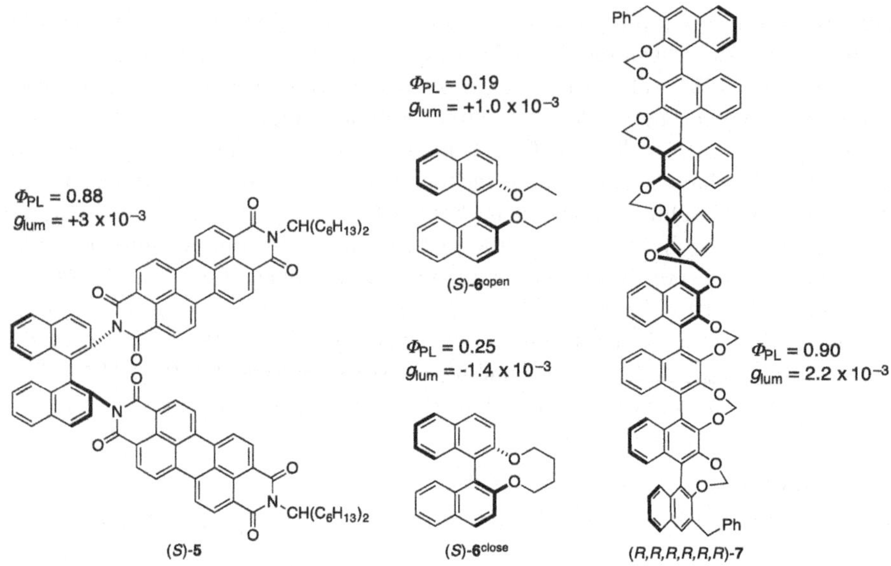

Fig. 10.3 Representative examples of oligoaryl-based CPL emitters with axial chirality

Biaryl-based axial chirality is an easily accessible chiral scaffold for CPL emission. Figure 10.3 shows the binaphthyl-based molecules. In 2007, perylenebiscarboxydiimide-containing binaphtylene (*S*)-**5** was prepared (Kawai et al. 2007). This compound showed intense emission with Φ_{PL} of 0.88, and the g_{lum} was estimated to be approximately 3×10^{-3}. Simple biaryl-based molecules (open- and closed-type molecules (*S*)-**6**^{open} and (*S*)-**6**^{close}, respectively) were synthesized (Kimoto et al. 2012). Despite their similar photoluminescence behavior and identical absolute configuration, these molecules showed different CPL signals. The values for (*S*)-**6**^{open} and (*S*)-**6**^{close} were calculated to be $+1.0 \times 10^{-3}$ ($\Phi_{PL} = 0.19$) and -1.4×10^{-3} ($\Phi_{PL} = 0.25$), respectively. Optically active naphthalene oligomers were synthesized; the representative molecule (*R,R,R,R,R,R,R*)-**7** consisting of eight naphthyl moieties is also shown in Fig. 10.3 (Takaishi et al. 2017). This molecule showed bright CPL emission with $g_{lum} = +2.2 \times 10^{-3}$ and $\Phi_{PL} = 0.90$. Compound (*S*)-**5** emitted from the perylenediimide moieties, whereas (*S*)-**6** and **7** showed emission from the oligoaryl scaffolds.

Helical chirality is also employed for CPL emission. Generally, Φ_{PL} of a helical molecule is not very high; however, its g_{lum} value is very large, often on the order of 10^{-2}. Figure 10.4 shows the helicene derivatives (*M*)-**8** (Field et al. 2003) and (*M*)-**9** (Sawada et al. 2012) as representative examples. Molecule **8** is the first helicene emitting CPL in dilute solution. This molecule was obtained as a diastereomer due to the attachment of a camphanate unit as the chiral auxiliary. The $|g_{lum}|$ value was calculated to be 1.1×10^{-3}. A very large $|g_{lum}|$ value of 3.2×10^{-2} ($\Phi_{PL} = 0.30$) was obtained from molecule **9**, which was prepared by the Rh-catalyzed enantioselective [2 + 2 + 2] cycloaddition. Molecule **10** is a stimuli-responsive compound used for

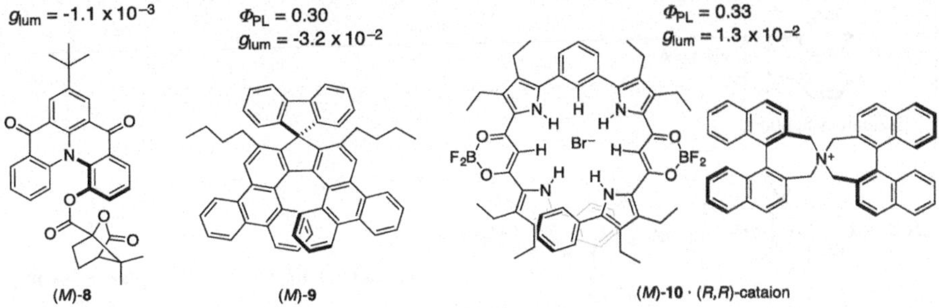

Fig. 10.4 Representative examples of helical CPL emitters

capturing halogen anions to form a helix (Haketa et al. 2012). The chiral counter cation is allowed for the introduction of one-handed helicity in dilute solution. The resulting ion pair was emissive, and the $|g_{lum}|$ value was calculated to be 1.3×10^{-2} with Φ_{PL} of 0.33.

Recently, planar chirality has also been applied to CPL emission. However, there are fewer CPL emitters based on planar chirality as compared to those based on axial and helical chiralities. Unique optically active molecules have been prepared by taking advantage of the planar chirality of a [2.2]paracyclophane scaffold, and large g_{lum} values, as well as high Φ_{PL} were achieved (Morisaki and Chujo 2019). Various interesting second-ordered structures were also constructed. This chapter focuses on planar chirality and CPL emission in the mono-dispersed state from the optically active small organic molecules based on the planar chiral [2.2]paracyclophanes.

10.2 Introduction: [2.2]Paracyclophane and Its Planar Chirality

Cyclophane is a cyclic compound containing at least one aromatic ring in the main chain skeleton. It is a well-known cyclic compound and have been used particularly in the field of synthetic organic chemistry and organometallic chemistry (Vögtle 1993; Gleiter and Hopf 2004). The typical cyclophane is [2.2]paracyclophane (Fig. 10.5), in which two phenylenes are stacked in proximity (the distance between two benzene rings is approximately 3.0 Å). [2.2]Paracyclophane was first synthesized in 1949, by pyrolysis of *para*-xylene (Brown and Farthing 1949); subsequently, a direct synthesis from 1,4-bis-bromomethylbenzene by Wurtz-type intramolecular cyclization was reported in 1951 (Cram and Steinberg 1951). Since then, various synthetic routes have been developed to prepare a wide variety of [2.2]paracyclophane molecules. [2.2]Paracyclophane has a unique π-stacked structure, and there have been many studies on its physical properties within the field of organic chemistry (Gleiter and Hopf 2004; Brown and Farthing 1949).

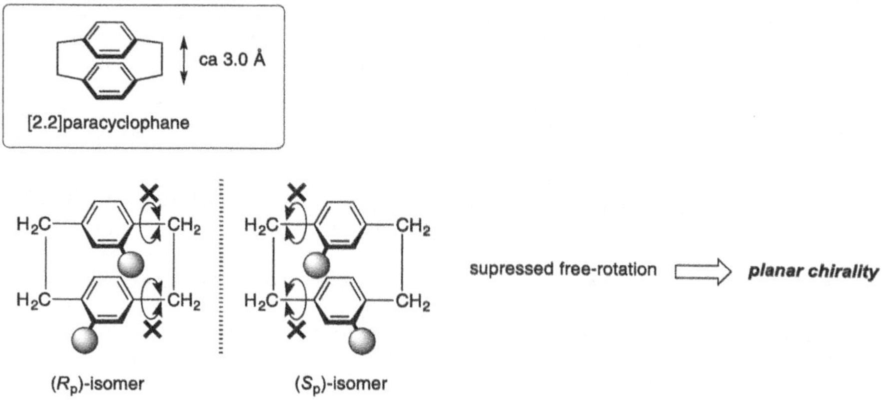

Fig. 10.5 Structure of [2.2]paracyclophane and the corresponding planar chirality

Although [2.2]paracyclophane is a molecule that has generally been used in the field of organic chemistry and organometallic chemistry, its utilization is not widespread in the fields of polymer chemistry and materials chemistry (Hopf 2008; Morisaki and Chujo 2006, 2008, 2009, 2011, 2012; Mizogami and Yoshimura 1985; Guyard and Audebert 2001; Guyard et al. 2002; Salhi et al. 2002; Salhi and Collard 2003; Jagtap and Collard 2010; Weiland et al. 2019). In terms of the through-space conjugated system, [2.2]paracyclophane-based π-stacked molecules have been systematically prepared, and their electronic communication between the stacked π-electron systems have been known since 1998 (Oldham et al. 1998; Bazan et al. 1998; Bartholomew and Bazan 2001; Bazan 2007). In 2001, thiophene-substituted [2.2]paracyclophanes were polymerized electrochemically; however, the polymer was deposited on the electrode and was insoluble in solvents (Guyard and Audebert 2001). In 2002, soluble π-stacked polymers consisting of [2.2]paracyclophane as a repeating unit in the main chain were synthesized (Morisaki and Chujo 2002). Since then, various π-stacked polymers have been prepared (Morisaki and Chujo 2006, 2008, 2009, 2011, 2012). The structures have been well-characterized, and the properties, such as optical properties, have been shown due to their solubility in organic solvents. Since their optical properties changed continuously depending on the number of the stacked π-electron systems, they were called "through-space conjugated polymers". Highly efficient unidirectional fluorescence resonance energy transfer (FRET) (Morisaki et al. 2013, 2014a, 2014b, 2017), as well as through-space electron transfer (Molina-Ontoria et al. 2011; Wielopolski et al. 2013) was achieved by precisely designing the stacked π-electron systems. [2.2]Paracyclophane-based through-space conjugated polymers and oligomers can act as single molecular wires.

As described above, [2.2]paracyclophane consists of two phenylene units fixed in proximity; therefore, the rotational movement of the benzene rings is suppressed. By introducing a substituent at appropriate positions on the benzene ring(s), the corresponding [2.2]paracyclophane becomes a planar chiral molecule (Fig. 10.5) (Cram and Allinger 1955; Rozenberg et al. 2004; Rowlands 2008; Gibson and Knight 2003;

Aly and Brown 2009; Paradies 2011). The planar chirality of [2.2]paracyclophane is well-known in the fields of organic chemistry and organometallic chemistry, and planar chiral [2.2]paracyclophanes have been utilized as chiral auxiliaries and chiral ligands. However, until recently, the planar chirality of [2.2]paracyclophane has been ignored in the fields of polymer chemistry and materials chemistry until recently. In 2012, a new optical resolution method was developed for pseudo-*ortho*-disubstituted [2.2]paracyclophane, and the transformation and polymerization of the enantiopure [2.2]paracyclophanes were reported (Morisaki et al. 2012a, b). The resulting optically active polymer emitted CPL (Morisaki et al. 2012).

This chapter highlights and introduces the recent results on the synthesis of enantiopure disubstituted and tetrasubstituted [2.2]paracyclophanes. The preparation of optically active molecules based on the [2.2]paracyclophanes for their application in the fields of polymer and materials chemistry as the CPL emitters is also described.

10.3 Synthesis of Enantiopure Disubstituted [2.2]Paracyclophane and Optically Active π-Stacked Molecules

Optical resolution routes of mono-substituted [2.2]paracyclophanes were developed, and various enantiopure *ortho*-, pseudo-*geminal*-, and *syn-latero*-disubstituted [2.2]paracyclophanes were prepared (Cram and Allinger 1955; Rozenberg et al. 2004; Rowlands 2008; Gibson and Knight 2003; Aly and Brown 2009; Paradies 2011). In addition, several synthetic routes to the syntheses of enantiopure disubstituted [2.2]paracyclophanes, e.g., pseudo-*ortho*-disubstituted [2.2]paracyclophanes, have been reported (Pye et al. 1997; Rossen et al. 1997; Zhuravsky et al. 2008; Jiang and Zhao 2004; Jones et al. 2003; Pamperin et al. 1997; Pamperin et al. 1998; Braddock et al. 2002), as shown in Fig. 10.6. A representative example is the synthesis of enantiopure pseudo-*ortho*-bis(diarylphosphino)[2.2]paracyclophane (Ph-PHANEPHOS) (Fig. 10.6a) (Pye et al. 1997). The resulting (S_p)- and (R_p)-Ph-PHANEPHOS are the commercially available chiral ligands for the transition metal-catalyzed asymmetric reactions. This PHANEPHOS makes it possible to produce enantioenriched pseudo-*ortho*-dibromo[2.2]paracyclophane (Fig. 10.6b) by kinetic resolution (Rossen et al. 1997). Synthesis of optically active 4-bromo-12-hydroxy[2.2]paracyclophane (Zhuravsky et al. 2008), pseudo-*ortho*-dihydroxy[2.2]paracyclophane (PHANOL) (Jiang and Zhao 2004), and pseudo-*ortho*-dihydroxymethyl[2.2]paracyclophane (Jones et al. 2003) were successfully synthesized by using chiral camphanic acid chloride as the chiral auxiliary; for example, Fig. 10.6c shows the optical resolution of PHANOL racemate. The enzyme-promoted kinetic resolutions of pseudo-*ortho*-disubstituted [2.2]paracyclophanes have also been developed (Pamperin et al. 1997, 1998; Braddock et al. 2002).

A practical route to the optical resolution of pseudo-*ortho*-dibromo[2.2]paracyclophane was reported in 2012, in which (1*R*,2*S*,5*R*)-(-)-menthyl

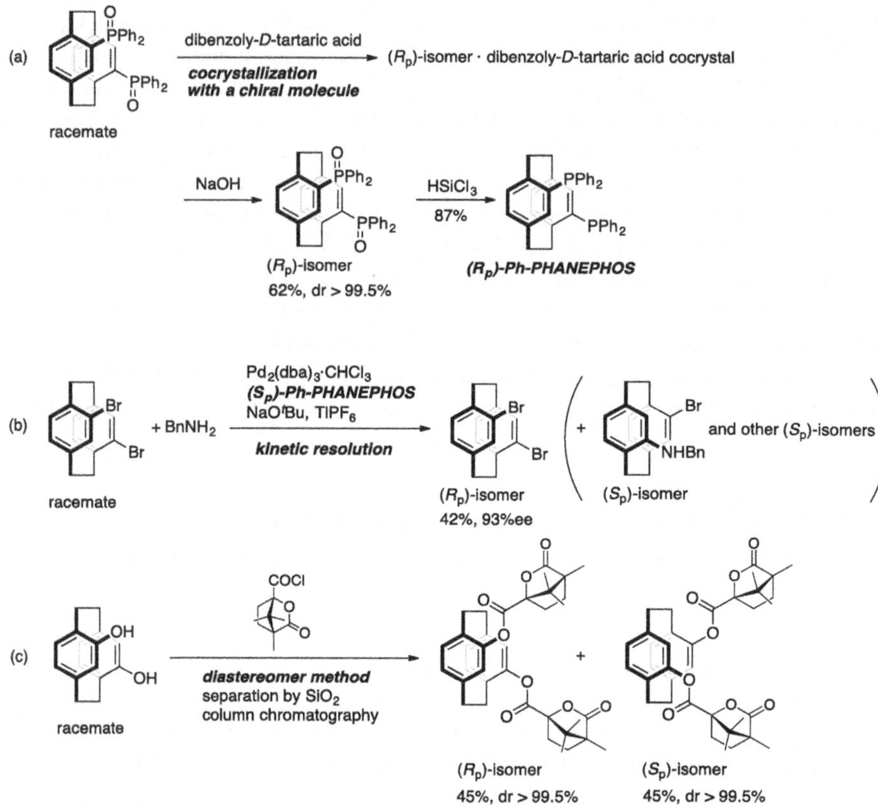

Fig. 10.6 Examples of optical resolution of pseudo-*ortho*-disubstituted [2.2]paracyclophanes

(*S*)-*p*-toluenesulfinate was used as the chiral auxiliary (Fig. 10.7) (Morisaki et al. 2012). Racemic pseudo-*ortho*-dibromo[2.2]paracyclophane *rac*-**11** was reacted with *n*-butyllithium (*n*-BuLi). Then, the reaction with (1*R*,2*S*,5*R*)-(-)-menthyl-*p*-toluenesulfinate afforded the diastereomers (R_p,*S*)- and (S_p,*S*)-**12**, which could be separated by using SiO_2 column chromatography. The isolated diastereomers were reacted with *t*-BuLi to form dilithiated intermediate **13**; not only the lithium-halogen exchange reaction, but also the lithium-sulfur exchange reaction occurred (Rowlands 2008; Clayden 2002; Hitchcock et al. 2005; Parmar et al. 2010). The subsequent reaction with various electrophiles produced enantiopure pseudo-*ortho*-disubstituted [2.2]paracyclophanes (Fig. 10.7). (R_p)- and (S_p)-Diformyl[2.2]paracyclophanes **14** could be separated by chiral column chromatography; thus, their chromatographic optical resolution was also possible (Morisaki et al. 2012). This synthetic route can be used to produce new PHANEPHOS; for example, cyclohexyl groups could be introduced on phosphorus atoms to produce Cy-PHANEPHOS **16** (Fig. 10.7).

Pseudo-*ortho*-diformyl[2.2]paracyclophane **14** could be converted to the corresponding diethynyl[2.2]paracyclophane **17** using the Ohira-Bestmann reagent

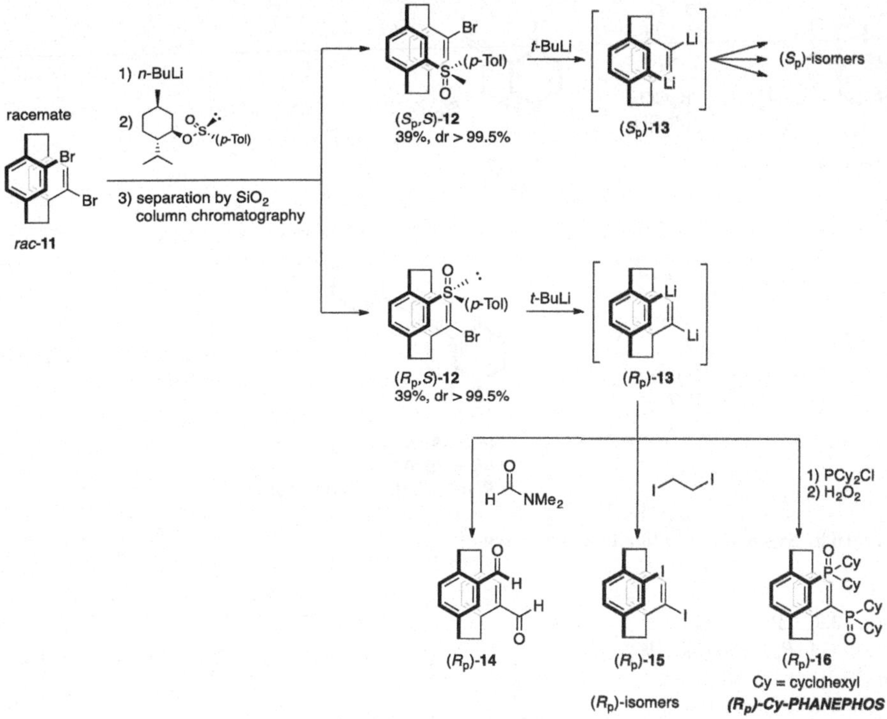

Fig. 10.7 Optical resolution of pseudo-*ortho*-disubstituted [2.2]paracyclophane and transformations

(Fig. 10.8) (Ohira 1989; Müller et al. 1996), which was used as a monomer to synthesize poly(*p*-arylene-ethynylene)s (poly-PAEs) via Sonogashira-Hagihara coupling (Tohda et al. 1975; Sonogashira 2002) polymerization. The treatment of **17** with 1,4-diiodobenzene derivative **18** affords the optically active through-space conjugated polymers (R_p)- and (S_p)-**19** (Morisaki et al. 2012). Chiroptical properties (chiral optical properties) of **9** showed the intense CPL in the emission region (Fig. 10.9). In addition, its absolute photoluminescence (PL) quantum efficiency (Φ_{PL}) was approximately 80% and the $|g_{lum}|$ value was 2.2×10^{-3} in dilute $CHCl_3$ solution. These results indicated that the optically active polymer **19** is a good CPL emitter.

The chiroptical properties of PAE oligomers were investigated (Morisaki et al. 2014). As shown in Fig. 10.10, the optically active PAE-type π-stacked dimer (R_p)-**20**, trimer (R_p)-**21**, tetramer (R_p)-**22**, and cyclic trimer (R_p)-**23** were prepared from (R_p)-**15** (shown in Fig. 10.7) and (R_p)-**17** as the key chiral building blocks.

Figure 10.10 includes the optical and chiroptical data. The absolute anisotropic factors of absorbance (g_{abs} values = [(molar absorption coefficient of left-handed circular polarized light) – (molar absorption coefficient of right-handed circular polarized light)]/(molar absorption coefficient) at $\lambda_{abs,max}$) were calculated from the

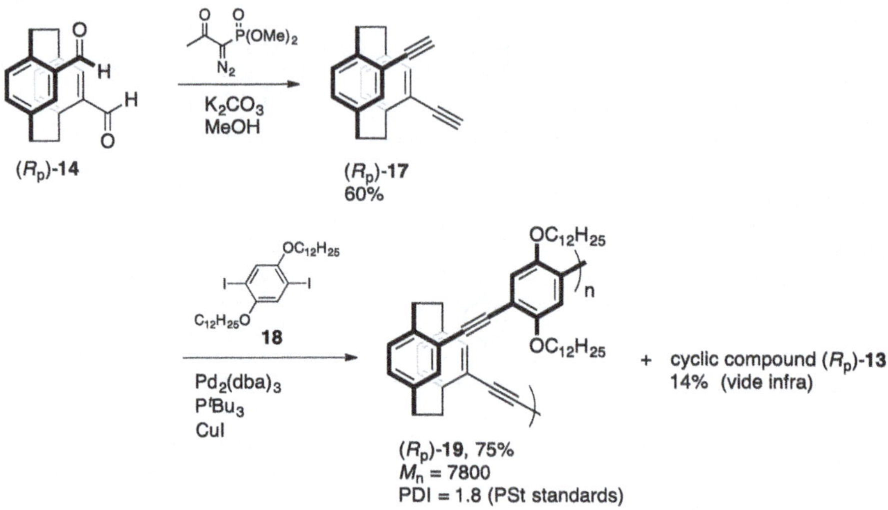

Fig. 10.8 Synthesis of optically active polymer

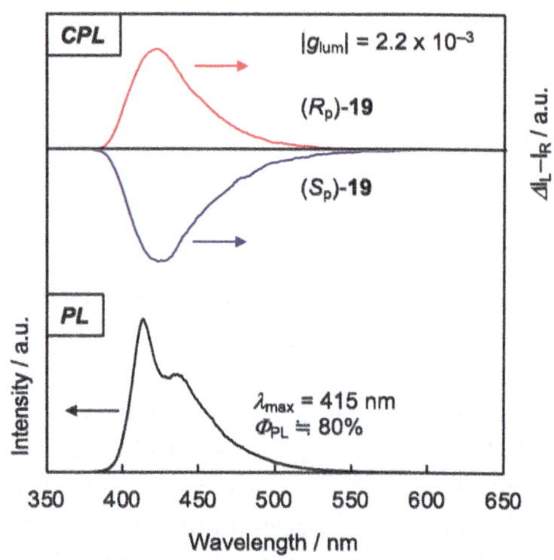

Fig. 10.9 CPL and PL spectra of (R_p)- and (S_p)-**19** in $CHCl_3$ (10×10^{-5} M) excited at 290 nm for CPL and $\lambda_{abs,max}$ for PL

CD spectra of the optically active oligomers, and the $|g_{lum}|$ values at $\lambda_{PL,max}$ calculated from CPL spectra. The $|g_{abs}|$ values of **20–22** were substantially constant at 3.0 $\times$ 10^{-3}, regardless of the number of stacked π-electron systems. Linearly π-stacked structures of **20–22** in the ground state can adopt various conformations in dilute solution, leading to the constant g_{abs} values. In other words, the constant g_{abs} values reflect only the chirality of optically active V-shaped skeleton of the cyclophane moiety. A g_{abs} value is a normalized value; therefore, the g_{abs} values of **20–22** were constant independent of the number of stacked π-electron systems.

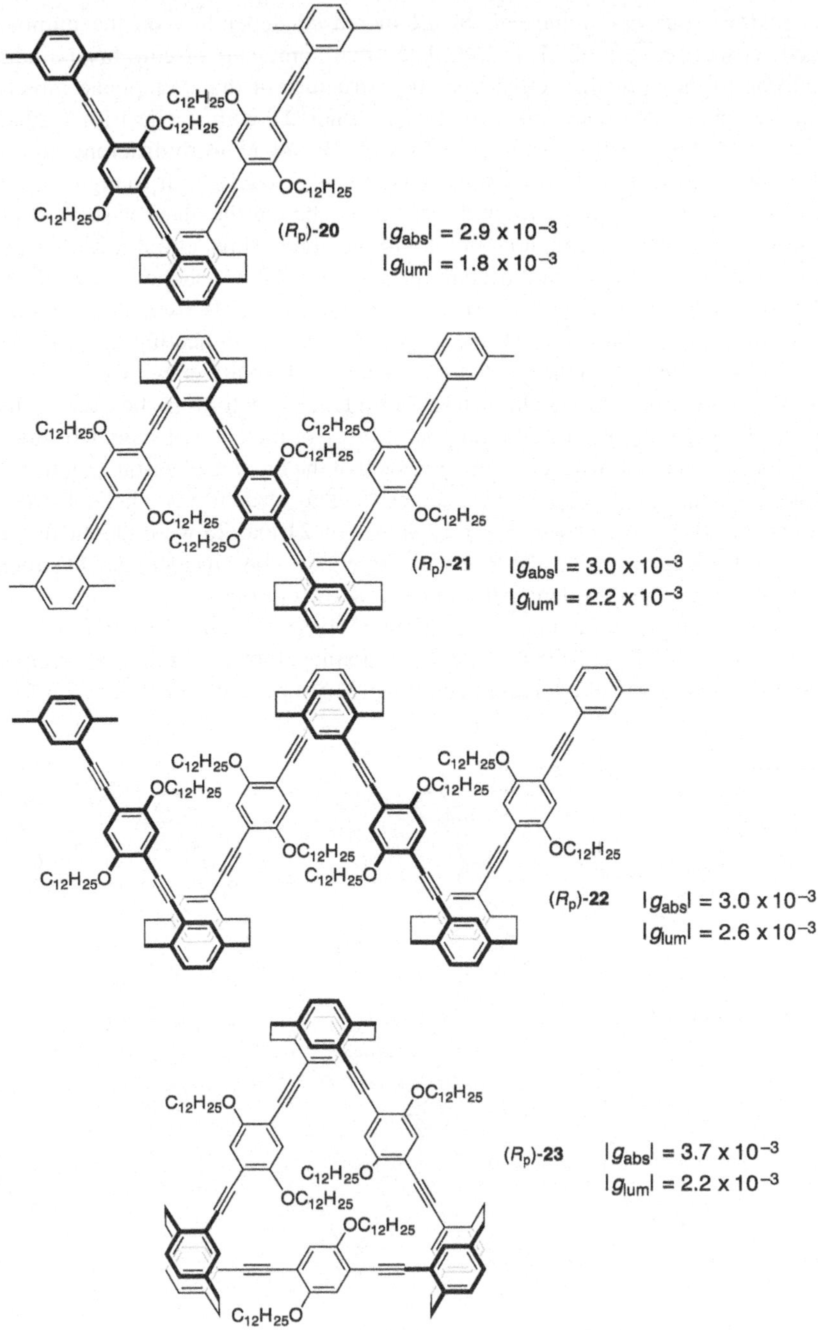

Fig. 10.10 Optically active PAE-type π-stacked oligomers (R_p)-**20–22** and cyclic trimer (R_p)-**23**

In contrast, the g_{lum} values of **20–22** increased depending on the number of stacked π-electron systems. It is speculated that some sort of chirality is induced in addition to the optically active V-shaped structure of the cyclophane moiety in the excited state. The comparison of linear trimer **21** with cyclic trimer **23** gave the answer to this question. The g_{abs} values of **21** and **23** were different, and their g_{lum} values were identical. The cyclic structure is a fixed chiral triangle structure (fixed optically active second-ordered structure), whereas the linear π-stacked structure adopts various conformations in the ground state. Thus, chiralities of both the second-ordered structure (chiral triangle structure) and the V-shaped structure affect the larger g_{abs} value of **23** (3.7×10^{-3}) more than that of **21** (3.0×10^{-3}). On the other hand, the g_{lum} values of **21** and **23** were identical, indicating that **21** and **23** adopt a similar structure in the excited state. PAEs, poly(*p*-arylenevinylene)s (PAVs), and poly(*p*-arylene)s (PAs) are known to form planar structures in the excited states, because of the contribution of the quinoid moiety in the excited state. Considering the conformation of **21** in the excited state, each of the three π-electron systems of **21** assumes a planar structure, producing either a zigzag structure or a helical structure (foldamer) in the excited state. The g_{lum} values of **21** and **23** were identical; therefore, **21** forms a structure similar to a cyclic one, i.e., a one-handed helical structure (optically active second-ordered structure), as shown in Fig. 10.11.

Optically active through-space conjugated polymer **26** consisting of enantiopure planar chiral 4,12-disubstituted [2.2]paracyclophane and quaterthiophene was synthesized by Suzuki-Miyaura coupling (Miyaura et al. 1979) between **24** and

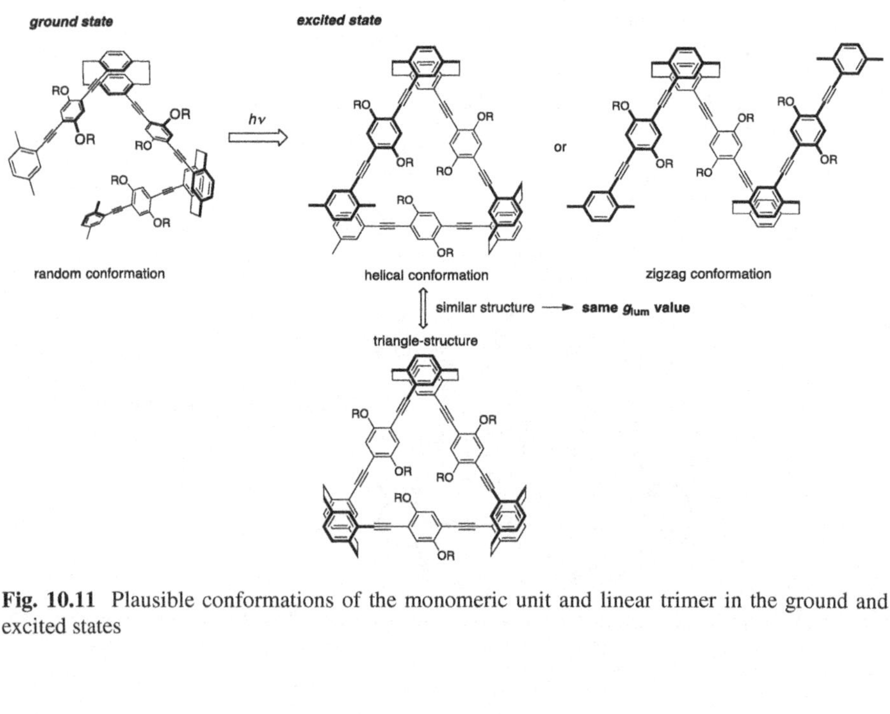

Fig. 10.11 Plausible conformations of the monomeric unit and linear trimer in the ground and excited states

25 (Fig. 10.12) (Morisaki et al. 2015). Moderate CPL was observed (Fig. 10.13), and the Φ_{PL} and $|g_{lum}|$ values were estimated to be 0.07 and 5×10^{-4}, respectively.

Chromatographic optical resolutions of pseudo-*ortho*-disubstituted [2.2]paracyclophanes were reported by Lützen (Meyer-Eppler et al. 2013) and Hasegawa (Kobayakawa et al. 2014; Hasegawa et al. 2017, 2019; Ishioka et al. 2019), and several enantiopure π-stacked molecules have been produced. Figure 10.14 shows the (R_p)-isomers prepared by Hasegawa and coworkers (Kobayakawa et al. 2014; Hasegawa et al. 2017; Hasegawa et al. 2019; Ishioka et al. 2019). Although the CPL properties of these compounds were not reported, their interesting chiroptical and electrochemical properties were shown. Compound **27** was used as a chiral dopant for nematic liquid crystals with a helical twisting power of around 10 m^{-1} (Kobayakawa et al. 2014), and **28** (Hasegawa et al. 2017) and **29** (Hasegawa et al. 2019) exhibited

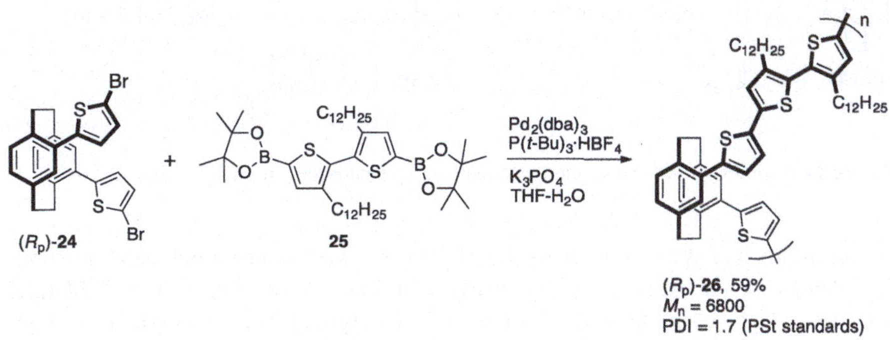

Fig. 10.12 Synthesis of optically active polymer consisting of [2.2]paracyclophane and quaterthiophene

Fig. 10.13 CPL and PL spectra of (R_p)- and (S_p)-**26** in $CHCl_3$ (10×10^{-5} M) excited at 380 nm for CPL and $\lambda_{abs,max}$ for PL

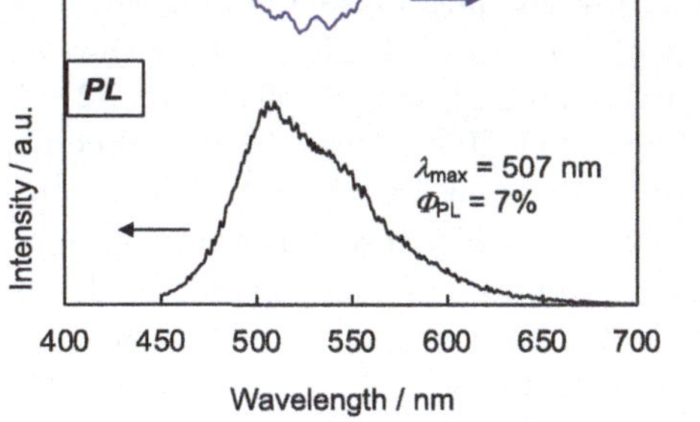

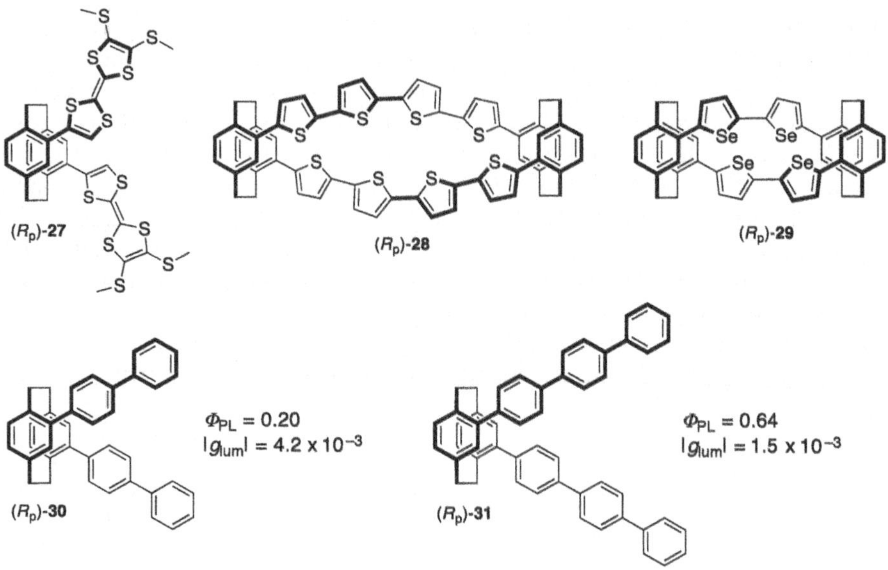

Fig. 10.14 Pseudo-*ortho*-disubstituted [2.2]paracyclophane-based chiral molecules

unique redox behaviors owing to the closely stacked quaterthiophenes and biselenophenes, respectively. Recently, PA-type π-stacked molecules **30** and **31** have been reported, in which terphenylenes and quaterphenylenes were stacked at the terminal benzene rings (Ishioka et al. 2019), respectively, as shown in Fig. 10.14. Intense CPL was observed from **30** with Φ_{PL} of 0.20 and $|g_{lum}|$ of 4.2×10^{-3} as well as **31** with Φ_{PL} of 0.64 and $|g_{lum}|$ of 1.5×10^{-3}.

10.4 Synthesis of Enantiopure 4,7,12,15-Tetrasubstituted [2.2]Paracyclophane and Optically Active π-Stacked Molecules

In 2014, an optical resolution method of 4,7,12,15-tetrasubstituted [2.2]paracyclophane was reported (Fig. 10.15) (Morisaki et al. 2014). The racemate 4,7,12,15-tetrabromo[2.2]paracyclophane *rac*-**32** was synthesized by Chow and coworkers (Chow et al. 2005). One of the bromo groups of *rac*-**32** was converted to a hydroxy group to obtain *rac*-**33**, and the reaction with (1*S*,4*R*)-(-)-camphanic chloride as a chiral auxiliary resulted in the diastereomers (R_p,1S,4*R*)-**34** and (S_p,1*S*,4*R*)-**34**. The absolute configuration could be determined by X-ray crystallography. Separation of diastereomers was readily carried out in gram-scale by simple SiO_2 column chromatography.

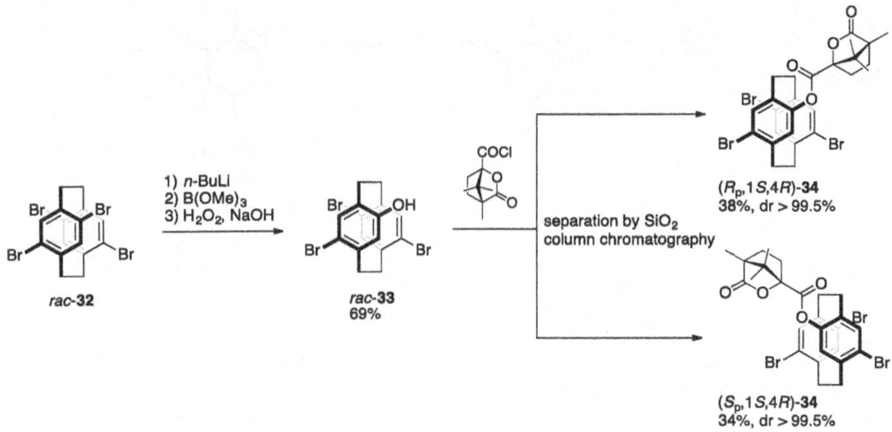

Fig. 10.15 Optical resolution of 4,7,12,15-tetrasubstituted [2.2]paracyclophane

The chiral unit was easily removed by hydrolysis to afford the optically active phenol **33** (Fig. 10.16). The treatment of **33** with trifluoromethanesulfonic anhydride afforded the optically active 4,7,12,15-tetrasubstituted [2.2]paracyclophane **35**. Subsequent Sonogashira-Hagihara coupling between the optically active cyclophane **35** with trimethylsilyl (TMS) acetylene proceeded smoothly to afford triyne **36** with good yield. Interestingly, $Pd_2(dba)_3/P^tBu_3/CuI$ catalyst system (dba = dibenzylideneacetone and P^tBu_3 = tri(*t*-butyl)phosphine) led to chemoselective coupling with bromo groups. Then, the remaining trifluoromethylsulfonyl group could be reacted with TMS acetylene using a $PdCl_2$(dppf) catalyst (dppf = 1,1′-bis(diphenylphosphino)ferrocene) to give the optically active tetrayne **37**. The TMS groups were removed to give the optically active 4,7,12,15-tetraethynyl[2.2]paracyclophane **38**.

Optically active cyclic molecules **39–41** were synthesized with **38** serving as a chiral building block (Fig. 10.17), and their chiroptical properties were revealed. The large chirality was induced in both the ground state and the excited state. For example, the specific rotation of **39** was approximately 1,500, and the molar ellipticity reached 3,000,000 deg cm^2 dmol^{-1}. Cyclic molecule **39** exhibited excellent CPL profiles (Fig. 10.18), and the $|g_{lum}|$ value at $\lambda_{PL,max}$ was on the order of 10^{-2} with Φ_{PL} of 0.45. In addition to the planar chirality of the [2.2]paracyclophane skeleton, the chiral second-ordered structure, namely, the chiral two blades propeller-shaped structure contributed to the chiral induction in the excited state. The propeller-shaped molecules **40** (Gon et al. 2015) and **41** (Gon et al. 2016) showed a large molar absorption coefficient, good Φ_{PL}, and large CPL g_{lum} value. Especially, propeller-shaped molecule **40** was also an excellent CPL emitter with a large g_{lum} value and good Φ_{PL} of 0.60.

Optically active X-shaped molecules **42–44** were synthesized from 4,7,12,15-tetrasubstituted [2.2]paracyclophane **38** (Fig. 10.19) (Gon et al. 2015). Various aromatic units such as benzene, naphthalene, and anthracene could be introduced to

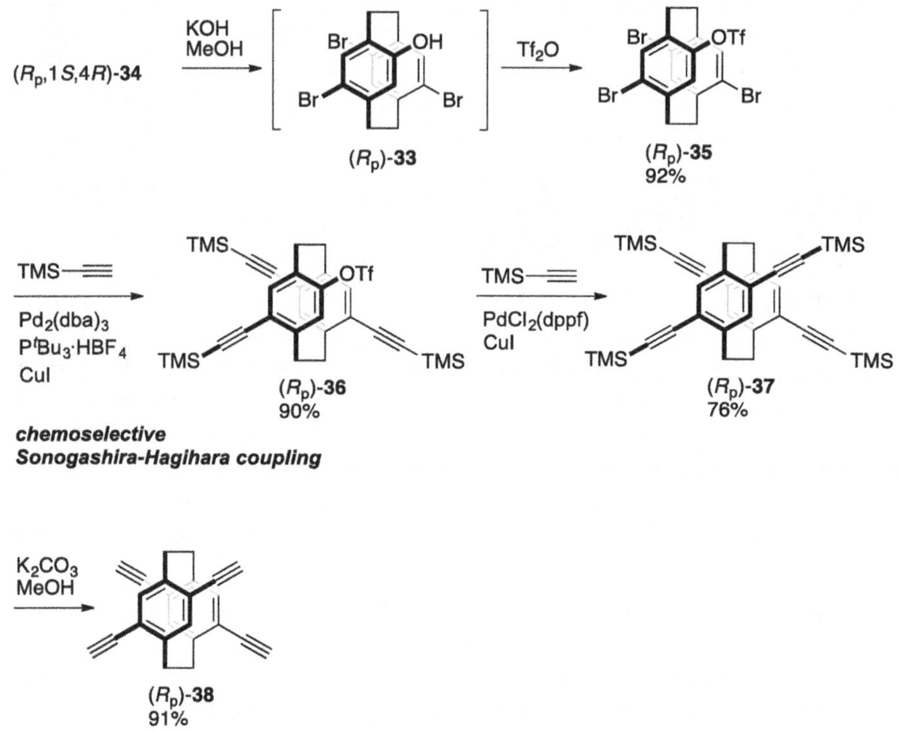

Fig. 10.16 Synthesis of optically active 4,7,12,15-tetrasubstituted [2.2]paracyclophane building blocks

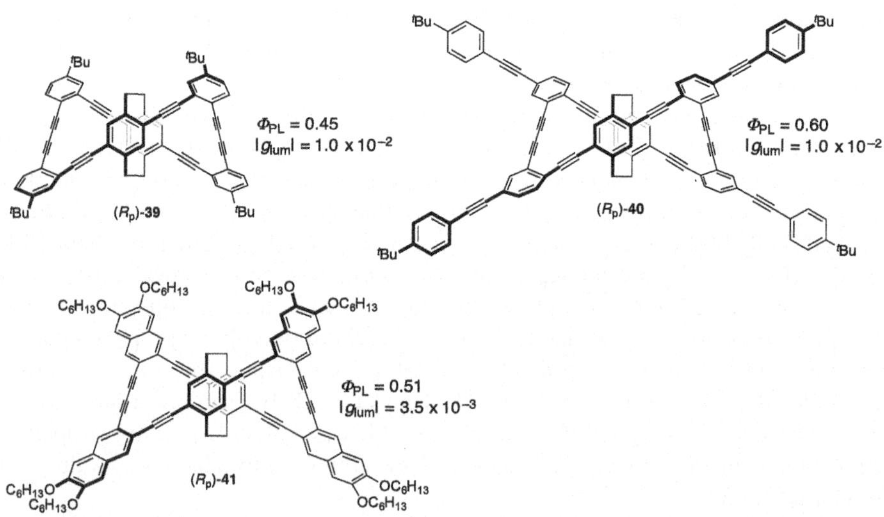

Fig. 10.17 Chiral propeller-shaped molecules based on optically active 4,7,12,15-tetrasubstituted [2.2]paracyclophane

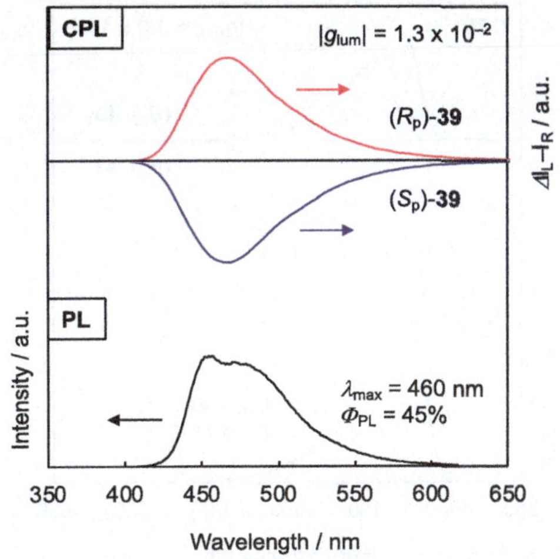

Fig. 10.18 CPL and PL spectra of (R_p)- and (S_p)-**39** in $CHCl_3$ (10×10^{-6} M) excited at 314 nm for CPL and $\lambda_{abs,max}$ for PL

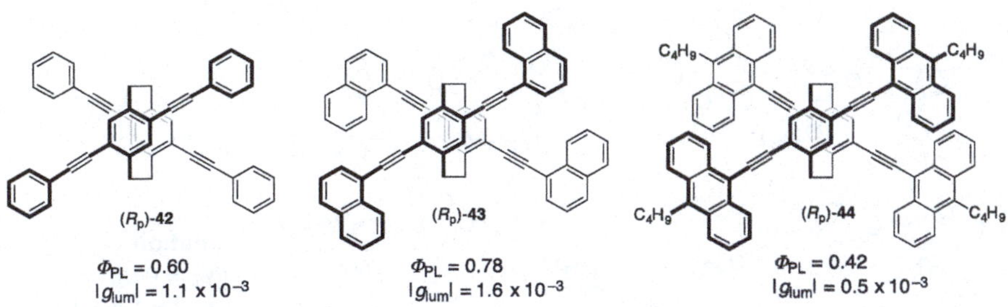

Fig. 10.19 Structures of chiral X-shaped molecules

the 4,7,12,15-tetrasubstituted [2.2]paracyclophane skeleton. All compounds emitted CPL with good Φ_{PL}, and large CPL g_{lum} value; as a representative example, PL and CPL spectra of **43** are shown in Fig. 10.20.

Figure 10.21 shows the 3rd generation chiral dendrimer (Gon et al. 2016) consisting of a chiral X-shaped core and Fréchet-type dendrons that were prepared by the convergent method (Hawker and Fréchet 1990; Fréchet 1994). The 2nd to 4th generation dendrimers exhibited good film-forming ability, and the thin films could be obtained by the spin-coating approach. The thin film emitted bright CPL from the chiral core unit upon excitation of the benzene rings of dendrons. Energy transfer occurred from the dendrons to the chiral core, and intense PL was emitted. Due to the dendrons, aggregation-caused quenching was completely suppressed, and thus, the bright CPL was observed; CPL and PL spectra are shown in Fig. 10.22. The Φ_{PL} of the thin films were estimated to be 65%, and the g_{lum} value at $\lambda_{PL,max}$ was approximately 2×10^{-3}.

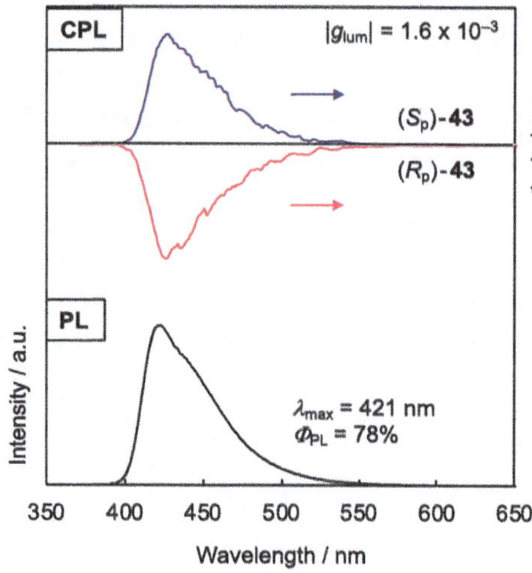

Fig. 10.20 CPL and PL spectra of (R_p)- and (S_p)-**43** in $CHCl_3$ (10×10^{-5} M) excited at 300 nm for CPL and $\lambda_{abs,max}$ for PL

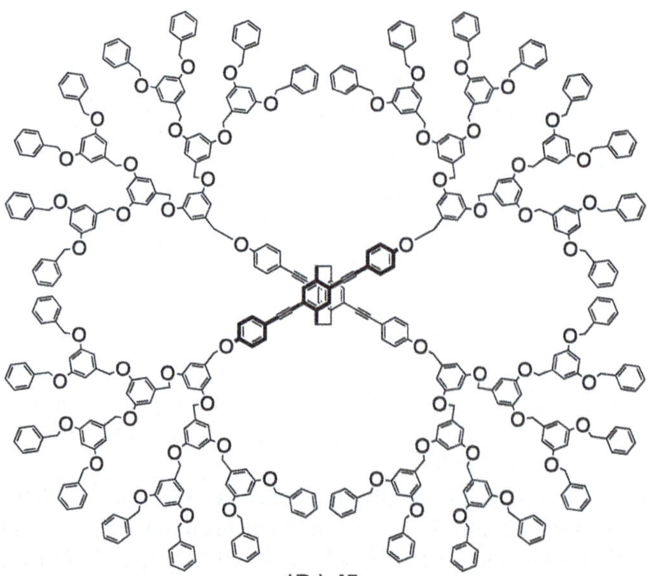

(R_p)-**45**

Fig. 10.21 Structure of chiral dendrimer with a chiral X-shaped [2.2]paracyclophane core

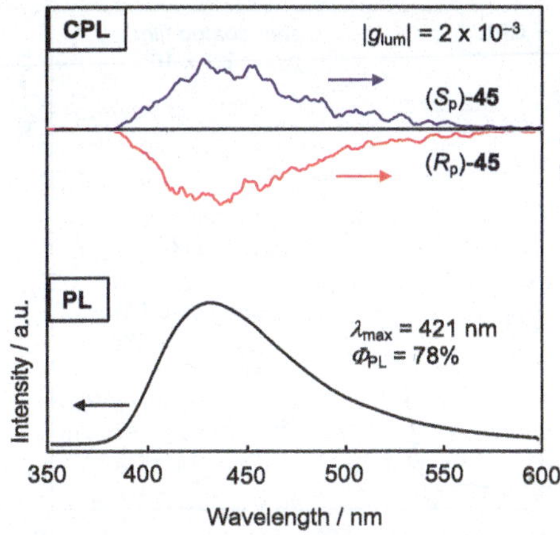

Fig. 10.22 CPL and PL spectra of the thin film of (R_p)- and (S_p)-**45** excited at 279 nm for CPL and $\lambda_{abs,max}$ for PL

An optically active X-shaped molecule **46** consisting of a more extended π-conjugation system was prepared from **38** as a chiral building block (Fig. 10.23) (Gon et al. 2017). The X-shaped molecule **46** emitted CPL with Φ_{PL} of 87% in the dilute solution with a good $|g_{lum}|$ value of 1.2×10^{-3}. In addition, this molecule showed good film formability. The g_{lum} value of the thin film formed by the spin-coating method exhibited the positive CPL signal with the g_{lum} value of $+2.1 \times 10^{-2}$ (Fig. 10.24), which was larger by one order of magnitude compared with that the dilute solution. The annealed spin-coated film exhibited a negative g_{lum} value of

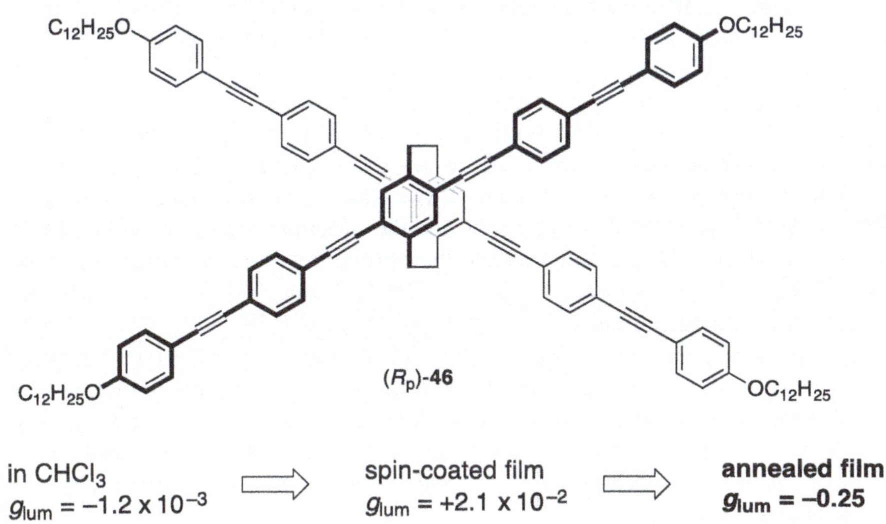

Fig. 10.23 Structure of chiral X-shaped molecule (R_p)-**46** and the CPL profiles

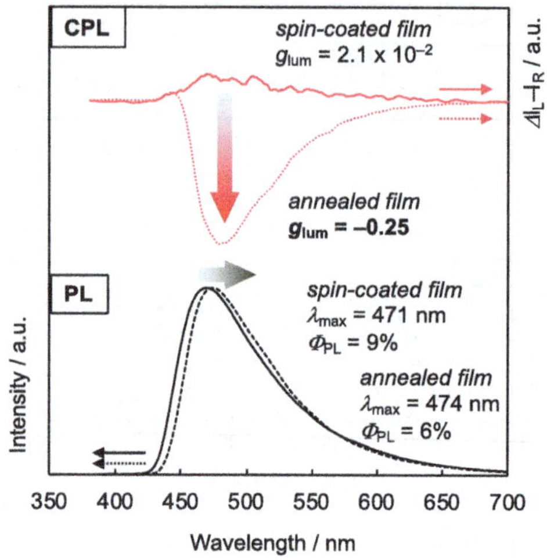

Fig. 10.24 CPL and PL spectra of the spin-coated and the annealed film of (R_p)-**46**

–0.25 (Fig. 10.24), which was a very large value on the order of 10^{-1}. The optically active higher-ordered structure was constructed in the thin film by π-π interactions among the extended π-conjugation systems, as well as the van der Waals force of the $C_{12}H_{25}$ chains. It seems that the thermodynamically stable higher-ordered structure was formed by the heating protocol.

Chemoselective Sonogashira-Hagihara coupling using $Pd_2(dba)_3/P^tBu_3$ catalyst system was utilized to stack different π-electron systems (Sasai et al. 2018). The triyne **36** mentioned above was used as the substrate to produce chiral X-shaped π-stacked dimers **48** and **50** consisting of different π-electron systems as shown in Fig. 10.25. Normal X-shaped molecule **51** was also prepared from **36** via **37** (Gon et al. 2017; Kikuchi et al. 2019). The optical, as well as chiroptical properties of 48 and 50, were almost the same as those of the corresponding X-shaped molecule **51** (Table 10.1). The number of methoxy groups increased, spectra were redshifted and Φ_{PL} and g_{lum} values were increased. Thus, they were excellent CPL emitters.

Optically active *p*-arylene-vinylenes (PAVs)-stacked X-shaped molecules **52** and **53** were also prepared (Gon et al. 2017), and the structures are shown in Fig. 10.26. Racemic molecule **52** was prepared and their properties were investigated in detail (Bazan et al. 1998; Bartholomew and Bazan 2001; Bazan 2007; Morisaki and Chujo 2002). Molecule **52** exhibited CPL with a good Φ_{PL} of 0.78 and $|g_{lum}|$ value of 3.7 × 10^{-3}, while the PL of the aggregates quenched to exhibit Φ_{PL} of 0.03 with good $|g_{lum}| = 4.3 \times 10^{-3}$. Molecule **53** showed moderate PL profiles both in solution (Φ_{PL} = 0.58) and in the aggregated state (Φ_{PL} = 0.24), and moderate CPL properties with $|g_{lum}|$ in the order of 10^{-4} in solution, as well as in the aggregation state were observed. The optical and chiroptical properties varied drastically by attaching phenyl groups to the ethene moieties.

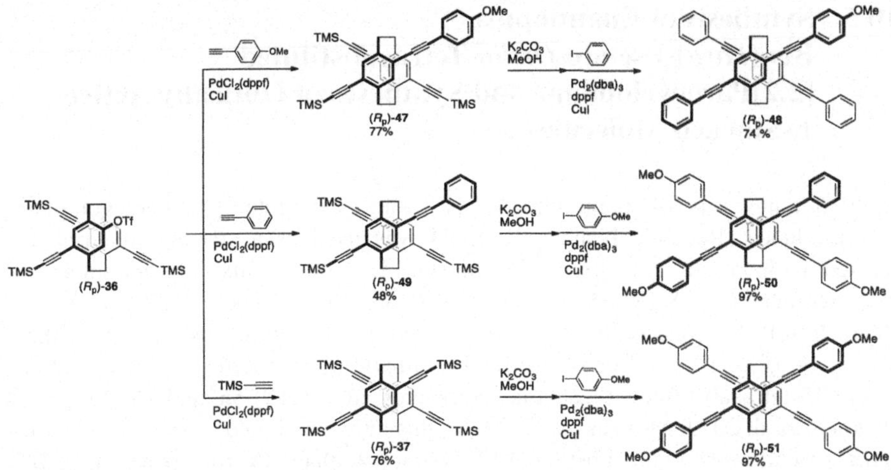

Fig. 10.25 Synthesis of X-shaped π-stacked dimers

Table 10.1 Optical and chiroptical data of **48**, **50**, and **51** in the diluted solutions

Compound	UV λ_{max}/nm ($\varepsilon/10^5$ cm^{-1} M^{-1})	PL λ_{max}/nm	Φ_{PL}	CPL$\lvert g_{lum}\rvert$
48	352 (0.60)	420	0.71	1.5×10^{-3}
50	357 (0.56)	424	0.73	1.7×10^{-3}
51	361 (0.68)	427	0.75	1.7×10^{-3}

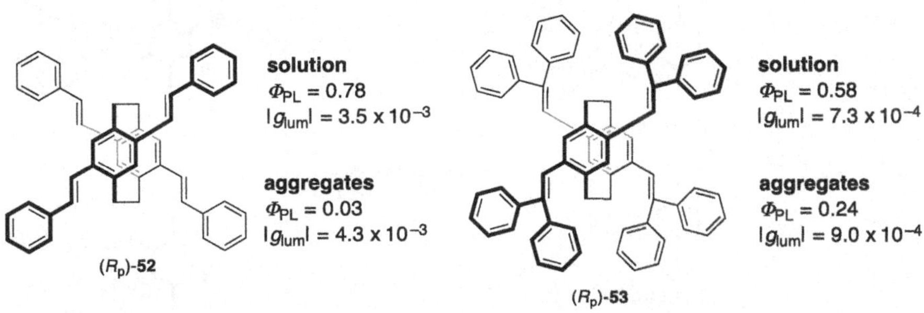

Fig. 10.26 Structures of chiral X-shaped molecules consisting of PAVs

10.5 Synthesis of Enantiopure Bis-(*Para*)-Pseudo-*Ortho*-Tetrasubstituted [2.2]Paracyclophane and Syntheses of Optically Active π-Stacked Molecules

In 2016, a new type of enantiopure 4,7,12,15-tetrasubstituted [2.2]paracyclophane was produced. Racemic bisphenol *rac*-**54** was used as a starting material and reacted with (1*S*,4*R*)-(-)-camphanic chloride as a chiral auxiliary to obtain the diastereomers (R_p,1*S*,4*R*)-**55** and (S_p,1*S*,4*R*)-**55** (Figs. 10.6c, 10.27) (Jiang and Zhao 2004). The diastereomers were reacted with bromine using iron without separation (Fig. 10.27). Regioselective bromination proceeded with iron to obtain 4,7,12,15-tetrasubstituted [2.2]paracyclophanes (R_p,1*S*,4*R*)-**56** and (S_p,1*S*,4*R*)-**56** (Kikuchi et al. 2019; Morisaki et al. 2016); this tetrasubstituted isomer is called bis-(*para*)-pseudo-*ortho*-tetrasubstituted [2.2]paracyclophane (Vorontsova et al. 2008). The diastereomers were separated by simple column chromatography using SiO_2. Removal of the chiral auxiliary groups with KOH formed chiral bisphenol **57**, and the successive reaction with trifluoromethanesulfonic anhydride resulted in enantiopure bis-(*para*)-pseudo-*ortho*-type tetrasubstituted [2.2]paracyclophane building blocks **58** (Morisaki et al. 2016).

Chemoselective Sonogashira-Hagihara coupling was available for **58**. The treatment of **58** with TMS acetylene in the presence of a catalytic amount of

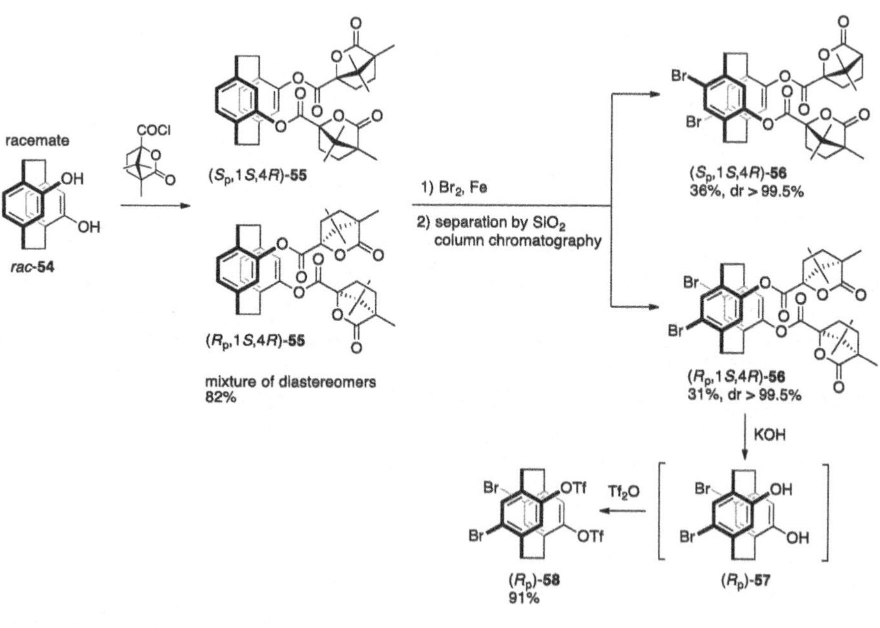

Fig. 10.27 Optical resolution of 4,7,12,15-bis-(*para*)-pseudo-*ortho*-type tetrasubstituted [2.2]paracyclophane and the transformations

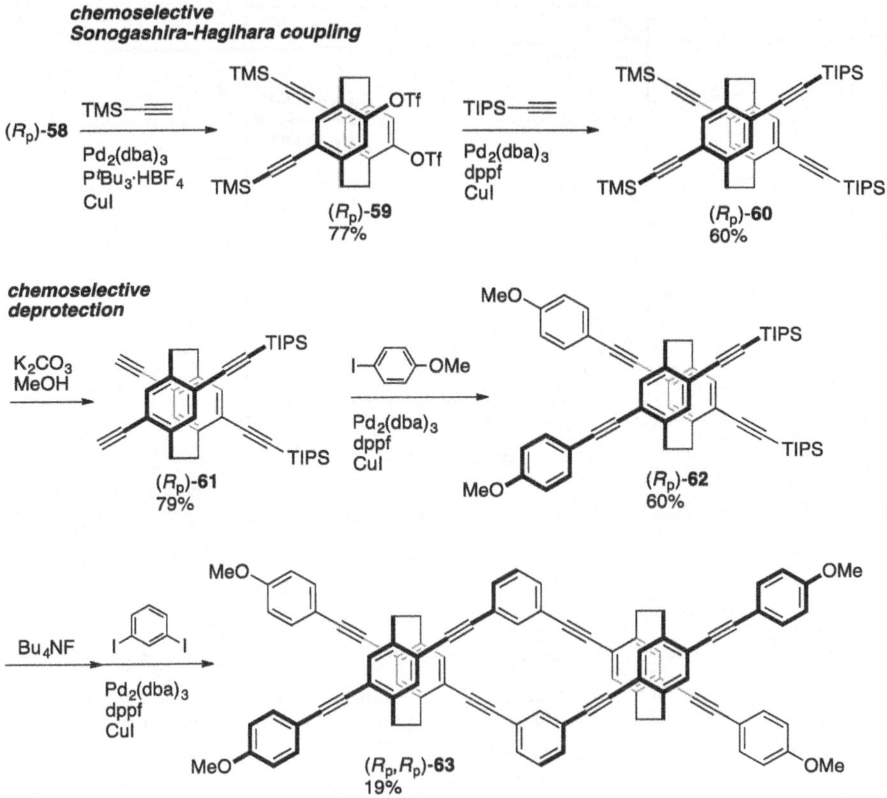

Fig. 10.28 Synthesis of one-handed double helical molecule

$Pd_2(dba)_3/P^tBu_3/CuI$ afforded the diyne **59** selectively, as shown in Fig. 10.28. The coupling reaction between **59** with triisopropylsilyl (TIPS) acetylene using a $Pd_2(dba)_3$/dppf catalyst system enabled us to obtain the corresponding tetrayne **60**. K_2CO_3/MeOH allowed the chemoselective removal of the TMS group from **60** to obtain diyne **61**. Then, Sonogashira-Hagihara coupling of **60** with *p*-iodoanisole produced **62**. The TIPS group was easily removed by Bu_4NF, and the reaction with *m*-diiodobenzene provided enantiopure cyclic molecule **63** consisting of two optically active cyclophanes, in which boomerang-shaped arylene-ethynylene containing five benzene rings are stacked at the second and fourth phenylene moieties. This cyclic molecule forms a one-handed double helical structure; for example, the (R_p,R_p)-isomer construct left-handed double helices. This double helical molecule also emitted intense CPL with Φ_{PL} of 0.62 and $|g_{lum}| = 1.6 \times 10^{-3}$; the PL and CPL spectra are shown in Fig. 10.29.

Enantiopure bisphenol **57** was reacted with K_2CO_3 and MeI to obtain a new chiral building block (R_p)-**64** in high yield (Fig. 10.30) (Kikuchi et al. 2019). The reaction of (R_p)-**64** with TMS acetylene in the presence of $Pd_2(dba)_3/P^tBu_3/CuI$

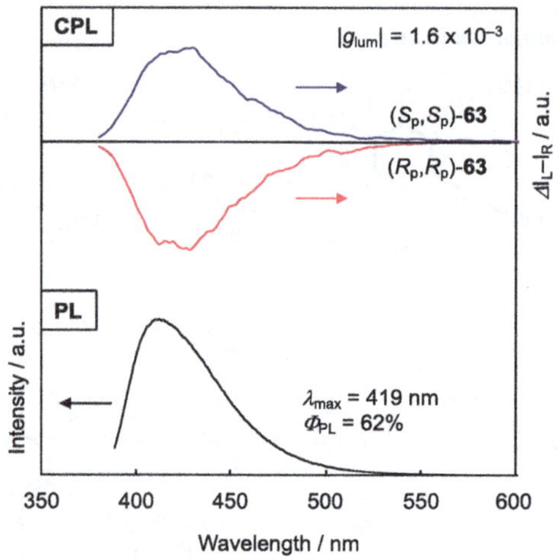

Fig. 10.29 CPL and PL spectra of (R_p)- and (S_p)-**63** in $CHCl_3$ (10×10^{-5} M) excited at 300 nm for CPL and $\lambda_{abs,max}$ for PL

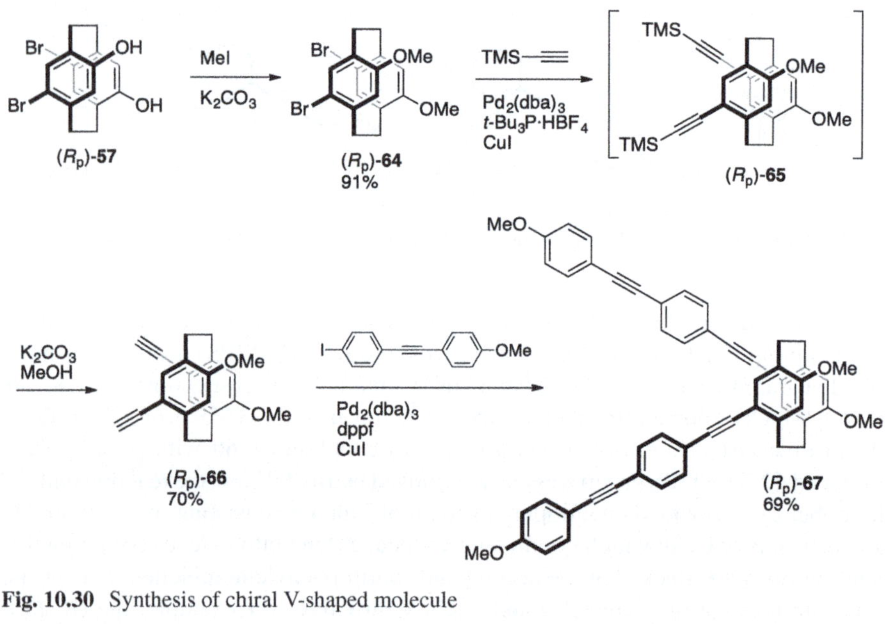

Fig. 10.30 Synthesis of chiral V-shaped molecule

catalyst system proceeded smoothly to obtain the cross-coupling product (R_p)-**65**. The TMS groups of (R_p)-**65** were easily removed by K_2CO_3/MeOH to obtain diyne (R_p)-**66**. The Sonogashira-Hagihara reaction of (R_p)-**66** with iodotoluene derivative resulted in the corresponding V-shaped molecule (R_p)-**67** (Fig. 10.30). In molecule **67**, π-electron systems are stacked at the terminal benzene rings, whereas the same

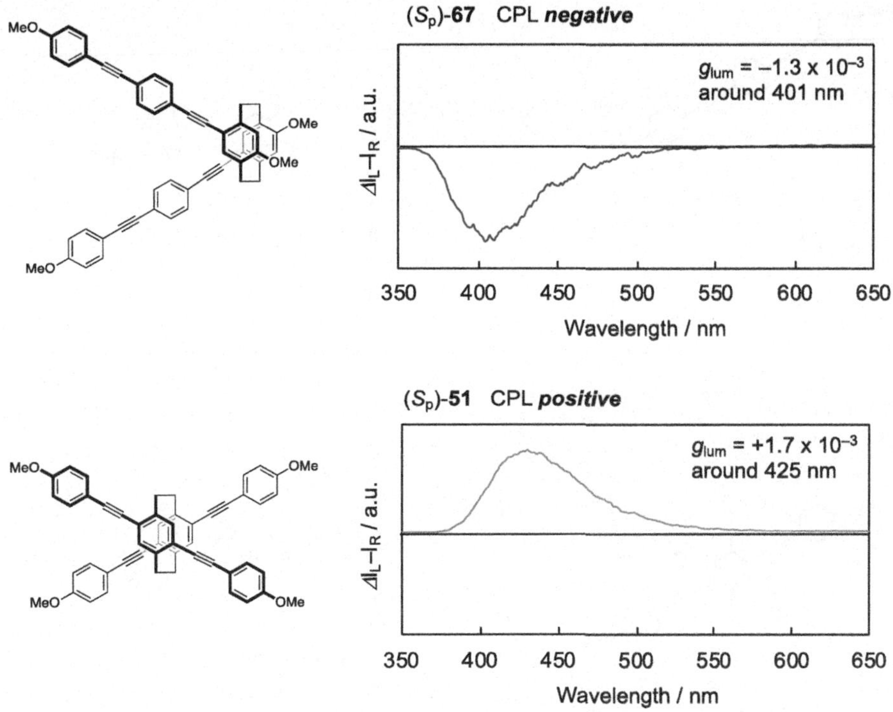

Fig. 10.31 CPL spectra of (S_p)-**67** and **51** in $CHCl_3$ (10×10^{-5} M)

π-electron systems are stacked at the central benzene rings in the X-shaped molecule **51** (Fig. 10.25). Their CPL spectra are shown in Fig. 10.31, indicating the opposite CPL signs in spite of having the same absolute configuration. Incidentally, Fig. 10.31 shows the spectra of (S_p)-isomers, because the simulation was carried out for them (vide infra).

The electronic transition dipole moments and magnetic transition dipole moments were simulated for V-shaped **67** and X-shaped **51** (Fig. 10.32) (Kikuchi et al. 2019); the simulations were carried out for (S_p)-isomers. Theoretically, a g_{lum} value is expressed in the following equation: $4|\mu||\mathbf{m}|\cos\theta/(|\mu|^2 + |\mathbf{m}|^2)$ as mentioned. The μ and **m** represent electric and magnetic transition dipole moments, respectively, and the θ represents the angle between the μ and **m**. The angle θ between μ and **m** of (S_p)-**67** was estimated to be 144°, and that of (S_p)-**51** was estimated to be 87°. Simulation results predicted opposite CPL signs for **67** (negative due to $\theta = 144°$) and **51** (positive due to $\theta = 87°$), which was supported by the experimental results (Fig. 10.31). Thus, the appropriate construction of the intermolecular orientations of π-electron systems is important not only for CPL intensity, but also for CPL signs. In other words, negative and positive signs of CPL can be controlled by the orientation of the π-electron systems.

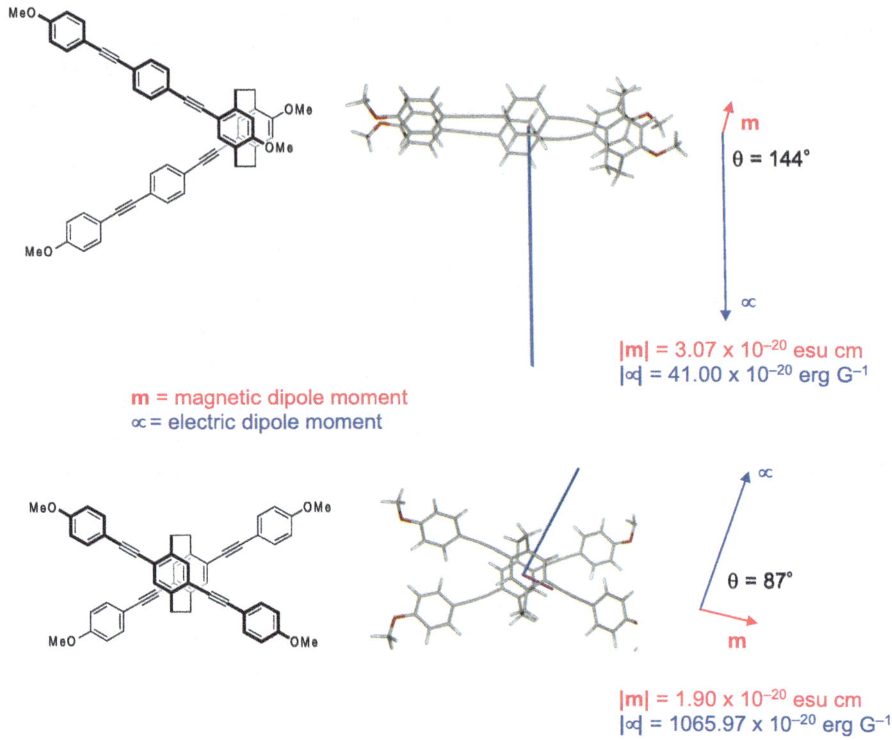

Fig. 10.32 Simulation results: the transition dipole moments of (S_p)-**67** and (S_p)-**51** in the excited states by using the TD-DFT calculations at the ωB97XD/6-31G(d,p) level of theory with following options; TD = (NStates = 10, Root = 1), SCRF = (SOLVENT = Chloroform). Lengths of the dipole moments on the structures are based on the atomic units

10.6 Synthesis of Enantiopure Bis-(*Para*)-Pseudo-*Meta*-Tetrasubstituted [2.2]Paracyclophane

In 2018, enantiopure bis-(*para*)-pseudo-*meta*-tetrasubstituted [2.2]paracyclophane derivatives were prepared as new chiral building blocks by means of diastereomeric optical resolution (Sawada et al. 2018). On the other hand, Lützen and coworkers had already reported the chromatographic optical resolution of disubstituted analogs in 2014 (Meyer-Eppler et al. 2014). Racemic 4,7,12,15-tetrasubstituted [2.2]paracyclophane *rac*-**33** was used as the starting material, which was reacted with BuLi to form the dilithiated intermediate *rac*-**68** (Fig. 10.33). After the formation of lithium phenoxide, the next lithium-halogen exchange occurred at pseudo-*meta*-position due to the electronic effect, leading to bis-(*para*)-pseudo-*meta*-type dilithiated [2.2]paracyclophane. Addition of $B(OMe)_3$ to the intermediate and the treatment with H_2O_2/NaOH resulted in dibromodihydroxy[2.2]paracyclophane *rac*-**69**.

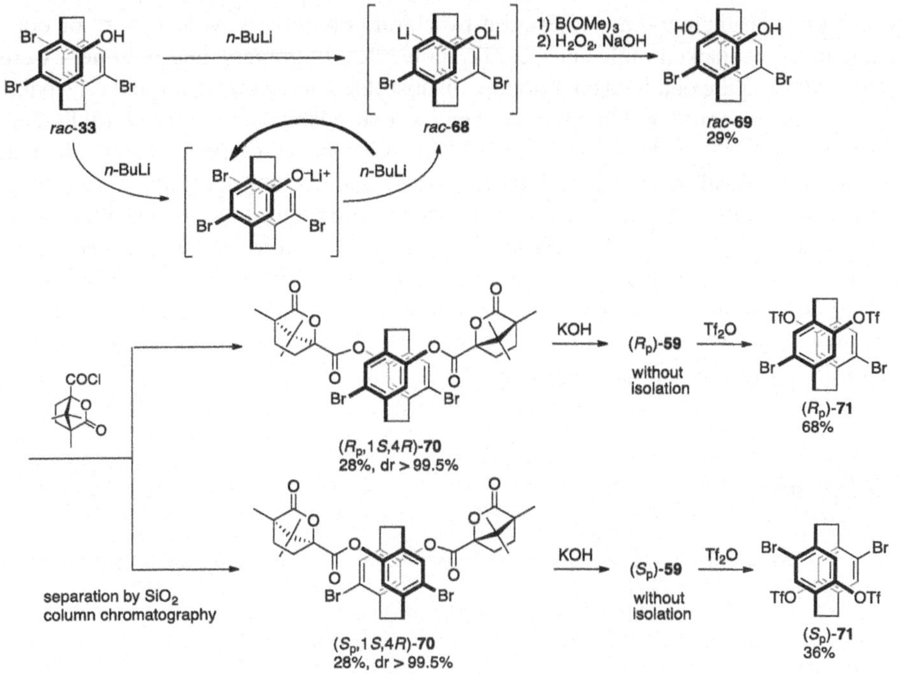

Fig. 10.33 Optical resolution of 4,7,12,15-bis-(*para*)-pseudo-*meta*-type tetrasubstituted [2.2]paracyclophane and the transformations

Diastereomer approach using (1*S*,4*R*)-(-)-camphanic chloride enabled the optical resolution of the bis-(*para*)-pseudo-*meta*-type tetrasubstituted [2.2]paracyclophane (Fig. 10.33). The resulting diastereomers (R_p,1*S*,4*R*)-**70** and (S_p,1*S*,4*R*)-**70** were separated by simple column chromatography using SiO_2. The reaction of **70** with KOH afforded the bisphenol and the next reaction with trifluoromethanesulfonic anhydride produced enantiopure bis-(*para*)-pseudo-*meta*-type tetrasubstituted [2.2]paracyclophanes **71**. These enantiomers are also used as the chiral building blocks for various CPL emitters.

10.7 Conclusion

In summary, the authors have focused on the planar chirality of [2.2]paracyclophane and the intense CPL emission from the enantiopure [2.2]paracyclophane-based molecules in this chapter. Their optical resolution methods basically consisted of diastereomer resolution methods using the common chiral auxiliaries such as (1*R*,2*S*,5*R*)-(-)-menthyl-*p*-toluenesulfinate and (1*S*,4*R*)-(-)-camphanic chloride instead of chromatographic resolution by chiral columns. A wide variety of optically active π-stacked molecules, cyclic molecules, oligomers, and polymers were prepared using the obtained enantiopure disubstituted and tetrasubstituted [2.2]paracyclophanes as chiral building blocks. They were used to construct the chiral higher-ordered structures such as V-shaped, X-shaped, propeller-shaped, triangle-shaped, and double helical structures both in the ground and excited states due to the structurally stable planar chiral cyclophanes. This is important to show excellent CPL performance. [2.2]Paracyclophane-based molecules exhibit good PL property with high Φ_{PL}. Their ε values are large because of the extended π-electron systems. Generally, it is difficult to produce the materials emitting intense CPL with high Φ_{PL} and large g_{lum}. The planar chiral [2.2]paracyclophane skeleton seems to be an ideal scaffold for developing materials exhibiting strong CPL.

References

Aly AA, Brown AB (2009) Asymmetric and fused heterocycles based on [2.2]paracyclophane. Tetrahedron 65:8055–8089

Amako T, Nakabayashi K, Mori T, Inoue Y, Fujiki M, Imai Y (2014) Sign inversion of circularly polarized luminescence by geometry manipulation of four naphthalene units introduced into a tartaric acid scaffold. Chem Commun 50:12836–12839

Bartholomew GP, Bazan GC (2001) Bichromophoric paracyclophanes: models for interchromophore delocalization. Acc Chem Res 34:30–39

Bazan GC (2007) Novel organic materials through control of multichromophore interactions. J Org Chem 72:8615–8635

Bazan GC, Oldham WJ, Lachicotte RJ, Tretiak S, Chernyak V, Mukamel S (1998) Stilbenoid dimers: dissection of a paracyclophane chromophore. J Am Chem Soc 120:9188–9204

Berova N, Nakanishi K, Woody RW (eds) (2000) Circular dichroism, 2nd edn. Wiley-VCH, Toronto

Braddock DC, MacGilp ID, Perry BG (2002) Improved synthesis of (±) -4,12-dohydroxy[2.2]paracyclophane and its enantiomeric resolution by enzymatic methods: plana chiral (*r*)- and (*s*)-phanol. J Org Chem 67:8679–8681

Brown CJ, Farthing AC (1949) Preparation and structure of di-p-xylylene. Nature 164:915–916

Chem N, Yan B (2018) Recent theoretical and experimental progress in circularly polarized luminescence of small organic molecules. Molecules 23:3376/1–32

Chow HF, Low KH, Wong KY (2005) An improved method for the regiospecific synthesis of polysubstituted [2.2]paracyclophanes. Synlett 2130–2134

Clayden J (2002) In organolithiums: selectivity for synthesis. Pergamon, Oxford, pp 141–142

Cram DJ, Allinger NL (1955) Macro rings. xii. stereochemical consequences of steric compression in the smallest paracyclophane. J Am Chem Soc 77:6289–6294

Cram DJ, Steinberg H (1951) Macro rings. i. preparation and spectra of the paracyclophanes. J Am Chem Soc 73:5691–5704

Emeris CA, Oosterhoff LJ (1967) Emission of circularly-polarized radiation by optically-active compounds. 1:129–132

Field JE, Muller G, Riehl JP, Venkataraman D (2003) circularly polarized luminescence from bridgedtriarylamine helicenes. J Am Chem Soc 125:11808–11809

Fréchet JMJ (1994) Functional polymers and dendrimers: reactivity, molecular architecture, and interfacial energy. Science 263:1710–1715

Gibson SE, Knight JD (2003) [2.2]Paracyclophane derivatives in asymmetric catalysis. Org Biomol Chem 1:1256–1269

Gleiter R, Hopf H (2004) Modern cyclophane chemistry. Wiley-VCH, Weinheim
Gon M, Morisaki Y, Chujo Y (2015a) Optically active cyclic compounds based on planar chiral [2.2]paracyclophane: extension of the conjugated systems and chiroptical properties. J Mater Chem C 3:521–529
Gon M, Morisaki Y, Chujo Y (2015) Highly emissive circularly polarized luminescence from optically active conjugated dimers consisting of planar chiral [2.2]paracyclophane. Eur J Org Chem 7756–7762
Gon M, Kozuka H, Morisaki Y, Chujo Y (2016a) Optically active cyclic compounds based on planar chiral [2.2]paracyclophane: extension of the π-surface with naphthalene units. Asian J Org Chem 5:353–359
Gon M, Morisaki Y, Sawada R, Chujo (2016) Synthesis of optically active x-shaped conjugated compounds and dendrimers based on planar chiral [2.2]paracyclophane, leading to highly emissive circularly polarized luminescence materials. Chem Eur J 22:2291–2298
Gon M, Sawada R, Morisaki Y, Chujo (2017) Enhancement and controlling the signal of circularly polarized luminescence based on a planar chiral tetrasubstituted [2.2]paracyclophane framework in aggregation system. Macromolecules 50:1790–1802
Gon M, Morisaki Y, Chujo (2017) optically active phenylethene dimers based on planar chiral tetrasubstituted [2.2]paracyclophane. Chem Eur J 23:6323–6329
Guyard L, Audebert P (2001) Synthesis and electrochemical polymerization of bis-dithienyl cyclophane. Electrochem Commun 3:164–167
Guyard L, Audebert P, Dolbier WR Jr, Duan JX (2002) Synthesis and electrochemical polymerization of new oligothiophene functionalized fluorocyclophanes. J Electroanal Chem 537:189–193
Haketa Y, Bando Y, Takaishi K, Uchiyama M, Muranaka A, Naito M, Shibaguchi H, Kawai T, Maeda H (2012) Asymmetric induction in the preparation of helical receptor-anion complexes: ion-pair formation with chiral cations. Angew Chem Int Ed 51:7967–7971
Hasegawa M, Kobayakawa K, Matsuzawa H, Nishinaga T, Hirose T, Sako K, Mazaki Y (2017) Chem Eur J 23:3267–3271
Hasegawa M, Kobayakawa K, Nojima Y, Mazaki Y (2019) Org Biomol Chem 17:8822–8826
Hawker CJ, Fréchet JMJ (1990) Preparation of polymers with controlled molecular architecture. a new convergent approach to dendritic macromolecules. J Am Chem Soc 112:7638–7647
Hitchcock PB, Rowlands GJ, Parmar R (2005) The synthesis of enantiomerically pure 4-substituted [2.2]paracyclophane derivatives by sulfoxide–metal exchange. Chem Commun 4219–4221
Hopf H (2008) [2.2]paracyclophane in polymer chemistry and materials chemistry. Angew Chem Int Ed 47:2–7
Ishioka S, Hasegawa M, Hara N, Sasaki H, Nojima Y, Imai Y, Mazaki Y (2019) Chiroptical properties of oligophenylenes anchoring with stereogenic [2.2]paracyclophane 48:640–643
Jagtap SP, Collard DM (2010) Multitiered 2D π-stacked conjugated polymers based on pseudo-geminal disubstituted [2.2]paracyclophane. J Am Chem Soc 132:12208–12209
Jiang B, Zhao XL (2004) A simple and efficient resolution of (±)-4,12-dihydroxy[2.2]paracyclophane. Tetrahedron Asym 15:1141–1143
Jones PG, Hillmer J, Hopf H (2003) (S)-4,16-dihydroxymethyl-[2.2]paracyclophane bis-(1s)-camphanoate. Acta Cryst E59:o24–o25
Kawai T, Kawamura K, Tsumatori H, Ishikawa M, Naito M, Fujiki M, Nakashima T (2007) Circularly polarized luminescence of a fluorescent chiral binaphthylene-perylenebiscarboxydiimide dimer. Chem Phys Chem 8:1465:1468
Kikuchi K, Nakamura J, Nagata Y, Tsuchida H, Kakuta T, Ogoshi T, Morisaki Y (2019) Control of circularly polarized luminescence by orientation of stacked π-electron systems. Chem Asian J 14:1681–1685
Kimoto T, Tajima N, Fujiki M, Imai Y (2012) Control of circularly polarized luminescence by using open- and closed-type binaphthyl derivatives with the same axial chirality. Chem Asian J 7:2836–2841

Kobayakawa K, Hasegawa M, Sasaki H, Endo J, Matsuzawa H, Sako K, Yoshida J, Mazaki Y (2014) Dimeric tetrathiafulvalene linked to pseudo-*ortho*-[2.2]paracyclophane: chiral electrochromic properties and use as a chiral dopant. Chem Asian J 9:2751–2754

Lunkley JL, Shirotani D, Yamanari K, Kaizaki S, Muller G (2008) Extraordinary circularly polarized luminescence activity exhibited by cesium tetrakis(3-heptafluoro-butylryl-(+)-camphorato) Eu(III) complexes in EtOH and $CHCl_3$ Solutions. J Am Chem Soc 130:13814–13815

Maeda H, Bando Y (2013) Recent progress in research on stimuli-responsive circularly polarized luminescence based on π-conjugated molecules. Pure Appl Chem 85:1967–1978

Meyer-Eppler G, Vogelsang E, Benkhäuser C, Schneider A, Schnakenburg G, Lützen A (2013) Synthesis, chiral resolution, and absolute configuration of dissymmetric 4,12-difunctionalized [2.2]paracyclophane. Eur J Org Chem 4523–4532

Meyer-Eppler G, Sure R, Schneider A, Schnakenburg G, Grimme S, Lützen A (2014) Synthesis, chiral resolution, and absolute configuration of dissymmetric 4,15-difunctionalized [2.2]paracyclophanes. J Org Chem 79:6679–6687

Miyaura N, Yamada K, Suzuki A (1979) A new stereospecific cross-coupling by the palladium-catalyzed reaction of 1-alkenylboranes with 1-alkenyl or 1-alkynyl halides. Tetrahedron Lett 20:3437–3440

Mizogami S, Yoshimura S (1985) Synthesis of a new crystalline polymer: polymetacyclophane. J Chem Soc Chem Commun 1736–1738

Molina-Ontoria A, Wielopolski M, Gebhardt J, Gouloumis A, Clark T, Guldi DM, Martín N (2011) [2,2']paracyclophane-based π-conjugated molecular wires reveal molecular-junction behavior. J Am Chem Soc 133:2370–2373

Morisaki Y, Chujo Y (2002) Synthesis of novel π-conjugated polymers having [2.2]paracyclophane skeleton in the main chain. extension of π-conjugated length via the through-space. Macromolecules 35:587–589

Morisaki Y, Chujo Y (2006) Through-space conjugated polymers based on cyclophanes. Angew Chem Int Ed 45:6430–6437

Morisaki Y, Chujo Y (2008) cyclophane-containing polymers. Prog Polym Sci 33:346–364

Morisaki Y, Chujo Y (2009) synthesis of π-stacked polymers on the basis of [2.2]paracyclophane. Bull Chem Soc Jpn 82:1070–1082

Morisaki Y, Chujo Y (2011) Through-space conjugated polymers consisting of [2.2]paracyclophane. Polym Chem 2:1249–1257

Morisaki Y, Chujo Y (2012) π-electron-system-layered polymers based on [2.2]paracyclophane. Chem Lett 41:840–846

Morisaki Y, Chujo Y (2019) Planar chiral [2.2]paracyclophanes: optical resolution and transformation to optically active π-stacked molecules. Bull Chem Soc Jpn 92:265–274

Morisaki Y, Hifumi R, Lin L, Inoshita K, Chujo Y (2012a) T practical optical resolution of planar chiral pseudo-*ortho*-disubstituted [2.2]paracyclophane. Chem Lett 41:990–992

Morisaki Y, Hifumi R, Lin L, Inoshita K, Chujo Y (2012b) Through-space conjugated polymers consisting of planar chiral pseudo-*ortho*-linked [2.2]paracyclophane. Polym Chem 3:2727–2730

Morisaki Y, Kawakami N, Nakano T, Chujo Y (2013) Energy transfer properties of a [2.2]paracyclophane-based through-space dimer. Chem Eur J 19:17715–17718

Morisaki Y, Kawakami N, Shibata S, Chujo Y (2014a) Through-space conjugated molecular wire consisting of three π-electron systems. Chem Asian J 9:2891–2895

Morisaki Y, Kawakami N, Shibata S, Chujo Y (2014b) Synthesis and properties of a through-space conjugated dimer. Chem Lett 43:426–428

Morisaki Y, Inoshita K, Chujo Y (2014c) Planar chiral through-space conjugated oligomers: synthesis and characterization of chiroptical properties. Chem Eur J 20:8386–8390

Morisaki Y, Gon M, Sasamori T, Tokitoh N, Chujo Y (2014d) Planar chiral tetrasubstituted [2.2]paracyclophane: optical resolution and functionalization. J Am Chem Soc 136:3350–3353

Morisaki Y, Inoshita K, Shibata S, Chujo Y (2015) Synthesis of optically active through-space conjugated polymers consisting of planar chiral [2.2]paracyclophane and quaterthiophene. Polym J 47:278–281

Morisaki Y, Sawada R, Gon M, Chujo Y (2016) New type of planar chiral [2.2]paracyclophanes and construction of one-handed double helices. Chem Asian J 11:2524–2527

Morisaki Y, Shibata S, Chujo Y (2017) [2.2]paracyclophane-based single molecular wire consisting of four π-electron systems can. J Chem 95:424–431

Muller G (2014) In luminescence of lanthanide ions in coordination compounds and nanomaterials. In: de Bettencourt-Dias A (ed). John Wiley & Sons, Chichester, U.K., pp 77–124

Müller S, Liepold B, Roth GJ, Bestmann HJ (1996) An improved one-pot procedure for the synthesis of alkynes from aldehydes. Synlett 1996:521522

Ohira S (1989) Methanolysis of dimethyl (1-diazo-2-oxopropyl) phosphonate: generation of dimethyl (diazomethyl) phosphonate and reaction with carbonyl compounds. Synth Commun 19:561–564

Oldham WJ, Miao YJ, Lachicotte RJ, Bazan GC (1998) Stilbenoid dimers: effect of conjugation length and relative chromophore orientation. J Am Chem Soc 120:419–420

Pamperin D, Hopf H, Syldatk C, Pietzsch M (1997) Synthesis of Planar Chiral [2.2]Paracyclophanes by Biotransformations: Kinetic Resolution of 4-Formyl[2.2]paracyclophane by Asymmetric Reduction. Tetrahedron Asym 8:319–325

Pamperin D, Ohse B, Hopf H, Pietzsch M (1998) Synthesis of planar-chiral [2.2]paracyclophanes by biotransformations: screening for hydrolase activity for the kinetic resolution of 4-acetoxy-[2.2]paracyclophane. J Mol Cat B Enzymatic 5:317–319

Paradies J (2011) [2.2]Paracyclophane derivatives: synthesis and application in catalysis. Synthesis 3749–3766

Parmar R, Coles MP, Hitchcock PB, Rowlands GJ (2010) Towards a flexible strategy for the synthesis of enantiomerically pure [2.2]paracyclophane derivatives: the chemistry of 4-tolylsulfinyl[2.2]paracyclophane. Synthesis 4177–4187

Pye PJ, Rossen K, Reamer RA, Tsou NN, Volante RP, Reider PJ (1997) A new planar chiral bisphosphine ligand for asymmetric catalysis: highly enantioselective hydrogenations under mild conditions. J Am Chem Soc 119:6207–6208

Riehl JP, Muller F (2012) Comprehensive chiroptical spectroscopy. Wiley and Sons, New York

Riehl JP, Richardson FS (1986) Circularly polarized luminescence spectroscopy. Chem Rev 86:1–16

Rossen K, Pye PJ, Maliakal A, Volante RP (1997) Kinetic resolution of rac-4,12-dibromo[2.2]paracyclophane in a palladium [2.2]phanephos catalyzed amination. J Org Chem 62:6462–6463

Rowlands GJ (2008) The synthesis of enantiomerically pure [2.2]paracyclophane derivatives. Org Biomol Chem 6:1527–1534

Rozenberg V, Sergeeva E, Hopf H (2004) In Gleiter R, Hopf H (eds) Modern cyclophane chemistry. Wiley-VCH, Weinheim, Germany, pp 435–462

Salhi F, Collard DM (2003) π-stacked conjugated polymers: the influence of paracyclophane π-stacks on the redox and optical properties of a new class of broken conjugated polythiophenes. Adv Mater 15:81–85

Salhi F, Lee B, Metz C, Bottomley LA, Collard DM (2002) Influenece of π-stacking. on the redox properties of oligothiophenes: (α-alkyloligo-thienyl)para[2.2]cyclophane. Org Lett 4:3195–3198

Sánchez-Carnerero EM, Agarrabeitia AR, Moreno F, Maroto BL, Muller G, Ortiz MJ, de la Moya S (2015) Circularly polarized luminescence from simple organic molecules. Chem Eur J 21:13488–13500

Sasai Y, Tsuchida H, Kakuta T, Ogoshi T, Morisaki Y (2018) Synthesis of optically active π-stacked compounds based on planar chiral tetrasubstituted [2.2]paracyclophane. Mater Chem Front 2:791–795

Sato S, Yoshii a, Takahashi S, Furumi S, Takauchi M, Isobe H (2017) Chiral intertwined spirals and magnetic transition dipole moments dictated by cylinder helicity. PNAS 114:13097–13101

Sawada Y, Furumi S, Takai A, Takeuchi M, Noguchi K, Tanaka K (2012) Rhodium-catalyzed Enantioselective synthesis, crystal structures, and photophysical properties of helically chiral 1,1'-bitriphenylenes. J Am Chem Soc 134:4080–4083

Sawada R, Gon M, Nakamura J, Morisaki Y, Chujo Y (2018) Synthesis of enantiopure planar chiral bis-(*para*)-pseudo-*meta*-type [2.2]paracyclophanes. Chirality 30:1109–1114

Sheng Y, Ma J, Liu S, Wang Y, Zhu C, Cheng Y (2016) Strong and reversible circularly polarized luminescence emission of a chiral 1,8-naphthalimide fluorophore induced by excimer emission and orderly aggregation. Chem Eur J 22:9519–9522

Sonogashira K (2002) In: Negishi E (ed) Handbook of organopalladium chemistry for organic synthesis. Wiley-Interscience, New York, pp 493–529

Takaishi K, Yamamoto S, Hinoide S, Ema T (2017) Helical oligonaphthodioxepins showing intense Circularly Polarized Luminescence (CPL) in solution and in the solid state. Chem Eur J 23:9249–9252

Tanaka H, Inoue Y, Mori T (2018) Circularly polarized luminescence and circular dichroisms in small organic molecules: correlation between excitation and emission dissymmetry factors. ChemPhotoChem 2:386–402

Tohda Y, Sonogashira K, Hagihara N (1975) A convenient synthesis of acetylenes: catalytic substitutions of acetylenic hydrogen with bromoalkenes, iodoarenes and bromopyridines. Tetrahedron Lett 16:4467–4470

Vögtle F (1993) cyclophane chemistry: synthesis, structures and reactions. John Wiley & Sons, Chichester

Vorontsova NV, Rozenberg VI, Sergeeva EV, Vorontsov EV, Starikova ZA, Lyssenko KA, Hopf H (2008) Symmetrically tetrasubstituted [2.2]paracyclophanes: their systematization and regioselective synthesis of several types of bis-bifunctional derivatives by double electrophilic substitution. Chem Eur J 14:4600–4617

Weiland KJ, Gallego A, Mayor M (2019) Beyond simple substitution patterns—symmetrically tetrasubstituted [2.2]paracyclophanes as 3D functional materials. Eur J Org Chem 3073–3085

Wielopolski M, Molina-Ontoria A, Schubert C, Margraf JT, Krokos E, Kirschner J, Gouloumis A, Clark T, Guldi DM, Martín N (2013) Blending through-space and through-bond π-π-coupling in [2,2']-paracyclophane-oligophenylenevinylene molecular wires. J Am Chem Soc 135:10372–10381

Zhuravsky R, Starikova Z, Vorontsov E, Rozenberg V (2008) Novel Strategy for the synthesis of chiral pseudo-ortho-substituted hydrxy[2.2]paracyclophane-based ligands from the resolved 4-bromo-12-hydroxy[2.2]paracyclophane as a parent compound. Tetrahedron Asym 19:216–222

Part III
Applications of Functional Dyes

Chapter 11
Fluorescent Chemosensors

Yosuke Niko and Shigeru Watanabe

Abstract Chemosensors are molecules capable of monitoring changes in the concentration, structure, or location of chemical species based on a detectable physical signal and can therefore be used in quantitative analysis or for the monitoring and/or visualization of targeted analytes (ions, biomolecules, organelles, etc.). Besides the analyte itself, chemosensors can be used indirectly to observe chemical reactions, biological events, or specific phases in materials where the analyte appears (e.g., the concentration of reactive oxygen species in mitochondria is related to apoptosis). As functional dyes show inherent optical properties such as photon absorption/emission at distinct wavelengths, they are potential optical signal indicators in chemosensors. In fact, in the more than 150 years since F. Goppelsröder invented a morin-based chemosensor to detect aluminum anion using fluorescence signal (Wu et al. in Chem Soc Rev 46:7105–7123, 2017), a number of such "fluorescent chemosensors" have been developed. Some of them are widely used in practical scientific fields such as biology, physiology, pharmacology, food chemistry, and environmental chemistry, as well as in industrial and military/defense fields. This chapter describes the representative principles and molecular designs of fluorescent chemosensors, and several historically important progresses are introduced. The reason to limit the discussion to fluorescence is that fluorescence-based techniques generally exhibit superior sensitivity to absorption-based ones, and therefore many dyes have been reported for the development of fluorescent chemosensors.

Keywords Photo-induced electron transfer · Förster resonance energy transfer · Excimer · Aggregation · Intramolecular charge transfer · Excited state intramolecular proton transfer

---

Y. Niko (✉) · S. Watanabe
Research and Education Faculty, Multidisciplinary Science Cluster, Interdisciplinary Science Unit, Kochi University, 2-5-1, Akebono-cho, Kochi-shi, Kochi 780-8520, Japan
e-mail: y.niko@kochi-u.ac.jp

Y. Ooyama and S. Yagi (eds.), *Progress in the Science of Functional Dyes*,
https://doi.org/10.1007/978-981-33-4392-4_11

11.1 Definition of "Fluorescent Chemosensors"

"Fluorescent chemosensors" and "fluorescent probes" are often used as interchangeable terms, although some readers may associate different types of observation object, available instrument, and sensing mechanism with each term. For instance, some people regard fluorescent chemosensors to be molecules for sensing ions or other chemical species using spectroscopic methods, whereas fluorescent probes are tools for visualizing inter/intracellular biomolecules, cell organelles, or tissues by means of microscopy. A similar terminological confusion comes from the function of fluorescent molecules. For some people, fluorescent chemosensors should be able to vary their fluorescent properties when the sensor captures the analyte. While this might be the general definition of fluorescent chemosensors (Rice et al. 2002), the sensors may be called "fluorescent chemodosimeters" instead when their binding to the analyte is irreversible. On the other hand, some fluorescent molecules simply bind to the analyte without any specific change in their fluorescence properties. While these molecules certainly have their uses, they tend to be excluded from the category of fluorescent chemosensors and termed instead as fluorescent probes, markers, and so on.

To avoid the terminological confusion, here the authors would like to include what may be known as fluorescent chemodosimeters, fluorescent probes, etc. in the discussion of fluorescent chemosensors. That is, all fluorescent molecules capable of quantitative/qualitative analysis, monitoring, or visualization of their target objects will be regarded as fluorescent chemosensors in this chapter.

11.2 Molecular Structure of Fluorescent Chemosensors

11.2.1 Components

Basically, fluorescent chemosensor consists of a **fluorophore** as a signal indicator and a **binding site** to capture the analyte of interest. When the fluorescent function of the sensor is activated by host–guest interaction (vide infra, see Sect. 11.4.1), its binding site may be called a "receptor". Moreover, if the binding is selective for the receptor in proteins, the binding molecule will be called a "ligand". To avoid this terminological complexity, the phrases "fluorophore" and "binding site" are used as much as possible in this chapter.

Generally, fluorophores are selected from functional dyes as described in Chaps. 1, 2, 3, 4, 5 and 6. On the other hand, there are many different types of the binding site. For instance, the receptors (chelators) in host–guest chemistry, typified by the crown ether family (Wu et al. 2017; Rice et al. 2002), ligands for receptors in peptides (Kubota and Hamachi 2015), aptamers (Famulok et al. 2007), antibodies (Urano et al. 2009), and others (Wu et al. 2015; Sapsford et al. 2013; Zhu et al. 2016), have been intensively used as binding sites in fluorescent chemosensors. However, it should be kept in mind that any molecule exhibiting high reactivity or affinity to the analyte (even the dye itself) may act as a binding site. For example, cationic and/or planar fluorophores tend to bind to DNA despite having no specific binding sites. Thus, various molecular structural characteristics, such as the polarity, rigidity, and hydrophobicity/hydrophilicity may direct the fluorophore to the analyte.

11.2.2 Representative Structures

The typical structure of existing fluorescent chemosensors is described in Fig. 11.1a, where a fluorophore is covalently connected to the binding site through a spacer. Most fluorescent chemosensors introduced in this chapter fall within this framework. In general, fluorescent chemosensors with such a structure are expected to change their fluorescent properties when their binding site interacts with the analyte through a signal transduction mechanism that will be described in Sect. 11.4. However, this does not always occur for the so-called "fluorescent probes (or markers)", as mentioned in Sect. 11.1.

In the past two decades, new structures have been employed in fluorescent chemosensors that use non-covalent intermolecular interactions (hydrogen bond, electrostatics, π-π stacking, Van der Waals, or hydrophilic/hydrophobic interaction). As represented in Fig. 11.1b, the fluorophore is initially non-covalently connected to a binding site. In the presence of analyte, it separates from the original binding site, resulting in a change in the fluorescence properties. Although this so-called

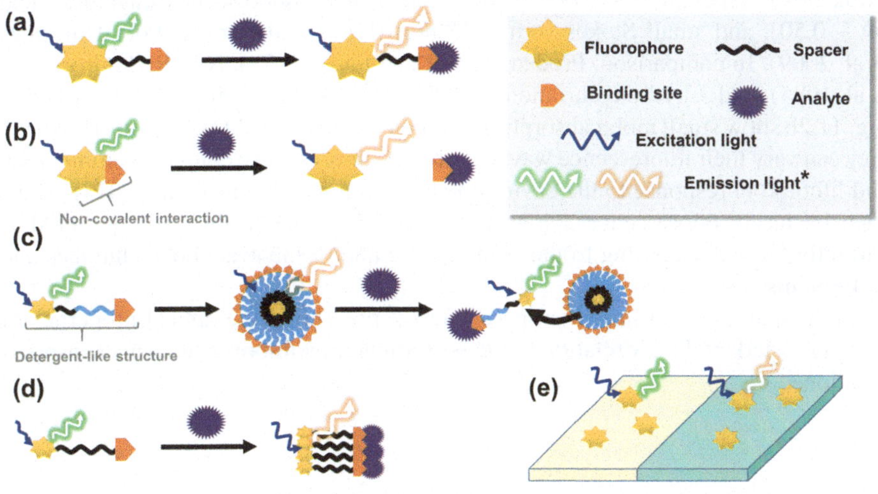

Fig. 11.1 Representative structures and emitting behaviors of fluorescent chemosensors. Different colors in the emission arrows denote change in the fluorescence color or intensity

supramolecular approach (ensemble approach) is not yet a mature technique, theoretically such fluorescent chemosensors can be prepared by simply mixing a fluorophore and binding site. Thus, this method could expand the available kinds of fluorophores and simplify the molecular synthesis. This type of sensor is skipped in this chapter; however, they are well summarized in the review by Kim and co-workers (Wu et al. 2015).

Another type of supramolecular approach is described in Fig. 11.1c, d. Here, the assembly/disassembly of a fluorophore in the presence/absence of analyte is strategically used as a trigger to induce variation in the fluorescent properties of the fluorophore. Relevant examples will be shown in Sects. 11.4.6 and 11.4.7.

Moreover, as explained in Sect. 11.2.1, some fluorescent chemosensors do not contain a specific molecule as binding site. One example is when the sensor is required to detect not specific chemical species but the "phase" in a certain system, as shown in Fig. 11.1e. Environment-sensitive dyes such as Prodan and Nile Red (Klymchenko 2017) should be good candidates for such an application. The relevant examples will be shown in Sects. 11.4.8 and 11.4.9.

11.3 Dye Selection in Fluorescent Chemosensors

The choice of fluorophore is probably the most important issue in developing novel fluorescent chemosensors. Before this chapter, several small organic dyes such as rhodamines, cyanines, and BODIPY shown in Fig. 11.2a were introduced in detail. Those dyes are generally known to exhibit large molar absorption coefficients (molar extinction coefficient $\varepsilon \sim 100{,}000\ M^{-1}\ cm^{-1}$), high fluorescence quantum yield ($\Phi \sim 0.50$), and small Stokes shift (~25 nm) (Lavis and Raines 2008; Umezawa et al. 2009). In comparison, Prodan (Weber and Farris 1979), Nile Red (Greenspan et al. 1985), and 3-hydroxychromones (3HC) (Demchenko et al. 2009) described in Fig. 11.2b show small molar absorption coefficients and large Stokes shifts. However, they can vary their fluorescence wavelength (i.e., fluorescence color), quantum yield, and lifetime in response to the environmental condition (Klymchenko 2017). Thus, each dye has its fluorescence characteristics, and therefore the fluorophore should be carefully chosen according to the intended use and/or the available techniques and instruments.

In this part, several important parameters representing the optical properties of dyes are listed, and their relation to the performance as fluorescent chemosensors is mentioned.

11.3.1 Absorption/Fluorescence Wavelengths

Each dye exhibits inherent wavelengths for photon absorption/fluorescence. The first consideration is determining what to observe and which analytical instruments and

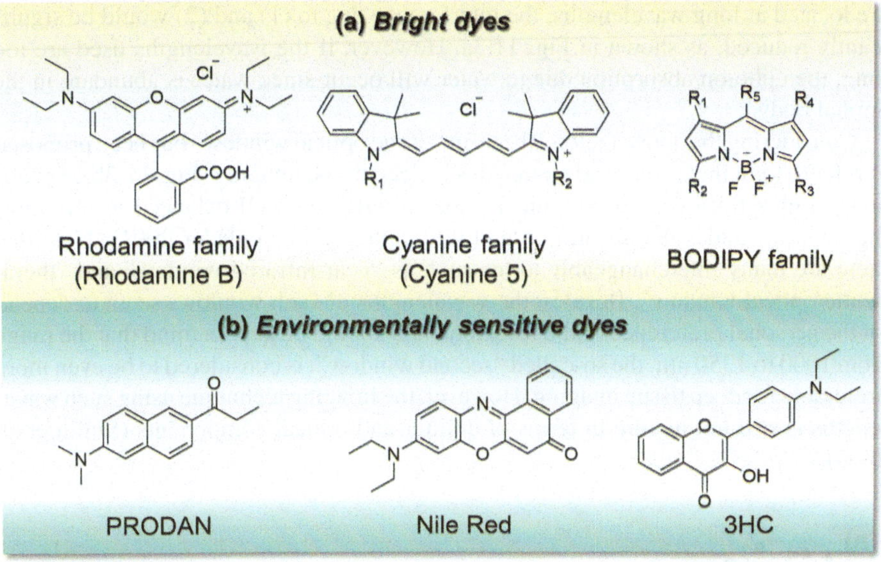

Fig. 11.2 Molecular structures of representative organic dyes used in fluorescent chemosensors (Lavis and Raines 2008; Umezawa et al. 2009; Weber and Farris 1979; Greenspan et al. 1985; Demchenko et al. 2009)

techniques are available. Then, dyes exhibiting the desired absorption/fluorescence wavelengths could be considered as candidate fluorophores.

In cell imaging using fluorescence microscopy, dyes absorbing visible light tend to be favored. In fact, commercial microscopes such as the wide field and confocal types are generally equipped with the optical system for such dyes. This is because the photo-excitation of dyes by UV light would cause unwanted photon absorption from inherent molecules such as NADPH and flavins, which produce auto-fluorescence that interferes with observing the target of interest (Monici 2005). However, for dyes exhibiting fluorescence at longer wavelength or with a large Stokes shift, the auto-fluorescence would be less of a problem.

In the case of in vivo imaging using animals, fluorophores with absorption and emission at wavelengths longer than visible light, i.e., near-infrared (NIR) light are highly valuable. There are two reasons for this. (1) When using UV or visible light for excitation, the abovementioned undesired absorption due to both extra- and intracellular components in tissues such as collagen, elastin, hemoglobin, NADH, riboflavin, retinol, and folic acid is much more pronounced than in living cells (Monici 2005). (2) Furthermore, the scattering of excitation/emitting light in tissue imaging is much more crucial than in cell imaging (Jacques 2013). In fact, (1) and (2) are synergistic in hampering the penetration of excitation light into the dyes in the animal body. As a result, both the excitation of the fluorophore and the observation of fluorescence from the dye are largely restricted. If the dye's excitation and emission wavelengths are located at long wavelengths, the interference due to (1) and (2) would be significantly reduced, as shown in Fig. 11.3a. However, if the wavelengths used are too long, then photon absorption due to water will occur since water is abundant in the animal body.

Considering the above factors, the term "tissue optical window" has been proposed to refer to light that is more advantageous in deep tissue imaging (Fig. 11.3b). Specifically, lights from 650 to 900 nm (far-red to NIR) are well-balanced to minimize the effect of undesired absorption and the scattering (Weissleder 2001). Note that there are many interchangeable terms such as "near-infrared window" and "therapeutic/optical window". Besides, the wavelengths of such windows seem to depend on the personal preference of the researchers. It should be kept in mind that the range from 1000 to 1350 nm, the so-called "second window", is considered to be even more preferable for deep tissue imaging. However, the imaging technique using such wavelengths is still not mature in terms of emitter and optical components (Smith et al. 2009).

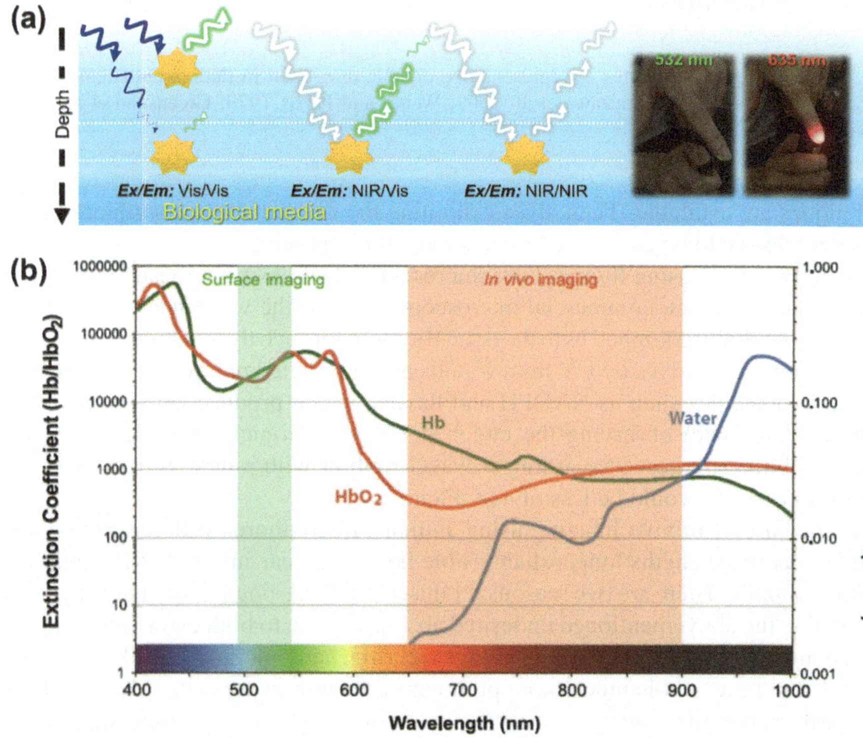

Fig. 11.3 **a** Schematic illustration of the advantages of NIR (red-shifted) excitation/fluorescence. **b** Extinction coefficients of water and oxy- and deoxyhemoglobin plotted from visible to NIR wavelengths. Panel **b** is adapted with permission from Urano et al. (2009). Copyright 2010, American Chemical Society

High-performance liquid chromatography (HPLC) coupled with fluorescence detection is often used for the detection, characterization, and analysis of trace substances (vide infra, see Sect. 11.4.5) (Schuster 1988). Such detectors usually use the UV/Vis region, therefore dyes with absorption/fluorescence around these wavelengths are available as fluorophores. Similarly, when fluorescence spectroscopy is the main instrument, dyes that work in the UV/Vis region might be required. This is because spectrofluorometers that can function from the UV/Vis to NIR region are relatively expensive.

11.3.2 Brightness

Many researchers previously seemed to regard the fluorescence quantum yield as the emitting ability of the dye. However, considering that fluorescence occurs upon photon excitation of the dye, the product of fluorescence quantum yield and molar extinction coefficient ($\Phi\varepsilon$) should be a more practical parameter to represent the dye's ability to emit. This parameter is called "theoretical brightness", and its values for representative fluorophores are summarized in Table. 11.1 (Lavis and Raines 2008). Highly bright dyes have significant advantages (Reisch and Klymchenko 2016). For instance, they allow the power of excitation laser in microscopy to be decreased while maintaining the high signal to noise (S/N) ratio. This is quite valuable in suppressing the dye bleaching, damage to biomolecules, temperature change in the matrix of interest, and so on. Moreover, several techniques often require a high brightness, such as single-molecule tracking and the observation of deep tissue. In not only fluorescence microscopy but also other methods such as spectroscopy and HPLC, bright dyes are expected to provide high sensitivity for detecting the analyte.

The theoretical brightness is conventionally obtained using a spectrofluorometer equipped with xenon lamp. Alternatively, the brightness can be directly estimated from a fluorescence microscope equipped with laser, by counting the number of photons emitted by the dye in unit time. Note that the brightness values obtained by these two methods are not always identical. This is because when excited by a high-power laser (10–1000 W cm^{-2}), more of the dye suffers fast photo-bleaching and/or generates non-fluorescent and long-lived triplet excited state as compared to excitation with the xenon lamp (~1 mW cm^{-2}), resulting in fewer emitted photons

Table. 11.1 Brightness and photostability of representative organic/inorganic dyes

	Organic dyes		Quantum dots (Reisch and Klymchenko 2016)
	Rhodamine 6G (Soper et al. 1993)	Nile Red (Niko et al. 2015)	
Brightness ($\Phi\varepsilon$)	1.1×10^5	2.6×10^4	10^5–10^6
Photostability[a]	1.9×10^6	1.9×10^5	10^7–10^8

[a]The total number of photons emitted from a single dye molecule or particle before photo-bleaching

in a given time. On the other hand, the brightness estimated from microscopy is apparently more reliable, therefore this method is recommended for bioimaging. In any case, it should be noted that the illumination power also tends to affect the brightness of the dyes (Reisch and Klymchenko 2016).

An important consideration comes from the above discussion: between two dyes with the same theoretical brightness, the one exhibiting higher fluorescence quantum yield might be more useful than the other, because it is less likely to generate non-fluorescent species during laser excitation.

11.3.3 Photostability

The main drawback of organic dyes is that they are easily photo-bleached compared to inorganic dyes. This often becomes an obstacle in bioimaging, which usually uses a high-power laser as the excitation source. Several mechanisms have been proposed for photo-bleaching. The main one is the generation of excited triplet state of the dye via intersystem crossing after photo-excitation, which results in photochemical reaction like the Norrish type (IUPAC 1997a, b). In addition, dye in the triplet excited state may transfer its excitation energy to molecular oxygen, thereby generating highly reactive singlet oxygen that can attack several kinds of molecules including the dye itself (Zheng et al. 2014). Another mechanism is photo-isomerization, typified by the isomerization between *trans*-stilbene (fluorescent) and *cis*-stilbene (non-fluorescent) (Görner and Kuhn 2007).

Photostable dyes have the advantage of enabling long-term imaging. Moreover, dyes with high photostability are indispensable for some techniques such as single-molecule tracking (Zheng et al. 2014) and super resolution imaging (e.g., stimulated emission depletion microscopy (STED)) (Wang et al. 2017). In addition, the photo-stability would be important for storing the dye. On the other hand, extreme photo-stability is not always useful in fluorescence imaging, because in one technique the dye is intentionally bleached.

Photostability of the dye is often characterized by a spectrofluorometer equipped with xenon lamp. In this method, the dye is continuously excited for a given time while the decay of fluorescence intensity is monitored. Photostability of the dye can be characterized by the decay constant, which is easily estimated from the analysis of the fluorescence decay curve, by assuming that photodegradation of the excited dye is a first-order reaction. This method is applicable when laser is the excitation source. Alternatively, photostability of the dye can be estimated using a microscope equipped with laser, by counting the total number of photons emitted by the dye until photo-bleaching (see Table 11.1). It is known that small organic dyes generally deliver between 10^5 and 10^6 photons for periods of 10–100 s under the strong illumination required for their imaging ($\approx$1 kW cm^{-2}). Note that photostability of the dye obtained from spectrofluorometer and microscopy are not always identical, like the case of brightness characterization (Reisch and Klymchenko 2016). This might be related to the excess accumulation of long-lived excited triplet state under high-power excitation, which accelerates the photo-bleaching.

11.3.4 Two-Photon Absorption Cross Section

Typically, an organic dye absorbs a "single" photon to become excited and then returns to the ground state via emission (fluorescence), thermal deactivation (internal conversion and subsequent vibrational relaxation), or intersystem crossing to triplet state and subsequent thermal deactivation or phosphorescence. However, when using an extremely intense excitation light, such as femtosecond pulse laser, the dye may absorb two photons simultaneously. This phenomenon is called "two-photon absorption" or "two-photon excitation" of the dye (Fig. 11.4a).

After either one-photon or two-photon excitation, the dye molecule decays to the ground state via photophysical processes. Usually, the wavelength used in two-photon excitation is approximately twice that used in the corresponding single-photon excitation to maintain the same total excitation energy. This means that two-photon absorption can occur under significantly red-shifted light compared to single-photon excitation. As described in Sect. 11.3.1, this gives many advantages in fluorescence imaging. For instance, two-photon excitation should be very valuable for deep tissue imaging that requires excitation at long wavelengths (tissue optical window). Moreover, two-photon absorption usually occurs at the focal point of the laser where the photon flux is extremely high. Therefore, spatially selective excitation of the dye can be realized.

The efficiency of two-photon absorption by the dye is described by "two-photon absorption cross section", which can be estimated by the absolute method (z-scan method) or the relative method (two-photon excited fluorescence method) (Pawlicki et al. 2009). The product of two-photon absorption cross section and fluorescence quantum yield is called the "two-photon action cross section", which can be regarded as the theoretical brightness under two-photon excitation.

It should be kept in mind that the wavelength for two-photon excitation is just as important as the two-photon action cross section. In in vivo imaging, an excitation wavelength over 900 nm is more preferable than that below 900 nm where several biomolecules would be excited (Fig. 11.4b, c) (Ahn et al. 2015).

11.3.5 Sensitivity to Environmental Condition

Some dyes vary their fluorescence properties (intensity, color, lifetime, etc.) in response to environmental conditions (e.g., polarity, viscosity, and pH) (Klymchenko 2017). As might be expected, the utility of such dyes largely depends on their sensitivity (the extent of fluorescence property variations) to the changing in environmental condition.

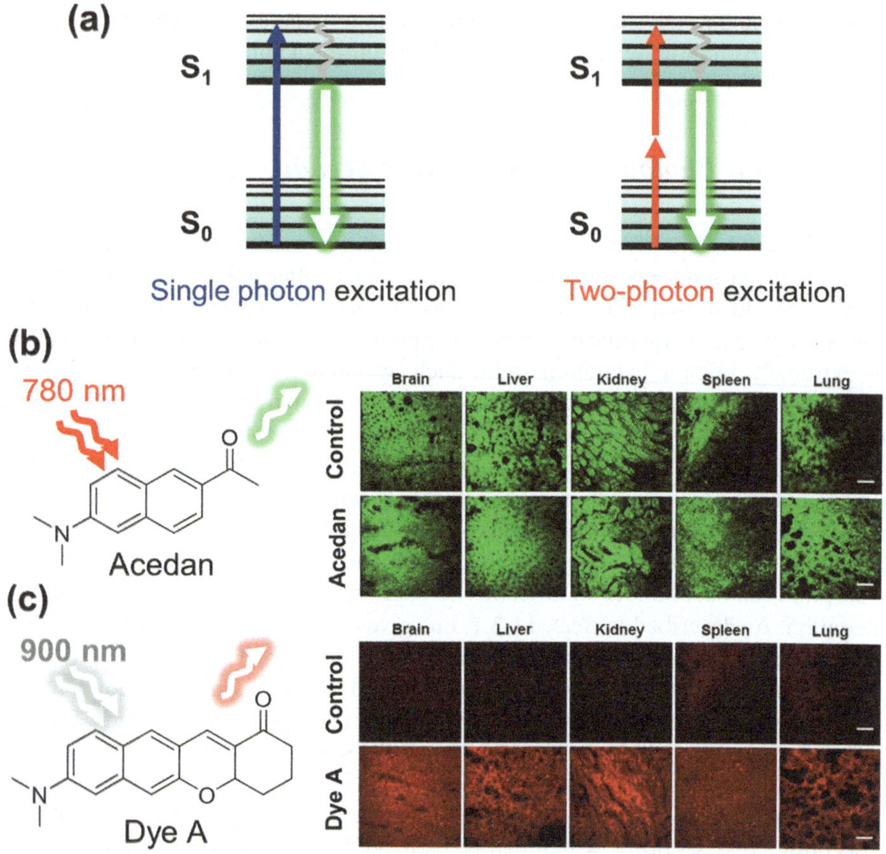

Fig. 11.4 **a** Schematic illustration of single and two-photon excited fluorescence. **b**, **c** Two-photon fluorescence microscopic imaging of tissues of different mouse organs, acquired with acedan (excited at 780 nm) and dye A (excited at 900 nm), respectively, at a middle depth (~20 μm) of the sectioned tissue (thickness ~50 μm). Images adapted with permission from Ahn et al. (2015). Copyright 2015, American Chemical Society

Dyes such as Prodan and Nile Red change their fluorescence color in response to the surrounding polarity, and this phenomenon and the dyes are called fluorescence solvatochromism and solvatochromic dyes, respectively. The sensitivity of fluorescence wavelength to environmental polarity is usually evaluated by introducing parameters such as orientation polarizability and/or $E_T(30)$ (Reichardt 2005) if the fluorescence solvatochromism is based on intramolecular charge transfer (ICT) (vide infra, see Sect. 11.4.8).

Orientation polarizability is a theoretical polarity parameter of solvent derived from the Onsager model, and it is expressed by the dielectric constant and refractive index of solvents. By plotting the Stokes shift versus the orientation polarizability of dye in given solvents, the difference in dye dipole moment between the ground and excited states can be estimated (Lippert-Mataga equation) (Mataga et al. 1956, 2005; Lippert 1957). If the difference is large, then the dye has strong ICT characteristics and therefore exhibits large Stokes shift, broadened fluorescence, and strong fluorescence solvatochromism.

$E_T(30)$ is a solvent polarity parameter experimentally obtained using Reichart's dye as a reference. Specifically, the slope of the fluorescence peak wavenumber versus $E_T(30)$ plot of the solvatochromic dye of interest in given solvents directly shows dye sensitivity.

Another class of environment-sensitive dyes is typified by 9-(dicyanovinyl)julolidine (DCVJ) (Klymchenko 2017; Loutfy 2007; Kung and Reed 1986), which varies its fluorescence intensity in response to the surrounding viscosity. The sensitivity of such dyes is generally evaluated by using Förster-Hoffmann equation (Loutfy 2007; Kung and Reed 1986).

In many cases, the abovementioned dyes change in their fluorescence lifetime as well as fluorescence intensity/color. These properties are also useful for fluorescence lifetime imaging (FLIM), and therefore their evaluation may expand the utility of the dyes (vide infra, see Sect. 11.4.8).

11.3.6 Other Considerations

The hydrophilicity/hydrophobicity, cell permeability, and cell toxicity of the newly developed fluorescent chemosensors may be significant factors in their practical use. Unfortunately, these considerations are beyond the focus of this chapter due to text length limit.

11.4 Operation Principles of Fluorescent Chemosensors

As mentioned in Sect. 11.2, fluorescent chemosensors generally capture the analyte at first and then change in their fluorescence intensity, color, or lifetime through certain signal transduction mechanisms. Such changes in fluorescence properties could be observed by a spectrofluorometer, fluorescence microscope, etc. to allow the detection, quantitation, monitoring, and visualization of the analyte and related phenomena. On the other hand, it should be kept in mind again that fluorescent chemosensors that selectively bind to the analyte without changing their fluorescence properties are also useful makers to visualize the object of interest.

In this chapter, we present representative operation principles by which fluorescent chemosensors change their fluorescence properties, along with the related photophysical mechanism. It should be noted that multiple principles can function simultaneously in a single fluorescent chemosensor. In addition, as mentioned in Sect. 11.1, the binding site itself in fluorescent chemosensors does not have to directly induce change in the fluorescence signal. That is, even when the binding site is used to simply capture a specific analyte, the fluorescent chemosensor may vary its fluorescence properties via certain mechanisms (vide infra, see Sect. 11.4.7). Unfortunately, details about the significance of the detection of each analyte and the underlying theoretical basis (e.g., equations to calculate the kinetic constants such as the association/disassociation rate constants) are skipped due to limitation in the length. The reader is referred to the corresponding references.

11.4.1 "Primordial" Fluorescent Chemosensors Based on Host–Guest Chemistry

First of all, the primordial fluorescent chemosensors developed by Sousa and co-worker should be introduced (Sousa and Larson 1977), because these sensors opened the door to today's intensive development of fluorescent chemosensors based on host–guest chemistry. In sensor **1** shown in Fig. 11.5, a crown ether and a naphthalene moiety are covalently connected and act as the binding site and the fluorophore, respectively. Once the crown ether forms a coordination complex with an alkali metal ion, the fluorescence intensity of the naphthalene fluorophore at 77 K is decreased whereas the phosphorescence intensity is increased. Contrarily, the fluorescence intensity of sensor **2** is increased when it is coordinated with potassium

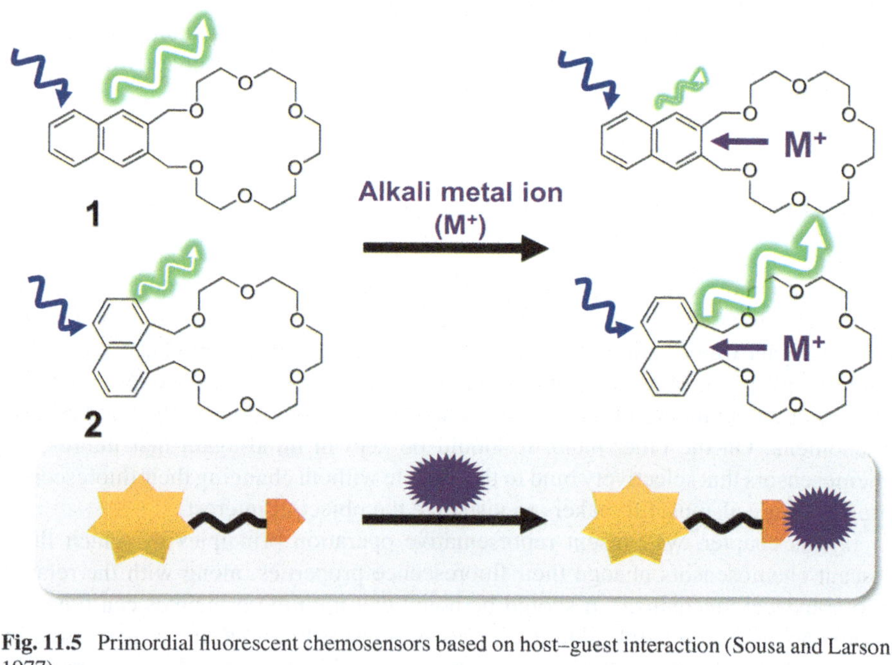

Fig. 11.5 Primordial fluorescent chemosensors based on host–guest interaction (Sousa and Larson 1977)

(K^+), rubidium (Rb^+), or cesium (Ce^+) ions. The successful design of **1** and **2** definitely implies that host–guest interaction would be useful for developing fluorescent chemosensors.

In the host–guest-based metal ion sensor, the metal coordination itself may affect the photophysical properties of fluorophore in several ways (Rice et al. 2002). For instance, the rigidity of the sensor molecule might be increased along with metal ion coordination, resulting in decreased thermal deactivation of the excited fluorophore, i.e., an increase in the fluorescence quantum yield. In addition, the energy difference between excited first singlet state and its adjacent excited triplet state of the fluorophore would change upon coordination, leading to significant change in the efficiency of intersystem crossing (the transition from excited singlet state to triplet state). Moreover, metal ion coordination may provoke the "heavy atom effect" (IUPAC 1997c), which is known to facilitate intersystem crossing by enhancing the spin–orbit interaction. Thus, the observed change in fluorescence intensity of **1** and **2** upon complexation might result from one or more of the abovementioned effects. In later research, host–guest interaction has been used much more strategically, and the representative approaches will be described in Sects. 11.4.2 and 11.4.3.

11.4.2 Photo-Induced Electron Transfer

Among fluorescent chemosensors based on host–guest chemistry, recently the most powerful and frequently used operation principle is "photo-induced electron transfer (PET)" (Silva et al. 2009). Generally, organic dyes are in the lowest excited singlet state upon photo-excitation. Fluorescence is the light emitted when such a dye molecule returns to the ground state (Fig. 11.6a). When an electron donor molecule (i.e., PET donor) is adjacent to the excited dye while its highest occupied molecular orbital (HOMO) is between energies of the two single occupied molecular orbitals (SOMO) formed by the excited dye, electron transfer would occur from HOMO of the PET donor to lower energy SOMO of the dye (Fig. 11.6b). This phenomenon is called reductive electron transfer. Contrarily, electron transfer from a high-energy

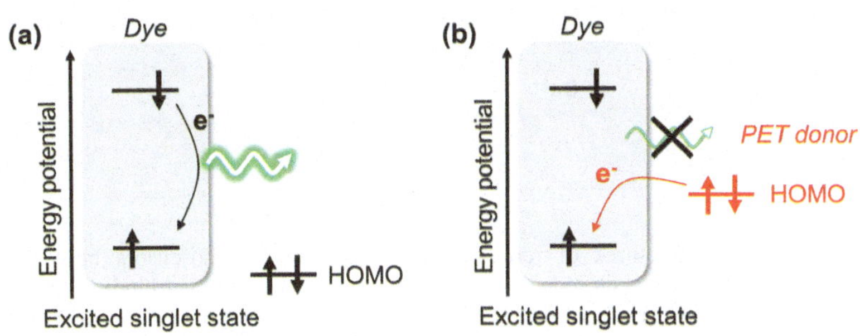

Fig. 11.6 Schematic illustration of PET

SOMO of the dye to the lowest unoccupied molecular orbital (LUMO) of an electron acceptor molecule (i.e., PET acceptor) is also possible and called oxidative electron transfer. These phenomena are collectively called PET, which causes fluorescence quenching of the dye (PET quenching).

In the developed fluorescent chemosensors based on PET mechanism, the fluorophore is initially connected to the PET donor or acceptor molecule so that it is temporarily non-fluorescent (OFF state). Typically, the PET donor or acceptor moiety also acts as a binding site. Specifically, when the PET donor as a binding site captures the analyte, its HOMO level would be significantly decreased. This results in the suppression of PET quenching, i.e., enhanced fluorescence intensity of the fluorophore (ON state). Thus, PET-based fluorescent chemosensors enable quantitative analysis by switching their fluorescence to OFF/ON states in the absence/presence of the analyte.

A milestone in the history of PET-based fluorescent chemosensors is 1,2-bis (2-aminophenoxy) ethane-*N*,*N*,*N'*,*N'*-tetraacetic acid (BAPTA) (Fig. 11.7 for Ca^{2+}), the binding site to capture calcium ion (Ca^{2+}) developed by Roger and co-workers in 1980 (Tsien 1980). BAPTA contains tertiary amino groups whose HOMO energy is between the two SOMO of the fluorophore. The fluorophore connected to BAPTA itself is quenched through the PET mechanism (OFF state). On the other hand, BAPTA can capture Ca^{2+} to form a chelate complex, then the HOMO level of its tertiary amino groups is significantly decreased because lone pairs of the tertiary amino group participate in the coordination. As a result, the initial PET quenching is suppressed and the fluorophore becomes fluorescent (ON state). Using such OFF–ON switching, high-quality sensing of Ca^{2+} can be realized. What should be noted here is that BAPTA exhibits remarkable selectivity towards Ca^{2+}, although typical sensors could not distinguish between Ca^{2+} and magnesium ion (Mg^{2+}). Efforts to

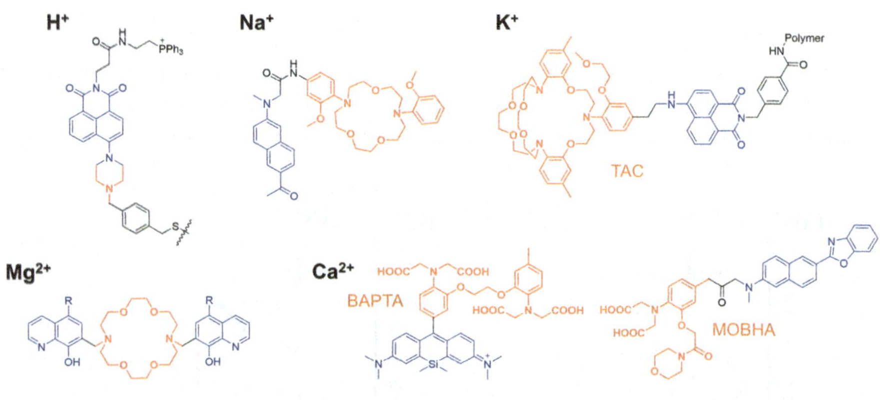

Fig. 11.7 Molecular structures of representative fluorescent chemosensors based on PET quenching. Blue- and orange-colored structures denote the fluorophore and binding sites, respectively. H^+: Lee et al. (2014), Na^+: Kim et al. (2010), K^+: Mortellaro et al. (2003), Mg^{2+}: Sale et al. (2006), and Ca^{2+}: Koide et al. (2011) for BAPTA and Kim et al. (2010) for MOBHA

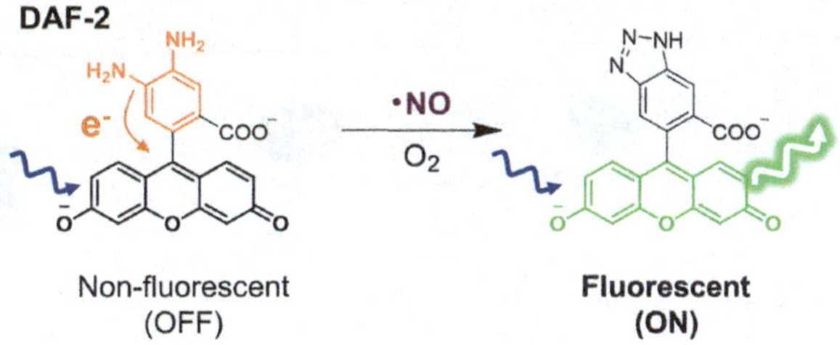

Fig. 11.8 Molecular structure of DAF-2, a PET-based fluorescent chemosensor activated by the reaction with nitric oxide (Hirata et al. 2002)

develop other fluorescent sensors bearing BAPTA have been made, and currently there is one that could realize in vivo imaging (Yin et al. 2015).

As is apparent from the example of BAPTA, the binding site has a significant role in PET-based fluorescent chemosensors. The binding site as PET donor (or acceptor) must have the ability to quench the fluorophore and also possess high selectivity for the analyte. The development of PET-based fluorescent chemosensors so far was accompanied by the search for superior binding sites. Some exemplar binding sites for PET-based fluorescent chemosensors are listed in Fig. 11.7, which reveals that many binding sites consist of tertiary amino group with a chelator like crown ether. Therefore, host–guest interaction is very often used in PET-based fluorescent chemosensors, because crown ether exhibits good selectivity to targeted metal ion. On the other hand, it should be kept in mind that the crown ether-based binding site is often hard to prepare, and several highly accessible and selective alternatives have been proposed (Watanabe et al. 1998, 2001).

Note that PET-based fluorescent sensors do not always use host–guest interactions. A representative example is DAF-2 shown in Fig. 11.8 (Hirata et al. 2002). DAF-2 is initially PET-quenched due to electron transfer from the phenyl moieties containing two amino groups to the xanthene fluorophore (OFF state). However, two amino groups located at the ortho position can react in living cells with nitric oxide (NO), which plays important roles in many biological processes, resulting in the formation of a triazole structure and subsequent ON state of the fluorophore. That is, DAF-2 works as an OFF/ON sensor for the detection of NO through the "reaction" between diamino groups and NO. In this type of sensor, the moiety exhibiting reactivity to the analyte is equivalent to the binding site.

11.4.3 Heavy Atom Effect

A sophisticated use of metal ion coordination, i.e., host–guest interaction is fluorescent chemosensors activated by the heavy atom effect. As shown in Fig. 11.9, sensor **3** (Madhu et al. 2014), which consists of the BODIPY fluorophore and Schiff base as a binding site, is fluorescent at first. Next, it is intentionally coordinated with the mercury ion (Hg^{2+}) to be in the OFF state due to heavy atom effect. Hg^{2+} is known to bond to chloride (Cl^{-}), and so once the sensor in OFF state meet Cl^{-}, it releases

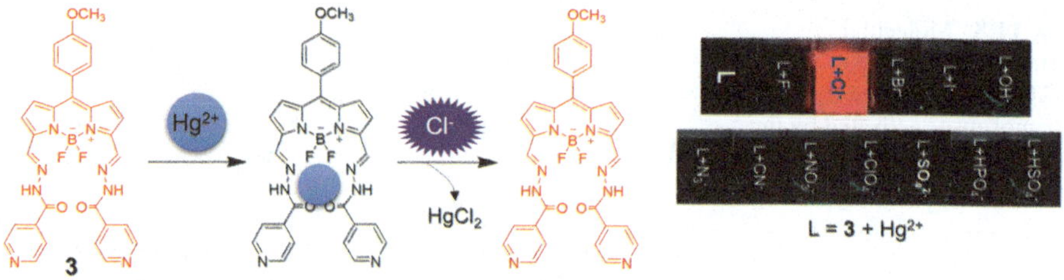

Fig. 11.9 Molecular structure of chloride sensor **3** designed to use heavy atom effect. Image adapted with permission from Madhu et al. (2014). Copyright 2014, Royal Society of Chemistry

Hg^{2+} and becomes fluorescent (ON state). That is, **3** works as an OFF–ON switching sensor for Cl^-. As is apparent from this example, this type of sensor often targets anions that exhibit strong interaction with heavy metal ions.

11.4.4 Förster Resonance Energy Transfer

Sometimes, the fluorescence from dye B is observed through photo-excitation of dye A located in the vicinity of B. This phenomenon usually results from energy transfer from A to dye B, which is generally explained by two mechanisms, i.e., Förster or Dexter mechanisms. The former is very frequently used in fluorescent chemosensors, therefore it should be the focus here. Note that the Förster type energy transfer, i.e., Förster resonance energy transfer (FRET) is often called "fluorescence resonance transfer" instead, even though the former is recommended by IUPAC (1997d).

Here, two fluorescent dyes are involved, namely, the FRET donor (D) and acceptor (A) (see Fig. 11.10). When the overlap between the fluorescence spectrum of D and the absorption spectrum of A is large and the distance between D and A is short

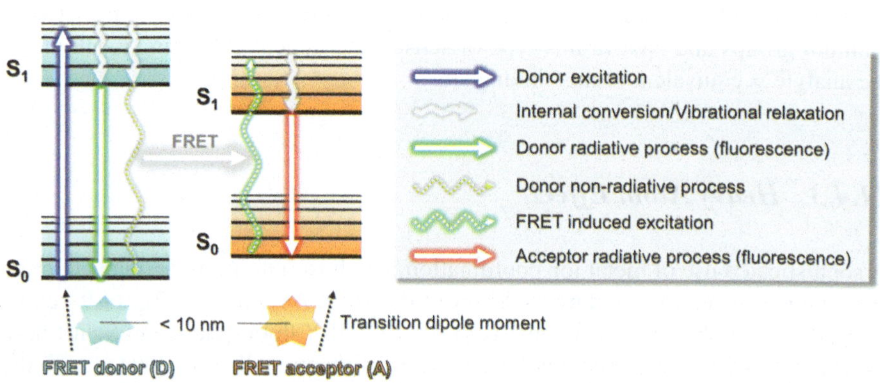

Fig. 11.10 Schematic diagram of Förster resonance energy transfer (FRET)

(usually less than 10 nm), the excitation energy of D could transfer to A, therefore fluorescence from A would be observed. Such FRET process is often explained by analogy to the resonance observed between a vibrating and a non-vibrating tuning fork. A large spectral overlap and a shorter distance between D and A result in highly efficient energy transfer. Thus, if the D and A species are fixed, the efficiency of FRET is dependent on their distance, which allows the measurement of distance between two positions in biomolecules where D and A are attached (Sahoo 2011; Ma et al. 2014). Additionally, the orientation of transition dipole moments between D and A may also be important. This is because FRET theoretically occurs when the dipolar oscillation of D in the excited state satisfies the condition to resonate with that of A in ground state, and so FRET will be forbidden if the two transition dipole moments are perpendicularly oriented. This orientation factor may cause difficulty in precisely measuring the distance between D and A bound to macromolecules like proteins, where the dyes cannot freely rotate and so that the orientation factor is not averaged.

It should be kept in mind that FRET is based on resonance, which means A does not absorb the fluorescence of D (i.e., reabsorption). Rather, an energy transfer process based on light reabsorption is called radiative energy transfer, whereas FRET is called non-radiative (radiationless) energy transfer. These two cases can be distinguished by comparing the fluorescence lifetime of a single D molecule (without energy transfer) to that of D with A (with energy transfer), because the fluorescence lifetime of D before and after FRET should be different. Such a distinction is sometimes required for precise fluorescence lifetime imaging.

Apparently, fluorescent chemosensors based on FRET basically exhibit the FRET ON–OFF (or OFF–ON) switching in the presence/absence of the analyte. A recent pioneering example is shown in Fig. 11.11, where the fluorescence chemosensor **4** consists of a rhodamine as D and silica-rhodamine as A to induce the FRET ON state (Kamiya et al. 2016). The silica-rhodamine part can react with glutathione, causing a blue shift in its absorption spectrum along with change in the π-conjugation system. As a result, the initial FRET is quenched (OFF state) and therefore the ratio of fluorescence intensities due to D and A is strongly varied, which allows several

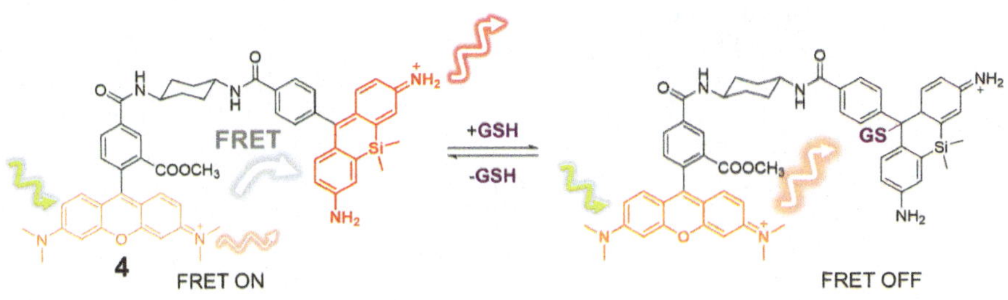

Fig. 11.11 Molecular structure of FRET-based fluorescent chemosensor **4** and its FRET ON–OFF behavior (Kamiya et al. 2016)

quantitative ways to analyze glutathione using fluorescence lifetime imaging and ratiometric imaging using microscopy.

11.4.5 Formation of Emissive π-System via Chemical Reaction

The fluorescence properties of organic fluorescent dyes are critically associated with their π-conjugation structure. Even if the π-system is initially non-fluorescent, it might be made fluorescent through chemical reactions such as transition metal-catalyzed cross-coupling, Witting reaction, or Knoevenagel condensation, all of them are used to extend the π-conjugation of aromatic molecules. Here, fluorescent chemosensors related to their π-conjugation structures will be presented.

First, the sensor **5** developed by Czarnic and co-worker in 1997 is shown in Fig. 11.12 (Dujols et al. 1997). Initially, the sensor molecule containing hydrazide as a binding site is not fluorescent. However, the hydrazide moiety can selectively react with copper ion (Cu^{2+}) as the analyte, and then the cyclic structure in the sensor is opened along with the formation of a large π-conjugated structure as fluorophore (Rhodamine B). After this pioneering fluorescent chemosensor driven by ring-opening reaction, a number of similar sensors have been developed. In particular, the principle is well compatible with species that induce specific chemical reactions, e.g., d-block metal ions (Cu^{2+}, Zn^{2+}, and Hg^{2+}), reactive oxygen species (ROS), reactive sulfur species (RSS), and reactive nitrogen species (RNS) (Wu et al. 2017). In addition, sensors of this type also change their

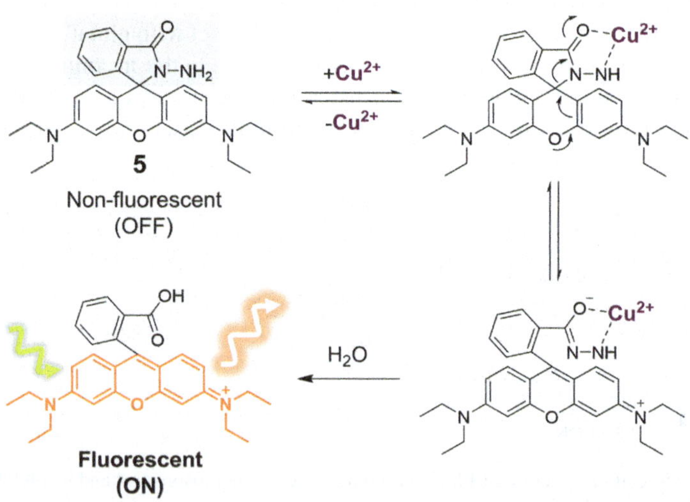

Fig. 11.12 Molecular structure of fluorescent chemosensor **5** and its turn-on behavior based on the formation of a large π-conjugation through ring-opening reaction with Cu^{2+} (Dujols et al. 1997)

HOMO/LUMO energies, so that this principle can be hybridized with sensors based on FRET.

It might be valuable to mention *o*-phthalaldehye (OPA) shown in Fig. 11.13a, a traditional sensor for the detection of amino acids (Schuster 1988). Although OPA is basically non-fluorescent, it forms a pyrrole ring via reaction with primary amine in the presence of thiol to form a π-extended fluorescent product. The naphthalene analog of OPA, called NDA, is also developed and known to exhibit brighter fluorescence than OPA. While neither OPA nor NDA has the selectivity to distinguish the kinds of amino acids, they have been well-used in conjunction with HPLC technique in several fields such as medical and food industries (Fig. 11.13b).

As can be seen in OPA and NDA, a π-system (such as benzene, naphthalene, and pyrene) that is attached to aldehyde or ketone group tends to become less fluorescent due to intersystem crossing. It is known as El-Sayed rule, that the n-orbital on carbonyl group of aldehyde/ketone creates n-π* excited states that efficiently mix

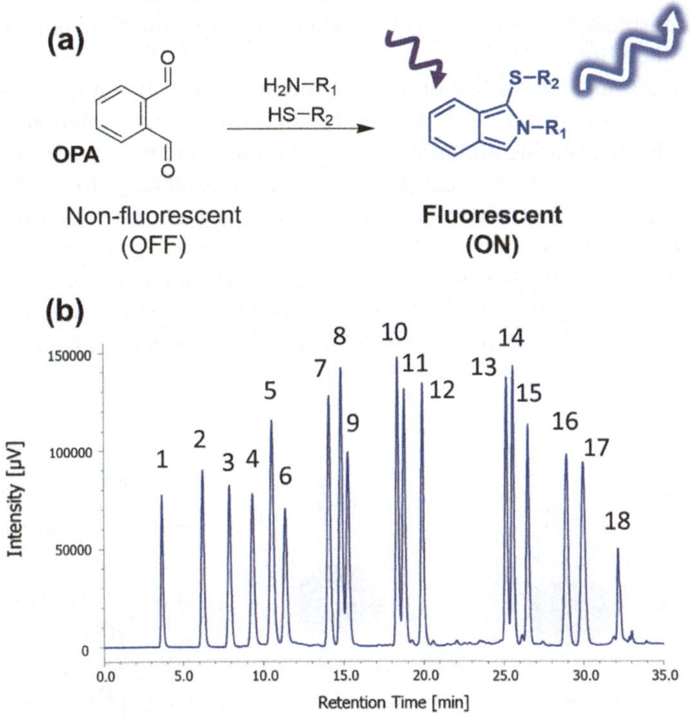

Fig. 11.13 **a** Reaction of OPA with a primary amine in the presence of thiol (Schuster 1988). **b** HPLC chromatogram of amino acids labeled by OPA. 1: Aspartic acid, 2: Glutamic acid, 3: Asparagine, 4: Histidine, 5: Serine, 6: Glutamine, 7: Arginine, 8: Glycine, 9: Threonine, 10: Taurine, 11: Alanine, 12: Tyrosine, 13: Methionine, 14: Valine, 15: Phenylalanine, 16: Isoleucine, 17: Leucine, 18: Lysine. The chromatogram was kindly provided by JASCO Corporation (LC application data No. 430029H, 2016)

with π-π* excited states in different spin multiplicity (IUPAC 1997e). That is, the aldehyde and ketone may induce greater spin–orbit interaction, resulting in inter-system crossing. In the cases of OPA and NDA, the two formyl groups disappear after the reaction with thiol and primary amine, and larger π-systems due to the pyrrole ring are formed. This is the photophysical basis for realizing the fluorescent OFF–ON system in OPA and NDA.

11.4.6 Excimer Emission

When a dye molecule in the ground state is located near another dye in the excited state, a so-called excited complex might be formed between the two. If this occurs between two molecules of the same species, the dimer is called an "excimer (excited dimer)", and otherwise an "exciplex". Pyrene, one of the polyaromatic hydrocarbons, is a representative chromophore that forms excimer with a variation in its fluorescence color from purplish blue (monomer) to greenish blue (excimer), making it a useful fluorescent chemosensor.

The excimer must be formed during the lifetime of the dye in the excited state, therefore it is normally observed in highly concentrated dye solutions, the solid state, and molecule bearing multiple dyes. For example, the introduction of multiple pyrenes into a cyclic molecule such as calix arene causes the dye to exhibit excimer emission because the introduced pyrene rings are very close to each other. From the example of sensor **6** shown in Fig. 11.14, if such a dye has a binding site that captures the analyte and also enlarges the distance between pyrenes, it can work as a fluorescent chemosensor based on excimer ON–OFF switching (Lee et al. 2004). A number of such sensors have been developed for sensing halogen anions or metal cations (Gutsche 2016).

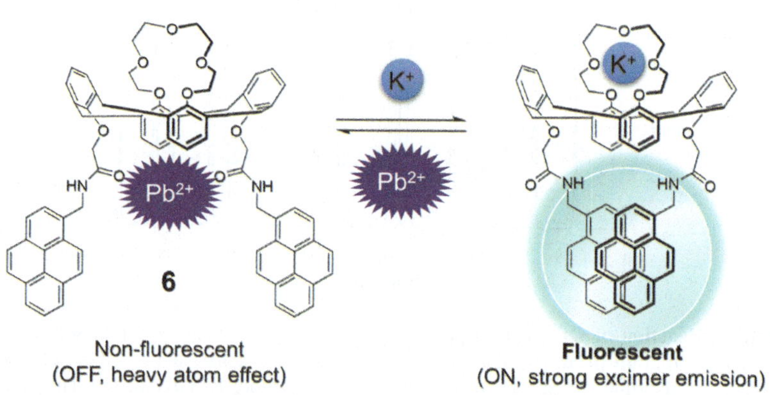

Fig. 11.14 Molecular structure of pyrene-tethered fluorescent chemosensor **6** and its fluorescence turn ON–OFF behavior in the presence of analyte (Lee et al. 2004)

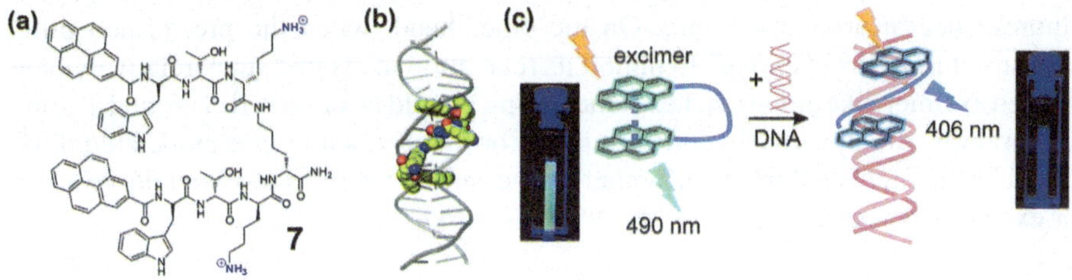

Fig. 11.15 Molecular structure and function of pyrene-based molecular beacon. Image adapted with permission from Wu et al. (2012). Copyright 2012, American Chemical Society

Another example using pyrene's structural characteristic is the pyrene-based peptide beacon reported by Schmuck and co-workers, shown in Fig. 11.15 (Wu et al. 2012). The pyrene moieties in sensor **7** produce excimer emission in water solution, because pyrene forms aggregates due to its high hydrophobicity. On the other hand, the pyrene switches its fluorescence from excimer to monomer emission in the presence of DNA. This is because the highly planar pyrene moieties in the dye intercalate into the double helix, resulting in the discrimination of excimer formation. In this dye, pyrene acts as both the fluorophore and the binding site.

Here, it might be worth mentioning the terminology for excimers. The classical definition of excimer by Birks (1963), as mentioned above, refers to an excited complex formed two identical dye molecules in the excited state and ground state. According to the review written by Winnik, excimers are often separated into two types: "dynamic excimer" and "static excimer" (see Fig. 11.16) (Winnik 1993). The former is the classical excimer. In the static excimer, the two dyes already form a dimer in the ground state. This dimer in excited state (D*, partially overlapped) undergoes structural relaxation to form a more stable structure (E*, sandwich type), which is equal to that formed by the dynamic excimer, and then it exhibits excimer

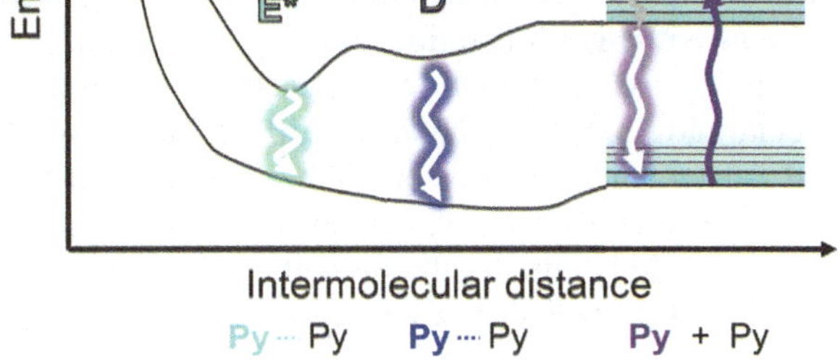

Fig. 11.16 Schematic illustration of pyrene's excimer formation. Image adapted with permission from Winnik (1993). Copyright 1993, American Chemical Society

fluorescence at around 480 nm. On the other hand, when the pre-formed dimer was structurally fixed enough to inhibit its relaxation in excited state, relatively blue-shifted excimer-like emission, i.e., static excimer could be observed at around 420 nm. Importantly, the dynamic excimer emission or that formed by pre-associated dimer could be distinguished from each other by measuring the time-resolved fluorescence or excitation spectra.

11.4.7 Assembly- or Disassembly-Induced Emission

Typical organic fluorescent dyes are known to be less or non-fluorescent when they aggregate, which is called aggregation-caused quenching (ACQ) (Hong et al. 2009). ACQ could be explained by several ways, for example, the formation of non-fluorescent "H-aggregates" through dipole–dipole interaction between two dyes, and/or consecutive energy transfer within aggregates (HOMO-FRET) that gradually diminishes the photon output (fluorescence) against the photon input (excitation). On the other hand, aggregation does not always quench the fluorescence from a dye. In some cases, molecules existing as H-aggregates in the ground state may still form fluorescent dimer (e.g., excimer, see Sect. 11.4.6) through structural relaxation in the excited state. In addition, some J-aggregates are known to exhibit red-shifted absorption and fluorescence in contrast to the H-aggregates (Fig. 11.17a) (Spano and Introduction 2009). Furthermore, Tang proposed that dyes with propeller-like structures such as hexaphenylsilole and tetraphenyl ethene are basically less or non-fluorescent due to efficient thermal deactivation, whereas they become fluorescent in the aggregated state due to suppressed molecular motion. This phenomenon is known as "aggregation-induced emission" (AIE) (Fig. 11.17b) (Hong et al. 2009). In summary, organic dyes vary their fluorescence behaviors in response to the formation of aggregation/disaggregation states, which is a useful characteristic in the design of fluorescent chemosensors.

An example of fluorescent chemosensors based on aggregation/disaggregation is shown in Fig. 11.18. Hamachi and co-workers developed a fluorescent chemosensor **8** that forms non-fluorescent aggregates in water (OFF state) (Mizusawa et al. 2012). This sensor contains the binding site to capture specific proteins on the cell surface (so this binding site could be called a "ligand"). As a result, some of the dye can leave the aggregates to bind to the protein, exhibiting a recovery of fluorescence (ON state, i.e., "turn-on"). The concept of this sensor is very useful for imaging non-enzymatic proteins with high S/N ratio.

Another sensor based on aggregation/disaggregation was developed by Klymchenko and co-workers (Niko et al. 2014). The sensor **9** has Nile Red as a fluorophore, and a detergent structure containing long alkyl chain as the hydrophobic part, and ornithine and quarternalized amine as the hydrophilic part. As shown in Fig. 11.19a, this molecule is initially fluorescent but becomes quenched upon forming micelles in water (OFF state). These micelles are stabilized by polymerization using a reagent

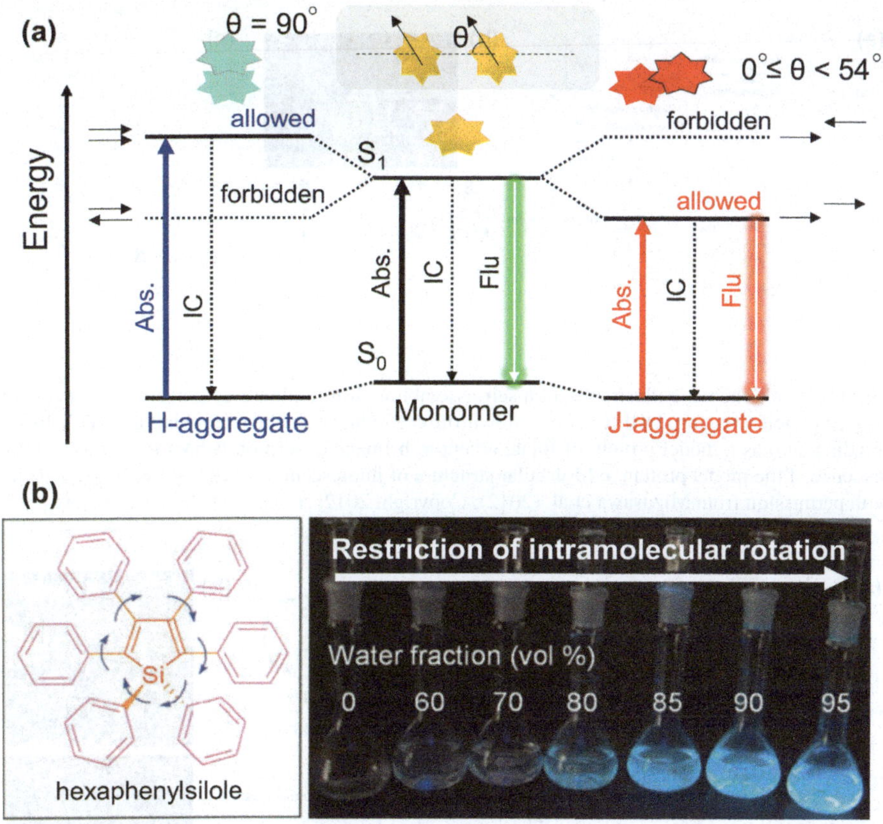

Fig. 11.17 **a** Schematic illustration of the energy change when the dye forms H- or J-aggregates. Small arrows denote transition dipole moment of the dye. **b** Molecular structure of hexaphenylsilole and its AIE behavior. Image adapted with permission from Hong et al. (2009). Copyright 2009, Royal Society of Chemistry

containing disulfide bond (DSP), which is known to be cleaved by intracellular reduction species such as glutathione. The polymerized micelles can be internalized by living cells through endocytosis, and then they gradually become fluorescent due to deaggregation (ON state) driven by cleavage of the cross-linker (Fig. 11.19b). That is, this sensor can fluoresce only in living cells, which might be useful for the so-called theranostics (therapy + diagnostics).

Furthermore, examples of fluorescent chemosensors based on AIE are shown in Fig. 11.20. Sensors **10** and **11** consist of tetraphenylethene as an AIE-active fluorophore and a chelator as the binding site to capture Hg^{2+} or Ag^{+} ions (Liu et al. 2008). As mentioned above, the sensors are basically non-fluorescent (OFF state) in the free state, but become fluorescent upon the formation of aggregates driven by coordination to Hg^{2+} or Ag^{+} (ON state).

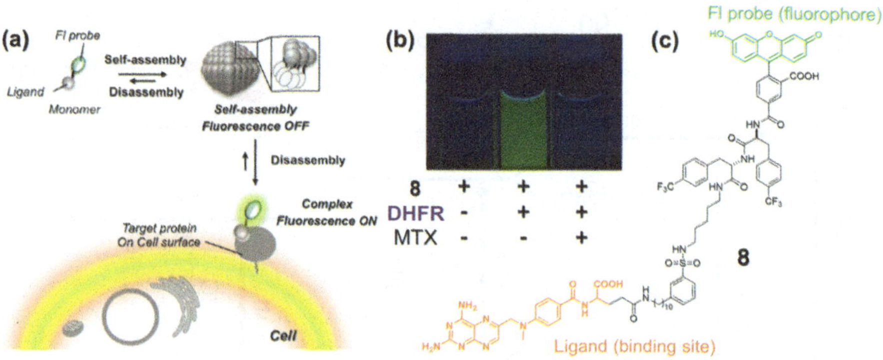

Fig. 11.18 **a** Schematic illustration of a self-assembling turn-on fluorescent chemosensor for cell surface protein imaging. DHFR denotes dihydrofolate reductase, which can bind the MTX ligand (binding site) as a model protein of folate receptor. **b** Image of turn-on behavior of sensor in the presence of the model protein. **c** Molecular structure of fluorescent chemosensor **8**. Image adapted with permission from Mizusawa et al. (2012). Copyright 2012, American Chemical Society

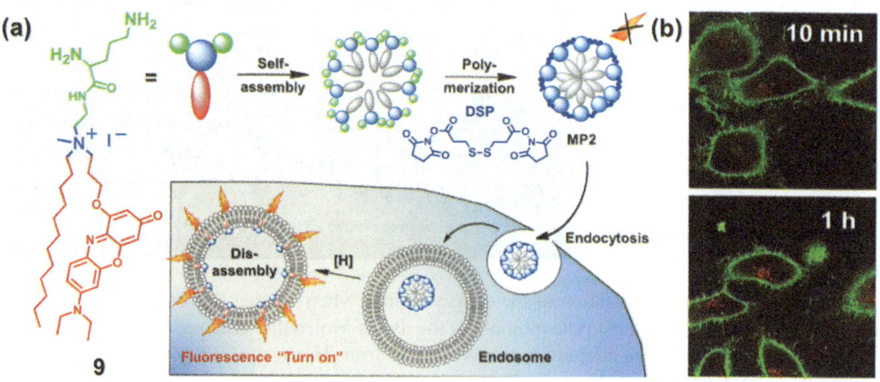

Fig. 11.19 **a** Schematic illustration of fluorescence turn-on behavior of polymerized micellar fluorescent chemosensor under intracellular stimuli. **b** Fluorescence imaging of HeLa cells incubated with the micellar sensor. Images adapted with permission from Niko et al. (2014). Copyright 2014, WILEY–VCH Verlag GmbH & Co

11.4.8 Intramolecular Charge Transfer

Section 11.3.5 mentioned that some organic dyes vary their fluorescence color in response to the surrounding polarity (fluorescence solvatochromism) (Klymchenko 2017). The typical mechanism of this phenomenon is intramolecular charge transfer (ICT), which is generally observed in the so-called "push–pull dye" (Fig. 11.21). As typified by Prodan and Nile Red, push–pull dyes consist of electronic donor moiety such as diarylamino, dialkylamino, or alkoxy groups and electron acceptor moiety such as aldehyde, ketone, or cyano groups.

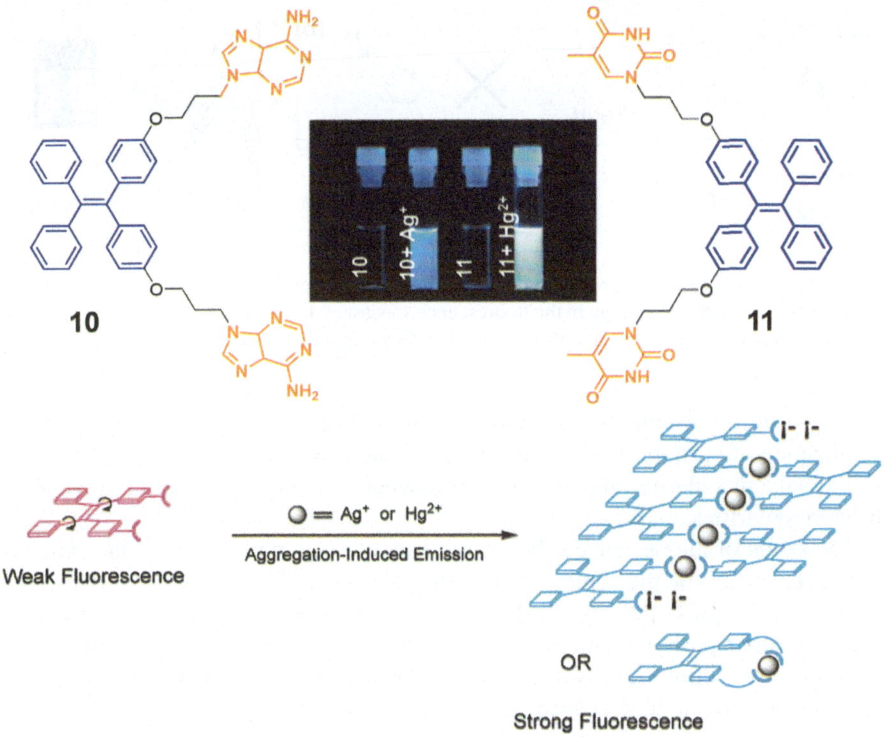

Fig. 11.20 Molecular structures of AIE-active fluorescent chemosensors **10** and **11**, and a schematic illustration of their fluorescence turn-on behavior in the presence of analyte. Image adapted with permission from Liu et al. (2008). Copyright 2008, American Chemical Society

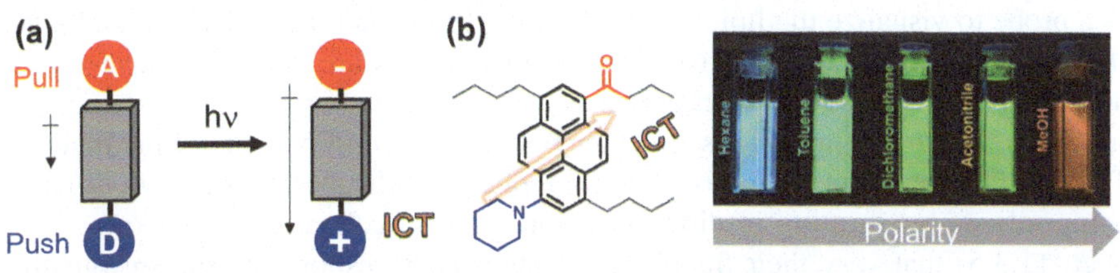

Fig. 11.21 **a** Schematic illustration of photo-induced intramolecular charge transfer. Arrows denote dipole moment of the dye. **b** A typical push–pull dye and its solvatochromism. Image adapted with permission from Niko et al. (2013). Copyright 2013, WILEY–VCH Verlag GmbH & Co

In general, organic dyes exhibit larger dipole moments in the excited state than in the ground state, except for ionic dyes like Reichart's dye (Reichardt 2005). Such a difference in dipole moment is especially pronounced for the push–pull dyes due to their ICT characteristic. Therefore, their excited state is largely stabilized by the surrounding solvent through dipole–dipole interaction along with conformational

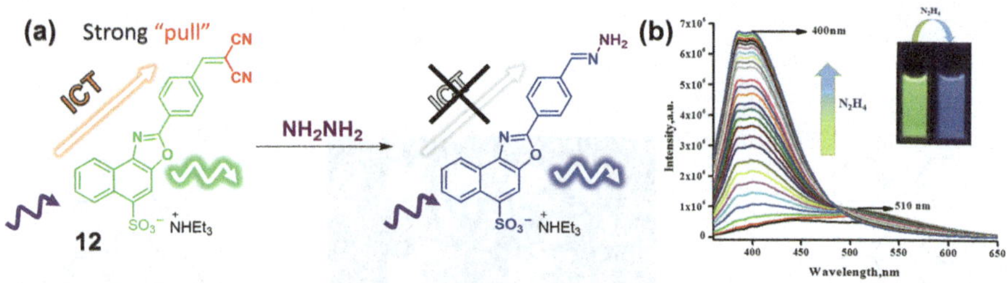

Fig. 11.22 **a** Push–pull structured fluorescent chemosensor **12** and its structural change due to reaction with NH_2NH_2. **b** The change in the fluorescence spectra of **12** in the presence of analyte. Image adapted with permission from Shweta et al. (2016). Copyright 2016, Royal Society of Chemistry

change, resulting in fluorescence with a large Stokes shift. Moreover, the extent of stabilization of push–pull dyes in the excited state in a solvent is strongly dependent on the solvent polarity, and so the fluorescence color can change in solvents of different polarities.

The extent of fluorescence solvatochromism of push–pull dyes, i.e., the extent of their ICT depends on the combination of the push and pull moieties in the dyes. This means that the push–pull dyes exhibit different fluorescent colors upon structural change in their push–pull moieties even in the same solvent. Thus, in fluorescent chemosensors exhibiting solvatochromism, the sensing is often performed by monitoring change in the fluorescence color caused by structural change in the push–pull moieties when the sensor reacts with analyte, as can be seen in sensor **12** shown in Fig. 11.22 (Shweta et al. 2016).

As mentioned in Sect. 11.2.2, the fluorescent solvatochromic dyes are used for not only detecting molecular analytes but also visualizing specific "phase" in a certain substance. For instance, a fluorescent solvatochromic pyrene derivative **PA** is used as a probe to visualize the lipid order between plasma and intracellular membranes (Fig. 11.23) (Didier et al. 2016). The cell membranes are known to have locally different compositions (lipids, cholesterol, etc.) with different intra- and intercellular functions, and the compositions affect the polarity and rigidity in their lipid bilayers. This enables the use of PA to visualize local phases of the membranes with different compositions by emitting in different colors. The so-called molecular rotors (see Sect. 11.3.5) that vary their fluorescence intensity in response to the surrounding viscosity can be used for similar applications, although the details are omitted here.

11.4.9 Intramolecular Proton Transfer

Some dyes exhibit tautomerization through "excited intramolecular proton transfer" (ESIPT) in the excited state. In this phenomenon, a proton in the dye moves from one specific position to another in the structure (usually from a hydroxyl proton donor to a carbonyl proton acceptor), resulting in the formation of an excited tautomer (T*)

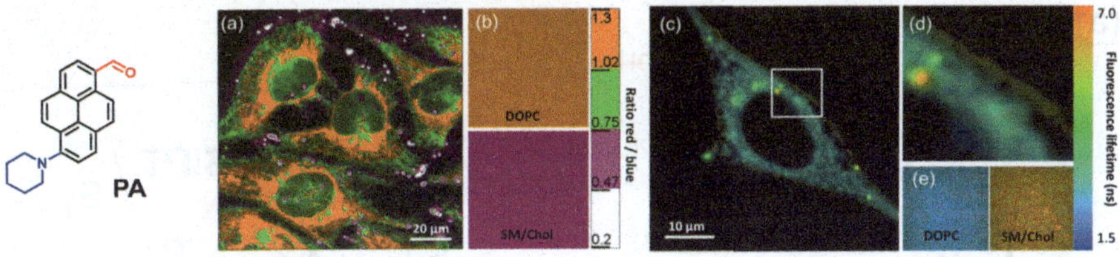

Fig. 11.23 Ratiometric (**a**, **b**) and FLIM (**c**–**e**) imaging of lipid order in living HeLa cells using solvatochromic probe PA. The pseudo colors represent the ratio of the long- to short-wavelength emission channels (550–700 to 470–550 nm). **b** Calibration of the ratio using suspension of lipid vesicles composed of DOPC (Ld phase) and sphingomyelin/cholesterol (SM/Cho, Lo phase). **c** FLIM image of HeLa cells and **d** zoom on the region of interest. Calibration images of suspensions of DOPC and SM/Chol **e** vesicles were recorded with the same instrumental settings. Images adapted with permission from Klymchenko (2017) and Didier et al. (2016). Copyright 2016, Nature Publishing Group and Copyright 2018, American Chemical Society, respectively

(see Fig. 11.24a, b). T* is more stable than the normal excited species (N*), and therefore fluorescence derived from it is observed at longer wavelength than that from N*. Remarkably, as typified by 3-hydroxychromone (3HC) derivatives, dyes exhibiting ESIPT can show dual emission from N* and T* by carefully controlling the energy barrier and energy level difference between N* and T* (Fig. 11.24c) (Klymchenko 2017; Demchenko et al. 2009). Importantly, the ratio of fluorescence intensities from N* and T* depends on the dielectric constant and H-bond donor ability of the surrounding solvent, and this property is a useful characteristic in fluorescent chemosensors as well as in fluorescence solvatochromism based on ICT. Furthermore, the N* fluorescence of 3HC is sensitive to polarity due to its ICT characteristics. That is, 3HC exhibits ESIPT-guided dual emission along with fluorescence solvatochromism based on ICT, meaning that it provides more parameters to analyze its environmental condition than the dyes discussed in Sect. 11.3.5. In fact, the interaction between 3HC-labeled peptide and biomolecules (nucleic acid and antibodies) can be very clearly observed (Fig. 11.24d) (Zamotaiev et al. 2014).

11.5 Outlook: Fluorescent Nanoparticles as Next-Generation Fluorophores

In this chapter, the authors have described the structures of various representative fluorescent chemosensors, the roles of their fluorophore and binding sites, and several photophysical phenomena as their working principles. Although almost all fluorophores introduced here are small organic molecules, currently several kinds of fluorescent nanoparticles (including quantum dots, dye-doped silica nanoparticles, polymer precipitates, nano-lipids, etc.) have attracted considerable attention as new ultra-bright emitters that could expand the use of fluorescent chemosensors (Reisch

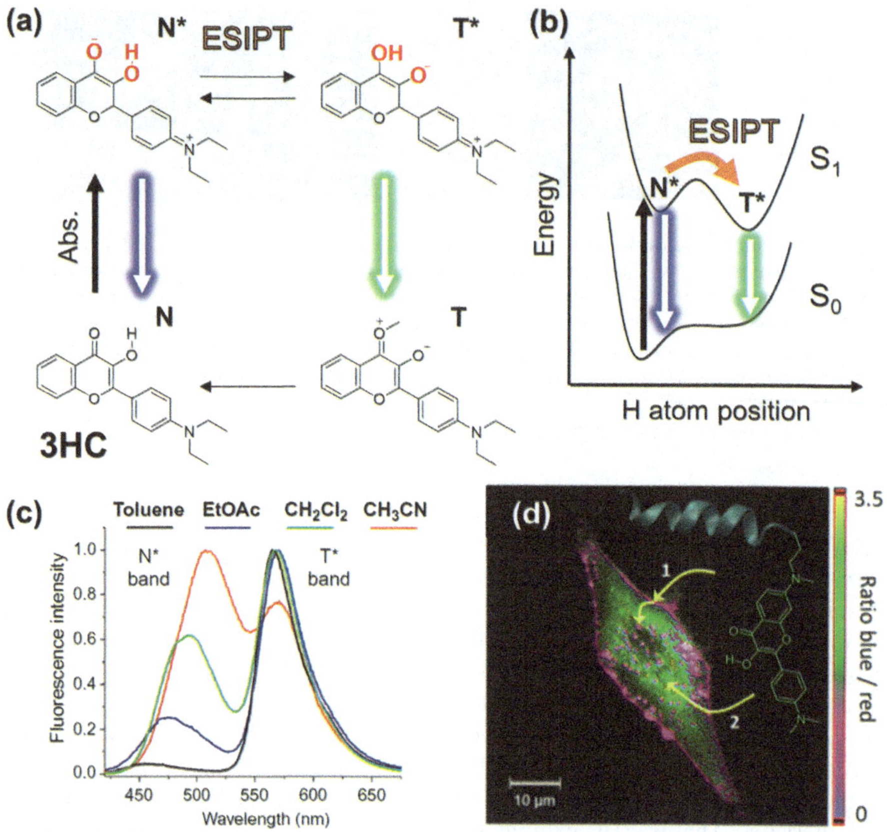

Fig. 11.24 **a** Schematic illustration of the occurrence of ESIPT. **b** Representative energy diagram of the dye exhibiting ESIPT. **c** Fluorescence spectra of 3HC in solvents of different polarity. **d** Ratiometric imaging of internalization of penetration into HeLa cells using a 3HC label. Image and spectroscopic data adapted with permission from Zamotaiev et al. (2014) and Klymchenko (2017). Copyright 2014 Royal Society of Chemistry and Copyright 2018, American Chemical Society, respectively

and Klymchenko 2016). In particular, dye-loaded polymer nanoparticles and nanolipids are known to be less toxic and 10–1000 times brighter than conventional small-molecule dyes, and even brighter than inorganic nanoparticles (Sekida et al. 2018). Therefore, they may function as a new class of fluorophores in the detection/visualization of new analytes and related phenomena. On the other hand, to the best of the authors' knowledge, there exist only very limited reports on the technique to control the collective fluorescent properties (e.g., OFF–ON switching and fluorescence color change in the presence/absence of analyte) of the fluorescent nanoparticles. As a result, further development of fluorescent nanoparticles is needed before they can be used practically in fluorescent chemosensors on par with conventional small-molecule dyes.

References

Ahn KH, Kim D, Kim KH, Park BS, Singha S, Jun YW, Baik SH, Wang T, Mook-Jung I, Moon H, Jung J (2015) Two-photon absorbing dyes with minimal autofluorescence in tissue imaging: application to in vivo imaging of amyloid-β plaques with a negligible background signal. J Am Chem Soc 137:6781–6789. https://doi.org/10.1021/jacs.5b03548

Birks J (1963) Excimer fluorescence spectra of pyrene derivatives. Spectrochim Acta 19:401–410. https://doi.org/10.1016/0371-1951(63)80051-X

De Silva AP, Moody TS, Wright GD (2009) Fluorescent PET (Photoinduced Electron Transfer) sensors as potent analytical tools. Analyst 134:2385–2393. https://doi.org/10.1039/b912527m

Demchenko AP, Mély Y, Duportail G, Klymchenko AS (2009) Monitoring biophysical properties of lipid membranes by environment-sensitive fluorescent probes. Biophys J 96:3461–3470. https://doi.org/10.1016/j.bpj.2009.02.012

Didier P, Mely Y, Konishi G, Klymchenko AS, Niko Y (2016) Bright and photostable push-pull pyrene dye visualizes lipid order variation between plasma and intracellular membranes. Sci Rep 6:1–9. https://doi.org/10.1038/srep18870

Dujols V, Ford F, Czarnik AW (1997) A long-wavelength fluorescent chemodosimeter selective for Cu(II) ion in water. J Am Chem Soc 119:7386–7387. https://doi.org/10.1021/ja971221g

Famulok M, Hartig JS, Mayer G (2007) Functional aptamers and aptazymes in biotechnology, diagnostics, and therapy. Chem Rev 107:3715–3743. https://doi.org/10.1021/cr0306743

Görner H, Kuhn HJ (2007) Cis-trans photoisomerization of stilbenes and stilbene-like molecules advances in photochemistry, pp 1–117

Greenspan P, Mayer EP, Fowler SD (1985) Nile red: a selective fluorescent stain for intracellular lipid droplets. J Cell Biol 100:965–973

Gutsche CD (2016) Fluorescent calixarene hosts. Calixarenes and beyond chapter, vol 28, pp 401–410. https://doi.org/10.1007/978-3-319-31867-7_28

Hirata Y, Kawahara S, Kojima H, Nagano T, Kikuchi K, Nagoshi H, Nakatsubo N, Kirino Y (2002) Detection and imaging of nitric oxide with novel fluorescent indicators: diaminofluoresceins. Anal Chem 70:2446–2453. https://doi.org/10.1021/ac9801723

Hong Y, Lam JWY, Tang BZ (2009) Aggregation-induced emission: Phenomenon, mechanism and applications. Chem Commun 4332–4353. https://doi.org/10.1039/b904665h

IUPAC (1997a) McNaught AD, Wilkinson A. Compendium of chemical terminology (the "Gold Book"), 2nd edn. Blackwell Scientific Publications, Oxford. In: Nic M, Jirat J, Kosata B (2006) XML on-line corrected version. https://goldbook.iupac.org. Updated by Jenkins A. ISBN 0-9678550-9-8. https://doi.org/10.1351/goldbook.N04219.

IUPAC (1997b) McNaught AD, Wilkinson A. Compendium of chemical terminology (the "Gold Book"), 2nd edn. Blackwell Scientific Publications, Oxford. In: Nic M, Jirat J, Kosata B (2006) XML on-line corrected version. https://goldbook.iupac.org. Updated by Jenkins A. ISBN 0-9678550-9-8. https://doi.org/10.1351/goldbook.N04218

IUPAC (1997c) McNaught AD, Wilkinson A. Compendium of chemical terminology (the "Gold Book"), 2nd edn. Blackwell Scientific Publications, Oxford. In: Chalk SJ (2019) Online. ISBN 0-9678550-9-8. https://doi.org/10.1351/goldbook.H02756

IUPAC (1997d) McNaught AD, Wilkinson A. Compendium of chemical terminology (the "Gold Book"), 2nd edn. Blackwell Scientific Publications, Oxford. In: Chalk SJ (2019) Online. ISBN 0-9678550-9-8. https://doi.org/10.1351/goldbook.FT07381

IUPAC (1997e) McNaught AD, Wilkinson A. Compendium of chemical terminology (the "Gold Book"), 2nd edn. Blackwell Scientific Publications, Oxford. In: Chalk SJ (2019) Online. ISBN 0-9678550-9-8. https://doi.org/10.1351/goldbook.ET07369

Jacques SL (2013) Optical properties of biological tissues: a review. Phys Med Biol 58:R37–R61. https://doi.org/10.1088/0031-9155/58/11/R37

Kamiya M, Yamasoba T, Urano Y, Yoshida M, Umezawa K (2016) Rational design of reversible fluorescent probes for live-cell imaging and quantification of fast glutathione dynamics. Nat Chem 9:279–286. https://doi.org/10.1038/nchem.2648

Kim HJ, Han JH, Kim MK, Lim CS, Kim HM, Cho BR (2010) Dual-color imaging of sodium/calcium ion activities with twophoton fluorescent probes. Angew Chemie Int Ed 49:6786–6789. https://doi.org/10.1002/anie.201002907

Klymchenko AS (2017) Solvatochromic and fluorogenic dyes as environment-sensitive probes: design and biological applications. Acc Chem Res 50:366–375. https://doi.org/10.1021/acs.accounts.6b00517

Koide Y, Takahashi N, Ikegaya Y, Terai T, Matsuki N, Egawa T, Ueno T, Komatsu T, Hanaoka K, Nagano T, Ujita S (2011) Development of a far-red to near-infrared fluorescence probe for calcium ion and its application to multicolor neuronal imaging. J Am Chem Soc 133:14157–14159. https://doi.org/10.1021/ja205809h

Kubota R, Hamachi I (2015) Protein recognition using synthetic small-molecular binders toward optical protein sensing in vitro and in live cells. Chem Soc Rev 44:4454–4471. https://doi.org/10.1039/c4cs00381k

Kung CE, Reed JK (1986) Microviscosity measurements of phospholipid bilayers using fluorescent dyes that undergo torsional relaxation. Biochemistry 25:6114–6121. https://doi.org/10.1021/bi00368a042

Lavis LD, Raines RT (2008) Bright ideas for chemical biology—supporting information. ACS Chem Biol 3:142–155. https://doi.org/10.1021/cb700248m

Lee JY, Kim JS, Bartsch RA, Kim SK, Lee SH, Lee JY (2004) An excimer-based, binuclear, on−off switchable calix[4]crown chemosensor. J Am Chem Soc 126:16499–16506. https://doi.org/10.1021/ja045689c

Lee MH, Park N, Yi C, Han JH, Hong JH, Kim KP, Kang DH, Sessler JL, Kang C, Kim JS (2014) Mitochondria-immobilized ph-sensitive off-on fluorescent probe. J Am Chem Soc 136:14136–14142. https://doi.org/10.1021/ja506301n

Lippert VE (1957) Spektroskopische bestimmung des dipolmomentes aromatischer verbindungen im ersten angeregten singulettzustand. Ber Bunsenges Phys Chem 61:962–975. https://doi.org/10.1002/bbpc.19570610819

Liu L, Zhang G, Xiang J, Zhang D, Zhu D (2008) Fluorescence "turn on" chemosensors for Ag+ and Hg2+ based on tetraphenylethylene motif featuring adenine and thymine moieties. Org Lett 10:4581–4584. https://doi.org/10.1021/ol801855s

Loutfy RO (2007) Fluorescence probes for polymer free-volume. Pure Appl Chem 58:1239–1248. https://doi.org/10.1351/pac198658091239

Ma L, Yang F, Zheng J (2014) Application of fluorescence resonance energy transfer in protein studies. J Mol Struct 1077:87–100. https://doi.org/10.1016/j.molstruc.2013.12.071

Madhu S, Kalaiyarasi R, Basu SK, Jadhav S, Ravikanth M (2014) A boron-dipyrrin-mercury(ii) complex as a fluorescence turn-on sensor for chloride and applications towards logic gates. J Mater Chem C 2:2534–2544. https://doi.org/10.1039/c3tc32188f

Mataga N, Kaifu Y, Koizumi M (1956) Solvent Effects upon fluorescence spectra and the dipolemoments of excited molecules. Bull Chem Soc Jpn 29:465–470. https://doi.org/10.1246/bcsj.29.465

Mataga N, Chosrowjan H, Taniguchi S (2005) Ultrafast charge transfer in excited electronic states and investigations into fundamental problems of exciplex chemistry: Our early studies and recent developments. J Photochem Photobiol C Photochem Rev 6:37–79. https://doi.org/10.1016/j.jphotochemrev.2005.02.003

Mizusawa K, Takaoka Y, Hamachi I (2012) Specific cell surface protein imaging by extended self-assembling fluorescent turn-on nanoprobes. J Am Chem Soc 134:13386–13395. https://doi.org/10.1021/ja304239g

Monici M (2005) Cell and tissue autofluorescence research and diagnostic applications. Biotechnol Annu Rev 11:227–256. https://doi.org/10.1016/S1387-2656(05)11007-2

Mortellaro MA, Fraatz RJ, He H, Tusa JK, Leiner MJP (2003) A fluorescent sensor with high selectivity and sensitivity for potassium in water. J Am Chem Soc 125:1468–1469. https://doi.org/10.1021/ja0284761

Niko Y, Kawauchi S, Konishi GI (2013) Solvatochromic pyrene analogues of prodan exhibiting extremely high fluorescence quantum yields in apolar and polar solvents. Chem A Eur J 19:9760–9765. https://doi.org/10.1002/chem.201301020

Niko Y, Arntz Y, Mely Y, Konishi GI, Klymchenko AS (2014) Disassembly-driven fluorescence turn-on of polymerized micelles by reductive stimuli in living cells. Chem A Eur J 20:16473–16477. https://doi.org/10.1002/chem.201405040

Niko Y, Sasaki S, Narushima K, Sharma DK, Vacha M, Konishi GI (2015) 1-, 3-, 6-, and 8-tetrasubstituted asymmetric pyrene derivatives with electron donors and acceptors: high photostability and regioisomer-specific photophysical properties. J Org Chem 80:10794–10805. https://doi.org/10.1021/acs.joc.5b01987

Pawlicki M, Collins HA, Denning RG, Anderson HL (2009) Two-photon absorption and the design of two-photon dyes. Angew Chemie Int Ed 48:3244–3266. https://doi.org/10.1002/anie.200805257

Reichardt C (2005) Solvatochromic dyes as solvent polarity indicators. Chem Rev 94:2319–2358. https://doi.org/10.1021/cr00032a005

Reisch A, Klymchenko AS (2016) Fluorescent polymer nanoparticles based on dyes: seeking brighter tools for bioimaging. Small 12:1968–1992. https://doi.org/10.1002/smll.201503396

Rice TE, McCoy CP, Gunaratne HQN, de Silva AP, Rademacher JT, Huxley AJM, Gunnlaugsson T (2002) Signaling recognition events with fluorescent sensors and switches. Chem Rev 97:1515–1566. https://doi.org/10.1021/cr960386p

Sahoo H (2011) Förster resonance energy transfer—a spectroscopic nanoruler: principle and applications. J Photochem Photobiol C Photochem Rev 12:20–30. https://doi.org/10.1016/j.jphotochemrev.2011.05.001

Sale P, Montalti M, Farruggia G, Wolf FI, Prodi L, Zaccheroni N, Savage PB, Iotti S, Trapani V (2006) 8-hydroxyquinoline derivatives as fluorescent sensors for magnesium in living cells. J Am Chem Soc 128:344–350. https://doi.org/10.1021/ja056523u

Sapsford KE, Algar WR, Berti L, Gemmill KB, Casey BJ, Oh E, Stewart MH, Medintz IL (2013) Functionalizing nanoparticles with biological molecules: developing chemistries that facilitate nanotechnology. Chem Rev 113:1904–2074. https://doi.org/10.1021/cr300143v

Schuster R (1988) Determination of amino acids in biological, pharmaceutical, plant and food samples by automated precolumn derivatization and high-performance liquid chromatography. J Chromatogr B Biomed Sci Appl 431:271–284. https://doi.org/10.1016/S0378-4347(00)83096-0

Sekida S, Kameyama T, Koga T, Hadano S, Watanabe S, Niko Y (2018) Highly lipophilic and solid emissive N-annulated perylene bisimide synthesis for facile preparation of bright and far-red excimer fluorescent nano-emulsions with large Stokes shift. J Photochem Photobiol A Chem 364:16–21. https://doi.org/10.1016/j.jphotochem.2018.05.023

Shweta KA, Neeraj ASK, Prakash A, Roy JK, Tiwari I, Upadhyay KK (2016) A highly sensitive naphthaoxazole-based cell-permeable ratiometric chemodosimeter for hydrazine. RSC Adv 6:94959–94966. https://doi.org/10.1039/c6ra15081k

Smith AM, Mancini MC, Nie S (2009) Bioimaging: second window for in vivo imaging. Nat Nanotechnol 4:710–711. https://doi.org/10.1038/nnano.2009.326

Soper SA, Nutter HL, Keller RA, Davis LM, Shera EB (1993) The photophysical constants of several fluorescent dyes pertaining to ultrasensitive fluorescence spectroscopy. Photochem Photobiol 57:972–977. https://doi.org/10.1111/j.1751-1097.1993.tb02957.x

Sousa LR, Larson JM (1977) Crown ether model systems for the study of photoexcited state response to geometrically oriented perturbers. The effect of alkali metal ions on emission from naphthalene derivatives. J Am Chem Soc 99:307–310. https://doi.org/10.1021/ja00443a084

Spano FC, Introduction I (2009) The spectral signatures of frenkel polarons in. Acc Chem Res 43:429–439. https://doi.org/10.1021/ar900233v

Tsien RY (1980) new calcium indicators and buffers with high selectivity against magnesium and protons: design, synthesis, and properties of prototype structures. Biochemistry 19:2396–2404. https://doi.org/10.1021/bi00552a018

Umezawa K, Matsui A, Nakamura Y, Citterio D, Suzuki K (2009) Bright, color-tunable fluorescent dyes in the Vis/NIR region: establishment of new "tailor-made" multicolor fluorophores based on borondipyrromethene. Chem A Eur J 15:1096–1106. https://doi.org/10.1002/chem.200801906

Urano Y, Choyke PL, Alford R, Kobayashi H, Ogawa M (2009) New strategies for fluorescent probe design in medical diagnostic imaging. Chem Rev 110:2620–2640. https://doi.org/10.1021/cr900263j

Wang C, Taki M, Sato Y, Fukazawa A, Higashiyama T, Yamaguchi S (2017) Super-photostable phosphole-based dye for multiple-acquisition stimulated emission depletion imaging. J Am Chem Soc 139:10374–10381. https://doi.org/10.1021/jacs.7b04418

Watanabe S, Onogawa O, Komatsu Y, Yoshida K (1998) Luminescent metalloreceptor with a neutral bis(acylaminoimidazoline) binding site: optical sensing of anionic and neural phosphodiesters. J Am Chem Soc 120:229–230. https://doi.org/10.1021/ja973263a

Watanabe S, Ikishima S, Matsuo T, Yoshida K (2001) A luminescent metalloreceptor exhibiting remarkably high selectivity for Mg2+ over Ca2+ [1]. J Am Chem Soc 123:8402–8403. https://doi.org/10.1021/ja010931q

Weber G, Farris FJ (1979) Synthesis and spectral properties of a hydrophobic fluorescent probe: 6-propionyl-2-(dimethylamino)naphthalene. Biochemistry 18:3075–3078. https://doi.org/10.1021/bi00581a025

Weissleder R (2001) A clearer vision for in vivo imaging biological imaging toward the phosphoproteome. Nat Biotechnol 19:316–317. https://doi.org/10.1038/86684

Winnik FM (1993) Photophysics of preassociated pyrenes in aqueous polymer solutions and in other organized media. Chem Rev 93:587–614. https://doi.org/10.1021/cr00018a001

Wu J, Zou Y, Li C, Sicking W, Piantanida I, Yi T, Schmuck C (2012) A molecular peptide beacon for the ratiometric sensing of nucleic acids. J Am Chem Soc 134:1958–1961. https://doi.org/10.1021/ja2103845

Wu J, Kwon B, Liu W, Anslyn EV, Wang P, Kim JS (2015) Chromogenic/fluorogenic ensemble chemosensing systems. Chem Rev 115:7893–7943. https://doi.org/10.1021/cr500553d

Wu D, Sedgwick AC, Gunnlaugsson T, Akkaya EU, Yoon J, James TD (2017) Fluorescent chemosensors: the past, present and future. Chem Soc Rev 46:7105–7123. https://doi.org/10.1039/c7cs00240h

Yin J, Hu Y, Yoon J (2015) Fluorescent probes and bioimaging: alkali metals, alkaline earth metals and pH. Chem Soc Rev 44:4619–4644. https://doi.org/10.1039/c4cs00275j

Zamotaiev OM, Postupalenko VY, Shvadchak VV, Pivovarenko VG, Klymchenko AS, Mély Y (2014) Monitoring penetratin interactions with lipid membranes and cell internalization using a new hydration-sensitive fluorescent probe. Org Biomol Chem 12:7036–7044. https://doi.org/10.1039/c4ob01242a

Zheng Q, Juette MF, Jockusch S, Wasserman MR, Zhou Z, Altman RB, Blanchard SC (2014) Ultra-stable organic fluorophores for single-molecule research. Chem Soc Rev 43:1044–1056. https://doi.org/10.1039/c3cs60237k

Zhu H, Fan J, Du J, Peng X (2016) Fluorescent probes for sensing and imaging within specific cellular organelles. Acc Chem Res 49:2115–2126. https://doi.org/10.1021/acs.accounts.6b00292

Chapter 12
White-Light Emissive Materials Based on Supramolecular Approach

Yuji Kubo

Abstract In material science, organic compound-based white-light emission (WLE) systems that emit in the entire visible region have received much attention due to low-cost, lightweight, easy fabrication of thin film, and so on. Synthesis of single molecules with WLE function has been still a challenge because of constraints imposed by Kasha's rule, whereas it is easier to fabricate multicomponent-based emissive systems. In the latter case, supramolecular organization of the related components is promising, where well-tailored intermolecular interactions between them would tune their interplay on the physical properties to give WLE. Toward this end, the use of Förester resonance energy transfer (FRET) is benefitable for the interplay between the emissive components. Significant overlap between the emission spectrum of donor dye and absorption band of adjacent acceptor one is indispensable. This chapter describes supramolecular ensembles capable of tuning the primary colors, i.e., red, green, and blue, or two complementary colors, e.g., cyan and orange, with dynamic and reversible non-covalent bonds involving hydrogen bonds, metal coordination, electrostatic interactions, hydrophobic interactions, and π–π stacking interactions. Emissive systems other than FRET behavior are also introduced in this chapter.

Keywords White-light emission · Supramolecular chemistry · Energy transfer · Self-assembly

12.1 Introduction

Design and development of white-light emission (WLE) materials have become one of the emerging research areas in solid-state lighting, flat-panel display devices, and backlights for liquid crystalline display (Forrest and Thompson 2007; Xiao et al. 2011; Xia et al. 2016; Su et al. 2016; Farinola and Ragni 2011; Wu et al. 2009; Reineke et al. 2009). White-light is composed of the three primary colors:

Y. Kubo (✉)
Department of Applied Chemistry, Graduate School of Urban Environmental Sciences, Tokyo Metropolitan University, 1-1 Minami-Osawa, Hachioji, Tokyo 192-0397, Japan
e-mail: yujik@tmu.ac.jp

Y. Ooyama and S. Yagi (eds.), *Progress in the Science of Functional Dyes*,
https://doi.org/10.1007/978-981-33-4392-4_12

red (R), green (G), and blue (B). To avoid subjective color perception, the International Commission on Illumination (Commission Internationale de l'Éclairage, CIE) coordinates in the chromaticity diagram (known as CIE 1931) have been used as the colorimetric standard (Jones 1943). All colors are represented by two color coordinates (x and y) in the system; accordingly, white light is defined as (0.33, 0.33). Thus, white light can refer to any color with coordinates in a region with energy equal to that of white light. The diagram also indicates that white light can be obtained by mixing two complementary colors for which the connection line of their color coordinates crosses the white light region. The quality of white light is evaluated by the color rendering index (CRI) and correlated color temperature (CCT) (Xu et al. 2018). For organic materials, great effect has been devoted to prepare a single molecule-based white-light emission system owing to the potent application to white organic light-emitting devices (Lili and Michael 2014). However, there is still a challenge to provide systems with "panchromatic emission" because of Kasha's rule (Turro 1991), which states that photon emission occurs in an appreciable yield from the lowest excitation state. To overcome such difficulty, plausible dyes have been designed to have at least two forms, such as an emissive tautomer, to entirely cover the visible region. As a typical example, excited state intramolecular proton transfer (ESIPT) has great promise to provide the desired white-light emitters (Liu et al. 2017a; Sakai et al. 2016), whose molecules have a stable enol form in the ground state, which is switchable to the corresponding keto form in the excited state through intramolecular proton transfer (Zhao et al. 2012a). Such a unique photophysical feature leads to a large Stokes shift of the emission, providing a very useful methodology to cover the longer wavelengths of the visible region. Aggregation-induced emission (AIE) is also a promising way toward this end (Mei et al. 2015). A propeller-like organic compounds involving tetraphenylethylene (TPE) (La et al. 2018), hexaphenylsilole (Hong et al. 2009), and tetraphenylpyrazole (TPP) (Chen et al. 2015) have been developed as AIE luminogens. Those compounds are nonemissive in solution but emit in the aggregation state, which is the exact opposite of aggregation-caused quenching (ACQ) phenomena. Such a unique feature is responsible for the restriction of intramolecular rotation. In this context, Tang, et al. prepared single molecules with dual fluorescent modes containing TPE and pyrene fluorophore through covalent bond formation to realize WLE (Feng et al. 2018a). The structure-property relationship was investigated to understand the origin of WLE. Thermally activated delayed fluorescence (TADF) (Uoyama et al. 2012) is known as an effective tool to obtain a high photoluminescence efficiency in the development of materials for OLEDs. For the strategy, it should be considered that common TADF molecules can easily aggregate through π–π interactions, resulting in exciton concentration quenching. The drawback was overcome by Zheng, Chi, and co-workers who developed single organic material composed of carbazoly- and phenothiazinyl-substituted benzophenone. The compound exhibited an aggregation-induced emission delayed fluorescence (Xie et al. 2015).

Facile way to tune fluorescence involving WLE is to control the Förester resonance energy transfer (FRET) (van der Meer 2013) from donor dye to acceptor one in a well-organized microenvironment, which makes chemists fabricate plural dye-containing systems. Subsequently, supramolecular self-assembly has great promise for the generation of WLE materials. Efficacy of the energy transfer (*E*) is defined by the following Eq. (12.1) (Lakowicz 2006):

$$E = \frac{k_{ET}}{k_{ET} + \tau_D^{-1}} \tag{12.1}$$

where k_{ET} is the rate constant for the energy transfer process, being given by Eq. (12.2)

$$k_{ET} = \tau_{DA}^{-1} - \tau_D^{-1} \tag{12.2}$$

in which τ_{DA} and τ_{D} are averaged lifetime values of the donor in the presence and absence of the acceptor, respectively. The *E* value is also given by using the Förster distance (R_0) and donor/acceptor distance (*r*). Accordingly, the donor/acceptor distance can be experimentally evaluated using the following Eq. (12.3) when FRET efficiency and the Förster distance are obtained. In this context, we should consider the FRET efficiency affected by the acceptor concentration (Gartzia-Rivero et al. 2014).

$$E = \frac{R_0^6}{R_0^6 + r^6} \tag{12.3}$$

In principle, the donor-to-acceptor energy transfer significantly depends on the following parameters: distance, orientation factor, and overlap integral of donor and acceptor. Based on such FRET behavior, the sophisticated engineering between three primary colors (a) and complementary color (b) in Fig. 12.1 allows for production of WLE. In this case, it is highly desired to circumvent self-quenching in the emission such as high concentration- or aggregation-induced quenching (ACQ).

This chapter describes the preparation of white-light emissive materials using a rational supramolecular approach through hydrogen bonding, metallo-coordination, host-guest interaction, self-assembly, and dynamic covalent bond, which mainly reviews how to control FRET behavior to give WLE in molecule-based systems.

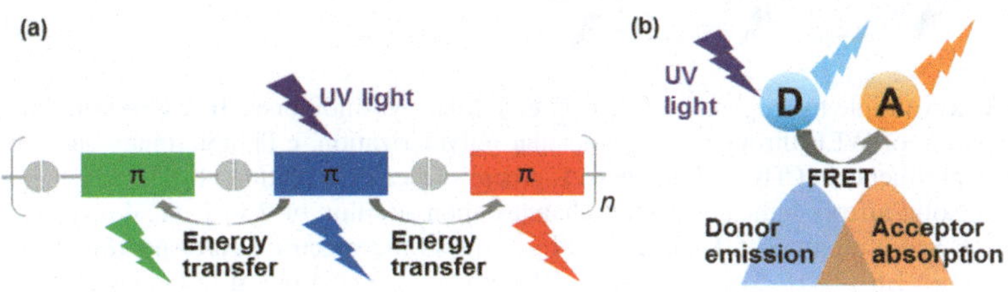

Fig. 12.1 Schematic drawing of RGB-based **a** or two complementary color-based **b** white-light emission; energy transfer is used to cover whole visible region

12.2 Hydrogen Bonding

Hydrogen bonding has been greatly used for the construction of supramolecules (Jeffry 1997), which is essential in the formation of double strand DNA and protein folding. A fascinating feature of interaction mode is the directionality of the hydrogen bond, allowing for precise design of the targeted complexes and control of the geometry of supramolecules. The pioneering work for the production of supramolecule-based WLE materials was conducted by Meijer, Schenning, and co-workers (Abbel et al. 2009). Self-complementary hydrogen bonding interactions between ureido-pyrimidinone (UPy) units (Fig. 12.2a) were used for linkages to connect three primary color emissive units. As shown in Fig. 12.2b, a blue-emitting oligofluorene **1** with quantum yield (Φ) of 0.50, green-emitting oligo(phenylene vinylene) **2** ($\Phi = 0.45$), and a red-emitting perylene bisimide **3** ($\Phi = 0.55$) were synthesized, where UPy

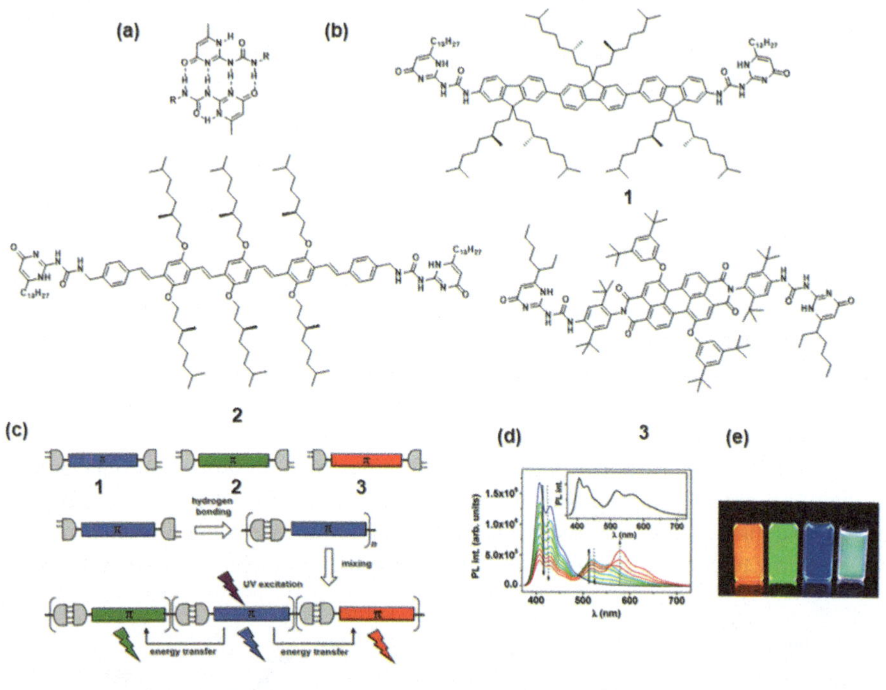

Fig. 12.2 **a** Multiple hydrogen bonding between uride-pyrimidinones. **b** A schematic illustration of the creation of WLE through supramolecular polymerization. **c** Photoluminescence spectra of titration experiments in $CHCl_3$ (blue: pure **1**, green: successive addition of **2**, red: further addition of **3**). The solid arrows indicate spectral changes upon addition of **2** to **1**, the dotted arrow upon addition of **3** to a mixture of **1** and **2**. The inset shows the spectrum corresponding to a ratio of 59:33:8. [**1**] = 1.6×10^{-6} M excitation wavelength (λ_{ex}) = 364 nm. **d** Solutions of pure di-UPy chromophores and a WLE mixture in $CHCl_3$ under UV irradiation (λ_{ex} = 365 nm). Reproduced with permission of The American Chemical Society from Abbel et al. (2009). Copyright 2009 American Chemical Society

units were introduced at both ends of each component. In fluorescence titrations of **1** with either **2** or **3**, the emission of the donor was quenched when excited at 364 nm, which is consistent with the maximum wavelength of **1**, whereas the emission intensity of these acceptors increased simultaneously. High Stern-Volmer constants suggest efficient energy transfer in the solution through UPy-based hydrogen interactions between the components. Supramolecular copolymers containing all three RGB components were prepared by simultaneous incorporation of **2** and **3** into the chains of **1**. Fine-tuning of the component ratio allowed for a partial energy transfer (Fig. 12.2d), resulting in the appearance of WLE at a certain ratio (**1/2/3** = 59:33:8) as shown in Fig. 12.2e. Interestingly, WLE could not been achieved using the mixture of oligo(fluorene), oligo(phenylene vinylene), and perylene bisimide chromophores that lacked the UPy units.

Similar approach using self-complementary hydrogen bonding was also reported by Bassani et al. (Tseng et al. 2011). Three kinds of π-conjugated fluorene-, bis(fluorenyl)benzothiadiazole-, and bis(thiophenyl)benzothiadiazole-derivatives with biuret groups at the terminal groups were synthesized to emit red (R), green (G), and blue (B), respectively. It is interesting to note that highly emissive RGB organic nanospheres formed spontaneously when dropcast from dilute THF solutions onto a substrate, where energy transfer process occurred in the tricomponent spherical aggregates.

The investigation of co-crystal or multicomponent crystals has been interested from the standpoint of advanced optical and electrical materials (Yan and Evans 2014). Ono and Hisaeda reported three-, four-, and five-component co-crystals with flexile color tuning capability involving WLE (Ono and Hisaeda 2019). Toward this end, supramolecular host was prepared by complexation of *N*,*N*-dipyrid-3-yl-1,4,5,8-naphthalenediimide (**NDI**) with two tris(pentafluorophenyl)borane (**TPFB**) linked by boron–nitrogen dative bonds as Lewis acid–base pairs, which spontaneously formed upon mixing the components (Fig. 12.3a). Initially, three-component crystals with 14 representative aromatic guests (**4–17**) were evaluated by single-crystal X-ray structure analysis, where various intermolecular interactions such as C–H···F hydrogen bonding interactions, π–π stacking, charge-transfer interactions, and inclusion phenomena were of significance for crystal formation, although co-crystals cannot be prepared with **C18**.

The photophysical properties of the three-component crystals **C4–C17** (Fig. 12.3a), as well as guest-free crystal (**GF**), were investigated by means of UV-vis diffuse-reflectance spectroscopy measurements with an integrating sphere. Although **GF** and **C4** showed similar absorption spectra, **C5–C17** showed a new absorption band in the visible region from 400 to 600 nm, which was dependent on the guest molecules included. Those co-crystals (**C4–C17**) also showed unique photoluminescence characters upon excitation at 330–380 nm, which reflected the nature of the guest molecules. It was found that color variation over the range of electron-deficient guests such as difluorobenzene (**C5–C7**), from visible-to-NIR luminescence, was achieved. The emission occurred with high quantum yields of 16.7% for **C9**, 18.9% for **C10**, 28.3% for **C11**, 17.3% for **C13**, 26.2% for **C14**, and 10.8% for **C15**, in the crystalline state. This result is attributable to the crystal structure and the

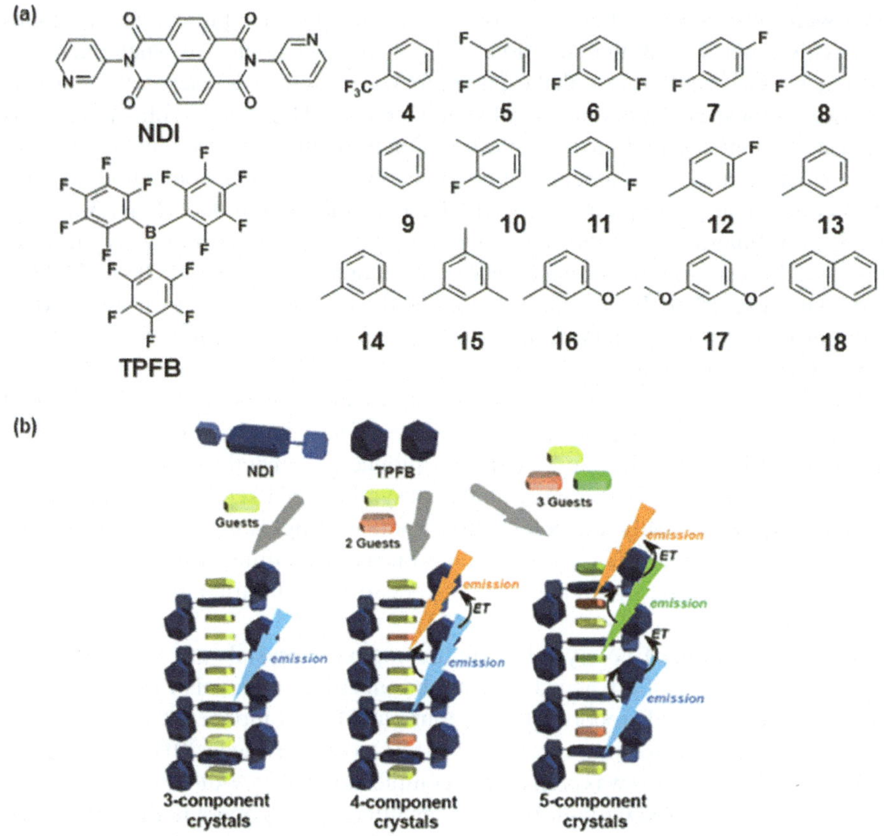

Fig. 12.3 a Chemical structures of the multicomponent crystal motifs used in this study. **b** Schematic illustration of three-, four-, and five-component crystalline materials. Reproduced from Ono and Hisaeda (2019) by permission of The Royal Society of Chemistry

electronic properties of the guest molecules, whereas the strong electron-donating properties of *m*-methylanisole and 1,3-dimethoxybenene might quench the CT-based photoluminescence. The electron transfer process from *m*-methylanisole and 1,3-dimethoxybenzene to **NDI** deactivates the excited state by non-radiative decay. Such successful findings led to the investigation of multi guests-accommodation by the host complex (**NDI/TPFB**) to see flexible multi-light emission. As a typical example, a four-component crystal **MC** composed of a binary host including *m*-fluorotoluene **11**/naphthalene **18** (96.7/3.3) showed WLE, which was subjected to fluorescence time-resolved spectroscopy (TCSPC). As time passed, the short wavelength emission bands (420–520 nm) gradually decreased and red-shifted, whereas the long wavelength emission (520–720 nm) continuously increased and reached its maximum intensity at around 600 nm (Fig. 12.4a). The delayed growth strongly suggests

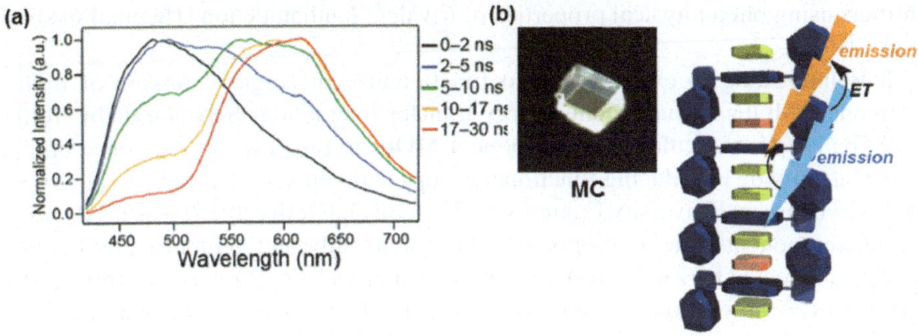

Fig. 12.4 **a** Time-resolved emission spectra of **MC** at different times after the pulse (365 nm). Excitation at 365 nm. **b** Schematic representation of the energy transfer in **MC**. The photograph shows fluorescence microscopy images of WLE of **MC**. Reproduced from Ono and Hisaeda (2019) by permission of The Royal Society of Chemistry

that energy transfer from the higher (blue emission) to the lower (orange emission) energy sites occurred. The relatively longer lifetime may arise from CT states. The possible energy transfer process is shown in Fig. 12.4b, where an appropriate distance between the donor and acceptor was reached due to a trace of naphthalene molecules well-dispersed in the *m*-fluorotoluene/naphthalene binary guest systems.

12.3 Metallo-coordination

Use of coordination bond between organic ligand with metal ions is profitable because not only plural coordination geometry but also tremendous kinds of properties of metal ions used can lead to versatile supramolecular systems (Cook et al. 2013; Bentz and Cohen 2018). Notably, metal ions have the opportunity to participate in electron transfer process, catalysis, or redox process to play a significant role in the concomitant functionality of the system. Metal-organic frameworks (MOFs) are a type of ideal crystalline material with rigid reticular structures (Kirchon et al. 2018), being useful for the construction of metallo-coordinated WLE systems (Pan et al. 2018). In this context, the use of lanthanide ions is effective owing to the high luminescent efficiency and narrowband, applicable to biomedical analyses and imaging fields (Bünzli 2010). Making use of Eu^{3+} and Tb^{3+} ions being strong red and green emissive activators, respectively (Tang et al. 2014), a MOF system with white-light emission can be prepared by trapping lanthanide ions, such as Eu^{3+} and Tb^{3+}, to sensitize their emission properties in the solid-state through a Ln^{3+}-exchanged process (Ma et al. 2014). For the approach, however, sensitization of the luminescence was attained through Dexter's theory, not employed by the FRET mechanism (Feng et al. 2015). The readers can see the comprehensive review paper about solid-state white-light emitters using photophysical properties of trivalent lanthanide ions (SeethaLekshmi et al. 2017).

It is interesting to explore how to functionalize the organic linkers of multicomponent MOFs to manipulate energy transfer interactions between them. Telfer et al., built up isoreticular multicomponent MOFs to tune the luminescence output systematically by introducing functional groups to the linkers (Cornelio et al. 2018). MUF-77 (Massey University Framework-77) (Liu and Telfer 2015) was used, being a multicomponent system comprising three chemically and geometrically distinct ligands connected by Zn_4O clusters. For such approach, $Zn_4O(hxtt)_{4/3}(gua)_{1/2}(bdc\text{-}NH_2)_{1/2}$ (x = alkyl groups) were prepared (Fig. 12.5), where hxtt, gua, and bdc-NH_2 were alkyl-substituted truxene-2,7,12 tricarboxylate, guanidine-functionalized 4,4′-bipenylcarboxyale with yellow emission (λ_{em} = 570 nm), amino-substituted 1,4-benzenedicarboxylate with blue emission (λ_{em} = 427 nm), respectively. The resultant MOFs (**19**, **20**, **21**, **22**) incorporate both blue (bdc-NH_2) and yellow (gua) emissive units together with the four triopic truxene linkers. As shown in Fig. 12.6a, an interplay between all three components endowed frameworks with tunable emission profiles. Subsequently, energy transfer interactions between those components in **21** led to the production of WLE with CIE coordinates (0.3218, 0.3590) and CCT of 5935 K close to that of pure white light (Fig. 12.6b).

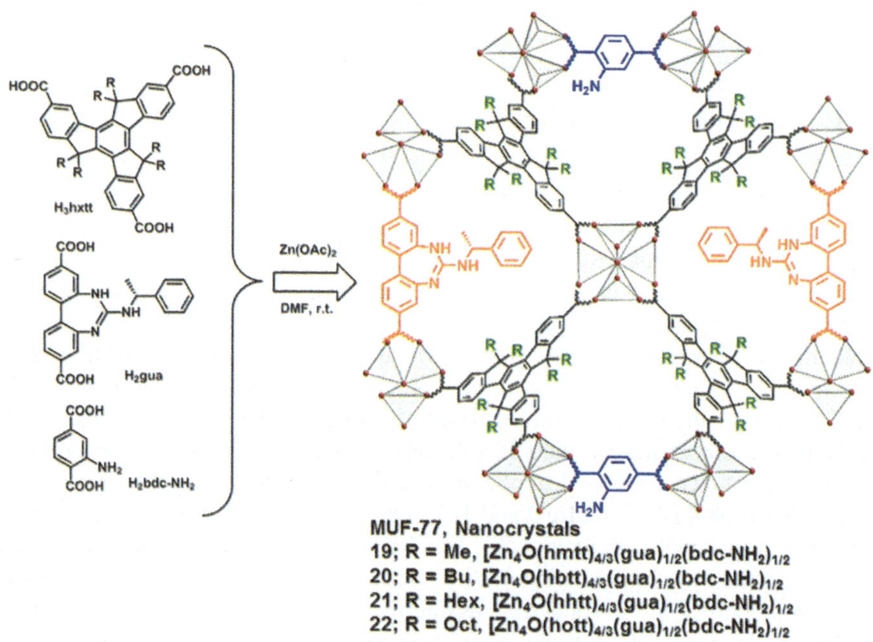

Fig. 12.5 Preparation of MUF-77 nanocrystals **19–22**. Reproduced with permission of The American Chemical Society from Cornelio et al. (2018). Copyright 2018 American Chemical Society

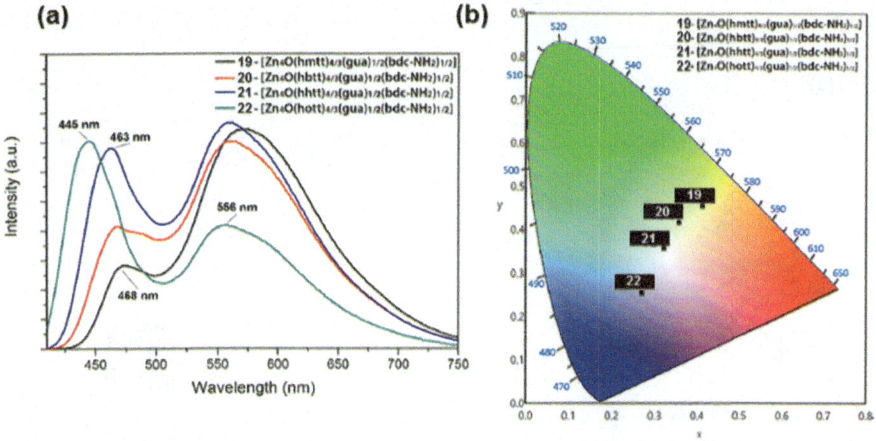

Fig. 12.6 **a** Emission spectra of MUF-77 nanocrystals **19–22**, suspended in DMF, with an excitation of 390 nm, **b** CIE diagram for the emission spectra of **19–22**. Reproduced with permission of The American Chemical Society from Cornelio et al. (2018). Copyright 2018 American Chemical Society

Isoreticular feature of MOF can serve as a host framework for encapsulating guest fluorescent dyes through noncovalent interactions with the guest. Bu et al. prepared a Zn^{2+}-based MOF with a carbazole-derived triangular carboxylic ligand, which had a high porosity a blue light emission based on the carbazole segment (Xia et al. 2018). The structural and photophysical characteristics made it an ideal energy transfer platform for the sensitization of red and green emissive dyes accommodated in the MOF. As a result, a WLE luminophore was obtained after carefully controlling their contents. These studies demonstrated that MOFs are among the most superior frameworks for application in WLE diodes.

Stang et al., prepared platinum-coordination driven macrocycle **23** that has an orange emission at 563 nm with AIE property (Zhang et al. 2017). On another front, fluorine-based bisammonium **24** was synthesized, which showed two emission bands at 391 nm and 411 nm, corresponding to blue emission. Those chemical structures are shown in Fig. 12.7a. Concentration-dependent ^{1}H NMR measurements were conducted to study the formation of supramolecular oligomers (Fig. 12.7b). When the concentration of a stoichiometric 1:1 mixture of **23** and **24** was increased from 1.0 to 40.0 mM, noticeable chemical shifts were observed for both compounds. Not only benzyl protons (H_t), but also methylene protons (H_u) of **24** were upfield-shifted, whereas ethylene protons arising from crown ether unit of **23** shifted downfield. The results indicated the formation of supramolecular oligomers through host-guest complexation, which was supported by two-dimensional diffusion-ordered NMR experiments. The emission properties of the supramolecular ensemble were investigated. It is interesting to note that the assembles formed by **23** and **24** with a 1:1 molar ratio showed concentration-dependent fluorescence in a wide color change. At high concentration (>0.5 mM) an orange emission was observed, whereas a blue

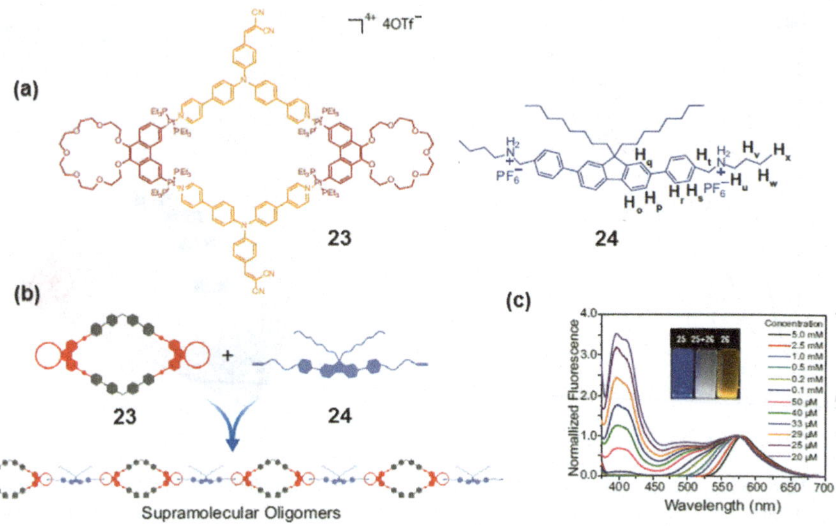

Fig. 12.7 **a** Chemical structures of **23** and **24**. **b** Cartoon representation of the formation of supramolecular oligomers. **c** Emission spectra of equal molar **23** and **24** at different concentrations. Reproduced with permission (Zhang et al. 2017). Copyright 2017, National Academy Sciences

emission was detected at low concentration (<25 μM). Given that macrocycle **23** and linker **24** act as AIE luminogen and ACQ fluorophore, respectively, the AIE luminogen plays a significant role at high concentration, whereas at low concentration the ACQ fluorophore dominates fluorescent feature to emit blue light. As a result, the assemble displayed WLE (CIE chromaticity coordinate: 0.30, 0.34) when set up the concentration of 29 μM. Under such conditions, both the orange and blue emission bands were observed, endowing the solution with an overall WLE upon fine-tuning the concentration. Although this approach was not due to FRET control, the precise manipulation of the AIE and ACQ properties of composed emitters in the self-assembles would be a potent way to provide WLE systems.

Tetraphenylethylene (TPE) and its derivatives are among the most promising AIE luminogens (Zhao et al. 2012b; Yang et al. 2018). Their superior synthetic diversity has allowed for a number of applications such as organic light-emitting diodes, chemosensors, bioprobes, and so on. Notably, the ability to modify them in a simple manner has led to their use in organic light-emitting diodes (Jadhav et al. 2017), bioprobes (Han et al. 2016), and chemosensors for molecular recognition (Noguchi et al. 2014; Park and Hong 2010; Zhang et al. 2015; Khandare et al. 2015), among others (Sick et al. 2018; Hoshi et al. 2018). Yang, Li, and co-workers developed TPE-incorporated metallo-supramolecules to build luminescent materials based on AIE behavior (Yin et al. 2018). To this end, ligands (**L1**, **L2**, and **L3**) were synthesized by introducing 2,2′;6′,2″-terpyridine (TPY) onto TPE core through several steps of Suzuki or/and Sonogashira couplings. Those ligands were assembled with $Cd(NO_3)_2$ in exact stoichiometric ratios in $CHCl_3$/MeOH mixture to form three generations

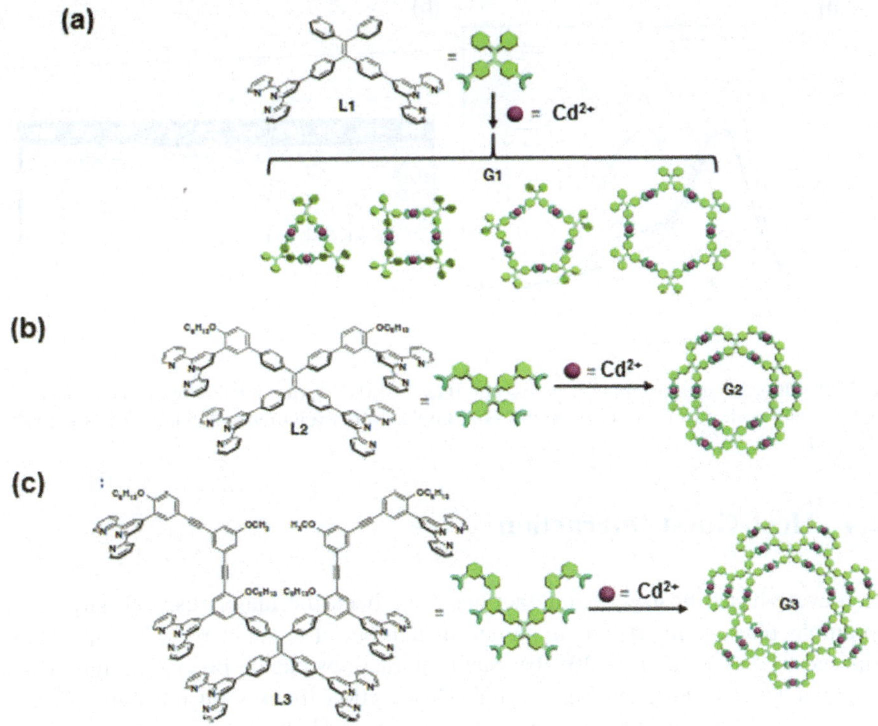

Fig. 12.8 Formation of supramolecular rosettes **G1** (**a**), **G2** (**b**), and **G3** (**c**) assembled with Cd^{2+}

of supramolecular rosettes (**G1–G3**) (Fig. 12.8). The self-assembly of **L1** gave a mixture of macrocycles, whereas the use of **L2** and **L3** produced discrete hexamer (**G2**) and heptamer (**G3**) through complexation with Cd^{2+}, respectively. Emission properties of the assembles with rosettes structures were investigated in both solution and aggregation states. In solution, a remarkable increase in Φ_F values of **G2** and **G3** was observed compared to the corresponding **L2** and **L3**, due to restriction of intramolecular rotation, whereas **G1** in solution remained weak emission. Among them, the optical feature of **G2** in solution and aggregation state allowed for room temperature WLE. The emission spectrum of **G2** in CH_3CN displayed two main bands centered at 450 nm and 580 nm, being ascribable to the local excited (LE) state of TPE and intramolecular charge transfer (ICT) character from TPE-based electron donor to the electron acceptor, TPY, moieties, respectively. Further enhanced intensity of the ICT band and decrease in the LE band was observed with increasing water fraction in a mixture of CH_3CN/water (Fig. 12.9a). In the case of 60% water fraction, **G2** emitted pure white light, as shown in Fig. 12.9b, with coordination (0.325, 0.355) in 1931 International Commission on Illumination (Commission Internationale de l'Éclairage, CIE) coordinates in the chromaticity diagram.

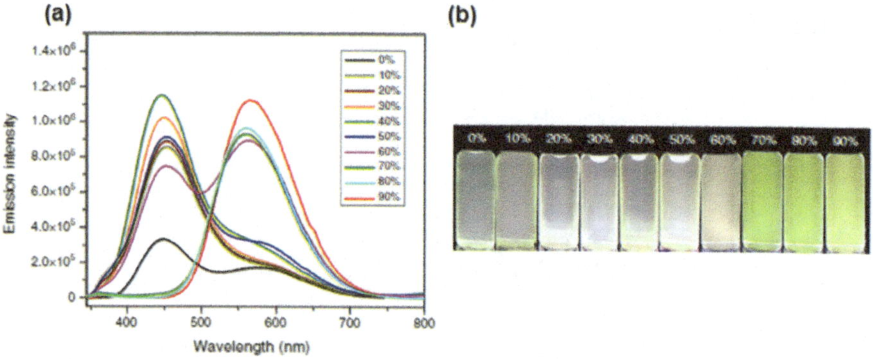

Fig. 12.9 Fluorescence spectra ($\lambda_{ex} = 320$ nm, [**G2**] = 1.0 μM) (**a**) and photographs (**b**) of **G2** in CH_3CN/water with various water fractions. **G2** samples were excitation at 365 nm on 298 K ([**G2**] = 1.0 μM)

12.4 Host-Guest Interaction

Macrocycle-based host-guest interactions have been the major research targets in supramolecular chemistry, where numerous number of host macrocycles have been proposed to date (Liu et al. 2017b). Such interactions enable precise design at the molecular level to tune emission color. This section focuses on intriguing examples of WLE through host-gust interactions. Liu et al., focused on the host-guest interactions between cucurbit[*n*]urils (n = 7, 8) (**CB[*n*]**s) and styrylpyridinium dyes (**SP**s) for color tuning (Li et al. 2017). As shown in Fig. 12.10a, **CB[*n*]**s are among the most popular macrocycles, which consist of methylene bridge-linked glycoluril units (Freeman et al. 1981). Carbonyl-enriched cavity is suitable for encapsulation of cations through ion-dipole interactions. The cavity size can be tuned by the number of the glycoluril unit. On the other hand, **SP**s are typical dipolar ones (Fig. 12.10a), being easy access to different structures with various emission colors and interact with **CB[*n*]** (Barrow et al. 2015). Host **CB[8]** was allowed to include 2 molecules of **B0** as a head-to-tail pattern (Fig. 12.10b) to show fluorescence with λ_{em} of 495 nm and 536 nm arising from *anti*-excimer, whereas **B3–B5** tended to form the *syn*-excimers in

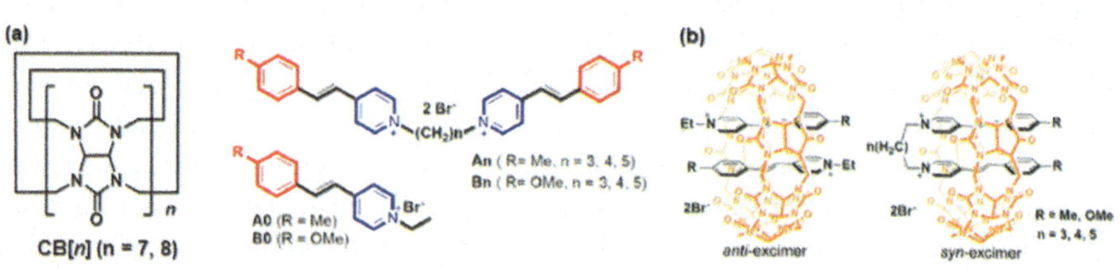

Fig. 12.10 **a** Chemical structures of cucurbit[n]urils **CB[*n*]**s and **SP**s. **b** Different inclusion patten of **SP** in **CB[8]**. Reproduced with permission of The American Chemical Society from Li et al. (2017). Copyright 2017 American Chemical Society

CB[8] through a stoichiometric 1:1 complexation, showing more redshifted emission than that of *anti*-excimer mode. Similarly, addition of 1 equiv. of **CB[8]** to **A3–A5** could enhance emissions at different levels. Also, emissive color-tunable properties of **An** took place on the basis of different ratios of monomer and *syn*-excimer emissions. Further, reversibly tunable emission was investigated by a supramolecular method. Figure 12.11a, b summarized change in the emissive color of **A3–A5** in the presence of **CB[7]**. This result was explained on the basis of **CB[7]** that could separate two **SP** rings through significant interactions between **CB[7]** and **SP** segment (Fig. 12.11c). In this context, bis-SP (**A3**) preferred to adopt *syn*-excimer formation where the **SP** segments intramolecularly stacked as head-to-head patterns in aqueous solution, leading to an emission at 607 nm. However, upon addition of **CB[7]** to an aqueous solution of **A3,** the *syn*-excimer fluorescence at 607 nm was weakened, whereas monomer emission at 460 nm was increased. Concomitantly, the emissive color changed from orange to blue, which included a near white-light emissive region. On the other hand, addition of 1-adamantaamine hydrochloride (**ADA**) with a strong affinity to **CB[7]** made **A3** release from the cavity of **CB[7]** competitively and recover the *syn*-excimer. This study demonstrated a facile method of luminescent color tuning by supramolecular assembly in which host-guest interaction can play a significant role in tuning of emission color.

Organic materials with room temperature phosphorescence (RTP) property have received lots of attention due to long afterglow lifetimes, which have opened the way to new possible applications (Forni et al. 2018). However, it is not so easy to prepare metal-free RTP materials since the energy of the excited triplet energy would be easily lost by thermally vibrational and collisional process or oxygen. Therefore, the study to develop pure organic RTP compounds has mainly focused on phosphorescence in the crystalline state. Accordingly, noncrystalline metal-free small compounds with

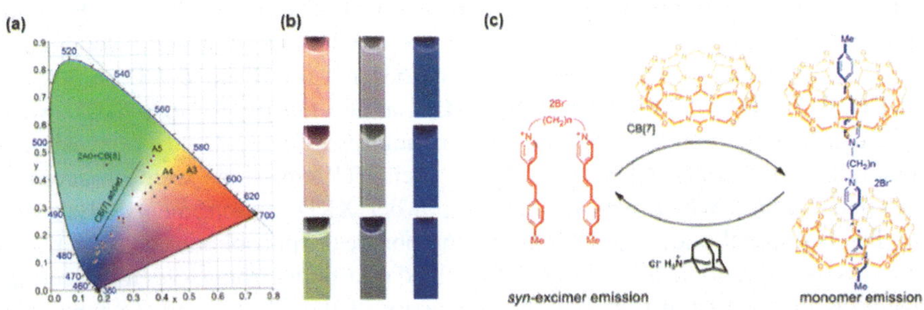

Fig. 12.11 **a** CIE chromaticity coordinate changes of **A3** (10.0 μM), **A4** (10.0 μM), and **A5** (10.0 μM) in the presence of **CB[7]** (0.0, 0.2, 0.4, 0.6, 0.8, 1.0, 1.2, 1.4, 1.6, 2.0 equiv.). **b** Photographs of fluorescent examples with CIE coordinate (from left up to right bottom): **A3** (10.0 μM) with 0.0, 1.0, and 2.0 equiv. of **CB[7]** (0.47, 0.41; 0.32, 0.29; 0.17, 0.16), **A4** (10.0 μM) with 0.0, 0.8, and 2.0 equiv of **CB[7]** (0.39, 0.40; 0.32, 0.32; 016, 0.13), **A5** (10.0 μM) with 0.0, 0.8, and 2.0 equiv of **CB[7]** (0.38, 0.48; 0.30, 0.34; 015, 0.10). **c** Schematic representation of reversibly tuning emissions from monomer to *syn*-excimer. Reproduced with permission of The American Chemical Society from Li et al. (2017). Copyright 2017 American Chemical Society

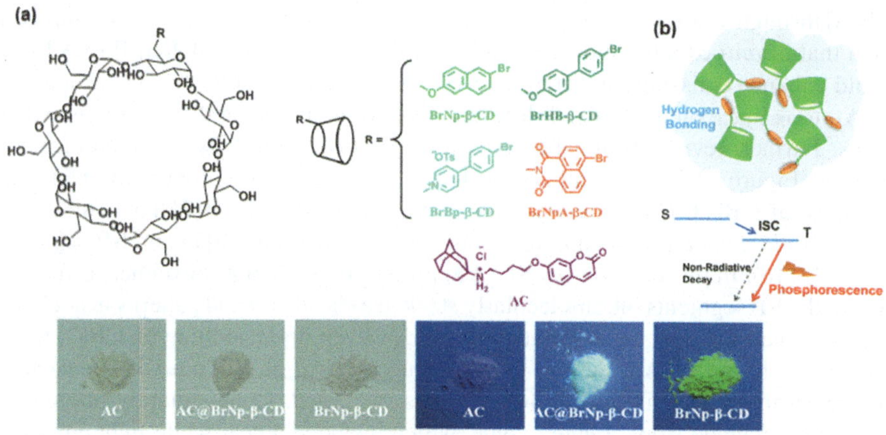

Fig. 12.12 **a** Molecular structures of RTP emissive cyclodextrin derivatives (**BrNp-*β*-CD**, **BrHB-*β*-CD**, **BrBp-*β*-CD**, and **BrNpA-*β*-CD**) and fluorescent guest molecule **AC**. **b** Schematic illustration of the phosphorescence emission. **c** Images of the solids under daylight and UV irradiation. Reproduced with permission of The American Chemical Society from Li et al. (2018). Copyright 2018 American Chemical Society

efficient RTP emission has been an intriguing target to be studied. In 2018, a series of amorphous organic small compounds were developed, which have efficient RTP emission capability to conveniently modifying phosphor moieties to β-cyclodextrin (β-CD) as shown in Fig. 12.12 (Li et al. 2018). CDs are cyclic oligosaccharides composed of six, seven, or eight D-(+)-glucopyranose units that are linked by *α*-(1-4)-glycoside linkage, which are named *α*, *β*, and *γ*-cyclodextrin, respectively (Wenz et al. 2006). Phosphors employed in this study are illustrated in Fig. 12.12a were connected at the primary OH group of *β*-CD. The introduction of Br atom as a heavy one into those systems led to a strong RTP emission in the solid-state; the emission peaks were detected at 512 and 537 nm for **BrNp-*β*-CD**, at 500 nm for **BrHB-*β*-CD**, at 500 nm for **BrBp-*β*-CD**, at 585 nm for **BrNpA-*β*-CD** when excited at 285 nm, 300 nm, 350 nm, and 365 nm, respectively. Unlike conventional organic RTP materials (Chai et al. 2017), such RTP emissions were observed in amorphous states, which was verified by powder XRD measurements. Hydrogen bonding between the adjacent CDs might immobilize the phosphorescent moieties to suppress the vibrational quenching and shielded phosphors from oxygen to a certain extent, as inferred from Fig. 12.12b. The cavity of the CD derivative enables the accommodation of guest molecules. Taking into account adamantine that can form a stable complex with β-CD (Liu et al. 2008), alkyl chain-linked adamantane/coumarin conjugate (**AC**) was synthesized and used as guest molecules to tune emission color in the system. Subsequently, a host-guest complex formation of **AC@BrNp-*β*-CD** in a 1:1 binding stoichiometry was confirmed by UV/Vis absorption spectroscopy, ^{1}H NMR techniques involving 2D ROESY measurement, and MALDI-TOF-MS spectrometry. It was found that the photoluminescence spectra of the solid type of

AC@BrNp-β-CD complex showed noticeable broad dual emission peaks, which was assignable to the emission of coumarin group at 400 nm (blue-purple band) and the BrNp group emission at 512 and 537 nm (yellow-green band). It is interesting to note that almost pure WLE with the CIE coordinates calculated as (0.29, 0.33) was realized when **BrNp-β-CD** was incorporated with 0.1 equiv. of **AC** under excitation at 295 nm.

Recently, an amphiphilic AIE cage system with tetraphenylpyrazole (TPP) was synthesized by stepwise nucleophilic aromatic substitution (S_nAr) reaction with **TPP-4OH** (Fig. 12.13a) (Feng et al. 2018b). The **TTP-cage** emitted strong blue fluorescence at 397 nm with photoluminescence quantum yield of 34.4% in THF solution under 365 nm UV irradiation due to the restriction of intramolecular rotation of the TTP unit. The emission remained unchanged when up to 60% of water was added to the THF solution. The information of the structural feature came from UV and CD measurements. The UV spectrum in THF exhibited an absorption maximum at 360 nm due to π–π* transition. It is interesting to note that Cotton effect was observed in the absorption regions of the phenol rings and TPP units. It indicated that the propeller-like TPP adopted either clockwise (*P*) or anticlockwise (*M*) rotational patterns. Crystal structure of **TPP-Cage** analog indicated that **TPP-Cage** could adopt a cubic prism structure, where the distance between the top and the bottom pyrazine cores was about 4.50 Å, and the distance between the two triazine ring nitrogen atoms was about 11.23 Å. The size of the cavity was large enough to encapsulate many guest molecules. In addition, π–π stacking, C–H–π, and C–H–N interaction allowed **TPP-Cage** to pack to form a grid-like porous structure. Given the blue emission of **TPP-Cage**, **DPP** was used as a guest molecule that emits a yellow and its molecular size enabled it to thread into the cavity of the cage system, forming a host-guest complex (Fig. 12.13b). WLE was easily obtained by adding water to the complex. As shown in Fig. 12.13c, **TPP-Cage** exhibited a deep blue emission at 397 nm when excited at 365 nm, which shifted to 430 nm upon adding water, whereas the glaring yellow emission of **DPP** was quenched by water addition due to the ACQ effect. However, the amphiphilic nature of the cage compound by introducing 2-[2-(2-methoxyethoxy)ethoxy]ethyloxy group allowed it to serve as a light harvesting platform to encapsulate **DPP** in its hydrophobic cavity in an aqueous solution, as evidenced from ^{1}H NMR and ^{1}H-^{1}H 2D NOESY NMR measurement. Subsequently, aggregation-induced quenching of **DPP** was avoided through such encapsulation. Indeed, a new peak assignable to the emission band of **DPP** encapsulated in the cavity was observed at 555 nm, with increasing water fraction in THF (Fig. 12.13c). Interestingly, the complementary emission colors of **TPP** and **DPP** enabled the complex to emit white light in the aggregation state.

12.5 Self-assembly

Organic assemblies from solution to solid substrates emerge as one of the most facile and promising way to retain the pure WLE, which is relevant for device

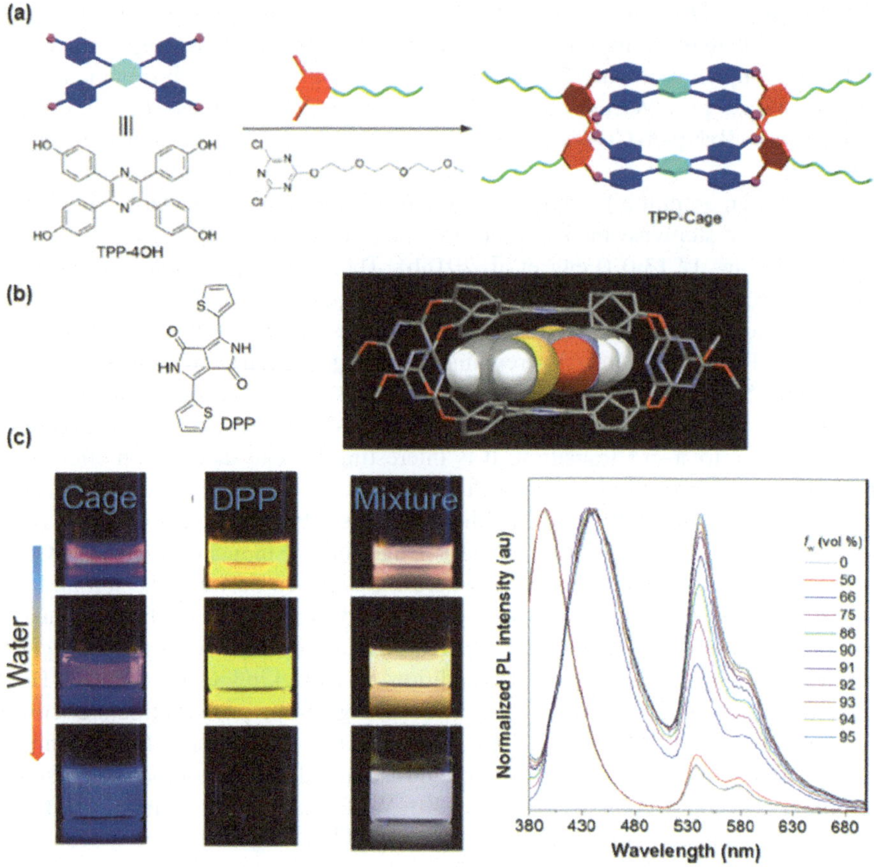

Fig. 12.13 **a** Schematic drawing of construction of **TPP-cage**. **b** Side view simulation of encapsulation of **DPP** in the cavity of **TPP-Cage** analog. **c** Photos of **TPP-Cage, DPP**, and their mixture (1:1, w/w) in THF (top) and THF/water mixtures (middle and bottom) with 75% (middle) and 95% (bottom) water fractions taken under 365 nm UV irradiation. **d** Normalized PL spectra of **TPP-Cage/DPP** mixture in THF and THF/water mixtures with different water fractions (f_W). [**TPP-Cage**] = [**DPP**] = 2.5×10^{-5} M; excitation wavelength: 365 nm. Reproduced with permission of The American Chemical Society from Feng et al. (2018b). Copyright © 2018, American Chemical Society

fabrication (Rao et al. 2013). Considering the environmentally benign construction of desired systems, the use of eco-friendly solvents such as water has been desirable. Würthner et al. prepared water-soluble and self-assembled nanocapsules composed of perylene bisimide vesicles (Zhang et al. 2009). For this approach, a small amount of THF solution of amphiphilic **PBI1** and **PBI2** (Fig. 12.14a) was added slowly to an aqueous solution of guest **bispyrene**. Subsequently, these loaded vesicles were stabilized by in situ photopolymerization using 2,2-dimethoxy-2-pneylacetophenone under 350 nm ultraviolet irradiation. The characterization was

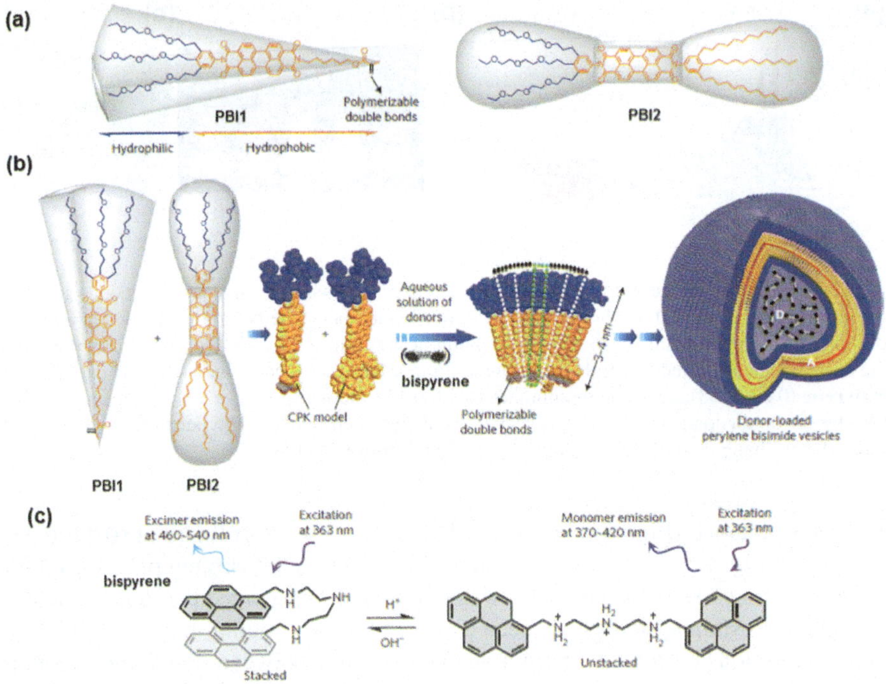

Fig. 12.14 **a** Chemical structures of amphiphilic perylene bisimide **PBI1** and **PBI2**. **b** Preparation of nanocapsule with bispyrene (**bispyrene**) guest incorporated. **c** Chemical structures of bispyrene donors with unstacked or stacked conformation. Reproduced with permission of Springer Nature from Zhang et al. (2009). Copyright © 2009, Springer Nature

conducted by infrared spectroscopy, TEM observation, and dynamic light scattering measurements. Bis(pyrene)-based donor and bilayer perylene acceptor membranes displayed well-separated blue-green fluorescence at 350–540 nm and red fluorescence at 600–850 nm in aqueous solution, respectively. The resultant nanocapsules (Fig. 12.14b) were stable in aqueous conditions, which had the energy donor molecules in the interior and bilayer perylene bisimide acceptors as the vesicular wall. Interestingly, a pH-dependent change in emission color was observed in the vesicular system. The vesicle showed blue-purple fluorescence at 370–420 nm excitation at 363 nm under acidic conditions, being ascribable to monomer emission of protonated bispyrene with unstacked conformation (Fig. 12.14c). However, with increasing pH, the blue–purple emission decreased, whereas the blue-green emission at 460–540 nm initially increased, being due to the excimer emission of bispyrene with π–π stacked conformation. And then, the red fluorescence arising from the perylene bisimide acceptor membrane appeared at 670 nm (Fig. 12.15a). It means that efficient FRET occurred under basic conditions from the stacked bispyrene to the perylene bisimide acceptor because the absorption band of the acceptor unit was overlapped with the band of excimer emission of the bispyrene donor. At pH = 9.0

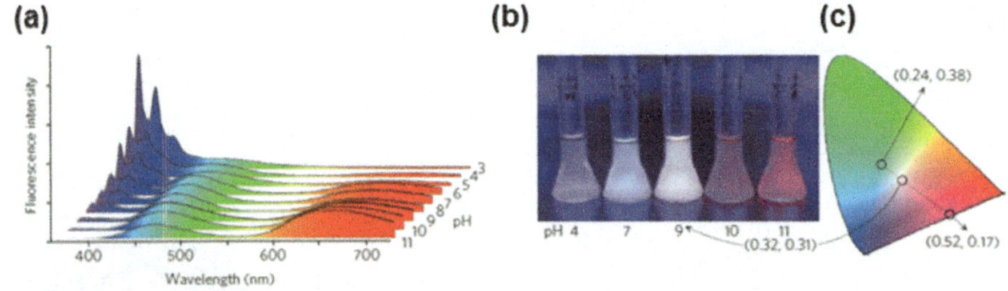

Fig. 12.15 **a** Fluorescence spectra of donor-loaded perylene bisimide vesicles in aqueous solution ranging from pH 3.0–11.0. **b** Photographs of donor-loaded perylene bisimide vesicles in aqueous solution at different pH under an ultraviolet lamp at 366 nm. **c** CIE 1931 chromaticity diagram. The three points indicated by circles signify the fluorescence color coordinates for the excimers of **bispyrene** (0.24, 0.38), perylene membranes (0.52, 0.17) and white fluorescence coordinate (0.32, 0.31) for the donor-loaded polymerized vesicles at pH 9.0. Reproduced with permission of Springer Nature from Zhang et al. (2009). Copyright © 2009, Springer Nature

WLE was obtained (Fig. 12.15b) with CIE chromaticity coordinate of (0.32, 0.31) as shown in Fig. 12.15c. In this way, such a FRET tuning by varying pH allowed for the production of WLE behavior for water-soluble donor-loaded perylene vesicle.

Light harvesting systems have been explored for the generation of multiple fluorescent emissions by tuning energy transfer efficiency between donor and acceptor components. Guo et al., developed an amphiphilic calixarene-built in platform toward this end (Fig. 12.16) (Xu et al. 2016). The phenol-derived cyclic oligomer is called "calixarenes" (Kim et al. 2012). The amphiphilic derivatives (**AmC*n*A**s, n = 4 and 5) in this study were composed of hydrophilic choline groups at the upper rim and hydrophobic alkyl chains at the lower rim. Prior to the energy transfer experiment, the assembly properties of **AmC*n*A**s were evaluated. Both **AmC4A** and **AmC5A** have similar critical aggregation concentration of 5.2×10^{-6} and 5.5×10^{-6} M, respectively, indicating unique superiority in self-assembly and host-guest interaction. Indeed, dynamic light scattering (DLS) measurements exhibited that **AmC4A** and **AmC5A** formed large-sized aggregates with averaged diameters of 149 and 79 nm, respectively. It was found that the vesicular aggregate arising from **AmC5A** was stable without agglomeration and precipitation. To fabricate an efficient light harvesting system based on calixarene amphiphile platform, 1-anilino-8-naphthalenesulfonate (**1,8-ANS**) and 4,7-bis(thien-2-yl)-2,1,3-benzo-thiazole (**DBT**) were used (Fig. 12.16a). Subsequently, **1,8-ANS** was favorably encapsulated to the cavity of the calixarene (Fig. 12.16b) with emission enhancement at ca. 475 nm (λ_{ex} = 354 nm). On another front, the hydrophobic feature of **DBT** made it to be entrapped into the hydrophobic bilayer. It is worthy to note that the absorption band of **1,8-ANS** overlaps with the emission band of **DBT** to promise substantial FRET behavior. As a result, the noncovalent loading of **1,8-ANS** donor encapsulated in the calixarene-based cavity and **DBT** convolved in the hydrophobic alkyl chain led to the production of tunable colorful emission through the FRET process. Subsequently, the emission of **AmC5A/1,8-ANS/DBT** vesicles

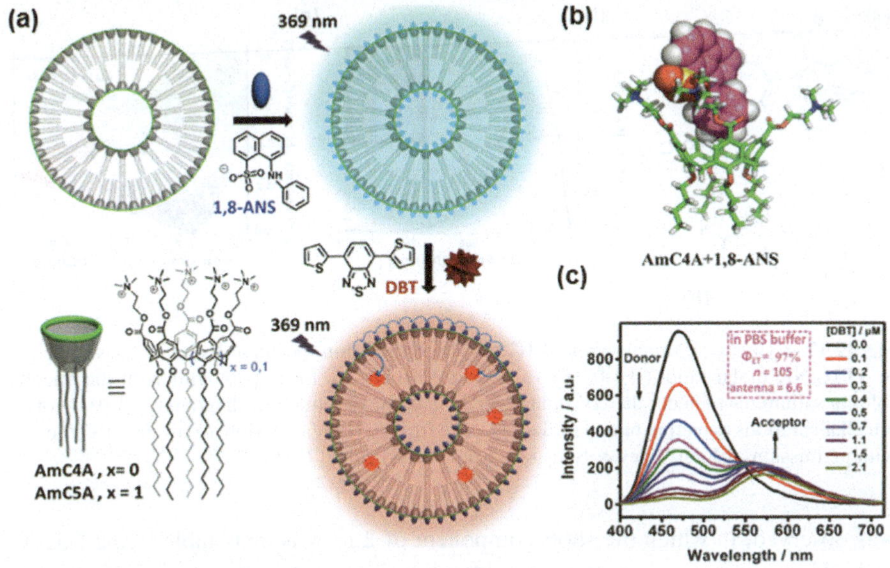

Fig. 12.16 **a** Schematic illustration of light harvesting vesicular system. **b** Molecular simulation of **AmC4A/1,8-ANS**, where the long alkyl chains of **AmC4A** were omitted for clarity. **c** Fluorescence spectra of **AmC5A/1,8-ANS** in PBS buffer with different concentration of **DBT**, $\lambda_{ex} = 369$ nm, [**1,8-ANS**] = [**AmC5A**] = 2.0×10^{-5} M. Reproduced with permission of John Wiley & Sons Ltd. from Xu et al. (2016)

in phosphate-buffered saline (PBS) was tuned by varying donor-to-acceptor ratios (Fig. 12.16c) to attain pure WLE with color coordinates (0.31, 0.34) with color temperature of 6581 K.

It has been known that an amphiphilic building block self-assembles in water to yield a morphologically distinct micelle (Bales et al. 1998). Pioneer work to produce FRET-based WLE nanosystem was conducted by Würthner et al., in which perylene dye micelles with red emission were prepared (Zhang et al. 2013). When loaded blue emissive 1,3-di-*N*-carbazolepropanes into the micelles to give WLE from the two fluorescent colors. However, taking into account facile, cost-effective, and environmentally benign protocol for the fabrication of WLE micelle, the use of commercially available surfactants is a very effective way. Patra et al., synthesized hydrophobic fluorophore, 6,7,8,9-tetrapropylpyrido[1,2-a]indole-10-carbaldehyde (**TPIC**) (Fig. 12.17a) and carried out screening with a series of cationic (CTAB and Adogen 464), anionic (SDS and sodium taurocholate), and nonionic surfactants (Triton X-45, Triton X-100, Tween 20, and Tween 80) (Pallavi et al. 2018). Subsequently, the solubility of **TPIC** was significantly enhanced in an aqueous solution of Triton X-45 to induce fluorescence enhancement of **TPIC** ($\lambda_{em} = 485$ nm, $\lambda_{ex} = 410$ nm). The fluorescence quantum yield was also found to be 33% in aqueous Triton X-45 dispersion. The photophysical properties of **TPIC** were probed by using the time-correlated single-photon counting (TCSPC). Bi-exponential decay kinetics

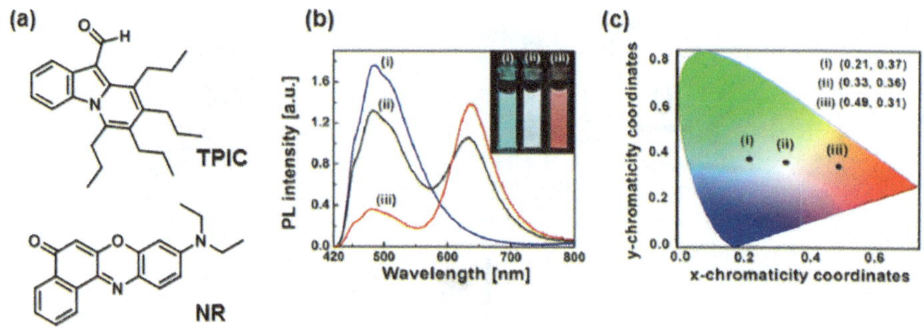

Fig. 12.17 **a** Chemical structures of **TPIC** and **NR**. **b** Emission spectra in Triton X-45 as a function of **TPIC**:**NR** molar ratio: (i) 1:0, (ii) 1:0.05, and (iii) 1:0.25. Inset: photographs of the micellar solutions illuminated under the UV light at 365 nm. **c** The chromaticity diagram (CIE 1931) of the micellar solutions depicting tuning of fluorescence from: (i) blue, (ii) white to (iii) red. Reproduced with permission of John Wiley & Sons Ltd. from Pallavi et al. (2018)

was observed, in which the short component of 2 ns was assignable to the fraction of TPIC residing at the micelle-water interface. The long decay time of 9.5 ns originated from the molecules present at the core of the micelles. These results indicated hydrophobic **TPIC** were protected from the aqueous environment to circumvent the ACQ effect in the nonionic micelles. To explore the possibility of multicolor emission through a rationalized FRET pair, Nile red (**NR**) (Fig. 12.17a) was chosen as a FRET acceptor because the emission of **TPCI** was considerably overlapped with the absorbance of **NR**. Figure 12.17b shows fluorescent spectra of **TPIC** and **NR** as a function of **TPIC**:**NR** molar ratio of 1:0 (i), 1:0.05 (ii), and 1:0.25 (iii). When set up **TPIC**:**NR** molar ratio of 1:0.05 (ii), excitation of 410 nm led to WLE emission with displaying similar peak intensities at 485 nm and 640 nm, covering the entire visible region (Fig. 12.17b). The CIE coordination indicated $x = 0.33$, $y = 0.36$, being close to the pure white (Fig. 12.17c). Easy-to-make micelle-induced WLE using water as environmentally friendly solvent promises potential applications for lighting and display devices.

12.6 Supramolecular Gels

Gels have been well-recognized as supramolecular nanosystems. Indeed, the dynamic nature of gels stemming from low molecular weight gelators (LMWGs) have received much attention over the past few decades because of their possible use as smart materials in a wide range of applications, including catalysis, crystal growth, ion sensing, drug delivery, data storage, and biomaterial replacement (Babu et al. 2014; Jones and Steed 2016; Segarra-Maset et al. 2013; Yu et al. 2013; Sun et al. 2014; Zhang and Su 2013; Duan et al. 2014; Zhang et al. 2016). Given that the FRET process from donor to acceptor plays a significant role for light harvesting to emit white light, gel as a platform is so available that it provides a molecular organization that could help efficient FRET process (Praveen et al. 2014). As a typical example, a co-assembled hydrogel of melamine was developed, where 6,7-dimethoxy-2,4[1*H*,3*H*]-quinazolinedione (**Q**), riboflavin (**R**), and rhodamine B (**RhB**) were mixed (Fig. 12.18a) (Bairi et al. 2013). Melamine (**M**) is a useful supramolecular synthon that can bind emissive molecules through hydrogen bonding interactions, resulting in the formation of self-assembled gels. On photo-excitation at 297 nm, **Q** emitted at 384 nm in the aqueous solution ($\Phi = 0.53$), whereas **R** exhibited absorption peaks at 373 nm corresponding to the π–π^* transition coupled with the n–π^* transition and at 444 nm due to the π–π^* transition of the flavin moiety, respectively. And **R** emitted at 532 nm in aqueous solution ($\Phi = 0.27$) when excited at 373 nm. Such a good spectral overlap of the absorption spectra of **R** and fluorescence of **Q** led to the investigation of the interplay of the components in an aqueous solution by applying the Stern-Volmer plot. The fluorescence titration of **Q** was conducted upon adding incremental amounts of **R**, in which the plot of I_0/I versus [**R**] plot showed a nonlinear Stern-Volmer curve. It suggested that not only the static quenching occurred for the complex formation at the ground state, but the

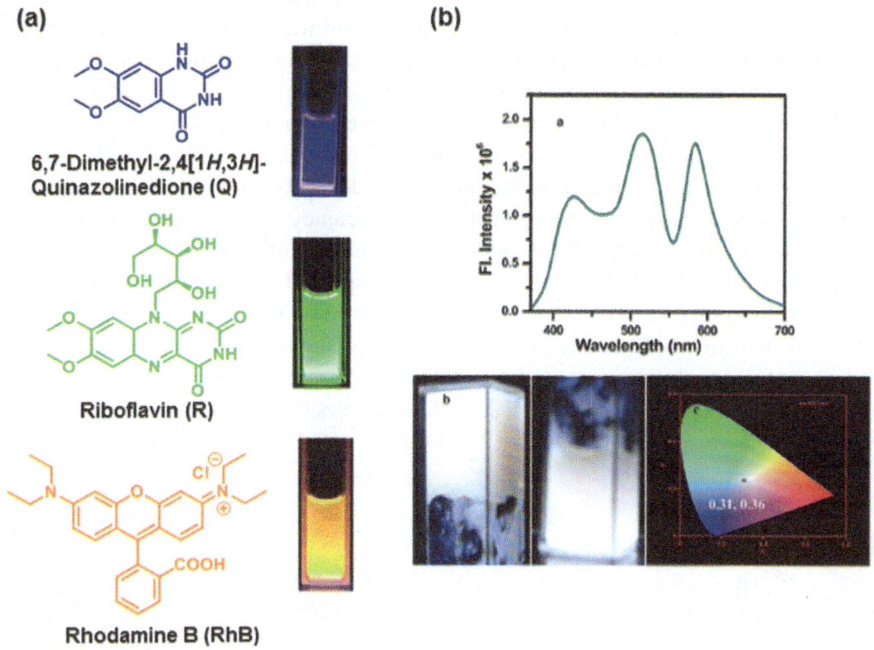

Fig. 12.18 **a** Chemical Structures of **Q**, **R**, and **RhB**, **b** Emission spectra of WLE gel having molar ratio **Q**:**M**:**R**:**RhB** 100:100:0.5:0.02 for excitation at 360 nm "a", photograph of under UV light irradiation at 365 nm "b", and CIE coordinate for the WLE gel "c". Reproduced with permission of The American Chemical Society from Bairi et al. (2013). Copyright 2013 American Chemical Society

dynamic quenching also took place due to the energy transfer at the excited state. On the other hand, the fluorescence quenching of **R** with **RhB** showed a straight line of the Stern-Volmer plot of I_0/I versus [**RhB**]. Taking into account the lifetime of the **R** + **RhB** mixture decreasing from 5.05 ns of pure **R** to 3.81 ns indicating the dymaic quenching and suggesting the excited state energy transfer from **R** to **RhB**. However, no energy transfer from **Q** to **RhB** occurred in the solution state. Alternatively, the closer proximity between the donor (**Q**) and acceptor (**RhB**) in the gel state enabled it to cause an appreciable energy transfer between them. WLE gel was obtained by varying the **RhB** and **R** concentration in the 1:1 **MQ** gel, which had a molar ratio of **M**:**Q**:**R**:**RhB** = 100:100:0.5:0.02 on excitation at 360 nm (Fig. 12.18b). The CIE coordinates of 0.31 and 0.36, the values being very close to those of white light.

Two-component WLE organogel was reported by Yi et al., where a naphthalimide-based gelator (**Naph**) served as a donor moiety for the energy transfer process (Fig. 12.19) (Cao et al. 2015). Compound **Naph** in CH_3CN emitted blue light at 430 nm (Φ= 26.7%) under excitation at 360 nm, which gelated many kinds of organic solvents because of the intermolecular π–π, hydrogen bonding, and hydrophobic interactions. On the other hand, the Ir^{3+} complex ($Ir(bt)_2(acac)$) (bt = 2-phenylbenzothiazole, acac = acetylacetone), **Ir**, absorbed in a broad wavelength region ranging from 375 nm to 525 nm with extinction coefficient (ε) ≈ 10^3 M^{-1} cm^{-1}, being assignable to the mixed singlet and triplet metal-to-ligand charge-transfer (MLCT) transition (dp(Ir)–p*(bt)) (Li et al. 2011). Simultaneously, the Ir^{3+} complex **Ir** shows a phosphorescent band with λ_{em} value of 563 nm and a shoulder at 603 nm with a 1.6% luminescence quantum yield when excited at 356 nm. Such spectral features allowed for envisioning the FRET process from **Naph** to **Ir**. Indeed, the emission spectra of **naph** decreased upon adding **Ir** to CH_3CN (Fig. 12.20a), resulting in an emission of white light. The efficiency of the energy transfer was investigated by means of Förster theory. Time-resolved fluorescence measurement of **naph** in a diluted CH_3CN solution (10^{-5} M) showed a single exponential decay with a lifetime of $\tau = 2.05$ ns, which was shortened to 1.29 ns with the addition of 0.5 equivalent of **naph**. It was indicative of a FRET process between donor **naph** and acceptor **Ir**. However, the FRET efficiency was evaluated to be 15.5–36.6% according to Förster theory (Fig. 12.20b). The Förster distance was calculated to be 6.95–8.42 nm on the basis of (12.3) equation (vide supra). A relatively larger

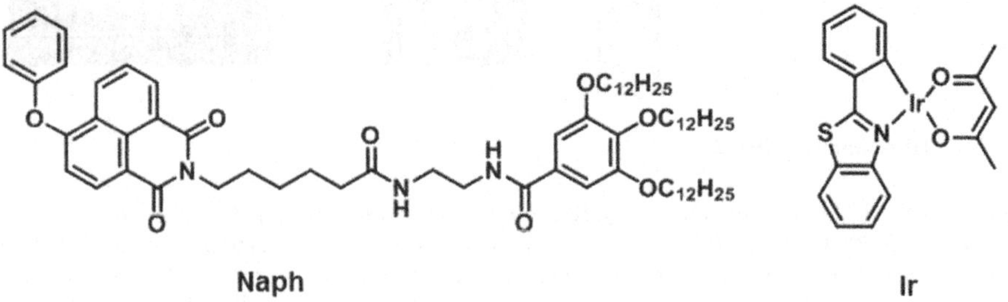

Fig. 12.19 Chemical structures of compound **27** and Ir^{3+}-complex **Ir**

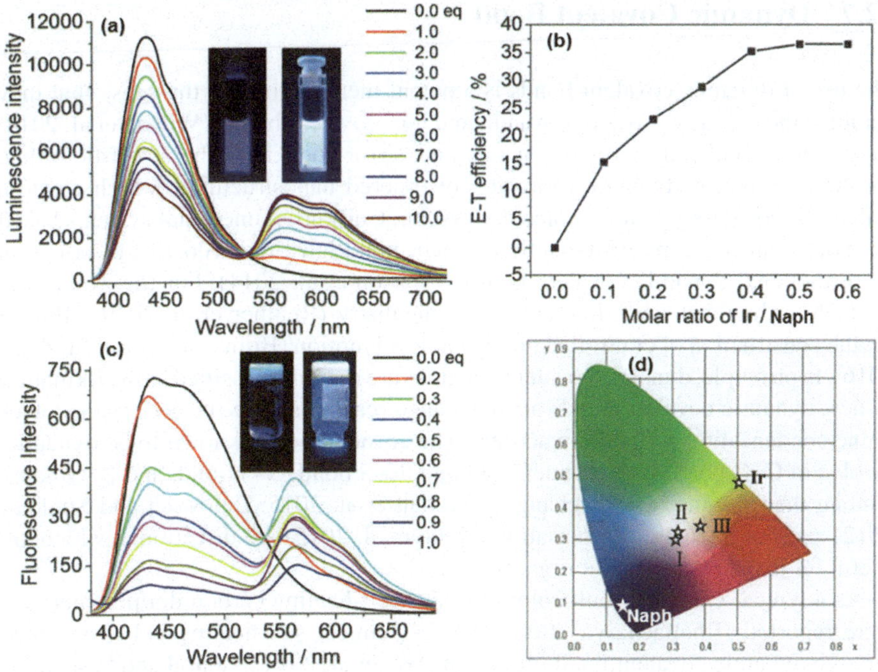

Fig. 12.20 **a** Emission spectral changes of **naph** with the incremental addition of **Ir** ([**naph**] = 1.0 × 10^{-5} M and λ_{ex} = 356 nm); **b** Change in energy transfer efficiency with the addition of different amounts of **Ir** to **naph** in CH_3CN ([**naph**] = 1.0 × 10^{-5} M); **c** Emission spectra of gel **naph** and gel **naph** – **Ir** with different molar ratios of **Ir/naph** (λ_{ex} = 356 nm, [**naph**] = 25 mg mL^{-1}); **d** CIE 1931 chromaticity diagram of the Ir^{3+} solution, gel **naph** and gel **naph** –**Ir** (I: 0.3 eq.; II: 0.5 eq.; III: 1.0 eq. of **Ir**) (λ_{ex} = 374 nm). The insets in (**a**) and (**c**) are luminescence images of **naph** (left) and **naph** –**Ir** (right) in solution and gel, respectively. Reproduced from Cao et al. (2015) with permission from the PCCP Owner Societies

amount of **Ir** was needed for WLE in CH_3CN (Fig. 12.20a). On another front, in the gel state, the energy transfer efficiency was found to be 41.1% (Fig. 12.20c), being higher than that in the solution. The distance between the donor and acceptor in the gel state was evaluated to be 6.57 nm. Subsequently, more efficient FRET property reduced the amount of Ir^{3+} complex, although 'aggregation-caused quenching' effect in the gel led to decrease emission intensity. The morphological structure and mechanical properties of the gel were investigated to obtain insight that the addition of **Ir** did not affect the mechanical properties or the surface wettability of the gel. As a result, the **naph-Ir** gel with the addition of 0.5 equivalents of **Ir** could emit white light with CIE coordinates of (0.33, 0.31) with excitation at 374 nm (Fig. 12.20d).

12.7 Dynamic Covalent Bond

The use of dynamic covalent bonds is a potent methodology in the sense that they combine the characteristics of covalent and non-covalent bonds (Wilson et al. 2014; Rowan et al. 2002). Its reversible nature and "error-checking" characteristics allow for the easy thermodynamic formation of ordered nanostructures, which create a stable and covalently bound molecular system. Currently, functional systems based on more than one type of dynamic covalent bond have received lots of attention as "orthogonal dynamic covalent bonds" (Wilson et al. 2014). For the expansion, "tunable orthogonal reversible covalent" chemistry (Reuther et al. 2019) allowed for the construction of controlled molecular level motion (Bruns and Fraser Stoddart 2016). In principle, dynamic covalent reactions are roughly classified as the formation of new dynamic covalent bonds and exchange reactions. The former case involves imine condensation and aldol reactions. The covalent bonds known to be dynamic, involve in C–C bond, C–N bond, C–O bond, C–S bond, S–S bond, and B–O bond. Among them, imine bond/exchange (Vardhan et al. 2015; Belowich and Stoddart 2012) and boronic acid condensation (Kubo et al. 2015) are directional, which are useful for the supramolecular organization.

As a typical example, multicolor emissive self-healing carbon dot/polymer gels were reported (Bhattacharya et al. 2019). Chemical self-healing gels assembled through dynamic covalent bond formation (Wei et al. 2014; Amaral and Pasparakis 2017) are well-known in the field of soft materials. In this study, Jelinek et al., utilized carbon dots (C-dots) as luminescent nanoparticles, whose emission properties can be tuned by the different carbonaceous precursors used (Miao et al. 2018). Figure 12.21a shows the synthetic schemes and assembly of emissive C-dot/polymer gel. The three aldehyde building blocks and experimental conditions gave the differently colored C-dots; green emissive C-dots (**G-C-dots**) were prepared using glutaraldehyde at 150 °C for 2 h. Blue emissive C-dots (**B-C dots**) and yellow C-dots (**CoAP-C-dots**) were obtained from benzaldehyde and polymeric cyclooctadiene-aldehyde, respectively. The synthetic strategy of the targeted gel was depicted in Fig. 12.21b. Mixing the C-dots and polyethylhennkoueneimine (PEI) led to Schiff base reaction between the aldehydes on their surfaces of the C-dots and a number of amines in the branched PEI framework to give imine bonds, resulting in the formation of C-dots/PEI gel. The dynamic covalent imine bonds are core structural elements that are responsible for the self-healing properties (Fig. 12.19c). For the characterization, the high-resolution transmission electron microscopy (HR-TEM) was used to reveal well-resolved lattice planes confirming the formation of sp^2 graphitic cores. The average C-dot particle sizes were determined to be 6 ± 3 nm, 3 ± 1 nm, and 7 ± 2 nm for **G-C-dots**, **B-C-dots**, and **CoAP-C-dots**, respectively. It is noteworthy that no aggregation-induced fluorescence quenching was widely observed in C-dot systems in the solid phase, strongly suggesting immobilization of the cross-linked C-dots within the gel matrix to prevent their close proximity and concomitant self-quenching. The formation of imine bonds was evidenced by FT IR and XPS measurements. A significant decrease in 1700 cm^{-1} due to C=O was observed, whereas the generation of imine bonds was

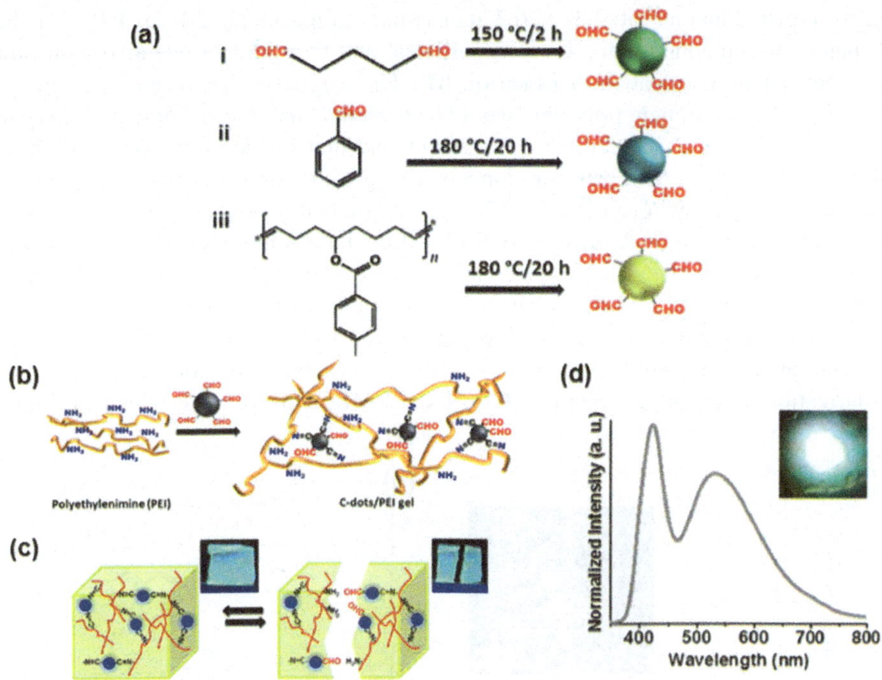

Fig. 12.21 **a** Synthesis of aldehyde-containing C-dots. **b** Cartoon of assembly of the C-dot/PEI gel through Schiff base reaction between the aldehyde units upon the C-dots' surface and the amines within the PEI framework. **c** Self-healing properties of the gel are attained through reversible imine bond formation. The inset representative fluorescence images of the self-healing phenomenon are shown. **d** Emission spectra and optical images of **B-C-dot/G-C-dot**/PEI plus **CoAP-Cdot**/PEI films under illumination of 403 nm-LED. Reproduced with permission of The American Chemical Society from Bhattacharya et al. (2019). Copyright © 2019, American Chemical Society

observed at 1650 cm^{-1}. Similarly, a new species attributable to the –C=N peak was observed in the XPS analysis. The imine bond formation in the gel was supported by UV/Vis absorption spectra.

The self-healing properties of the aldehyde-C-dot/PEI gels were investigated to show high sensitivity to external stimuli. Figure 12.21c highlights such properties attained through reversible imine bond formation. When one cuts it in the middle, the two separated pieces of the gel reattached upon attaining a physical contact, demonstrating the rapid room temperature reconstitution of the imine bonds at the interface between the two films. Notably, the border between the two fused pieces could hardly be checked in the recovered film. The multicolor properties of the gels were applied to WLE materials. Subsequently, the **B-C-dot/G-C-dot**/PEI and **CoAP-Cdot**/PEI films stacked horizontally on opposite sides of a quartz glass slide led to the generation of white light under irradiation with a 403 nm LED.

Kubo, et al., investigated the sequential dehydration benzene-1,4-diboronic acid with pentaerythritol induced mono-dispersed flower-like microparticles (**BPs**) with an averaged diameter of 2.3 ± 0.3 μm (Matsushima et al. 2012). **BP**s can be dispersed in aqueous media as microparticles, and they have a negative value for the zeta potential (ξ) at the surface of **BP**. This suggests that terminal hydroxyl groups of the component polymers would be present at the microscopic periphery of the microparticles (Vijayakumar et al. 2011), and that surface functionalization by grafting is possible. As a proof-to-concept, dye grafting on the surface was investigated using 1-pyreneboronic acid (**25**). For the procedure, **BP** (10 mg) was soaked in a MeOH solution of **25** (1.0×10^{-5} M, 10 mL) at room temperature for 24 h to give emissive particles. Fluorescence microscopy revealed dispersed blue emissive nanoparticles (Fig. 12.22a), which were assignable to monomeric pyrenyl emissions and the quantum yield (Φ) was determined to be 0.74. Direct evidence for grafting pyrene through boronate esterification was subject to solid-state ^{13}C cross-polarization magic angle spinning (CP-MAS) NMR measurement, which exhibited

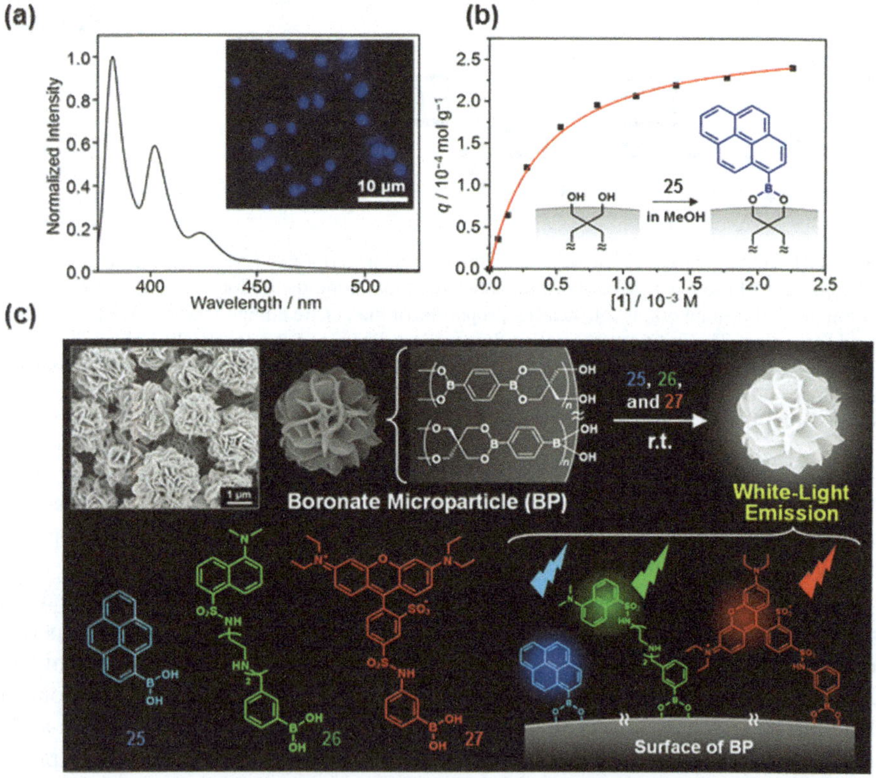

Fig. 12.22 **a** Fluorescence spectrum of pyrene-grafted **BP** in 5 mM HEPES buffer (pH 7.0). Inset: fluorescence microscopic image of pyrene-grafted-**BP**. **b** The graft of boronic acid-appended pyrene onto the surface of **BP** as a function of concentration of **25** in MeOH where q denotes the amounts of dye grafted. **c** Preparation of water-dispersible white-light emitting boronate microparticles. Reproduced from Nishiyabu et al. (2013) by permission of The Royal Society of Chemistry

signals ($\delta = 121.5$–129.5 ppm) assignable to the aromatic C–H carbons of the pyrenyl moiety. Surface analysis was also conducted by using a Langmuir-type adsorption isotherm to provide an estimated theoretical maximum value of 45% for the fraction of surface coverage by boronate pyrene molecules (Fig. 12.22b). To fabricate nanoparticles with WLE, dansyl-appended boronic acid (**26**) and a rhodamine B derivative (**27**) was grafted on the surface in a similar manner. The resultant particles were emitted at 515 nm ($\Phi = 0.52$) and 600 nm ($\Phi = 0.34$) in the 4-(2-hydroxyethyl)-1-piperazineethanesulfonic acid (HEPES) buffer (pH 7.0), respectively. According to CIE coordinates in the chromaticity diagram, color tuning of the particles was achieved by adjusting the relative concentrations of the blue-, green-, and red-emissive dihydroxylboryl dyes to produce the desired **Sens-BP** microparticles with white-light emissive properties (Fig. 12.22c) (Nishiyabu et al. 2013).

It is worthy to note that **Sens-BP** microparticles served as a WLE chemosensors in water because the green component grafted by **26** has a preferable binding site for metal ions. As shown in Fig. 12.23a, the addition of Cu^{2+} into the HEPES buffer (pH 7.0) led to an efficient decrease of the emission intensity in the wide range of the visible spectrum (450–625 nm). This suggests that complex formation between the dansyldiethylenetriamine moiety and Cu^{2+} could induce fluorescence quenching of dansyl fluorophore and decrease the fluorescence intensity of the rhodamine B unit because of the reduced FRET from dansyl to the rhodamine B units. This synergistic quenching upon adding Cu^{2+} led to a notable change of the emission color, which was easily detected by the naked eye (Fig. 12.23a). The Cu^{2+} selectivity and reusability for chemosensor applications were also investigated (Fig. 12.23b), where

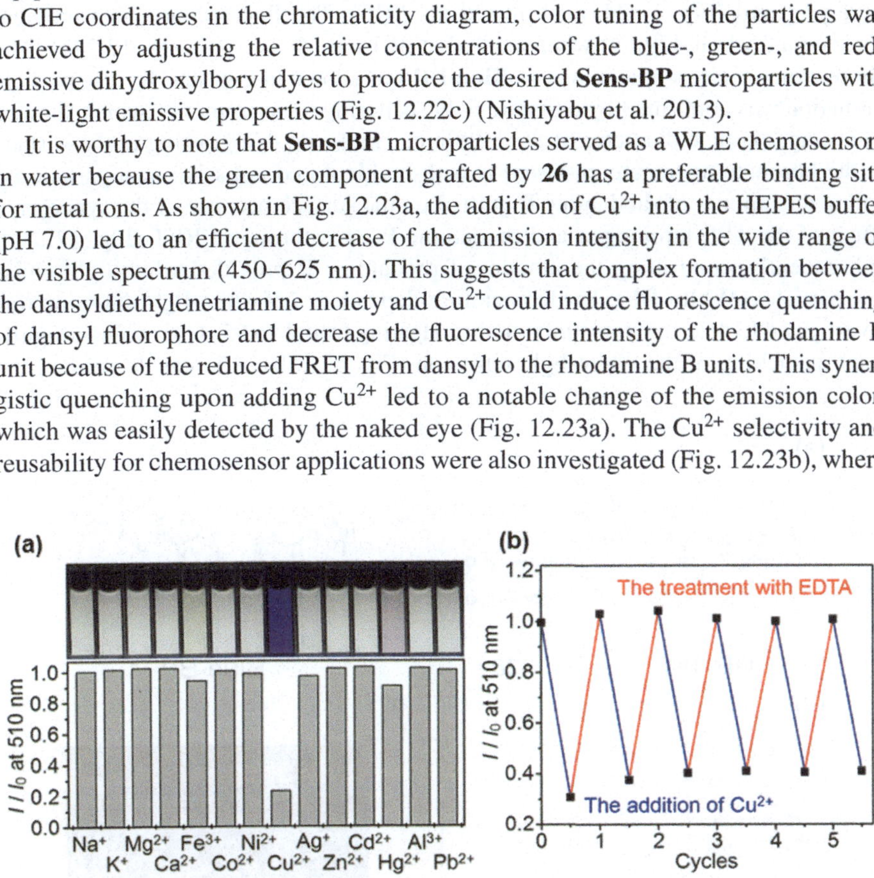

Fig. 12.23 **a** Emission colors and changes in the fluorescence intensity of **Sen-BP** (1.0 mg mL^{-1}) after immersion in 5 mM HEPES buffer solutions (pH 7.0) of Na^+, K^+, Mg^{2+}, Ca^{2+}, Fe^{3+}, Co^{2+}, Ni^{2+}, Cu^{2+}, Ag^+, Zn^{2+}, Cd^{2+}, Hg^{2+}, Al^{3+}, and Pb^{2+}, respectively. $[M^{n+}] = 3.0 \times 10^5$ M, $\lambda_{ex} = 365$ nm, $\lambda_{em} = 510$ nm. **b** Recyclability of **Sen-BP** (1.0 mg mL^{-1}) for Cu^{2+} detection. The fluorescence quenching caused upon addition of Cu^{2+} (3.0×10^5 M) in 5 mM HEPES buffer (pH 7.0) and the fluorescence was recovered by treatment with EDTA (1.0×10^3 M): $\lambda_{ex} = 365$ nm, $\lambda_{em} = 510$ nm. Reproduced from Nishiyabu et al. (2013) by permission of The Royal Society of Chemistry

their recyclability as chemosensors was demonstrated by using ethylenediaminetetraacetic acid (EDTA). Taken together, a WLE chemosensor workable in water was proposed by the surface modification of boronate microparticles.

Two-component approach to emit white light was applied to boronate microparticle's functionalization. In this context, the incorporation of AIE-active luminogens into boronate assembly was investigated by using a sequential boronate esterification of di(boronic acid)-appended tetraphenylethylene (TPE) (**DB-TPE**) with pentaerythritol (**28**). As a result, blue emissive boronate nanoparticles **Blue-BP**s were prepared (Fig. 12.24a) (Ozawa et al. 2015). The fluorescence measurement detected a peak at 479 nm when excited at 365 nm ($\Phi = 21\%$; Fig. 12.24b). The fluorescence microscopic image of **Blue-BP** was shown in Fig. 12.24c. The particle formation was confirmed by dynamic light scattering measurements, which showed an average particle diameter of 86 ± 28 nm. The facile dye grafting at the surface of the boronate aggregate endowed it with multicolored emissive nanoparticle by tuning the amount of rhodamine B grafted on the surface because of the synergistic effect of the TPE luminogen and rhodamine B units through FRET (Fig. 12.25a). Subsequently, controlling the amount of rhodamine B provided white-light emissive particles (**White-BP**). The temperature-dependence of white-light emission was investigated to determine its functionality. Heating from 5 to 65 °C provided a significant quenching of the white-light emission (Fig. 12.25b), whereas decreasing the

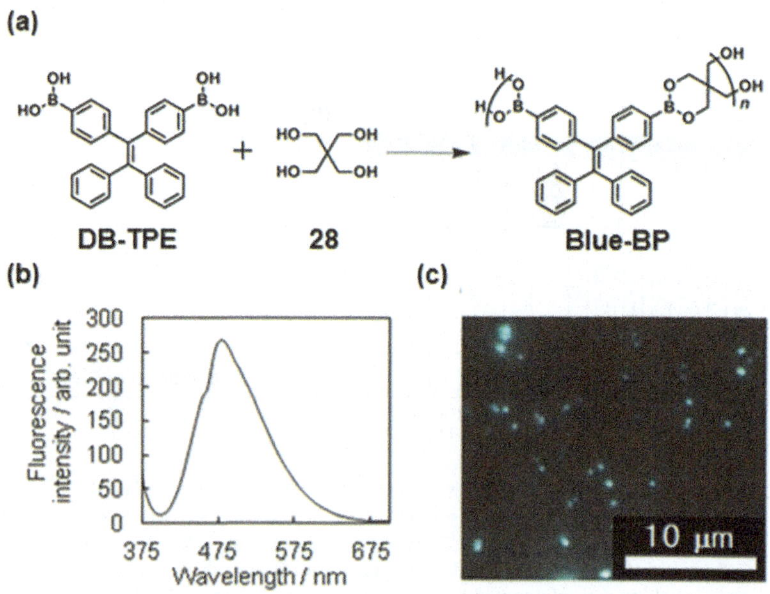

Fig. 12.24 **a** Sequential boronate esterification of **DB-TPE** with **28** in methanol to produce blue emissive nanoparticles **Blue-BP**. **b** The fluorescence spectrum of **Blue-BP** in water at 25 °C, λ_{ex} = 365 nm. **c** The fluorescence microscopy image of **Blue-BP**. Reproduced from Ozawa et al. (2015) by permission of The Royal Society of Chemistry

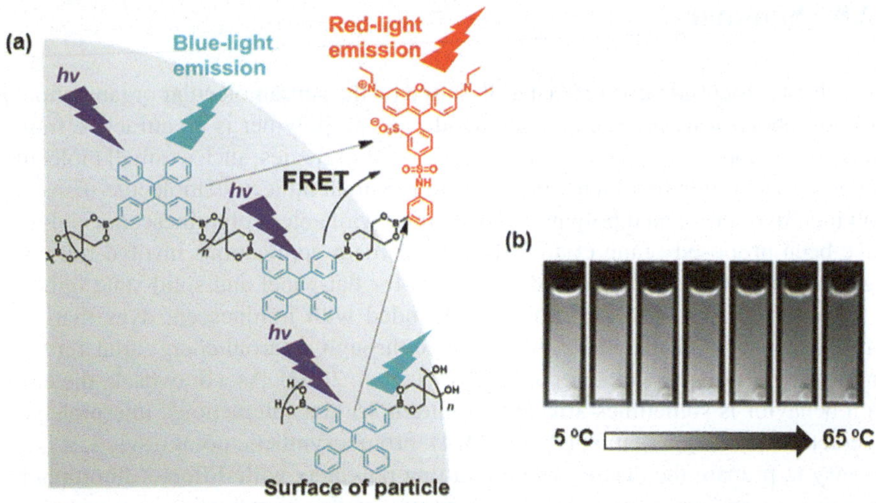

Fig. 12.25 **a** WLE of TPE- and rhodamine B-containing boronate nanoparticles **White-BP**. **b** Photographs of aqueous dispersion of **White-BP** (0.1 mg mL^{-1}) under irradiation with UV light (365 nm) from 5 to 65 °C

temperature led to an increased emission. The heating/cooling cycle showed that the particle responded linearly to the range of temperatures studied, and a small hysteresis was observed. Thus, the temperature sensitivity (S) over the temperature range was defined according to the following Eq. (12.4):

$$S = \frac{\Delta Q}{Q_T \Delta T} \times 100\% \quad (12.4)$$

in which Q_T is the fluorescence intensity at a low temperature, ΔQ corresponds to the quenching of fluorescence and is equal to the change in the fluorescence intensity, ΔT is the temperature range, and S is given in % K^{-1} (Brites et al. 2012). The value of S for **White-BP** was 1.1% K^{-1}, which is consistent with that of CdTe luminescence thermometry (Wang et al. 2002). In this way, **White-BP** could serve as a white-light emissive nanothermometer with a temperature sensitivity of 1.1% K^{-1} and be operated at physiological temperatures in water. Reversible thermo-switching of fluorescence was successfully achieved due to their stability in water.

12.8 Summary

This chapter focused on WLE materials based on the supramolecular organization of bi-fluorophors or tri-fluorophors. On another front, polymer is an attractive framework to develop organic WLE materials; versatile strategies, such as small molecule-doped polymer, polymer blend, polymer-doped small molecule, molecule-dispersed polymer, dye-terminated polymer, and excimer white electroluminescence polymer have been proposed (Tang et al. 2013). Their main applications involve full color light-emitting diodes (LEDs) and backlights for flat-panel and solid-state lighting. A common strategy involves polymers blended with luminescent dyes that emit simultaneously over the whole visible range through a partial energy transfer from energy donor polymers to acceptor dyes (Wu et al. 2007). As a drawback, the emission behavior is sometimes affected by intramolecular interactions, intermolecular interactions, and aggregation (Reiter 2014). From a synthetic point of view, it is not so easy to prepare the desired white-emitting polymers with different fluorophores in a precise mixing ratio. Therefore, supramolecular organization is a promising way because of its facile processability and precise tuning. Furthermore, noncovalent interactions can strengthen the relationship between the components for energy transfer. However, it should be pointed out that supramolecular systems would be in some cases unstable due to competitive interactions under employed conditions. Besides, in terms of practical use, such supramolecules are highly desired for WLE in an aqueous solution and in solid-state. With such a subject in mind, more effort should be carried out.

References

Abbel R, Grenier C, Pouderoijen MJ, Stouwdam JW, re PELGL, Sijbesma RP, Meijer EW, Schenning APHJ (2009) White-light emitting hydrogen-bonded supramolecular copolymers based on π-conjugated oligomers. J Am Chem Soc 131:833–843

Amaral AJR, Pasparakis G (2017) Stimuli responsive self-healing polymers: gels, elastomers and membranes. Polym Chem 8(42):6464–6484. https://doi.org/10.1039/C7PY01386H

Babu SS, Praveen VK, Ajayaghosh A (2014) Functional π-gelators and their applications. Chem Rev 114(4):1973–2129. https://doi.org/10.1021/cr400195e

Bairi P, Roy B, Chakraborty P, Nandi AK (2013) Co-assembled white-light-emitting hydrogel of melamine. ACS Appl Mater Interfaces 5(12):5478–5485. https://doi.org/10.1021/am4013566

Bales BL, Messina L, Vidal A, Peric M, Nascimento OR (1998) Precision relative aggregation number determinations of SDS micelles using a spin probe. a model of micelle surface hydration. J Phys Chem B 102 (50):10347–10358. https://doi.org/10.1021/jp983364a

Barrow SJ, Kasera S, Rowland MJ, del Barrio J, Scherman OA (2015) Cucurbituril-based molecular recognition. Chem Rev 115(22):12320–12406. https://doi.org/10.1021/acs.chemrev.5b00341

Belowich ME, Stoddart JF (2012) Dynamic imine chemistry. Chem Soc Rev 41(6):2003–2024. https://doi.org/10.1039/C2CS15305J

Bentz KC, Cohen SM (2018) Supramolecular metallopolymers: from linear materials to infinite networks. Angew Chem Int Ed 57(46):14992–15001. https://doi.org/10.1002/anie.201806912

Bhattacharya S, Phatake RS, Nabha Barnea S, Zerby N, Zhu J-J, Shikler R, Lemcoff NG, Jelinek R (2019) Fluorescent self-healing carbon dot/polymer gels. ACS Nano. https://doi.org/10.1021/acsnano.8b07087

Brites CDS, Lima PP, Silva NJO, Millan A, Amaral VS, Palacio F, Carlos LD (2012) Thermometry at the nanoscale. Nanoscale 4:4799–4829. https://doi.org/10.1039/C2NR30663H

Bruns CJ, Fraser Stoddart J (2016) The nature of the mechanical bond: from molecules to machines. Wiley, Hoboken

Bünzli J-CG (2010) Lanthanide luminescence for biomedical analyses and imaging. Chem Rev 110:2729–2755

Cao X, Lan H, Li Z, Mao Y, Chen L, Wu Y, Yi T (2015) White light emission from a two-component hybrid gel via an energy transfer process. PCCP 17(48):32297–32303. https://doi.org/10.1039/C5CP05232G

Chai Z, Wang C, Wang J, Liu F, Xie Y, Zhang Y-Z, Li J-R, Li Q, Li Z (2017) Abnormal room temperature phosphorescence of purely organic boron-containing compounds: the relationship between the emissive behavior and the molecular packing, and the potential related applications. Chem Sci 8(12):8336–8344. https://doi.org/10.1039/C7SC04098A

Chen M, Li L, Nie H, Tong J, Yan L, Xu B, Sun JZ, Tian W, Zhao Z, Qin A, Tang BZ (2015) Tetraphenylpyrazine-based AIEgens: facile preparation and tunable light emission. Chem Sci 6(3):1932–1937. https://doi.org/10.1039/C4SC03365E

Cook TR, Zheng Y-R, Stang PJ (2013) Metal-organic frameworks and self-assembled supramolecular coordination complexes: comparing and contrasting the design, synthesis, and functionality of metal-organic materials. Chem Rev 113(1):734–777. https://doi.org/10.1021/cr3002824

Cornelio J, Zhou T-Y, Alkaş A, Telfer SG (2018) Systematic tuning of the luminescence output of multicomponent metal-organic frameworks. J Am Chem Soc 140(45):15470–15476. https://doi.org/10.1021/jacs.8b09887

Duan P, Cao H, Zhang L, Liu M (2014) Gelation induced supramolecular chirality: chirality transfer, amplification and application. Soft Matter 10(30):5428–5448. https://doi.org/10.1039/c4sm00507d

Farinola GM, Ragni R (2011) Electroluminescent materials for white organic light emitting diodes. Chem Soc Rev 40(7):3467–3482. https://doi.org/10.1039/C0CS00204F

Feng C, Sun JW, Yan PF, Li YX, Liu TQ, Sun QY, Li GM (2015) Color-tunable and white-light emission of one-dimensional L-di-2-thenoyltartaric acid mixed-lanthanide coordination polymers. Dalton Trans 44(10):4640–4647. https://doi.org/10.1039/c4dt03457k

Feng X, Qi C, Feng H-T, Zhao Z, Sung HHY, Williams ID, Kwok RTK, Lam Jacky WY, Qin A, Tang BZ (2018a) Dual fluorescence of tetraphenylethylene-substituted pyrenes with aggregation-induced emission characteristics for white-light emission. Chem Sci 9(25):5679–5687. https://doi.org/10.1039/C8SC01709C

Feng H-T, Zheng X, Gu X, Chen M, Lam JWY, Huang X, Tang BZ (2018b) White-light emission of a binary light-harvesting platform based on an amphiphilic organic cage. Chem Mater 30(4):1285–1290. https://doi.org/10.1021/acs.chemmater.7b04703

Forni A, Lucenti E, Botta C, Cariati E (2018) Metal free room temperature phosphorescence from molecular self-interactions in the solid state. J Mater Chem C 6(17):4603–4626. https://doi.org/10.1039/C8TC01007B

Forrest SR, Thompson ME (2007) Introduction: organic electronics and optoelectronics. Chem Rev 107(4):923–925. https://doi.org/10.1021/cr0501590

Freeman WA, Mock WL, Shih NY (1981) Cucurbituril. J Am Chem Soc 103(24):7367–7368. https://doi.org/10.1021/ja00414a070

Gartzia-Rivero L, Cerdán L, Bañuelos J, Enciso E, López Arbeloa Í, Costela Á, García-Moreno I (2014) Förster resonance energy transfer and laser efficiency in colloidal suspensions of dye-doped nanoparticles: concentration effects. J Phys Chem C 118(24):13107–13117. https://doi.org/10.1021/jp503218z

Han A, Wang H, Kwok RTK, Ji S, Li J, Kong D, Tang BZ, Liu B, Yang Z, Ding D (2016) Peptide-Induced AIEgen self-assembly: a new strategy to realize highly sensitive fluorescent light-up probes. Anal Chem 88(7):3872–3878. https://doi.org/10.1021/acs.analchem.6b00023
Hong Y, Lam JWY, Tang BZ (2009) Aggregation-induced emission: phenomenon, mechanism and applications. Chem Commun 29:4332–4353. https://doi.org/10.1039/B904665H
Hoshi M, Kaneko N, Nishiyabu R, Kubo Y (2018) Formation of emissive nanoparticles from tetraphenylethylene-containing boronate macrocycles: preparation, characterization and functionalization. J Mater Chem C 6(41):11052–11062. https://doi.org/10.1039/C8TC03748E
Jadhav T, Choi JM, Shinde J, Lee JY, Misra R (2017) Mechanochromism and electroluminescence in positional isomers of tetraphenylethylene substituted phenanthroimidazoles. J Mater Chem C 5(24):6014–6020. https://doi.org/10.1039/C7TC00950J
Jeffry GA (1997) An introdution to hydrogen bonding. Oxford University Press, Oxford
Jones LA (1943) The historical background and evolution of the colorimetry report. J Opt Soc Am 33(10):534–543. https://doi.org/10.1364/JOSA.33.000534
Jones CD, Steed JW (2016) Gels with sense: supramolecular materials that respond to heat, light and sound. Chem Soc Rev. https://doi.org/10.1039/C6CS00435K
Khandare DG, Joshi H, Banerjee M, Majik MS, Chatterjee A (2015) Fluorescence turn-on chemosensor for the detection of dissolved CO_2 based on ion-induced aggregation of tetraphenylethylene derivative. Anal Chem 87(21):10871–10877. https://doi.org/10.1021/acs.analchem.5b02339
Kim HJ, Lee MH, Mutihac L, Vicens J, Kim JS (2012) Host–guest sensing by calixarenes on the surfaces. Chem Soc Rev 41(3):1173–1190. https://doi.org/10.1039/C1CS15169J
Kirchon A, Feng L, Drake HF, Joseph EA, Zhou H-C (2018) From fundamentals to applications: a toolbox for robust and multifunctional MOF materials. Chem Soc Rev 47(23):8611–8638. https://doi.org/10.1039/C8CS00688A
Kubo Y, Nishiyabu R, James TD (2015) Hierarchical supramolecules and organization using boronic acid building blocks. Chem Commun 51(11):2005–2020. https://doi.org/10.1039/C4CC07712A
La DD, Bhosale SV, Jones LA, Bhosale SV (2018) Tetraphenylethylene-based AIE-active probes for sensing applications. ACS Appl Mater Interfaces 10(15):12189–12216. https://doi.org/10.1021/acsami.7b12320
Lakowicz JR (2006) Princiles of fluorescnece spectroscopy, 3rd edn. Springer, New York
Li C, Yu M, Sun Y, Wu Y, Huang C, Li F (2011) A nonemissive iridium(III) complex that specifically lights-up the nuclei of living cells. J Am Chem Soc 133:11231–11239. https://doi.org/10.1021/ja202344c
Li S-H, Xu X, Zhou Y, Zhao Q, Liu Y (2017) Reversibly tunable white-light emissions of styrylpyridiniums with cucurbiturils in aqueous solution. Org Lett 19(24):6650–6653. https://doi.org/10.1021/acs.orglett.7b03377
Li D, Lu F, Wang J, Hu W, Cao X-M, Ma X, Tian H (2018) Amorphous metal-free room-temperature phosphorescent small molecules with multicolor photoluminescence via a host-guest and dual-emission strategy. J Am Chem Soc 140(5):1916–1923. https://doi.org/10.1021/jacs.7b12800
Lili B, Michael DH (2014) A review of single white-light emitters: the quest for picture perfect dyes in the next generation of single layer WOLED displays. Curr Org Chem 18(6):740–772. https://doi.org/10.2174/15701794113106660071
Liu L, Telfer SG (2015) Systematic ligand modulation enhances the moisture stability and gas sorption characteristics of quaternary metal-organic frameworks. J Am Chem Soc 137(11):3901–3909. https://doi.org/10.1021/jacs.5b00365
Liu Y, Yu Z-L, Zhang Y-M, Guo D-S, Liu Y-P (2008) Supramolecular architectures of β-cyclodextrin-modified chitosan and pyrene derivatives mediated by carbon nanotubes and their DNA condensation. J Am Chem Soc 130(31):10431–10439. https://doi.org/10.1021/ja802465g
Liu H, Cheng X, Zhang H, Wang Y, Zhang H, Yamaguchi S (2017a) ESIPT-active organic compounds with white luminescence based on crystallization-induced keto emission (CIKE). Chem Commun 53(55):7832–7835. https://doi.org/10.1039/C7CC03758A

Liu Z, Nalluri SKM, Stoddart JF (2017b) Surveying macrocyclic chemistry: from flexible crown ethers to rigid cyclophanes. Chem Soc Rev 46(9):2459–2478. https://doi.org/10.1039/C7CS00185A

Ma M-L, Qin J-H, Ji C, Xu H, Wang R, Li B-J, Zang S-Q, Hou H-W, Batten SR (2014) Anionic porous metal–organic framework with novel 5-connected *vbk* topology for rapid adsorption of dyes and tunable white light emission. J Mater Chem C 2(6):1085–1093. https://doi.org/10.1039/c3tc32001d

Matsushima Y, Nishiyabu R, Takanashi N, Haruta M, Kimura H, Kubo Y (2012) Boronate self-assemblies with embedded Au nanoparticles: Preparation, characterization and their catalytic activities for the reduction of nitroaromatic compounds. J Mater Chem 22:24124–24131. https://doi.org/10.1039/c2jm34797k

Mei J, Leung NLC, Kwok RTK, Lam JWY, Tang BZ (2015) Aggregation-induced emission: together we shine, united we soar! Chem Rev 115(21):11718–11940. https://doi.org/10.1021/acs.chemrev.5b00263

Miao X, Qu D, Yang D, Nie B, Zhao Y, Fan H, Sun Z (2018) Synthesis of carbon dots with multiple color emission by controlled graphitization and surface functionalization. Adv Mater 30(1):1704740. https://doi.org/10.1002/adma.201704740

Nishiyabu R, Sugino Y, Kubo Y (2013) White-light emitting boronate microparticles for potential use as reusable bright chemosensors in water. Chem Commun 49:9869–9871. https://doi.org/10.1039/c3cc45739g

Noguchi T, Roy B, Yoshihara D, Tsuchiya Y, Yamamoto T, Shinkai S (2014) Cyclization-Induced turn-on fluorescence system applicable to dicarboxylate sensing. Chem Eur J 20(2):381–384. https://doi.org/10.1002/chem.201304031

Ono T, Hisaeda Y (2019) Flexible-color tuning and white-light emission in three-, four-, and five-component host/guest co-crystals by charge-transfer emissions as well as effective energy transfers. J Mater Chem C 7(10):2829–2842. https://doi.org/10.1039/C8TC06165C

Ozawa A, Shimizu A, Nishiyabu R, Kubo Y (2015) Thermo-responsive white-light emission based on tetraphenylethylene- and rhodamine B-containing boronate nanoparticles. Chem Commun 51:118–121. https://doi.org/10.1039/c4cc07405j

Pallavi P, Sk B, Ahir P, Patra A (2018) Tuning the Förster resonance energy transfer through a self-assembly approach for efficient white-light emission in an aqueous medium. Chem Eur J 24(5):1151–1158. https://doi.org/10.1002/chem.201704437

Pan M, Liao W-M, Yin S-Y, Sun S-S, Su C-Y (2018) Single-phase white-light-emitting and photoluminescent color-tuning coordination assemblies. Chem Rev 118(18):8889–8935. https://doi.org/10.1021/acs.chemrev.8b00222

Park C, Hong J-I (2010) A new fluorescent sensor for the detection of pyrophosphate based on a tetraphenylethylene moiety. Tetrahedron Lett 51(15):1960–1962. https://doi.org/10.1016/j.tetlet.2010.02.009

Praveen VK, Ranjith C, Armaroli N (2014) White-light-emitting supramolecular gels. Angew Chem Int Ed 53(2):365–368. https://doi.org/10.1002/anie.201306787

Rao KV, Datta KKR, Eswaramoorthy M, George SJ (2013) Highly pure solid-state white-light emission from solution-processable soft-hybrids. Adv Mater 25(12):1713–1718. https://doi.org/10.1002/adma.201204407

Reineke S, Lindner F, Schwartz G, Seidler N, Walzer K, Lüssem B, Leo K (2009) White organic light-emitting diodes with fluorescent tube efficiency. Nature 459:234. https://doi.org/10.1038/nature08003. https://www.nature.com/articles/nature08003#supplementary-information

Reiter G (2014) Some unique features of polymer crystallisation. Chem Soc Rev 43(7):2055–2065. https://doi.org/10.1039/C3CS60306G

Reuther JF, Dahlhauser SD, Anslyn EV (2019) Tunable orthogonal reversible covalent (TORC) bonds: dynamic chemical control over molecular assembly. Angew Chem Int Ed 58(1):74–85. https://doi.org/10.1002/anie.201808371

Rowan SJ, Cantrill SJ, Cousins GRL, Sanders JKM, Stoddart JF (2002) Dynamic covalent chemistry. Angew Chem Int Ed 41:898–952

Sakai K-i, Tsuchiya S, Kikuchi T, Akutagawa T (2016) An ESIPT fluorophore with a switchable intramolecular hydrogen bond for applications in solid-state fluorochromism and white light generation. J Mater Chem C 4(10):2011–2016. https://doi.org/10.1039/C5TC04290A

SeethaLekshmi S, Ramya AR, Reddy MLP, Varughese S (2017) Lanthanide complex-derived white-light emitting solids: a survey on design strategies. J J Photochem Photobiol C 33:109–131. https://doi.org/10.1016/j.jphotochemrev.2017.11.001

Segarra-Maset MD, Nebot VJ, Miravet JF, Escuder B (2013) Control of molecular gelation by chemical stimuli. Chem Soc Rev 42(17):7086–7098. https://doi.org/10.1039/C2CS35436E

Sick T, Hufnagel AG, Kampmann J, Kondofersky I, Calik M, Rotter JM, Evans A, Döblinger M, Herbert S, Peters K, Böhm D, Knochel P, Medina DD, Fattakhova-Rohlfing D, Bein T (2018) Oriented films of conjugated 2D covalent organic frameworks as photocathodes for water splitting. J Am Chem Soc 140(6):2085–2092. https://doi.org/10.1021/jacs.7b06081

Su L, Zhang X, Zhang Y, Rogach AL (2016) Recent progress in quantum dot based white light-emitting devices. Top Curr Chem 374. https://doi.org/10.1007/s41061-016-0041-3

Sun Z, Huang Q, He T, Li Z, Zhang Y, Yi L (2014) Multistimuli-responsive supramolecular gels: design rationale, recent advances, and perspectives. ChemPhysChem 15(12):2421–2430. https://doi.org/10.1002/cphc.201402187

Tang C, Liu X-D, Liu F, Wang X-L, Xu H, Huang W (2013) Recent progress in polymer white light-emitting materials and devices. Macromol Chem Phys 214(3):314–342. https://doi.org/10.1002/macp.201200305

Tang Q, Liu S, Liu Y, He D, Miao J, Wang X, Ji Y, Zheng Z (2014) Color tuning and white light emission via in situ doping of luminescent lanthanide metal-organic frameworks. Inorg Chem 53(1):289–293. https://doi.org/10.1021/ic402228g

Tseng K-P, Fang F-C, Shyue J-J, Wong K-T, Raffy G, DellGuerzo A, Bassani DM (2011) Spontaneous generation of highly emissive RGB organic nanospheres. Angew Chem Int Ed 50(31):7032–7036. https://doi.org/10.1002/anie.201101945

Turro NJ (1991) Modern molecular photochemistry, 1st ed. University Science Books, Sausalito

Uoyama H, Goushi K, Shizu K, Nomura H, Adachi C (2012) Highly efficient organic light-emitting diodes from delayed fluorescence. Nature 492(7428):234–238. https://doi.org/10.1038/nature11687

van der Meer BW (2013) Förster theory. In: FRET—Förster resonance energy transfer: from theory to applications, pp 23–62. https://doi.org/10.1002/9783527656028.ch03

Vardhan H, Mehta A, Nath I, Verpoort F (2015) Dynamic imine chemistry in metal–organic polyhedra. RSC Adv 5(82):67011–67030. https://doi.org/10.1039/C5RA10801B

Vijayakumar C, Sugiyasu K, Takeuchi M (2011) Oligofluorene-based electrophoretic nanoparticles in aqueous medium as a donor scaffold for fluorescence resonance energy transfer and white-light emission. Chem Sci 2:291–294. https://doi.org/10.1039/C0SC00343C

Wang S, Westcott S, Chen W (2002) Nanoparticle luminescence thermometry. J Phys Chem B 106:11203–11209

Wei Z, Yang JH, Zhou J, Xu F, Zrínyi M, Dussault PH, Osada Y, Chen YM (2014) Self-healing gels based on constitutional dynamic chemistry and their potential applications. Chem Soc Rev 43(23):8114–8131. https://doi.org/10.1039/C4CS00219A

Wenz G, Han B-H, Müller A (2006) Cyclodextrin rotaxanes and polyrotaxanes. Chem Rev 106(3):782–817. https://doi.org/10.1021/cr970027+

Wilson A, Gasparini G, Matile S (2014) Functional systems with orthogonal dynamic covalent bonds. Chem Soc Rev 43:1948–1962. https://doi.org/10.1039/C3CS60342C

Wu W-C, Lee W-Y, Pai C-L, Chen W-C, Tuan C-S, Lin J-L (2007) Photophysical and electroluminescent properties of fluorene-based binary and ternary donor–acceptor polymer blends. J Polym Sci Part B: Polym Phys 45:67–78. https://doi.org/10.1002/polb.20997

Wu H, Ying L, Yang W, Cao Y (2009) Progress and perspective of polymer white light-emitting devices and materials. Chem Soc Rev 38(12):3391–3400. https://doi.org/10.1039/B816352A

Xia Z, Xu Z, Chen M, Liu Q (2016) Recent developments in the new inorganic solid-state LED phosphors. Dalton Trans 45(28):11214–11232. https://doi.org/10.1039/c6dt01230b

Xia Y-P, Wang C-X, An L-C, Zhang D-S, Hu T-L, Xu J, Chang Z, Bu X-H (2018) Utilizing an effective framework to dye energy transfer in a carbazole-based metal–organic framework for high performance white light emission tuning. Inorg Chem Front 5(11):2868–2874. https://doi.org/10.1039/C8QI00747K

Xiao L, Chen Z, Qu B, Luo J, Kong S, Gong Q, Kido J (2011) Recent progresses on materials for electrophosphorescent organic light-emitting devices. Adv Mater 23(8):926–952. https://doi.org/10.1002/adma.201003128

Xie Z, Chen C, Xu S, Li J, Zhang Y, Liu S, Xu J, Chi Z (2015) White-light emission strategy of a single organic compound with aggregation-induced emission and delayed fluorescence properties. Angew Chem Int Ed 54(24):7181–7184. https://doi.org/10.1002/anie.201502180

Xu Z, Peng S, Wang Y-Y, Zhang J-K, Lazar AI, Guo D-S (2016) Broad-spectrum tunable photoluminescent nanomaterials constructed from a modular light-harvesting platform based on macrocyclic amphiphiles. Adv Mater 28(35):7666–7671. https://doi.org/10.1002/adma.201601719

Xu C, Guan H, Song Y, An Z, Zhang X, Zhou X, Shi Z, Sheng Y, Zou H (2018) Novel highly efficient single-component multi-peak emitting aluminosilicate phosphors co-activated with Ce^{3+}, Tb^{3+} and Eu^{2+}: luminescence properties, tunable color, and thermal properties. PCCP 20(3):1591–1607. https://doi.org/10.1039/C7CP07108F

Yan D, Evans DG (2014) Molecular crystalline materials with tunable luminescent properties: from polymorphs to multi-component solids. Mater Horiz 1(1):46–57. https://doi.org/10.1039/C3MH00023K

Yang Z, Chi Z, Mao Z, Zhang Y, Liu S, Zhao J, Aldred MP, Chi Z (2018) Recent advances in mechano-responsive luminescence of tetraphenylethylene derivatives with aggregation-induced emission properties. Mater Chem Front 2(5):861–890. https://doi.org/10.1039/C8QM00062J

Yin G-Q, Wang H, Wang X-Q, Song B, Chen L-J, Wang L, Hao X-Q, Yang H-B, Li X (2018) Self-assembly of emissive supramolecular rosettes with increasing complexity using multitopic terpyridine ligands. Nat Commun 9(1):567. https://doi.org/10.1038/s41467-018-02959-w

Yu G, Yan X, Han C, Huang F (2013) Characterization of supramolecular gels. Chem Soc Rev 42(16):6697–6722. https://doi.org/10.1039/c3cs60080g

Zhang J, Su CY (2013) Metal-organic gels: from discrete metallogelators to coordination polymers. Coord Chem Rev 257(7–8):1373–1408. https://doi.org/10.1016/j.ccr.2013.01.005

Zhang X, Rehm S, Safont-Sempere MM, Würthner F (2009) Vesicular perylene dye nanocapsules as supramolecular fluorescent pH sensor systems. Nature Chem 1:623. https://doi.org/10.1038/nchem.368. https://www.nature.com/articles/nchem.368#supplementary-information

Zhang X, Görl D, Würthner F (2013) White-light emitting dye micelles in aqueous solution. Chem Commun 49(74):8178–8180. https://doi.org/10.1039/C3CC44875D

Zhang G, Hu F, Zhang D (2015) Manipulation of the aggregation and deaggregation of tetraphenylethylene and silole fluorophores by amphiphiles: emission modulation and sensing applications. Langmuir 31(16):4593–4604. https://doi.org/10.1021/la5029367

Zhang L, Jin Q, Liu M (2016) Enantioselective recognition by chiral supramolecular gels. Chem Asian J 11(19):2642–2649. https://doi.org/10.1002/asia.201600441

Zhang M, Yin S, Zhang J, Zhou Z, Saha ML, Lu C, Stang PJ (2017) Metallacycle-cored supramolecular assemblies with tunable fluorescence including white-light emission. Proc Natl Acad Sci USA 114(12):3044–3049. https://doi.org/10.1073/pnas.1702510114

Zhao J, Ji S, Chen Y, Guo H, Yang P (2012a) Excited state intramolecular proton transfer (ESIPT): from principal photophysics to the development of new chromophores and applications in fluorescent molecular probes and luminescent materials. PCCP 14(25):8803–8817. https://doi.org/10.1039/C2CP23144A

Zhao Z, Lam JWY, Tang BZ (2012b) Tetraphenylethene: a versatile AIE building block for the construction of efficient luminescent materials for organic light-emitting diodes. J Mater Chem 22(45):23726–23740. https://doi.org/10.1039/C2JM31949G

Chapter 13
Photodynamic Therapy

Keiichi Imato and Yousuke Ooyama

Abstract Photosensitizing dyes with the ability to produce singlet oxygen ($^1O_2$) under light irradiation can be used for photodynamic therapy (PDT), a treatment of early-stage cancer with less stress on bodies, where the photogenerated $^1O_2$ destroys cancer cells. For PDT use, photosensitizing dyes require several capabilities such as strong photoabsorption in the phototherapeutic window (650–900 nm), high $^1O_2$ generation quantum yield (Φ_{Δ}), good water solubility, and low toxicity without light irradiation. Therefore, the effects of chemical structures and substituents of photosensitizing dyes on these properties have been investigated, which enables to create new photosensitizing dyes with excellent performance for PDT. This chapter overviews recent studies and developments in photosensitizing dyes with a focus on porphyrin, phthalocyanine, boron-dipyrromethene (BODIPY), xanthene, phenothiazinium, heteropolycycle, pyrylium, azinium, squalin, and transition metal (Ru, Ir, Pt) complex skeletons.

Keywords Photosensitizing dyes · Photodynamic therapy · Singlet oxygen · Intersystem crossing · Photoabsorption

13.1 Introduction

Photodynamic therapy (PDT) using photosensitizing dyes with the ability to produce singlet oxygen ($^1O_2$) and low-power laser light has attracted much attention as a treatment for cancer in an early stage with less stress on bodies (Fig. 13.1a) (DeRosa and Crutchley 2002; Patrice 2003; Pawlicki et al. 2009; Lovell et al. 2010; Dąbrowski and Arnaut 2015; Zhou et al. 2016; Callaghan and Senge 2018). The mechanism of PDT is shown in Fig. 13.1b. Photosensitizing dyes (^{1}Dye) adsorbed to cancer cells are excited to photoexcited singlet states (^{1}Dye*) by light irradiation, undergo intersystem crossing (ISC) to be excited triplet states (^{3}Dye*), and generate reactive $^1O_2$ via ISC and energy transfer from ^{3}Dye* to ground-state triplet oxygen ($^3O_2$) (Type

K. Imato · Y. Ooyama (✉)
Hiroshima University, Higashihiroshima, Japan
e-mail: yooyama@hiroshima-u.ac.jp

Y. Ooyama and S. Yagi (eds.), *Progress in the Science of Functional Dyes*,
https://doi.org/10.1007/978-981-33-4392-4_13

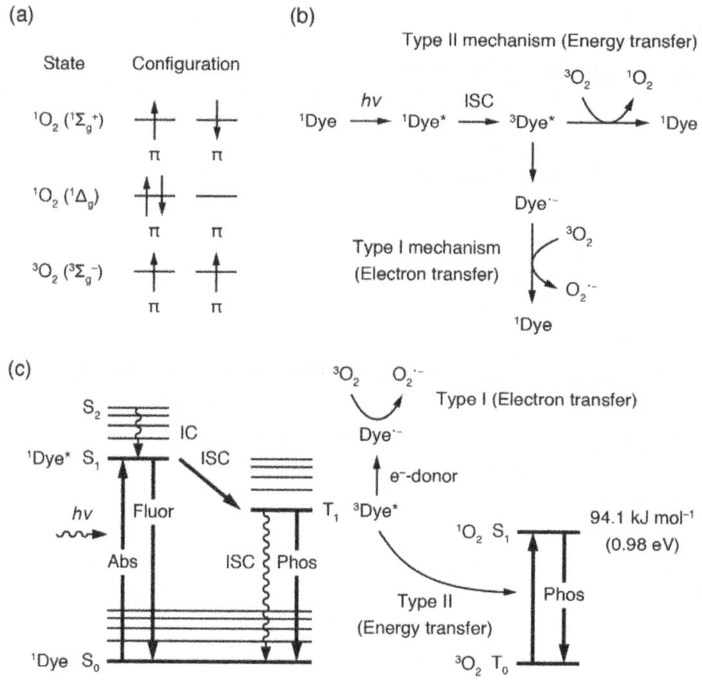

Fig. 13.1 **a** Electron configurations of ground and excited oxygen molecules and **b** production of reactive oxygen species by photosensitizing dyes and **c** its energy state diagram (Abs: absorption, Fluor: fluorescence, Phos: phosphorescence, IC: internal conversion, ISC: intersystem crossing)

II mechanism). In other words, PDT is a treatment method in which the reactive $^1O_2$ destroys cancer cells. $^1O_2$-generating photosensitizing dyes for PDT require the following properties: (1) strong absorption of light in the phototherapeutic window (650–900 nm), which can penetrate deep into living tissues, (2) high triplet quantum yield (Φ_T) due to highly efficient ISC from ^{1}Dye* to ^{3}Dye* (from S_1 to T_1), (3) high $^1O_2$ generation quantum yield (Φ_Δ) due to highly efficient ISC between ^{3}Dye* and $^3O_2$, (4) good solubility in water, (5) specific accumulation in tumor tissues (high affinity for tumor), (6) no induction of skin inflammation and rashes by photosensitivity of photosensitizing dyes, (7) low toxicity without light irradiation. On the other hand, superoxide ($O_2^{\bullet-}$) is generated via Type I mechanism involving electron transfer from radical anion of dye (Dye$^{\bullet-}$), which is formed by electron transfer between ^{3}Dye* and surrounding electron donors including solvent and substrate, to $^3O_2$. Similar to $^1O_2$, $O_2^{\bullet-}$ is a reactive oxygen species (ROS), but considered to show less anticancer activity than $^1O_2$ (Fig. 13.1b).

Porphyrin dyes, phthalocyanine dyes, phenothiazine dyes, xanthene dyes, and fullerene (C_{60}) derivatives are known as representative photosensitizing dyes that can produce $^1O_2$ (Bonnett 1995; Redmond and Gamlin 1999; Mroz et al. 2007; Ethirajan et al. 2011). Currently, some porphyrin dyes are approved and practically used as PDT dyes. However, after PDT using these porphyrin dyes, patients are forced to live in a dark room to avoid direct sunlight for several weeks to one month, because the dyes cause skin irritation and rashes due to their photosensitivity. In addition, the porphyrin dyes have a disadvantage of low PDT efficiency due to the weak photoabsorption (small molar absorbance coefficient, ε) in the phototherapeutic window. In recent years, a number of BODIPY (abbreviation of boron-dipyrromethene) based photosensitizing dyes have been developed and extensively studied as PDT dyes by evaluation of the $^1O_2$ generation efficiency and destruction rate of cancer cells (Awuah and You 2012; Kamkaew et al. 2013). More recently, photosensitizing dyes with high brightness (brightness: $BT = \varepsilon \times \Phi_{fl}$ (fluorescence quantum yield)) and high phototoxicity (phototoxic power: $PP = \varepsilon \times \Phi_{\Delta}$) have been attracted scientific interest to achieve both fluorescence imaging of cancer cells and PDT. This chapter outlines the effects of chemical structures and substituents of photosensitizing dyes on their photoabsorption properties and $^1O_2$ generation efficiency and highlights recent developments in photosensitizing dyes.

13.2 Methods for Evaluation of Singlet Oxygen $^1O_2$ Generation

$^1O_2$ produced by photosensitizing dyes behaves as an electrophile and causes (A) Diels–Alder [4 + 2] cycloaddition reaction with dienes, (B) [2 + 2] cycloaddition reaction with olefins, and (C) Schenck-Ene reaction with olefins having a hydrogen atom at the allylic position, as shown in Fig. 13.2a (Prein and Adam 1996; Greer 2006). In particular, the scavenger method is generally used as an indirect spectroscopic method for the detection of $^1O_2$ using the reaction A. In the method, $^1O_2$

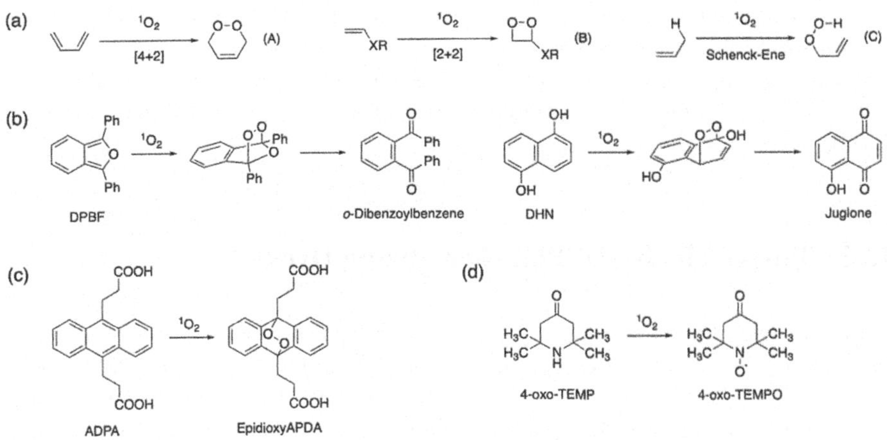

Fig. 13.2 **a** Reaction of singlet oxygen with olefins, **b**, **c** singlet oxygen scavengers (DPBF, DHN, and ADPA), and **d** reaction of a spin trapping agent (4-oxo-TEMP) with singlet oxygen

generation can be indirectly estimated from changes in photoabsorption spectra of scavengers such as 1,3-diphenylisobenzofuran (DPBF, $\lambda^{abs}{}_{max} = 410$–$420$ nm) (Gollnick and Griesbeck 1985) and 1,5-dihydroxynaphthalene (DHN, $\lambda^{abs}{}_{max} = 300$ nm) (Takizawa et al. 2011), which are oxidized by $^1O_2$ via [4 + 2] cycloaddition reaction (Fig. 13.2b). To quantify $^1O_2$ generation by the scavenger method, a relative quantum yield method is commonly used. In the method, $\Phi_{\Delta sam}$ of a target photosensitizing dye (sam) can be calculated by comparing changes in photoabsorption spectra of a scavenger during its reaction with $^1O_2$ generated using the target photosensitizing dye and spectral changes using a standard photosensitizing dye (ref) represented by methylene blue, as well as comparison with $\Phi_{\Delta ref}$ as follows (Eq. (13.1)):

$$\Phi_{\Delta sam} = \Phi_{\Delta ref} \times [(m_{sam}/m_{ref}) \times (L_{ref}/L_{sam})] \quad (13.1)$$

where $\Phi_{\Delta sam}$ and $\Phi_{\Delta ref}$ are $^1O_2$ generation quantum yields of target and standard photosensitizing dyes, respectively, m_{sam} and m_{ref} are slopes in changes of absorbance at the photoabsorption maximum wavelength ($\lambda^{abs}{}_{max}$) of a scavenger against photoirradiation time using target and standard photosensitizing dyes, respectively, and L_{sam} and L_{ref} are light harvesting efficiencies of target and standard photosensitizing dyes, respectively, at photoirradiation wavelength ($L = 1 - 10^{-A}$, "A" is absorbance at the photoirradiation (photoexcitation) wavelength).

Alternatively, the evaluation of changes in fluorescence spectra of a fluorescent scavenger such as anthracene-9,10-dipropionic acid (ADPA, $\lambda^{fl}{}_{max} = 430$ nm) during its reaction with $^1O_2$ via [4 + 2] cycloaddition is also well known as another method to estimate $^1O_2$ generation (Fig. 13.2c) (Aubry et al. 1995).

In addition, the estimation of 4-oxo-TEMPO radicals formed via the reaction of 4-oxo-TEMP, a spin trapping agent, with $^1O_2$ by electron paramagnetic resonance (EPR) spectroscopy enables to indirectly detect and quantify photogenerated $^1O_2$ (Fig. 13.2d) (Yamakoshi et al. 2003; Oriana et al. 2013).

As a direct method for detection and quantification of $^1O_2$, a spectroscopic method for measuring the phosphorescence of $^1O_2$ [$^1O_2 \rightarrow {}^3O_2 + h\nu$ (1270 nm, 0.98 eV)] generated by photosensitizing dyes is available but requires an expensive spectrophotometer to detect near-infrared fluorescence (Fig. 13.1c) (Yamakoshi et al. 2003).

13.3 Porphyrin-Based Photosensitizing Dyes

Porphyrin dyes have a strong Soret absorption band at 400–450 nm ($\varepsilon = -10^6$ M^{-1} cm^{-1}) and a weak Q absorption band at 500–700 nm ($\varepsilon = -10^4$ M^{-1} cm^{-1}), and therefore, have been expected as excellent PDT dyes with the photosensitizing ability in the phototherapeutic window since the early days. Water-soluble porphyrin-based photosensitizing dyes can be obtained by introducing sulfonic acid, sulfonate, carboxylic acid, carboxylate, or *N*-alkylpyridinium groups into the porphyrin skeleton. Currently, in Japan, porfimer sodium (photofrin) and talaporfin sodium (Laserphyrin) are approved by the Ministry of Health, Labor and Welfare and practically used for PDT, as porphyrin-based photosensitizing dyes (Fig. 13.3a) (Ethirajan et al. 2011). Porfimer sodium, a first-generation photosensitizer, has a photoabsorption band ($\lambda^{abs}{}_{max}$ = 630 nm) in a wavelength region relatively favorable to tissue penetration. On the other hand, talaporfin sodium, a second-generation photosensitizer, has a chlorin skeleton and photoabsorption band ($\lambda^{abs}{}_{max}$ = 650 nm) in a wavelength region more tissue-permeable than that of porphyrin-based porfimer sodium.

Furthermore, because talaporfin sodium shows highly specific accumulation in tumor tissues and can modestly suppress photosensitivity, it is used as an excellent photosensitizer for PDT.

So far, much knowledge about the effects of the central metal and substituents of porphyrin-based photosensitizing dyes on the efficiency of $^1O_2$ generation has been accumulated (Fig. 13.3a) (DeRosa and Crutchley 2002; Bonnett 1995). Tetraphenylporphyrin (**H2TPP**) without a central metal shows a Φ_Δ value of ca. 0.6–0.7 and is used as a standard photosensitizing dye for porphyrin dyes (Fig. 13.3a). Porphyrin dyes having diamagnetic metal ions including Zn^{2+}, Pd^{2+}, and Cd^{2+} as the central metal exhibit high Φ_Δ values. In contrast, Φ_Δ values of porphyrin dyes with paramagnetic metal ions such as Mn^{2+}, Co^{2+}, and Cu^{2+} are extremely low. Φ_Δ correlates with triplet quantum yield (Φ_T) and triplet state lifetime (τ_T); for example, **ZnTPP** ($\Phi_\Delta = 0.83$) and **PdTPP** ($\Phi_\Delta = 0.88$) have high Φ_T (0.8–1.0) and long-lived τ_T ($10^3 - 10^4$ μs), whereas **CoTPP** and **CuTPP** ($\Phi_\Delta = < 0.01$) exhibit extremely low Φ_T and short-lived τ_T. On the other hand, **PtTPPS** with Pt^{2+} as the central metal has a low Φ_Δ value (0.06), but **Pt4TPyP** having four pyridyl groups coordinated with Pt^{4+} shows a relatively high Φ_Δ value (0.50) (Naik et al. 2014).

It is known as an interesting substituent effect that the Φ_Δ values of sulfonated **H_2TPP** (**H_2TPPS**, $\Phi_\Delta = 0.58$) and carboxylated **H_2TPP** (**H_2TPPC**, $\Phi_\Delta = 0.56$) are improved by introduction of silyl groups into **H_2TPPS** (**H_2TPPSSi**, $\Phi_\Delta = 0.66$) and **H_2TPPC** (**H_2TPPCSi**, $\Phi_\Delta = 0.72$) (Fig. 13.3a) (Horiuchi et al. 2014). It has been reported that the Φ_Δ value (0.92) of diazaporphyrin (**H_2DAP**) is higher than that of **H2TPP** (Shinmori et al. 2005).

Porphycene, a structural isomer of porphyrin, is expected as a second-generation photosensitizer because of the larger ε value (–50,000 M^{-1} cm^{-1}) than porphyrin in the red region above 630 nm. Compared to porphycenes without the central metal (**H_2TPPo**, $\Phi_T = 0.52$, $\Phi_\Delta = 0.23$) and with the central Cu^{2+} (**CuTPPo**, $\Phi_T = 0.35$, $\Phi_\Delta = 0.24$), a porphycene with Pd^{2+} as the central metal (**PdTPPo**) has a higher Φ_T (0.78) and Φ_Δ (0.78) values (Rubio et al. 2005). In addition, bromine-substituted porphycenes (**H2TPrPoBr1–4**) show high Φ_Δ values (0.49–0.95) (Shimakoshi et al. 2008), resulting from the increased spin–orbit interaction due to the heavy electron effect of bromine atom (decreased Φ_{fl}) and enhanced Φ_T (Fig. 13.3b).

Rubilin is a porphyrin analogue, and a Se-substituted rubilin, selenium rubilin (**RubSe**), has a strong Soret absorption band at 633 nm (ε = 221,000 M^{-1} cm^{-1}) and two Q absorption bands at 835 nm (ε = 42,300 M^{-1} cm^{-1}) and 1156 nm (ε = 121,000 M^{-1} cm^{-1}) (Fig. 13.3b) (Tian et al. 2013). **RubSe** exhibits pH-dependent

$^1O_2$ generation characteristics, that is, low (0.06) and high (0.69) Φ_Δ values at pH 7.4 and 5, respectively.

Optical, electrochemical, and photosensitizing properties of porphyrin, chlorin, and bacteriochlorin skeletons have been compared. Bacteriochlorin-based dyes show Q absorption bands in a 90 nm-longer wavelength region (ca. 740 nm) but moderately lower Φ_Δ values than those of porphyrin and chlorine-based dyes (Fig. 13.3b, **H2THPP, H2THPC, H2THPB**) (Arnaut et al. 2014).

In recent years, porphyrin-based photosensitizing dyes that have strong two-photon absorption cross section (δ_2/GM) have been developed to overcome the disadvantage of porphyrin dyes, i.e., weak Q absorption bands in a short-wavelength region. Two-photon absorption (2PA) is a phenomenon in which a molecule is excited by simultaneous absorption of two photons; for example, excitation of a dye by simultaneous absorption of two 1000 nm photons is comparable to that by absorption of one 500 nm photon, which has a double energy gap. Therefore, two-photon-absorbable dyes enable the use of tissue-permeable near-infrared light and promise two-photon PDT, in which cancer cells deep in tissues can be selectively killed. For example, dendronized **DenH$_2$TPP** with enhanced light collection efficiency (290 GM @λ^{2PA} = 790 nm), **TPA-ZnP** with a triphenylamine group (251 GM @λ^{2PA} = 830 nm), and **DPP-ZnP** with a diketopyrrolopyrrole moiety (2000 GM @λ^{2PA} = 910 nm) have been designed and synthesized as porphyrin-based photosensitizing dyes exhibiting two-photon absorption characteristics (Fig. 13.3c) (Yao et al. 2015; Hammerer et al. 2014; Schmitt et al. 2015).

As described above, the molecular design and synthesis of new porphyrin-based photosensitizing dyes have been extensively studied, and the effects of central metals and substituents on photochemical and electrochemical properties and $^1O_2$ generation efficiency have been also investigated.

13.4 Phthalocyanine-Based Photosensitizing Dyes

Phthalocyanine dyes have a weak Soret absorption band at 300–400 nm and a strong Q absorption band above 650 nm (ε = ~ 10^5 M^{-1} cm^{-1}). Porphyrin dyes show high light collection efficiency in the Soret absorption bands, whereas phthalocyanine dyes show high light collection efficiency in the Q absorption bands. Therefore, phthalocyanine dyes are expected to be second-generation photosensitizers for PDT with excellent photosensitizing abilities in the phototherapeutic window. To render phthalocyanine dyes water-soluble, the introduction of sulfonic acid, sulfonate, carboxylic acid, and carboxylate into the phthalocyanine skeleton is an effective way. On the other hand, phthalocyanine dyes form aggregates in solution such as dimers and multimers stacked due to their strong π-π interaction and hydrophobic interaction. The formation of aggregates not only shifts the Q absorption bands to a short-wavelength region but also deactivates the excited states (energy deactivation between dyes), thereby reducing the light collection efficiency, Φ_T, and Φ_Δ. The aggregation can be suppressed by introduction of bulky water-soluble substituents such as glucose, galactose, and cyclodextrin into the phthalocyanine skeleton (Fig. 13.4a) (DeRosa and Crutchley 2002; Li et al. 2015; Mori et al. 2015; Lau et al. 2011).

Similar to porphyrin-based photosensitizing dyes, phthalocyanine (Pc) dyes with diamagnetic metal ions such as Zn^{2+} and Al^{3+} as the central metal exhibit high Φ_{Δ} values, whereas those with paramagnetic metal ions including Co^{2+} and Cu^{2+} show significantly low Φ_{Δ} values, compared to the metal-free Pc (Fig. 13.4a,b) (DeRosa and Crutchley 2002). This is due to the long-lived triplet states of Pc-based photosensitizing dyes with a diamagnetic metal ($\tau_T = 10^2$ μs). **ZnPc** having Zn^{2+} as the central metal has a Φ_{Δ} value of ca. 0.5–0.6 and is used as a standard photosensitizing dye for Pc dyes.

Tetrapyrazinoporphyrazine-based dyes are aza analogues of Pc and have both fluorescence and $^1O_2$ generation characteristics. Therefore, recently, the effects

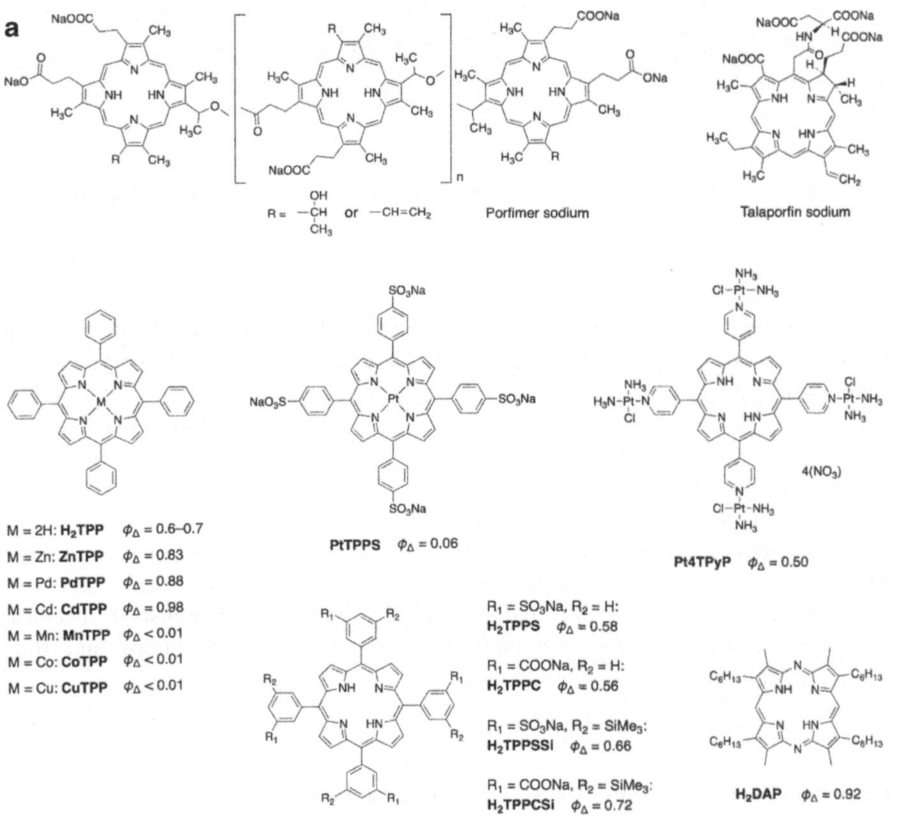

Fig. 13.3 **a** Talaporfin sodium, porfimer sodium, metalloporphyrin, and azaporphyrin photosensitizing dyes. **b** Photosensitizing dyes having porphycene, rubilin, porphyrin, chlorin, and bacteriochlorin skeletons. **c** Porphyrin-based photosensitizing dyes with two-photon absorption characteristics

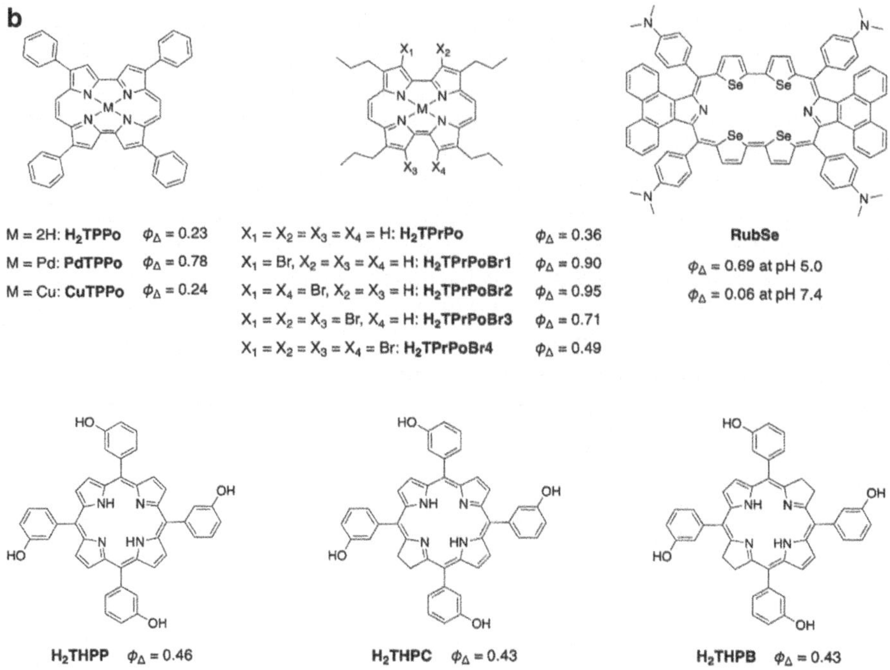

Fig. 13.3 (continued)

of central metals and substituents on the photochemical and electrochemical properties and $^1O_2$ generation efficiency have attracted considerable interest. Tetrapyrazinoporphyrazine-based dyes having a bulky phenol group exhibit optical properties derived from the free dyes without aggregation in solution (Fig. 13.4c). Compared to the metal-free **TPyzPzsPO** ($\lambda^{abs}{}_{max} = 672$ nm, $\Phi_{fl} = 0.036$, $\Phi_\Delta = 0.056$), **ZnTPyzPzsPO** ($\lambda^{abs}{}_{max} = 651$ nm) with Zn^{2+} as the central metal has high Φ_{fl} (0.28) and Φ_Δ (0.58) (Novakova et al. 2015).

Thus, phthalocyanine-based photosensitizing dyes are expected to be practically applied as second-generation photosensitizers for PDT, and the correlation between Pc structures and $^1O_2$ generation characteristics has been studied intensively, because the dyes have strong Q absorption bands in the phototherapeutic window and show relatively high $^1O_2$ generation efficiency.

13.5 BODIPY-Based Photosensitizing Dyes

Boron-dipyrromethene (BODIPY) dyes show excellent photostability, strong photoabsorption in red and near-infrared regions, and highly fluorescent properties, and therefore, promise to be applied to optoelectronic devices such as

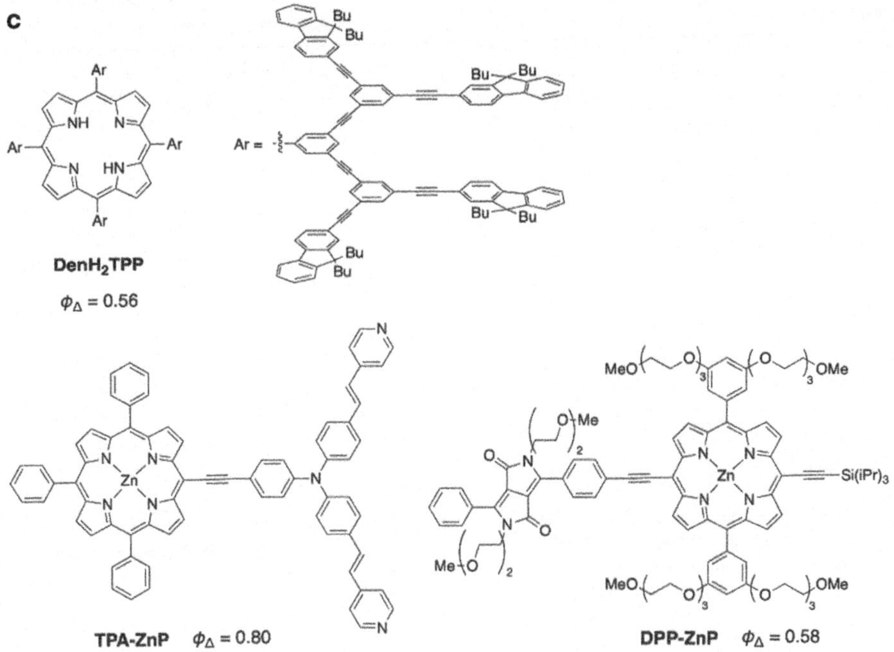

Fig. 13.3 (continued)

organic light-emitting diode (OLED) and dye-sensitized solar cells (DSSC), optical sensors, and photosensitizing dyes for PDT. Their photophysical, electrochemical, photosensitizing, and photoelectric conversion characteristics have been extensively studied.

Most BODIPY dyes have photoabsorption maxima around 500–550 nm ($\varepsilon = -10^5$ M^{-1} cm^{-1}) and are unavailable in the phototherapeutic window (650–900 nm). Thus, the effects of substituents on the photoabsorption maximum wavelengths of BODIPY dyes have been investigated (Awuah and You 2012; Kamkaew et al. 2013). The introduction of a styryl group at the 3rd or 5th position of the BODIPY skeleton causes a bathochromic shift of the photoabsorption maximum wavelength by ca. 100 nm (Fig. 13.5a; **BODIPY-P1**, $\lambda^{abs}{}_{max} = 498$ nm; **Sty-BODIPY-P1**, $\lambda^{abs}{}_{max} =$ 599 nm) (Coskun and Akkaya 2004). The photoabsorption maximum wavelength is bathochromically shifted by 30 or 50 nm by introducing bromine or iodine atoms at the 2nd and 6th positions of the BODIPY skeleton, respectively (Fig. 13.5a, **I2-BODIPY-P1**, $\lambda^{abs}{}_{max} = 529$ nm) (Li et al. 2013). Although BODIPY dyes have high Φ_{fl} values and thereby low Φ_T values, which is unsuitable for PDT, the introduction of heavy atoms including bromine and iodine atoms on the BODIPY skeleton can enhance the spin–orbit interaction due to the heavy atom effect, decrease Φ_{fl} values, and dramatically improve Φ_T and Φ_Δ values (Fig. 13.5a; **BODIPY-P1**, $\Phi_{fl} = 0.65$, $\Phi_\Delta \approx 0$; **I2-BODIPY-P1**, $\Phi_{fl} = 0.02$, $\Phi_\Delta = 0.83$). It has been reported

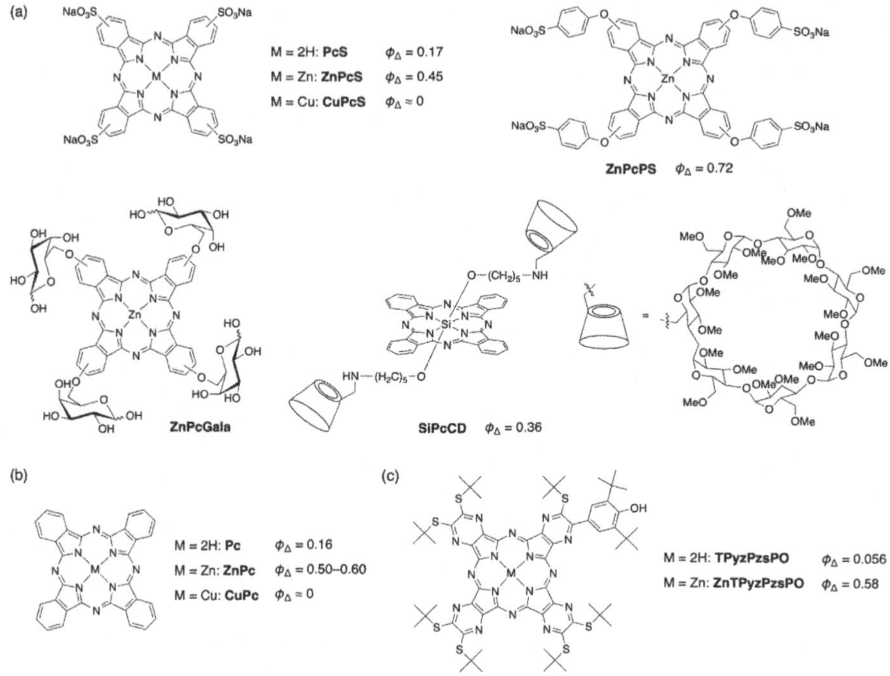

Fig. 13.4 a Water-soluble phthalocyanines, **b** metal phthalocyanines, and **c** tetrapyrazinoporphyrazine-based photosensitizing dyes

that **TMBODIPY-D28** and **TMBODIPY-D88** obtained by dimerization of 1,3,5,7-tetramethyl **TMBODIPY-1** at the 2,8′-positions and 8,8′-positions, respectively, can exhibit high Φ_Δ values compared to the corresponding monomers (Cakmak et al. 2011). Photoabsorption spectrum of a thienopyrrole-fused BODIPY dye (**T2-BODIPY**, $\lambda^{abs}{}_{max}$ = 571 nm) is bathochromically shifted by ca. 40 nm compared to a non-fused BODIPY dye (**BODIPY-T2**, $\lambda^{abs}{}_{max}$ = 529 nm), and some fused BODIPY dyes show photoabsorption maxima around 700 nm (Fig. 13.5b) (Ji et al. 2015). In addition, the introduction of bromine atoms into fused BODIPY structures decreases the Φ_{fl} values but shifts the photoabsorption maxima to longer wavelength regions and significantly enhances the Φ_Δ values (Fig. 13.5b) (Yang et al. 2013); for example, **T2-BODIPY-T2Br2** shows the lower Φ_{fl} value (0.04) but higher Φ_Δ value (0.63) than those of **T2-BODIPY-T2** ($\Phi_{fl} \approx 0.2$, $\Phi_\Delta \approx 0$).

As mentioned above, although the enhanced spin–orbit interaction due to the heavy atom effect of bromine and iodine atoms introduced into the BODIPY core structure can lower the Φ_{fl} values and improve the Φ_T values, the presence of bromine and iodine atoms renders the dyes toxic under no light irradiation. On the other hand, recently, photosensitizing dyes with high brightness (BT = $\varepsilon \times \Phi_{fl}$) and high phototoxicity (PP = $\varepsilon \times \Phi_\Delta$) have attracted much interest to achieve both fluorescence imaging of cancer cells and PDT. D'Souza and You et al. reported that a

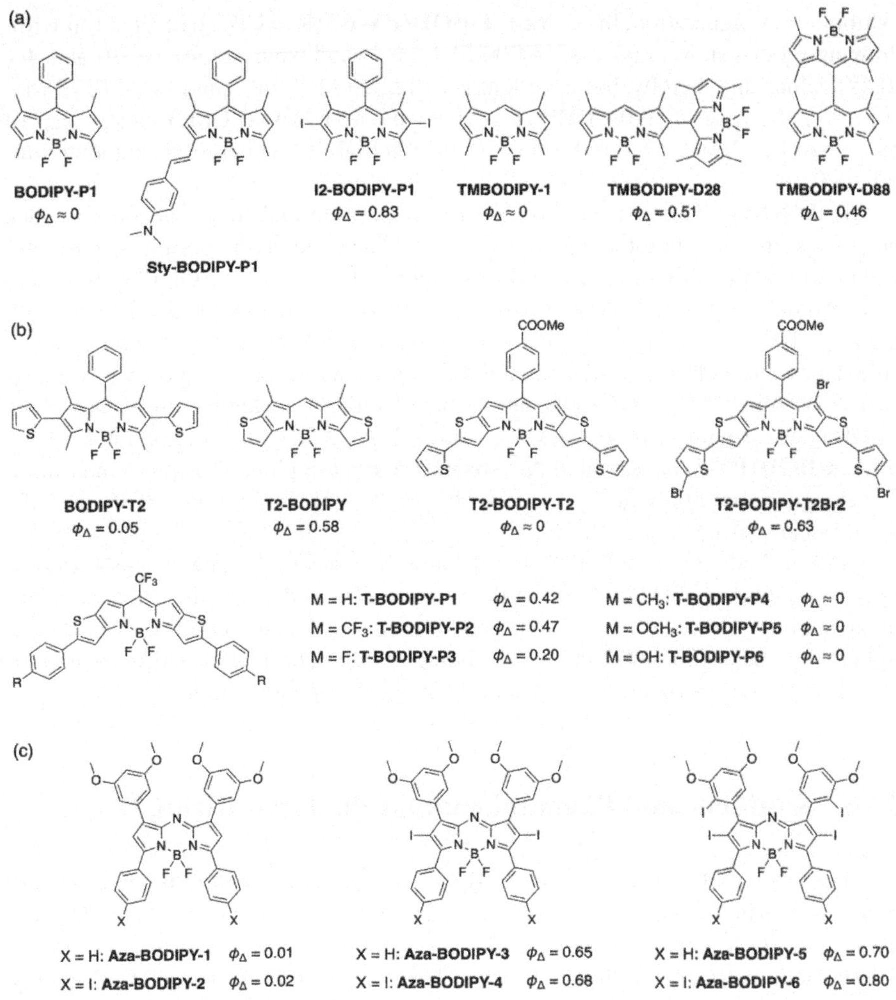

Fig. 13.5 **a** Halogen-substituted BODIPY and BODIPY dimer, **b** condensed BODIPY, and **c** aza-BODIPY photosensitizing dyes

thienopyrrole-fused BODIPY dye, **T-BODIPY-P1** (R = H), without any heavy atom substitution shows a high ε (120,000 M^{-1} cm^{-1}) around 690 nm, a moderate Φ_{fl} value (0.22), and relatively high Φ_{Δ} (0.42), BT (26,400 M^{-1} cm^{-1}), and PP (50,400 M^{-1} cm^{-1}) values (Fig. 13.5b) (Watley et al. 2015). In addition, the study on the influence of electron withdrawing and electron-donating groups on the $^{1}O_2$ generation properties of a series of **T-BODIPY-P** dyes ($\lambda^{abs}{}_{max}$ = 688–738 nm, ε = 120,000–287,000 M^{-1} cm^{-1}) demonstrated that the electron withdrawing groups increase the HOMO-LUMO energy gaps and Φ_{Δ} values. **T-BODIPY-P4** (R = CH_3), **T-BODIPY-P5** (R = OCH_3), and **T-BODIPY-P6** (R = OH) having electron-donating groups exhibit no $^1O_2$ generation. In contrast, **T-BODIPY-P2** ($R = CF_3$) with electron withdrawing groups shows high ε (211,000 M^{-1} cm^{-1}) and balanced Φ_{fl} (0.39) and Φ_Δ (0.47) values and thereby, has excellent BT (82,290 M^{-1} cm^{-1}) and PP (99,170 M^{-1} cm^{-1}). In the case of **T-BODIPY-P** derivatives, a HOMO–LUMO energy gap of more than 1.5 eV is necessary for them to exhibit high $^1O_2$ generation characteristics ($\Phi_\Delta > 0.2$).

Aza-BODIPY dyes exhibit strong photoabsorption and moderate fluorescence properties in the phototherapeutic window. Their photoabsorption spectra are bathochromically shifted by ca. 100 nm compared to those of BODIPY dyes, and their photoabsorption maximum wavelengths are observed around 650–700 nm with high ε (-10^5 M^{-1} cm^{-1}) in the spectra. In addition, aza-BODIPY dyes show remarkable heavy atom effects similar to BODIPY dyes; the introduction of heavy atoms such as bromine and iodine atoms onto the aza-BODIPY skeleton reduces Φ_{fl} values and dramatically improve Φ_T and Φ_Δ values (Fig. 13.5c). For example, in the case of **Aza-BODIPY-1–6**, iodinated **Aza-BODIPY-6** has a photoabsorption maximum wavelength around 680 nm (ε = ca. 50,000 M^{-1} cm^{-1}) and high Φ_T (0.86) and Φ_Δ (0.80) values (Adarsh et al. 2012).

As described above, BODIPY dyes promise to construct a new group of photosensitizing dyes for PDT because of their high ε, Φ_{fl}, and Φ_Δ values, and excellent photostability. However, most of the previously developed BODIPY dyes absorb light mainly below 500–600 nm, which should be bathochromically shifted to exhibit excellent photosensitizing properties in the phototherapeutic window.

13.6 Xanthene and Phenothiazinium Photosensitizing Dyes

Xanthene dyes including rose bengal (RB), erythrosine B, eosin blue, fluorescein, and rhodamine (Rhod), and phenothiazinium dyes including methylene blue (MB) show photoabsorption maximum wavelengths at ca. 550–650 nm ($\varepsilon = -10^5$ M^{-1} cm^{-1}). These xanthene and phenothiazinium dyes have excited triplet energy levels (energy difference between their ground states, ^{1}Dye, and excited triplet states, ^{3}Dye*) close to the energy difference between $^1O_2$ ($^1\Delta_g$) and $^3O_2$ ($^3\Sigma_g^-$) (94.1 kJ $mol^{-1} \approx 1270$ nm ($^1O_2$ phosphorescence wavelength) ≈ 0.98 eV, Fig. 13.1c), and thereby, show high Φ_Δ values due to the efficient energy transfer between ^{3}Dye* and $^1O_2$ (Fig. 13.6a) (DeRosa and Crutchley 2002; Ronzani et al. 2013). The Φ_Δ values of RB and MB are ca. 0.7–0.8 and 0.5–0.6, respectively, and they are used as standard photosensitizing dyes for xanthene and phenothiazinium dyes.

Many studies on the photochemical and electrochemical properties of xanthene and phenothiazinium dyes have revealed that the introduction of bromine and iodine atoms into the xanthene skeleton bathochromically shifts the photoabsorption maximum wavelength, and the heavy atom effect of halogen atoms improves the Φ_T and Φ_Δ values. Therefore, tetraiodide-substituted RB and erythrosine B are efficient photosensitizing dyes (Fig. 13.6a) (Pal et al. 1996). Detty et al. synthesized **TMR-E**, analogues of rhodamine/rosamine dyes in which the bridge atoms are displaced by

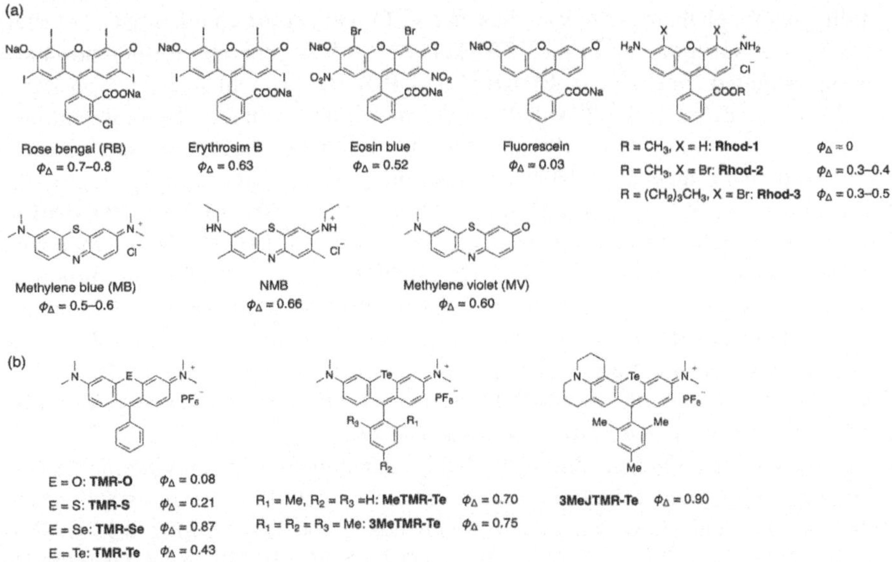

Fig. 13.6 **a** Xanthene and phenothiazinium, and **b** chalcogen atom (S, Se, and Te)-substituted rhodamine/rosamine photosensitizing dyes

S, Se, and Te (Fig. 13.6b) (Kryman et al. 2014), investigated the effects of chalcogen atoms on the optical properties and $^1O_2$ generation efficiency, and reported that the Φ_{fl} values decrease in the order of O (0.84) < S (0.44) < Se (0.009) < Te (<0.005), and the Φ_Δ values increase in the order of O (0.08) < S (0.21) < Te (0.43) < Se (0.87). This result is in good agreement with the heavy atom effect of chalcogen atoms, and they ascribe the relatively low Φ_Δ value of **TMR-Te** to the short-lived τ_T. In addition, they investigated the substituent effects of rosamin dyes and reported that the introduction of bulky phenyl groups at the 9-position (**MeTMR-Te** and **3MeTMR-Te**) and fusion with the julolidine ring (**3MeJTMR-Te**) suppress the free rotation of substituents that causes non-radiative deactivation from the photoexcited states (S_1), resulting in highly efficient ISC (from S_1 to T_1) and high Φ_Δ values (Kryman et al. 2013).

As described above, most xanthene and phenothiazinium dyes have high Φ_Δ values but cannot absorb light above 650 nm. Thus, the improvement of photoabsorption in the phototherapeutic window is required for the use in PDT.

13.7 Heteropolycyclic Photosensitizing Dyes

Perylene diimide (PDI) has been widely used as dyes for optical sensors, biosensors, and optoelectronics due to its photoabsorption maximum wavelength around 500–600 nm ($\varepsilon = 10^4$ M^{-1} cm^{-1}) and high chemical durability and photo/thermal stability. Although there are few studies on the $^1O_2$ generation characteristics of PDI, the effects of substituents on the PDI skeleton on $^1O_2$ generation efficiency have been investigated (Fig. 13.7a). **PDI-Br2** and **PDI-Br4**, in which the 1,7-positions or 2,5,8,11-positions of the PDI skeleton are substituted with 2 or 4 bromine atoms, show high Φ_Δ values (0.23 and 0.85, respectively) due to the heavy atom effect of bromine atoms (Yang et al. 2016). The substitution with phenylethenyl groups at the 1,7-positions or 2,5,8,11-positions of the PDI skeleton (**PDI-PhE2** and **PDI-PhE4**) also enhances the Φ_Δ values. In particular, the Φ_Δ value of **PDI-PhE4** modified with four phenylethenyl groups reaches ca. 0.6. **Ph-PDI** ($\lambda^{abs}{}_{max}$ = ca. 530 nm) modified with para-substituted aryl groups at the 2,5,8,11-positions shows higher Φ_T and Φ_Δ values than unsubstituted PDI ($\lambda^{abs}{}_{max}$ = 525 nm) (Yu et al. 2016). The Φ_Δ value of **MeSPh-PDI** having methylthioxyphenyl groups is extremely high (0.8). The Φ_T values increase in the order of **Ph-PDI** (0.08) < **MeOPh-PDI** (0.54) < **MeSPh-PDI** (0.86), which is attributed to the increased spin–orbit interaction.

A series of naphthalene diimides (NDIs) substituted with alkylamino groups containing a quaternary ammonium salt at one or both of the 2,6-positions have high water solubility and show photoabsorption bands at 500–600 nm (Fig. 13.7a, **NDI-n2N3Br**, **NDI-n3N3Br**, **NDI-n2N4**, **NDI-n3N4**). **NDI-n2N3Br** and **NDI-n3N3Br** exhibit high Φ_Δ values (ca. 0.4–0.6) due to the heavy atom effect of a bromine atom. **NDI-n3N4**, on the other hand, has no bromine atoms but exhibits a moderate Φ_Δ value (0.3) (Doria et al. 2013).

In aromatic carbonyl compounds such as benzophenone, highly efficient ISC (from S_1 to T_1) from ^{1}Dye* to ^{3}Dye* is observed because of the low energy levels of n-π* excited states. The reasons for the highly efficient ISC are that (1) the transition from ^{1}(n-π)* to ^{3}(π-π)* is allowed although that from ^{1}(π-π)* to ^{3}(π-π)* is forbidden and (2) the small electron exchange energy for the n-π* transition due to the spatial separation of the n and π orbitals renders the energy gaps between S_1 and T_1 of aromatic carbonyl compounds small (Fig. 13.1b). Zhao et al. reported that a series of ketocoumarin dyes (**KCOU-1–5**) have photoabsorptions in a short-wavelength region ($\lambda^{abs}{}_{max}$ = ca. 450 nm, ε = ca. 10,000 M^{-1} cm^{-1}, Fig. 13.7b), but their triplet states are long-lived, which contributes to moderate $^1O_2$ generation characteristics (Φ_Δ = 0.28–0.48) (Huang et al. 2013).

Water-soluble bisarylidenecycloalkanone dyes (**AC-1** and **AC-2**) have photoabsorption maxima in a short-wavelength region around 480 nm (ε = ca. 60,000 M^{-1} cm^{-1}) but show two-photon absorption characteristics (**AC-1**, ca. 900 GM @λ^{2PA} = 820 nm; **AC-2**, ca. 1100 GM @λ^{2PA} = 820 nm) and moderate Φ_Δ values (**AC-1**, 0.26; **AC-2**, 0.14; Fig. 13.7c) (Zou et al. 2015).

As described above, heteropolycyclic photosensitizing dyes have relatively good $^1O_2$ generation characteristics. However, most of them show photoabsorption bands in short-wavelength regions below the phototherapeutic window (650–900 nm). To be used as photosensitizing dyes for PDT, the photoabsorption of heteropolycyclic photosensitizing dyes should be improved in long-wavelength regions.

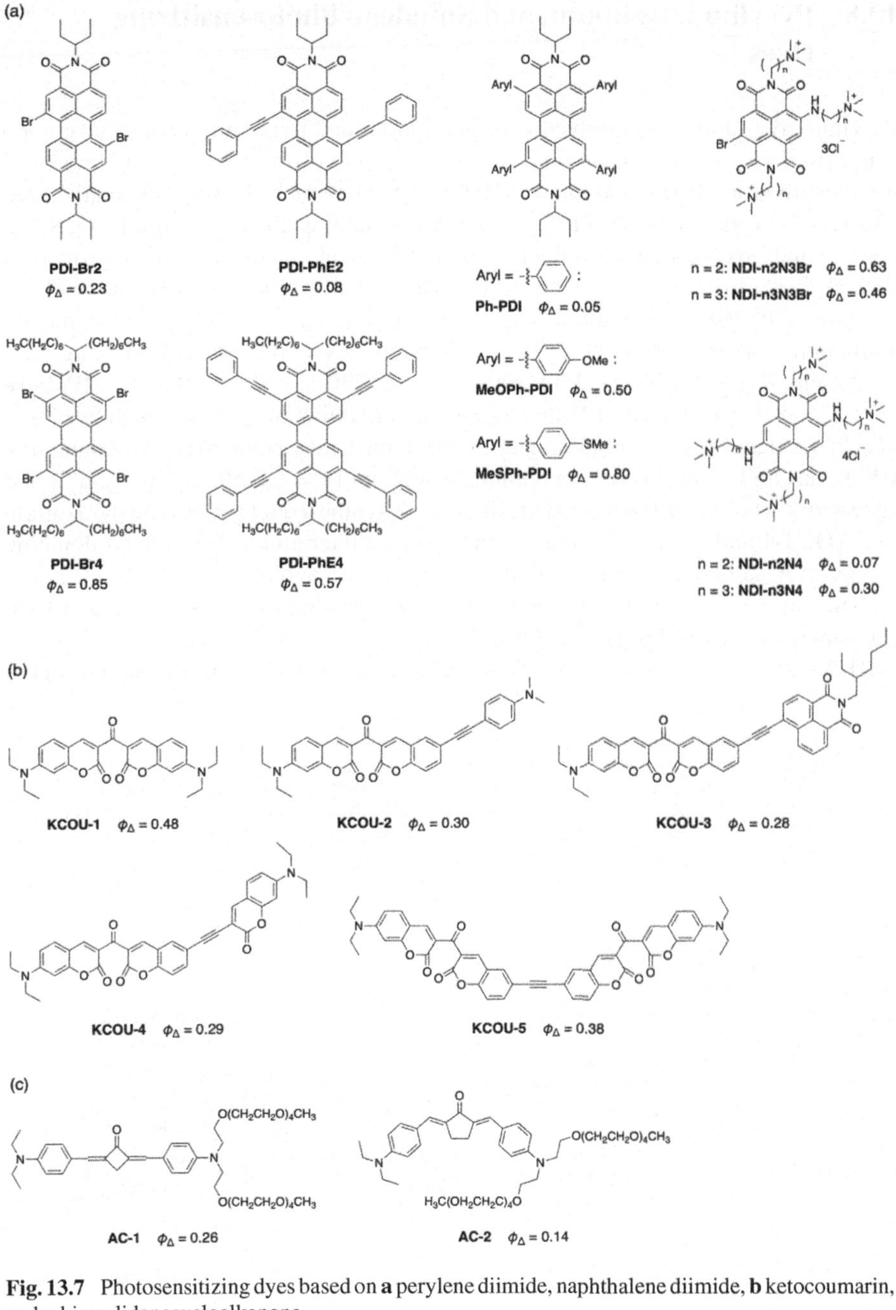

Fig. 13.7 Photosensitizing dyes based on **a** perylene diimide, naphthalene diimide, **b** ketocoumarin, and **c** bisarylidenecycloalkanone

13.8 Pyrylium, Azinium, and Squalene Photosensitizing Dyes

Pyrylium dyes and azinium dyes with pyridinium and pyrazinium rings have good water solubility and tumor affinity (adsorption) and absorb light in a wide range of wavelength around 600–800 nm ($\varepsilon = 10^4$ to 10^5 M^{-1} cm^{-1}). Detty et al. synthesized a series of pyrylium dyes (**PYP-E** and **PY-E**) containing chalcogen atoms (O, S, Se, and Te) and investigated the influence of chalcogen atoms on the optical properties and $^1O_2$ generation efficiency (Fig. 13.8a) (Detty et al. 1990; Leonard et al. 1999). In the case of **PYP-E**, their photoabsorption maximum wavelengths bathochromically shifts in the order of **PYP-O** ($\lambda^{abs}{}_{max} = 593$ nm, $\Phi_\Delta = 0.0004$) < **PYP-S** ($\lambda^{abs}{}_{max} = 685$ nm, $\Phi_\Delta = 0.0006$) < **PYP-Se** ($\lambda^{abs}{}_{max} = 730$ nm, $\Phi_\Delta = 0.014$) < **PYP-Te** ($\lambda^{abs}{}_{max} = 810$ nm, $\Phi_\Delta = 0.12$), and Se and Te-containing pyrylium dyes show higher Φ_Δ values than O and S-containing ones due to the heavy atom effect. Te-containing **PY-Te** has a photoabsorption maximum at 672 nm (ε = ca. 55,000 M^{-1} cm^{-1}) and shows $\Phi_\Delta = 0.037$. Ooyama et al. designed and synthesized D-π-A type pyrazinium dyes (**OEJ-1** and **OEJ-2**) having Br^- or I^- as a counter anion (D: electron-donating group, π: π skeleton, A: electron withdrawing group), and investigated the effects of halogen atoms on the optical properties and $^1O_2$ generation efficiency (Fig. 13.8b) (Ooyama et al. 2016). **OEJ-1** and **OEJ-2** absorb light in a wide range of 500–700 nm. **OEJ-2** with I^- shows a photoabsorption in a longer wavelength region and higher

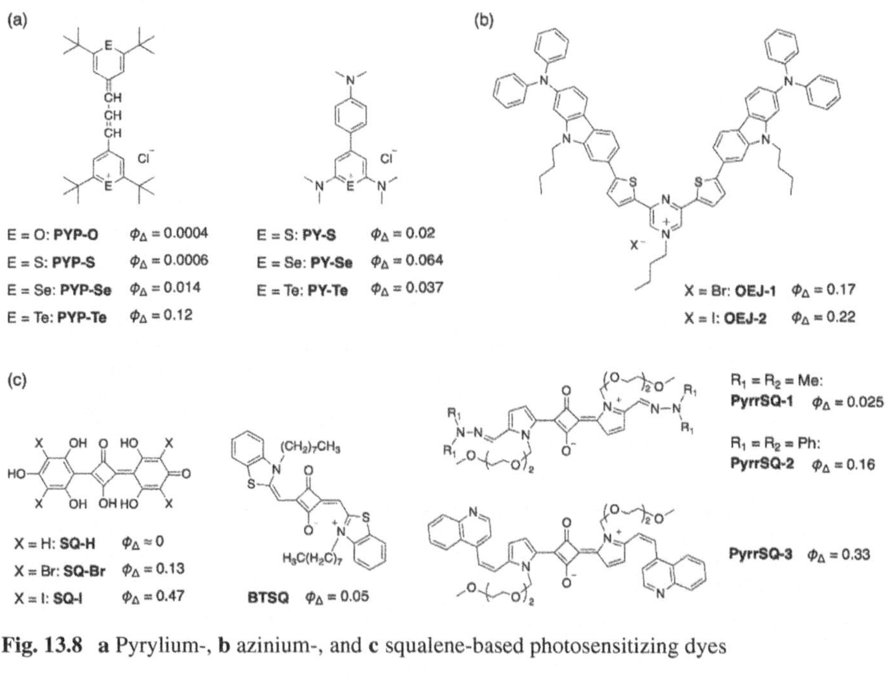

Fig. 13.8 **a** Pyrylium-, **b** azinium-, and **c** squalene-based photosensitizing dyes

Φ_Δ value (0.22) than **OEJ-1** with Br^- ($\Phi_\Delta = 0.17$). This result is attributed to the superior heavy atom effect of I^- to B^-.

Squalene dyes have been applied to optoelectronic devices including information recording, OLED, and organic solar cells as red and near-infrared absorbing dyes since the first synthesis in 1965. **BTSQ** ($\lambda^{abs}{}_{max} = 682$ nm, $\varepsilon = 295{,}000$ M^{-1} cm^{-1}), a squaraine dye developed as a photosensitizing dye for PDT, shows $\Phi_\Delta = 0.05$ (Salice et al. 2010). **PyrrSQ-1–3** with two-photon absorption characteristics have photoabsorption maximum wavelengths around 700 nm and show Φ_Δ values of 0.025–0.33 (Fig. 13.8c) (Beverina et al. 2008). It has been also reported that the heavy atom effect of halogen atoms improves the Φ_Δ values (Fig. 13.8c, **SQ-H**, **SQ-Br**, **SQ-I**) (Ramaiah et al. 1997).

Thus, pyrylium-, azinium-, and squalene-based photosensitizing dyes show low $^1O_2$ generation efficiency, although they have excellent water solubility, tumor affinity, and good photoabsorption properties in the phototherapeutic window. To improve the problem, the relationship between the modification of dye skeletons and $^1O_2$ generation efficiency is required to be elucidated.

13.9 Photosensitizing Dyes Based on Transition Metal (Ru, Ir, Pt) Complex

Photosensitizing dyes based on complexes with transition metals such as Ru, Ir, and Pt promise the use in PDT due to the broad photoabsorptions derived from metal-to-ligand charge transfer (MLCT) in a visible region (350–550 nm), highly efficient ISC (from S_1 to T_1) by the heavy transition metals, long-lived ^{3}MLCT excited states, and phosphorescence characteristics.

$^1O_2$ generation properties of trisdiimine Ru(II) complexes (**[Ru^{II}(N^N)$_3$]$^{2+}$**) have been extensively studied (Fig. 13.9a) (Fresnadillo et al. 1996; Stacey and Pope 2013). Most Ru(II) complexes have photoabsorption bands derived from MLCT around 440–550 nm ($\varepsilon = -10^4$ M^{-1} cm^{-1}). The Φ_Δ values of Ru(II) complexes are significantly affected by the ligands. The Φ_Δ values of tris(2,2′-bipyridyl)ruthenium (II) complex (**[Ru(bpy)$_3$]$^{2+}$**) are 0.22 in deuterated water and 0.73 in deuterated methanol, while those of tris(4,7-diphenyl-1,10-phenanthroline)ruthenium (II) complex (**[Ru(dpp)$_3$]$^{2+}$**) are 0.42 in deuterated water and 0.97 in deuterated methanol. Ru (II) complex with benzenesulfonic acid sodium salts at the 4,7-positions of the 1,10-phenanthroline ligands (**[Ru(dpds)$_3$]$^{2+}$**) has good water solubility and shows $\Phi_\Delta = 0.42$ in deuterated water and $\Phi_\Delta =$ ca. 1.0 in deuterated methanol.

Recently, the $^1O_2$ generation characteristics of Ir(III) complexes and development of new Ir(III) complex-based photosensitizing dyes have been extensively studied (Ashen-Garry and Selke 2014). In acetonitrile, the Φ_Δ value of tris(2,2′-bipyridyl)iridium(III) complex (**[Ir(bpy)$_3$]$^{3+}$**) is ca. 0.1, which is lower than that of **[Ru(bpy)$_3$]$^{2+}$** ($\Phi_\Delta = 0.57$) (Fig. 13.9b). On the other hand, Thompson et al. reported that dicyclometal-diketonated Ir(III) complexes (**[Ir(C^N)$_2$(O^O)]**) exhibit

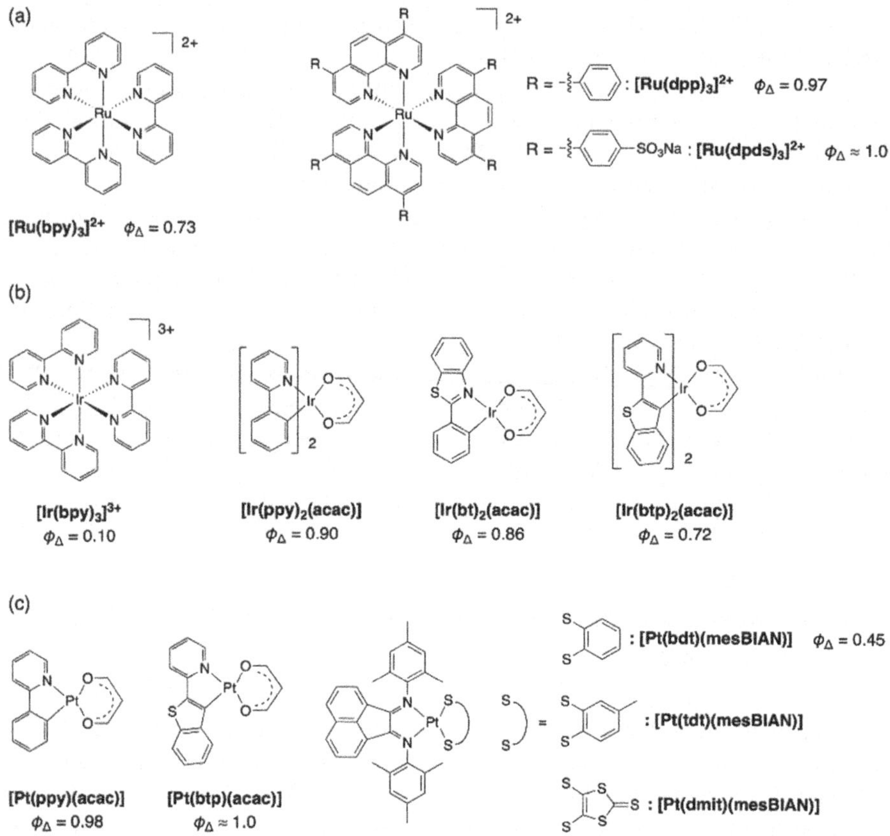

Fig. 13.9 Photosensitizing dyes based on transition metal complexes: **a** Ru, **b** Ir, and **c** Pt metal complexes

long-lived τ_T and high Φ_Δ values (Fig. 13.9b, **[Ir(ppy)$_2$(acac)]**, **[Ir(bt)$_2$(acac)]**, **[Ir(btp)$_2$(acac)]**) (Stacey and Pope 2013; Djurovich et al. 2007). For example, the Φ_Δ value of a cyclometalated Ir(III) complex with 2-phenylpyridine (ppy) and acetylacetone (acac) (**[Ir(ppy)$_2$(acac)]**) is 0.9.

Pt(II) complexes show excellent $^1O_2$ generation characteristics, although some of them are photochemically unstable (Fig. 13.9c). Cyclometal-diketonated Pt(II) complexes with monoanionic ligands (C^N) (**[Pt(C^N)(O^O)]**) have broad absorption bands derived from MLCT around 350–450 nm (ε = 2000–6000 M^{-1} cm^{-1}). For example, **[Pt(ppy)(acac)]** and **[Pt(btp)(acac)]** (acac: acetylacetone) with 2-phenylpyridine (ppy) and 2,2′-benzothienopyridine (btp) as cyclometalated ligands, respectively, have photoabsorption bands in short-wavelength regions but show Φ_Δ values close to 1 in deuterated methanol (Djurovich et al. 2007). The relatively strong MLCT photoabsorption bands of diimine-aryldithiolate platinum(II) complexes (**[Pt(S^S)(mesBIAN)]**) (mesBIAN: bismesitylbiazanaphthenequinone) are observed around 750 nm ($\varepsilon = \sim 10^4$ M^{-1} cm^{-1}), and **[Pt(bdt)(mesBIAN)]** exhibits a moderate Φ_Δ value (0.45) (Fig. 13.9c, **[Pt(bdt)(mesBIAN)]**, **[Pt(tdt)(mesBIAN)]**, **[Pt(dmit)(mesBIAN)]**) (Adams et al. 2007).

As described above, transition metal-based photosensitizing dyes show high Φ_Δ values, but their photoabsorption remains below 500 nm. To use them efficiently in the phototherapeutic window, photoabsorption in a longer wavelength region is required. In addition, their water solubility and low toxicity under no light irradiation should be improved.

13.10 Conclusions and Perspectives

In this chapter, the photoabsorption properties, $^1O_2$ generation characteristics, and water solubility of photosensitizing dyes for PDT are organized by dye skeleton, and the advantages and disadvantages of each dye skeleton are highlighted. The introduction of heavy atoms such as bromine and iodine into the dye skeletons promotes ISC (from S_1 to T_1) by enhanced spin–orbit interaction (while decreasing fluorescence emission), leading to high Φ_T values. However, since the introduction of bromine or iodine atoms causes toxicity under no light irradiation, the development of halogen atom-free photosensitizing dyes is desired. In recent years, photosensitizing dyes with high brightness ($BT = \varepsilon \times \Phi_{fl}$) and strong phototoxicity ($PP = \varepsilon \times \Phi_\Delta$) have attracted much interest to achieve both fluorescence imaging and PDT of cancer cells. In particular, BODIPY dyes promise to be a new group of photosensitizing dyes for PDT because those without halogen atoms have high BT and PP characteristics. However, the photosensitizing dyes that efficiently generate $^1O_2$ shown in this chapter are based on the existing porphyrin and BODIPY skeletons with photoabsorption in the phototherapeutic window (650–900 nm). At present, the molecular design and development of new dye skeletons for PDT are stagnant. In the future, to promote the improvement and spread of early cancer treatment by PDT, addressing to the construction of a new group of photosensitizing dyes with high BT and PP characteristics and exploring the possibilities for PDT use are desired.

References

Adams CJ, Fey N, Parfitt M, Pope SJA, Weinstein JA (2007) Synthesis, structures and properties of a new series of platinum–diimine–dithiolate complexes. Dalton Trans 111:4446–4456. https://doi.org/10.1039/b709252k

Adarsh N, Shanmugasundaram M, Avirah RR, Ramaiah D (2012) Aza-BODIPY derivatives: enhanced quantum yields of triplet excited states and the generation of singlet oxygen and their role as facile sustainable photooxygenation catalysts. Chem Eur J 18:12655–12662. https://doi.org/10.1002/chem.201202438

Arnaut LG, Pereira MM, Dąbrowski JM, Silva EFF, Schaberle FA, Abreu AR, Rocha LB, Barsan MM, Urbańska K, Stochel G, Brett CMA (2014) Photodynamic therapy efficacy enhanced by

dynamics: the role of charge transfer and photostability in the selection of photosensitizers. Chem Eur J 20:5346–5357. https://doi.org/10.1002/chem.201304202
Ashen-Garry D, Selke M (2014) Singlet oxygen generation by cyclometalated complexes and applications. Photochem Photobiol 90:257–274. https://doi.org/10.1111/php.12211
Aubry J-M, Mandard-Cazin B, Rougee M, Bensasson RV (1995) Kinetic studies of singlet oxygen [4+2]-cycloadditions with cyclic 1,3-dienes in 28 solvents. J Am Chem Soc 117:9159–9164. https://doi.org/10.1021/ja00141a006
Awuah SG, You Y (2012) Boron dipyrromethene (BODIPY)-based photosensitizers for photodynamic therapy. RSC Adv 2:11169–11183. https://doi.org/10.1039/c2ra21404k
Beverina L, Crippa M, Landenna M, Ruffo R, Salice P, Silvestri F, Versari S, Villa A, Ciaffoni L, Collini E, Ferrante C, Bradamante S, Mari CM, Bozio R, Pagani GA (2008) Assessment of water-soluble π-extended squaraines as one- and two-photon singlet oxygen photosensitizers: design, synthesis, and characterization. J Am Chem Soc 130:1894–1902. https://doi.org/10.1021/ja075933a
Bonnett R (1995) Photosensitizers of the porphyrin and phthalocyanine series for photodynamic therapy. Chem Soc Rev 24:19–33. https://doi.org/10.1039/cs9952400019
Cakmak Y, Kolemen S, Duman S, Dede Y, Dolen Y, Kilic B, Kostereli Z, Yildirim LT, Dogan AL, Guc D, Akkaya EU (2011) Designing excited states: theory-guided access to efficient photosensitizers for photodynamic action. Angew Chem Int Ed 50:11937–11941. https://doi.org/10.1002/anie.201105736
Callaghan S, Senge MO (2018) The good, the bad, and the ugly—controlling singlet oxygen through design of photosensitizers and delivery systems for photodynamic therapy. Photochem Photobiol Sci 17:1490–1514. https://doi.org/10.1039/C8PP00008E
Coskun A, Akkaya EU (2004) Difluorobora-s-diazaindacene dyes as highly selective dosimetric reagents for fluoride anions. Tetrahedron Lett 45:4947–4949. https://doi.org/10.1016/j.tetlet.2004.04.130
Dąbrowski JM, Arnaut LG (2015) Photodynamic therapy (PDT) of cancer: from local to systemic treatment. Photochem Photobiol Sci 14:1765–1780. https://doi.org/10.1039/C5PP00132C
DeRosa MC, Crutchley RJ (2002) Photosensitized singlet oxygen and its applications. Coord Chem Rev 233–234:351–371. https://doi.org/10.1016/S0010-8545(02)00034-6
Detty MR, Merkel PB, Hilf R, Gibson SL, Powers SK (1990) Chalcogenapyrylium dyes as photochemotherapeutic agents. 2. Tumor uptake, mitochondrial targeting, and singlet-oxygen-induced inhibition of cytochrome c oxidase. J Med Chem 33:1108–1116. https://doi.org/10.1021/jm00166a005
Djurovich PI, Murphy D, Thompson ME, Hernandez B, Gao R, Hunt PL, Selke M (2007) Cyclometalated iridium and platinum complexes as singlet oxygen photosensitizers: quantum yields, quenching rates and correlation with electronic structures. Dalton Trans 298:3763–3770. https://doi.org/10.1039/b704595f
Doria F, Manet I, Grande V, Monti S, Freccero M (2013) Water-soluble naphthalene diimides as singlet oxygen sensitizers. J Org Chem 78:8065–8073. https://doi.org/10.1021/jo401347z
Ethirajan M, Chen Y, Joshi P, Pandey RK (2011) The role of porphyrin chemistry in tumor imaging and photodynamic therapy. Chem Soc Rev 40:340–362. https://doi.org/10.1039/B915149B
Fresnadillo DG, Georgiadou Y, Orellana G, Braun AM, Oliveros E (1996) Singlet-oxygen ($^1\Delta_g$) production by ruthenium(II) complexes containing polyazaheterocyclic ligands in methanol and in water. Helv Chim Acta 79:1222–1238. https://doi.org/10.1002/hlca.19960790428
Gollnick K, Griesbeck A (1985) Singlet oxygen photooxygenation of furans: isolation and reactions of (4+2)-cycloaddition products (unsaturated sec.-ozonides). Tetrahedron 41:2057–2068. https://doi.org/10.1016/S0040-4020(01)96576-7
Greer A (2006) Christopher foote's discovery of the role of singlet oxygen [$^1O_2$ ($^1\Delta_g$)] in photosensitized oxidation reactions. Acc Chem Res 39:797–804. https://doi.org/10.1021/ar050191g
Hammerer F, Garcia G, Chen S, Poyer F, Achelle S, Fiorini-Debuisschert C, Teulade-Fichou M-P, Maillard P (2014) Synthesis and characterization of glycoconjugated porphyrin triphenylamine

hybrids for targeted two-photon photodynamic therapy. J Org Chem 79:1406–1417. https://doi.org/10.1021/jo402658h

Horiuchi H, Hosaka M, Mashio H, Terata M, Ishida S, Kyushin S, Okutsu T, Takeuchi T, Hiratsuka H (2014) Silylation improves the photodynamic activity of tetraphenylporphyrin derivatives in vitro and in vivo. Chem Eur J 20:6054–6060. https://doi.org/10.1002/chem.201303120

Huang D, Sun J, Ma L, Zhang C, Zhao J (2013) Preparation of ketocoumarins as heavy atom-free triplet photosensitizers for triplet–triplet annihilation upconversion. Photochem Photobiol Sci 12:872–882. https://doi.org/10.1039/c3pp25416j

Ji S, Ge J, Escudero D, Wang Z, Zhao J, Jacquemin D (2015) Molecular structure–intersystem crossing relationship of heavy-atom-free BODIPY triplet photosensitizers. J Org Chem 80:5958–5963. https://doi.org/10.1021/acs.joc.5b00691

Kamkaew A, Lim SH, Lee HB, Kiew LV, Chung LY, Burgess K (2013) BODIPY dyes in photodynamic therapy. Chem Soc Rev 42:77–88. https://doi.org/10.1039/C2CS35216H

Kryman MW, Schamerhorn GA, Yung K, Sathyamoorthy B, Sukumaran DK, Ohulchanskyy TY, Benedict JB, Detty MR (2013) Organotellurium fluorescence probes for redox reactions: 9-aryl-3,6-diaminotelluroxanthylium dyes and their telluroxides. Organometallics 32:4321–4333. https://doi.org/10.1021/om400467s

Kryman MW, Schamerhorn GA, Hill JE, Calitree BD, Davies KS, Linder MK, Ohulchanskyy TY, Detty MR (2014) Synthesis and properties of heavy chalcogen analogues of the texas reds and related rhodamines. Organometallics 33:2628–2640. https://doi.org/10.1021/om500346j

Lau JTF, Lo P-C, Fong W-P, Ng DKP (2011) Preparation and photodynamic activities of silicon(IV) phthalocyanines substituted with permethylated β-cyclodextrins. Chem Eur J 17:7569–7577. https://doi.org/10.1002/chem.201100621

Leonard KA, Nelen MI, Anderson LT, Gibson SL, Hilf R, Detty MR (1999) 2,4,6-Triarylchalcogenopyrylium dyes related in structure to the antitumor agent AA1 as in vitro sensitizers for the photodynamic therapy of cancer. J Med Chem 42:3942–3952. https://doi.org/10.1021/jm990134r

Li W, Li L, Xiao H, Qi R, Huang Y, Xie Z, Jing X, Zhang H (2013) Iodo-BODIPY: a visible-light-driven, highly efficient and photostable metal-free organic photocatalyst. RSC Adv 3:13417–13421. https://doi.org/10.1039/c3ra40932e

Li X-S, Ke M-R, Huang W, Ye C-H, Huang J-D (2015) A pH-responsive layered double hydroxide (LDH)-phthalocyanine nanohybrid for efficient photodynamic therapy. Chem Eur J 21:3310–3317. https://doi.org/10.1002/chem.201404514

Lovell JF, Liu TWB, Chen J, Zheng G (2010) Activatable photosensitizers for imaging and therapy. Chem Rev 110:2839–2857. https://doi.org/10.1021/cr900236h

Mori S, Yoshiyama H, Tokunaga E, Iida N, Hayashi M, Obata T, Tanaka M, Shibata N (2015) Design, synthesis, spectral investigations and biological activity of fluorinated phthalocyanine conjugated with galactose and comparison to its non-fluorinated counterpart. J Fluor Chem 174:137–141. https://doi.org/10.1016/j.jfluchem.2014.11.003

Mroz P, Tegos GP, Gali H, Wharton T, Sarna T, Hamblin MR (2007) Photodynamic therapy with fullerenes. Photochem Photobiol Sci 6:1139–1149. https://doi.org/10.1039/b711141j

Naik A, Rubbiani R, Gasser G, Spingler B (2014) Visible-light-induced annihilation of tumor cells with platinum-porphyrin conjugates. Angew Chem Int Ed 53:6938–6941. https://doi.org/10.1002/anie.201400533

Novakova V, Lásková M, Vavřičková H, Zimcik P (2015) Phenol-substituted tetrapyrazinoporphyrazines: pH-dependent fluorescence in basic media. Chem Eur J 21:14382–14392. https://doi.org/10.1002/chem.201502533

Ooyama Y, Enoki T, Ohshita J (2016) Development of a D–π–A pyrazinium photosensitizer possessing singlet oxygen generation. RSC Adv 6:5428–5435. https://doi.org/10.1039/C5RA26647E

Oriana S, Aroua S, Söllner JOB, Ma X-J, Iwamoto Y, Yamakoshi Y (2013) Water-soluble C_{60}– and C_{70}–PVP polymers for biomaterials with efficient $^1O_2$ generation. Chem Commun 49:9302–9304. https://doi.org/10.1039/c3cc45501g

Pal P, Zeng H, Durocher G, Girard D, Li T, Gupta AK, Giasson R, Blanchard L, Gaboury L, Balassy A, Turmel C, Laperrière A, Villeneuve L (1996) Phototoxicity of some bromine-substituted rhodamine dyes: synthesis, photophysical properties and application as photosensitizers. Photochem Photobiol 63:161–168. https://doi.org/10.1111/j.1751-1097.1996.tb03008.x

Patrice T (2003) Photodynamic therapy. Royal Society of Chemistry, London

Pawlicki M, Collins HA, Denning RG, Anderson HL (2009) Two-photon absorption and the design of two-photon dyes. Angew Chem Int Ed 48:3244–3266. https://doi.org/10.1002/anie.200805257

Prein M, Adam W (1996) The Schenck ene reaction: diastereoselective oxyfunctionalization with singlet oxygen in synthetic applications. Angew Chem Int Ed 35:477–494. https://doi.org/10.1002/anie.199604771

Ramaiah D, Joy A, Chandrasekhar N, Eldho NV, Das S, George MV (1997) Halogenated squaraine dyes as potential photochemotherapeutic agents: synthesis and study of photophysical properties and quantum efficiencies of singlet oxygen generation. Photochem Photobiol 65:783–790. https://doi.org/10.1111/j.1751-1097.1997.tb01925.x

Redmond RW, Gamlin JN (1999) A compilation of singlet oxygen yields from biologically relevant molecules. Photochem Photobiol 70:391–475. https://doi.org/10.1111/j.1751-1097.1999.tb08240.x

Ronzani F, Trivella A, Arzoumanian E, Blanc S, Sarakha M, Richard C, Oliveros E, Lacombe S (2013) Comparison of the photophysical properties of three phenothiazine derivatives: transient detection and singlet oxygen production. Photochem Photobiol Sci 12:2160–2169. https://doi.org/10.1039/c3pp50246e

Rubio N, Prat F, Bou N, Borrell JI, Teixidó J, Villanueva Á, Juarranz Á, Cañete M, Stockert JC, Nonell S (2005) A comparison between the photophysical and photosensitising properties of tetraphenyl porphycenes and porphyrins. New J Chem 29:378–384. https://doi.org/10.1039/B415314F

Salice P, Arnbjerg J, Pedersen BW, Toftegaard R, Beverina L, Pagani GA, Ogilby PR (2010) Photophysics of squaraine dyes: role of charge-transfer in singlet oxygen production and removal. J Phys Chem a 114:2518–2525. https://doi.org/10.1021/jp911180n

Schmitt J, Heitz V, Sour A, Bolze F, Ftouni H, Nicoud J-F, Flamigni L, Ventura B (2015) Diketopyrrolopyrrole-porphyrin conjugates with high two-photon absorption and singlet oxygen generation for two-photon photodynamic therapy. Angew Chem Int Ed 54:169–173. https://doi.org/10.1002/anie.201407537

Shimakoshi H, Baba T, Iseki Y, Aritome I, Endo A, Adachi C, Hisaeda Y (2008) Photophysical and photosensitizing properties of brominated porphycenes. Chem Commun 25:2882–2884. https://doi.org/10.1039/b802730g

Shinmori H, Kodaira F, Matsugo S, Kawabata S, Osuka A (2005) Photosensitizing properties of diazaporphyrin derivatives for singlet oxygen generation. Chem Lett 34:322–323. https://doi.org/10.1246/cl.2005.322

Stacey OJ, Pope SJA (2013) New avenues in the design and potential application of metal complexes for photodynamic therapy. RSC Adv 3:25550–25564. https://doi.org/10.1039/c3ra45219k

Takizawa S-Y, Aboshi R, Murata S (2011) Photooxidation of 1,5-dihydroxynaphthalene with iridium complexes as singlet oxygen sensitizers. Photochem Photobiol Sci 10:895–903. https://doi.org/10.1039/c0pp00265h

Tian J, Ding L, Xu H-J, Shen Z, Ju H, Jia L, Bao L, Yu J-S (2013) Cell-specific and pH-activatable rubyrin-loaded nanoparticles for highly selective near-infrared photodynamic therapy against cancer. J Am Chem Soc 135:18850–18858. https://doi.org/10.1021/ja408286k

Watley RL, Awuah SG, Bio M, Cantu R, Gobeze HB, Nesterov VN, Das SK, D'Souza F, You Y (2015) Dual functioning thieno-pyrrole fused BODIPY dyes for NIR optical imaging and photodynamic therapy: singlet oxygen generation without heavy halogen atom assistance. Chem Asian J 10:1335–1343. https://doi.org/10.1002/asia.201500140

Yamakoshi Y, Umezawa N, Ryu A, Arakane K, Miyata N, Goda Y, Masumizu T, Nagano T (2003) Active oxygen species generated from photoexcited fullerene (C60) as potential medicines: O2-• versus 1O2. J Am Chem Soc 125:12803–12809. https://doi.org/10.1021/ja0355574

Yang Y, Guo Q, Chen H, Zhou Z, Guo Z, Shen Z (2013) Thienopyrrole-expanded BODIPY as a potential NIR photosensitizer for photodynamic therapy. Chem Commun 49:3940–3942. https://doi.org/10.1039/c3cc40746b

Yang W, Zhao J, Sonn C, Escudero D, Karatay A, Yaglioglu HG, Küçüköz B, Hayvali M, Li C, Jacquemin D (2016) Efficient intersystem crossing in heavy-atom-free perylenebisimide derivatives. J Phys Chem C 120:10162–10175. https://doi.org/10.1021/acs.jpcc.6b01584

Yao D, Hugues V, Blanchard-Desce M, Mongin O, Paul-Roth CO, Paul F (2015) Dendritic molecular assemblies for singlet oxygen generation: meso-tetraphenylporphyrin-based biphotonic sensitizers with remarkable luminescence. New J Chem 39:7730–7733. https://doi.org/10.1039/C5NJ01381J

Yu Z, Wu Y, Peng Q, Sun C, Chen J, Yao J, Fu H (2016) Accessing the triplet state in heavy-atom-free perylene diimides. Chem Eur J 22:4717–4722. https://doi.org/10.1002/chem.201600300

Zhou Z, Song J, Nie L, Chen X (2016) Reactive oxygen species generating systems meeting challenges of photodynamic cancer therapy. Chem Soc Rev 45:6597–6626. https://doi.org/10.1039/C6CS00271D

Zou Q, Zhao H, Zhao Y, Fang Y, Chen D, Ren J, Wang X, Wang Y, Gu Y, Wu F (2015) Effective two-photon excited photodynamic therapy of xenograft tumors sensitized by water-soluble bis(arylidene)cycloalkanone photosensitizers. J Med Chem 58:7949–7958. https://doi.org/10.1021/acs.jmedchem.5b00731

Chapter 14
Photoenergy Conversion (Dye-Sensitized Solar Cells)

Yousuke Ooyama

Abstract Organic solar (photovoltaic) cell is one of the most promising new renewable photovoltaic cells. In particular, dye-sensitized solar cells (DSSCs) based on dye photosensitizers adsorbed on photoelectrodes (oxide semiconductor electrodes such as TiO_2, ZnO, and NiO) have received considerable attention from the viewpoint of the fascinating construction and operational principles, decorative natures, low cost of fabrication, and high power conversion efficiency. For improvement of the photovoltaic performances of DSSCs, it is essential to create new efficient dye photosensitizers. Therefore, a variety of ruthenium (Ru) complex dyes and organic dyes that possess excellent light-harvesting properties over the wide spectral region of sunlight and enhanced electron communication between the dye and photoelectrode and between the dye and electrolyte have been designed and developed. To date, the solar energy-to-electricity conversion yield (η) of up to 13% has been achieved in DSSCs.

Keywords Dye-Sensitized solar cells · Dye sensitizers · TiO_2 electrode · ZnO electrode · NiO electrode

14.1 Introduction

For the 2030 agenda for Sustainable Development Goals (SDGs) adopted by all United Nations Member States in 2015 that provides a shared blueprint for peace and prosperity for people and the planet now and into the future, the development of new renewable and sustainable energy sources is an issue of global concern to be addressed. Solar (photovoltaic) cells have attracted much attention among sustainable energy sources. Solar cells based on silicon (crystalline, multicrystalline, microcrystalline, and amorphous silicon), compound semiconductor (e.g., cadmium telluride (CdTe) and chalcopyrite compounds including copper indium gallium

Y. Ooyama (✉)
Department of Applied Chemistry, Graduate School of Engineering, Hiroshima University, 1-4-1 Kagamiyama, Higashi-Hiroshima 739-8527, Japan
e-mail: yooyama@hiroshima-u.ac.jp

Y. Ooyama and S. Yagi (eds.), *Progress in the Science of Functional Dyes*,
https://doi.org/10.1007/978-981-33-4392-4_14

selenide (CIGS)) are representative inorganic solar cells, which have achieved the solar energy-to-electricity conversion efficiency of up to 15–20%.

Organic solar cells based on dye photosensitizers (organic dyes and metal complex dyes) or polymer blend films (e.g., poly(3-hexylthiophene): P3HT) with fullerene derivatives (e.g., [6,6]-phenyl-C61-butyric acid methyl ester: PCBM) are known as one of the third-generation photovoltaic cells (O'Regan and Grätzel 1991; Hagfeldt et al. 2010; Kim et al. 2007; Günes et al. 2007).

Dye-sensitized solar cells (DSSCs), in which dye photosensitizers are adsorbed on photoelectrodes (e.g., TiO_2 and ZnO as n-type oxide semiconductor electrode or NiO as p-type oxide semiconductor electrode) have particularly received considerable attention from the viewpoint of their interesting construction and operational principles, decorative natures, high power conversion efficiency, and low cost of fabrication (O'Regan and Grätzel 1991; Hagfeldt et al. 2010; Nazeeruddin et al. 1993, 1997, 1999; 2001, 2003; Pashaei et al. 2016; Robertson 2006; Chen et al. 2007; Xie and Guo 2007; Imahori et al. 2009; Ooyama and Harima 2009; Ning and Tian 2009; Mishra et al. 2009; Ning et al. 2010; Yen et al. 2012; Odobel et al. 2010, 2012; Ooyama and Harima 2012; Yum et al. 2011; Nikolaou et al. 2017; Li and Diau 2013; Ladomenou et al. 2014; Manfredi et al. 2014; Zhang and Cole 2015; Lee et al. 2015; Li et al. 2017; Urbani et al. 2014; Higashino 2015; Brogdon et al. 2018; Ji et al. 2018; Mathew et al. 2014; Yella et al. 2014). n-TiO_2- or n-ZnO-based DSSCs generally consist of a dye-adsorbed nanocrystalline TiO_2 or ZnO electrode as a photoanode and a Pt-coated glass plate as a photoinert cathode sandwiching an electrolyte solution containing an I_3^-/I^- ion couple as a redox mediator, as shown in Fig. 14.1, which is illuminated through the TiO_2 or n-ZnO film side.

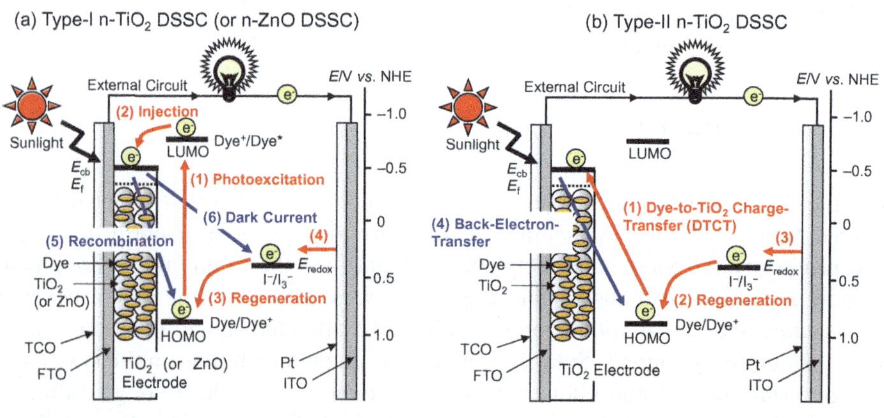

Fig. 14.1 Schematic representation of the construction and operational principles of **a** type-I n-TiO_2-based or n-ZnO-based DSSC (an electron is excited from the HOMO to the LUMO of the dye, followed by injection into the CB of TiO_2 or ZnO) and **b** type-II n-TiO_2-based DSSC (an electron is injected directly from the HOMO of the dye into the CB of TiO_2 or ZnO upon photoexcitation). The red and blue arrows represent the processes required for the generation of photocurrent and the recombination processes (electron-loss processes), respectively. ITO: indium tin oxide, FTO: fluorine-doped tin oxide, TCO: transparent conductive oxide

n-TiO_2 or n-ZnO-based DSSCs can be actually classified into two types, I and II, depending on the electron-injection pathway from the dye to the conduction band (CB) of the TiO_2 or ZnO electrode (Fig. 14.1). In type-I DSSCs, photoexcitation of the local band of the dye adsorbed on the TiO_2 or ZnO surface occurs, followed by injection of an electron from the photoexcited dye to the CB of the TiO_2 or ZnO electrode (Fig. 14.1a). In other words, an electron is excited from the highest occupied molecular orbital (HOMO) to the lowest unoccupied molecular orbital (LUMO) of the dye sensitizer by sunlight illumination, followed by injection of the LUMO electron into the CB of TiO_2 or ZnO electrode. Thus, this pathway can also be called the "two-step" or "indirect" electron-injection pathway. The resultant oxidized dye is subsequently reduced back to its original neutral state by electron donation from the I^- ions in the redox mediator. The injected electrons move through the interconnected network of TiO_2 or ZnO nanoparticles to reach the fluorine-doped tin oxide (FTO) and then pass through the external circuit to the cathode (Pt-coated glass). The I^- ion is regenerated by reduction of the I_3^- ion at the cathode through electron donation from the external circuit, completing the circuit. To achieve efficient electron injection from the excited dye to the CB of the TiO_2 or ZnO electrode, the LUMO energy level of the dye in a type-I DSSC must be higher (more negative) than the energy level (E_{cb}) of the CB (–0.5 V versus the normal hydrogen electrode (NHE)) of the TiO_2 or ZnO electrode. Specifically, some researchers have proposed that an energy gap of over 0.2–0.3 V is necessary for efficient electron injection. In order to achieve efficient regeneration of the oxidized state by electron transfer from the I_3^-/I^- ion couple in the electrolyte, the HOMO energy level of the dye must be lower (more positive) than that of the I_3^-/I^- redox potential (0.4 V vs. NHE). Specifically, some researchers have proposed that an energy gap of over 0.2–0.3 V is necessary for efficient regeneration. Thus, electron injection and dye regeneration must be thermodynamically feasible.

In contrast to type-I DSSCs, the electron injection is direct in type-II DSSCs, i.e., "one-step" electron injection occurs from the ground state of the dye to the CB of the TiO_2 electrode by photoexcitation of the dye-to-TiO_2 charge transfer (DTCT) band (Fig. 14.1b). In other words, the HOMO electron of the dye is directly injected into the CB of the TiO_2 electrode upon photoexcitation. Therefore, in type-II dye sensitizers, a strong new photoabsorption band is observed in a longer wavelength region due to the DTCT. The light-harvesting capability over the wide spectral range of sunlight is one of the most advantageous aspects in type-II DSSCs over type-I DSSCs because the direct electron injection enables the creation of the broad DTCT bands and the easing of restrictions on the LUMO energy levels of the dye sensitizers relative to the indirect electron injection in type-I DSSCs. Meanwhile, the photoabsorption coefficient of the DTCT band observed in type-II dye sensitizers is much lower than that of the photoabsorption band originating from the π–π^* transition in conventional dye sensitizers because of the lower overlap between the ground (molecular-localized) and the excited state (Ti-localized) wave functions.

In type-I DSSCs, there are electron-loss processes, which lower the photovoltaic performances: recombination of the electrons injected into the TiO_2 or ZnO electrode with the oxidized dyes (recombination process) or with the I_3^- ions at the TiO_2 or ZnO

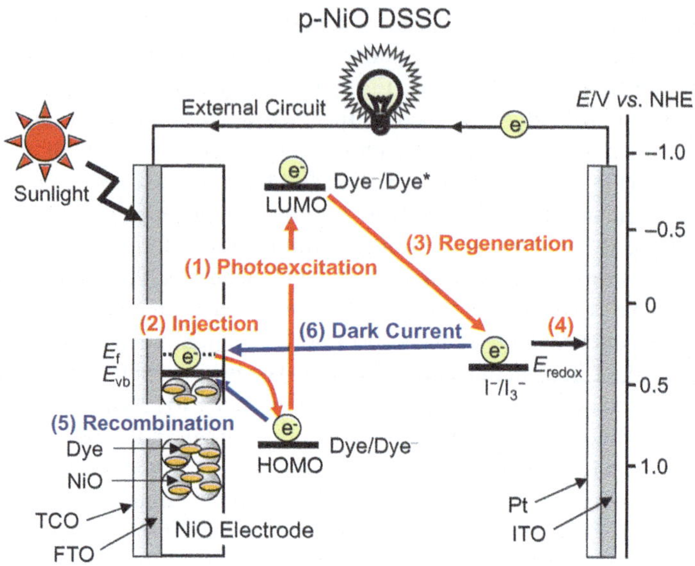

Fig. 14.2 Schematic representation of the construction and operational principles of p-NiO-based DSSC. The red and blue arrows represent the processes required for the generation of photocurrent and the recombination processes (electron-loss processes), respectively

surface (dark current process). The recombination process almost always competes with the regeneration of the oxidized dyes by the I^- ions. For the generation of greater photocurrents and photovoltages, the electron injection and dye regeneration processes must be kinetically more favorable than the electron-loss recombination and dark current processes.

p-NiO-based DSSCs are composed of a dye-adsorbed NiO electrode as a photocathode and a Pt-coated glass plate as a photoinert anode sandwiching an electrolyte solution containing an I_3^-/I^- ion couple as a redox mediator, as shown in Fig. 14.2, which is illuminated through the NiO film side (Hagfeldt et al. 2010; Mishra et al. 2009; Ning et al. 2010; Odobel et al. 2010; Yum et al. 2011; Nikolaou et al. 2017). First, the dye sensitizer absorbs sunlight, exciting an electron from its HOMO to LUMO. Then, the photogenerated hole is injected from the HOMO of the dye to the valence band (VB) of the NiO electrode. The resulting reduced dye is subsequently oxidized back to its original neutral state by electron transfer from the reduced dye to the I_3^- ion in the redox mediator. The injected holes diffuse in the NiO electrode, reach the FTO, and then, pass through the external circuit to the anode (Pt-coated glass). The I_3^- ion is regenerated by oxidation of the I^- ion at the anode through the hole donation, completing the circuit. In order to achieve efficient hole injection from the excited dye to the NiO VB, the HOMO energy level of the dye must be lower (more positive) than that of the VB (E_{vb}, 0.54 V vs. NHE) of the NiO electrode. An energy gap of over 0.6 V is necessary for the efficient hole injection, according to a few previous reports. Moreover, the LUMO energy level of the dye must be higher (more negative) than the redox potential of the I_3^-/I^- ion couple (0.4 V vs. NHE), in order to achieve efficient dye regeneration by electron transfer from the reduced dye to the I_3^- ion in the redox mediator. An energy gap of over 0.5 V is necessary for efficient regeneration, according to a few previous reports. Clearly, the LUMO energy level must be sufficiently lower than the CB energy level of the NiO electrode (E_{cb}, −3.0 V vs. NHE). However, there are also electron-loss processes, which lower the photovoltaic performances of DSSCs: recombination of the holes injected into the VB of the NiO electrode with the reduced dyes (geminate recombination) or with the I^- ions at the NiO surface (dark current). The geminate recombination process almost always competes with the oxidation of the reduced dyes by the I_3^- ions. Thus, for the generation of greater photocurrents and photovoltages, the hole injection and dye regeneration processes must be kinetically more favorable than the electron-loss geminate recombination and dark current processes.

Consequently, innovative ideas for molecular design of efficient dye sensitizers are required to develop high-performance DSSCs. In this context, a variety of ruthenium (Ru) complex dyes and organic dyes have been designed and developed. To date, the solar energy-to-electricity conversion yield (η) of up to 13% has been achieved in DSSCs (O'Regan and Grätzel 1991; Hagfeldt et al. 2010; Nazeeruddin et al. 1993, 1997, 1999, 2001, 2003; Pashaei et al. 2016; Robertson 2006; Chen et al. 2007; Xie and Guo 2007; Imahori et al. 2009; Ooyama and Harima 2009; Ning and Tian 2009; Mishra et al. 2009; Ning et al. 2010; Yen et al. 2012; Ooyama et al. 2012; Ooyama and Harima 2012; Odobel et al. 2010; Yum et al. 2011; Nikolaou et al. 2017; Li and Diau 2013; Ladomenou et al. 2014; Manfredi et al. 2014; Zhang and Cole 2015; Lee et al. 2015; Li et al. 2017; Urbani et al. 2014; Higashino 2015; Brogdon et al. 2018; Ji et al. 2018; Mathew et al. 2014; Yella et al. 2014).

14.2 Dye Sensitizers for N-TiO_2-Based Type-I DSSCs

Most of the conventional dye sensitizers developed for n-TiO_2-based type-I DSSCs have at least one carboxyl group (–COOH) as anchoring groups to bind to the TiO_2 electrode by the formation of bidentate linkage between the carboxyl group of the dye and the Brønsted acid site, hydroxyl group (Ti–OH), on the TiO_2 surface (Fig. 14.3). Ru complexes (O'Regan and Grätzel 1991; Hagfeldt et al. 2010; Nazeeruddin et al. 1993, 1997, 1999, 2001, 2003; Pashaei et al. 2016), porphyrin dyes (Imahori et al. 2009; Li and Diau 2013; Urbani et al. 2014; Higashino 2015; Brogdon et al. 2018; Ji et al. 2018; Mathew et al. 2014; Yella et al. 2014; Wang et al. 2005a; Tanaka et al. 2007; Hayashi et al. 2008; Eu et al. 2008; Campbell et al. 2007; Lee et al. 2009; Lu et al. 2009; Hsieh et al. 2010; Bessho et al. 2010; Yella et al. 2011; Higashino et al. 2015; Liu et al. 2016), phthalocyanine dyes (He et al. 2001; Aranyos et al. 2001; Nazeeruddin et al. 1999a; He et al. 2002; Reddy et al. 2007a, b; Giribabu et al. 2007; Mori et al. 2010; Kimura et al. 2012; Cid et al. 2007, 2009; Yum et al. 2008; García-Iglesias et al. 2011; Ragoussi et al. 2012; Ince et al. 2012; Yamamoto et al. 2016), and organic dyes (Hagfeldt et al. 2010; Robertson 2006; Chen et al. 2007; Xie and

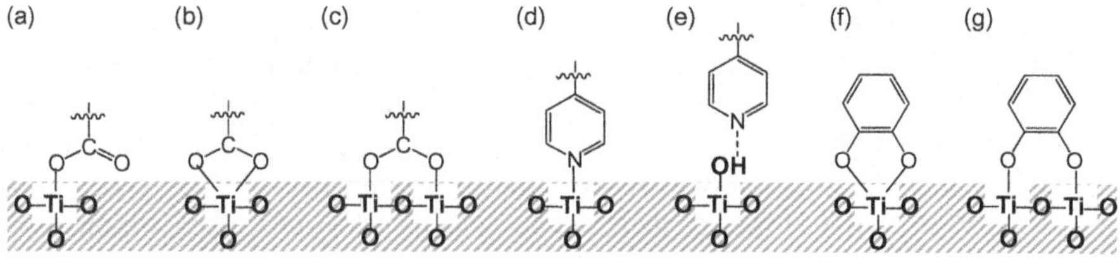

Fig. 14.3 Possible binding modes of carboxyl group, pyridyl group, and catechol on the TiO_2 surface: **a** ester linkage, **b** chelating linkage, **c** bidentate bridging linkage for carboxyl group, **d** coordinate bonding linkage, **e** hydrogen bonding linkage for pyridyl group, **f** bidentate mononuclear chelating linkage, and **g** bidentate dinuclear bridging linkage for catechol

Guo 2007; Ooyama and Harima 2009; Ning and Tian 2009; Mishra et al. 2009; Ning et al. 2010; Yen et al. 2012; Ooyama et al. 2012; Ooyama and Harima 2012) bearing carboxyl groups can inject electrons into the TiO_2 electrode, according to the type-I pathway. Sulfo group ($-SO_3H$), phosphonate group ($-H_2PO_3$), trialkoxysilyl group ($-Si(OR)_3$), and pyridyl group are also known as other anchoring groups. In order to further improve photovoltaic performances of n-TiO_2-based type-I DSSCs, much effort has been made toward the development of a variety of organic dye sensitizers in addition to Ru complex dyes. They possess the following characteristics: (i) good light-harvesting property over the wide spectral region of sunlight, (ii) high electron injection efficiency from the photoexcited dyes to the CB of the TiO_2 electrode, (iii) high surface loading and coverage of the dyes on the TiO_2 electrode, (iv) efficient reductive regeneration of the original neutral dyes by electron transfer from the I^- ions in the electrolyte containing the I_3^-/I^- redox couple, (v) suppressed charge recombination between the injected electrons in the CB of the TiO_2 electrode and the oxidized dyes or the I_3^- ions in the electrolyte, and (vi) suppressed dye aggregation on the TiO_2 surface.

Donor–acceptor–π-conjugated (D–π–A) dyes are expected to be one of the most promising classes of organic dye sensitizers bearing carboxyl groups for type-I DSSCs (Fig. 14.4), because the wide variety and facile modification of the chemical structures expand the possibility of molecular design, for example, the introduction of various substituents onto the dye skeletons, which allows easy control of not only their photophysical and electrochemical properties (HOMO and LUMO energy levels) but also their stereochemical structures. D–π–A dyes are constructed of an electron donor (D) such as diphenyl- and dialkylamino groups, an electron acceptor (A) with the anchoring role to bind to the TiO_2 electrode such as carboxylic acid, cyanoacrylic acid, and rhodanine-3-acetic acid moieties, and a π-conjugated bridge between D and A such as oligoene, coumarin, and oligothiophene. What is noteworthy about the structural features of D–π–A dyes is that the HOMO is localized over the π-conjugated system close to the D moiety and in contrast, the LUMO is localized over the A moiety (Ooyama and Harima 2009; Ning and Tian 2009; Ning et al. 2010; Ooyama and Harima 2012; Manfredi et al. 2014). Therefore, in terms of the concept of D–π–A structure, the extended π-conjugation and the enhanced

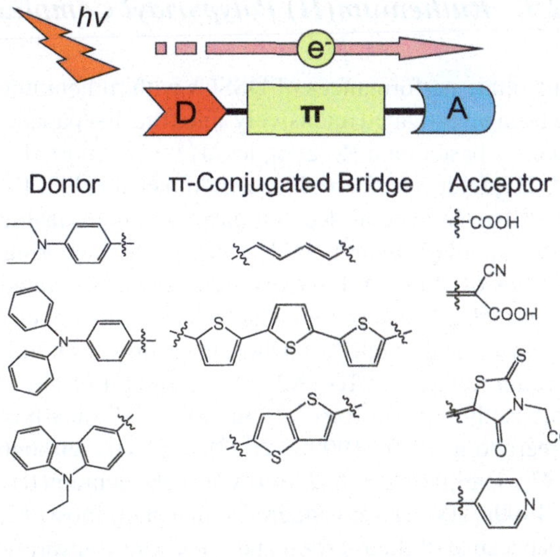

Fig. 14.4 Schematic representation of a D–π–A organic dye sensitizer for type-I TiO_2 or ZnO DSSCs

electron-donating and electron-accepting abilities of the D and A moieties, respectively, cause the decreased energy gap between the HOMO and LUMO. Moreover, the photoabsorption properties of D–π–A dyes are closely associated with intramolecular charge transfer (ICT) excitation from the D to A moieties in the dyes, which allows efficient electron transfer from the photoexcited dye to the CB of the TiO_2 electrode through the acceptor moiety (carboxyl group). Therefore, the D–π–A dye structure can facilitate the ICT and subsequent separation of the photogenerated charges between the D and A moieties in the excited dye, rapidly inject electrons into the CB of TiO_2, and effectively retard the charge recombination due to the spatial separation of the cationic charge on the D moiety apart from the TiO_2 surface with the injected electrons by the π-conjugated bridge. Indeed, the n-TiO_2-based type-I DSSCs with D–π–A dyes have achieved η values as high as 10–13%, which are comparable to those of Ru complex dyes.

Unfortunately, these D–π–A dyes have a major drawback of the aggregate formation by strong π–π stacking between dye molecules on the TiO_2 surfaces, which can reduce yields of electron injection from the dyes to the CB of TiO_2 due to intermolecular energy transfer between the stacked dye molecules. In order to prevent dye molecules on the TiO_2 surfaces from aggregating, coadsorption of deoxycholic acid (DCA) with D–π–A dyes has been used in a number of studies. Additionally, the introduction of sterically hindered substituents (bulky groups) including long alkyl chains and aromatic units into dye structures promises to suppress the dye aggregation efficiently, due to a disturbance of the π-stacking. Such hydrophobic substituents on the dye skeleton can also act as barriers to prevent the hydrophilic I_3^- ions from approaching the TiO_2 surface, thereby suppressing the charge recombination (dark current) of injected electrons with the I_3^- ions.

14.2.1 Ruthenium(II) Polypyridyl Complex Dyes

Photovoltaic performances of DSSCs with ruthenium(II) polypyridyl complex dyes have been investigated intensively since the first paper on high solar cell performances of DSSCs based on a Ru complex dye was reported in 1991 (O'Regan and Grätzel 1991; Hagfeldt et al. 2010; Nazeeruddin et al. 1993, 1997, 1999, 2001, 2003; Pashaei et al. 2016). In general, Ru complex dye photosensitizers are composed of a Ru(II) ion and ancillary ligands with at least one anchoring group. The photoabsorption properties of Ru complex dyes are closely associated with metal-to-ligand charge transfer (MLCT) excitation, which causes efficient electron transfer to the CB of the TiO_2 electrode through the anchoring group. *cis*-Dithiocyanobis(4,4′-dicarboxy-2,2′-bipyridine)ruthenium(II) (**N3**) (Nazeeruddin et al. 1993), ditetrabutylammonium-*cis*-bis(isothiocyanato)bis(2,2′-bipyridyl-4,4′-dicarboxylato) ruthenium(II) (**N719**) (Nazeeruddin et al. 1999b, 2003), and *tris*-tetrabutylammonium trithiocyanato-4,4′,4″-tricarboxy-2,2′:6′,2″-terpyridineruthenium(II) (**Black dye**) (Nazeeruddin et al. 1999a, 2001) are considered as one of the most efficient Ru(II) polypyridyl sensitizers because of their strong and wide-range absorption of visible light (Fig. 14.5). The TiO_2 DSSC based on **N3** exhibited a power-conversion efficiency (η) value of up to 10%, a short-circuit photocurrent density (J_{sc}) value of 18.2 mA cm^{-2}, an open-circuit photovoltage (V_{oc}) value of 720 mV, and a fill factor (*ff*) value of 0.73 under standard air mass (AM) 1.5 simulated sunlight (100 mW cm^{-2}). The incident photon-to-current conversion efficiency (IPCE) exceeded 80% in the visible region. The TiO_2 DSSC based on **N719**, which is commonly used as a standard sample among many research groups, achieved maximal η values over 11%.

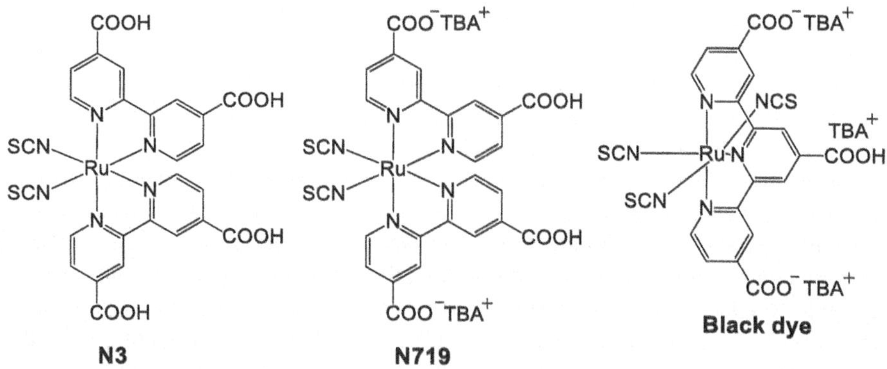

Fig. 14.5 Chemical structures of Ru(II) polypyridyl complex sensitizers **N3**, **N719**, and **Black dye** for DSSCs

14.2.2 Polyene Dyes

Polyene dyes with one of the simplest D–π–A structures composed of an amino moiety as an electron-donating part and a methine (–CH=CH–) unit connecting to a cyanoacrylic acid moiety as an electron-accepting part have been systematically studied by a number of researchers (Fig. 14.6) (Hara et al. 2003a, 2005a; Hagberg et al. 2006, 2007, 2008; Johansson et al. 2007; Tian et al. 2008; Kim et al. 2008a; Im et al. 2010). Expansion of the π-conjugated systems by the introduction of a methine unit caused a red-shift of the ICT photoabsorption band due to destabilization of the HOMO energy level and/or stabilization of the LUMO energy level, narrowing the HOMO–LUMO band gap. The photoabsorption maximum was shifted bathochromically by 30 nm per methine unit, which was possible up to four methine units. However, the expansion of the π-conjugated systems by a large number of methine units caused aggregation of the dyes on the TiO_2 surface by the formation of intermolecular π–π interactions between the dye molecules, which resulted in low yields of electron injection. A TiO_2 DSSC based on a polyene dye **TA-DM-CA** showed an η value of up to 9.67% with a J_{sc} value of 20.85 mA cm^{-2}, a V_{oc} value of 697 mV, and a *ff* value of 0.666 under AM 1.5 simulated sunlight (100 mW cm^{-2}). The IPCE value of 80% was achieved in the wavelength range of 440–510 nm (Im et al. 2010).

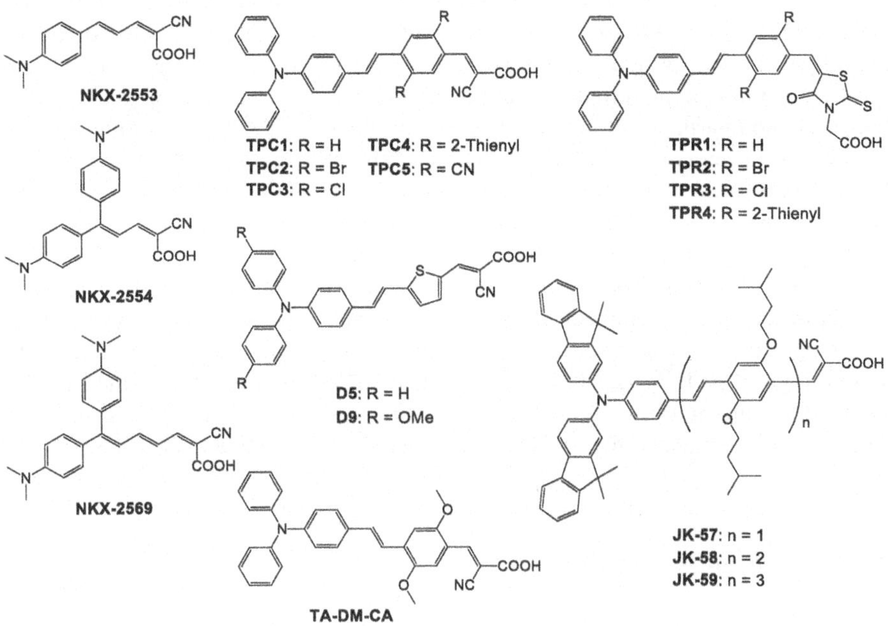

Fig. 14.6 Chemical structures of polyene dye sensitizers for TiO_2 DSSCs

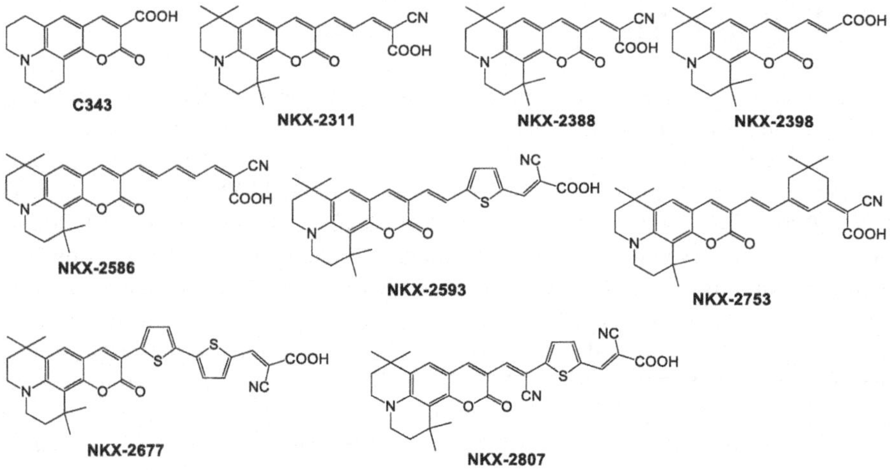

Fig. 14.7 Chemical structures of coumarin dye sensitizers for TiO_2 DSSCs

14.2.3 Coumarin Dyes

Coumarin dyes are one of the most promising classes of dye sensitizers in the D–π–A dyes and have been employed in DSSCs so far (Fig. 14.7) (Hara et al. 2001, 2003a, b, c, 2004, 2005b, c; Furube et al. 2005; Wang et al. 2005b, 2007a, b, 2008a). A series of new coumarin dye sensitizers have been designed to possess a different number of thiophene and/or methine units as a π-conjugated bridge between a coumarin skeleton with an amino group as an electron-donating part and a cyanoacrylic acid moiety as an electron-accepting part, based on the concept of the D–π–A structure. Their photovoltaic performances for DSSCs were systematically studied. The photoabsorption spectra of these new coumarin dyes were significantly shifted to a long-wavelength visible region, compared to the spectrum of the typical coumarin dye **C343**. The introduction of a methine unit (–CH=CH–) into the coumarin skeleton to connect both the cyano and carboxyl groups extended the π-conjugated system of the dye, thus resulting in a wide photoabsorption in the visible region. A TiO_2 DSSC based on **NKX-2677** exhibited an η value of 7.7% with a J_{sc} value of 14.3 mA cm^{-2}, a V_{oc} value of 730 mV, and a *ff* value of 0.74. The IPCE value reached nearly 80% in the visible region (Hara et al. 2003d).

14.2.4 Thiophene-Based Dyes

Thiophene- or oligothiophene-based dyes have also huge potential as dye sensitizers for DSSC use owing to their high chemical and environmental stabilities in addition to their strong photoabsorption properties and electronic tunability (Fig. 14.8) (Koumura et al. 2006; Wang et al. 2008b; Choi et al. 2007, 2008; Kim et al. 2006,

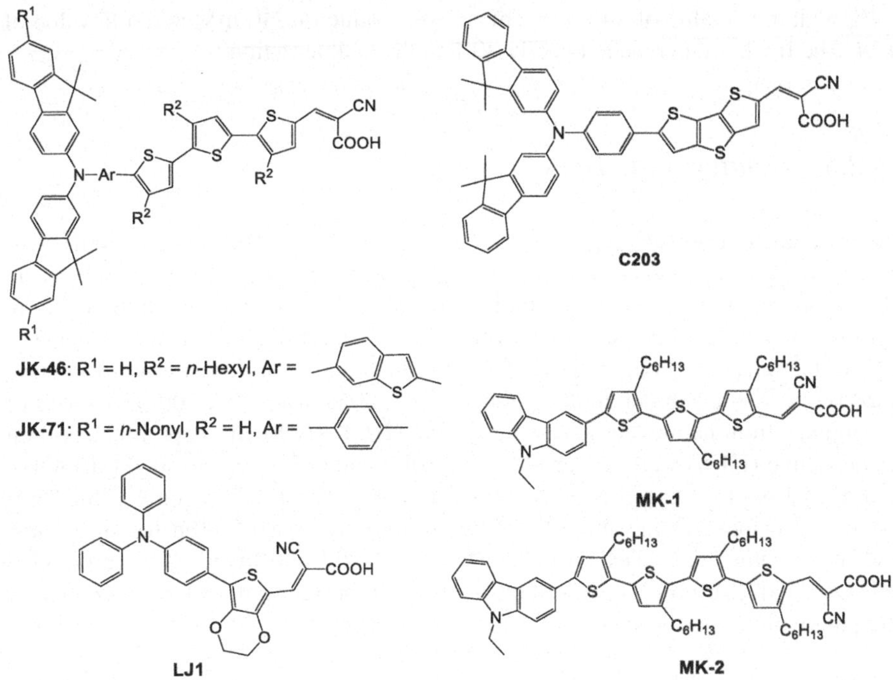

Fig. 14.8 Chemical structures of thiophene- or oligothiophene-based dye sensitizers for TiO_2 DSSCs

2008b, c; Liu et al. 2008; Thomas et al. 2008; Qin et al. 2008a). Expansion of the π-conjugated systems by the introduction of thiophene, dithienothiophene, and 3,4-ethylenedioxythiophene (EDOT) units largely shifted the ICT photoabsorption band to a long-wavelength region. In some D–π–A dyes, the red-shift of the photoabsorption maximum by 20 nm per thiophene unit was possible up to four thiophene units. However, the oligothiophene-based dyes were liable to form aggregation by strong intermolecular π–π interactions, which hampers photocurrent generation. In order to overcome this disadvantage, new oligothiophene-based dyes **MK-1** and **MK-2** were designed and synthesized (Koumura et al. 2006; Wang et al. 2008b). **MK-1** and **MK-2** are constructed of a carbazole moiety as an electron donor, an alkyl-functionalized oligothiophene as a π-conjugated linkage, and a cyanoacrylic acid unit as both electron acceptor and anchoring group. The charge recombination between the injected electrons with the dye cations and the I_3^- ions and the aggregation of the dye molecules on the TiO_2 surface can be diminished by the structural modification. What is noteworthy about the structural features of the dyes is that the steric hindrance of *n*-hexyl substituents on the oligothiophene linkage prevents the electrolyte (i.e., I_3^- ions) from approaching the TiO_2 surface and also the dye molecules from aggregating. The TiO_2 DSSC based on **MK-2** showed an η value of

7.7% with a J_{sc} value of 14.0 mA cm^{-2}, a V_{oc} value of 740 mV, and a *ff* value of 0.74. The IPCE value reached nearly 80% in the visible region.

14.2.5 Hemicyanine Dyes

Hemicyanine dye sensitizers, which are efficient in TiO_2 DSSCs, are cationic dyes based on D–π–A structures composed of a *p*-dialkylaminophenyl group as an electron-donating part, a cationic moiety including benzo- and naphthothiazolium, pyridinium, and indolium salts as a strong electron accepting part, and a methane (–CH=CH–) unit as a π-conjugated bridge between the electron-donating and accepting parts (Fig. 14.9) (Wang et al. 2000; Chen et al. 2005a; Yao et al. 2003). A series of benzothiazolium hamicyanine dye sensitizers (**HC1–HC5**) and naphthothiazolium hemicyanine (**HC6**) were designed to bear sulfonate (–SO_3^-), carboxyl (–COOH), and/or hydroxyl (–OH) groups as anchoring parts. The IPCE spectra of the TiO_2 DSSCs using these hemicyanine dyes were strongly affected by the anchoring groups, and the η values decreased in the order of carboxyl and hydroxyl > carboxyl > sulfonate and hydroxyl. This result indicates the importance of anchoring groups for the photovoltaic performances of DSSCs. The highest η value of 5.2% and IPCE

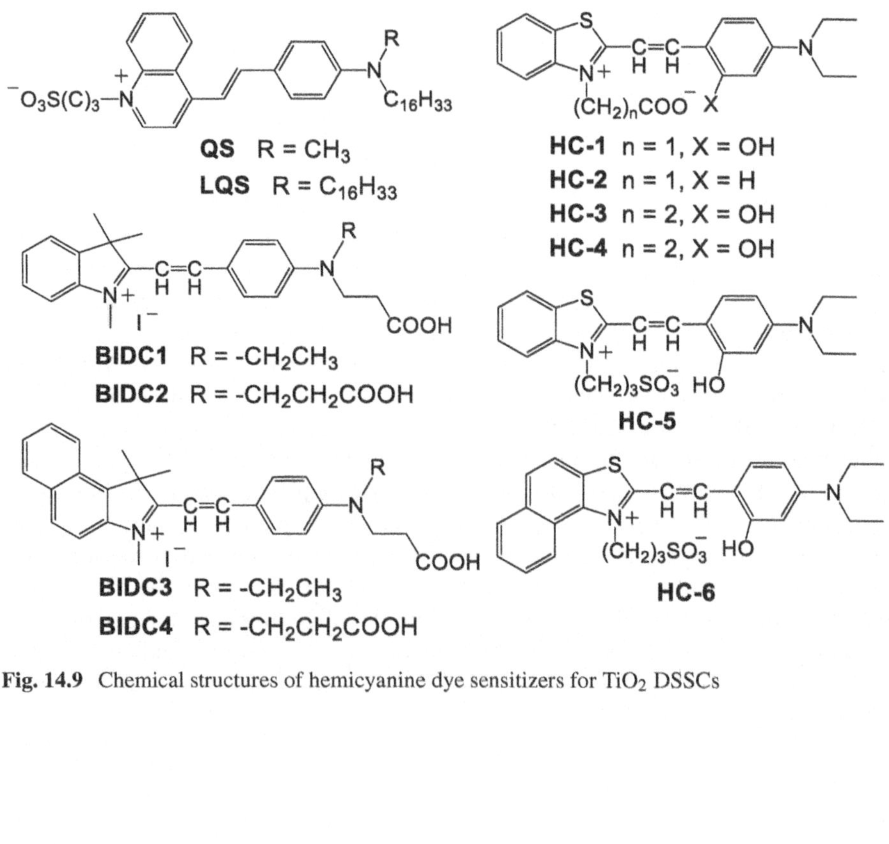

Fig. 14.9 Chemical structures of hemicyanine dye sensitizers for TiO_2 DSSCs

value of 74% were achieved in the TiO_2 DSSC based on **HC-1** under AM 1.5 simulated sunlight. **HC2–HC5** showed the η values of 3.5%, 4.4%, 3.1%, 2.0%, and 2.6%, respectively (Chen et al. 2005a).

14.2.6 Indoline Dyes

Indoline dye sensitizers are also one of the most promising candidates as organic sensitizers in TiO_2 DSSCs. The newly designed indoline dye sensitizers **DYE1–5**, **D102**, and **D205** (Fig. 14.10) are composed of three important parts: (1) rhodanine rings as an electron-accepting part at one end of the dye structures (the photoabsorption maximum is gradually red-shifted with increasing the number of rhodanine rings up to three), (2) an indoline moiety as a moderate electron-donating part, and (3) aromatic units on the indoline skeleton, which extends the π-conjugated systems of the dyes, causes a red-shift of the photoabsorption maximum and enhances the molar absorption coefficient (ε) (Horiuchi et al. 2003; Horiuchi et al. 2004a, b; Schmidt-Mende et al. 2005; Ito et al. 2006; Howie et al. 2008; Kuang et al. 2008; Ito et al. 2008). Furthermore, the modifications of the rhodanine ring with sterically hindered substituents represented by *n*-octyl groups and the indoline moiety with a 4-(2,2-diphenylvinyl) phenyl group are effective in suppressing dye aggregation and charge recombination (dark current) between the I_3^- ions and injected electrons in the TiO_2 electrode. The photovoltaic performance of the DSSC using **D205**, which has an *n*-octyl group, was superior to that of the DSSC using **DYE2**, which has an ethyl group, indicating that the improved V_{oc} and J_{sc} values of **D205** are attributed to the longer alkyl chain on the terminal rhodanine moiety. The TiO_2 DSSC based

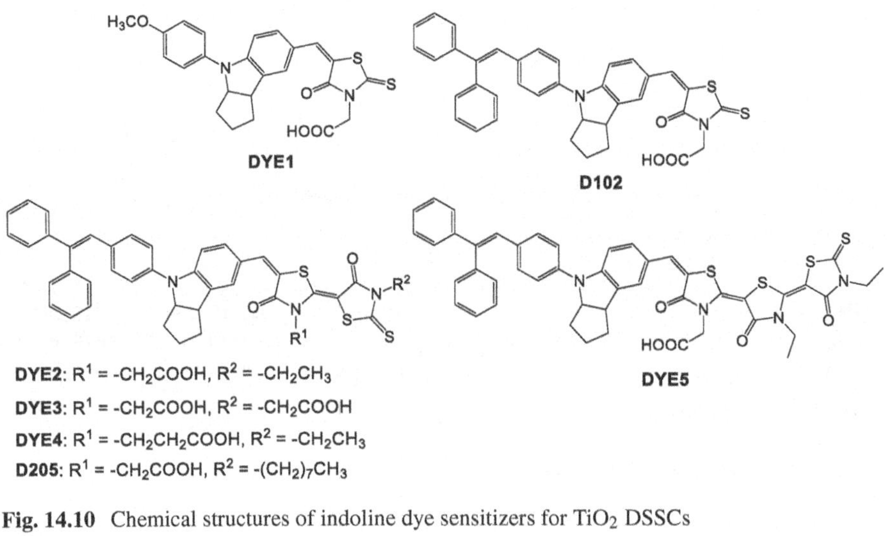

Fig. 14.10 Chemical structures of indoline dye sensitizers for TiO_2 DSSCs

on **D205** showed an η value of 9.5% with a J_{sc} value of 18.6 mA cm^{-2}, a V_{oc} value of 720 mV, and a *ff* value of 0.72. The IPCE value reached nearly 80% in the visible region (Kuang et al. 2008; Ito et al. 2008).

14.2.7 Heteropolycyclic Dyes

A series of D–π–A-type heteropolycyclic fluorescent dyes **OH1**, **OH2**, **OH4**, **OH7**, and **OH17** were designed as heteropolycyclic dye sensitizers to bear carboxyl groups at different positions on a chromophore skeleton (Fig. 14.11) (Ooyama et al. 2007a, b, 2009, 2011a). In **OH1**, the carboxyl group acts as not only an anchoring group to bind to the TiO_2 surface but also an electron acceptor. On the other hand, in **OH2**, **OH4**, and **OH7**, the carboxy groups act as anchoring groups, but the electron acceptor is the cyano group. **OH17** has two carboxyl groups, one of which is located at the same position as that of **OH1** and the other is located at the same position as that of **OH7**. The J_{sc} and η values of the DSSC using **OH2**, different parts of which play the roles as an anchoring group or an electron acceptor moiety, were comparable to those using **OH1** and higher than those using **OH3**, **OH4**, or **OH17**. Furthermore, **OH5–OH7**, **OH18**, and **OH19** with different lengths of non-conjugated alkyl chains and a carboxyl group at the end of each alkyl chain, were designed and synthesized. The J_{sc} and η values of the TiO_2 DSSCs based on **OH5–OH7** or **OH19** were comparable to or slightly smaller than those using **OH2**. In terms of the molecular structures, it is considered that the electron-accepting cyano group in **OH2**, **OH5–OH7**, and **OH19** can be located close to the TiO_2 surface without any

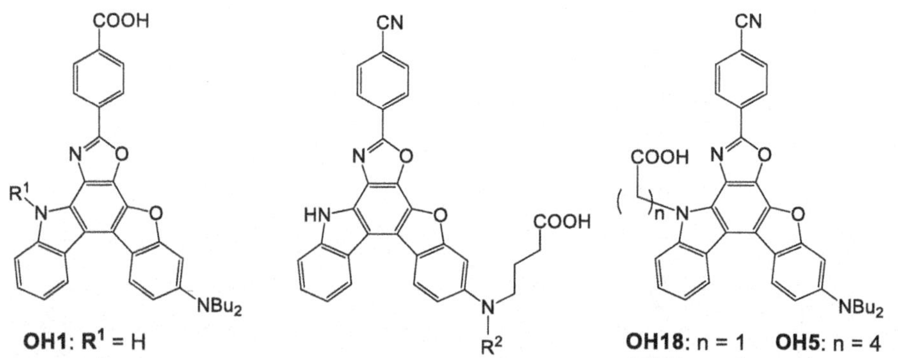

OH1: **R¹** = H
OH8: **R¹** = *n*-Butyl
OH9: **R¹** = Benzyl
OH10: **R¹** = 5-Nonyl
OH16: **R¹** = -$(CH_2)_3COOH$
OH17: **R¹** = -$(CH_2)_6COOH$

OH3: **R²** = *n*-Butyl
OH4: **R²** = -$(CH_2)_3COOH$

OH18: n = 1 **OH5**: n = 4
OH19: n = 2 **OH6**: n = 5
OH2: n = 3 **OH7**: n = 6

Fig. 14.11 Chemical structures of heteropolycyclic dye sensitizers for TiO_2 DSSCs

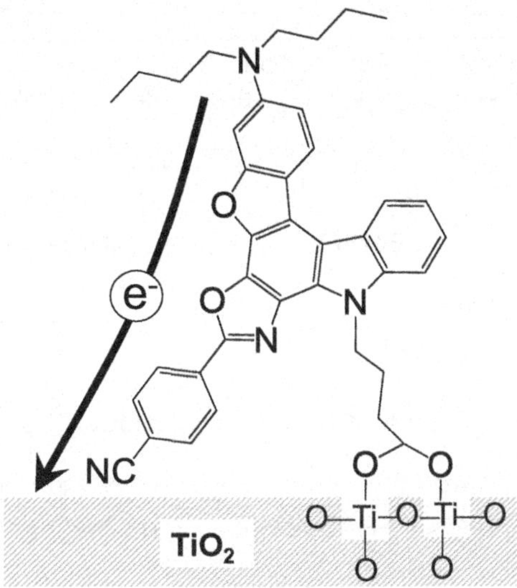

Fig. 14.12 Plausible configuration of a D–π–A dye sensitizer **OH**$_2$, which is functionally separated into an anchoring group (–COOH) and an electron-accepting group (–CN), on the TiO_2 surface

relation to the length of alkyl chains, probably due to their flexibility of alkyl chains (Fig. 14.12). Therefore, good electron communication between the dyes and the TiO_2 surface can be established, and **OH2**, **OH5–OH7**, and **OH19** can efficiently inject electrons into the CB of the TiO_2 electrode through the cyano group. Based on these results, a strong interaction between the acceptor moiety of dye sensitizers and the TiO_2 surface should be essential to efficient D–π–A dye sensitizers for TiO_2 DSSCs. In fact, several efficient D–π–A dye sensitizers with functionally separated parts of a strong electron-accepting moiety and an anchoring group have been designed and developed, according to the molecular design strategy (Ooyama et al. 2010a, b; Hao et al. 2009; Cong et al. 2012).

14.2.8 Xanthene Dyes

Xanthene dyes including eosin Y, mercurochrome, fluorescein, Rose Bengal, and rhodamine 6G are commercially available at low cost and were used as dye sensitizers in the early TiO_2 DSSCs (Fig. 14.13) (Sayama et al. 1998; Hara et al. 2000a; Hattori et al. 2004; Mann et al. 2008). However, the photovoltaic performances of the TiO_2 DSSCs based on the xanthene dyes were low due to the narrow photoabsorption bands and instabilities of the dyes. The η values of the TiO_2 DSSCs based on eosin Y or mercurochrome were 1.3% and 1.4%, respectively. A series of chalcogenoxanthylium dye sensitizers (**1-E**, **2-E**, and **4-Se**) were also synthesized. The H-aggregated dyes on the TiO_2 surface showed increased light-harvesting efficiencies and IPCE values. The IPCE values of the TiO_2 DSSCs based on these dyes were determined to be from 70% to 84% (Cong et al. 2012).

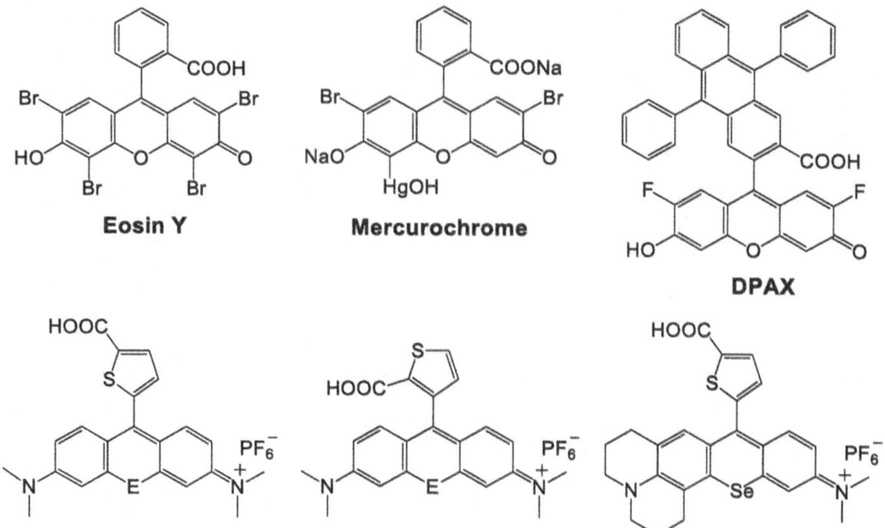

Fig. 14.13 Chemical structures of xanthene dye sensitizers for TiO_2 DSSCs

14.2.9 Perylene Dyes

Perylene dye sensitizers are also useful in DSSCs due to their outstanding chemical, thermal, and photochemical stabilities and high molar absorption coefficients in the long-wavelength region around 600 nm. The early studies on TiO_2 DSSCs using perylenedicarboxylic acid anhydride and its derivatives showed that the introduction of an amino group at the 9-position shifted the photoabsorption maxima bathochromically and improved the power conversion efficiency because of the electron-donating ability of the amino substitution (Ferrere and Gregg 2001, 2002). However, the adsorption of perylenedicarboxylic acid anhydride onto TiO_2 caused a blue-shift of the photoabsorption, because the anhydride group of the perylene underwent ring-opening to be two carboxylates and formed strong chemical interactions with the TiO_2 surface and effective electronic coupling.

A series of perylenedicarboxylic acid anhydrides **ID17**, **ID22**, **ID28**, **ID34**, and **ID35** were designed to show ICT characteristics due to the substituents of diphenylamino, *p*-diphenylaminophenyl, and *p*-(dimethylamino)phenylethynyl with different electron-donating abilities (Fig. 14.14) (Edvinsson et al. 2007). The experimental studies and time-dependent density functional theory calculations demonstrated that their ICT characteristics increased in the order of **ID17** < **ID35** < **ID22** < **ID34** and **ID28**. The η values of the TiO_2 DSSCs base on these dyes also increased in the order of **ID17** (1.4%) < **ID35** (2.0%) < **ID22** (2.4%) < **ID34** and **ID28** (3.2% and 3.9%, respectively), in accordance with the ICT characteristics. These electron donors work in the following two ways: (i) generally increasing the LUMO energy level and thereby, injecting electrons from the excited dye to the CB of TiO_2 more

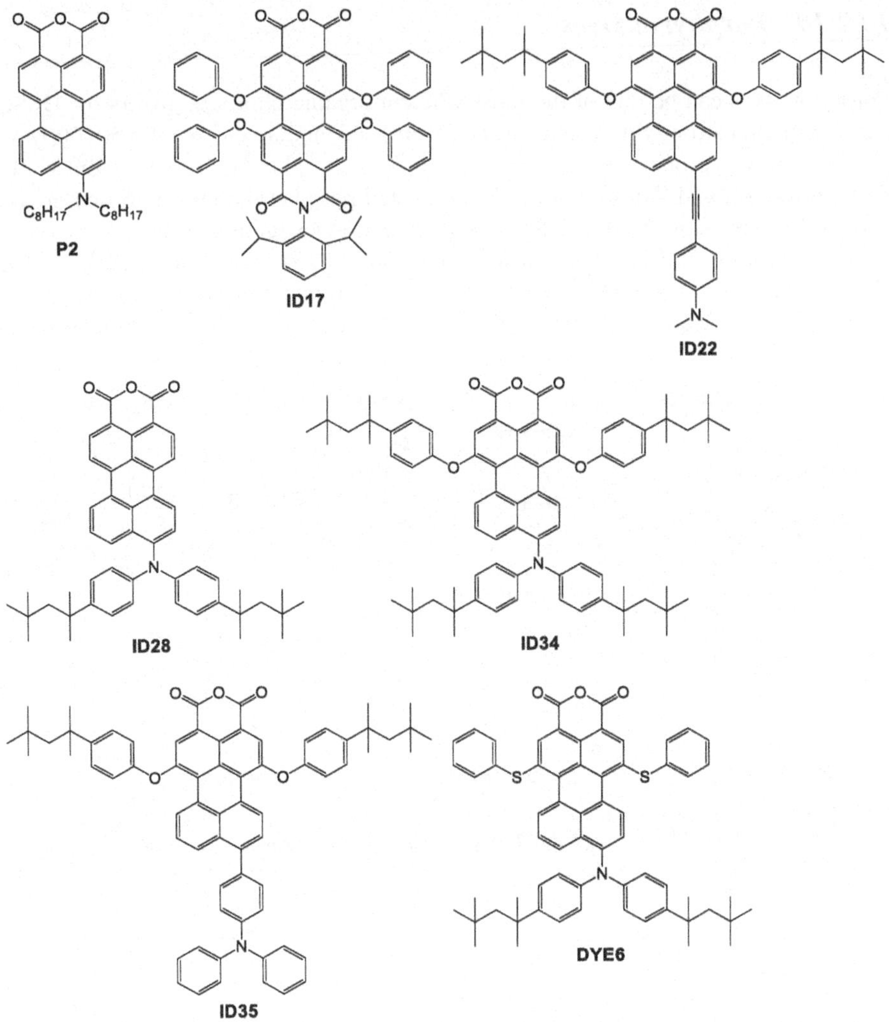

Fig. 14.14 Chemical structures of perylene dye sensitizers for TiO_2 DSSCs

efficiently and (ii) enhancing the intramolecular charge separation in the dye and retarding charge recombination of the injected electrons with dye cations. Moreover, a TiO_2 DSSC based on a perylene sensitizer **DYE6** bearing two thiophenol groups exhibited an η value of 6.8% with an IPCE value of 87%, a J_{sc} value of 12.60 mA cm^{-2}, a V_{oc} value of 728 mV, and a *ff* value of 0.74 under AM 1.5 simulated sunlight (Li et al. 2008).

14.2.10 Porphyrin Dyes

Porphyrin dyes can be one of the most efficient organic photosensitizers for DSSC because of their strong Soret band (400–450 nm) and moderate Q band (550–600 nm) photoabsorption properties. A series of zinc (Zn) tetraphenylporphyrins **Zn-3**, **Zn-5**, **Zn-8**, **Zn-11**, and **Zn-13** bearing a conjugated peripheral chain with one or two carboxyl groups were synthesized, and a detailed investigation of the photovoltaic performances of their DSSCs was reported (Fig. 14.15) (Wang et al. 2005a). The TiO_2 DSSC based on **Zn3** showed the following remarkable performance: J_{sc} = 13.5 mA cm^{-2}, V_{oc} = 560 mV, and η = 5.2%. A series of porphyrin sensitizers

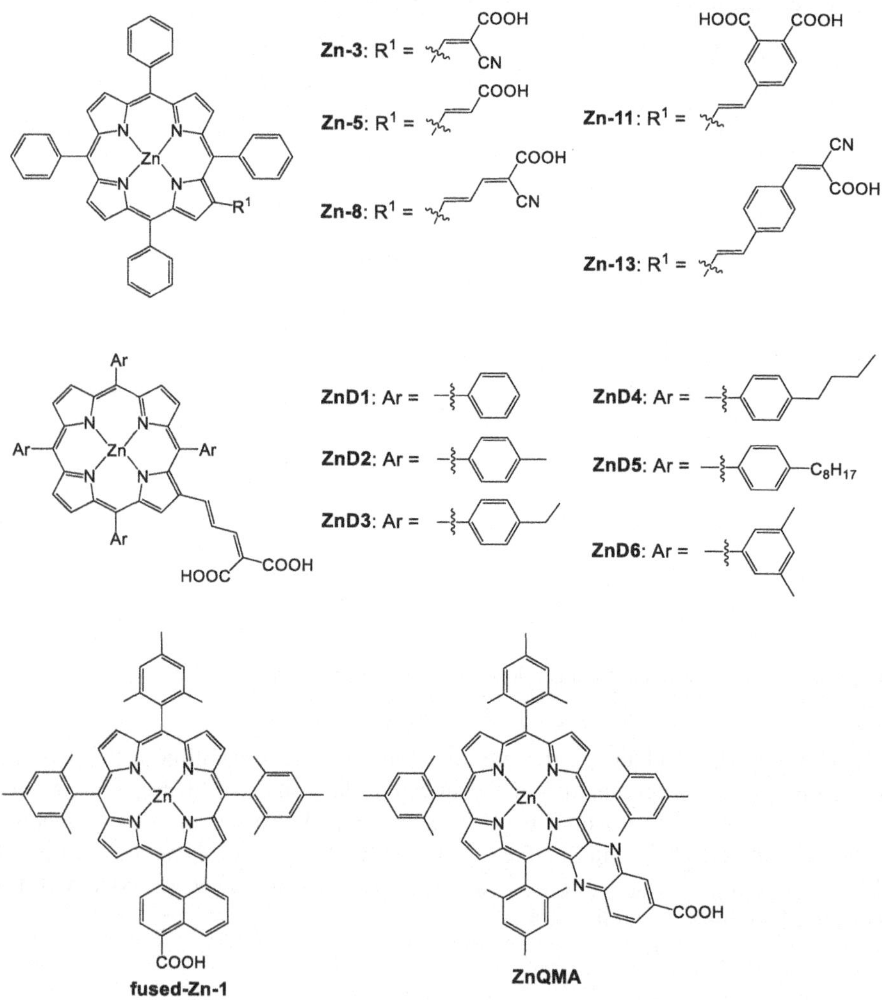

Fig. 14.15 Chemical structures of porphyrin dye sensitizers for TiO_2 DSSCs

ZnD1–6 (Campbell et al. 2007), *meso*- and β-naphthalene-fused porphyrincarboxylic acid **fused-Zn-1** (Tanaka et al. 2007; Hayashi et al. 2008), and β,β'-edge-fused **ZnQMA** with a quinoxaline moiety (Eu et al. 2008) were designed and developed as highly efficient push–pull-type Zn-porphyrin dye sensitizers for TiO_2 DSSCs. In these dyes, the aryl groups act as electron donors, and the carboxyl and malonic acid anchoring groups act as electron acceptors. The fused naphthyl moiety in the porphyrin ring of **fused-Zn-1** can shift the photoabsorption band (the Soret and Q bands) bathochromically. A carboxyl group is attached to the fused naphthyl moiety to facilitate electron injection from the photoexcited singlet state of **fused-Zn-1** to the CB of the TiO_2 electrode. Bulky mesityl groups are also introduced at the three *meso* positions of the porphyrin ring to suppress the aggregation of the **fused-Zn-1** molecules on the TiO_2 surface. The TiO_2 DSSC based on **fused-Zn-1** exhibited an η value of 4.1%, while those based on **ZnD1–6** showed η values of >5%. Therefore, it has been shown that the porphyrin π-system strongly interacts with the olefin-bridged electron acceptor substituted at the β-pyrrolic position of the porphyrin ring and that the malonic acid group binds strongly to the TiO_2 surface with a consequent improvement in the electronic coupling of the dye.

A new series of D–π–A porphyrin dyes with good photoabsorption in the red/near-infrared (NIR) region have been successfully designed and developed to improve the performances of DSSCs. A series of D–π–A Zn-porphyrin dyes **YD2**, **YD6**, and **YD7**, which possess a diarylamino group as an electron-donating part and an ethynylbenzoic acid moiety as an electron-accepting part linked through a porphyrin chromophore as a π-bridge, showed good photoabsorption properties (bathochromic shift and broadening of the Soret and Q bands) in the red/NIR region (~700 nm) of the solar spectrum and unidirectional flow of electrons in the photoexcited states (Fig. 14.16) (Higashino 2015; Brogdon et al. 2018; Ji et al. 2018; Mathew et al. 2014; Yella et al. 2014; Wang et al. 2005a; Tanaka et al. 2007; Hayashi et al. 2008; Eu et al. 2008; Campbell et al. 2007; Lee et al. 2009; Lu et al. 2009; Hsieh et al. 2010; Bessho et al. 2010). The TiO_2 DSSC based on **YD2** exhibited an η value of 11% with a J_{sc} value of 18.6 mA cm^{-2}, a V_{oc} value of 770 mV, a *ff* value of 0.764, and a maximum IPCE value of 90% at 675 nm under standard AM 1.5 simulated sunlight (100 mW cm^{-2}), which are comparable to those of ruthenium complexes (Bessho et al. 2010). Moreover, a D–π–A Zn porphyrin dye **YD2-*o*-C8** was designed and developed as a judiciously tailored variant of **YD2** by modification with long alkyl chains to attenuate interfacial back electron transfer reaction and dye aggregation (Yella et al. 2011). **YD2-*o*-C8** absorbed light over the whole visible range. The DSSC based on **YD2-*o*-C8** with the $Co^{II/III}$*tris*(2,2′-bipyridine) electrolyte exhibited impressively high IPCE values over the whole visible wavelength range, exceeding 80% from 450 to 680 nm except a narrow dip around 530 nm. High η values of nearly 11.9% have been achieved in the DSSC, which showed a V_{oc} value of 965 mV, a J_{sc} value of 17.3 mA cm^{-2}, and a *ff* value of 0.71 under standard AM 1.5 sunlight at 100 mW cm^{-2} intensity. This result is attributed to the fact that the molecular design of **YD2-*o*-C8**, particularly, the presence of the four octyloxy groups, greatly retards the rate of interfacial back electron transfer from the CB of the TiO_2 electrode to

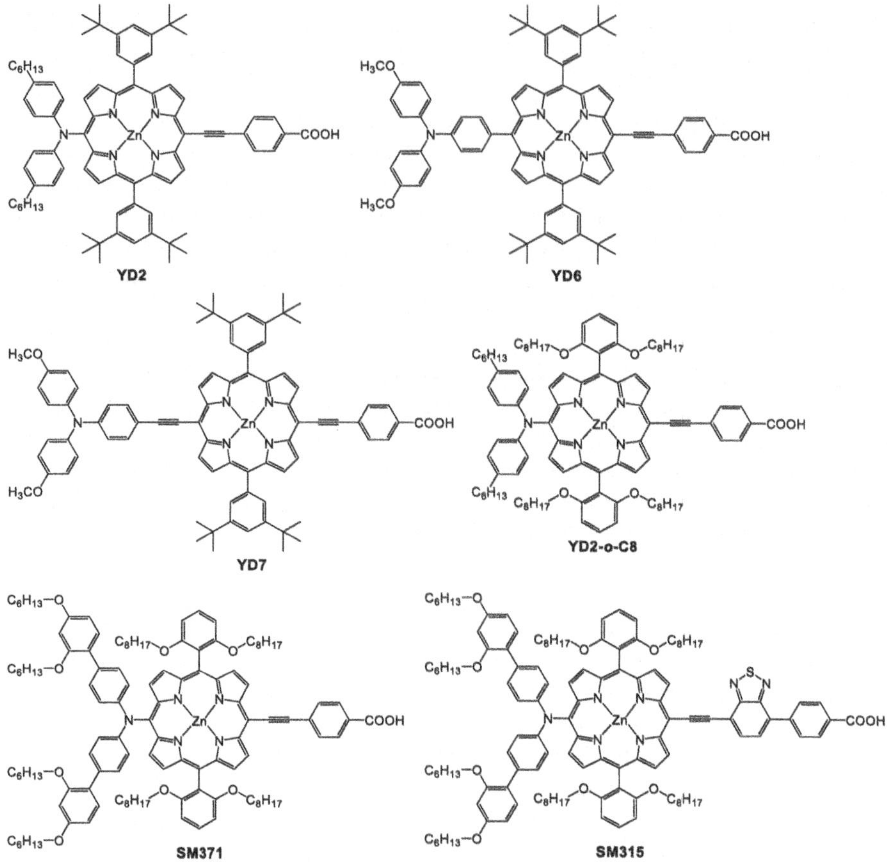

Fig. 14.16 Chemical structures of D–π–A porphyrin dye sensitizers for TiO_2 DSSCs

the oxidized cobalt mediator (Co^{III}*tris*(bipyridyl)), which enabled the strikingly high photovoltages of nearly 1.0 V.

Grätzel and co-workers further reported D–π–A porphyrins **SM371** with a bulky bis(2′,4′-bis(hexyloxy)-[1,1′-biphenyl]-4-yl)amine donor and a 4-ethynylbenzoic acid group and **SM315** with a more strongly electron-withdrawing benzothiadiazole (BTD) unit, which was used to improve the light-harvesting property and electrolyte compatibility (Fig. 14.16) (Mathew et al. 2014; Yella et al. 2014). Indeed, the introduction of the BTD electron-accepting unit significantly broadened the Soret and Q bands of **SM315** compared to those of **SM371** and improved the light-harvesting property in both green (500–600 nm) and red (up to 800 nm) regions. A broader and split Soret band (λ_{max} = 440 and 454 nm), a broader photoabsorption in the 500–600 nm region, and a more intense and red-shifted Q band (λ_{max} = 668 nm, ε = 53000 M^{-1} cm^{-1}) were observed in the spectrum of **SM315** in comparison with those observed in the spectrum of **SM371** (Q band: λ_{max} = 646 nm, ε = 29000 M^{-1}

cm^{-1}). The DSSC based on **SM371** with the $Co^{II/III}$*tris*(2,2′-bipyridine) electrolyte yielded a high V_{oc} value of 960 mV, a J_{sc} value of 15.9 mA cm^{-2}, and a *ff* value of 0.79 and achieved an overall η value of 12.0%, which is slightly improved compared to a similar porphyrin sensitizer **YD2-*o*-C8**. The DSSC based on **SM315** with the $Co^{II/III}$*tris*(2,2′-bipyridine) electrolyte showed a high J_{sc} value of 18.1 mA cm^{-2}, a *ff* value of 0.78, and an η value of 13%, outperforming **SM371** in spite of a slightly lower V_{oc} value of 910 mV, as a result of the improvement of the light-harvesting property in the visible and NIR regions. In fact, the DSSC using **SM315** demonstrated impressively high IPCE values across the whole visible wavelength region, 80% from 450 to 750 nm, demonstrating the utility of the BTD-functionalized anchoring moiety to improve light-harvesting properties in the visible and NIR regions.

14.2.11 Merocyanine Dyes

Merocyanine dyes can form J- and H-aggregates on the TiO_2 electrode surface. The photoabsorption of J-aggregates is observed in a longer wavelength region while that of the H-aggregates observed in a shorter wavelength region, with reference to the photoabsorption of the monomeric form. The relationship between dye aggregation on the TiO_2 surface and the photoelectrochemical properties has been investigated using a series of benzothiazole merocyanine dyes (**Mc[m, n]**) with alkyl chains (carbon number: $m = 2$, 5, 10, 18, and 20) attached to the benzothiazole ring and different lengths of methylene chains (carbon number: $n = 1$, 3, and 5) between a carboxyl group at the end and the dye chromophore (Fig. 14.17) (Khazraji et al. 1999; Sayama et al. 2000, 2002, 2003). **Mc[m, n]** formed J-aggregates on the TiO_2 surface. The formation of J-aggregates occurred more readily with increasing the length of alkyl chains (m). In the DSSCs using **Mc[m, n]**, the η and IPCE values also increased with increasing the alkyl chain length (m). In addition, the IPCE values significantly decreased with increasing the methylene length (n). It was suggested that a shorter distance between the dye chromophore and the TiO_2 surface generated a larger photocurrent because of the efficient charge transfer. The highest η value of

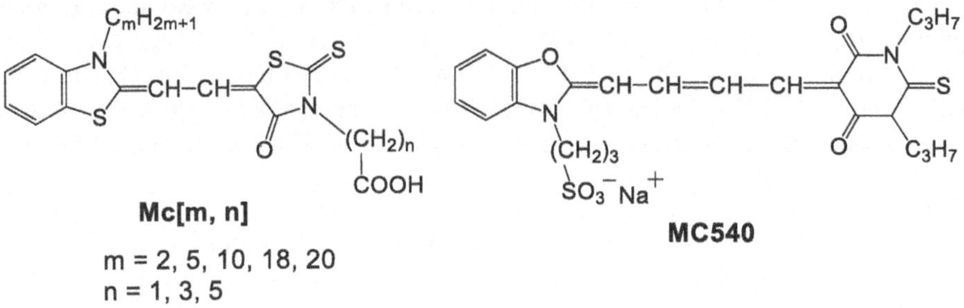

Fig. 14.17 Chemical structures of merocyanine dye sensitizers for TiO_2 DSSCs

4.5% was obtained in the TiO_2 DSSC based on **Mc[18, 1]** dye. Since the distance between the chromophore and the TiO_2 surface affected the IPCE values, the control of configuration and aggregation of merocyanine dyes is significantly important to improve photovoltaic performances of DSSCs based on merocyanine dyes.

14.2.12 Squaraine Dyes

A successful strategy for improvement of the light-harvesting efficiency of organic dyes would be the use of NIR dye skeletons, which have good photoabsorption properties in the red/NIR region of the solar spectrum because the red/NIR radiation (600–1000 nm) accounts for about 25% of the total energy of sunlight reaching the earth surface, while the visible radiation (350–700 nm) accounts for about 45%. Squaraine dyes are known to absorb the red/NIR radiation well and therefore, have been widely studied as dye sensitizers for DSSCs. However, these cationic squaraine dyes have low LUMO energy levels and also reduced electron injection yields from the excited dyes to the CB of the TiO_2 electrode due to the formation of H-aggregates (Qin et al. 2013; Chen et al. 2005b; Li et al. 2005, 2010a; Tatay et al. 2007; Burke et al. 2007, 2008; Alex et al. 2005; Yum et al. 2007; Park et al. 2012; Kuster et al. 2010; Choi et al. 2010; Shi et al. 2011; Paek et al. 2011; Bisht et al. 2017; Jradi et al. 2015; Alagumalai et al. 2016).

A series of symmetric (**Sq1–4**) and asymmetric (**Sq5–7**) squaraine dyes were designed and synthesized, and the photovoltaic performances of the TiO_2 DSSCs based on these dyes were investigated (Fig. 14.18) (Alex et al. 2005). The asymmetric squaraine dyes showed much higher performances as sensitizers than those of the symmetric squaraine dyes. Considering the molecular design of D–π–A dye sensitizers, an asymmetric squaraine dye **SQ1** with a carboxylic acid group directly attached to the chromophore was developed. **SQ1** also showed much higher performances than those of symmetric squaraine dyes, as a result of unidirectional flow of electrons toward the indoline moiety with the carboxylic acid anchoring group in the photoexcited state and consequently enhanced charge separation (Yum et al. 2007; Park et al. 2012). The TiO_2 DSSC using **SQ1** achieved an η value of up to 4.5%. Additionally, an asymmetric bis(indoline) squaraine dye **YR6** showed good photoabsorption property in the red/NIR region (~700 nm) of the solar spectrum and unidirectional flow of electrons in the photoexcited state (Shi et al. 2011). The TiO_2 DSSC based on **YR6** yielded a J_{sc} value of 14.8 mA cm^{-2}, a V_{oc} value of 642 mV, and a *ff* value of 0.71, corresponding to an η value of 6.74%, under standard AM 1.5 simulated sunlight.

Paek et al. developed a new series of stable asymmetric squaraine NIR sensitizers **JK-216** and **JK-217** with both thiophenyl pyrrolyl and indolium groups. **JK-216** and **JK-217** exhibited an photoabsorption band at 669 nm ($\varepsilon = 93400$ M^{-1} cm^{-1}) and 672 nm ($\varepsilon = 77900$ M^{-1} cm^{-1}), respectively (Paek et al. 2011). The DSSC using **JK-216** yielded a J_{sc} value of 13.92 mA cm^{-2}, a V_{oc} value of 610 mV, and a *ff* value of 0.74, corresponding to an η value of 6.29%, while that using **JK-217** yielded an

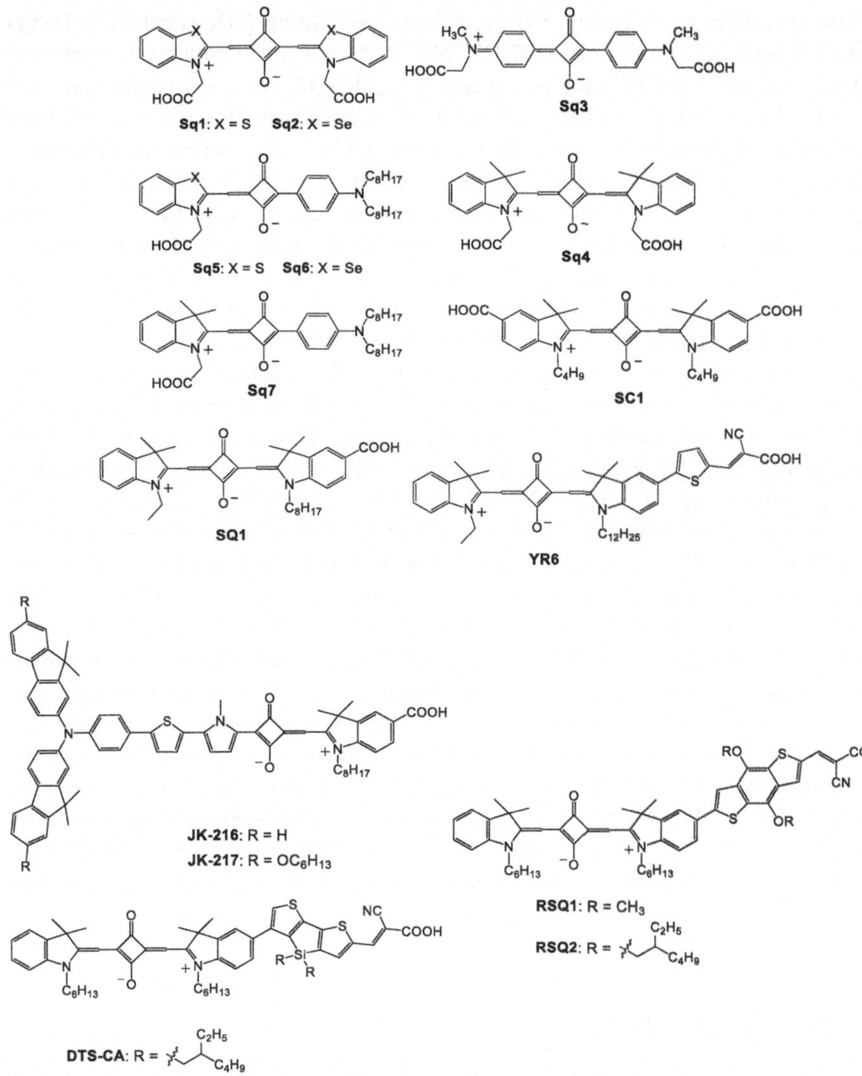

Fig. 14.18 Chemical structures of squaraine dye sensitizers for TiO_2 DSSCs

η value of 5.54% with a J_{sc} value of 13.73 mA cm^{-2}, a V_{oc} value of 580 mV, and a *ff* value of 0.70. The IPCE value of **JK-216** exceeded 55% in a broad spectral region from 550 to 780 nm and showed a profound dip at 550 nm. The DSSC based on **JK-217** showed an excellent stability under light irradiation at 60 °C for 1000 h. These dyes successfully suppressed their aggregation and also stabilized long-term device performances by the incorporation of a nonplanar dimethyl fluoreneaniline part and an asymmetric squaraine moiety into the structures.

Nithyanandhan et al. developed asymmetric squaraine NIR sensitizers **RSQ1** and **RSQ2** with benzodithiophene (BDT) as a π-spacer and cyanoacrylic acid as an electron acceptor. The BDT moieties of **RSQ1** and **RSQ2** were modified with methyl and 2-ethylhexyl groups, respectively, in order to investigate the effect of alkylated π-spacers on the dye aggregation on the TiO_2 surface and charge recombination reactions at TiO_2/dye/electrolyte interfaces (Bisht et al. 2017). The DSSC using **RSQ2** worked well with higher V_{oc} and *ff* values than those of the DSSC using **RSQ1** in spite of a comparable J_{sc} value. The DSSC using **RSQ2** exhibited an η value of 6.72% with a J_{sc} value of 18.53 mA cm^{-2}, a V_{oc} value of 538 mV, and a *ff* value of 0.67, without any coadsorbent. The higher V_{oc} and *ff* values using **RSQ2** were attributed to attenuation of the charge recombination.

El-Sayed et al. studied a series of squaraine NIR sensitizers with asymmetric push–pull D–π–A structures composed of cyanoacetic acid or cyanophosphonic acid as an electron acceptor and anchoring group and a π-bridge of 4-hexyl-4H-dithieno[3,2-*b*:2′,3′-*d*]pyrrole (DTP), dithieno[3,2-b:2′,3′-d]thiophene (DTT), or 4,4-bis(2-ethylhexyl)-4Hsilolo[3,2-b:4,5-b′]dithiophene (DTS) (Jradi et al. 2015). The squaraine dyes with a DTS bridge achieved the highest efficiencies, which is attributed to high J_{sc} and V_{oc} values due to suppressed dye aggregation and slow charge recombination, respectively; the η values were 5.0% and 8.9% in the DSSCs based on **DTS-CA** with cyanoacetic acid and **DTS-PA** with cyanophosphonic acid, respectively. The charge injection efficiencies from the squaraine dyes with cyanoacetic acid to the TiO_2 electrode were higher than those from the squaraine dyes with cyanophosphonic acid. On the other hand, it was possible that the squaraine dyes with cyanophosphonic acid enabled slower charge recombination between the injected electrons into the TiO_2 electrode and the I_3^- ions in the electrolyte than the squaraine dyes with cyanoacetic acid. Moreover, this study suggested the possibility that the two out-of-plane bulky (2-ethylhexyl) substituents on the DTS π-bridge decreased the dye aggregation and charge recombination, consequently, increased the J_{sc} and V_{oc} values of the DSSCs and enhanced the DSSC performances.

14.2.13 Cyanine Dyes

Cyanine dyes have significantly high absorption extinction coefficients (ca. 10^5 M^{-1} cm^{-1}), with intense and broad photoabsorption bands in the visible and NIR regions. Cyanine dyes adsorbed on the TiO_2 surface can form J- or H-aggregates. To establish a guidance for designing efficient cyanine dye sensitizers, the photovoltaic performances of the TiO_2 DSSCs based on various cyanine dyes **Cn-D** ($n = 0$–3), **Cn-N** ($n = 1$–3), **C1'**, **C1-D#**, **C2-Ns**, and **C2-ND** have been systematically investigated (Fig. 14.19) (Sayama et al. 2001). The IPCE values of the DSSCs based on these cyanine dyes increased with decreasing the distance between the dye skeleton and the TiO_2 surface because of efficient charge transfer between them. Additionally, although the increase of the number of methine unit (n) shifted the photoabsorption maxim to longer wavelength regions by ca. 100 nm per methine unit, it suppressed

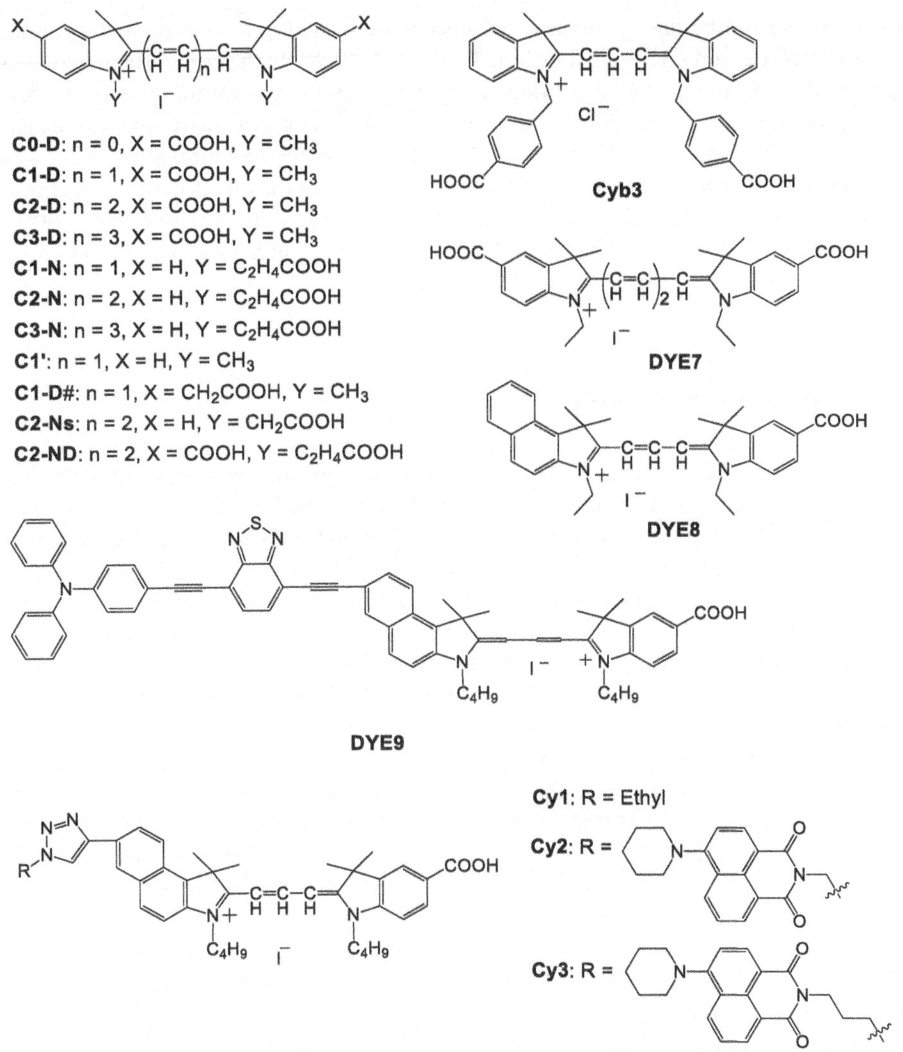

Fig. 14.19 Chemical structures of cyanine dye sensitizers for TiO_2 DSSCs

electron transfer from the excited dyes to the CB of TiO_2 and dramatically decreased the IPCE values.

Symmetric cyanine dyes **Cyb3** and **DYE7** and an asymmetric cyanine dye **DYE8** were also designed and synthesized as sensitizers for TiO_2 DSSCs (Fig. 14.19) (Guo et al. 2005; Chen et al. 2005c). The cyanine dyes aggregated on the TiO_2 surface and efficiently harvested light. The TiO_2 DSSCs using **Cyb3, DYE7**, and **DYE8** yielded η values of 1.01%, 1.3%, and 2.9%, respectively. The electron density was distributed on the indole moiety with a carboxylic acid anchoring group in the asymmetric cyanine dye unlike the symmetric cyanine dyes. In addition, asymmetric cyanine dyes **Cy1**, **Cy2**, and **Cy3** were developed (Wu et al. 2008). In the comparison between **Cy2** and **Cy3**, the J_{sc} and V_{oc} values increased with increasing the length of alkyl chains between the naphthalimide and triazole groups; consequently, the η value also increased. The long alkyl chain is considered to prevent the recombination processes of injected electrons with the dye cations and I_3^- ions in the electrolyte. In the three cyanine dyes, the TiO_2 DSSC using **Cy3** yielded the highest η value of 4.80%. Furthermore, a TiO_2 DSSC based on **DYE9** with a triphenylamine moiety exhibited an η value of 7.62% (Ma et al. 2008).

14.2.14 Phthalocyanine Dyes

Phthalocyanine (Pc) dyes are also one of the most promising classes of NIR dye sensitizers because of their intense photoabsorption in the UV/blue (Soret band) and red/NIR (Q band at around 700 nm) spectral regions, in addition to their electrochemical, photochemical, and thermal stabilities. However, the power conversion efficiencies of TiO_2 DSSCs based on Pcs are much lower than those using other organic dye sensitizers. The low efficiencies are attributed to aggregation of Pc dyes, lack of electron transfer directionality in the excited states, and poor electronic coupling between their LUMO and the CB of TiO_2. Therefore, to minimize the formation of dye aggregates and also improve the solubility in organic solvents, bulky substituents such as *tert*-butyl groups have been introduced into the phthalocyanine skeleton (He et al. 2001, 2002; Aranyos et al. 2001; Nazeeruddin et al. 1999b). An asymmetric phthalocyaninatozinc sensitizer (**PCH001**) with three *tert*-butyl and two carboxylic acid groups as "push" and "pull" groups, respectively, achieved efficient electron transfer from the excited dye to the TiO_2 electrode because of the good electronic coupling between the dye's LUMO and the TiO_2 CB energy levels and suppressed the formation of dye aggregates (Fig. 14.20) (Reddy et al. 2007a; Giribabu et al. 2007). In addition, other efficient push–pull-type ZnPc dye sensitizers with bulky electron-donating groups have been developed, including **PcS** series with 2,6-diphenylphenoxy or 2,6-diphenyl-4-methoxyphenoxy groups (Mori et al. 2010; Kimura et al. 2012), **TT** series with *tert*-butyl or 2,6-diphenylphenoxy groups (Reddy et al. 2007b; Cid et al. 2007, 2009; Yum et al. 2008; García-Iglesias et al. 2011; Ragoussi et al. 2012), **ZnPc1** with triarylamine-terminated bisthiophene groups (Ince et al. 2012), and **ZnPc2** with hexylbisthiophene groups (Yamamoto et al. 2016) (Fig. 14.20), and the photovoltaic performances of their DSSCs were evaluated. The TiO_2 DSSC based on **TT1** with three *tert*-butyl and one carboxylic acid groups directly attached to the Pc skeleton yielded an η value of 3.52%. In highly sterically hindered asymmetric ZnPcs (**PcS6** and **PcS15**) with 2,6-diphenylphenoxy or 2,6-diphenyl-4-methoxyphenoxy groups, respectively, the formation of dye aggregates onto the TiO_2 surface was inhibited almost completely. The η values of the TiO_2 DSSCs using **PcS6** or **PcS15** are 4.6% and 5.3%, respectively (Mori et al. 2010; Kimura et al. 2012). Furthermore, a TiO_2 DSSC based on a low-symmetry

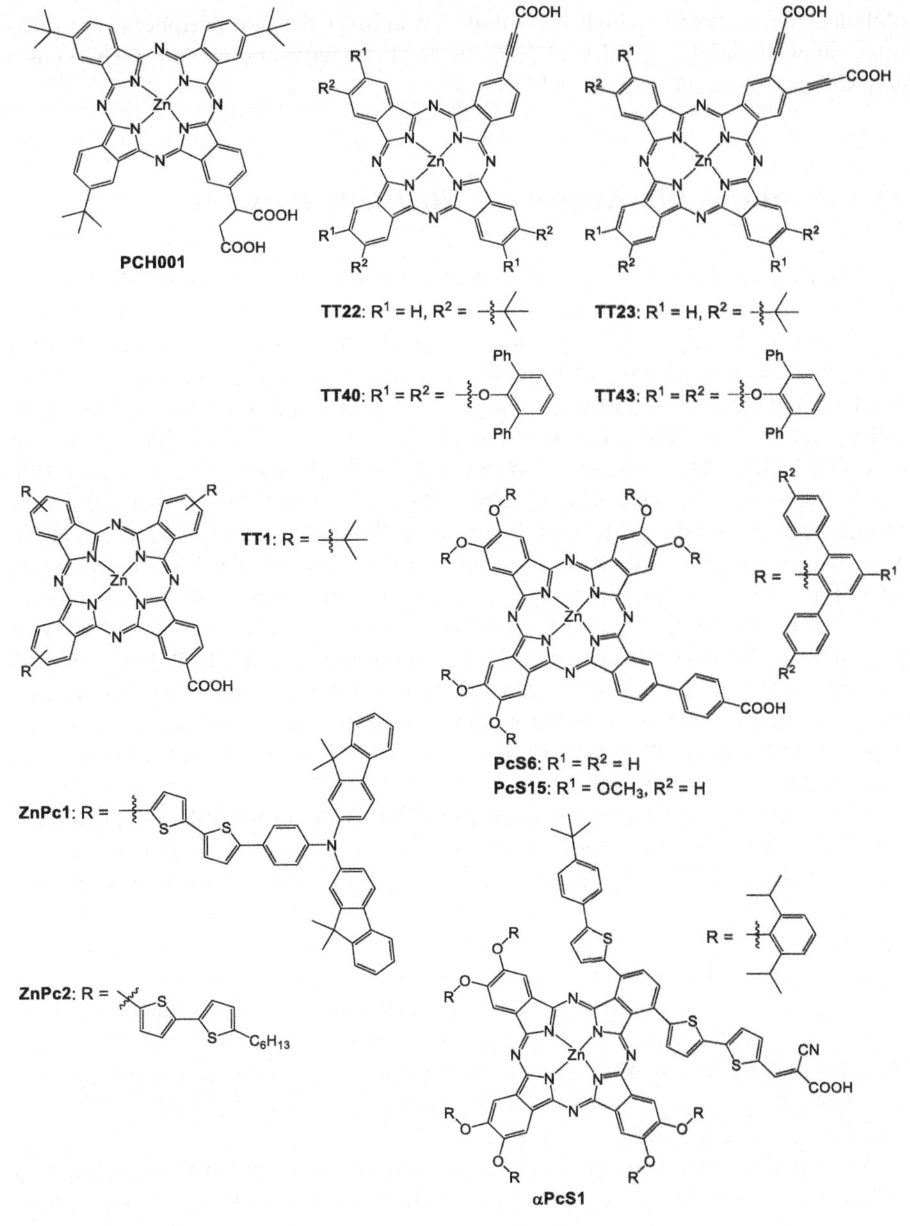

Fig. 14.20 Chemical structures of phthalocyanine dye sensitizers for TiO_2 DSSCs

phthalocyanine **αPcS1**, which has thiophene units at the non-peripheral (α) positions, showed a higher η value of 5.5% using chenodeoxycholic acid (CDCA) as a coadsorbent (Yamamoto et al. 2016).

14.2.15 Boron Dipyrromethene (BODIPY) Dyes

Boron dipyrromethene (BODIPY) dyes are fascinating sensitizers for DSSCs because they have high chemical stability, photostability, and solubility in organic solvents, in addition to large photoabsorption coefficients in the visible and NIR regions, strong fluorescence property, and ease of electrochemical modification with electron-donating and electron-accepting groups onto the BODIPY core (Singh and Gayathri 2014; Hattori et al. 2005; Kumaresan et al. 2009; Erten-Ela et al. 2008; Kolemen et al. 2010, 2011; Mao et al. 2012; Wang et al. 2012). A series of D–π–A BODIPY dye sensitizers **DYE10**, **DYE11**, and **PS-1–PS-4**, which have two triphenylamine or trialkoxyphenyl groups as electron donors at the 3- and 5-positions and a carboxylic acid or cyanoacrylic acid group as an electron acceptor at the 8-position on the BODIPY core, were developed (Fig. 14.21) (Erten-Ela et al. 2008; Kolemen et al. 2010, 2011). The BODIPY dyes **PS-1–PS-4** without two methyl groups at the 1- and 7-positions were designed, because orthogonality between the BODIPY core and the carboxyphenyl moiety, an electron-accepting and anchoring group, can reduce the π-conjugation and electron flow from the electron donor to the electron acceptor. The TiO_2 DSSC using **PS-1** yielded an η value of 2.46% with a J_{sc} of 9.17 mA cm^{-2}, a V_{oc} of 430 mV, and a *ff* of 0.62.

A series of D–π–A BODIPY dye sensitizers **B1–B3** and **PB1–PB4**, which possess triphenylamine or 1H-phenanthro[9,10-*d*]imidazole as an electron donor at the 6-position and cyanoacrylic acid as an electron acceptor at the 2-position, were also designed and synthesized (Fig. 14.21) (Mao et al. 2012; Wang et al. 2012). The introduction of a substituent at 8-position on the BODIPY core can suppress dye aggregation and charge recombination between the injected electrons in TiO_2 and the I_3^- ions, increasing in both J_{sc} and V_{OC} values. In fact, the η value of the TiO_2 DSSC based on **B3** with a hexyl group at the 8-position is 1.83%, which is higher than that of **B1** with a methyl group at the 8-position ($\eta = 1.28\%$). The TiO_2 DSSC based on **PB4** yielded an η value of 2.26% with a J_{sc} value of 5.10 mA cm^{-2}, a V_{oc} value of 610 mV, and a *ff* value of 0.72.

Kubo et al. designed and developed a series of boron–dibenzopyrromethene dyes **DB1–3** with thienyl-cyanoacrylic acid units (Kubo et al. 2014). The dyes exhibited intense absorption bands in a long wavelength region with λ_{max} values of 647 nm (ε = 157000 M^{-1} cm^{-1}) for **DB1** containing two anchoring units, 660 nm (ε = 109000 M^{-1} cm^{-1}) for regioisomer **DB2**, and 644 nm (ε = 139000 M^{-1} cm^{-1}) for **DB3** containing one anchoring unit, although these dyes have no strong electron-donating groups, such as arylamines. The excellent light-harvesting abilities of the dyes allowed high J_{sc} values in their DSSCs. As a result, the DSSC with butterfly-shaped **DB2** showed a J_{sc} value of 19.02 mA cm^{-2}, a V_{oc} value of 540 mV, and

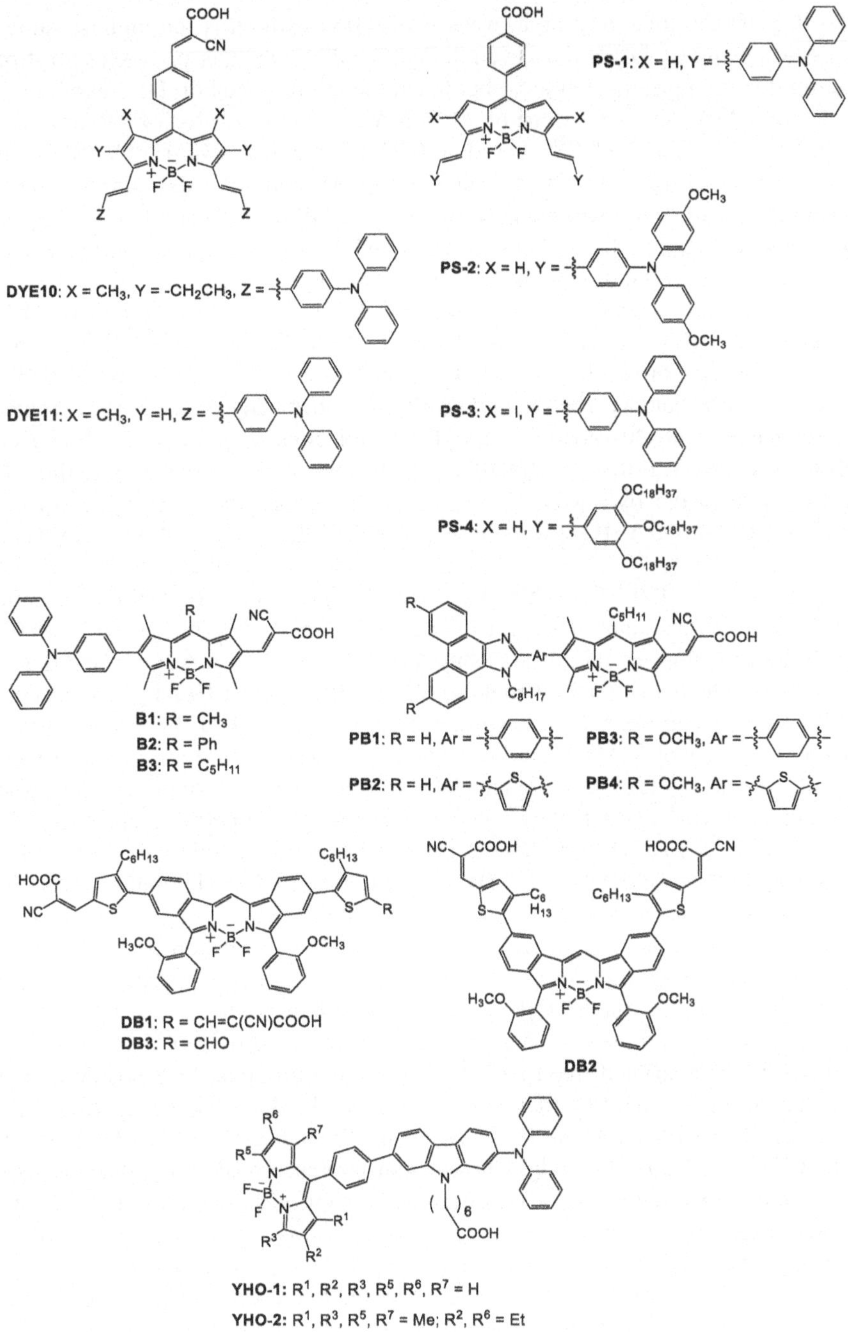

Fig. 14.21 Chemical structures of boron dipyrromethene (BODIPY) dye sensitizers for TiO_2 DSSCs

a *ff* value of 0.59, resulting in an η value of 6.06%, which is the highest value in the DSSCs using BODIPY dyes even though the dye sensitizer possesses no strong electron-donating groups. On the other hand, the DSSC based on **DB1** exhibited an η value of 5.24% with a J_{sc} value of 19.24 mA cm^{-2}, a V_{oc} value of 530 mV, and a *ff* value of 0.514, and the DSSC based on **DB3** exhibited an η value of 5.48% with a J_{sc} value of 17.21 mA cm^{-2}, a V_{oc} value of 550 mV, and a *ff* value of 0.579. These results indicate that photovoltaic performance of a DSSC is affected by the position of the thienyl-cyanoacrylic acid unit on the isoindole ring and by the resultant change in the orientation of the dye on the TiO_2 surface.

Furthermore, Ooyama et al. designed and developed a nonalkylated BODIPY dye sensitizer **YHO-1** and a hexa-alkylated BODIPY dye sensitizer **YHO-2**, which have a diphenylamino-carbazole moiety as an electron donor at the 8-position on the BODIPY core and a carboxyhexyl group as an anchoring group on the carbazole ring (Ooyama et al. 2013a). In terms of the molecular structures of **YHO-1** and **YHO-2**, it is assumed that the BODIPY core locates in close proximity to the TiO_2 surface, when the two dyes were adsorbed on the TiO_2 surface. The strong absorption bands of **YHO-1** and **YHO-2** were observed at 501 nm ($\varepsilon = 54800$ M^{-1} cm^{-1}) and 524 nm ($\varepsilon = 50600$ M^{-1} cm^{-1}), respectively, which are attributed to the $S_0 \rightarrow S_1$ transition of the BODIPY core. **YHO-2** ($\lambda_{em} = 535$ nm, $\Phi_f = 0.62$) exhibited a much higher fluorescence quantum yield (Φ_f) than **YHO-1** ($\lambda_{em} = 532$ nm, $\Phi_f = 0.06$). The J_{sc} and η values of the DSSC using **YHO-2** were 2.33 mA cm^{-2} and 0.65%, respectively, which are greater than those of the DSSC using **YHO-1** (1.24 mA cm^{-2} and 0.34%, respectively). The maximum IPCE value of **YHO-2** exceeded 30% at 530 nm, while that of **YHO-1** was 11% at 517 nm. These results indicate that **YHO-2** can inject electrons efficiently from the BODIPY core to the CB of the TiO_2 electrode in comparison with **YHO-1**. Based on the spectroscopic properties and photovoltaic performances of **YHO-1** and **YHO-2**, the relatively low photovoltaic performance of the nonalkylated BODIPY dye **YHO-1** is attributed to the radiationless relaxation of the photoexcited state due to the free rotation of the phenyl group at the 8-position on the BODIPY core and/or due to the fluorescence deactivation from the twisted intramolecular charge transfer (TICT) state, and the resulting suppression of direct electron injection from the BODIPY core to the CB of the TiO_2 electrode. In contrast, the higher photovoltaic performance of the hexa-alkylated BODIPY dye **YHO-2** is attributed to the restricted free rotation of the phenyl group at the 8-position by the hexa-alkylation onto the BODIPY core. Consequently, this study demonstrated that fluorescent BODIPY dyes can inject electrons efficiently from the BODIPY core to the CB of the TiO_2 electrode, whereas non-fluorescent BODIPY dyes generate low photocurrents due to the radiationless relaxation of the photoexcited state. On the other hand, the V_{oc} values of BODIPY dye sensitizers are lower than those of conventional organic dye sensitizers. The relatively low V_{oc} values (400–430 mV) of the DSSCs using **YHO-1** or **YHO-2** are attributed to the fast charge recombination between the injected electrons in the CB of TiO_2 and the I_3^- ions in the electrolyte due to the approach of the I_3^- ions to the TiO_2 surface by the electrostatic interactions between the BODIPY core and the I_3^- ion.

14.2.16 Phenothiazine and Phenoxazine Dyes

The phenothiazine skeleton with the electron-rich nitrogen and sulfur atoms possesses a stronger electron-donating character than other amines such as triphenylamine, tetrahydroquinoline, carbazole, iminodibenzyl, and many other *N*-heterocycles. The introduction of an electron-accepting group such as cyanoacrylic acid on the phenothiazine skeleton can produce a D–π–A system, which exhibits an ICT band in the visible light region (Fig. 14.22) (Huang et al. 2016; Wan et al. 2012; Baheti et al. 2015; Wu et al. 2010; Yang et al. 2012). Indeed, Thomas et al. designed and synthesized two types of fluorene-based sensitizers with a phenothiazine donor and a cyanoacrylic acid acceptor, that is, the phenothiazine unit was functionalized at either the nitrogen or the 3-position carbon to obtain a T-shape dye **4a** and a rod-like dye **7a**, respectively (Baheti et al. 2015). It is evident that the phenothiazine and fluorene units are perpendicular to one another in the T-shaped dye **4a**, while they are in a nearly coplanar arrangement in the rod-like dye **7a**. The rod-like dye **7a** showed a broader and red-shifted photoabsorption band (λ_{max} =

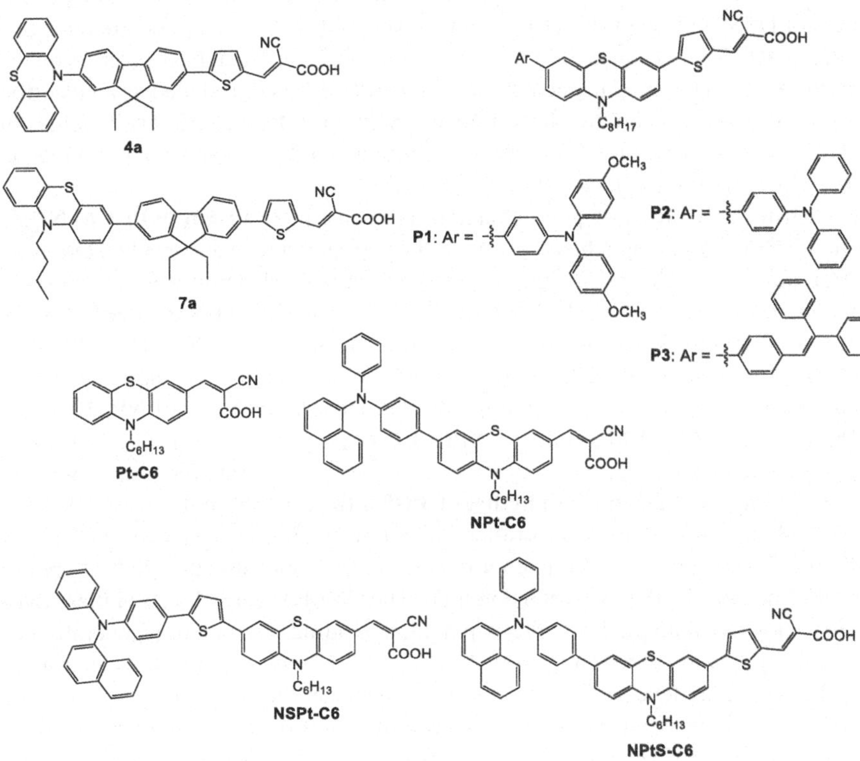

Fig. 14.22 Chemical structures of phenothiazine dye sensitizers for TiO_2 DSSCs

441 nm) compared to the T-shaped dye **4a** (λ_{max} = 429 nm) due to the longer conjugation channel in **7a**. This bathochromic shift in **7a** probably points to the extended π-conjugation into the phenothiazine ring via the 3-position. In other words, the electronic coupling between the phenothiazine-donor and fluorine-acceptor moieties is poor in the T-shaped dye **4a** due to the perpendicular orientation of phenothiazine and the conjugation pathway. The DSSC based on the T-shaped dye **4a** exhibited a high IPCE value (>80%) compared to that of the rod-like dye **7a**. Although **7a** showed a broader and intense photoabsorption band in the visible region than **4a**, the photovoltaic performance (J_{sc} = 4.54 mA cm^{-2}, V_{oc} = 595 mV, *ff* = 0.63, and η = 1.7%) of the DSSC based on **7a** is lower than that (J_{sc} = 9.36 mA cm^{-2}, V_{oc} = 690 mV, *ff* = 0.69, and η = 4.47%) of **4a**. The low photovoltaic performance of the DSSC based on the rod-shaped dye **7a** is probably due to the dye aggregation on the TiO_2 surface, the resulting enhancement of the excited state quenching pathways, and the reduction of the electron injection efficiency, because rod-shaped dyes are more likely to aggregate on the TiO_2 surface compared with T-shaped dyes.

To date, much effort has been put into the development of various phenothiazine dye sensitizers for DSSCs, and there has been a gradual accumulation of information about the relationship between the chemical structures and photovoltaic performances of DSSCs. Tian et al. designed and developed D–D–π–A phenothiazine dye sensitizers **P1–3** with cyanoacetic acid as an electron acceptor at the 3-position on the phenothiazine skeleton. A triphenylamine substituent was introduced as an electron donor at the 7-position on the phenothiazine skeleton in **P1** and **P2**. The η values of the TiO_2 DSSCs based on **P1** and **P2** are 3.78% and 4.41%, respectively (Wu et al. 2010), which are higher than that of **P3** (2.48%).

Furthermore, a series of phenothiazine dye sensitizers **Pt-C6**, **NPt-C6**, **NPtS-C6**, and **NSPt-C6** having a hexyl group at the nitrogen and cyanoacrylic acid as an electron acceptor at the 3-position on the phenothiazine skeleton were synthesized by Chow et al. (Yang et al. 2012). A triarylamine substituent was attached as an electron donor at the 7-position on the phenothiazine skeleton in **NPt-C6**, **NPtS-C6**, and **NSPt-C6**. In addition, **NPtS-C6** and **NSPt-C6** have a thiophene unit at different positions on the phenothiazine skeleton. The photoabsorption bands of **NPt-C6** (λ_{max} = 453 nm, ε = 18200 M^{-1} cm^{-1}), **NPtS-C6** (λ_{max} = 463 nm, ε = 22700 M^{-1} cm^{-1}), and **NSPt-C6** (λ_{max} = 442 nm, ε = 19300 M^{-1} cm^{-1}) are red-shifted with an increase in the ε value compared to that of **Pt-C6** (λ_{max} = 438 nm, ε = 14900 M^{-1} cm^{-1}) without a triarylamine substituent. The introduction of an electron-donating triarylamine substituent on the phenothiazine skeleton elevated the HOMO energy level, thus decreasing the oxidation potential. The HOMO energy levels of these dyes are more positive than the I_3^-/I^- redox potential, leading to efficient regeneration of the oxidized dyes by electron transfer from the I_3^-/I^- ion couple in the electrolyte. The LUMO energy levels of these dyes are higher than the energy level (E_{cb}) of the CB of TiO_2, which suggests that the electron injection from the photoexcited dyes into the CB of the TiO_2 electrode is thermodynamically feasible. The DSSC based on **Pt-C6** showed an η value of 4.49% with a J_{sc} value of 10.73 mA cm^{-2}, a V_{oc} value of 650 mV, and a *ff* value of 0.64. It is worth mentioning that inserting a thiophene unit between the phenothiazine moiety and the acceptor group reduces the photovoltaic

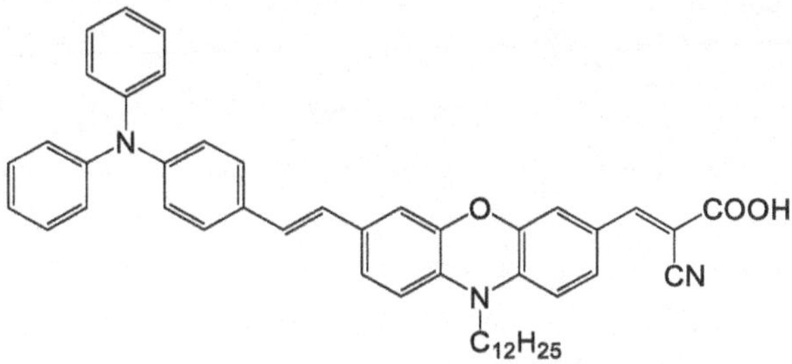

TH305

Fig. 14.23 Chemical structure of a phenoxazine dye sensitizer **TH305** for TiO_2 DSSCs

performance of a DSSC, whereas inserting it between the phenothiazine moiety and the triarylamine donor group enhances the photovoltaic performance. The η values of the DSSCs based on the dyes with or without a thiophene moiety are in the order of **NSPt-C6** (6.22%) > **NPt-C6** (5.60%) > **NPtS-C6** (2.40%). Consequently, the best performance was observed in the DSSC based on **NSPt-C6**, which exhibited a J_{sc} value of 14.42 mA cm^{-2}, a V_{oc} value of 690 mV, and a *ff* value of 0.63, corresponding to an η value of 6.22%.

On the other hand, a few phenoxazine dyes have been synthesized as dye sensitizers for TiO_2 DSSCs (Fig. 14.23). The η value of a TiO_2 DSSC based on **TH305** is 7.7% (Tian et al. 2009).

14.2.17 Cyclopentadithiophene and Dithienosilole (DTS) Dyes

Compared to cyclopentadithiophene, the extended π-conjugated system of dithienosilole (DTS) effectively causes a large red-shift of the photoabsorption band (ICT band) as a result of destabilization of the HOMO energy level and/or stabilization of the LUMO energy level, which narrows the HOMO–LUMO band gap (Fig. 14.24). Consequently, the HOMO–LUMO band gap of DTS is smaller than that of 2,2'-bithiophene because of the low LUMO energy level of DTS. The extended π-conjugation is enabled by σ*–π* conjugation through the interaction of a silicon σ* orbital and a bithiophene π* orbital. A TiO_2 DSSC using **C219** with a DTS moiety reached an impressive efficiency over 10% (Zeng et al. 2010).

14.2.18 Polymeric Dyes

Conjugated polymers such as polypyrrole, polyaniline, poly(*p*-phenyleneethynylene), and polythiophene have been investigated for TiO_2 DSSCs by evaluation of their photovoltaic performances (Hong et al. 2002; Hao et al.

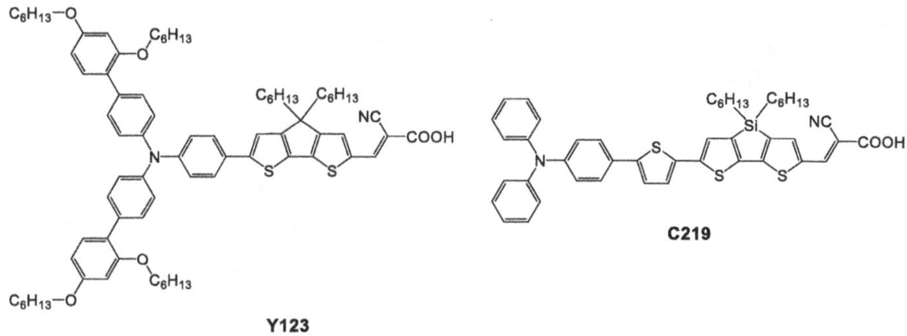

Fig. 14.24 Chemical structures of a cyclopentadithiophene dye sensitizer **Y123** and a dithienosilole (DTS) dye sensitizer **C219** for TiO_2 DSSCs

1998; Mwaura et al. 2006; Kim et al. 2003; Senadeera et al. 2003, 2005; Fang et al. 2011; Zhang et al. 2009; Chang et al. 2011). In general, conjugated polymers have large absorption coefficients, and their HOMO and LUMO energy levels can be fine-tuned. Among them, polythiophenes are particularly promising candidates as polymeric dye photosensitizers because of their thermal and environmental stabilities, high conductivities, reversible transitions between redox and neutral states, and easily controllable solubility and photophysical and electrochemical properties by the introduction of side chains. A TiO_2 DSSC using poly(3-thiopheneacetic acid) **P3TAA** showed $\eta \approx 1.6\%$ (Fig. 14.25) (Kim et al. 2003; Senadeera et al. 2005). Moreover, conjugated polymers (**P2** and **RCP-1**) with the D–π–A architecture were synthesized for TiO_2 DSSCs. **P2** and **RCP-1** are composed of triphenylamine or carbazole, respectively, as an electron-donating backbone and cyanoacrylic acid as an electron-accepting side chain connected through a conjugated thiophene unit as a π-bridge. The η value of the TiO_2 DSSC using **P2** is 3.39% under AM 1.5 simulated sunlight (100 mW cm^{-2}) (Zhang et al. 2009). The TiO_2 DSSC using **RCP-1** yielded a J_{sc} value of 8.43 mA cm^{-2}, a V_{oc} value of 670 mV, and a *ff* value of 0.73, corresponding to an η value of 4.11% (Chang et al. 2011), which is the highest conversion efficiency in TiO_2 DSSCs based on polymeric dyes.

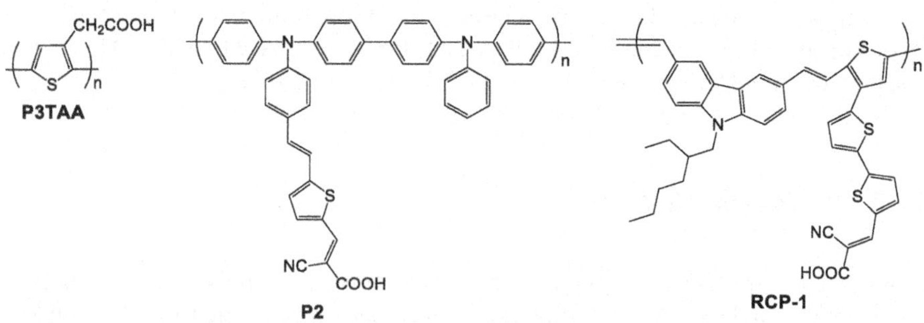

Fig. 14.25 Chemical structures of polymeric dye sensitizers for TiO_2 DSSCs

14.2.19 *D–π–a Dyes with Pyridine, Pyrazine, and Triazine as Anchoring Group*

In the above-mentioned TiO_2 DSSCs, –COOH, $–SO_3H$,$–H_2PO_3$, or $-Si(OR)_3$ have been generally incorporated into conventional dye sensitizers as anchoring groups, by which the dyes are adsorbed on the TiO_2 surface to form a bidentate bridge (dye-O-Ti) at Brønsted acid sites of surface-bound hydroxyl groups (Ti–OH) on the TiO_2 surface (Fig. 14.3c). Particularly, various kinds of D–π–A dye sensitizers with a carboxyl group (–COOH) as an electron-withdrawing and anchoring group have been reported to establish good electronic communication between the dyes and TiO_2 by the formation of the strong bidentate bridge, as described above.

On the other hand, Ooyama et al. designed and developed a new type of D–π–A carbazole-pyridine dye sensitizers **NI3–NI6** containing a pyridine ring as an electron-withdrawing, electron-injecting, and anchoring groups (Fig. 14.26) (Ooyama et al. 2011b, c). In addition, functionally separated D–π–A dye sensitizers **NI7** and **NI8** having a carboxyl group as an anchoring group and a pyridine ring as an electron-accepting group were also synthesized. The FTIR spectra of **NI3–NI8** adsorbed on the TiO_2 nanoparticles indicate that **NI3–NI6** formed the coordination bond [dye-(nitrogen of pyridine)-Ti] between their pyridine rings and Lewis acid sites (exposed Ti^{n+} cations) on the TiO_2 surface (Fig. 14.3d) and **NI7** and **NI8** formed both bidentate bridge and coordination bond. The TiO_2 DSSCs using the new-type D–π–A dye sensitizers **NI3–NI6** showed higher J_{sc} and η values than those of conventional D–π–A dye sensitizers **NI1** and **NI2** with a carboxyl group as an electron-withdrawing and anchoring group. Consequently, the formation of the coordination bond between the pyridine ring of **NI3–NI6** and the Lewis acid sites on the TiO_2 surface enabled efficient electron injection through the good electron communication between them, rather than the formation of the bidentate bridge between the carboxyl group of **NI1** and **NI2** and the Brønsted acid sites of the TiO_2 surface.

High-performance DSSCs based on porphyrin dyes or D–π–A dyes bearing pyridyl groups have been reported (Fig. 14.27) (Zhang et al. 2013; Lu et al. 2013; Daphnomili et al. 2012; Stangel et al. 2014; Mai et al. 2015; Ooyama et al. 2013b, c; Daphnomili et al. 2013; Wang et al. 2013; Sakurada et al. 2014; Mao et al. 2014; Ikeuchi et al. 2015; Bolisetty et al. 2015; Franchi et al. 2014; Verbitskiy et al. 2014; Jia et al. 2016; Angaridis et al. 2016; Meinhardt et al. 2016). Zhou et al. reported that a DSSC using a D–π–A dye **TTC105** with a pyridyl group showed an η value of 1.88% (Zhang et al. 2013). Wang et al. reported that a DSSC using a D–π–A porphyrin dye **LW11** with a pyridyl group showed an η value of 3.96% (Lu et al. 2013). Goutsolelos et al. developed porphyrin dyes **P1–3** with pyridyl groups, and the DSSC using **P2** with four pyridyl groups yielded an η value of 3.9% (Daphnomili et al. 2012). However, the amounts ($<5.0 \times 10^{16}$ molecules per cm^2) of these porphyrin dye sensitizers with pyridyl groups adsorbed on the TiO_2 electrode were much lower than those of porphyrin dye sensitizers with carboxyl groups. The low loading of the dye sensitizers led directly to poor surface coverage of the

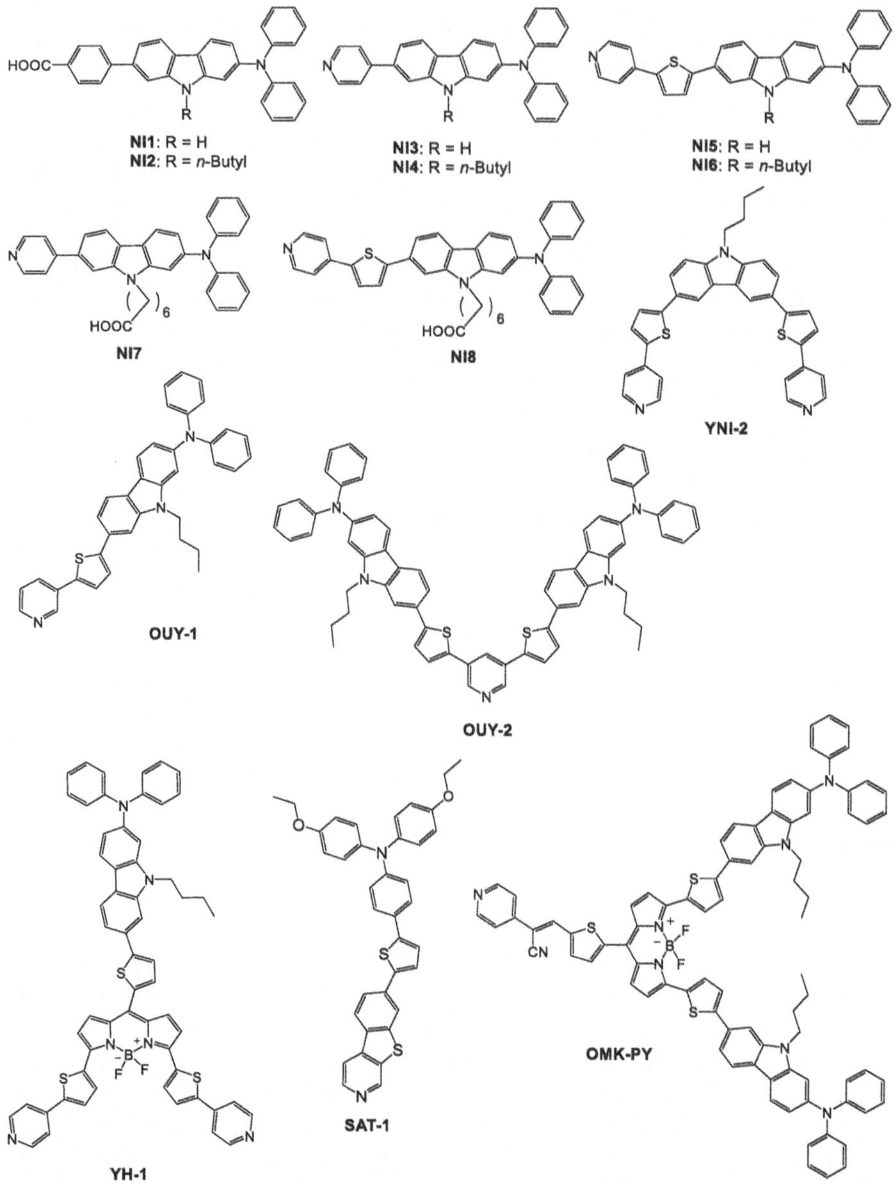

Fig. 14.26 Chemical structures of conventional D–π–A dye sensitizers **NI1** and **NI2**, D–π–A carbazole-pyridine dye sensitizers **NI3**–**NI6**, **OUY-1**, **OUY-2**, **SAT-1**, **YH-1**, and **OMK-PY**, and functionally separated D–π–A carbazole-pyridine dye sensitizers **NI7** and **NI8** reported by Ooyama et al.

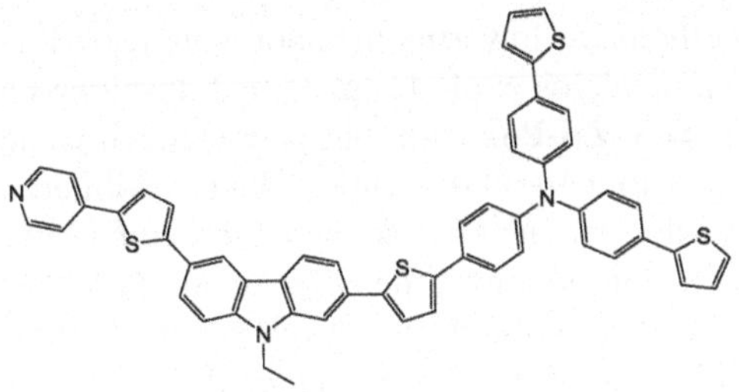

TTC105

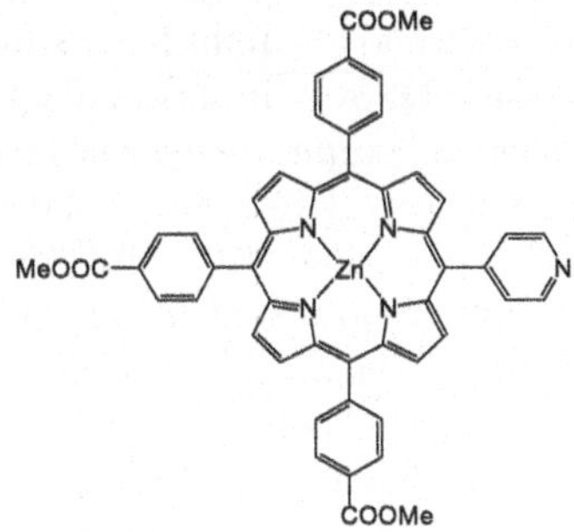

P1

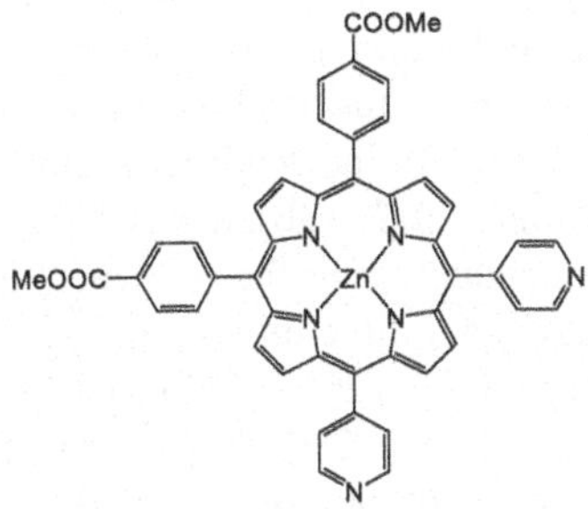

P2

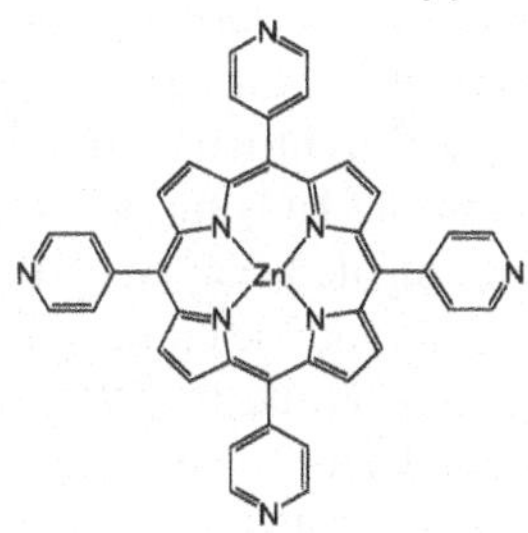

P3

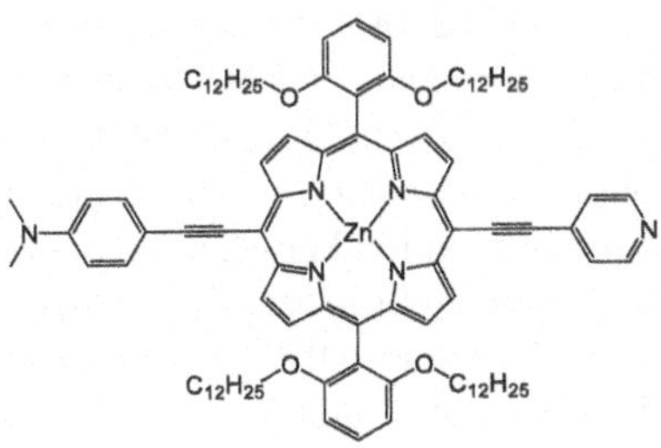

LW11

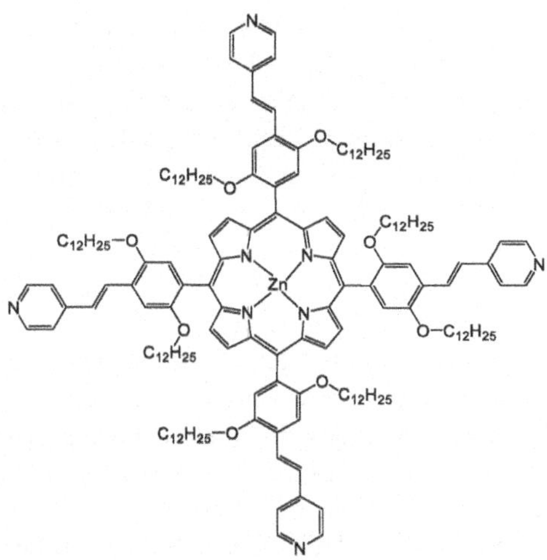

1PV-Zn-Por

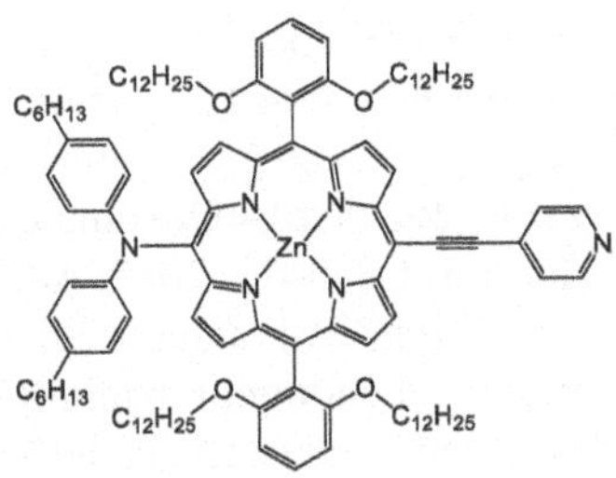

MH-3

Fig. 14.27 Chemical structures of conventional D–π–A carbazole-pyridine dye sensitizer **TTC105** and porphyrin-pyridine dye sensitizers **P1–3**, **LW11**, **MH-3**, and **1PV-Zn-Por** for TiO_2 DSSCs

TiO_2 electrode and low light-harvesting efficiency, lowering the photovoltaic performances of the DSSCs. In this context, Goutsolelos et al. designed and developed a "spider-shaped" porphyrin dye sensitizer **1PV-Zn-Por** with four phenylenevinylene groups, eight long dodecyloxy chains, and four pyridyl groups, which exhibited a high dye loading value (1.9×10^{17} molecules per cm^2) due to the increased basicity of the pyridyl groups (Stangel et al. 2014). Consequently, the DSSC using **1PV-Zn-Por** yielded an η value of 5.12%. Grätzel and co-workers also developed the highest performance DSSC with an η value of 8.2% based on **MH-3** (Mai et al. 2015).

On the other hand, Ooyama et al. designed and developed a D–π–A-type BODIPY dye sensitizer **YH-1** with two pyridyl groups as electron-withdrawing and anchoring groups at the end of the 3- and 5-positions and a carbazole–diphenylamine moiety as an electron-donating group at the 8-position on the BODIPY core (Fig. 14.26). **YH-1** showed a good light-harvesting efficiency in the red/NIR region and a good adsorption ability on the TiO_2 surface (Ooyama et al. 2013b). It was demonstrated that the extended π-conjugated system by the introduction of the two thienylpyridines at the 3- and 5-positions and the carbazole–diphenylamine moiety through a thiophene unit at the 8-position on the BODIPY core can cause red-shift and broadening of the photoabsorption band in the red/NIR region. The DSSC using **YH-1** yielded an incident photon-to-current conversion efficiency of ca. 10% over 500–700 nm with an onset at 800 nm. Using a coadsorbent of CDCA to prevent dye aggregation on the TiO_2 surface, the DSSC based on **YH-1** showed a J_{sc} value of 1.85 mA cm^{-2} and an η value of 0.39%. The relatively low photovoltaic performance of **YH-1** is attributed to not only the formation of strongly π-stacked aggregates, the low LUMO energy level, and the radiationless relaxation of the photoexcited state, which reduce the electron injection yield but also the fast charge recombination between the injected electrons and the I_3^- ions, which decreases the V_{oc} value (320–350 mV). Moreover, Ooyama et al. designed and synthesized a $(D)_2$–π–A-type BODIPY dye **OMK-PY** with two diphenylamine–thienylcarbazole moieties at the 3- and 5-positions and a pyridyl group at the end of the 8-position on the BODIPY core (Ooyama et al. 2017a). The DSSC based on **OMK-PY** showed a J_{sc} value of 1.08 mA cm^{-2}, a V_{oc} value of 335 mV, and a *ff* value of 0.53, corresponding to an η value of 0.19%.

Furthermore, Ooyama et al. designed and developed a new-type of D–π–A dye sensitizers with azine rings such as pyridine, pyrazine, triazine, and pyridinium ring, as electron-withdrawing and anchoring groups (Figs. 14.26, 14.28) (Ooyama et al. 2010c, 2013c, 2014a, b, 2015a, b, c, d, 2016a, b, c; Harima et al. 2013, 2015; Adachi et al. 2016). It was revealed that the D–π–A dye sensitizers with azine rings can adsorb onto the TiO_2 surface by the formation of hydrogen bonding or azinium ion at the Brønsted acid sites or coordination bonding at the Lewis acid sites on the TiO_2 surface, which enables efficient electron injection from the dyes to the CB of the TiO_2 electrode. As a result, azine rings were demonstrated to be one of the most promising components of dye sensitizers for DSSCs.

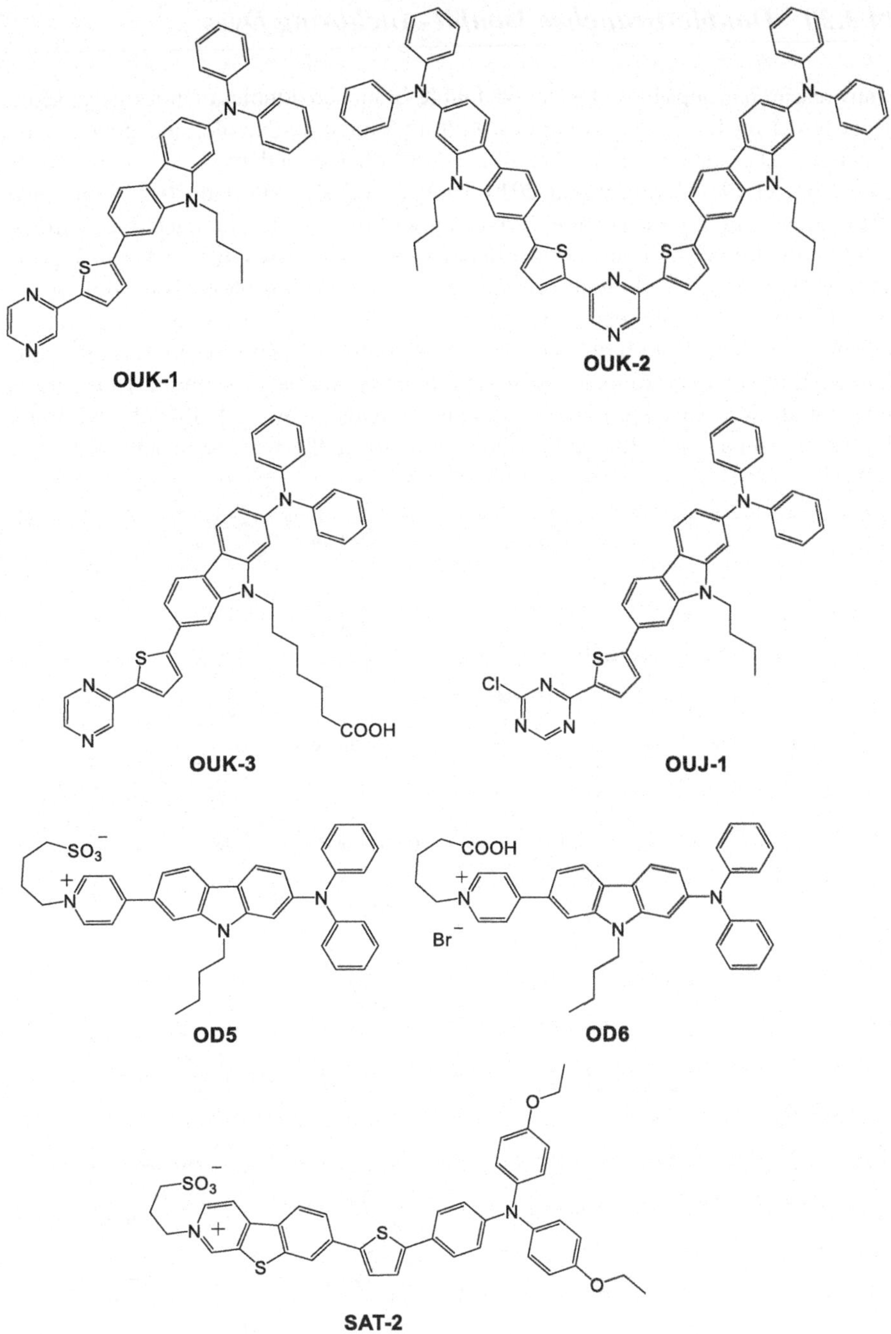

Fig. 14.28 Chemical structures of D–π–A carbazole-pyrazine dye sensitizers **OUK-1**, **OUK-2**, and **OUK-3**, D–π–A carbazole-triazine dye sensitizer **OUJ-1**, D–π–A carbazole-pyridinium dye sensitizers **OD-5** and **OD-6**, and D–π–A carbazole benzo[4,5]thieno[2,3-*c*]pyridinium dye sensitizer **SAT-2** reported by Ooyama et al.

14.2.20 Double-Branched Double-Anchoring Dyes

Many research groups have developed double-branched double-anchoring dye sensitizers based on D–(π–A)$_2$ structures with arylamine, carbazole, and phenoxazine moieties as D groups, and the photovoltaic performances of their DSSCs were investigated (Fig. 14.29) (Manfredi et al. 2014; Hung et al. 2015; Liu et al. 2011; Gupta et al. 2014; Hung et al. 2014). The main advantages of the double-branched dyes with D–(π–A)$_2$ structures over conventional linear D–π–A dyes are enhanced optical properties such as broader photoabsorption spectra, red-shifted photoabsorption peaks, lower band gaps, and higher molar extinction coefficients. A further intrinsic strategic advantage of the double-branched dyes is the control of their intermolecular aggregation. However, disadvantages such as poor molecular packing on the TiO_2 surface have not allowed very high photovoltaic performances of the DSSCs based on the double-branched dyes, although η values of up to ca. 8% have been achieved.

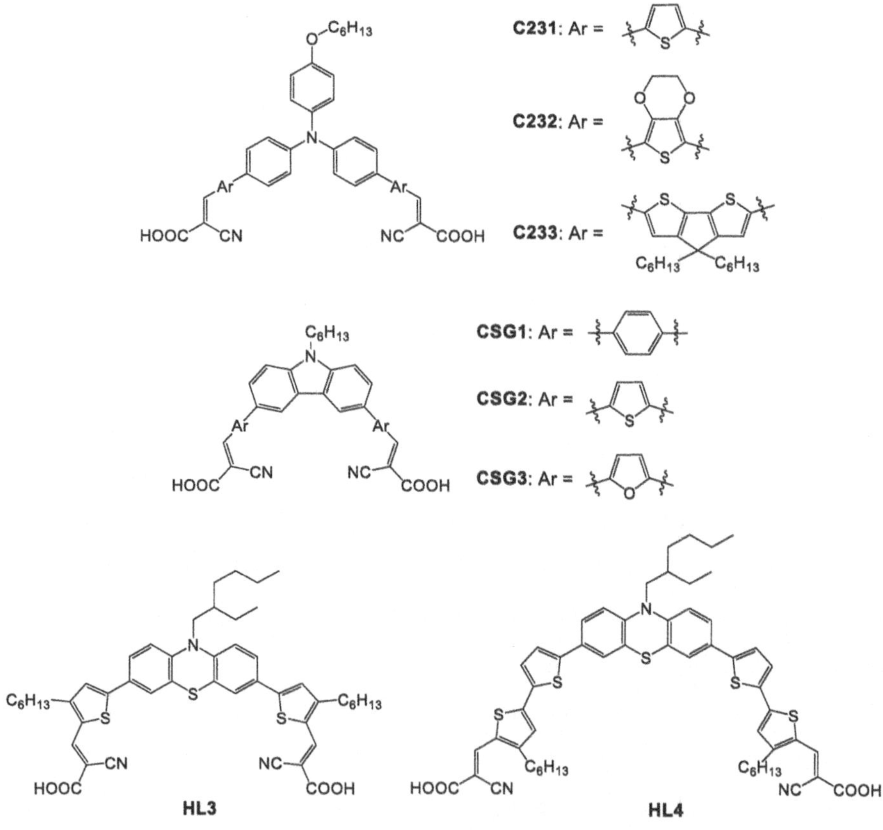

Fig. 14.29 Chemical structures of double-branched double-anchoring dyes with D–(π–A)$_2$ structures

Double-branched (D–π–A)$_2$L dyes with two D–π–A units, are separately located and linked by a bridge of a saturated chain or a ring system (non-conjugated linker: L), are highly promising sensitizers for DSSCs (Fig. 14.30) (Meier et al. 2017). The advantages of (D–π–A)$_2$L dyes compared to D–π–A dyes are the broad and intense photoabsorption properties, the lower dye aggregation tendency, and the higher dye loading density on the TiO_2 electrode. In addition, the intramolecular transfer of the excitation energy can enhance the contact time of the adsorbed dyes in the excited state S_1 and thereby, reduce the undesirable charge recombination (dark current). Photovoltaic performance of a DSSC using a (D–π–A)$_2$L dye is higher than that of the corresponding single branched D–π–A dye sensitizer. Indeed, the highest η value of 8.1% was achieved in the DSSC based on a (D–π–A)$_2$L dye **D1** with two 3,4-ethylenedioxythiophene (EDOT) units and two triphenylamino groups (Jiang et al. 2014).

On the other hand, type-I/type-II hybrid dye sensitizers in DSSCs can inject electrons into the CB of the TiO_2 electrode by both the ICT excitation (type-I pathway)

Fig. 14.30 Chemical structure of a double-branched double-anchoring dye **D1** with (D–π–A)$_2$L structure (non-conjugated linker: L)

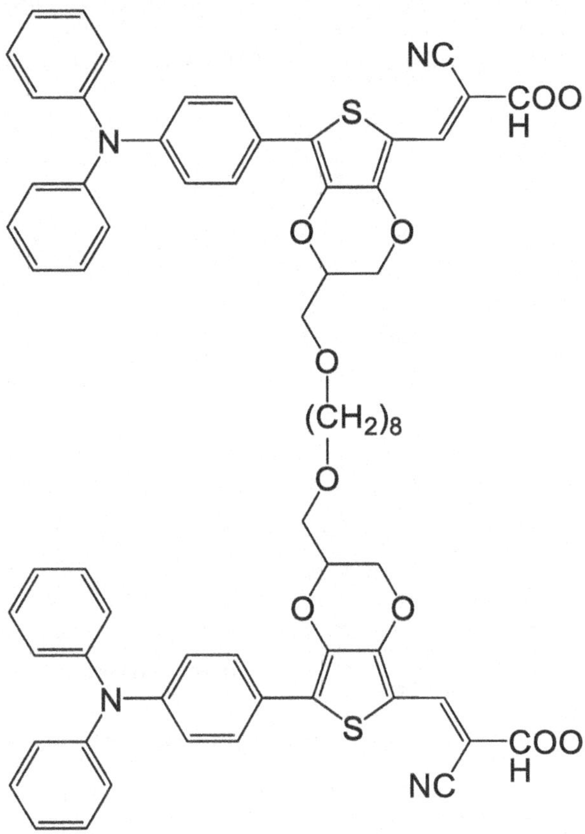

D1

and the photoexcitation of the DTCT band (type-II pathway). Therefore, Ooyama et al. designed and synthesized a dye sensitizer **OF-Py-Cat** with a pyridyl group and a catechol unit as anchoring groups to adsorb on the TiO_2 electrode through both coordination bond at the Lewis acid sites and bidentate binuclear bridge at the Brønsted acid sites on the TiO_2 surface (Fig. 14.31) (Ooyama et al. 2016b). Moreover, a type-I dye sensitizer **OF-Car-Car** with two carboxyl anchoring groups, a type-I dye sensitizer **OF-Py-Py** with two pyridyl anchoring groups, a type-I dye sensitizer **OF-Py-Car** with a pyridyl anchoring group and a carboxyl anchoring group, and a type-II dye sensitizer **OF-Cat-Cat** with two catechol anchoring groups were also synthesized. **OF-Py-Cat** adsorbed on the TiO_2 surface showed a broad photoabsorption band originating from both ICT and DTCT characteristics, i.e., **OF-Py-Cat** worked as a type-I/type-II hybrid dye sensitizer. However, the photovoltaic performance of the type-I/type-II DSSC using **OF-Py-Cat** was lower than those of the type-I DSSCs using **OF-Car-Car**, **OF-Py-Py**, or **OF-Py-Car** with carboxyl groups and pyridyl groups but comparable with that of the type-II DSSC using **OF-Cat-Cat** with catechol groups. The low photovoltaic performance of the DSSC using **OF-Py-Cat** is attributed to much faster back-electron transfer from the electrons injected into the TiO_2 electrode to the oxidized dye, despite the efficient electron injection from the pyridyl group to the CB of the TiO_2 electrode.

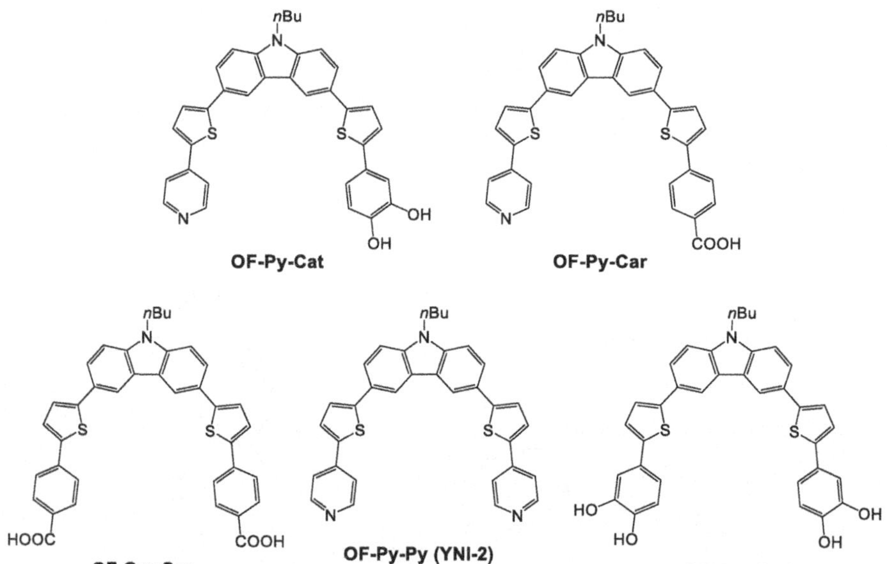

Fig. 14.31 Chemical structures of a type-I/type-II hybrid dye sensitizer **OF-Py-Cat** with a pyridyl group and a catechol unit, a type-I dye sensitizer **OF-Car-Car** with two carboxyl groups, a type-I dye sensitizer **OF-Py-Py** with two pyridyl groups, a type-I dye sensitizer **OF-Py-Car** with a pyridyl group and a carboxyl group, and a type-II dye sensitizer **OF-Cat-Cat** with two catechol units

14.2.21 Co-Sensitized DSSC Employing Two Dyes

In order to further improve photovoltaic performances of DSSCs, co-sensitization using two kinds of organic dyes, e.g., combinations of phthalocyanine and organic dyes, porphyrin and organic dyes, and Ru-complex and organic dyes, has been expected as an effective approach, because it can improve the surface dye coverage on the TiO_2 electrode and the light-harvesting efficiency (LHE) and suppress dye aggregation and charge recombination between the injected electrons in the TiO_2 electrode and the I_3^- ions in the electrolyte. In fact, Grätzel et al. reported a high-performance co-sensitized DSSC using two kinds of dyes with a carboxyl group and a $Co^{II/III}$tris(bipyridyl)-based redox electrolyte. The co-sensitization of the D–π–A zinc porphyrin dye **YD2-o-C8** (Fig. 14.16) and an organic D–π–A dye **Y123** (Fig. 14.24) yielded a J_{sc} value of 17.66 mA cm^{-2}, a V_{oc} value of 935 mV, and a *ff* value of 0.74, corresponding to an η value of 12.3% (Yella et al. 2011). However, in such co-sensitization, both carboxyl-anchoring dyes often compete to adsorb at the Brønsted acid sites on the TiO_2 surface, which decreases the amount of adsorbed dyes on the TiO_2 electrode, lowers the LHE, and reduces the power conversion efficiency. Thus, co-sensitized DSSCs based on the combinations of the Ru-complex dye (**black dye**, Fig. 14.5) with carboxyl groups and a D–π–A dye (**NI5** or **YNI-2**, Fig. 14.26) with one or two pyridyl groups were fabricated to achieve site-selective dye adsorption on the TiO_2 surface and reduce the competitive adsorption between the two dyes (Shibayama et al. 2014). **Black dye** with carboxyl groups can adsorb at the Brønsted acid sites (surface-bound hydroxy groups, Ti–OH) on the TiO_2 surface, while **NI5** and **YNI-2** with pyridyl groups can adsorb at the Lewis acid sites (exposed Ti^{n+} cations) on the TiO_2 surface. Therefore, the competitive adsorption is expected not to occur between carboxyl-anchoring dyes and pyridine-anchoring dyes. Indeed, the photovoltaic performances of the co-sensitized DSSCs using the combinations of **black dye** and **NI-5** ($\eta = 10\%$) or **black dye** and **YNI-2** ($\eta = 9.8\%$) were higher than that of the DSSC using only **black dye** ($\eta = 9.5\%$), which successfully demonstrated the effective co-sensitization of two dyes bearing different anchoring groups by their site-selective adsorption without competitive adsorption.

Moreover, as an innovative approach that enables effective and convenient co-sensitization with an enhanced dye coverage on the TiO_2 electrode, Ooyama et al. developed a new co-sensitization method using two kinds of D–π–A dyes with pyridyl groups capable of adsorbing at both Brønsted acid sites and Lewis acid sites on the TiO_2 surface, i.e., one-step co-adsorption of **SAT-1** and **NI-6** or **YNI-2** to the TiO_2 electrode using their mixed solution (Fig. 14.26) (Ooyama et al. 2015d). **SAT-1** with benzo[4,5]thieno[2,3-*c*]pyridine can predominantly adsorb on the TiO_2 electrode through hydrogen bonding at the Brønsted acid sites on the TiO_2 surface, while **NI6** and **YNI-2** adsorb at the Lewis acid sites. The co-sensitization of **SAT-1** and **NI-6** or **YNI-2** on the TiO_2 electrode successfully demonstrated their site-selective adsorption without competitive adsorption and resulting enhancement of the dye coverage on the TiO_2 electrode. On the other hand, Ooyama et al. also reported another co-sensitization method for the site-selective adsorption using a D–π–A dye sensitizer (**NI-6** or **YNI-2**) with one or two pyridyl groups and a type-II D–π–Cat dye sensitizer (**YM-2**), which can adsorb at the Lewis acid sites and the Brønsted acid sites on the TiO_2 surface, respectively (Ooyama et al. 2015e). It was found that the most convenient and rapid co-adsorption of two different dye sensitizers on the TiO_2 electrode has been achieved by one-step co-adsorption using their mixed dye solution. Therefore, the co-sensitization method using two kinds of D–π–A dyes with different bonding abilities to the Lewis acid sites and the Brønsted acid sites on the TiO_2 surface was demonstrated to be one of the most promising strategies to enhance the dye coverage on the TiO_2 electrode and the light-harvesting efficiency of DSSCs.

14.3 N-ZnO-Based DSSCs

Dye sensitizers for n-ZnO-based DSSCs have been developed under a strategy similar to that in the molecular design for n-TiO_2-based DSSCs. As organic dye sensitizers for ZnO DSSCs, Ru complex dyes (Shin et al. 2007; Saito and Fujihara 2008; Nguyen et al. 2009; Tennakone et al. 1999; Keis et al. 2000, 2002; Kakiuchi et al. 2006; Quintana et al. 2007; Chou et al. 2007; Wu et al. 2007), xanthene dyes (Hara et al. 2000a, b; Yoshida et al. 2000, 2004; Hosono et al. 2004; Bahadur and Srivastava 2003; Guillén et al. 2008), indoline dyes (Hosono et al. 2008; Dentani et al. 2009; Matsui et al. 2009, 2010a, b, 2011; Sakuragi et al. 2010; Guillén et al. 2011; Ambade et al. 2011; Higashijima et al. 2011; Cheng and Hsieh 2010), perylene dyes (Dentani et al. 2007; Szarko et al. 2008), heptamethine–cyanine dyes (Mikroyannidis et al. 2009; Matsui et al. 2005; Otsuka et al. 2008; Dentani et al. 2008; Funabiki et al. 2011), and squaraine dyes (Otsuka et al. 2006; Matsui et al. 2006) have been employed. However, the photovoltaic performances of their ZnO DSSCs were lower than those of TiO_2 DSSCs because of the chemical instability of ZnO, which is considered to be mainly related to the high acidity of the carboxyl-anchoring groups in the dyes. The adsorption of acidic dyes onto the ZnO electrode can cause the dissolution of Zn^{2+} followed by the formation of Zn^{2+}/dye agglomerates, which lowers electron injection efficiency and/or dye regeneration efficiency. In order to avoid forming Zn^{2+}/dye agglomerates, it would be effective to reduce the acidity of the dye sensitizers by structural modification, including optimization (reduction) of the number of carboxyl groups, and replacement of the anchoring groups by weakly acidic anchoring groups such as carboxylate, carboxylic acid anhydride, and catechol groups. This hypothesis may be supported by the fact that organic dyes with the weakly acidic aliphatic carboxylic acid and rhodanine-3-acetic acid anchoring groups have been generally used as dye sensitizers for ZnO DSSCs, instead of the strongly acidic cyanoacrylic acid group. In fact, the ZnO DSSC using the indoline dye **D102** with rhodanine-3-acetic acid (Fig. 14.10) yielded higher J_{sc} and η values than those of the ZnO DSSC using the corresponding indoline dye **D131** with cyanoacrylic acid (Fig. 14.32) (Hosono et al. 2008; Dentani et al. 2009; Matsui et al. 2009, 2010a, b, 2011; Sakuragi et al. 2010; Guillén et al. 2011; Ambade et al. 2011; Higashijima et al. 2011).

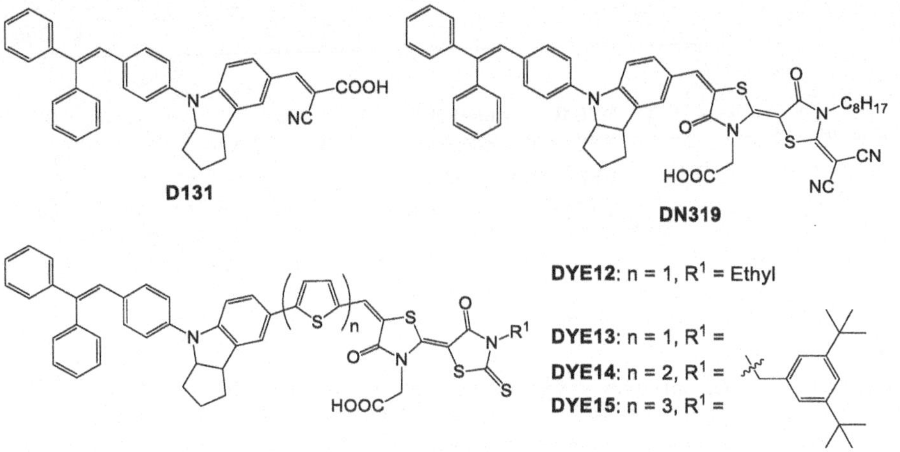

Fig. 14.32 Chemical structures of indoline dye sensitizers for ZnO DSSCs

14.3.1 Xanthene Dyes

Xanthene dyes including eosin Y, mercurochrome, fluorescein, Rose Bengal, and rhodamine 6G, are commercially available at low cost and were utilized as dye sensitizers in early studies on ZnO DSSCs (Fig. 14.13) (Hara et al. 2000a, b; Yoshida et al. 2000, 2004; Hosono et al. 2004; Bahadur and Srivastava 2003; Guillén et al. 2008). Particularly, the η values of ZnO DSSCs using eosin Y and mercurochrome were 2.3% and 2.5% (Hara et al. 2000a, b; Yoshida et al. 2004), respectively, which are higher than those of the corresponding TiO_2 DSSCs.

14.3.2 Indoline Dyes

Various D–π–A indoline dyes with cyanoacrylic acid or rhodanine-3-acetic acid groups have been developed as efficient dye sensitizers for ZnO DSSCs (Figs. 14.10, 14.32) (Hosono et al. 2008; Dentani et al. 2009; Matsui et al. 2009, 2010a, b, 2011; Sakuragi et al. 2010; Guillén et al. 2011; Ambade et al. 2011; Higashijima et al. 2011; Cheng and Hsieh 2010). The ZnO DSSC using the indoline–double rhodanine dye **D205** exhibited an η value of up to 5.34% with a J_{sc} value of 12.17 mA cm^{-2}, a V_{oc} value of 653 mV, and a *ff* value of 0.67 under AM 1.5 simulated sunlight (100 mW cm^{-2}). The η value is the highest conversion efficiency in ZnO DSSCs using organic dye sensitizers (Fig. 14.10) (Cheng and Hsieh 2010).

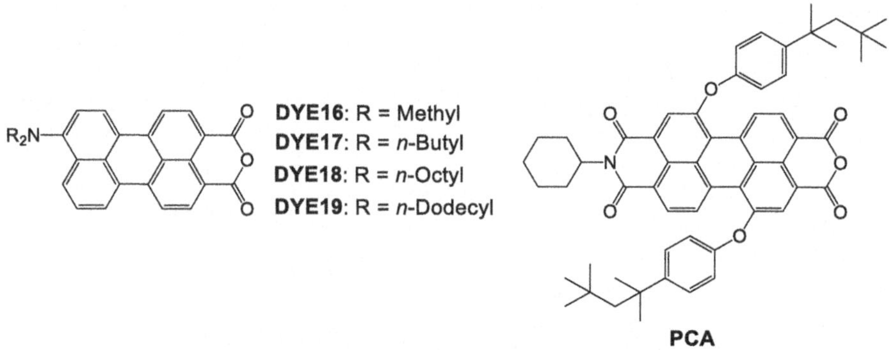

Fig. 14.33 Chemical structures of perylene dye sensitizers for ZnO DSSCs

14.3.3 Perylene Dyes

Perylene dyes **DYE16–19** (Dentani et al. 2007) and a perylene monoimide dye **PCA** (Mikroyannidis et al. 2009) with a carboxylic acid anhydride unit as an anchoring group have been developed as dye sensitizers for ZnO DSSCs (Fig. 14.33). The ZnO DSSC using **PCA** yielded an η value of 1.34% with a J_{sc} value of 4.22 mA cm^{-2}, a V_{oc} value of 610 mV, and a *ff* value of 0.52 under AM 1.5 simulated sunlight (100 mW cm^{-2}).

14.3.4 Heptamethine-Cyanine Dyes

Heptamethine–cyanine dyes **DYE20–26** (Matsui et al. 2005), **KFH-2**, and **KFH-3** (Funabiki et al. 2011) have been designed and synthesized as NIR-absorbing dye sensitizers for ZnO DSSCs (Fig. 14.34). The ZnO DSSC using **KFH-3** showed an η value of 1.23% with an IPCE value of 33% at 800 nm, a J_{sc} value of 3.34 mA cm^{-2}, a V_{oc} value of 490 mV, and a *ff* value of 0.76 under AM 1.5 simulated sunlight (100 mW cm^{-2}).

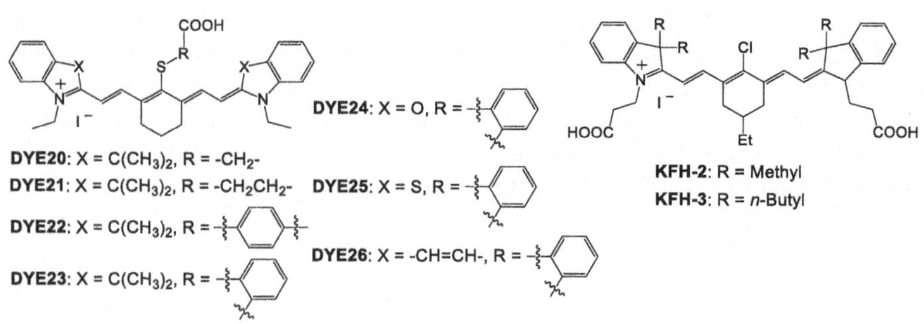

Fig. 14.34 Chemical structures of heptamethine–cyanine dye sensitizers for ZnO DSSCs

14.3.5 Squaraine Dyes

To date, only a few asymmetric squaraine dye sensitizers have been developed for ZnO DSSCs (Fig. 14.35). The η values of the ZnO DSSCs using **DYE27**, **DYE28**, and **DYE29** were 0.53%, 0.48%, and 0.51%, respectively (Otsuka et al. 2006; Matsui et al. 2006).

14.4 Catechol Dyes for N-TiO_2-Based Type-II DSSCs

As shown in Fig. 14.1b, the electron injection pathway in type-II DSSCs is direct, i.e., "one-step" electron injection from the ground state of the dye to the CB of the TiO_2 electrode by photoexcitation of the DTCT band. In other words, the HOMO electron of the dye is directly injected into the CB of the TiO_2 electrode upon photoexcitation. One of the most advantageous aspects in type-II DSSCs over type-I DSSCs is the light-harvesting capability over the wide spectral region of sunlight because the direct electron injection enables the creation of the broad DTCT bands and the easing of restrictions on the LUMO energy levels of the dye sensitizers relative to the indirect electron-injection in type-I DSSCs. Catechol (Cat) dyes including dopamine and fluorine, numerous natural pigments including bromopyrogallol red, and anthocyanins with Cat moieties are classified as type-II dye sensitizers and show a strong new photoabsorption band in a long-wavelength region due to the DTCT by binding to the TiO_2 surface (Fig. 14.36) (Frei et al. 1990; Moser et al. 1991; Cherepy et al. 1997; Tennakone et al. 1997, 1998; Jayaweera et al. 1999; Dai and Rabani 2002a, b; Hao et al. 2006; Kumara et al. 2006; Polo and Iha 2006; Agarkar et al. 2011; Sarker et al. 2012; Narayan 2012; Tae et al. 2005; An et al. 2010). Cat dyes are known to bind to the TiO_2 electrode through a bidentate mononuclear chelating linkage and/or a bidentate binuclear bridge at the Brønsted acid sites on the TiO_2 surface (Fig. 14.3f, g).

However, photovoltaic performances of type-II TiO_2 DSSCs using Cat dyes were significantly lower than those of type-I TiO_2 DSSCs using dye sensitizers with

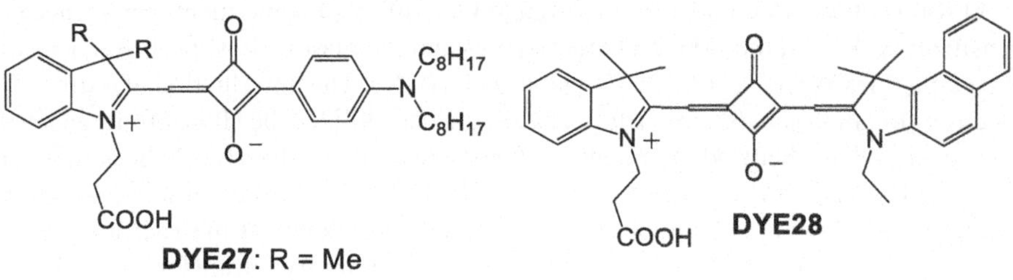

DYE27: R = Me
DYE29: R = *n*-butyl

DYE28

Fig. 14.35 Chemical structures of squaraine dye sensitizers for ZnO DSSCs

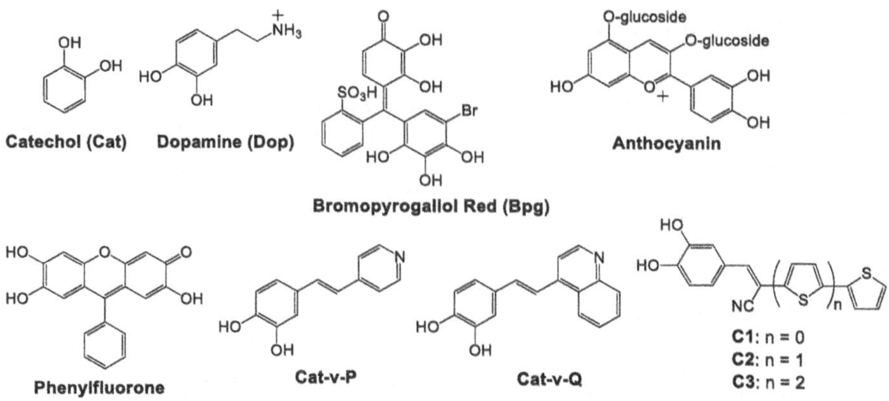

Fig. 14.36 Chemical structures of catechol dye sensitizers for type-II TiO_2 DSSCs

carboxyl groups because the back transfer (charge recombination) of the TiO_2-injected electrons to the oxidized dyes in the type-II pathway was much faster than that in the type-I pathway. In order to overcome this problem, Cat dyes **Cat-v-P** and **Cat-v-Q** were designed and synthesized (Fig. 14.36) (Tae et al. 2005). The modification of the Cat moiety with electron-donating groups, such as (pyridin-4-yl)vinyl (v-P) and (quinolin-4-yl)vinyl (v-Q), 2- and 2.7-fold increased the η value, respectively, driven by large increases in the J_{sc} value. The overall η values of the TiO_2 DSSCs using **Cat-v-P** or **Cat-v-Q** were 1.2% and 1.6%, respectively. This result clearly demonstrates that the modification of a primary electron-donating group (4-vinylcatechol unit) with a second electron-donating group (pyridine or quinoline unit) allows a dramatic increase in electron injection efficiency. The enhanced η values are attributed to the retardation of the back electron transfer by the consecutive charge shift from the secondary electron-donating group to the primary electron-donating group and the red-shift of the DTCT band caused by the increased electron-donating ability of the dye. Additionally, Cat-thiophene dyes **C1–C3** containing one, two, or three thiophene units were developed as type-II dye sensitizers (Fig. 14.36) (An et al. 2010). The η values for the TiO_2 DSSCs using **C1–C3** increased in the order of **C1** (0.61%) < **C2** (0.70%) < **C3** (0.76%).

Ooyama et al. systematically investigated the effect of substituents on a Cat dye sensitizer (**CAT**) on the DTCT characteristics and the photovoltaic performances of their type II DSSCs (Fig. 14.37) (Ooyama et al. 2015f). The introduction of a strongly electron-donating substituent to the Cat skeleton destabilized the HOMO energy level and generated the thermodynamically unfavorable oxidized dyes by electron transfer from the I^- ions in the electrolyte. On the other hand, the introduction of a strongly electron-withdrawing substituent to the Cat skeleton stabilized the LUMO energy level, decreased the DTCT characteristics, facilitated the back transfer of the TiO_2-injected electrons to the strongly electron-accepting oxidized dye, and lowered the photovoltaic performance of the DSSC. The introduction of a moderately electron-withdrawing substituent to the Cat skeleton not only increased the DTCT efficiency

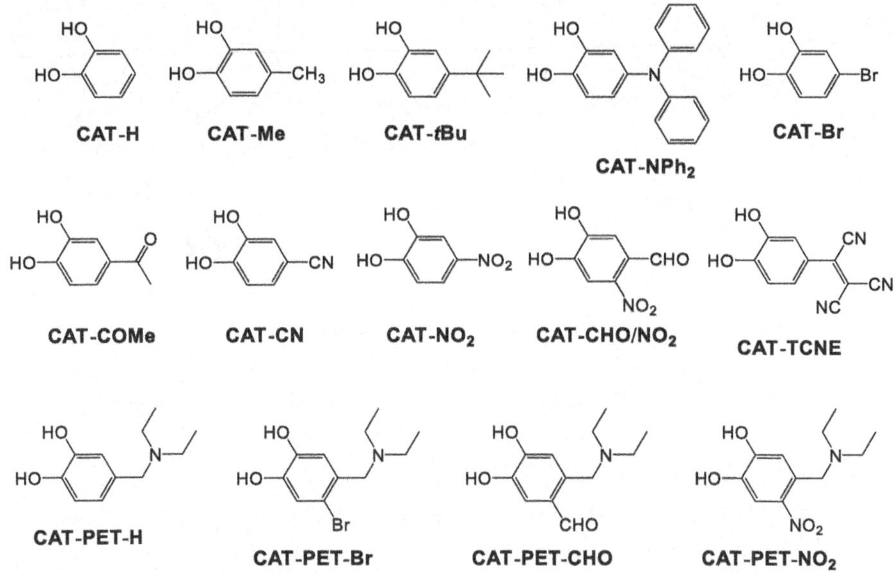

Fig. 14.37 Chemical structures of catechol dye sensitizers **CAT** and **CAT-PET** for type-II TiO_2 DSSCs reported by Ooyama et al.

but also retarded the back electron transfer, providing favorable conditions for the type-II electron injection pathway from the ground state of the dye to the CB of the TiO_2 electrode by the photoexcitation of the DTCT band.

Furthermore, to retard the back electron transfer in type-II DSSCs and increase the DTCT efficiency of Cat dyes, Ooyama et al. designed and synthesized Cat dye sensitizers (**CAT-PET** series) possessing photo-induced electron transfer (PET) characteristics. The **CAT-PET** dyes are composed of a Cat moiety with a diethylamino group as an electron-donating group linked by a methylene spacer (Fig. 14.37) (Ooyama et al. 2017b). The PET takes place from the nitrogen atom in the diethylamino group to the photoexcited Cat moiety and retards the back transfer of the TiO_2-injected electrons to the oxidized Cat moiety in type-II DSSCs. The photovoltaic performances of the DSSCs using the **CAT-PET** dyes were higher than those using the **CAT** dyes. This result indicates that the PET from the diethylamino group to the oxidized Cat dye adsorbed on the TiO_2 electrode efficiently retarded the back electron transfer and provided favorable conditions for the type-II electro -injection pathway from the ground state of the dye to the CB of the TiO_2 electrode by the photoexcitation of the DTCT band. Therefore, the introduction of PET characteristics to Cat dyes with a moderately electron-withdrawing substituent was demonstrated to be an effective molecular design for type-II dye sensitizers, because it increases the DTCT efficiency and retards the back electron transfer.

In order to establish a guideline for molecular design of efficient Cat dye sensitizers for type-II DSSCs, D–π–Cat fluorescent dyes **YM-1**, **YM-2**, and **YM-3** with fluorene, carbazole, and carbazole–terthiophene, respectively, as a π-conjugated moiety were

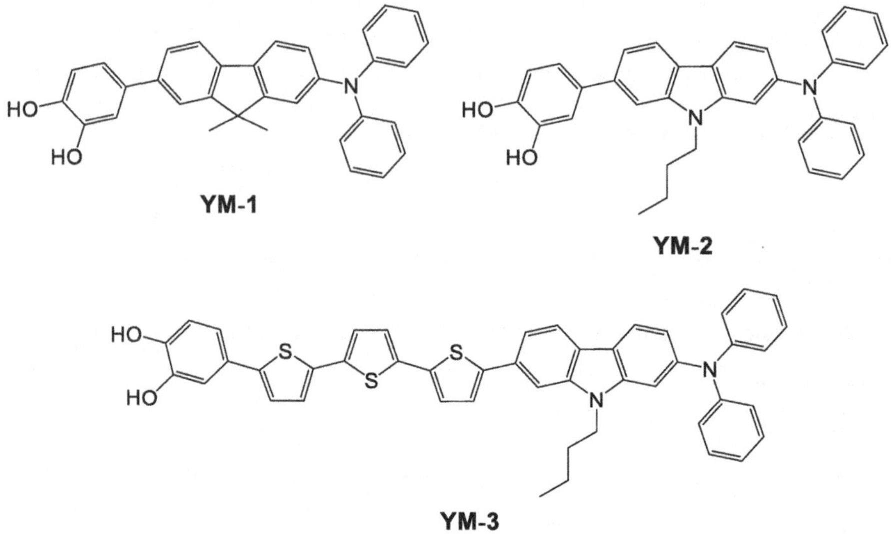

Fig. 14.38 Chemical structures of D–π–Cat dye sensitizers **YM-1**, **YM-2**, and **YM-3** for type-II TiO_2 DSSCs reported by Ooyama et al.

designed and developed (Fig. 14.38) (Ooyama et al. 2014c). **YM-1**, **YM-2**, and **YM-3** adsorbed on the TiO_2 film showed a broad photoabsorption band corresponding to the DTCT by the formation of Ti–Cat chelate complexes. The IPCE values corresponding to the DTCT bands in the DSSCs using **YM-1** and **YM-2** were higher than that using **YM-3**. Thus, the photocurrent in the DSSCs using **YM-1** and **YM-2** is considered to be mainly generated by the direct electron injection pathway from the Ti–Cat chelate complexes to the CB of the TiO_2 electrode, i.e., type-II pathway. On the other hand, the photocurrent in the DSSC using **YM-3** is considered to be mainly generated by an indirect electron-injection pathway from the excited dye to the CB of the TiO_2 electrode, i.e., type-I pathway. The stabilized LUMO energy level and the extended π-conjugated system by the introduction of a long π-bridge such as terthiophene to the Cat moiety increased the ICT excitation based on π–π* transition with the decreased DTCT characteristics, resulting in the indirect electron-injection pathway from the excited dye to the CB of the TiO_2 electrode by photoexcitation of the local band of the adsorbed dye on the TiO_2 electrode.

14.5 P-NiO-Based DSSCs

Organic dyes for p-NiO-based DSSCs also require at least one anchoring group, e.g., carboxyl, phosphonate, and catechol groups, to strongly adsorb onto the NiO photocathode surface and to establish good electron communication (Fig. 14.2) (Hagfeldt et al. 2010; Mishra et al. 2009; Ning et al. 2010; Odobel et al. 2010; Yum et al. 2011;

Nikolaou et al. 2017). Carboxyl groups have been widely employed as anchoring groups in dye sensitizers for NiO DSSCs, although they are not necessarily an optimal group for hole injection because they stabilize the LUMO rather than the HOMO. Therefore, electron-withdrawing anchoring groups including a carboxyl group are not appropriate for organic dyes to efficiently inject holes and suppress geminate recombination in NiO DSSCs. A study on photovoltaic performances of NiO DSSCs using ruthenium polypyridine complexes with carboxylic acids, biscarbodithioic acids, catechol, or methylphosphonic acids as anchoring groups suggested that methylphosphonic acid is a favorable anchoring group for a stable linkage with a metal oxide, while catechol is another promising anchoring group that can enhance electronic coupling with the NiO electrode because of the more rich electron density compared to carboxylic acid (Pellegrin et al. 2011). That is, the study proposed that catechol and methylphosphonic acid are promising anchoring groups for dye sensitizers to replace the conventional carboxyl group in NiO DSSCs.

The molecular design opposite to organic dyes for TiO_2 DSSCs and ZnO DSSCs is required for those used in NiO DSSCs: the organic dye sensitizer should be fixed on the NiO surface through an anchoring group connected to the electron-donating moiety (i.e., HOMO) of the dye, while the electron-accepting moiety (i.e., LUMO) of the dye should be located apart from the NiO surface. The HOMO energy level of the dye must be lower (more positive) than that of the VB of the NiO electrode, in order to achieve efficient hole injection from the excited dye to the VB of NiO. Moreover, the LUMO energy level of the dye must be higher (more negative) than the redox potential of the I_3^-/I^- ion couple, in order to achieve efficient dye regeneration of the reduced state through electron transfer from the reduced dyes to the I_3^- ions in the redox mediator. Obviously, the LUMO energy level must be sufficiently lower than the energy level of the NiO CB. Therefore, the hole injection and dye regeneration must be thermodynamically feasible.

A successful molecular design for further improvement of photovoltaic performances of NiO DSSCs is to enable dye sensitizers to efficiently facilitate hole injection and dye regeneration and retard geminate recombination between the reduced dyes and holes injected into the NiO electrode. An (anchoring group)–D–π–A and An–D (chromophore)–A dyes with an anchoring group on an electron-donating moiety are expected to be two of the most promising classes of dye sensitizers for NiO DSSCs that meet the above requirements (Fig. 14.39). In NiO DSSCs based on an An–D–π–A or An–D–A dye, photoexcitation can induce ICT from the D to A moieties of the dye, which generates a positive charge on the dye close to the NiO

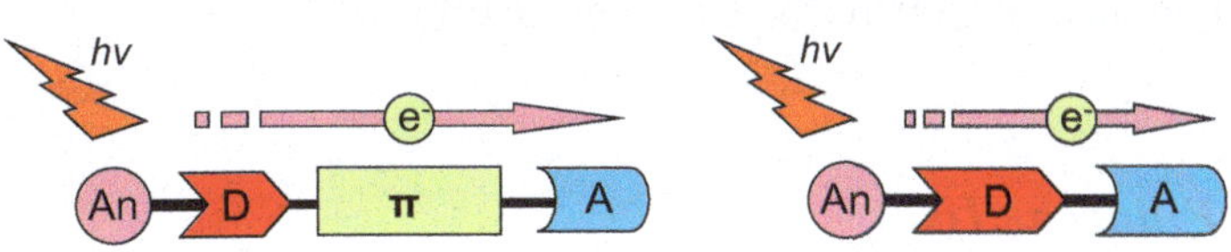

Fig. 14.39 Schematic representation of An–D–π–A and An–D (chromophore)–A organic dye sensitizers for p-NiO DSSCs

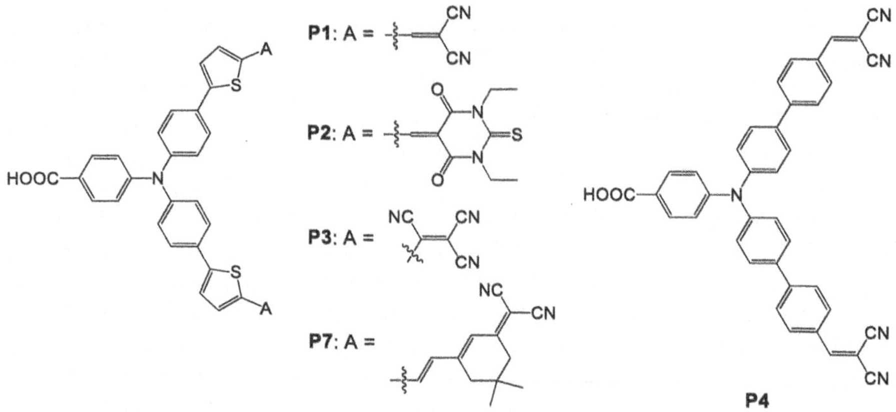

Fig. 14.40 Chemical structures of An–D–π–A triphenylamine dye sensitizers for NiO DSSCs

surface and a negative charge on the dye apart from the NiO surface. Therefore, the anchoring groups of An–D–π–A and An–D–A dyes for NiO DSSCs are attached to the electron-donating moieties, unlike D–π–A dyes use in TiO_2 DSSCs. The HOMO is delocalized on the π-conjugated bridge and electron-donating moiety including the anchoring group close to the NiO surface, resulting in efficient hole injection from the excited dye to the VB of NiO. On the other hand, the LUMO is mainly on the electron-withdrawing moiety, allowing efficient electron transfer from the dye to the I_3^- ions in the redox mediator.

14.5.1 Triphenylamine Dyes

In order to suppress not only geminate recombination but also dark current, An–D–π–A dye sensitizers with a carboxyl group as an anchoring group on the electron-donating moiety (triphenylamine) connecting to two electron-accepting groups (dicyanovinylene groups) through a thiophene or a phenyl linker were designed and developed (**P1**–**P4** and **P7**, Fig. 14.40) (Qin et al. 2008b, 2009, 2010; Li et al. 2010b). The maximum IPCE values of the NiO DSSCs using **P1** and **P4** were 35% and 44%, respectively, under AM 1.5 simulated sunlight (100 mW cm^{-2}). The η values of 0.08% and 0.09% were obtained in the NiO DSSCs based on **P1** and **P4**, respectively (Qin et al. 2008b, 2009; Li et al. 2010b).

14.5.2 Fluorene Dyes

In order to facilitate hole injection and suppress dark current, An–D–π–A fluorene dye sensitizers (**DYE30**–**32**, **DYE34**, and **DYE35**) with one or two carboxyl groups as anchoring groups on the electron-donating moiety (diphenylamine) connecting to an electron-accepting moiety (dicyanovinylene for **DYE30**, **DYE31**, and **DYE 34** or 1,3-diethyl-2-thioxodihydropyrimideine-4,6-dione for **DYE32** and **DYE 35**) through different π-linkers, and their related derivatives (**DYE33** and **O7**) were designed and developed (Fig. 14.41) (Yen et al. 2011; Ji et al. 2011). The NiO DSSCs using the dyes with two carboxyl groups showed higher V_{oc}, J_{sc}, and η values than those of the dyes with one carboxyl group. The relatively high V_{oc} values of the dyes with two carboxyl groups are attributed to efficient suppression of charge recombination (dark current) between the holes in the NiO VB and the I^- ions in the electrolyte. The NiO DSSCs using **O7** with two carboxyl groups on diphenylamine connecting to a dicyanovinyl group through an EDOT linker showed an η value of 0.060% with a J_{sc} value of 1.74 mA cm^{-2}, a V_{oc} value of 90 mV, and a *ff* value of 0.38, and a maximum IPCE value of 18% (Ji et al. 2011).

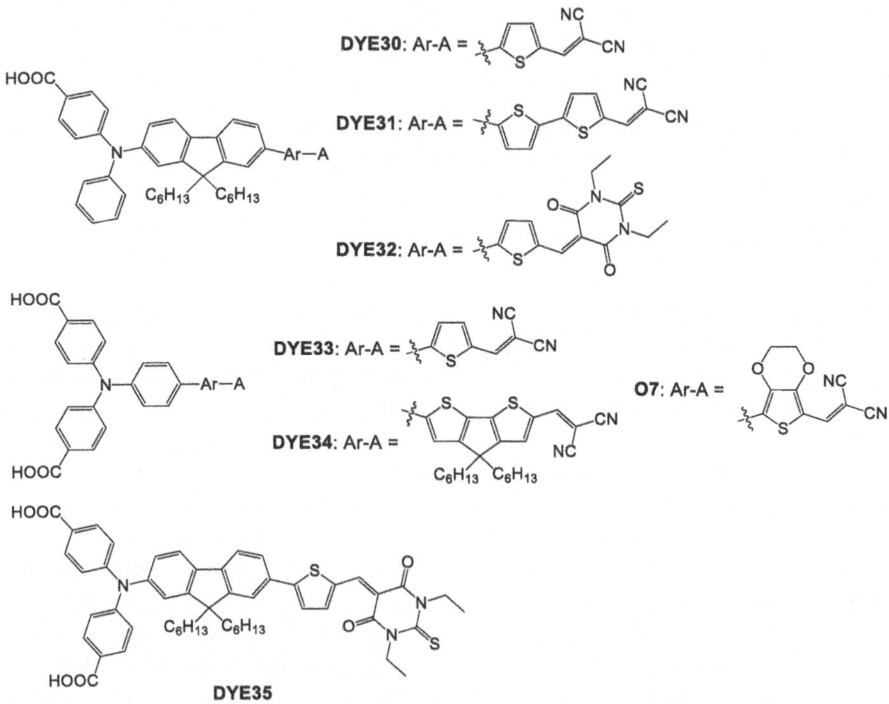

Fig. 14.41 Chemical structures of An–D–π–A fluorene dye sensitizers **DYE30–32**, **DYE34**, and **DYE35** and their related derivatives **DYE33** and **O7** for NiO DSSCs

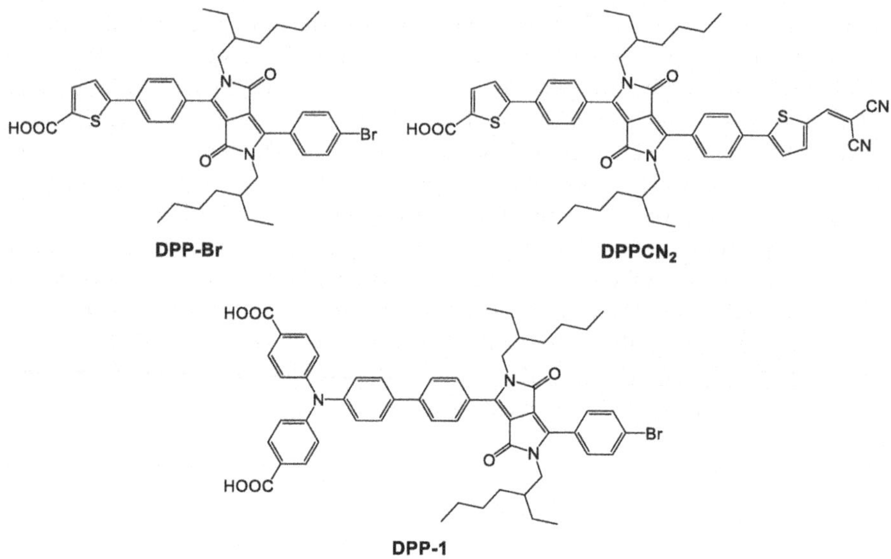

Fig. 14.42 Chemical structures of diketopyrrolopyrrole (DPP) dye sensitizers **DPP-Br** and **DPPCN$_2$** and an An–D–π–A DPP dye sensitizer **DPP-1** for NiO DSSCs

14.5.3 Diketopyrrolopyrrole Dyes

Diketopyrrolopyrrole (DPP) with electron-withdrawing ability shows a strong photoabsorption band around 500 nm and particularly high photostability. DPP-based dye sensitizers for p-type NIO DSSCs have been developed, and their photovoltaic performances have been investigated (Fig. 14.42). Odobel et al. developed DPP-based dyes with thienyl carboxylic acid as an anchoring group and with either bromine (**DPP-Br**) or dicyano vinyl (**DPPCN$_2$**). **DPP-Br** and **DPPCN$_2$** showed an intense photoabsorption band at 490 nm ($\varepsilon = 16900$ M^{-1} cm^{-1}) and 508 nm ($\varepsilon = 21400$ M^{-1} cm^{-1}), respectively, which is ascribed to a DPP-centered π–π* electron transition. The η values of the p-type NiO DSSCs based on **DPP-Br** and **DPPCN$_2$** were 0.03% and 0.07%, respectively (Favereau et al. 2013). On the other hand, Hammarström and co-workers reported that the NiO DSSCs based on an An–D–π–A DPP dye **DPP-1** with two carboxyl groups on diphenylamine yielded an η value of 0.063% with a J_{sc} value of 1.89 mA cm^{-2}, a V_{oc} value of 100 mV, and a *ff* value of 0.33, and a maximum IPCE value of 25% (Farré et al. 2016).

14.5.4 Squaraine Dyes

Lin et al. developed NIR squaraine dyes **p-SQ1** and **p-SQ2** with one and two anchoring groups as sensitizers for p-type NiO DSSCs (Fig. 14.43) (Chang et al.

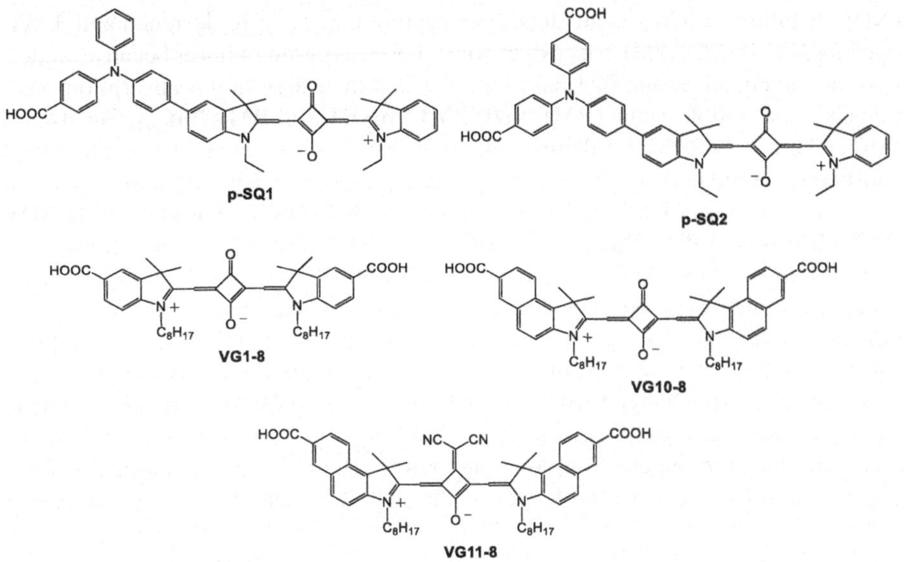

Fig. 14.43 Chemical structures of squaraine dye sensitizers **p-SQ1**, **p-SQ2**, **VG1-C8**, **VG10-C8**, and **VG11-C8** for NiO DSSCs

2012). The photovoltaic performance (J_{sc} = 1.92 mA cm^{-2}, V_{oc} = 140 mV, *ff* = 0.42, and η value = 0.113%) of the DSSC based on **p-SQ2** were higher than that (J_{sc} = 1.22 mA cm^{-2}, V_{oc} = 117 mV, *ff* = 0.371, and η value = 0.053%) based on **p-SQ1**, due to the high dye loading density of **p-SQ2** relative to **p-SQ1**. Barolo and co-workers designed and developed three symmetric squaraines **VG1-C8**, **VG10-C8**, and **VG11-C8** with octyl groups as *N*-substituents of the indolenine moieties (Bonomo et al. 2016). **VG11-C8** has a dicyano-vinyl substituent on the central squaric ring as an electron-withdrawing moiety. The dicyano-vinyl substituent is located apart from the two carboxylic anchoring groups, which connect the dye to the NiO surface. The p-type NiO DSSC based on **VG11-C8** showed the highest η value of 0.035% (η = 0.014% and 0.015% in the DSSCs based on **VG1-C8** and **VG10-C8**, respectively).

14.5.5 Perylene Monoimide Dyes

To generate long-lived charge-separated states and effectively retard geminate recombination, D–A-type dye sensitizers, "dyads," composed of a chromophore (electron donor) bound to the NiO surface and an electron acceptor located apart from the NiO surface have been studied. A series of An–D–A-type organic dyads with perylene monoimide (PMI) as an electron-accepting moiety connecting to naphthalene diimide

(NDI) or fullerene (C_{60}) as an electron-accepting moiety were designed and developed for NiO DSSCs. PMI-based dyes were chosen as chromophores because of their outstanding chemical and thermal properties and their high molar absorption coefficients in the visible range (**PMI-NDI, PMI-PhNDI**, and **PMI-PhC$_{60}$**, Fig. 14.44) (Morandeira et al. 2008; Le Pleux et al. 2009, 2011; Gibson et al. 2011). The *o*-aryl substituents could potentially prevent dye aggregation, which is well known to be an unfavorable factor in DSSCs. The η values of the NiO DSSCs based on **PMI-NDI**, **PMI-PhNDI**, and **PMI-PhC$_{60}$** were 0.073%, 0.074%, and 0.058%, respectively.

In addition, three An–D–A-type dyes **DYE36–38** with PMI as an electron-accepting moiety and oligothiophene coupled to triphenylamine as an electron-donating moiety were designed and synthesized (Fig. 14.45) (Nattestad et al. 2010; Zhang et al. 2011). In the dyes, the length of the regioregular oligo-3-hexylthiophene was systematically changed from bithiophene ($n = 1$, **DYE36**) to quaterthiophene ($n = 2$, **DYE37**) and sexithiophene ($n = 3$, **DYE38**), and carboxyl anchoring groups were introduced to the electron-donating moiety. Their charge recombination rates decreased with increasing the length of the oligothiophene linker. The enhanced

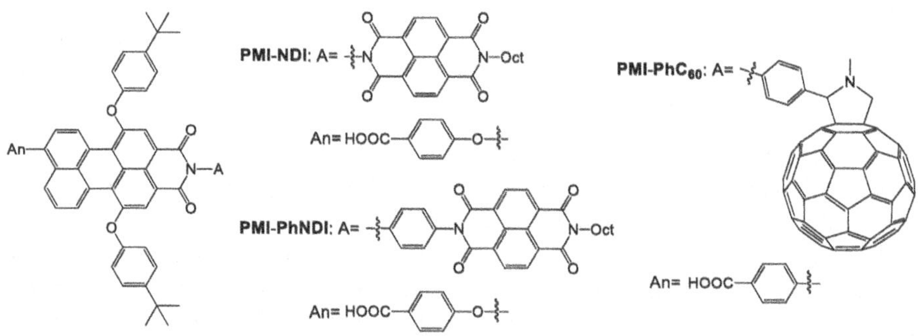

Fig. 14.44 Chemical structures of An–D–A perylene monoimide dye sensitizers **PMI-NDI, PMI-PhNDI**, and **PMI-PhC$_{60}$** for NiO DSSCs

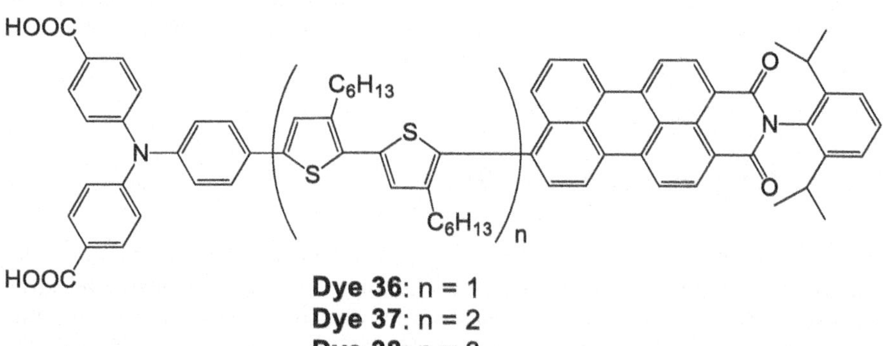

Fig. 14.45 Chemical structures of An–D–A perylene monoimide dye sensitizers **DYE36–38** for NiO DSSCs

hydrophobicity of the dyes with increasing the length of the oligothiophene linker prevented electrolytes from approaching the NiO surface and retarded interfacial charge recombination between the holes generated in the NiO electrode and the I^- ions in the electrolyte (dark current). The J_{sc}, V_{oc}, *ff*, and η values drastically increased with increasing the length of the oligothiophene linker in **DYE36–38**. The NiO DSSC using **DYE38** yielded a J_{sc} value of 5.35 mA cm^{-2}, a V_{oc} value of 218 mV, and a *ff* value of 0.35 under standard AM 1.5 simulated sunlight (100 mW cm^{-2}), corresponding to an η value of 0.41%, which is the highest conversion efficiency in NiO DSSCs. Moreover, the p-type NiO DSSC based on **DYE38** and tris(1,2-diaminoethane)cobalt(II)/(III) electrolytes exhibited a high η value of 1.3% (Powar et al. 2013).

14.6 Summary

As described in this chapter, exquisite molecular designs and synthetic strategies of organic dye sensitizers for various types of DSSCs have been developed so far to improve the photovoltaic performances and chemical and environmental stabilities. Additionally, there has been a gradual accumulation of useful knowledge about the effects of the photophysical and electrochemical properties and molecular structures of organic dye sensitizers on the DSSC performances, which has made a considerable contribution to research and development on DSSCs. However, the fact is that research on DSSCs using organic dyes has tended to stagnate in recent years. Obviously, it can be pointed out that DSSCs have not yet led to sufficient solar energy-to-electricity conversion over the wide spectral region of sunlight, i.e., from visible light to the red/NIR region, which is the major reason for the limited photovoltaic performances of DSSCs. Thus, an effective strategy to improve the photovoltaic performances is the use of panchromatic organic dye sensitizers with intense photoabsorption properties in the red/NIR region (600–1000 nm). We believe that the development of type-II organic dye sensitizers with efficient light-harvesting capabilities is a key to achieving the goal and achieving a breakthrough in the practical use of high-performance DSSCs exhibiting photosensitization over the whole region of sunlight because type-II dye sensitizers efficiently harvest sunlight particularly in the red region due to the DTCT band and eased restrictions on the LUMO energy levels. Such type-II DSSCs will attract further attention from chemists, physicists, and engineers because of the enormous scientific interest in their construction and operational principles, and much efforts are required for the practical use and creation of high-performance DSSCs.

References

Adachi Y, Ooyama Y, Shibayama N, Ohshita J (2016) Synthesis of organic photosensitizers containing dithienogermole and thiadiazolo[3,4-*c*]pyridine units for dye-sensitized solar cells. Dalton Trans 45:13817–13826

Agarkar SA, Kulkarni RR, Dhas VV, Chinchansure AA, Hazra P, Joshi SP, Ogale SB (2011) Isobutrin from butea monosperma (flame of the forest): a promising new natural sensitizer belonging to chalcone class. ACS Appl Mater Interfaces 3:2440–2444

Alagumalai A, Munavvar Fairoos MK, Vellimalai P, Sil MC, Nithyanandhan J (2016) Effect of out-of-plane Alkyl group's position in dye-sensitized solar cell efficiency: a structure–property relationship utilizing indoline-based unsymmetrical squaraine dyes. ACS Appl Mater Interfaces 8:35353–35367

Alex S, Santhosh U, Das S (2005) Dye sensitization of nanocrystalline TiO_2: enhanced efficiency of unsymmetrical versus symmetrical squaraine dyes. J Photochem Photobiol A: Chem 172:63–71

Ambade SB, Mane RS, Han S-H, Lee S-H, Sung M-M, Joo O-S (2011) Indoline-dye immobilized ZnO nanoparticles for whopping 5.44% light conversion efficiency. J Photochem Photobiol A 222:366–369

An B-K, Hu W, Burn PL, Meredith P (2010) New type II catechol-thiophene sensitizers for dye-sensitized solar cells. J Phys Chem C 114:17964–17974

Angaridis PA, Ferentinos E, Charalambidis G, Ladomenou K, Nikolaou V, Biswas S, Sharma GD, Coutsolelos AG (2016) Pyridyl vs. bipyridyl anchoring groups of porphyrin sensitizers for dye sensitized solar cells. RSC Adv 6:22187–22203

Aranyos V, Hjelm J, Hagfeldt A, Grennberg H (2001) Free-base tetra-arylphthalocyanines for dye-sensitised nanostructured solar cell applications. J Porphyrins Phthalocyanines 5:609–616

Bahadur L, Srivastava P (2003) Efficient photon-to-electron conversion with rhodamine 6G-sensitized nanocrystalline n-ZnO thin film electrodes in acetonitrile solution. Sol Energy Mater Sol Cells 79:235–248

Baheti AK, Thomas KRJ, Li C-T, Lee C-P, Ho K-C (2015) Fluorene-based sensitizers with a phenothiazine donor: effect of mode of donor tethering on the performance of dye-sensitized solar cells. ACS Appl Mater Interfaces **7**:2249–2262

Bessho T, Zakeeruddin SM, Yeh C-Y, Diau EW-G, Grätzel M (2010) Highly efficient mesoscopic dye-sensitized solar cells based on donor–acceptor-substituted porphyrins. Angew Chem 122:6796–6799 (2010); Angew Chem Int Ed 49:6646–6649

Bisht R, Munavvar Fairoos MK, Singh AK, Nithyanandhan J (2017) Panchromatic sensitizer for dye-sensitized solar cells: unsymmetrical squaraine dyes incorporating benzodithiophene π-spacer with alkyl chains to extend conjugation, control the dye assembly on TiO_2, and retard charge recombination. J Org Chem 82:1920–1930

Bolisetty MNKP, Li C-T, Thomas KRJ, Bodedla GB, Ho K-C (2015) Benzothiadiazole-based organic dyes with pyridine anchors for dye-sensitized solar cells: effect of donor on optical properties. Tetrahedron 71:4203–4212

Bonomo M, Barbero N, Matteocci F, Carlo AD, Barolo C, Dini D (2016) Beneficial effect of electron-withdrawing groups on the sensitizing action of squaraines for *p*-type dye-sensitized solar cells. J Phys Chem C 120:16340–16353

Brogdon P, Cheema H, Delcamp JH (2018) Near-infrared-absorbing metal-free organic, porphyrin, and phthalocyanine sensitizers for panchromatic dye-sensitized solar cells. Chem Sus Chem 11:86–103

Burke A, Schmidt-Mende L, Ito S, Grätzel M (2007) A novel blue dye for near-IR 'dye-sensitised' solar cell applications. Chem Commun 234–236

Burke A, Ito S, Snaith H, Bach U, Kwiatkowski J, Grätzel M (2008) The function of a TiO_2 compact layer in dye-sensitized solar cells incorporating "planar" organic dyes. Nano Lett 8:977–981

Campbell WM, Jolley KW, Wagner P, Wagner K, Walsh PJ, Gordon KC, Schmidt-Mende L, Nazeeruddin MK, Wang Q, Grätzel M, Officer DL (2007) Highly efficient porphyrin sensitizers for dye-sensitized solar cells. J Phys Chem C 111:11760–11762

Chang DW, Ko S-J, Kim JY, Park S-M, Lee HJ, Dai L, Baek J-B (2011) Multifunctional conjugated polymers with main-chain donors and side-chain acceptors for dye sensitized solar cells (DSSCs) and organic photovoltaic cells (OPVs). Macromol Rapid Commun 32:1809–1814

Chang C-H, Chen Y-C, Hsu C-Y, Chou H-H, Lin JT (2012) Squaraine-arylamine sensitizers for highly efficient p-type dye-sensitized solar cells. Org Lett 14:4726–4729

Chen Y-S, Li C, Zeng Z-H, Wang W-B, Wang X-S, Zhang B-W (2005a) Efficient electron injection due to a special adsorbing group's combination of carboxyl and hydroxyl: dye-sensitized solar cells based on new hemicyanine dyes. J Mater Chem 15:1654–1661

Chen Y, Zeng Z, Li C, Wang W, Wang X, Zhang B (2005b) Highly efficient co-sensitization of nanocrystalline TiO_2 electrodes with plural organic dyes. New J Chem 29:773–776

Chen X, Guo J, Peng X, Guo M, Xu Y, Shi L, Liang C, Wang L, Gao Y, Sun S, Cai S (2005c) Novel cyanine dyes with different methine chains as sensitizers for nanocrystalline solar cell. J Photochem Photobiol A: Chem 171:231–236

Chen Z, Li F, Huang C (2007) Organic d-pi-a dyes for dye-sensitized solar cell. Curr Org Chem 11:1241–1258

Cheng H-M, Hsieh W-F (2010) High-efficiency metal-free organic-dye-sensitized solar cells with hierarchical ZnO photoelectrode. Energy Environ Sci 3:442–447

Cherepy NJ, Smestad GP, Gr_tzel M, Zhang JZ (1997) Ultrafast electron injection: implications for a photoelectrochemical cell utilizing an anthocyanin dye-sensitized tio_2 nanocrystalline electrode. J Phys Chem B 101:9342–9351

Choi H, Lee JK, Song KH, Song K, Kang SO, Ko J (2007) Synthesis of new julolidine dyes having bithiophene derivatives for solar cell. Tetrahedron 63:1553–1559

Choi H, Baik C, Kang SO, Ko J, Kang M-S, Nazeeruddin MK, Grätzel M (2008) Highly efficient and thermally stable organic sensitizers for solvent-free dye-sensitized solar cells. Angew Chem 120:333–336 (2008); Angew Chem Int Ed 47:327–330

Choi H, Kim J-J, Song K, Ko J, Nazeeruddin MK, Grätzel M (2010) Molecular engineering of panchromatic unsymmetrical squaraines for dye-sensitized solar cell applications. J Mater Chem 20:3280–3286

Chou TP, Zhang Q, Cao G (2007) Effects of dye loading conditions on the energy conversion efficiency of ZnO and TiO_2 dye-sensitized solar cells. J Phys Chem C 111:18804–18811

Cid J-J, Yum J-H, Jang S-R, Nazeeruddin MK, Martinez-Ferrero E, Palomares E, Ko J, Grätzel M, Torres T (2007) Molecular cosensitization for efficient panchromatic dye-sensitized solar cells. Angew Chem 119:8510–8514 (2007); Angew Chem Int Ed 46:8358–8362

Cid J-J, García-Iglesias M, Yum J-H, Forneli A, Albero J, Martínez-Ferrero E, Vázquez P, Grätzel M, Nazeeruddin MK, Palomares E, Torres T (2009) Structure-function relationships in unsymmetrical zinc phthalocyanines for dye-sensitized solar cells. Chem Eur J 15:5130–5137

Cong J, Yang X, Liu J, Zhao J, Hao Y, Wang Y, Sun L (2012) Nitro group as a new anchoring group for organic dyes in dye-sensitized solar cells. Chem Commun 48:6663–6665

Dai Q, Rabani J (2002a) Unusually efficient photosensitization of nanocrystalline TiO_2 films by pomegranate pigments in aqueous medium. New J Chem 26:421–426

Dai Q, Rabani J (2002b) Photosensitization of nanocrystalline TiO_2 films by anthocyanin dyes. J Photochem Photobiol, A 148:17–24

Daphnomili D, Landrou G, Singh SP, Thomas A, Yesudas K, Sharma GD, Coutsolelos AG (2012) Photophysical, electrochemical and photovoltaic properties of dye sensitized solar cells using a series of pyridyl functionalized porphyrindyes. RSC Adv 2:12899–12908

Daphnomili D, Sharma GD, Biswas S, Justin TKR, Goutsolelos AG (2013) A new porphyrin bearing a pyridinylethynyl group as sensitizer for dye sensitized solar cells. J Photochem Photobiol, A 253:88–96

Dentani T, Funabiki K, Jin J-Y, Yoshida T, Minoura H, Matsui M (2007) Application of 9-substituted 3,4-perylenedicarboxylic anhydrides as sensitizers for zinc oxide solar cell. Dyes Pigm 72:303–307

Dentani T, Nagasaka K, Funabiki K, Jin J-Y, Yoshida T, Minoura H, Matsui M (2008) Flexible zinc oxide solar cells sensitized by styryl dyes. Dyes Pigm 77:59–69

Dentani T, Kubota Y, Funabiki K, Jin J, Yoshida T, Minoura H, Miura H, Matsui M (2009) Novel thiophene-conjugated indolinedyes for zinc oxide solar cells. New J Chem 33:93–101
Edvinsson T, Li C, Pschirer N, Schöneboom J, Eickemeyer F, Sens R, Boschloo G, Herrmann A, Müllen K, Hagfeldt A (2007) Intramolecular charge-transfer tuning of perylenes: spectroscopic features and performance in dye-sensitized solar cells. J Phys Chem C 111:15137–15140
Erten-Ela S, Yilmaz MD, Icli B, Dede Y, Icli S, Akkaya EU (2008) A panchromatic boradiazaindacene (BODIPY) sensitizer for dye-sensitized solar cells. Org Lett 10:3299–3302
Eu S, Hayashi S, Umeyama T, Matano Y, Araki Y, Imahori H (2008) Quinoxaline-fused porphyrins for dye-sensitized solar cells. J Phys Chem C 112:4396–4405
Fang Z, Eshbaugh AA, Schanze KS (2011) Low-bandgap donor–acceptor conjugated polymer sensitizers for dye-sensitized solar cells. J Am Chem Soc 133:3063–3069
Farré Y, Zhang L, Pellegrin Y, Planchat A, Blart E, Boujtita M, Hammarström L, Jacquemin D, Odobel F (2016) Second generation of diketopyrrolopyrrole dyes for NiO-based dye-sensitized solar cells. J Phys Chem C 120:7923–7940
Favereau L, Warnan J, Pellegrin Y, Blart E, Boujtita M, Jacquemin D, Odobel F (2013) Diketopyrrolopyrrole derivatives for efficient NiO-based dye-sensitized solar cells. Chem Commun 49:8018–8020
Ferrere S, Gregg BA (2001) Large increases in photocurrents and solar conversion efficiencies by UV illumination of dye sensitized solar cells. J Phys Chem B 105:7602–7605
Ferrere S, Gregg BA (2002) New J Chem 26:1155–1160
Franchi D, Calamante M, Reginato G, Zani L, Peruzzini M, Taddei M, de Biani FF, Basosi R, Sinicropi A, Colonna D, Di Carlo A, Mordin A (2014) A comparison of carboxypyridine isomers as sensitizers for dye-sensitized solar cells: assessment of device efficiency and stability. Tetrahedron 70:6285–6295
Frei H, Fitzmaurice DJ, Grätzel M (1990) Surface chelation of semiconductors and interfacial electron transfer. Langmuir 6:198–206
Funabiki K, Mase H, Hibino A, Tanaka N, Mizuhata N, Sakuragi Y, Nakashima A, Yoshida T, Kubota Y, Matsui M (2011) Synthesis of a novel heptamethine–cyanine dye for use in near-infrared active dye-sensitized solar cells with porous zinc oxide prepared at low temperature. Energy Environ Sci 4:2186–2192
Furube A, Katoh R, Hara K, Sato T, Murata S, Arakawa H, Tachiya M (2005) Lithium ion effect on electron injection from a photoexcited coumarin derivative into a TiO_2 nanocrystalline film investigated by visible-to-IR ultrafast spectroscopy. J Phys Chem B 109:16406–16414
García-Iglesias M, Yum J-H, Humphry-Baker R, Zakeeruddin SM, Péchy P, Vázquez P, Palomares E, Grätzel M, Nazeeruddin MK (2011) Effect of anchoring groups in zinc phthalocyanine on the dye-sensitized solar cell performance and stability. Chem Sci 2:1145–1150
Gibson EA, Smeigh AL, Pleux LL, Hammarstrcm L, Odobel F, Boschloo G, Hagfeldt A (2011) Cobalt polypyridyl-based electrolytes for p-type dye-sensitized solar cells. J Phys Chem C 115:9772–9779
Giribabu L, Kumar CV, Reddy VG, Reddy PY, Rao CS, Jang S-R, Yum J-H, Nazeeruddin MD, Grätzel M (2007) Unsymmetrical alkoxy zinc phthalocyanine for sensitization of nanocrystalline TiO_2 films. Sol Energy Mater Sol Cells 91:1611–1617
Guillén E, Gasanueva F, Anta JA, Vega-poot A, Oskam G, Alcántara R, Fernández-Lorenzo C, Martín-Calleja J (2008) Photovoltaic performance of nanostructured zinc oxide sensitised with xanthene dyes. J Photochem Photobiol, A 200:364–370
Guillén E, Azaceta E, Peter LM, Zukal A, Tena-Zaera R, Anta JA (2011) ZnO solar cells with an indoline sensitizer: a comparison between nanoparticulate films and electrodeposited nanowire arrays. Energy Environ Sci 4:3400–3407
Günes S, Neugebauer H, Sariciftci NS (2007) Conjugated polymer-based organic solar cells. Chem Rev 107:1324–1338
Guo M, Diao P, Ren Y-J, Meng F, Tian H, Cai S-M (2005) Photoelectrochemical studies of nanocrystalline TiO_2 co-sensitized by novel cyanine dyes. Sol Energy Mater Sol Cells 88:23–35

Gupta KSV, Suresh T, Singh SP, Islam A, Han L, Chandrasekharam M (2014) Carbazole based A-π-D-π-A dyes with double electron acceptor for dye-sensitized solar cell. Org Electron 15:266–275

Hagberg DP, Edvinsson T, Marinado T, Boschloo G, Hagfeldt A, Sun L (2006) A novel organic chromophore for dye-sensitized nanostructured solar cells. Chem Commun 2245–2247

Hagberg DP, Marinado T, Karlsson KM, Nonomura K, Qin P, Boschloo G, Brinck T, Hagfeldt A, Sun L (2007) Tuning the HOMO and LUMO energy levels of organic chromophores for dye sensitized solar cells. J Org Chem 72:9550–9556

Hagberg DP, Yum J-H, Lee H, Angelis FD, Marinado T, Karlsson KM, Humphry-Baker R, Sun L, Hagfeldt A, Grätzel M, Nazeeruddin MK (2008) Molecular engineering of organic sensitizers for dye-sensitized solar cell applications. J Am Chem Soc 130:6259–6266

Hagfeldt A, Boschloo G, Sun L, Kloo L, Pettersson H (2010) Dye-sensitized solar cells. Chem Rev 110:6595–6663

Hao Y, Yang M, Yu C, Cai S, Liu M, Fan L, Li Y (1998) Photoelectrochemical studies on acid-doped polyaniline as sensitizer for TiO_2 nanoporous film. Sol Energy Mater Sol Cells 56:75–84

Hao S, Wu J, Huang Y, Lin J (2006) Natural dyes as photosensitizers for dye-sensitized solar cell. Sol Energy 80:209–214

Hao Y, Yang X, Cong J, Tian H, Hagfeldt A, Sun L (2009) Efficient near infrared D–π–A sensitizers with lateral anchoring group for dye-sensitized solar cells. Chem Commun 4031–4033

Hara K, Horiguchi T, Kinoshita T, Sayama K, Sugihara H, Arakawa H (2000a) Highly efficient photon-to-electron conversion with mercurochrome-sensitized nanoporous oxide semiconductor solar cells. Sol Energy Mater Sol Cells 64:115–134

Hara K, Horiguchi T, Kinoshita T, Sayama K, Sugihara H, Arakawa H (2000) Highly efficient photon-to-electron conversion of mercurochrome-sensitized nanoporous ZnO solar cells. Chem Lett 316–317

Hara K, Sayama K, Ohga Y, Shinpo A, Suga S, Arakawa H (2001) A coumarin-derivative dye sensitized nanocrystalline TiO_2 solar cell having a high solar-energy conversion efficiency up to 5.6%. Chem Commun 569–570

Hara K, Kurashige M, Ito S, Shinpo A, Suga S, Sayama K, Arakawa H (2003a) Novel polyenedyes for highly efficient dye-sensitized solar cells. Chem Commun 252–253

Hara K, Tachibana Y, Ohga Y, Shinpo A, Suga S, Sayama K, Sugihara H, Arakawa H (2003b) Dye-sensitized nanocrystalline TiO_2 solar cells based on novel coumarin dyes. Sol Energy Mater Sol Cells 77:89–103

Hara K, Sato T, Katoh R, Furude A, Ohga Y, Shinpo A, Suga S, Sayama K, Sugihara H, Arakawa H (2003c) Molecular design of coumarin dyes for efficient dye-sensitized solar cells. J Phys Chem B 107:597–606

Hara K, Kurashige M, Dan-oh Y, Kasada C, Shinpo A, Suga S, Sayama K, Arakawa H (2003d) Design of new coumarindyes having thiophene moieties for highly efficient organic-dye-sensitized solar cells. New J Chem 27:783–785

Hara K, Dan-oh Y, Kasada C, Ohga Y, Shinpo A, Suga S, Sayama K, Arakawa H (2004) Effect of additives on the photovoltaic performance of coumarin-dye-sensitized nanocrystalline TiO_2 solar cells. Langmuir 20:4205–4210

Hara K, Sato T, Katoh R, Furube A, Yoshihara T, Murai M, Kurashige M, Ito S, Shinpo A, Suga S, Arakawa H (2005a) Novel conjugated organic dyes for efficient dye-sensitized solar cells. Adv Funct Mater 15:246–252

Hara K, Wang Z-S, Sato T, Furube A, Katoh R, Sugihara H, Dan-oh Y, Kasada C, Shinpo A, Suga S (2005b) Oligothiophene-containing coumarin dyes for efficient dye-sensitized solar cells. J Phys Chem B 109:15476–15482

Hara K, Miyamoto K, Abe Y, Yanagida M (2005c) Electron transport in coumarin-dye- sensitized nanocrystalline TiO_2 electrodes. J Phys Chem B 109:23776–23778

Harima Y, Fujita T, Kano Y, Imae I, Komaguchi K, Ooyama Y, Ohshita J (2013) Lewis-acid sites of TiO_2 Surface for adsorption of organic dye having pyridyl group as anchoring unit. J Phys Chem C 117:16364–16370

Harima Y, Kano Y, Fujita T, Imae I, Ooyama Y, Ohshita J (2015) Photoinduced electron injection from an organic dye having a pyridyl anchor to Lewis acid site of TiO_2 surfac. RSC Adv 5:71387–71392

Hattori S, Hasobe T, Ohkubo K, Urano Y, Umezawa N, Nagano T, Wada Y, Yanagida S, Fukuzumi S (2004) Enhanced energy and quantum efficiencies of a nanocrystalline photoelectrochemical cell sensitized with a donor−acceptor dyad derived from fluorescein. J Phys Chem B 108:15200–15205

Hattori S, Ohkubo K, Urano Y, Sunahara H, Nagano T, Wada Y, Tkachenko NV, Lemmetyineu H, Fukuzumi S (2005) Charge separation in a nonfluorescent donor-acceptor dyad derived from boron dipyrromethene dye, leading to photocurrent generation. J Phys Chem B 109:15368–15375

Hayashi S, Matsubara Y, Eu S, Hayashi H, Umeyama T, Matano Y, Imahori H (2008) Fused five-membered porphyrin for dye-sensitized solar cells. Chem Lett 37:846–847

He J, Hagfeldt A, Lindquist S-E, Grennberg H, Korodi F, Sun L, Åkermark B (2001) Phthalocyanine-sensitized nanostructured TiO_2 electrodes prepared by a novel anchoring method. Langmuir 17:2743–2747

He J, Benkö G, Korodi F, Polivka T, Lomoth R, Åkermark B, Sun L, Hagfeldt A, Sundström V (2002) Modified phthalocyanines for efficient near-ir sensitization of nanostructured TiO_2 electrode. J Am Chem Soc 124:4922–4932

Higashijima S, Miura H, Fujita T, Kubota Y, Funabiki K, Yoshida T, Matsui M (2011) Highly efficient new indoline dye having strong electron-withdrawing group for zinc oxide dye-sensitized solar cell. Tetrahedron 67:6289–6293

Higashino T, Imahori H (2015) Porphyrins as excellent dyes for dye-sensitized solar cells: recent developments and insights. Dalton Trans 44:448–463

Higashino T, Fujimori Y, Sugiura K, Tsuji Y, Ito S, Imahori H (2015) Tropolone as a high-performance robust anchoring group for dye-sensitized solar cells. Angew Chem Int Ed 54:9052–9056

Hong JS, Joo M, Vittal R, Kim K-J (2002) Improved photocurrent-voltage characteristics of Ru(II)-dye sensitized solar cells with polypyrrole-anchored TiO_2 films. J Electrochem Soc 149:E493–E496

Horiuchi T, Miura H, Uchida S (2003) Highly-efficient metal-free organic dyes for dye-sensitized solar cells. Chem Commun 3036–3037

Horiuchi T, Miura H, Uchida S (2004a) Highly efficient metal-free organic dyes for dye-sensitized solar cells. J Photochem Photobiol A: Chem 164:29–32

Horiuchi T, Miura H, Sumioka K, Uchida S (2004b) High efficiency of dye-sensitized solar cells based on metal-free indoline dyes. J Am Chem Soc 126:12218–12219

Hosono E, Fujihara S, Kimura T (2004) Synthesis, structure and photoelectrochemical performance of micro/nano-textured ZnO/eosin Y electrodes. Electrochim Acta 49:2287–2293

Hosono E, Mitsui Y, Zhou H (2008) Metal-free organic dye sensitized solar cell based on perpendicular zinc oxide nanosheet thick films with high conversion efficiency. Dalton Trans 5439–5441

Howie WH, Claeyssens F, Miura H, Peter LM (2008) Characterization of solid-state dye-sensitized solar cells utilizing high absorption coefficient metal-free organic dyes. J Am Chem Soc 130:1367–1375

Hsieh C-P, Lu H-P, Chiu C-L, Lee C-W, Chuang S-H, Mai C-L, Yen W-N, Hsu S-J, Diau EW-G, Yeh C-Y (2010) Synthesis and characterization of porphyrin sensitizers with various electron-donating substituents for highly efficient dye-sensitized solar cells. J Mater Chem 20:1127–1134

Huang Z-S, Meier H, Cao D (2016) Phenothiazine-based dyes for efficient dye-sensitized solar cells. J Mater Chem C 4:2404–2426

Hung W-I, Liao Y-Y, Hsu C-Y, Chou H-H, Lee T-H, Kao W-S, Lin JT (2014) High-performance dye-sensitized solar cells based on phenothiazine dyes containing double anchors and thiophene spacers. Chem Asian J 9:357–366

Hung W-I, Liao Y-Y, Lee T-H, Ting Y-C, Ni J-S, Kao W-S, Lin JT, Wei T-C, Yen Y-S (2015) Eugenic metal-free sensitizers with double anchors for high performance dye-sensitized solar cells. Chem Commun 51:2152–2155

Ikeuchi T, Agrawal S, Ezoe M, Mori S, Kimura M (2015) Enhanced charge separation efficiency in pyridine-anchored phthalocyanine-sensitized solar cells by linker elongation. Chem Asian J 10:2347–2351

Im H, Kim S, Park C, Jang S-H, Kim C-J, Kim K, Park N-G, Kim C (2010) High performance organic photosensitizers for dye-sensitized solar cells. Chem Commun 46:1335–1337

Imahori H, Umeyama T, Ito S (2009) Large π-aromatic molecules as potential sensitizers for highly efficient dye-sensitized solar cells. Acc Chem Res 42:1809–1818

Ince M, Cardinali F, Yum J-H, Martínez-Díaz MV, Nazeeruddin MK, Grätzel M, Torres T (2012) Convergent synthesis of near-infrared absorbing, "Push-Pull", bisthiophene-substituted, Zinc(II) phthalocyanines and their application in dye-sensitized solar cells. Chem Eur J 18:6343–6348

Ito S, Zakeeruddin SM, Humphry-Baker R, Liska P, Charvet R, Comte P, Nazeeruddin MK, Péchy P, Takata M, Miura H, Uchida S, Grätzel M (2006) High-Efficiency organic-dye-sensitized solar cells controlled by nanocrystalline-TiO_2 electrode thickness. Adv Mater 18:1202–1205

Ito S, Miura H, Uchida S, Takata M, Sumioka K, Liska P, Comte P, Péchy P, Grätzel M (2008) High-conversion-efficiency organic dye-sensitized solar cells with a novel indoline dye. Chem Commun 5194–5196

Jayaweera PM, Kumarasinghe AR, Tennakone T (1999) Nano-porous TiO_2 photovoltaic cells sensitized with metallochromic triphenylmethane dyes: [n-TiO_2/triphenylmethane dye/p-I –/I3–(or CuI)]. J Photochem Photobiol, A 126:111–115

Ji Z, Natu G, Huang Z, Wu Y (2011) Linker effect in organic donor–acceptor dyes for p-type NiO dye sensitized solar cells. Energy Environ Sci 4:2818–2821

Ji J-M, Zhou H, Kim HK (2018) Rational design criteria for D–π–A structured organic and porphyrin sensitizers for highly efficient dye-sensitized solar cells. J Mater Chem A 6:14518–14545

Jia H, Shen K, Ju XI, Zhanga M, Zheng H (2016) Enhanced performance of dye-sensitized solar cells with Y-shaped organic dyes containing di-anchoring groups. New J Chem 40:2799–2805

Jiang S, Fan S, Lu X, Zhou G, Wang Z-S (2014) Double D–π–A branched organic dye isomers for dye-sensitized solar cells. J Mater Chem A 2:17153–17164

Johansson EMJ, Edvinsson T, Odelius M, Hagberg DP, Sun L, Hagfeld A, Siegbahn H, Rensmo H (2007) Electronic and molecular surface structure of a polyene–diphenylaniline dye adsorbed from solution onto nanoporous TiO_2. J Phys Chem C 111:8580–8586

Jradi FM, Kang X, O'Neil D, Pajares G, Getmanenko YA, Szymanski P, Parker TC, El-Sayed MA, Marder SR (2015) Near-infrared asymmetrical squaraine sensitizers for highly efficient dye sensitized solar cells: the effect of π-bridges and anchoring groups on solar cell performance. Chem Mater 27:2480–2487

Kakiuchi K, Hosono E, Fujihara S (2006) Enhanced photoelectrochemical performance of ZnO electrodes sensitized with N-719. J Photochem Photobiol, A 179:81–86

Keis K, Lindgren J, Lindquist S-E, Hagfeldt A (2000) Studies of the adsorption process of ru complexes in nanoporous ZnO electrodes. Langmuir 16:4688–4694

Keis K, Magnusson E, Lindstrcm H, Lindquist S-E, Hagfeldt A (2002) A 5% efficient photoelectrochemical solar cell based on nanostructured ZnO electrodes. Sol Energy Mater Sol Cells 73:51–58

Khazraji AC, Hotchandani S, Das S, Kamat PV (1999) Controlling dye (merocyanine-540) aggregation on nanostructured TiO_2 films. An organized assembly approach for enhancing the efficiency of photosensitization. J Phys Chem B 103:4693–4700

Kim Y-G, Walker J, Samuelson LA, Kumar J (2003) Efficient light harvesting polymers for nanocrystalline TiO_2 photovoltaic cells. Nano Lett 3:523–525

Kim S, Lee JK, Kang SO, Ko J, Yum J-H, Frantacci S, Angelis FD, Censo DD, Nazeeruddin MK, Grätzel M (2006) Molecular engineering of organic sensitizers for solar cell applications. J Am Chem Soc 128:16701–16707

Kim JY, Lee K, Coates NE, Moses D, Nguyen T-Q, Dante M, Heeger AJ (2007) Efficient tandem polymer solar cells fabricated by all-solution processing. Science 317:222–225

Kim C, Choi H, Kim S, Baik C, Song K, Kang M-S, Kang SO, Ko J (2008a) Molecular engineering of organic sensitizers containing *p*-phenylene vinylene unit for dye-sensitized solar cells. J Org Chem 73:7072–7079

Kim S, Kim D, Choi H, Kang M-S, Song K, Kang SO, Ko J (2008b) Enhanced photovoltaic performance and long-term stability of quasi-solid-state dye-sensitized solar cellsvia molecular engineering. Chem Commun 4951–4953

Kim D, Kang M-S, Song K, Kang SO, Ko J (2008c) Molecular engineering of organic sensitizers containing indole moiety for dye-sensitized solar cells. Tetrahedron 64:10417–10424

Kimura M, Nomoto H, Masaki N, Mori S (2012) Dye molecules for simple co-sensitization process: fabrication of mixed-dye-sensitized solar cells. Angew Chem 124:4447–4450 (2012); Angew Chem Int Ed 51:4371–4374

Kolemen S, Cakmak Y, Erten-Ela S, Altay Y, Brendel J, Thelakkat M, Akkaya EU (2010) Solid-state dye-sensitized solar cells using red and near-IR absorbing bodipy sensitizers. Org Lett 12:3812–3815

Kolemen S, Bozdemir OA, Cakmak Y, Barin G, Erten-Ela S, Marszalek M, Yum J-H, Zakeeruddin SM, Nazeeruddin MK, Grätzel M, Akkaya EU (2011) Optimization of distyryl-Bodipy chromophores for efficient panchromatic sensitization in dye sensitized solar cells. Chem Sci 2:949–954

Koumura N, Wang Z-S, Mori S, Miyashita M, Suzuki E, Hara K (2006) Alkyl-functionalized organic dyes for efficient molecular photovoltaics. J Am Chem Soc 128:14256–14257

Kuang D, Uchida S, Humphry-Baker R, Zakeeruddin SK, Grätzel M (2008) Organic dye-sensitized ionic liquid based solar cells: remarkable enhancement in performance through molecular design of indoline sensitizers. Angew Chem 120:1949–1953 (2008); Angew Chem Int Ed 47:1923–1927

Kubo Y, Eguchi D, Matsumoto A, Nishiyabu R, Yakushiji H, Shigaki K, Kaneko M (2014) Boron–dibenzopyrromethene-based organic dyes for application in dye-sensitized solar cells. J Mater Chem A 2:5204–5211

Kumara GRA, Kaneko S, Okuya M, Onwona-Agyeman B, Konno A, Tennakone K (2006) Shiso leaf pigments for dye-sensitized solid-state solar cell. Sol Energy Mater Sol Cells 90:1220–1226

Kumaresan D, Thummel RP, Bura T, Ulrich G, Ziessel R (2009) Color tuning in new metal-free organic sensitizers (bodipys) for dye-sensitized solar cells. Chem Eur J 15:6335–6339

Kuster S, Sauvage F, Nazeeruddin MK, Grätzel M, Nüesh FA, Geiger T (2010) Unsymmetrical squaraine dimer with an extended π-electron framework: An approach in harvesting near infra-red photons for energy conversion. Dyes Pigm 87:30–38

Ladomenou K, Kitsopoulos TN, Sharma GD, Coutsolelos AG (2014) The importance of various anchoring groups attached on porphyrins as potential dyes for DSSC applications. RSC Adv 4:21379–21404

Le Pleux L, Chavillon B, Pellegrin Y, Blart E, Cario L, Jobic S, Odobel F (2009) Simple and reproducible procedure to prepare self-nanostructured NiO films for the fabrication of p-type dye-sensitized solar cells. Inorg Chem 48:8245–8250

Le Pleux L, Smeigh AL, Gibson E, Pellegrin Y, Blart E, Boschloo G, Hagfeldt A, Hammarstrcm L, Odobel F (2011) Synthesis, photophysical and photovoltaic investigations of acceptor-functionalized perylene monoimide dyes for nickel oxide p-type dye-sensitized solar cells. Energy Environ Sci 4:2075–2084

Lee C-W, Lu H-P, Lan C-M, Huang Y-L, Liang Y-R, Yen W-N, Liu Y-C, Lin Y-S, Diau EW-G, Yeh C-Y (2009) Novel zinc porphyrin sensitizers for dye-sensitized solar cells: synthesis and spectral, electrochemical, and photovoltaic properties. Chem Eur J 15:1403–1412

Lee C-P, Lin RY-Y, Lin L-Y, Li C-T, Chu T-C, Sun S-S, Lin JT, Ho K-C (2015) Recent progress in organic sensitizers for dye-sensitized solar cells. RSC Adv 5:23810–23825

Li L-L, Diau EW-G (2013) Porphyrin-sensitized solar cells. Chem Soc Rev 42:291–304

Li C, Wang W, Wang X, Zhang B, Cao Y (2005) Molecular design of squaraine dyes for efficient far-red and near-IR sensitization of solar cells. Chem Lett 34:554–555

Li C, Yum J-H, Moon S-J, Herrmann A, Eickemeyer F, Pschirer NG, Erk P, Schöneboom J, Müllen K, Grätzel M, Nazeeruddin MK (2008) An improved perylene sensitizer for solar cell applications. Chem Sus chem 1:615–618

Li J-Y, Chen C-Y, Lee C-P, Chen S-C, Lin T-H, Tsai H-H, Ho K-C, Wu C-G (2010a) Unsymmetrical squaraines incorporating the thiophene unit for panchromatic dye-sensitized solar cells. Org Lett 12:5454–5457

Li L, Gibson EA, Qin P, Boschloo G, Gorlov M, Hagfeldt A, Sun L (2010b) Double-layered NiO photocathodes for p-type DSSCs with record IPCE. Adv Mater 22:1759–1762

Li C-T, Lin RY-Y, Lin JT (2017) Sensitizers for aqueous-based solar cells. Chem Asian J 12:486–496

Liu W-H, Wu I-C, Lai C-H, Lai C-H, Chou P-T, Li Y-T, Chen C-L, Hsu Y-Y, Chi Y (2008) Simple organic molecules bearing a 3,4-ethylenedioxythiophene linker for efficient dye-sensitized solar cells. Chem Commun 5152–5154 (2008)

Liu J, Zhang J, Xu M, Zhou D, Jing X, Wang P (2011) Mesoscopic titania solar cells with the tris(1,10-phenanthroline)cobalt redox shuttle: uniped versus biped organic dyes. Energy Environ Sci 4:3021–3029

Liu Y-C, Chou H-H, Ho F-Y, Wei H-J, Wei T-C, Yeh C-Y (2016) A feasible scalable porphyrin dye for dye-sensitized solar cells under one sun and dim light environments. J Mater Chem A 4:11878–11887

Lu H-P, Tsai C-Y, Yen W-N, Hsieh C-P, Lee C-W, Yeh C-Y, Diau EW-G (2009) Control of dye aggregation and electron injection for highly efficient porphyrin sensitizers adsorbed on semiconductor films with varying ratios of coadsorbate. J Phys Chem C 113:20990–20997

Lu J, Xu X, Li Z, Cao K, Cui J, Zhang Y, Shen Y, LiY, Zhu J, Dai S, Chjen W, Cheng Y, Wang M (2013) Zinc porphyrins with a pyridine-ring-anchoring group for dye- sensitized solar cells. Chem Asian J 8:956–962 (2013)

Ma X, Hua J, Wu W, Jin Y, Meng F, Tian H (2008) A high-efficiency cyanine dye for dye-sensitized solar cells. Tetrahedron 64:345–350

Mai C-L, Moehl T, Hsieh C-H, Décoppet J-D, Zakeeruddin SM, Grätzel M, Yeh C-Y (2015) Porphyrin sensitizers bearing a pyridine-type anchoring group for dye-sensitized solar cells. ACS Appl Mater Interfaces 7:14975–14982

Manfredi N, Cecconi B, Abbotto A (2014) Multi-branched multi-anchoring metal-free dyes for dye-sensitized solar cells. Eur J Org Chem 7069–7086

Mann JR, Gannon MK, Fitzgibbons TC, Detty MR, Watson DF (2008) Optimizing the photocurrent efficiency of dye-sensitized solar cells through the controlled aggregation of chalcogenoxanthylium dyes on nanocrystalline titania films. J Phys Chem C 112:13057–13061

Mao M, Wang J-B, Xiao Z-F, Dai S-Y, Song Q-H (2012) New 2,6-modified BODIPY sensitizers for dye-sensitized solar cells. Dyes Pigm 94:224–232

Mao J, Wang D, Liu S-H, Hang Y, Xu Y, Zhang Q, Wu W, Chou P-T, Hua J (2014) Dye-sensitized solar cells based on functionally separated D-π-A dyes with 2-cyanopyridine as an electron-accepting and anchoring group. Asian J Org Chem 3:153–160

Mathew S, Yella A, Gao P, Humphry-Baker R, Curchod BFE, Ashari-Astani N, Tavernelli I, Rothlisberger U, Nazeeruddin MdK, Grätzel M (2014) Dye-sensitized solar cells with 13% efficiency achieved through the molecular engineering of porphyrin sensitizers. Nat Chem 6:242–247

Matsui M, Hashimoto Y, Funabiki K, Jin J-Y, Yoshida T, Minoura H (2005) Application of near-infrared absorbing heptamethine cyanine dyes as sensitizers for zinc oxide solar cell. Synth Met 148:147–153

Matsui M, Mase H, Jin J-Y, Funabiki K, Yoshida T, Minoura H (2006) Application of semisquaric acids as sensitizers for zinc oxide solar cell. Dyes Pigm 70:48–53

Matsui M, Ito A, Kotani M, Kubota Y, Funabiki K, Jin J, Yoshida T, Minoura H, Miura H (2009) The use of indoline dyes in a zinc oxide dye-sensitized solar cell. Dyes Pigm 80:233–238

Matsui M, Fujita T, Kubota Y, Funabiki K, Jin J, Yoshida T, Miura H (2010a) Substituent effects in a double rhodanine indoline dye on performance of zinc oxide dye-sensitized solar cell. Dyes Pigm 86:143–148

Matsui M, Asamura Y, Kubota Y, Funabiki K, Jin J, Yoshida T, Miura H (2010b) Highly efficient substituted triple rhodanine indoline dyes in zinc oxide dye-sensitized solar cell. Tetrahedron 66:7405–7410

Matsui M, Kotani M, Kubota Y, Funabiki K, Jin J, Yoshida T, Higashijima S, Miura H (2011) Comparison of performance between benzoindoline and indoline dyes in zinc oxide dye-sensitized solar cell. Dyes Pigm 91:145–152

Meier H, Huang Z-S, Cao D (2017) Double D–π–A branched dyes: a new class of metal-free organic dyes for efficient dye-sensitized solar cells. J Mater Chem C 5:9828–9837

Meinhardt U, Lodermeyer F, Schaub TA, Kunzmann A, Dral PO, Sale AC, Hampel F, Guldi DM, Costa RD, Kivala M (2016) *N*-Heterotriangulene chromophores with 4-pyridyl anchors for dye-sensitized solar cells. RSC Adv 6:67372–67377

Mikroyannidis JA, Stylianakis MM, Suresh P, Roy MS, Sharma GD (2009) Synthesis of perylene monoimide derivative and its use for quasi-solid-state dye-sensitized solar cells based on bare and modified nano-crystalline ZnO photoelectrodes. Energy Environ Sci 2:1293–1301

Mishra A, Fischer MKR, Bäuerle P (2009) Metal-free organic dyes for dye-sensitized solar cells: from structure: property relationships to design rules. Angew Chem 121:2510–2536 (2009); Angew Chem Int Ed 48:2474–2499

Morandeira A, Fortage J, Edvinsson T, Pleux LL, Blart E, Boscjloo G, Hagfeldt A, Hammarstrcm L, Odobel F (2008) Improved photon-to-current conversion efficiency with a nanoporous p-type NiO electrode by the use of a sensitizer-acceptor dyad. J Phys Chem C 112:1721–1728

Mori S, Nagata M, Nakahata Y, Yasuta K, Goto R, Kimura M, Taya M (2010) Enhancement of incident photon-to-current conversion efficiency for phthalocyanine-sensitized solar cells by 3D molecular structuralization. J Am Chem Soc 132:4054–4055

Moser J, Punchihewa S, Infelta PP, Grätzel M (1991) Surface complexation of colloidal semiconductors strongly enhances interfacial electron-transfer rates. Langmuir 7:3012–3018

Mwaura JK, Zhao X, Jiang H, Schanze KS, Reynolds JR (2006) Spectral broadening in nanocrystalline TiO_2 solar cells based on poly(*p*-phenylene ethynylene) and polythiophene sensitizers. Chem Mater 18:6109–6111

Narayan MR (2012) Review: dye sensitized solar cells based on natural photosensitizers. Renew Sustain Energy Rev 16:208–215

Nattestad A, Mozer AJ, Fischer MKR, Cheng Y-B, Mishra A, Bäuerle P, Bach U (2010) Highly efficient photocathodes for dye-sensitized tandem solar cells. Nat Mater 9:31–35

Nazeeruddin MK, Kay A, Rodicio I, Humphry-Baker R, Müller E, Liska P, Vlachopolous N, Grätzel M (1993) Conversion of light to electricity by *cis*-X2bis(2,2′-bipyridyl- 4,4′-dicarboxylate)ruthenium(II) charge-transfer sensitizers (X = Cl–, Br–, I–, CN–, and SCN–) on nanocrystalline titanium dioxide electrodes. J Am Chem Soc 115:6382–6390

Nazeeruddin MK, Pechy P, Grätzel M (1997) Efficient panchromatic sensitization of nanocrystalline TiO_2 films by a black dye based on a trithiocyanato–ruthenium complex. Chem Commun 1705–1706

Nazeeruddin MK, Zakeeruddin SM, Humphry-Baker R, Jirousek M, Liska P, Vlachopoulos N, Shklover V, Fischer C-H, Grätzel M (1999a) Acid−base equilibria of (2,2′-Bipyridyl-4,4′-dicarboxylic acid)ruthenium(II) complexes and the effect of protonation on charge-transfer sensitization of nanocrystalline titania. Inorg Chem 38:6298–6305

Nazeeruddin MD, Humphry-Baker R, Grätzel M, Wöhrle D, Schnurpfeil G, Schneider G, Hirth A, Trombach N (1999b) Efficient near-IR sensitization of nanocrystalline TiO_2 films by zinc and aluminum phthalocyanines. J Porphyrins Phthalocyanines 3:230–237

Nazeeruddin MK, Pechy P, Renouard T, Zakeeruddin SM, Humphry-Baker R, Comte P, Liska P, Cevey L, Costa E, Shklover V, Spiccia L, Deacon GB, Bignozzi CA, Grätzel M (2001) Engineering of efficient panchromatic sensitizers for nanocrystalline TiO_2-based solar cells. J Am Chem Soc 123:1613–1624

Nazeeruddin MK, Splivallo R, Liska P, Comte P, Grätzel M (2003) A swift dye uptake procedure for dye sensitized solar cells. Chem Commun 1456–1457

Nguyen H-M, Mane RS, Ganesh T, Han S-H, Kim N (2009) Aggregation-free ZnO nanocrystals coupled HMP-2 dye of higher extinction coefficient for enhancing energy conversion efficiency. J Phys Chem C 113:9206–9209

Nikolaou V, Charisiadis A, Charalambidis G, Coutsolelos AG, Odobel F (2017) Recent advances and insights in dye-sensitized NiO photocathodes for photovoltaic devices. J Mater Chem A 5:21077–21113

Ning Z, Tian H (2009) Triarylamine: a promising core unit for efficient photovoltaic materials. Chem Commun 5483–5495

Ning Z, Fu Y, Tian H (2010) Improvement of dye-sensitized solar cells: what we know and what we need to know. Energy Environ Sci 3:1170–1181

O'Regan B, Grätzel M (1991) A low-cost, high-efficiency solar cell based on dye-sensitized colloidal TiO_2 films. Nature 353:737–740

Odobel F, Pleux LL, Pellegrin Y, Blart E (2010) New photovoltaic devices based on the sensitization of p-type semiconductors: challenges and opportunities. Acc Chem Res 43:1063–1071

Ooyama Y, Harima Y (2009) Molecular designs and syntheses of organic dyes for dye- sensitized solar cells. Eur J Org Chem 2903–2934

Ooyama Y, Harima Y (2012) Photophysical and electrochemical properties, and molecular structures of organic dyes for dye-sensitized solar cells. Chem Phys Chem 13:4032–4080

Ooyama Y, Shimada Y, Kagawa Y, Imae I, Harima Y (2007a) Photovoltaic performance of dye-sensitized solar cells based on donor–acceptor π-conjugated benzofuro[2,3-*c*]oxazolo [4,5-*a*]carbazole-type fluorescent dyes with a carboxyl group at different positions of the chromophore skeleton. Org Biomol Chem 5:2046–2054

Ooyama Y, Shimada Y, Kagawa Y, Yamada Y, Imae I, Komaguchi K, Harima Y (2007b) Synthesis of new-type donor–acceptor π-conjugated benzofuro[2,3-*c*]oxazolo[4,5-*a*]carbazole fluorescent dyes and their photovoltaic performances of dye-sensitized solar cells. Tetrahedron Lett 48:9167–9170

Ooyama Y, Shimada Y, Ishii A, Ito G, Kagawa Y, Imae I, Komaguchi K, Harima Y (2009) Photovoltaic performance of dye-sensitized solar cells based on a series of new-type donor–acceptor π-conjugated sensitizer, benzofuro[2,3-*c*]oxazolo[4,5-*a*]carbazole fluorescent dyes. J Photochem Photobiol A 203:177–185

Ooyama Y, Inoue S, Asada R, Ito G, Kushimoto K, Komaguchi K, Imae I, Harima Y (2010a) Dye-sensitized solar cells based on a novel fluorescent dye with a pyridine ring and a pyridinium dye with the pyridinium ring forming strong interactions with nanocrystalline TiO_2 films. Eur J Org Chem 92–100 (2010)

Ooyama Y, Nagano T, Inoue S, Imae I, Komaguchi K, Harima Y (2010b) Dye-sensitized solar cells based on D–π–A fluorescent dyes with pyridine ring forming strong interaction with nanocrystalline TiO_2. Bull Chem Soc Jpn 83:1113–1121

Ooyama Y, Inoue S, Asada R, Ito G, Kushimoto K, Komaguchi K, Imae I, Harima Y (2010c) Dye-sensitized solar cells based on a novel fluorescent dye with a pyridine ring and a pyridinium dye with the pyridinium ring forming strong interactions with nanocrystalline TiO_2 films. Eur J Org Chem 92–100

Ooyama Y, Shimada Y, Inoue S, Nagano T, Fujikawa Y, Komaguchi K, Imae I, Harima Y (2011a) New molecular design of donor-π-acceptor dyes for dye-sensitized solar cells: control of molecular orientation and arrangement on TiO_2 surface. New J Chem 35:111–118

Ooyama Y, Inoue S, Nagano T, Kushimoto K, Ohshita J, Imae I, Komaguchi K, Harima Y (2011b) Dye-sensitized solar cells based on donor-acceptor π-conjugated fluorescent dyes with a pyridine ring as an electron-withdrawing anchoring group. Angew Chem 123:7567–7571 (2011); Angew Chem Int Ed 50:7429–7433

Ooyama Y, Nagano T, Inoue S, Imae I, Komaguchi K, Ohshita J, Harima Y (2011c) Dye-sensitized solar cells based on donor-π-acceptor fluorescent dyes with a pyridine ring as an electron-withdrawing-injecting anchoring group. Chem Eur J 17:14837–14843

Ooyama Y, Ohshita J, Harima Y (2012) Control of molecular arrangement and/or orientation of D–π–A fluorescent dyes for dye-sensitized solar cells. Chem Lett 41:1384–1396

Ooyama Y, Hagiwara Y, Mizumo T, Harima Y, Ohshita J (2013a) Synthesis of diphenylamino-carbazole substituted BODIPY dyes and their photovoltaic performance in dye-sensitized solar cells. RSC Adv 3:18099–18106

Ooyama Y, Hagiwara Y, Mizumo T, Harima Y, Ohshita J (2013b) Photovoltaic performance of dye-sensitized solar cells based on D–π–A type BODIPY dye with two pyridyl groups. New J Chem 37:2479–2485

Ooyama Y, Yamaguchi N, Imae I, Komaguchi K, Ohshita J, Harima Y (2013c) Dye-sensitized solar cells based on D–π–A fluorescent dyes with two pyridyl groups as an electron- withdrawing-injecting anchoring group. Chem Commun 49:2548–2550

Ooyama Y, Sato T, Harima Y, Ohshita J (2014a) Development of a D–π–A dye with benzothienopy-ridine as the electron-withdrawing anchoring group for dye-sensitized solar cells. J Mater Chem A 2:3293–3296

Ooyama Y, Uenaka K, Harima Y, Ohshita J (2014b) Development of D–π–A dyes with a pyrazine ring as an electron-withdrawing anchoring group for dye-sensitized solar cells. RSC Adv 4:30225–30228

Ooyama Y, Yamada T, Fujita T, Harima Y, Ohshita J (2014c) Development of D–π–Cat fluorescent dyes with a catechol group for dye-sensitized solar cells based on dye-to-TiO_2 charge transfer. J Mater Chem A 2:8500–8511

Ooyama Y, Uenaka K, Ohshita J (2015a) Development of D–π–A fluorescent dyes with a 3-pyridyl group as electron-withdrawing anchoring group for dye-sensitized solar cells. Eur J Org Chem 3713–3720

Ooyama Y, Uenaka K, Ohshita J (2015b) Synthesis, optical, electrochemical and photovoltaic properties of a D–π–A fluorescent dye with triazine ring as electron-withdrawing anchoring group for dye-sensitized solar cells. RSC Adv 5:21012–21018

Ooyama Y, Uenaka K, Ohshita J (2015c) Development of a functionally separated D–π-A fluorescent dye with a pyrazyl group as an electron-accepting group for dye-sensitized solar cells. Org Chem Front 2:552–559

Ooyama Y, Uenaka K, Sato T, Shibayama N, Ohshita J (2015d) Effective co-sensitization using D–π–A dyes with a pyridyl group adsorbing at Brønsted acid sites and Lewis acid sites on a TiO_2 surface for dye-sensitized solar cells. RSC Adv. 5:2531–2535

Ooyama Y, Uenaka K, Kanda M, Yamada T, Shibayama N, Ohshita J (2015e) A new co-sensitization method employing D–π–A dye with pyridyl group and D–π–Cat dye with catechol unit for dye-sensitized solar cells. Dyes Pigm 122:40–45

Ooyama Y, Kanda M, Uenaka K, Ohshita J (2015f) Effect of substituents in catechol dye sensitizers on photovoltaic performance of type II dye-sensitized solar cells. Chem Phys Chem 16:3049–3057

Ooyama Y, Uenaka K, Kamimura T, Ozako S, Kanda M, Koide T, Tani F (2016a) Dye- sensitized solar cell based on an inclusion complex of a cyclic porphyrin dimer bearing four 4-pyridyl groups and fullerene C_{60}. RSC Adv 6:16150–16158

Ooyama Y, Furue K, Enoki T, Kanda M, Adachi Y, Ohshita J (2016b) Development of type-I/type-II hybrid dye sensitizer with both pyridyl group and catechol unit as anchoring group for type-I/type-II dye-sensitized solar cell. Phys Chem Chem Phys 18:30662–30676

Ooyama Y, Yamaguchi N, Ohshita J, Harima Y (2016c) Impact of the molecular structure and adsorption mode of D–π–A dye sensitizers with a pyridyl group in dye-sensitized solar cells on the adsorption equilibrium constant for dye-adsorption on TiO_2 surface. Phys Chem Chem Phys 18:32992–32998

Ooyama Y, Kanda M, EnoKi T, Adachi Y, Ohshita J (2017a) Synthesis, optical and electrochemical properties, and photovoltaic performance of a panchromatic and near-infrared $(D)_2$–π–A type BODIPY dye with pyridyl group or cyanoacrylic acid. RSC Adv 7:13072–13081

Ooyama Y, Yamaji K, Ohshita J (2017b) Photovoltaic performances of type-II dye-sensitized solar cells based on catechol dye sensitizers: retardation of back-electron transfer by PET (photo-induced electron transfer). Mater Chem Front 1:2243–2255

Otsuka A, Funabiki K, Sugiyama N, Yoshida T, Minoura H, Matsui M (2006) Dye sensitization of ZnO by unsymmetrical squaraine dyes suppressing aggregation. Chem Lett 35:666–667

Otsuka A, Funabiki K, Sugiyama N, MAse H, Yoshida T, Minoura H, Matsui M (2008) Design and synthesis of near-infrared-active Heptamethine–Cyanine dyes to suppress aggregation in a dye-sensitized porous zinc oxide solar cell. Chem Lett 37:176–177

Paek S, Choi H, Kim C, Cho N, So S, Song K, Nazeeruddinc MK, Ko J (2011) Efficient and stable panchromatic squaraine dyes for dye-sensitized solar cells. Chem Commun 47:2874–2876

Park J, Barolo C, Sauvage F, Barbero N, Benzi C, Quagliotto P, Coluccia S, Censo DD, Grätzel M, Nazeeruddin MK, Viscardi G (2012) Symmetric vs. asymmetric squaraines as photosensitisers in mesoscopic injection solar cells: a structure–property relationship study. Chem Commun 48:782–2784

Pashaei B, Shahroosvand H, Graetzel M, Nazeeruddin MK (2016) Influence of ancillary ligands in dye-sensitized solar cells. Chem Rev 116:9485–9564

Pellegrin Y, Pleux LL, Blart E, Penaud A, Chavillon B, Szuwarski N, Boujtita M, Cario L, Jobic S, Jacquemin D, Odobel F (2011) Ruthenium polypyridine complexes as sensitizers in NiO based p-type dye-sensitized solar cells: Effects of the anchoring groups. J Photochem Photobiol, A 219:235–242

Polo AS, Iha NYM (2006) Blue sensitizers for solar cells: natural dyes from Calafate and Jaboticaba. Sol Energy Mater Sol Cells 90:1936–1944

Powar S, Daeneke T, Ma MT, Fu D, Duffy NW, Götz G, Weidelener M, Mishra A, Bäuerle P, Spiccia L, Bach U (2013) Highly efficient p-type dye-sensitized solar cells based on tris(1,2-diaminoethane)Cobalt(II)/(III) electrolytes. Angew Chem Int Ed 52:602–605

Qin H, Wenger S, Xu M, Gao F, Jing X, Wang P, Zakeeruddin SM, Grätzel M (2008a) An organic sensitizer with a fused dithienothiophene unit for efficient and stable dye- sensitized solar cells. J Am Chem Soc 130:9202–9203

Qin P, Zhu H, Edvinsson T, Boschloo G, Hagfeldt A, Sun L (2008b) Design of an organic chromophore for P-type dye-sensitized solar cells. J Am Chem Soc 130:8570–8571

Qin P, Linder M, Brinck T, Boschloo G, Hagfeldt A, Sun L (2009) High incident photon-to-current conversion efficiency of p-type dye-sensitized solar cells based on NiO and organic chromophores. Adv Mater 21:2993–2996

Qin P, Wiberg J, Gibson EA, Linder M, Li L, Brinck T, Hagfeldt A, Albinsson B, Sun L (2010) Synthesis and mechanistic studies of organic chromophores with different energy levels for p-type dye-sensitized solar cells. J Phys Chem C 114:4738–4748

Qin C, Wong W-Y, Han L (2013) Squaraine dyes for dye-sensitized solar cells: recent advances and future challenges. Chem Asian J 8:1706–1719

Quintana M, Edvinsson T, Hagfeldt A, Boschloo G (2007) Comparison of dye-sensitized ZnO and TiO_2 solar cells: studies of charge transport and carrier lifetime. J Phys Chem C 111:1035–1041

Ragoussi M-E, Cid J-J, Yum J-H, De La Torre G, Censo DD, Grätzel M, Nazeeruddin MK, Torres T (2012) Carboxyethynyl anchoring ligands: a means to improving the efficiency of phthalocyanine-sensitized solar cells. Angew Chem 124:4451–4454 (2012); Angew Chem Int Ed 51:4375–4378

Reddy PY, Giribabu L, Lyness C, Snaith HJ, Vijaykumar C, Chandrasekharam M, Lakshmikantam M, Yum J-H, Kalyanasundaram K, Grätzel M, Nazeeruddin MK (2007a) Efficient Sensitization of Nanocrystalline TiO_2 Films by a near-IR-absorbing Unsymmetrical Zinc Phthalocyanine. Angew Chem 119:377–380 (2007); Angew Chem Int Ed 46:373–376

Reddy PY, Giribabu L, Lyness C, Snaith HJ, Vijaykumar C, Chandrasekharam M, Lakshmikantam M, Yum J-H, Kalyanasundaram K, Grätzel M, Nazeeruddin MK (2007b) Efficient Sensitization of nanocrystalline TiO_2 films by a near-IR-absorbing unsymmetrical zinc phthalocyanine. Angew Chem 119:377–380 (2007); Angew Chem In Ed 46:373–376

Robertson N (2006) Optimizing dyes for dye-sensitized solar cells. Angew Chem 118: 2398–2405 (2006); Angew Chem Int Ed 45:2338–2345

Saito M, Fujihara S (2008) Large photocurrent generation in dye-sensitized ZnO solar cells. Energy Environ Sci 1:280–283

Sakurada T, Arai Y, Segawa H (2014) Porphyrins with β-acetylene-bridged functional groups for efficient dye-sensitized solar cells. RSC Adv 4:13201–13204

Sakuragi Y, Wang X-F, Miura H, Matsui M, Yoshida T (2010) Aggregation of indoline dyes as sensitizers for ZnO solar cells. J Photochem Photobiol, A 216:1–7

Sarker AK, Kang MG, Hong J-D (2012) A near-infrared dye for dye-sensitized solar cell: catecholate-functionalized zinc phthalocyanine. Dyes Pigm 92:1160–1165

Sayama K, Sugino M, Sugihara H, Abe Y, Arakawa H (1998) Photosensitization of porous TiO_2 semiconductor electrode with Xanthene Dyes. Chem Lett 27:753–754

Sayama K, Hara K, Mori N, Satsuki M, Suga S, Tsukagoshi S, Abe Y, Sugihara H, Arakawa H (2000) Photosensitization of a porous TiO_2 electrode with merocyanine dyes containing a carboxyl group and a long alkyl chain. Chem Commun 1173–1174

Sayama K, Hara K, Ohga Y, Shinpou A, Suga S, Arakawa H (2001) Significant effects of the distance between the cyanine dye skeleton and the semiconductor surface on the photoelectrochemical properties of dye-sensitized porous semiconductor electrodes. New J Chem 25:200–202

Sayama K, Tsukagoshi S, Hara K, Ohga Y, Shinpou A, Abe Y, Suga S, Arakawa H (2002) Photoelectrochemical properties of J aggregates of benzothiazole merocyanine dyes on a nanostructured TiO_2 film. J Phys Chem B 106:1363–1371

Sayama K, Tsukagoshi S, Mori T, Hara K, Ohga Y, Shinpo A, Abe Y, Suga S, Arakawa H (2003) Efficient sensitization of nanocrystalline TiO_2 films with cyanine and merocyanine organic dyes. Sol Energy Mater Sol Cells 80:47–71

Schmidt-Mende L, Bach U, Humphry-Baker R, Horiuchi T, Miura H, Ito S, Uchida S, Grätzel M (2005) Organic dye for highly efficient solid-state dye-sensitized solar cells. Adv Mater 17:813–815

Senadeera GKR, Nakamura K, Kitamura T, Wada Y, Yanagida S (2003) Fabrication of highly efficient polythiophene-sensitized metal oxide photovoltaic cells. Appl Phys Lett 83:5470–5472

Senadeera GKR, Kitamura T, Wada Y, Yanagida S (2005) Photosensitization of nanocrystalline TiO_2 films by a polymer with two carboxylic groups, poly (3-thiophenemalonic acid). Sol Energy Mater Sol Cells 88:315–322

Shi Y, Hill RBM, Yum J-H, Dualeh A, Barlow S, Grätzel M, Marder SR, Nazeeruddin MK (2011) A high-efficiency panchromatic squaraine sensitizer for dye-sensitized solar cells. Angew Chem 123:6749–6751 (2011); Angew Chem Int Ed 50:6619–6621

Shibayama N, Ozawa H, Abe M, Ooyama Y, Arakawa H (2014) A new cosensitization method using the Lewis acid sites of a TiO_2 photoelectrode for dye-sensitized solar cells. Chem Commun 50:6398–6401

Shin Y-J, Lee J-H, Park J-H, Park N-G (2007) Enhanced photovoltaic properties of SiO_2-treated ZnO nanocrystalline electrode for dye-sensitized solar cell. Chem Lett 36:1506–1507

Singh SP, Gayathri T (2014) Evolution of BODIPY dyes as potential sensitizers for dye-sensitized solar cells. Eur J Org Chem 4689–4707

Stangel C, Bagaki A, Angaridis PA, Charalambidis G, Sharma GD, Coutsolelos AG (2014) "Spider"-shaped porphyrins with conjugated pyridyl anchoring groups as efficient sensitizers for dye-sensitized solar cells. Inorg Chem 53:11871–11881

Szarko JM, Neubauer A, Bartelt A, Socaciu-Siebert L, Birkner F, Schwarzburg K, Hannappel T, Eichberger R (2008) The ultrafast temporal and spectral characterization of electron injection from perylene derivatives into ZnO and TiO_2 colloidal films. J Phys Chem C 112:10542–10552

Tae EL, Lee SH, Lee JK, Yoo SS, Kang EJ, Yoon KB (2005) A strategy to increase the efficiency of the dye-sensitized TiO_2 solar cells operated by photoexcitation of Dye-to-TiO_2 charge-transfer bands. J Phys Chem B 109:22513–22522

Tanaka M, Hayashi S, Eu S, Umeyama T, Matano Y, Imahori H (2007) Novel unsymmetrically π-elongated porphyrin for dye-sensitized $TiO{-}_2$ cells. Chem Commun 2069–2071

Tatay S, Haque SA, O'Regan B, Durrant JR, Verhees WJH, Kroon JM, Vidal-Ferran A, Gaviña P, Palomares E (2007) Kinetic competition in liquid electrolyte and solid-state cyanine dye sensitized solar cells. J Mater Chem 17:3037–3044

Tennakone K, Kumarasinghe AR, Kumara GRRA, Wijayantha KGU, Sirimanne PM (1997) Nanoporous TiO_2 photoanode sensitized with the flower pigment cyanidin. J Photochem Photobiol, A 108:193–195

Tennakone K, Kumara GRRA, Kottegoda IRM, Perera VPS, Weerasundara PSRS (1998) Sensitization of nano-porous films of TiO_2 with santalin (red sandalwood pigment) and construction of dye-sensitized solid-state photovoltaic cells. J Photochem Photobiol, A 117:137–142

Tennakone K, Kumara GRRA, Kottegoda IRM, Perera VPS (1999) An efficient dye-sensitized photoelectrochemical solar cell made from oxides of tin and zinc. Chem Commun 15–16

Thomas KRJ, Hsu Y-C, Lin JT, Lee K-M, Ho K-C, Lai C-H, Cheng Y-M, Chou P-T (2008) 2,3-disubstituted thiophene-based organic dyes for solar cells. Chem Mater 20:1830–1840

Tian H, Yang X, Chen R, Zhang R, Hagfeldt A, Sun L (2008) Effect of different dye baths and dye-structures on the performance of dye-sensitized solar cells based on triphenylamine dyes. J Phys Chem C 112:11023–11033

Tian H, Yang X, Cong J, Chen R, Liu J, Hao Y, Hagfeldt A, Sun L (2009) Tuning of phenoxazinechromophores for efficient organic dye-sensitized solar cells. Chem Commun 6288–6290

Urbani M, Grätzel M, Nazeeruddin MK, Torres T (2014) Meso-substituted porphyrins for dye-sensitized solar cells. Chem Rev 114:12330–12396

Verbitskiy EV, Cheprakova EM, Subbotina JO, Schepochkin AV, Slepukhin PA, Rusinov GL, Charushin VN, Chupakhin ON, Makarova NI, Metelitsa AV, Minkin VI (2014) Synthesis, spectral and electrochemical properties of pyrimidine-containing dyes as photosensitizers for dye-sensitized solar cells. Dyes Pigm 100:201–214

Wan Z, Jia C, Duan Y, Zhou L, Lin Y, Shi Y (2012) Phenothiazine–triphenylamine based organic dyes containing various conjugated linkers for efficient dye-sensitized solar cells. J Mater Chem 22:25140–25147

Wang Z-S, Li F-Y, Huang C-H, Wang L, Wei M, Jin L-P, Li N-Q (2000) Photoelectric conversion properties of nanocrystalline TiO_2 electrodes sensitized with hemicyanine derivatives. J Phys Chem B 104:9676–9682

Wang Q, Campbell WM, Bonfantani EE, Jolley KW, Officer DL, Walsh PJ, Gordon K, Humphry-Baker R, Nazeeruddin MK, Grätzel M (2005a) Efficient light harvesting by using green Zn-porphyrin-sensitized nanocrystalline TiO_2 films. J Phys Chem B 109:15397–15409

Wang Z-S, Hara K, Dan-oh Y, Kasada C, Shinpo A, Suga S, Arakawa H, Sugihara H (2005b) Photophysical and (Photo)electrochemical properties of a coumarin dye. J Phys Chem B 109:3907–3914

Wang Z-S, Cui Y, Hara K, Dan-oh Y, Kasada C, Shinpo A (2007a) A high-light-harvesting-efficiency coumarin dye for stable dye-sensitized solar cells. Adv Mater 19:1138–1141

Wang Z-S, Cui Y, Dan-oh Y, Kasada C, Shinpo A, Hara K (2007b) Thiophene-functionalized coumarin dye for efficient dye-sensitized solar cells: electron lifetime improved by coadsorption of deoxycholic acid. J Phys Chem C 111:7224–7230

Wang Z-S, Cui Y, Dan-oh Y, Kasada C, Shinpo A, Hara K (2008a) Molecular design of coumarin dyes for stable and efficient organic dye-sensitized solar cells. J Phys Chem C 112:17011–17017

Wang Z-S, Koumura N, Cui Y, Takahashi M, Sekiguchi H, Mori A, Kubo T, Furube A, Hara K (2008b) Hexylthiophene-functionalized carbazole dyes for efficient molecular photovoltaics: tuning of solar-cell performance by structural modification. Chem Mater 20:3993–4003

Wang J-B, Fang X-Q, Pan X, Dai S-Y, Song Q-H (2012) New 2, 6-modified bodipy sensitizers for dye-sensitized solar cells. Chem Asian J 7:696–700

Wang L, Yang X, Li S, Cheng M, Sun L (2013) A new type of organic sensitizers with pyridine-*N*-oxide as the anchoring group for dye-sensitized solar cells. RSC Adv 3:13677–13680

Wu J-J, Chen G-R, Yang H-H, Ku C-H, Lai J-Y (2007) Effects of dye adsorption on the electron transport properties in ZnO-nanowire dye-sensitized solar cells. Appl Phys Lett 90:213109

Wu W, Hua J, Jin Y, Zhan W, Tian H (2008) Photovoltaic properties of three new cyanine dyes for dye-sensitized solar cells. Photochem Photobiol Sci 7:63–68

Wu W, Yang J, Hua J, Tang J, Zhang L, Long Y, Tian H (2010) Efficient and stable dye-sensitized solar cells based on phenothiazine sensitizers with thiophene units. J Mater Chem 20:1772–1779

Xie P, Guo F (2007) Molecular engineering of ruthenium sensitizers in dye-sensitized solar cells. Curr Org Chem 11:1272–1286

Yamamoto S, Zhang A, Stillman MJ, Kobayashi N, Kimura M (2016) Low-symmetry Ω- shaped zinc phthalocyanine sensitizers with panchromatic light-harvesting properties for dye-sensitized solar cells. Chem Eur J 22:18760–18768

Yang C-J, Chang YJ, Watanabe M, Hon Y-S, Chow TJ (2012) Phenothiazine derivatives as organic sensitizers for highly efficient dye-sensitized solar cells. J Mater Chem 22:4040–4049

Yao Q-H, Meng F-S, Li F-Y, Tian H, Huang C-H (2003) Photoelectric conversion properties of four novel carboxylated hemicyanine dyes on TiO_2 electrode. J Mater Chem 13:1048–1053

Yella A, Lee H-W, Tsao HN, Yi C, Chandiran AK, Nazeeruddin MK, Diau EW-G, Yeh C-Y, Zakeeruddin SM, Grätzel M (2011) Porphyrin-sensitized solar cells with cobalt (II/III)–based redox electrolyte exceed 12 percent efficiency. Science 334:629–634

Yella A, Mai C-L, Zakeeruddin SM, Chang S-N, Hsieh C-H, Yeh C-Y, Grätzel M (2014) Molecular engineering of push-pull porphyrin dyes for highly efficient dye-sensitized solar cells: the role of benzene spacers. Angew Chem Int Ed 53:2973–2977

Yen Y-S, Chen W-T, Hsu C-Y, Chou H-H, Lin JT, Yeh M-CP (2011) Arylamine-based dyes for p-type dye-sensitized solar cells. Org Lett 13:4930–4933

Yen Y-S, Chou H-H, Chen Y-C, Hsu C-Y, Lin JT (2012) Recent developments in molecule-based organic materials for dye-sensitized solar cells. J Mater Chem 22:8734–8747

Yoshida T, Terada K, Schlettwein D, Oekermann T, Sugiura T, Minoura H (2000) Electrochemical self-assembly of nanoporous ZnO/Eosin Y thin films and their sensitized photoelectrochemical performance. Adv Mater 12:1214–1217

Yoshida T, Iwaya M, Ando H, Oekermann T, Nonomura K, Schlettwein D, Wchrle D, Minoura H (2004) Improved photoelectrochemical performance of electrodeposited ZnO/EosinY hybrid thin films by dye re-adsorptio. Chem Commun 400–401

Yum J-H, Walter P, Huber S, Rentsch S, Geiger T, Nüesch F, Angelis FD, Grätzel M, Nazeeruddin MK (2007) Efficient far red sensitization of nanocrystalline TiO_2 films by an unsymmetrical squaraine dye. J Am Chem Soc 129:10320–10321

Yum J-H, Jang S-R, Humphry-Baker R, Grätzel M, Cid J-J, Torres T, Nazeeruddin MD (2008) Effect of coadsorbent on the photovoltaic performance of zinc pthalocyanine-sensitized solar cells. Langmuir 24:5636–5640

Yum J-H, Baranoff E, Wenger S, Nazeeruddin MK, Grätzel M (2011) Panchromatic engineering for dye-sensitized solar cells. Energy Environ Sci 4:842–857

Zeng W, Cao Y, Bai Y, Wang Y, Shi Y, Zhang M, Wang F, Pan C, Wang P (2010) Efficient dye-sensitized solar cells with an organic photosensitizer featuring orderly conjugated ethylenedioxythiophene and dithienosilole blocks. Chem Mater 22:1915–1925

Zhang L, Cole JM (2015) Anchoring groups for dye-sensitized solar cells. ACS Appl Mater Interfaces 7:3427–3455

Zhang W, Feng Z, Su M, Saeys M, Liu B (2009) A Triphenylamine-based conjugated polymer with donor-π-acceptor architecture as organic sensitizer for dye-sensitized solar cells. Macromol Rapid Commun 30:1533–1537

Zhang XL, Huang F, Nattestad A, Wang K, Fu D, Mishra A, Bäuerle P, Bach U, Cheng Y-B (2011) Enhanced open-circuit voltage of p-type DSC with highly crystalline NiO nanoparticles. Chem Commun 47:4808–4810

Zhang M-D, Xie H-X, Ju X-H, Qin L, Yang Q-X, Zheng H-G, Zhou X-F (2013) D-D-π-A organic dyes containing 4,4'-di(2-thienyl)triphenylamine moiety for efficient dye-sensitized solar cells. Phys Chem Chem Phys 15:634–641

Chapter 15
π-Conjugated Polymers Incorporating Naphthalene-Based Nitrogen-Containing Heteroaromatics for Organic Photovoltaics

Itaru Osaka

Abstract Recently, a wide variety of π-conjugated polymers have been developed for the use as the active layer in organic photovoltaics (OPVs). This chapter will summarize the recent development of π-conjugated polymers with donor–acceptor motifs, specifically, based on naphthalene-based nitrogen-containing heteroaromatics as the acceptor unit. With the strong electron-deficient nature as well as the large π-conjugation system, incorporation of these heteroaromatics in the backbone allows us to create π-conjugated polymers having wide absorption range, i.e., narrow bandgap, deep HOMO energy levels, and high crystallinity, all of which are important for improving the efficiency of OPV cells. We describe the syntheses, properties, structural order in the thin films, and OPV performances of the π-conjugated polymers based on naphthobisthiadiazole and its analogues.

Keywords Π-conjugated polymers · Semiconducting polymers · Donor–acceptor polymers · Organic photovoltaics

15.1 Introduction

Over the past few decades, π-conjugated polymers have been widely studied as semiconducting materials owing to their good electrical and optical properties (Reynolds et al. 2019). An important feature of π-conjugated polymers is that they can be dissolved in organic solvents and thus can be solution-processed to form robust thin films on both rigid and flexible substrates. This feature is beneficial for applications to organic devices, including organic light-emitting diodes (OLEDs) (Adachi et al. 2019), organic field-effect transistors (OFETs)/organic thin-film transistors (OTFTs) (Bao and Locklin 2007), and organic photovoltaics (OPVs)/organic solar cells (OSCs) (Brabec et al. 2014). In particular, OPVs have recently been attracting more attention than other devices due to the pressing need for renewable energy sources.

I. Osaka (✉)
Department of Applied Chemistry, Hiroshima University, Higashihiroshima, Japan
e-mail: iosaka@hiroshima-u.ac.jp

Y. Ooyama and S. Yagi (eds.), *Progress in the Science of Functional Dyes*,
https://doi.org/10.1007/978-981-33-4392-4_15

Whereas regioregular poly(3-hexylthiophene) (P3HT), the most common π-conjugated polymer, has a simple 3-hexylthiophene ring as the lone building unit (McCullough 1998), π-conjugated polymers that are used as p-type organic semiconductors for OPVs typically comprise several heteroaromatic rings as building units in the backbone. More specifically, heteroaromatic rings having electron-rich (donor) and electron-poor (acceptor) characteristics are alternately incorporated into the π-conjugated polymer backbone (Boudreault et al. 2011; Biniek et al. 2012). This alternating donor–acceptor arrangement has resulted in molecular orbital mixing and thus intramolecular charge transfer (CT) interactions in the backbone, giving rise to a wide light absorption range, e.g., a narrow optical bandgap (E_g), which is essential for OPVs. Furthermore, by choosing the strengths of the donor and acceptor characteristics, the energy levels of the highest occupied molecular orbital (HOMO) and the lowest unoccupied molecular orbital (LUMO) can be easily tuned. When π-conjugated polymers are used as p-type organic semiconductors, their HOMO energy levels must be as deep as possible in order to afford a high open-circuit voltage (V_{OC}) for the OPV cell, because V_{OC} is proportional to the energy difference between the HOMO of the p-type organic semiconductor and the LUMO of the n-type organic semiconductor (Thompson and Fréchet 2008). Therefore, the combination of "weak" donor and "strong" acceptor building units is desirable for donor–acceptor π-conjugated polymers. In addition to their electronic features, donor–acceptor polymers can provide high charge carrier mobilities due to the enhanced intermolecular π–π stacking that originates in dipole–dipole interactions. Furthermore, the use of larger heteroaromatic fused rings as either donor or acceptor building units or both is also expected to provide improved π–π interactions.

Among a variety of heteroaromatic fused rings, naphthalene-based nitrogen-containing heterocycles (Fig. 15.1) have emerged as acceptor building units for π-conjugated polymers (Osaka and Takimiya 2017). Due to their large π-electron systems as well as strong electron deficiencies, these units offer relatively long-wavelength absorptions (narrow E_g), deep HOMO energy levels, and strong π–π interactions. Here, we summarize π-conjugated polymers incorporating naphthalene-based nitrogen-containing heterocycles and describe their syntheses, properties, ordering structures, and OPV performances.

Fig. 15.1 Chemical structure of naphthalene-based nitrogen-containing heteroaromatic rings

X = Y = O (**NOz**)
X = Y = S (**NTz**)
X = Y = Se (**NSz**)
X = Y = N-R (**TZNT**)
X = Y = C(R)=C(R) (**NPz**)

X = O, Y = S (**NOT**)
X = O, Y = Se (**NOS**)
X = S, Y = Se (**NTS**)
X = O, Y = C(R)=C(R) (**NOP**)
X = S, Y = C(R)=C(R) (**NTP**)
X = Se, Y = C(R)=C(R) (**NSP**)

15.2 Synthesis of Naphthalene-Based Nitrogen-Containing Heteroaromatic Rings

Naphtho[1,2-*c*:5,6-*c*′]bis[1,2,5]thiadiazole (NTz), the most well studied naphthalene-based nitrogen-containing heteroaromatic, was first synthesized by Mataka et al. (1991). They reported the reaction of tetrasulfur tetranitride (N4S4) with various naphthalenols and successfully obtained NTz and its isomers. However, due to the fact that N4S4 is an explosive compound (Villena-Blanco and Jolly 1967), this synthetic methodology is not scalable. An alternative synthesis of NTz was developed (Scheme 15.1) in which 1,5-dinitro-2,6-diaminonaphthalene (**1**) was reduced to give 1,2,5,6-tetraaminonaphthalene (**2**) (Imai et al. 1973), which was then reacted with thionyl chloride to afford NTz via cyclization. Furthermore, **2** could also undergo cyclization to form other naphthalene-based nitrogen-containing heteroaromatics. The use of seleninyl chloride, sodium nitrite, and α-diketones as the cyclization reagent for **2** provided naphtho[1,2-*c*:5,6-*c*′]bis[1,2,5] selenadiazole (NSz) (Kawashima et al. 2015a), naphtho[1,2-*c*:5,6-*c*′]bis-1*H*-[1,2,3]triazole (**3**)

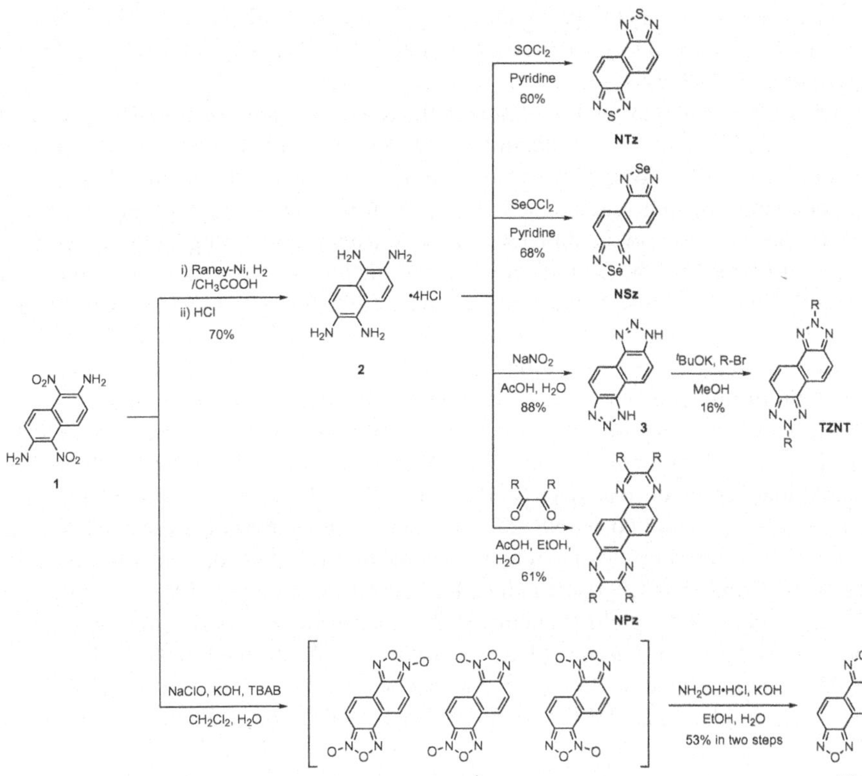

Scheme 15.1 Synthesis of naphthalene-based nitrogen-containing heteroaromatic rings

(Dong et al. 2013), and naphtho[1,2-*b*:5,6-*b*′]bisbispyrazine (NPz), respectively (Villena-Blanco and Jolly 1967; Mikie et al. 2017). Note that **3** was further converted into its alkylated derivatives (TZNT) (Dong et al. 2013). Interestingly, in the case of NPz, various α-diketones could be applied for the cyclization to afford NPz derivatives with different substituents at 2-, 3-, 8-, and 9-positions (Mikie and Osaka 2019; Yu et al. 2017). On the other hand, **1** was oxidized to form *N*-oxides of (naphtho[1,2-*c*:5,6-*c*′]bis[1,2,5]oxadiazole) NOz (**4**) and the following reduction of **1** gave NOz (Kawashima et al. 2015a).

These heteroaromatics could be further functionalized by bromine or boronate esters (Scheme 15.2). NTz, NSz, and NOz were easily brominated by *N*-bromosuccinimide (NBS) (Kawashima et al. 2015a) and TZNT was brominated by bromine to give dibrominated compounds (NTz-Br2, NSz-Br2, NOz-Br2, and TNTZ-Br2, respectively) in reasonably high yields (Scheme 15.2a, b). On the other hand, it was reported that NPz did not undergo bromination by either NBS or bromine. Thus, an alternative route was proposed in which NTz-Br2 was first converted into dibromo-tetraaminonaphthalene by the reduction/ring-opening reaction with zinc, and the following cyclization with an α-diketone gave dibrominated NPz (Scheme 15.2c) (Mikie et al. 2017). The introduction of boronate ester groups was only reported for NTz, in which the iridium-catalyzed direct C–H borylation reaction was employed to form 5,10-diborylated NTz (NTz-Bpin2) (Scheme 15.2d) (Kawashima et al. 2013).

Interestingly, whereas the use of zinc as the reducing agent for NTz-Br2 provided tetraaminonaphthalene, i.e., both thiadiazole moieties underwent a ring-opening reaction, the use of $NaBH_4/CoCl_2$ provided diamino-thiadiazolonaphthalene, i.e., a one-sided ring-opening reaction selectively occurred. Thus, by applying this reaction, asymmetric hybrid naphthalene-based nitrogen-containing heteroaromatics, in which two each of benzoxadiazole, benzothiadiazole, benzoselenadiazole, and benzopyrazine were fused, were synthesized (Scheme 15.3) (Mikie and Osaka 2018). However, because debromination occurred at the same time, NTz2T, in which both sides (5- and 10-positions) of NTz were substituted with alkylthiophenes instead of bromine groups, was employed. The use of diacetyl and seleninyl chloride as reagents for the ring-closing reaction provided [1,2,5]thiadiazolo[3′,4′-5,6]naphtho[1,2-*b*]pyrazine and [1,2,5]selenadiazolo[3′,4′:5,6]naphtho[1,2-*c*][1,2,5]thiadiazole derivatives (NTP2T and NTS2T) both in 90% yield. This reaction was also applied to NSz2T. The reduction of NSz2T by $NaBH_4/CoCl_2$ and the following ring-closing reaction using thionyl chloride and diacetyl also gave NTS2T and the [1,2,5]selenadiazolo[3′,4′-5,6]naphtho[1,2-*b*]pyrazine system (NSP2T) both in 92% yield (Scheme 15.3). Furthermore, this reaction was applicable to NOz2T, providing [1,2,5]oxadiazolo[3′,4′-5,6]naphtho[1,2-*b*]pyrazine (NOP2T), [1,2,5]thiadiazolo[3′,4′:5,6]naphtho[1,2-*c*][1,2,5]oxadiazole (NOT2T), and [1,2,5]selenadiazolo[3′,4′:5,6]naphtho[1,2-*c*][1,2,5]oxadiazole (NOS2T) systems, using diacetyl, thionyl chloride, and seleninyl chloride as reagents for the ring closure, respectively. All reactions had reasonably high yields of 70–80%.

All these naphthalene-based nitrogen-containing heteroaromatics and their derivatives are key compounds for the development of π-conjugated polymers. For

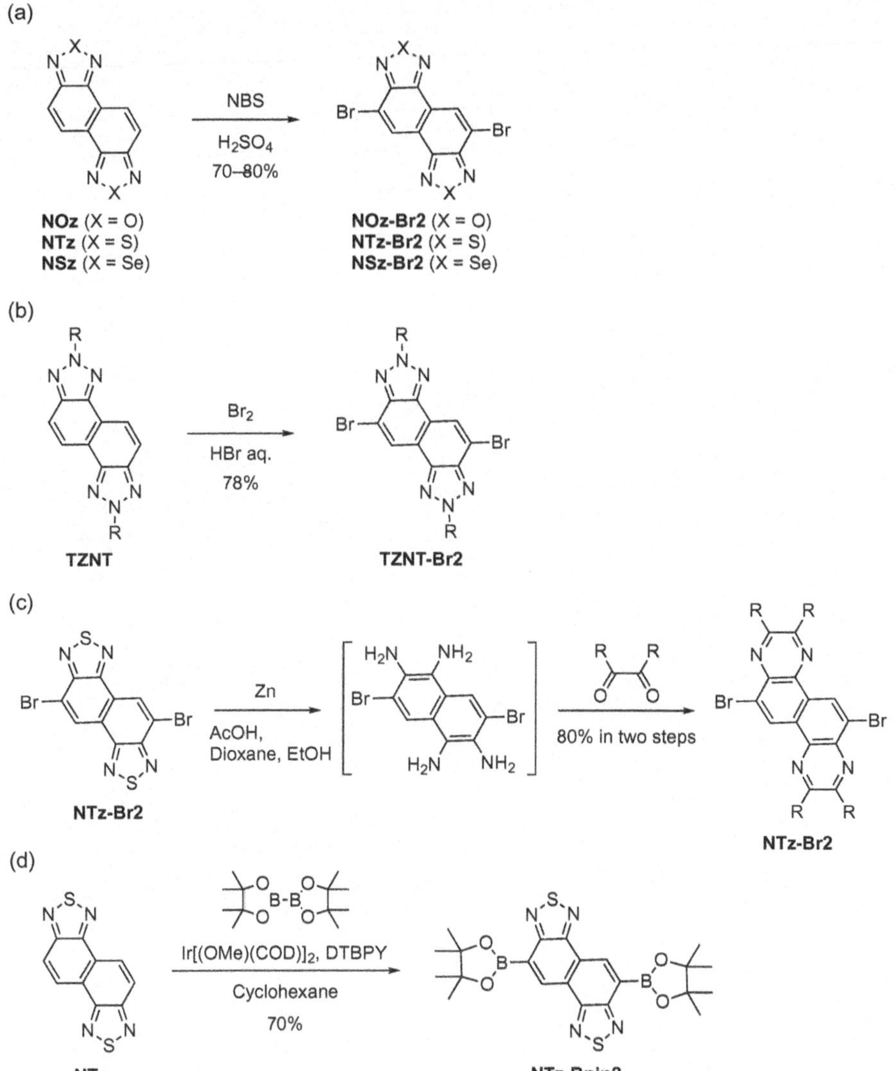

Scheme 15.2 Synthesis of naphthalene-based nitrogen-containing heteroaromatics with reactive groups

example, NTz-Br2 can be cross-coupled with stannylated alkylthiophene to give NTz2T, which can then be brominated into NTz2T-Br2. NTz2T-Br2 can be copolymerized with distannylated bithiophene (**5**) to yield a π-conjugated polymer (PNTz4T) (Scheme 15.4). As such, brominated and boronated naphthalene-based

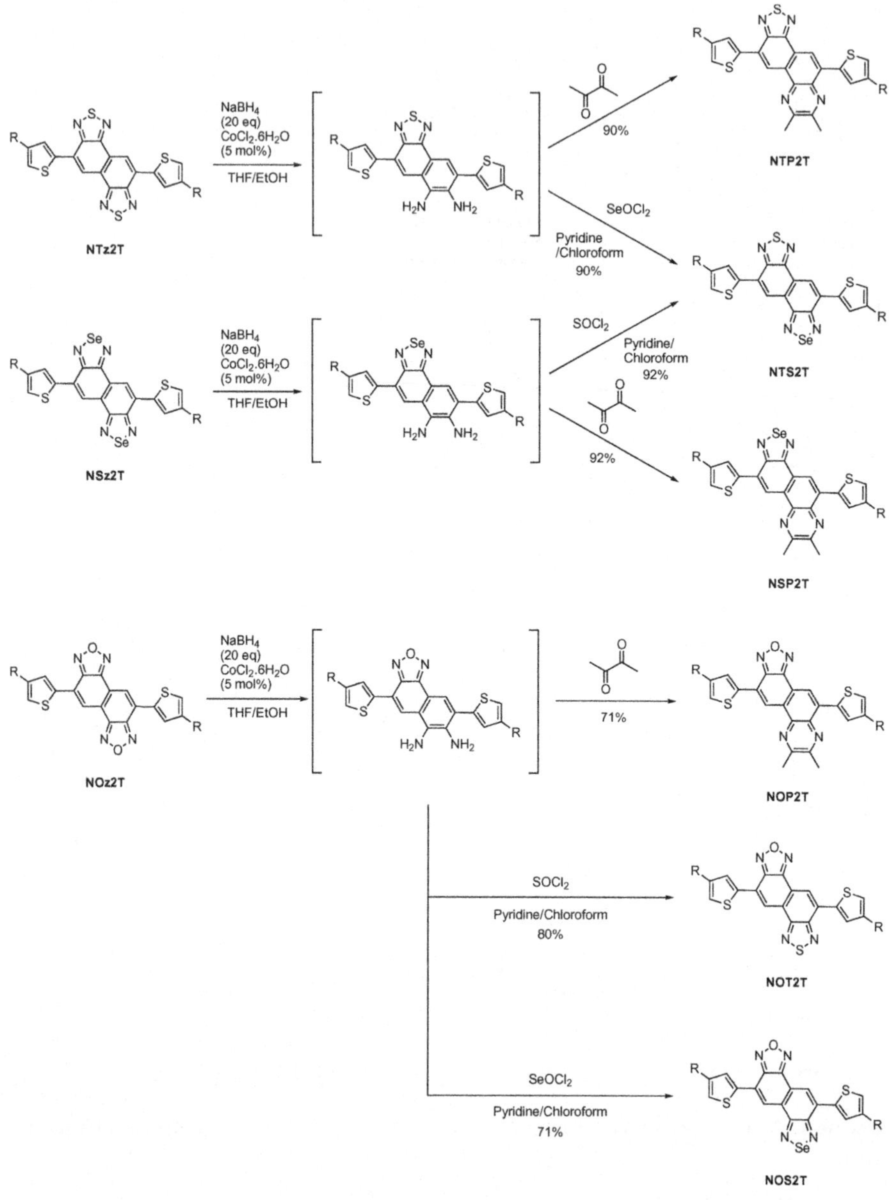

Scheme 15.3 Synthesis of hybrid naphthalene-based nitrogen-containing heteroaromatics

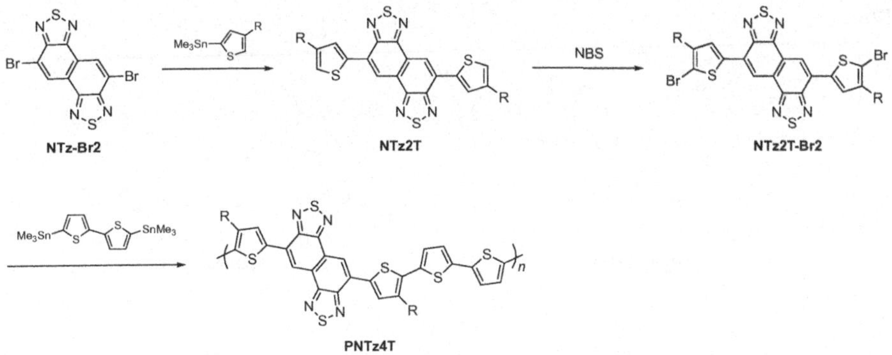

Scheme 15.4 An example of the synthesis of NTz-based polymers

heteroaromatics can be coupled with other heteroaromatic rings for use as co-monomers for the synthesis of π-conjugated polymers, or they can be directly polymerized with other co-monomers via cross-coupling reactions to give π-conjugated polymers.

15.3 π-Conjugated Polymers Incorporating Naphthalene-Based Nitrogen-Containing Heteroaromatic Rings

15.3.1 NTz-Based Polymers

Among the naphthalene-based nitrogen-containing heteroaromatics, NTz has been the most widely studied so far. Cao and co-workers reported for the first time the synthesis of an NTz-based polymer and its application to an OPV cell. They employed a benzodithiophene (BDT) moiety, one of the most well-known building units, as the donor unit for the copolymer. A BDT-NTz copolymer, namely, PBDT-DTNT (Fig. 15.2a), showed an E_g of 1.58 eV, which was narrower by 0.15 eV than that of its BTz counterpart, PBDT-DTBT (Fig. 15.2a) (Wang et al. 2011). Hole mobility evaluated by using the space-charge-limited current model was one order of magnitude higher for PBDT-DTNT (~10^{-5} cm^2 V^{-1} s^{-1}) than PBDT-DTBT (~10^{-6} cm^2 V^{-1} s^{-1}), and this was ascribed to the more pronounced aggregation of PBDT-DTNT as evidenced by absorption spectroscopy and X-ray diffraction study. Photovoltaic cells that had a conventional stacking structure (ITO/PEDOT:PSS/(polymer:$PC_{71}BM$)/Ca/Al) and used PBDT-DTNT with the p/n ratio of 1:1 showed as high as 6.0% power conversion efficiency (PCE) (J_{SC} = 11.7 mA cm^{-2}, V_{OC} = 0.80 V, FF = 0.61), which was higher than the PCEs of cells that used PBDT-DTBT (PCE = 2.1%; J_{SC} = 5.8 mA cm^{-2}, V_{OC} = 1.0 V, FF =

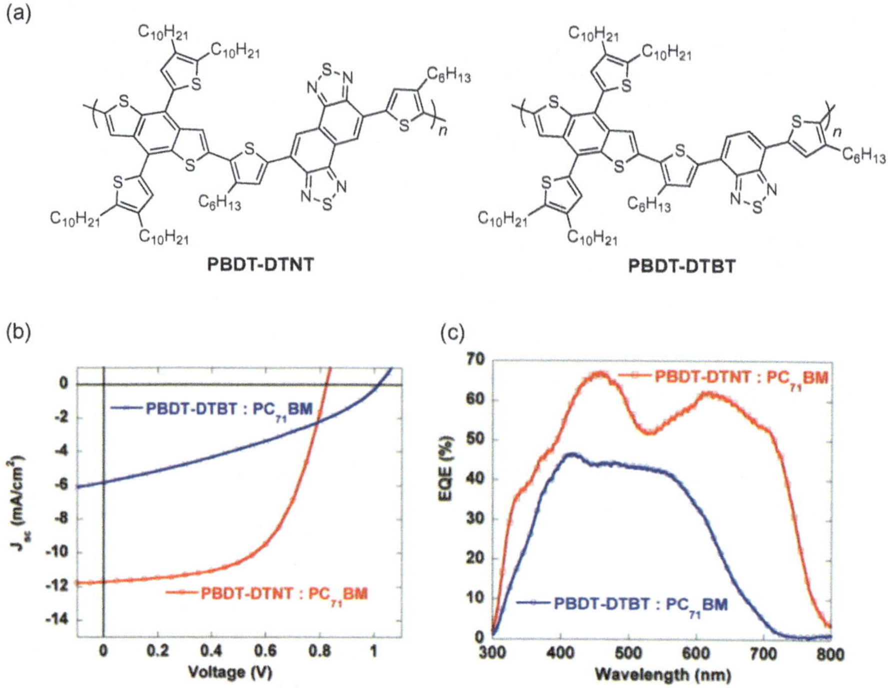

Fig. 15.2 **a** Chemical structures of BDT-NTz polymer (PBDT-DTNT) and its BTz counterpart (PBDT-DTBT). **b** Current density (*J*)–voltage (*V*) curves, and **c** external quantum efficiency (EQE) spectra of OPV cells that used these polymers. Reproduced with permission (Wang et al. 2011). Copyright 2011, American Chemical Society

0.35) (Fig. 15.2b, c). This indicated that NTz is a very promising building unit for π-conjugated polymers. Further, cells that had an inverted structure (ITO/ZnO/PFN-Br/(PBDT-DTNT:PC_{71}BM)/MoO_x/Ag) showed as high as 8.4% PCE (Yang et al. 2012).

Our group has reported the synthesis of a series of NTz-based polymers with quaterthiophene moieties as the donor unit, such as PNTz4T (Fig. 15.3a) (Osaka et al. 2012). In these polymers, a long branched alkyl group was used as the substituent to ensure sufficient solubility in solvents, which was attached to the thiophene rings neighboring the NTz moiety at β-positions heading the unsubstituted bithiophene moiety to avoid steric hindrance between NTz and alkylthiophene. PNTz4T had an E_g of 1.54 eV, which was 0.1 eV smaller than that of PBTz4T (Fig. 15.3a). Both the HOMO and LUMO energy levels were deeper for PNTz4T than PBTz4T. Hole mobility, evaluated by using transistor devices, for PNTz4T was as high as 0.56 cm^2 V^{-1} s^{-1} and approximately one order of magnitude higher than that for PBTz4T (~0.05 cm^2 V^{-1} s^{-1}), indicating the high potential of the NTz moiety. In photovoltaic cells with a conventional structure (ITO/PEDOT:PSS/(polymer/PC_{61}BM)/LiF/Al), whereas the PBTz4T cell showed a PCE of 2.6% (J_{SC} = 5.6 mA cm^{-2}, V_{OC} = 0.74 V,

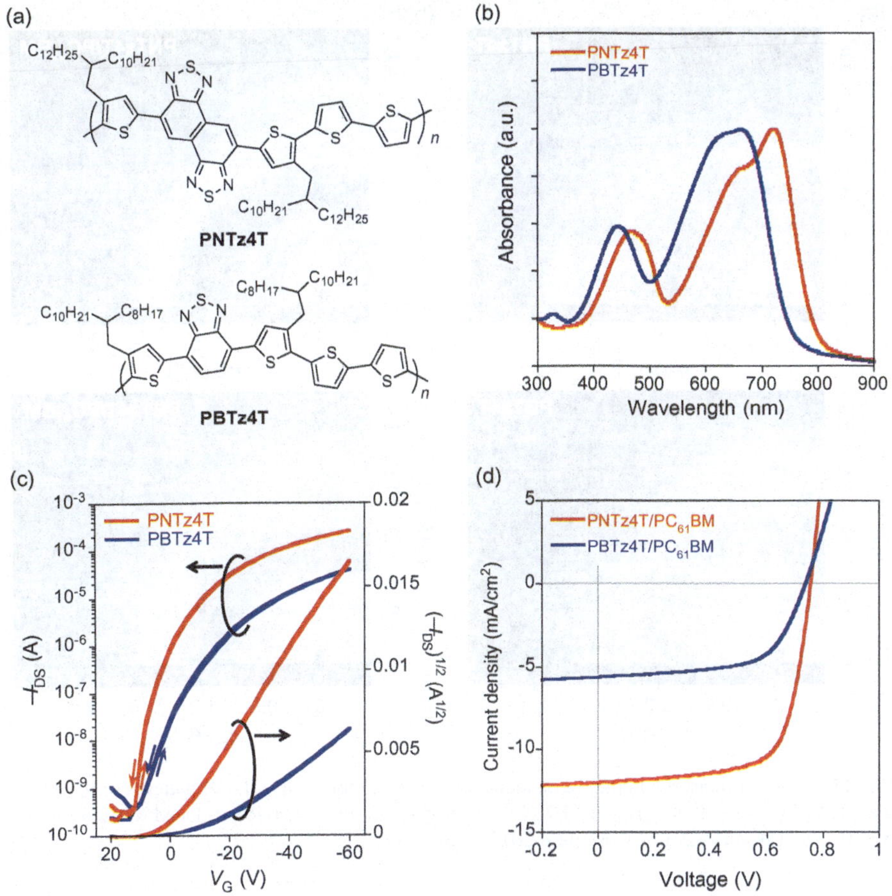

Fig. 15.3 **a** Chemical structures of NTz-quaterthiophene polymer (PNTz4T) and its BTz counterpart (PBTz4T). **b** *J–V* curves and **c** EQE spectra of OPV cells that used these polymers. Reproduced with permission (Osaka et al. 2012). Copyright 2012, American Chemical Society

FF = 0.63), the PNTz4T cell showed an approximately twofold higher PCE of 6.3% (J_{SC} = 12.0 mA cm^{-2}, V_{OC} = 0.76 V, FF = 0.69) (Fig. 15.3b, c).

The higher photovoltaic performance of PNTz4T relative to PBTz4T was understood by investigating the thin-film structure on the basis of grazing incidence X-ray diffraction (GIXD) patterns (Fig. 15.4). In the two-dimensional GIXD pattern for PNTz4T, diffractions assignable to the lamellar structure, (*h*00), and a diffraction assignable to the π − π stacking structures, (010), appeared on the quasi q_z ($\sim q_z \approx 0.25$ Å^{-1}) and q_{xy} ($q_{xy} \approx 1.7$ Å^{-1}) axes, respectively. This suggested that the backbones were predominantly formed with the edge-on orientation on the substrate surface. It was also noted that the lamellar diffractions appeared up to the fifth order,

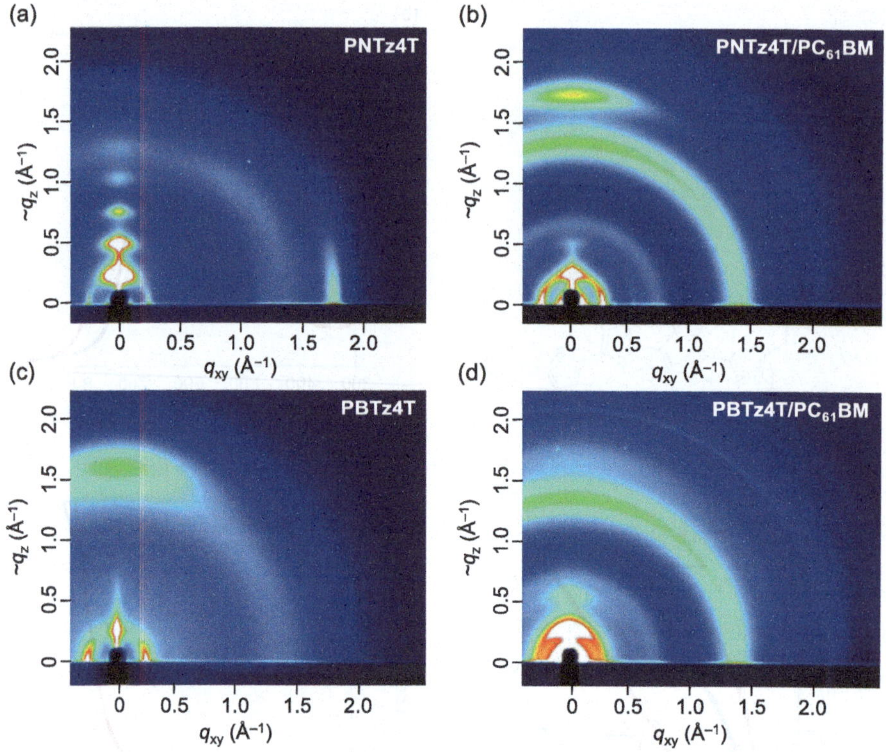

Fig. 15.4 Two-dimensional grazing incidence X-ray diffraction (2D GIXD) patterns of thin films. **a** PNTz4T, **b** PNTz4T/PC_{61}BM, **c** PBTz4T, and **d** PBTz4T/PC_{61}BM. Reproduced with permission (Osaka and Takimiya 2017). Copyright 2017, Wiley

indicating a highly ordered packing. The π–π stacking distance (d_π) was determined to be 3.5 Å, which was quite narrow for π-conjugated polymers. In contrast, PBTz4T showed lamellar and π–π stacking diffractions along the q_z and ~q_{xy} axes, indicating a face-on orientation. Further, the lamellar diffraction appeared only for the first order, indicating a less ordered structure. Interestingly, in the polymer/PC_{61}BM blend films, PNTz4T mainly oriented in a face-on manner as the π–π stacking diffraction appeared on the ~q_z axis. On the other hand, PBTz4T provided a much less crystalline feature in the blend film as there was no π–π stacking diffraction. The predominant face-on orientation as well as the highly crystalline structure with a narrow d_π for PNTz4T compared to PBTz4T explained well that the PNTz4T cell showed higher photovoltaic performance than the PBTz4T cell. Further, such structural features would enable high photovoltaic performance even with the thick active layer.

The photovoltaic performance of PNTz4T was further improved by the device optimization (Fig. 15.5) (Vohra et al. 2015). When the active layer thickness was increased to approximately 300 nm, J_{SC} improved markedly, resulting in the PCE of

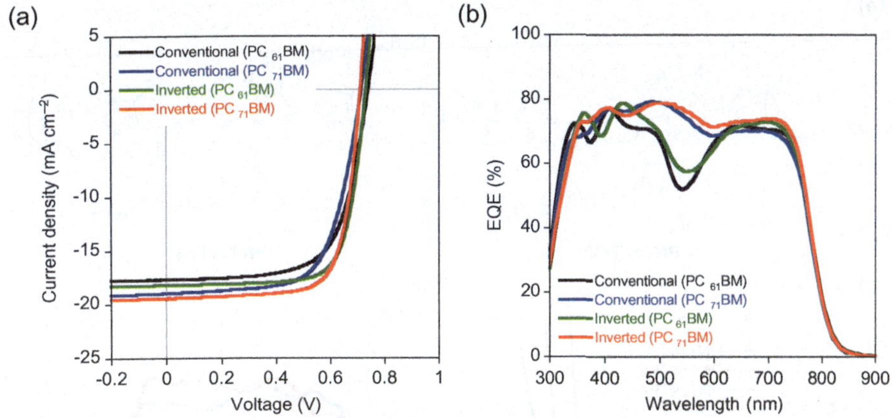

Fig. 15.5 **a** *J–V* curves and **b** EQE spectra of PNTz4T-based cells with conventional and inverted architectures. Reproduced with permission (Vohra et al. 2015). Copyright 2015, Nature Publishing Group

8.7%. Replacing $PC_{61}BM$ by $PC_{71}BM$ slightly improved PCE to 8.9% owing to the increased absorption at wavelengths of around 400–600 nm. The fact that maximum performance was achieved with a large active layer thickness of around 300 nm, which was approximately three times as large as that of typical OPVs, is quite important because a thick-layered cell is advantageous for light harvesting as well as for large-area printing (Peet et al. 2011; Osaka et al. 2014). Interestingly, use of the inverted structure (ITO/ZnO/(PNTz4T:PCBM)/MoO_x/Ag) resulted in a higher J_{SC} and thus the PCEs of 9.8% with $PC_{61}BM$ and 10.1% with $PC_{71}BM$. These improvements could be explained by the distribution of the backbone orientation (Vohra et al. 2015; Saito et al. 2018).

The fluorinated versions of PNTz4T (PNTz4TF2, PNTz4TF4) (Fig. 15.6a), in which fluorine atoms were introduced into the bithiophene moiety, were reported (Kawashima et al. 2016). Given its strong electron-withdrawing nature, the introduction of fluorine into the polymer backbone deepened the HOMO energy levels, thereby resulting in a large V_{OC}. In fact, the HOMO energy levels of PNTz4TF2 and PNTz4TF4 were −5.38 eV and −5.49 eV, respectively, which were deeper than that of PNTz4T (−5.14 eV), although the E_gs were slightly widened to 1.60 eV and 1.62 eV (E_g of PNTz4T was 1.56 eV). The inverted cells for the fluorinated polymers showed higher V_{OC}s of 0.82 V (PNTz4TF2) and 0.93 V (PNTz4TF4) than PNTz4T, consistent with the deeper HOMO energy levels (Fig. 15.6b). Although the PNTz4TF4 cell showed a low PCE of 6.5% (J_{SC} = 10.5 mA cm^{-2}, FF = 0.66) most likely owing to the increased recombination, the PNTz4TF2 cell exhibited an improved PCE of 10.5% (J_{SC} = 19.3 mA cm^{-2}, FF = 0.67), which was among the highest values for polymer/fullerene cells (Fig. 15.6b, c).

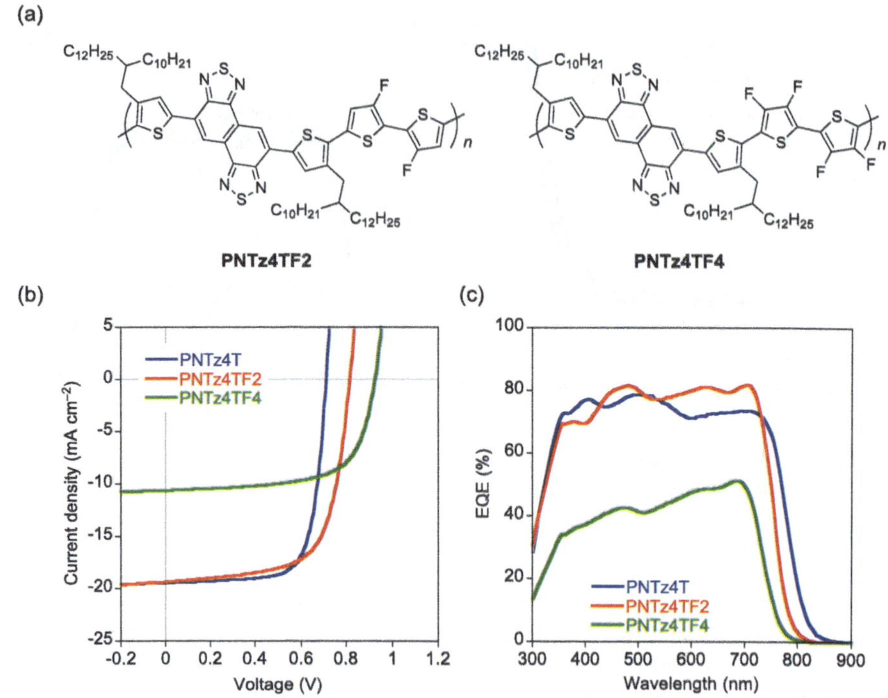

Fig. 15.6 **a** Chemical structures of fluorinated derivatives of PNTz4T (PNTz4TF2 and PNTz4TF4). **b** *J*–*V* curves and **c** EQE spectra of OPV cells that used PNTz4T, PNTz4TF2, and PNTz4TF4, in which $PC_{71}BM$ was used as the n-type material. Reproduced with permission (Kawashima et al. 2016). Copyright 2016, American Chemical Society

15.3.2 NOz-Based Polymers

Reports of NOz-based polymers have been very limited so far. Nevertheless, NOz has been proven to be a very important building unit for π-conjugated polymers. With the high electronegativity of oxygen, NOz possesses higher electron deficiency than NTz. Therefore, PNOz4T (Fig. 15.7a), a copolymer consisting of NOz and quaterthiophene, has HOMO and LUMO energy levels of −5.48 eV and −3.65 eV, respectively, both of which were deeper than those of PNTz4T by 0.2–0.3 eV. Further, E_g of PNOz4T was 1.52 eV, which was slightly narrower than that of PNTz4T (E_g = 1.56 eV). It was also noted that PNOz4T showed higher crystallinity than PNTz4T (Kawashima et al. 2015a), possibly because the smaller atomic radius of oxygen relative to sulfur allows the polymer backbones to pack in a more orderly fashion. Interestingly, PNOz4T showed ambipolar characteristics in OTFT devices with well-balanced hole and electron mobilities of up to 0.45 and 0.28 cm^2 V^{-1} s^{-1}, respectively, whereas PNTz4T showed unipolar p-type characteristics with hole mobility of ~0.22 cm^2 V^{-1} s^{-1} under the same conditions (Fig. 15.7b). The ambipolar characteristics of PNOz4T originated in the deep LUMO energy levels.

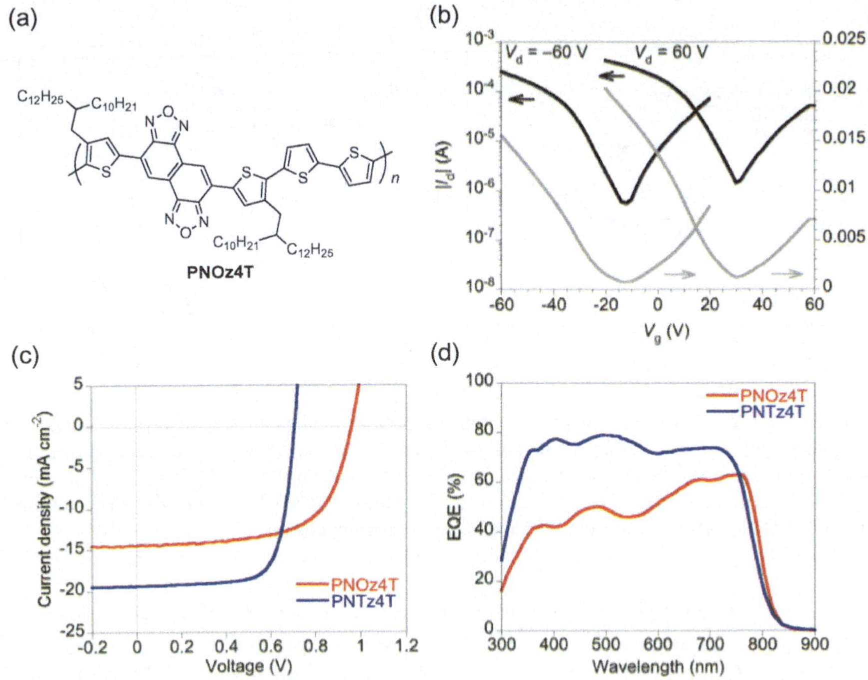

Fig. 15.7 **a** Chemical structure of NOz-based polymer (PNOz4T). **b** Transfer curves of the PNOz4T-based OFET device. **c** *J*–*V* curves and **d** EQE spectra of OPV cells that used PNOz4T and PNTz4T

The inverted cell that used PNOz4T (ITO/ZnO/PNOz4T:$PC_{71}BM/MoO_x$/Ag) afforded a relatively high PCE of 8.9% (J_{SC} = 14.5 mA cm^{-2}, V_{OC} = 0.96 V, FF = 0.64) (Fig. 15.7c, d) (Kawashima et al. 2015b). An important attribute of the PNOz4T cell was that it had a higher V_{OC} than the PNTz4T cell despite that the E_gs of both polymers were very similar. In other words, PNOz4T could significantly reduce photon energy loss (E_{loss}), which is defined by $E_g - eV_{OC}$, of OPVs. Thus, E_{loss} for the PNOz4T cell was 0.52–0.56 eV depending on the fabrication conditions, and was smaller by approximately 0.3 eV than that for the PNTz4T cell (0.80–0.85 eV). The HOMO energy level of PNOz4T was deeper by ca. 0.3 eV relative to that of PNTz4T, whereas the E_gs of both polymers were almost the same, agreeing well with the difference in E_{loss} between PNOz4T and PNTz4T.

The E_{loss} for OPVs is typically >0.8 eV (Wang et al. 2014), which is larger than those for inorganic and perovskite systems, which are around 0.4–0.5 eV (Fig. 15.8a) (King et al. 2011; Green et al. 2014). The large E_{loss} has hampered the improvement of V_{OC} and thus PCEs for OPVs (Fig. 15.8b). On the other hand, reducing E_{loss} would lead to loss of the driving force for charge separation in OPVs, which in turn would result in low J_{SC}s (Li et al. 2015). Therefore, with a small E_{loss} that is comparable to those in inorganic systems and a relatively high J_{SC} and thus the PCE of ~9%, PNOz4T is a highly unique π-conjugated polymer.

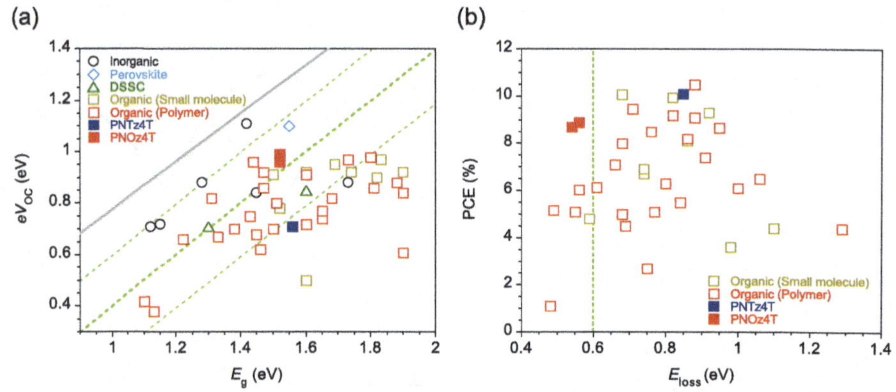

Fig. 15.8 **a** Plots of eV_{OC} against E_g, which was determined from the absorption onset, for various solar cell systems. The gray line and the green dotted line show E_{loss} calculated from the Shockley–Queisser (SQ) theory and $E_{loss} = 0.4$, 0.6, and 0.8 eV. **b** Plots of PCE against E_{loss} ($=E_g - eV_{OC}$) for organic solar cells (small molecules and polymers (PSCs)). Reproduced with permission (Kawashima et al. 2015b). Copyright 2015, Nature Publishing Group

Chou and co-workers also reported the synthesis of an NOz-based polymer (PSiNO) that incorporated benzobisthienosilolothiophene as the donor unit (Fig. 15.9a) (Zhang et al. 2015). In a similar manner to PNOz4T, PSiNO had a relatively narrow E_g of 1.56 eV and showed deep HOMO and LUMO energy levels of –5.50 and –3.73 eV, respectively. Reasonably high PCEs of around 8% were reported for both the conventional (ITO/PEDOT:PSS/PSiNO:$PC_{71}BM$/PFN/Ca/Al) and inverted (ITO/ZnO/PFN/PSiNO:$PC_{71}BM$/MoO_x/Ag) cells. It was noted that in both cases, a relatively high V_{OC} of 0.90 V was measured (Fig. 15.9b). As a result, E_{loss} for this system was 0.66 eV, which was relatively small for OPVs, indicating again that NOz is an interesting building unit for π-conjugated polymers.

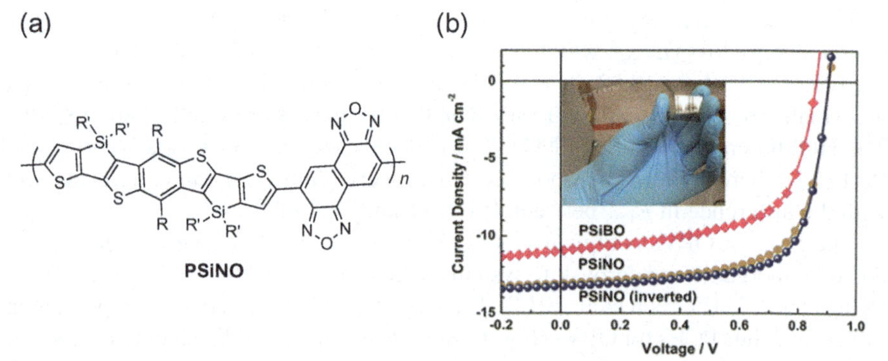

Fig. 15.9 **a** Chemical structure of PSiNO. **b** J–V curves of OPV cells that used PSiNO. Reproduced with permission (Zhang et al. 2015). Copyright 2015, Royal Society of Chemistry

15.3.3 TZNT-Based Polymers

TZNT is also an interesting building unit because it also allows the introduction of functional groups at *N*-positions, thereby leading to the high solubility of the corresponding polymers. Huang and co-workers reported the synthesis of TZNT-based polymers in which the BDT unit was incorporated as a donor unit (PBDT-TZNT) (Fig. 15.10a) (Dong et al. 2013). PBDT-TZNT had optical absorption at a

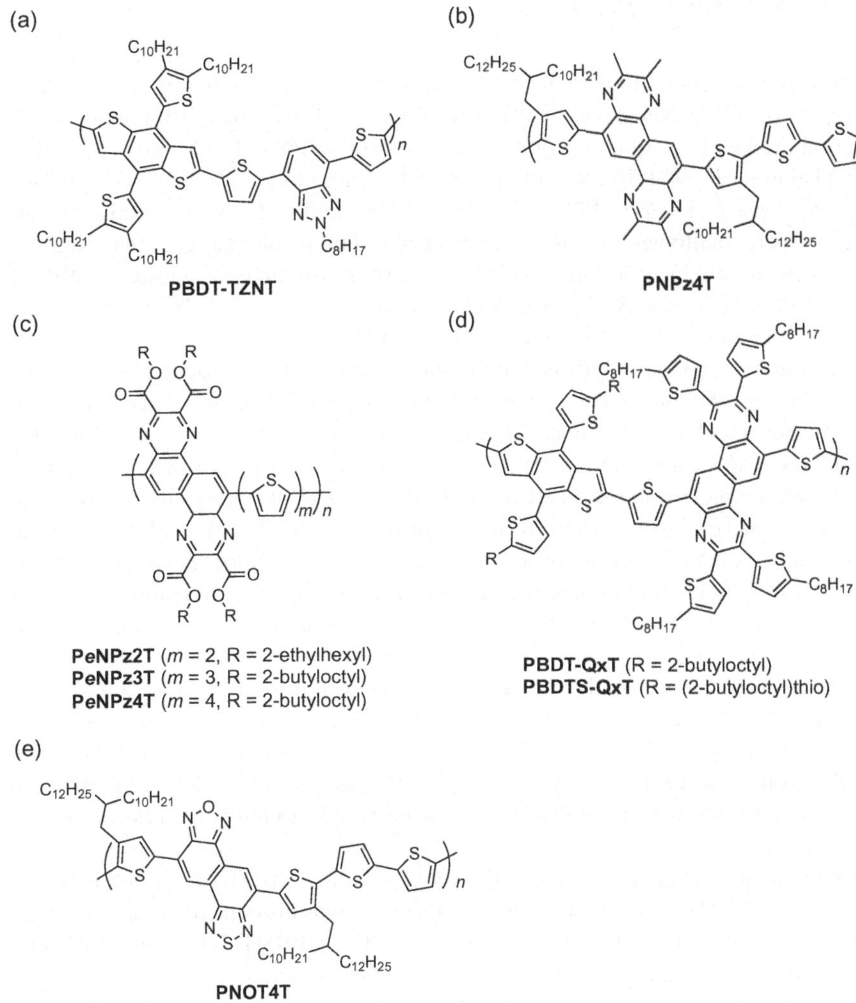

Fig. 15.10 Chemical structures of **a** a TZNT polymer (PBDT-TZNT), **b** an NPz-based polymer (PNPz4T), **c** ester-functionalized NPz-based polymers (PeNPz2T, PeNPz3T, and PeNPz4T), and **d** BDT-NPz polymers (PBDT-QxT and PBDTS-QxT)

wavelength range of 400–600 nm and E_g of approximately 1.9 eV. The absorption bands were very much blue-shifted compared to those of NTz-based polymers, which probably originated in the weak electron deficiency of the TZNT unit. The inverted OPV cells that used TZNT-based polymers had relatively high V_{OC}s of up to 0.92 V, which resulted from the relatively deep HOMO energy levels. PTZNT2 gave the highest PCE of 7.11% among these polymers.

15.3.4 NPz-Based Polymers

Our group has been involved in the synthesis of a series of NPz-based polymers. We first reported an NPz-based polymer (PNPz4T; Fig. 15.10b) in which tetramethyl NPz was combined with the quaterthiophene unit, similar to PNTz4T (Mikie et al. 2017). PNPz4T had an E_g of 1.76 eV, which was wider than that of PNTz4T. The HOMO and LUMO energy levels of PNPz4T were –5.05 eV and –3.02 eV, respectively, both of which were shallower than those of PNTz4T. The results suggest that NPz has weak electron deficiency relative to NTz. The photovoltaic cell that used PNPz4T (ITO/ZnO/PNPz4T:$PC_{61}BM/MoO_x$/Ag) showed a maximum PCE of 2.82% with J_{SC} of 7.9 mA cm^{-2}, V_{OC} of 0.59 V, and FF of 0.60.

One major advantage of NPz is that the substituents can be introduced at 2-, 3-, 8-, and 9-positions. We introduced one ester group to each of those positions, which was copolymerized with oligothiophene donor units (P*e*NPz2T, P*e*NPz3T, and P*e*NPz4T) (Fig. 15.10c) (Mikie and Osaka 2019). Owing to the electron-withdrawing property of the ester group, these polymers had deep HOMO and LUMO energy levels. Whereas the HOMO energy levels varied in the range of –5.1 to –5.4 eV with the oligothiophene unit, the LUMO energy levels were consistently approximately –3.5 eV. Further, it was interesting to note that the absorption range was red-shifted by more than 150 nm relative to that of PNPz4T with methyl or dodecyl substituents. E_g was around 1.50 eV. The polymers formed relatively highly crystalline structures with the face-on orientation, which were also favorable for OPVs, in both neat and blend films. In contrast, P*e*NPz2T with the bithiophene donor unit showed weak crystallinity, likely ascribed to the steric hindrance between the substituents. Although the OPV cells that used the polymers (ITO/ZnO/polymer:$PC_{61}BM/MoO_x$/Ag) had moderate PCEs that were as high as 6.3%, the narrow E_g of the polymers was indeed advantageous.

Peng and co-workers reported on BDT-NPz polymers PBDT-NQxT and PBDTS-NQxT (Fig. 15.10d), in which the octylthienyl group was introduced at each of the 2-, 3-, 8-, and 9-positions of the NPz unit (Yu et al. 2017). PBDT–NQxT and PBDTS–NQxT had HOMO energy levels of –5.24 eV and –5.31 eV, and E_gs of 1.80 eV and 1.81 eV, respectively. OPV cells were fabricated with the combination of $PC_{71}BM$ and a non-fullerene compound (ITIC) as the n-type material. The $PC_{71}BM$-based cell showed moderate PCEs of up to 7%, which were likely limited by the short photoresponse range below 700 nm. On the other hand, the ITIC-based cell had a photoresponse range of up to 800 nm, thereby resulting in the highest PCE of 11.5%.

15.3.5 NOT-Based Polymer

Among the hybrid naphthalene-based units, NOT was used as the acceptor unit for a polymer in combination with the quaterthiophene donor unit (PNOT4T) (Fig. 15.10e) (Mikie and Osaka 2018). The solubility of PNOT4T was significantly higher than that of the parent polymers, PNTz4T and PNOz4T, likely originating from the unsymmetrical structure of NOT that would lead to a regiorandom polymer backbone. Nevertheless, PNOT4T had high crystallinity with the edge-on orientation in the polymer-neat film, and with the face-on orientation in the blend film, similar to the parent polymers. The absorption range and the E_g (1.56 eV) were also similar to those of the parent systems. Interestingly, the HOMO and LUMO energy levels were in between those of PNTz4T and PNOz4T. The OPV cells with the inverted structure fabricated using $PC_{71}BM$ as the n-type material exhibited the maximum PCE of 8.9% ($J_{SC} = 17.7$ mA cm^{-2}, $V_{OC} = 0.77$ V, and FF = 0.65), which was reasonably high when the electronic properties and the film structures were considered.

15.4 Summary

This section highlighted a series of naphthalene-based nitrogen-containing heteroaromatics with symmetrical and unsymmetrical structures for π-conjugated polymers that are useful for organic electronic devices, in particular, OPVs. Due to the electron deficiency and the rigid structure, these heteroaromatics provide highly optically active π-conjugated polymers with broad absorption ranges, relatively deep HOMO and LUMO energy levels, and highly crystalline structures, all of which are very important for organic electronic applications. π-Conjugated polymers based on these building units have provided very high efficiencies of over 10% in OPVs when blended with both fullerene and non-fullerene materials.

References

Adachi C, Hattori R, Kaji H, Tsujimura T (2019) Handbook of organic light-emitting diodes. Springer

Bao Z, Locklin J (2007) Organic field-effect transistors. CRC Press

Biniek L, Schroeder BC, Nielsen CB, McCulloch I (2012) Recent advances in high mobility donor-acceptor semiconducting polymers. J Mater Chem 22:14803–14813

Boudreault P-LT, Najari A, Leclerc M (2011) Processable low-bandgap polymers for photovoltaic applications. Chem Mater 23:456–469

Brabec C, Scherf U, Dyakonov V (2014) Organic photovoltaics. Wiley-VCH

Dong Y, Hu X, Duan C, Liu P, Liu S, Lan L, Chen D, Ying L, Su S, Gong X, Huang F, Cao Y (2013) A series of new medium-bandgap conjugated polymers based on naphtho[1,2-*c*:5,6-*c*]bis(2-octyl-[1,2,3]triazole) for high-performance polymer solar cells. Adv Mater 25:3683–3688

Green MA, Ho-Baillie A, Snaith HJ (2014) The emergence of perovskite solar cells. Nat Photon 8:506–514

Imai K, Kurihara M, Mathias L, Wittmann J, Alston WB, Stille JK (1973) Synthesis and properties of thermally stable ladder polymers containing the 1,4-pyrazine ring obtained from polyheterocyclizations of tetramines and tetraketones in poly(phosphoric acid) and m-cresol. Macromolecules 6:158–162

Kawashima K, Miyazaki E, Shimawaki M, Inoue Y, Mori H, Takemura N, Osaka I, Takimiya K (2013) 5,10-diborylated naphtho[1,2-c:5,6-c′]bis[1,2,5]thiadiazole: a ready-to-use precursor for the synthesis of high-performance semiconducting polymers. Polym Chem 4:5224–5227

Kawashima K, Osaka I, Takimiya K (2015a) Effect of chalcogen atom on the properties of naphthobischalcogenadiazole-based π-conjugated polymers. Chem Mater 27:6558–6570

Kawashima K, Tamai Y, Ohkita H, Osaka I, Takimiya K (2015b) High-efficiency polymer solar cells with small photon energy loss. Nat Commun 6:10085

Kawashima K, Fukuhara T, Suda Y, Suzuki Y, Koganezawa T, Yoshida H, Ohkita H, Osaka I, Takimiya K (2016) Implication of fluorine atom on electronic properties, ordering structures, and photovoltaic performance in naphthobisthiadiazole-based semiconducting polymers. J Am Chem Soc 138:10265–10275

King RR, Bhusari D, Boca A, Larrabee D, Liu XQ, Hong W, Fetzer CM, Law DC, Karam NH (2011) Band gap-voltage offset and energy production in next-generation multijunction solar cells. Prog Photovolt 19:797–812

Li W, Hendriks KH, Furlan A, Wienk MM, Janssen RAJ (2015) High quantum efficiencies in polymer solar cells at energy losses below 0.6 eV. J Am Chem Soc 137:2231–2234

Mataka S, Takahashi K, Ikezaki Y, Hatta T, Tori-i A, Tashiro M (1991) Sulfur nitride in organic chemsitry. XIV, Selective formation of benzo-and benzobis [1, 2, 5] thiadiazole skeleton in the reaction of tetrasulfur tetranitride with naphthalenols and related compounds. Bull Chem Soc Jpn 64:68–73

McCullough RD (1998) The chemistry of conducting polythiophenes. Adv Mater 10:93–116

Mikie T, Osaka I (2018) Selective synthesis and properties of electron-deficient hybrid naphthalene-based π-conjugated systems. Chem Eur J 24:19228–19235

Mikie T, Osaka I (2019) Ester-Functionalized naphthobispyrazine as an acceptor building unit for semiconducting polymers: synthesis, properties, and photovoltaic performance. Macromolecules 52:3909–3917

Mikie T, Kawashima K, Komeyama K, Yoshida H, Osaka I (2017) Naphthobispyrazine as an electron-deficient building unit for π-conjugated polymers: efficient synthesis and polymer properties. Chem Lett 46:1193–1196

Osaka I, Takimiya K (2017) Naphthobischalcogenadiazole conjugated polymers: emerging materials for organic electronics. Adv Mater 39:1605218

Osaka I, Shimawaki M, Mori H, Doi I, Miyazaki E, Koganezawa T, Takimiya K (2012) Synthesis, characterization, and transistor and solar cell applications of a naphthobisthiadiazole-based semiconducting polymer. J Am Chem Soc 134:3498–3507

Osaka I, Saito M, Koganezawa T, Takimiya K (2014) Thiophene-thiazolothiazole copolymers: significant impact of side chain composition on backbone orientation and solar cell performances. Adv Mater 26:331–338

Peet J, Wen L, Byrne P, Rodman S, Forberich K, Shao Y, Drolet N, Gaudiana R, Dennler G, Waller D (2011) Bulk heterojunction solar cells with thick active layers and high fill factors enabled by a bithiophene-co-thiazolothiazole push-pull copolymer. Appl Phys Lett 98:043301

Reynolds JR, Thompson BC, Skotheim TA (2019) Handbook of conducting polymers, 4th ed. CRC Press

Saito M, Koganezawa T, Osaka I (2018) Correlation between distribution of polymer orientation and cell structure in organic photovoltaics. ACS Appl Mater Interfaces 10:32420–32425

Thompson BC, Fréchet JMJ (2008) Polymer-fullerene composite solar cells. Angew Chem Int Ed 47:58–77

Villena-Blanco M, Jolly WL (1967) Tetrasulfur tetranitride, S_4N_4. Inorganic syntheses, vol 9. Wiley, 98–102

Vohra V, Kawashima K, Kakara T, Koganezawa T, Osaka I, Takimiya K, Murata H (2015) Efficient inverted polymer solar cells employing favourable molecular orientation. Nat Photon 9:403–408

Wang M, Hu X, Liu P, Li W, Gong X, Huang F, Cao Y (2011) Donor-acceptor conjugated polymer based on naphtho[1,2-c:5,6-c]bis[1,2,5]thiadiazole for high-performance polymer solar cells. J Am Chem Soc 133:9638–9641

Wang M, Wang H, Yokoyama T, Liu X, Huang Y, Zhang Y, Nguyen TQ, Aramaki S, Bazan GC (2014) High open circuit voltage in regioregular narrow band gap polymer solar cells. J Am Chem Soc 136:12576–12579

Yang T, Wang M, Duan C, Hu X, Huang L, Peng J, Huang F, Gong X (2012) Inverted polymer solar cells with 8.4% efficiency by conjugated polyelectrolyte. Energy Environ Sci 5:8208–8214

Yu T, Xu X, Zhang G, Wan J, Li Y, Peng Q (2017) Wide bandgap copolymers based on quinoxalino[6,5-*f*]quinoxaline for highly efficient nonfullerene polymer solar cells. Adv Funct Mater 27:1701491

Zhang Z, Lin F, Chen HC, Wu HC, Chung CL, Lu C, Liu SH, Tung SH, Chen WC, Wong KT, Chou PT (2015) A silole copolymer containing a ladder-type heptacylic arene and naphthobisoxadiazole moieties for highly efficient polymer solar cells. Energy Environ Sci 8:552–557

Chapter 16
Luminescent Materials for Organic Light-Emitting Diodes

Shigeyuki Yagi

Abstract Nowadays, an organic light-emitting diode (OLED), consisting of nano-order luminescent and semiconductive organic thin films, attracts considerable interest from the viewpoint of application to flat panel displays and illumination devices. Among the constituent materials of OLEDs, emitting materials play an important role in determination of the device performance such as luminous efficiencies and chromaticity coordinates of electroluminescence. Here, luminescent dyes and related compounds used as emitters in OLEDs are reviewed in accordance with the mechanism of the exciton generation. First, fluorescent emitters based on organic dyes and π-conjugated polymers are focused on, and then the topic moves to phosphorescent organometallic emitters affording high device efficiencies. In the last part, organic TADF emitters are reviewed, which can realize the 100% internal quantum efficiency of electroluminescence in theory without the employment of precious metal-containing organometallic structures.

Keywords Organic light-emitting diode · Organic dye · Iridium complex · Platinum complex · Fluorescence · Phosphorescence · Thermally activated delayed fluorescence

16.1 Introduction

Organic light-emitting diode (OLED) is an electroluminescent device consisting of nano-order thin films of organic semiconducting and emitting materials and has attractive technological advantages such as surface self-light emission, quick on–off response, low voltage drive, thin and light device bodies, availability of flexible devices, and so on. Nowadays, OLEDs are applied to main light sources for large-area flat panel displays and lighting apparatuses. They are also used as image pixels of smart devices such as smartphones and wearable computers. In future, further

S. Yagi (✉)
Department of Applied Chemistry, Graduate School of Engineering, Osaka Prefecture University, 1-1 Gakuen-cho, Naka-ku Sakai, Osaka 599-8531, Japan
e-mail: yagi@chem.osakafu-u.ac.jp

Y. Ooyama and S. Yagi (eds.), *Progress in the Science of Functional Dyes*,
https://doi.org/10.1007/978-981-33-4392-4_16

demands for OLEDs are expected according to development of "Internet of Things (IoT)".

Emitting materials are one of the key constituents to determine the device performance of OLEDs, such as device efficiencies and chromaticity coordinates of electroluminescence (EL). The first report of EL from an organic compound was done in 1960s (Helfrich and Schneider 1965, 1966). Therein, although EL was obtained from a single crystalline anthracene, extremely high voltage was applied to it due to poor charge carrier injection. The prototype device structure of the current OLEDs was reported in 1987 by Tang and VanSlyke, which achieved high luminance of more than 1000 cd m^{-2} at less than 10 V (Tang and VanSlyke 1987). As shown in Fig. 16.1a, the device consisted of a stack of a hole-transporting diamine derivative (75 nm) and a fluorescent and electron-transporting aluminum–8-quinolinolate chelate complex Alq3 (60 nm) between an indium tin oxide (ITO) transparent anode and a Mg:Ag cathode, and all the constituent materials were vacuum deposited on ITO. At this point, the multilayered device structure was established to achieve high device performance. Next, Tang and coworkers fabricated a multilayer device including an emitting layer (EML) doped with a fluorescent dye, as shown in

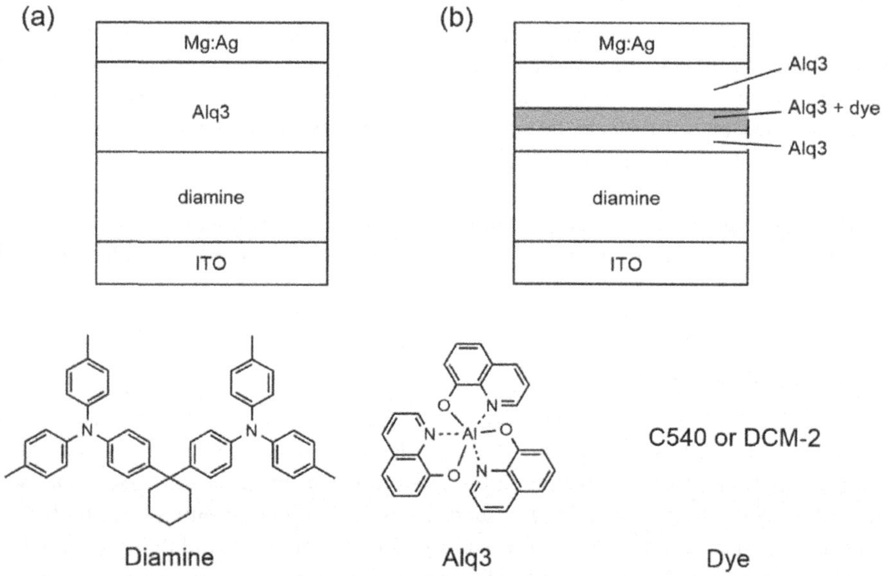

Fig. 16.1 Device structures of OLEDs reported by Tang and coworkers; **a** the undoped device (ITO (anode)/diamine/Alq3/Mg:Al (cathode)) and **b** the doped device (ITO (anode)/diamine/Alq3/Alq3 + 1 mol% dye/Alq3/Mg:Al (cathode)). The thickness of diamine layer was 75 nm, and the total thickness of the undoped Alq3 layer was 60 nm. The dye-doped Alq3 layer (5–60 and 4 nm for C540 and DCM-2, respectively) was inserted between the diamine/Alq3 interface (for C540) or into the undoped Alp3 layer (for DCM-2). In the case of the DCM-2-based device, the distance from the diamine/Alq3 interface was varied. The structures of C540 and DCM-2 are listed in Fig. 16.4

Fig. 16.1b (Tang and VanSlyke 1989). Employing a dye-doped EML led to achievement of an external quantum efficiency (η_{ext}) of 0.023–0.025, about two times larger than the undoped device (η_{ext}; 0.013). To obtain EL from the doped dye via energy transfer from the Alq3 host, it is essential to place the dye-doped layer in less than 5 nm from the diamine/Alq3 interface, indicating that the efficient energy transfer occurs in the charge recombination zone. After that, the emitter-doped device fabrication has been widely utilized as a general technique to achieve high-performance OLEDs. The importance of fabrication of the multilayer device structure was demonstrated by Saito and coworkers (Adachi et al. 1990). They fabricated a device consisting of stacked organic thin films of hole-transporting layer (HTL)/EML/electron-transporting layer (ETL) between the anode and the cathode. To obtain a high EL efficiency, it is important to achieve efficient exciton generation at EML through hole–electron recombination. In the case of the HTL/EML/ETL device structure, confinement of holes and electrons in EML by blocking at the EML/ETL and HTL/EML boundaries, respectively, leads to efficient charge carrier recombination.

At the early period of OLED development, sublimable materials were mainly used due to applicability to vacuum deposition (i.e., dry process). Thus, low-mass molecules have been widely developed. On the other hand, polymer materials have been also applied for OLED fabrication. Especially, π-conjugated polymers, represented by poly(*p*-phenylene vinylene)s (PPVs) such as MEH-PPV (EL maximum λ_{EL}; 625 nm) and MEH-PPDFV (λ_{EL}; 520 nm) (Losurdo et al. 2009) and poly(9,9-dialkyl-2,7-fluorene)s (PDAFs) (λ_{EL}; 436 nm, film) (Grice and Bradley 1998) (Fig. 16.2), have been used as emitting materials. Although these polymeric materials are not sublimable due to their high molecular masses, they are applied to OLED fabrication by coating and printing methods (i.e., wet process), due to their excellent film-forming ability. Nowadays, the wet process is attracting much attention as a reliable way for mass production of OLEDs.

The emitting materials for OLEDs have so far been classified into two; fluorescent and phosphorescent emitters. Fluorescent and phosphorescent emissions are obtained through the radiative decay from the lowest singlet and triplet excited states (S_1 and T_1), respectively, as shown in Fig. 16.3. In general, exciton generation in OLEDs is induced by electric current, where the singlet and triplet excitons of the emitting materials are spin-statistically generated in a ratio of 1:3. Taking the intersystem

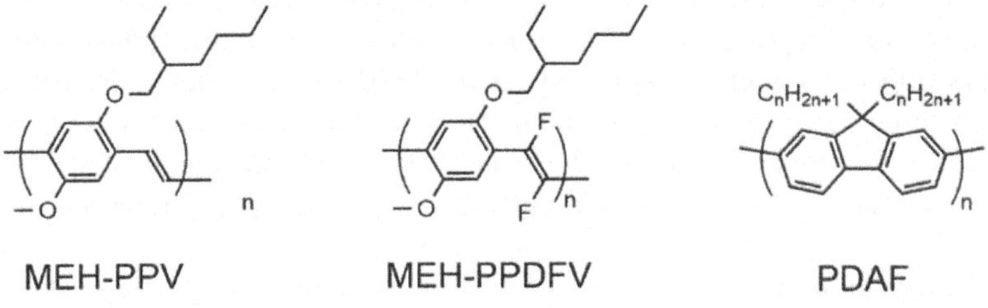

Fig. 16.2 Structures of π-conjugated polymers MEH-PPV, MEH-PPDFV, and PDAF

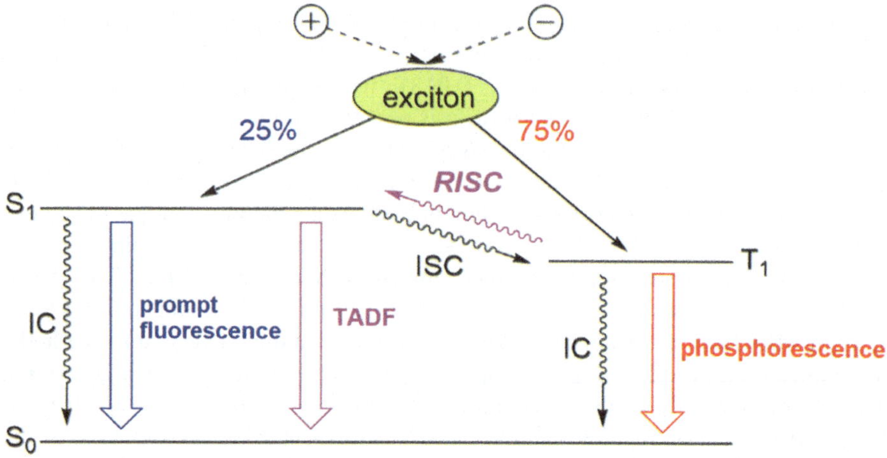

Fig. 16.3 Illustration of photophysical processes for emitting materials upon electric current excitation. S_0, S_1, and T_1 represent the singlet ground state, the lowest singlet excited state, and the lowest triplet state, respectively. ISC and RISC represent the intersystem crossing from S_1 to T_1 and its reverse process, respectively. IC represents non-radiative internal conversion

crossing from S_1 to T_1 into consideration, the exciton formation efficiency η_{ph} of the triplet exciton reaches as high as 1, although the singlet exciton generates with η_{ph} of 0.25 at most. The external quantum efficiency η_{ext} (%) of EL from OLED is defined as the following equation

$$\eta_{ext} = \eta_{out} \times \eta_{ph} \times \Phi_{PL} \times \gamma$$

where η_{out}, Φ_{PL}, and γ represent a light-extraction efficiency, a photoluminescence (PL) quantum yield, and a charge carrier balance factor, respectively. The typical value of η_{out} is *ca.* 0.2. According to this equation, OLEDs using phosphorescent emitters theoretically provide η_{ext} as high as 20%, whereas those using fluorescent ones do η_{ext} of 5% at most. Thus, phosphorescent OLEDs are superior in the device efficiency to fluorescent ones.

From the historical viewpoint, fluorescent emitters have been widely used since the early stage of development of the EL devices, because many of their basic skeletons have been derived from traditional organic fluorescent dyes. Thereafter, organometallic complexes emitting efficient room-temperature phosphorescence were developed in 2000s (Lamansky et al. 2001). Since then, phosphorescent organometallic complexes with platinum(II) and iridium(III) centers have been used to fabricate high-efficiency OLEDs. It is noteworthy that achievement of nearly 100% internal quantum efficiency was demonstrated by using a phosphorescent material (Adachi et al. 2001a). More recently, organic molecules exhibiting thermally activated delayed fluorescence (TADF) have been attracting much attention because the up-conversion process from the low-lying triplet state to the high-lying singlet excited one (i.e., reverse intersystem crossing) leads to the singlet exciton generation with η_{ph} more than the theoretical limit of conventional fluorescent OLEDs (Fig. 16.3) (Endo et al. 2011). Thus, TADF-based OLED can theoretically provide η_{ext} comparable to phosphorescent OLED. Nowadays, TADF materials are regarded as excellent candidates for development of high-efficiency OLEDs. It is noteworthy that TADF materials are beneficial in regards to production cost as well as influence on environment: they do not involve any precious metal elements, and thus can be developed relatively at low cost. Nowadays, from the viewpoint of historical background, fluorescent and phosphorescent materials are called the first and second generation emitters, respectively, and TADF materials are regarded as the third generation emitters.

This chapter reviews development of emitting materials for OLED applications, where the classification is based on the mechanism of photon generation; fluorescence, phosphorescence, and TADF materials. Recent advances of OLED emitters are also described, especially focusing on the aspects to overcome the technical problems such as optimization of chromaticity coordinates of emission colors, improvement of Φ_{PL}, and so on.

16.2 Fluorescent Materials: First-Generation Emitters

16.2.1 Fluorescent Emitters Based on Low-Mass Molecular Systems

At the early stage, traditional fluorescent dyes such as coumarins (e.g., C540 (or C6)) and merocyanines (e.g., DCM-1 and DCM-2) were used as emitters for OLEDs (Fig. 16.4) (Tang and VanSlyke 1989). Thereafter various types of fluorescent emitters have been developed to optimize the required properties such as Φ_{PL}s, emission colors, thermal stability, and so on. Many of fluorescent materials have relatively small Stokes shifts, and self-quenching usually occurs in high concentrations. In addition, the maximum η_{ext} of fluorescent OLED is 5% in theory, as described in Sect. 16.1. Thus, extremely high Φ_{PL}s are required in the solid state and solid matrix for non-doped and doped EMLs, respectively. As for emission colors, chromaticity coordinates consistent with three primary colors such as red, green, and blue are

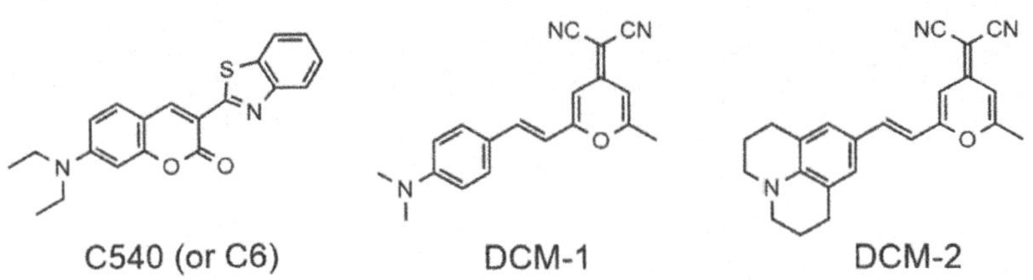

Fig. 16.4 Structures of coumarin dye C540 (or C6) and merocyanine dyes DCM-1 and DCM-2

required because OLEDs have been intended for application as picture elements for flat panel displays. To obtain a target emission color, one can tune the π-conjugation length of the chromophore. That is, if red-shifted emission is required, the extended π-conjugation system leading to a narrow HOMO–LUMO energy gap should be obtained to lower the energy level of S_1. On the contrary, shortening the π-conjugation system brings about the high-lying S_1, affording blue-shifted emission. Nowadays, with the help of progress of theoretical calculations, it is relatively easy to carry out molecular designs of fluorescent emitters focusing on the emission wavelengths. Here, the author summarizes representative fluorescent dyes for OLED applications.

As red fluorescent emitters, pyran-based merocyanine dyes like a DCM series have been intensely developed, which consist of the aminophenyl donor and dicyanomethylene acceptor groups. Although DCM-1 and DCM-2 were adopted by Tang and coworkers to fabricate OLEDs with a doped EML (Tang and VanSlyke 1989), they tend to suffer from self-aggregation lowering the device efficiency. To overcome this problem, the improved derivatives such as DCJTI and DCJTB (Fig. 16.5a) were developed (Chen et al. 2000), where branched alkyl groups are introduced to both of donor and acceptor moieties. When used as emitting dopants in OLEDs, these emitters exhibited saturated red EL, showing the Commission Internationale de L'éclairage (CIE) chromaticity coordinates of (0.626, 0.370) and (0.628, 0.367) for DCJTI and DCJTB, respectively. Besides the DCM-type dyes, various chromophoric systems have so far been reported as red fluorescent emitters. As shown in Fig. 16.5b, Yu and Shirota developed coumarin-based red fluorescent dyes ACY and QCY showing the CIE coordinates of (0.68, 0.32) and (0.70, 0.30) in their

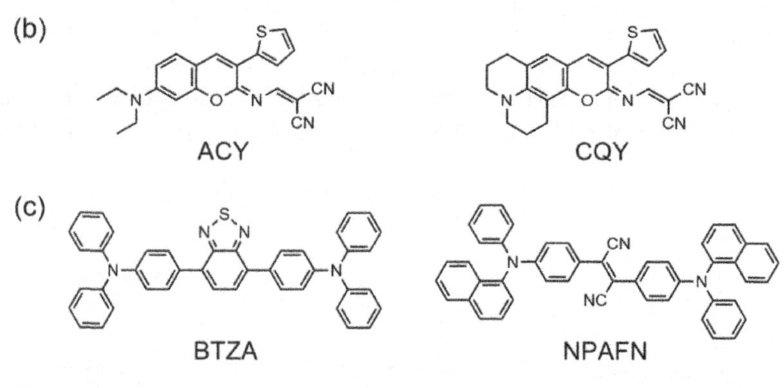

Fig. 16.5 Structures of red fluorescent emitters; **a** DCJTI and DCJTB, **b** ACY and CQY, and **c** BTZA and NPAFN

devices, closer to the National Television Standards Committee (NTSC) standard for red (defined as (0.67, 0.33)) than those obtained for DCJTI and DCJTB (Yu and Shirota 2002). Especially, the ACY-based device showed relatively efficient performance; a maximum luminance of 6400 cd m^{-2} at 15 V and η_{ext} of 0.8%. As other examples, 2,1,3-benzothiadiazole and fumaronitrile skeletons were incorporated to obtain BTZA (λ_{EL}, 626 nm; CIE (0.64, 0.36)) (Thomas et al. 2004) and NPAFN (λ_{EL}, 636 nm; CIE (0.64, 0.36)) (Yeh et al. 2003), respectively (Fig. 16.5c). Especially, the latter is highly emissive in the solid state, suitable for preparation of a non-doped EML.

For green fluorescent emitters, Alq3 and coumarin were used at the early stage (Tang and VanSlyke 1989). Then, lots of green fluorophores have been reported, as shown in Fig. 16.6. C545T is one of the commercially available green fluorescent emitters, which has been used in prototypical OLEDs. Chen and coworkers modified the structure of this dye to improve the device performance as well as thermal stability and photostability, namely C-545P (PL maximum λ_{PL}, 514 nm; Φ_{PL}, 0.99 in 1,2-dichloroethane) (Lee et al. 2004). Strategically placing methyl groups on the julolidyl rings, concentration quenching was more suppressed in comparison with C545, and the device efficiency of the C545P-based OLED was notably improved. Modification of C540 (or C6) and C545 series dyes with bulky substituents is also effective to improve the device efficiency of OLED as well as thermal stability, as seen in C6DB and C-545TB (Chen and Tang 2001). Quinacridone (QAD) and its derivatives are also typical green fluorescent emitters showing high thermal stability and photostability. Due to intermolecular strong hydrogen bonds, quinacridone is insoluble itself in any solvents. So, alkyl substituents are often introduced to the nitrogen atoms, as seen in DMQA (λ_{EL}, 544 nm; CIE (0.39, 0.59)) (Shi and Tang 1997). As other green fluorescent emitters, an anthracene-9,10-diamine derivative (TpAD) (Yu et al. 2011) and an oligoquinoxaline derivative (TPA-2PQx) (Hancock et al. 2006) have been reported. The former exhibits efficient green emission in dichloromethane (λ_{PL}, 546 nm; Φ_{PL}, 0.94), and λ_{EL} of its device is blue-shifted by 12 nm, showing the CIE coordinate of (0.35, 0.61). The latter exhibits an ambipolar character and allows for fabrication of an OLED with a non-doped EML showing η_{ext} of 3.0%. Green emitters for efficient non-doped OLEDs were also developed by Tang and coworkers, where they utilized aggregation-induced emission (AIE) behavior of 2,5-difuluororenyl-substituted silole derivatives (Chen et al. 2013). AIE is observed for certain dyes, where their emissions are enhanced by aggregation to suppress free bond rotations facilitating the non-radiative decay. So high Φ_{PL}s are obtained for AIE luminophores in the solid state rather than in solution. For example, in the case of MFMPS, the Φ_{PL} of the drop-cast film is 0.88, although that of the solution is just 0.026. The non-doped device with an MFMPS-based EML afforded green EL (CIE (0.36, 0.57)) with a high η_{ext} of 5.5% corresponding to the limit value for fluorescent OLEDs.

Blue fluorescent emitters are very important in OLED applications because they are still used as practical emitting materials for displays and illumination devices: as for blue OLEDs, the devices with fluorescent emitters are superior to those with phosphorescent and TADF emitters in duration and color purity, although η_{ext} is

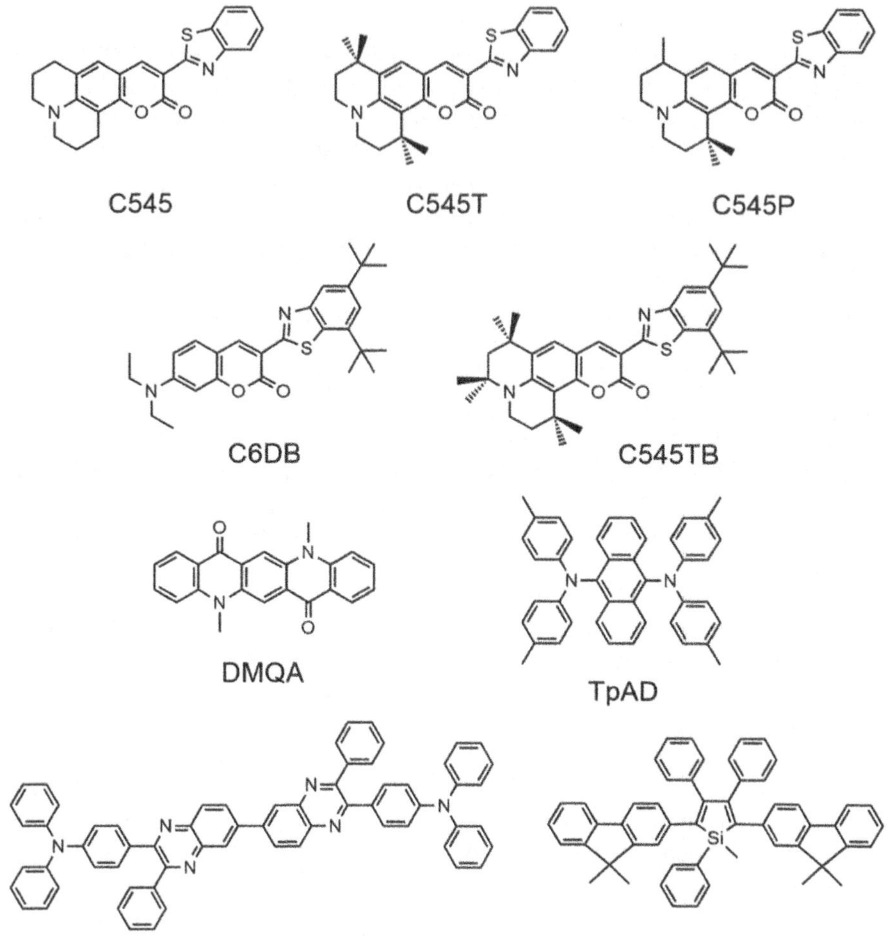

Fig. 16.6 Structures of green fluorescent emitters

inferior due to the spin statics on the exciton generation. Blue fluorophores consist of compact π-conjugation systems based on π–π^* electronic transition rather than intramolecular charge transfer (ICT) transition causing a low energy gap between the singlet ground state (S_0) and S_1. Representative blue fluorescent emitters are shown in Fig. 16.7. 2,5,8,11-Tetra-*tert*-butylperylene (TBPe) (Mi et al. 1999), TPB (Mikami et al. 2005), and DPVBi (Zheng et al. 2003) are commercially available. In the case of TBPe, the *tert*-butyl groups are essential to obtain stable color purity, suppressing the aggregate formation (Mi et al. 1999). The TBPe-based OLED showed blue EL at 460 nm, the CIE coordinate of which varied from (0.168, 0.273) to (0.175, 0.273) upon the increase in the doping level from 1 to 5%. On the other hand, the perylene-based device showed the CIE coordinate varying from (0.165, 0.196) to (0.178,

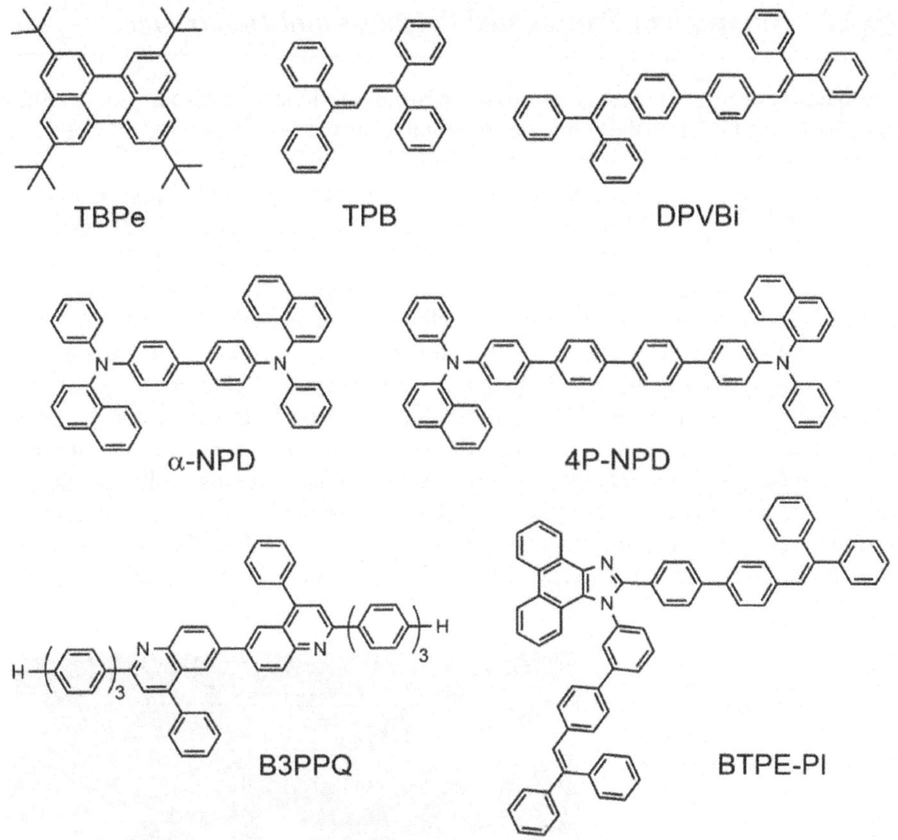

Fig. 16.7 Structures of blue fluorescent emitters

0.252). Although N^4,$N^{4\prime}$-di(naphthalen-1-yl)-N^4,$N^{4\prime}$-diphenyl-[1,1′-biphenyl]-4,4′-diamine (α-NPD) (Mikami et al. 2005) and N^4,$N^{4\prime\prime\prime}$-di(naphthalen-1-yl)-N^4,$N^{4\prime\prime\prime}$-diphenyl-[1,1′:4′,1″:4″,1‴-quaterphenyl]-4,4‴-diamine (4P-NPD) (Schwartz et al. 2007) are usually used as hole-transporting (i.e., p-type) host materials, they are also useful as blue emitters. They were used as deep blue emitters for fabrication of white OLEDs. An n-type blue emitter based on an oliogoquinoline π-conjugated structure (B3PPQ) was also reported by Jenekhe and coworkers, which emitted pure blue EL (λ_{EL}, 458 nm; η_{ext}, 2.8%; CIE (0.18, 0.21)) (Hancock et al. 2007). As a deep blue AIE emitter, BTPE-PI was reported by Tang and coworkers. The triphenylethane moieties are indispensable for generation of AIE, and non-doped OLED consisting of a VTPE-PI-based non-doped EML showed deep blue EL with CIE (0.15, 0.15), accompanied by a very high η_{ext} of 4.4% (Qin et al. 2015).

16.2.2 Fluorescent Conjugated Polymers and Dendrimers

Although emitting materials with high-mass polymeric structures are not used for dry-processed OLEDs, they are highly suitable to the wet-processed devices due to their excellent film-forming ability. Emitters bearing dendritic structures also come under this category. At the early stage of development of polymer-based OLEDs, PPVs and PDAFs were often used. MEH–PPV (Fig. 16.2), emitting orange-to-red EL, is used for fabrication of a flexible EL device (Fig. 16.8), which is a characteristic feature of "plastic"-based electronic devices (Gustafsson et al. 1992). Introduction of a cyano group to each vinyl moiety of PPV (PCTP, Fig. 16.9) led to a red shift of EL to the near-infrared region (λ_{EL}, 710 nm) (Greenham et al. 1993). As described in Sect. 16.1, PDAF is well known as a blue-emitting polymer (the structure shown in Fig. 16.2) (Ohmori et al. 1991). In this polymer, the alkyl groups are placed perpendicularly against the π-conjugation system of the fluorene unit to suppress the stacking interaction between neighboring polymer chains, and thus, self-quenching of emission is lowered. In addition to these polymers, various types of π-conjugated copolymers based on a 9,9-dialkyl-2,7-fluorenyl comonomer unit have been reported, as shown in Fig. 16.9. Varying the other

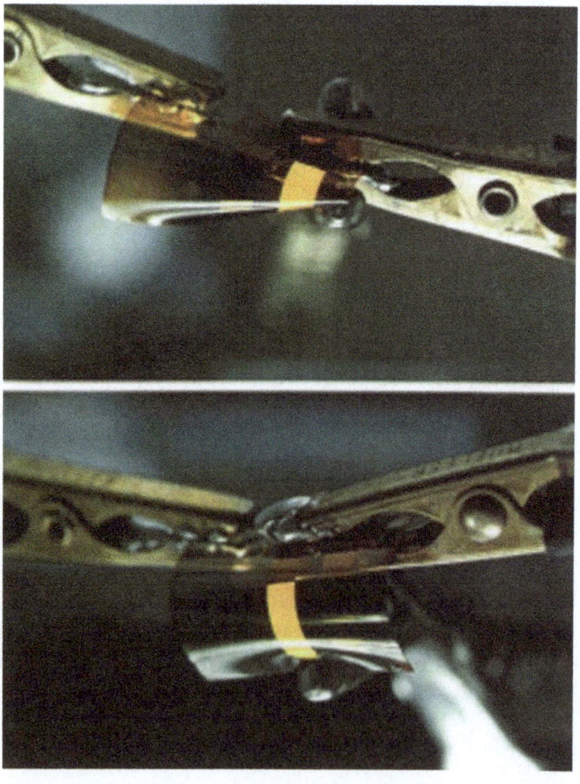

Fig. 16.8 Photographs of the flexible EL device with MEH–PPV as EML upon application of a bias voltage. The device was fabricated on a poly(ethylene terephthalate) (PET) substrate, where the device structure was PET/polyaniline (hole injection contact)/MEH-PPV (electroluminescent layer)/Ca (electron-injecting contact). Reprinted from Gustafsson et al. (1992) with permission from Springer Nature

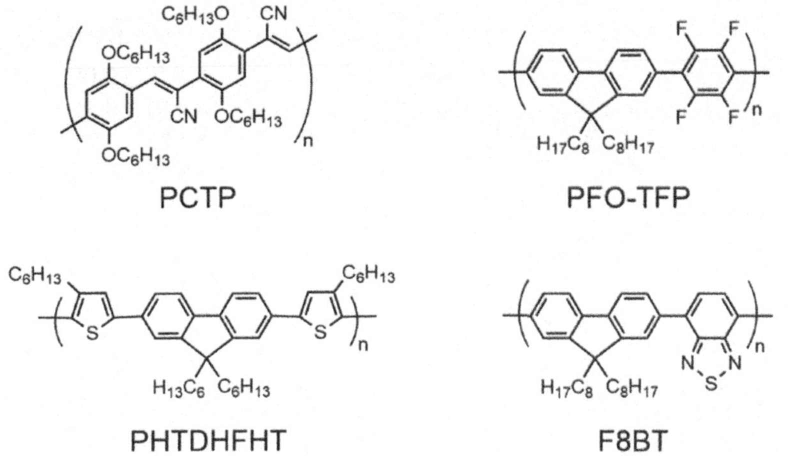

Fig. 16.9 Structures of polymer-based fluorescent emitters

comonomer unit, luminescent properties are tuned, and additional functions can be provided. The copolymer with an alternating fluorinated *p*-phenylene unit (PFO–TFP) was used as EML to fabricate highly efficient deep blue OLED (Giovanella et al. 2013). The λ_{EL} of the fabricated device was 405 nm (CIE (0.16, 0.05)), blue-shifted in comparison to a PFO-based device. When the multi-stacked device structure of ITO/PEDOT:PSS/PVCz/PFO-TFP/TPBi/Ba/Al (PEDOT:PSS, PVCz, and TPBi; poly(3,4-ethylenedioxythiophene):poly(4-stylenesulfonate), poly(9-vinylcarbazole), and 1,3,5-tris(1-phenyl-1*H*-benzo[*d*]imidazol-2-yl)benzene, respectively) was employed, a high η_{ext} (5.02%) was obtained, reaching a theoretical upper limit for fluorescent OLEDs. Employing a thiophen-2,5-diyl and a 2,1,3-benzothiadiazol-4,7-diyl comonomer unit, green fluorescent alternating copolymers PHTDHFHT (λ_{PL}, 493 nm with a shoulder at 515 nm, film; Φ_{PL}, 0.32, film) (Pei et al. 2000) and F8BT (λ_{EL}; 545 nm) (He et al. 1999) were obtained, respectively. As other examples, Ozturk and coworkers prepared a series of fluorene–dithienothiophene-*S*,*S*-dioxide copolymers having different ratios of the comonomer units (P1-X, P2-X, and P3-X; Fig. 16.10), demonstrating the tuning of the EL color from light blue to red (Osken et al. 2013).

Dendrimers are highly branched molecules with defined molecular masses, consisting of the core and the branched dendrons. They also come under the category of OLED polymer materials, and a variety of dendritic emitting materials consisting of a core emitter surrounded by covalently attached dendrons have been developed. The dendritic emitters have the great advantage of structural perfection afforded by a highly controlled synthesis, showing highly reliable structure–property relationships in comparison with conventional polymer materials. In addition, various functionalities are added to the core component by employing the functionalized dendrons. Low-mass molecules are not suitable for solution-processed device fabrication due to their poor film-forming ability, whereas dendritic molecules can overcome this

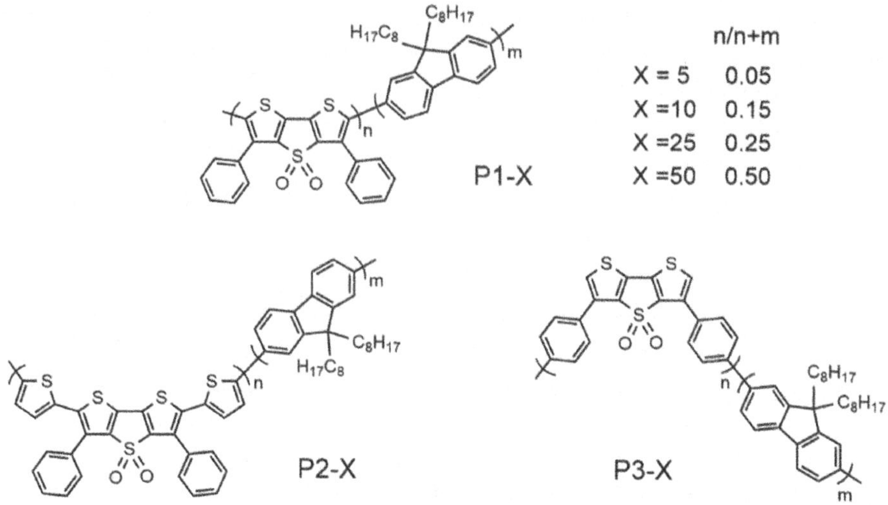

Fig. 16.10 Structures of fluorene–dithienothiophene-*S*,*S*-dioxide copolymers P1-X, P2-X, and P3-X

problem by attaching appropriate dendrons. In addition, attaching the dendrons brings about high glass transition temperature suppressing the deterioration of the device performance caused by polycrystallization. Thus, dendritic emitters are often used as EMLs in solution-processed non-doped OLEDs. It is noteworthy that the peripheral dendrons prevent the luminescent cores from intermolecular interactions facilitating self-quenching of emission. For example, pyrene-cored dendrimer PYGTPA (Fig. 16.11) was developed to fabricate a solution-processed deep blue OLED. Using this dendrimer as a non-doped EML, blue EL with CIE (0.153, 0.155) was obtained, where the maximum luminance and current efficiency were 9242 cd m^{-2} and 0.85 cd A^{-1}, respectively (Auer-Berger et al. 2016). It is well known that pyrene derivatives often form excimers between the excited and ground-state monomers causing red shifts of their emission spectra. In the case of PYGTPA, the intermolecular interaction between the pyrene cores is effectively suppressed, affording the blue monomer emission exclusively. Actually, the phosphazene-cored dendrimer PZPY (Fig. 16.11) in which the aminopyrenes are placed at each dendron (i.e., at the outer site of the dendritic structure) showed a red-shifted EL emission band along with the blue emission assignable to the monomeric aminopyrene, due to strong interaction between the pyrene fluorophores (Bolink et al. 2008). As another example, dendorimers with a blue emissive distylylbenzene (DSB) core were reported by Kwok and Wong (Fig. 16.12) (Kwok and Wong 2001). They prepared the dendrimers with the first generation and the second generation dendrons (DnDSB-1a–d and DnDSB-2a–d, respectively). These dendrimers exhibit blue emission in $CHCl_3$ with extremely high Φ_{PL}s more than 0.93. When the dendrimers were used as an emitting dopant for ITO/PVCz:DnDSB-1 (or DnDSB-2)/Al, EL corresponding to PL in solution and PVCz film is observed. The device performance for DnDSB-1 is superior to that

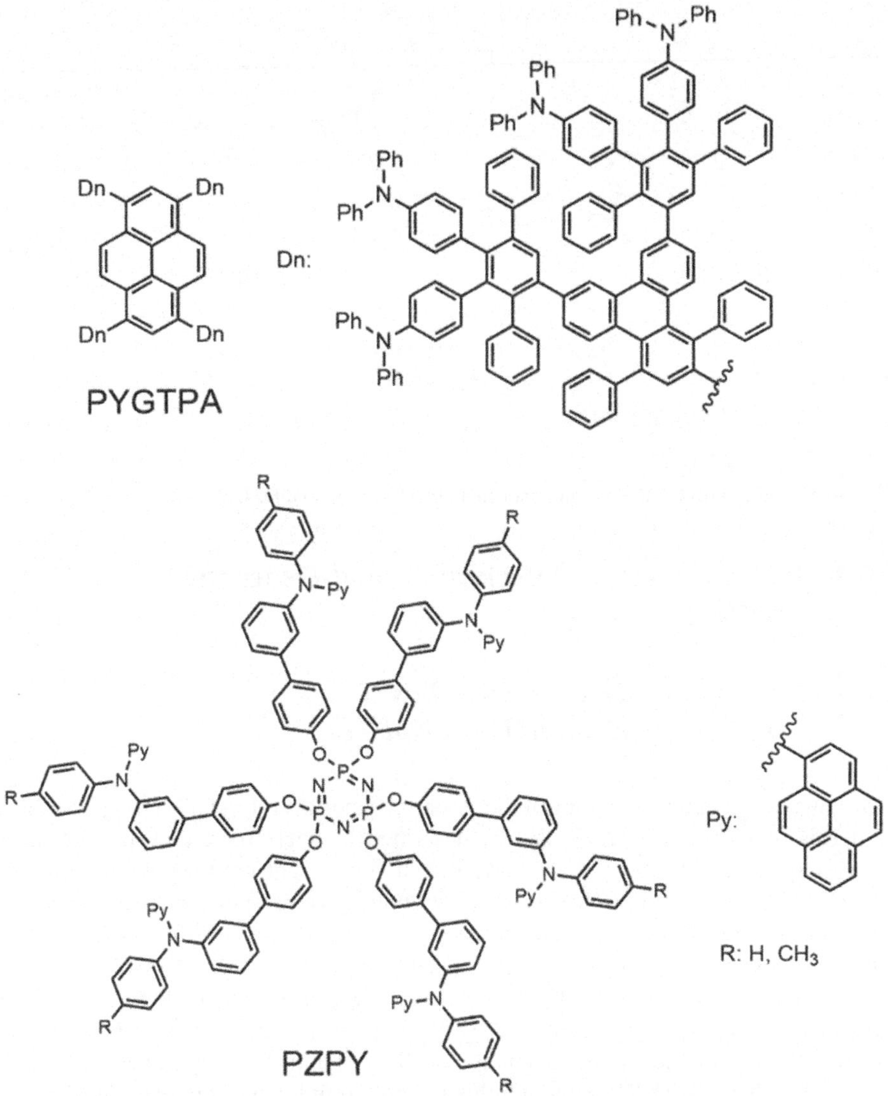

Fig. 16.11 Structures of fluorescent dendrimers PYGTPA and PZPY

for DnDSB-2, and DnDSB-1b, bearing peripheral propoxy groups showed the best performance.

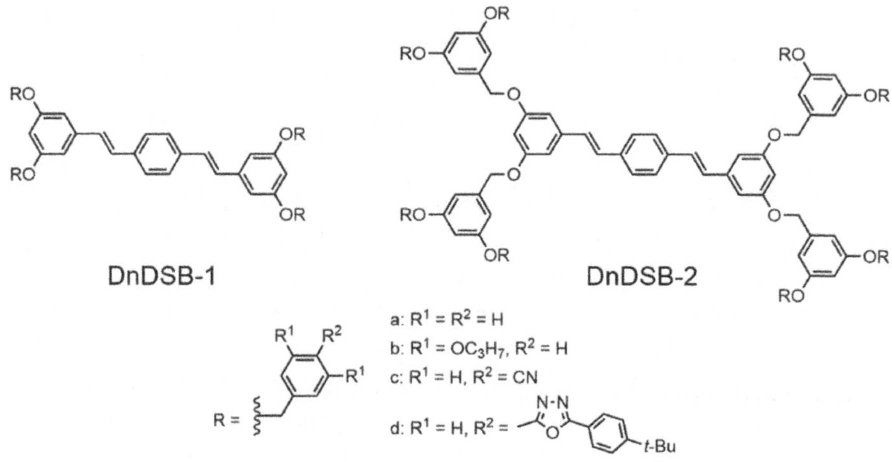

Fig. 16.12 Structures of fluorescent dendrimers DnDSB-1 and DnDSB-2

16.3 Phosphorescent Materials: Second-Generation Emitters

16.3.1 Phosphorescent Emitters Based on Organoiridium(III) Complexes

Phosphorescent emitters are one of the essential materials for today's OLED displays in terms of the device efficiency: as described in Sect. 16.1, they provide four times larger η_{ext} in theory than conventional fluorescent emitters. Especially, all the green and red emitters for the display application are phosphorescent ones at the present time. Among the phosphorescent emitters, organoiridium(III) emitters are most popular, represented by bis- (Lamansky et al. 2001a, 2001b) and tris-cyclometalated (Tamayo et al. 2003c) complexes, as shown in Fig. 16.13. The former complex typically consists of two 2-phenylpyridinate-type cyclometalated ligands (C^N) and one anionic bidentate ancillary ligand (LX), the general structure of which is Ir(C^N)$_2$(LX). Although this type of complex has geometrical isomers, the *N,N-trans, C,C-cis* configuration has been frequently reported (Chuang et al. 2009; Zhu et al. 2011a; Ikawa et al. 2012). As LX, β-diketonates (O^O) such as acetylacetonate (acac) and dipivaloylmethanate (dpm) is often used, as represented by Ir(ppy)$_2$(acac) (ppy; 2-phenylpyridinate). Other bidentate ancillary ligands such as picolinate and salicylimine (Lamansky et al. 2001a) have so far been reported. The latter complex is generally drawn as Ir(C^N)$_3$, the representative example of which is Ir(ppy)$_3$. This complex has two geometrical isomers such as facial (*fac*) and meridional (*mer*) isomers (Fig. 16.13), and the former isomer shows much higher Φ_{PL} because the non-radiative decay process is considerably facilitated in the latter one. For both of Ir(C^N)$_2$(LX) and Ir(C^N)$_3$, the iridium(III) center is

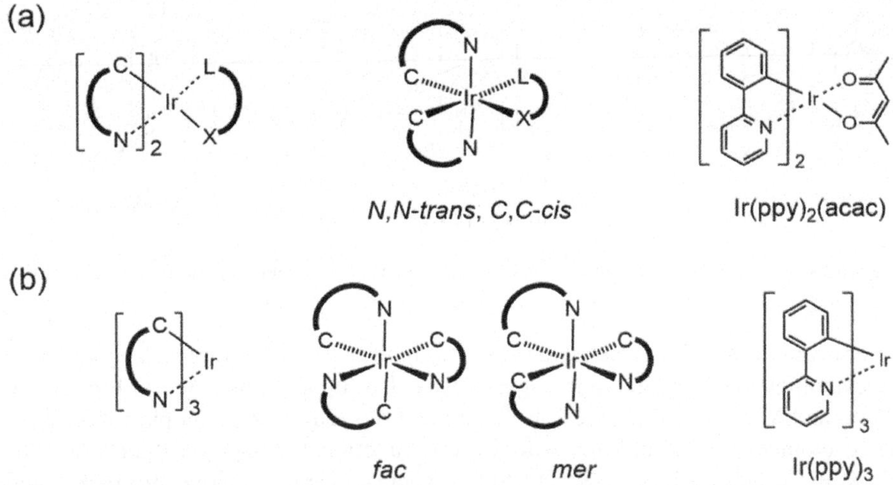

Fig. 16.13 General structures, isomeric configurations and typical examples of **a** bis- and **b** tris-cyclometalated iridium(III) complexes

essential to reduce the spin-forbidden nature of the complex through strong spin–orbit coupling, leading to efficient room-temperature phosphorescence. Thus, the efficient mixing of the singlet and triplet states shortens the photoluminescence lifetime down to microseconds, although typical phosphors often show the lifetimes from milliseconds to seconds. From the viewpoint of OLED applications, this photophysical feature is beneficial to suppress the triplet–triplet annihilation which reduce the device efficiency (Reineke et al. 2007). The cyclometalated structures are indispensable to obtain phosphorescence in the visible regions because the emission from the iridium(III) complexes is attributed to a mixture of triplet metal-to-ligand charge transfer (MLCT) and π–π* ligand states. Thus, C^N plays a significant role in determination of the triplet energy levels of ^{3}MLCT and $^3\pi$–π* responsible for the emission color. Here, organoiridium(III) complexes for OLED applications are reviewed, especially focusing on the emission colors such as red, green, and blue.

As mentioned above, the emission color tuning of organoiridium(III) emitters is carried out through designing C^N. It is relatively easy to obtain green phosphorescent organoiridium(III) complexes because the most typical complexes such as *fac*-Ir(ppy)$_3$ (λ_{PL}, 508 nm; Φ_{PL}, 0.97) (Sajoto et al. 2009) and Ir(ppy)$_2$(acac) (λ_{PL}, 516 nm; Φ_{PL}, 0.34) (Lamansky et al. 2001a) exhibit green emission. Thus, *fac*-Ir(ppy)$_3$ and Ir(ppy)$_2$(acac) are often used as benchmark materials of green phosphorescent emitters (Tanaka et al. 2007; Chen et al. 2009). For example, Kido and coworkers reported the *fac*-Ir(ppy)$_3$-based OLED showing green EL with extremely high η_{ext} of 26% at 1000 cd m^{-2}, where the value of η_{ext} was almost the theoretical limit for phosphorescent OLED (Tanaka et al. 2007a). Wong and coworkers used Ir(ppy)$_2$(acac) as an emitting dopant to yield green EL with the CIE coordinate of (0.33, 0.62), and η_{ext} of 17.5% was obtained (Chen et al. 2009a). As

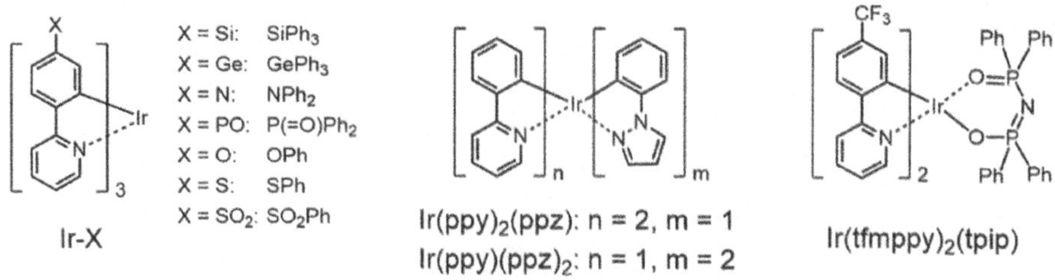

Fig. 16.14 Structures of green phosphorescent bis- and tris-cyclometalated iridium(III) complexes

other examples of green phosphorescent organoiridium(III) complexes, modulation by cyclometalated and ancillary ligands was also reported, as listed in Fig. 16.14. Wong and Ma reported a series of bluish green-to-yellowish green phosphorescent tris-cyclometalated complexes with various functional groups on their C^Ns (Ir-X; X = Si, Ge, N, PO, O, S, and SO_2), where λ_{EL} ranges from 496 to 532 nm (Φ_{PL}; 0.13–0.91 in degassed dichloromethane) (Zhou et al. 2008). Dedeian and coworkers reported the heteroleptic tris-cyclometalated complexes $Ir(ppy)_2(ppz)$ and $Ir(ppy)(ppz)_2$ (ppz; phenylpyrazolate-$N,C^{2'}$) exhibiting green phosphorescent emission at 504 and 498 nm with Φ_{PL}s of 0.72 and 0.68, respectively, although the application to OLED were not reported (Dedeian et al. 2005). Combination of 4′-trifluoromethyl-substituted penylpyridinate (tfmppy) with the tetraphenylimidodiphosphinate (tpip) ancillary ligand yielded the bis-cyclometalated complex $Ir(tfmppy)_2(tpip)$ exhibiting green phosphorescent emission (Zhu et al. 2011b). Although $Ir(tfmppy)_2(tpip)$ itself showed low Φ_{PL} of 0.12 in solution, the electroluminescent device yielded the high maximum current efficiency ($\eta_{j\,max}$) of 67.95 cd A^{-1} and the high power efficiency ($\eta_{p\,max}$) of 69.90 lm W^{-1}.

Since red emitters are very important materials as one of primary colors for displays, development of red phosphorescent materials has been one of hot research fields. To obtain organoiridium(III) complexes exhibiting red-shifted phosphorescence emission, extension of the π-conjugation system of C^N is effective because it brings about narrow energy gaps of MLCT and ligand-centered π–π* transitions. Representative examples of red phosphorescent organoiridium(III) complexes are shown in Fig. 16.15. The pioneering works are $Ir(piq)_3$ (λ_{PL}, 620 nm; Φ_{PL}, 0.26) and $Ir(btp)_2(acac)$ (λ_{PL}, 612 nm; Φ_{PL}, 0.21), reported by Tsuboyama et al. (2003) and Lamansky et al. (2001b), respectively (piq, 1-phenylisoquinolinate; btp, 2-(benzo[*b*]thiophen-2-yl)pyridinate). Especially, $Ir(piq)_3$ in OLED shows highly efficient red electroluminescence with a good CIE chromaticity coordinate of (0.68, 0.32) and a relatively high η_{ext} (10.3%), although it is not so emissive in solution. Thereafter, lots of organoiridium(III) complexes bearing π-extended C^Ns have been reported. In this context, considerable efforts have been devoted by many chemists to achieve high Φ_{PL}s, because lowering the triplet energy level tends to facilitate the non-radiative decay relative to the radiative decay, according to the energy gap law (Caspar et al. 1982). Yagi and coworkers developed a highly red luminescent bis-cyclometalated complex which consists of 1-(dibenzo[*c,d*]furan-4-yl)isoquinolinate

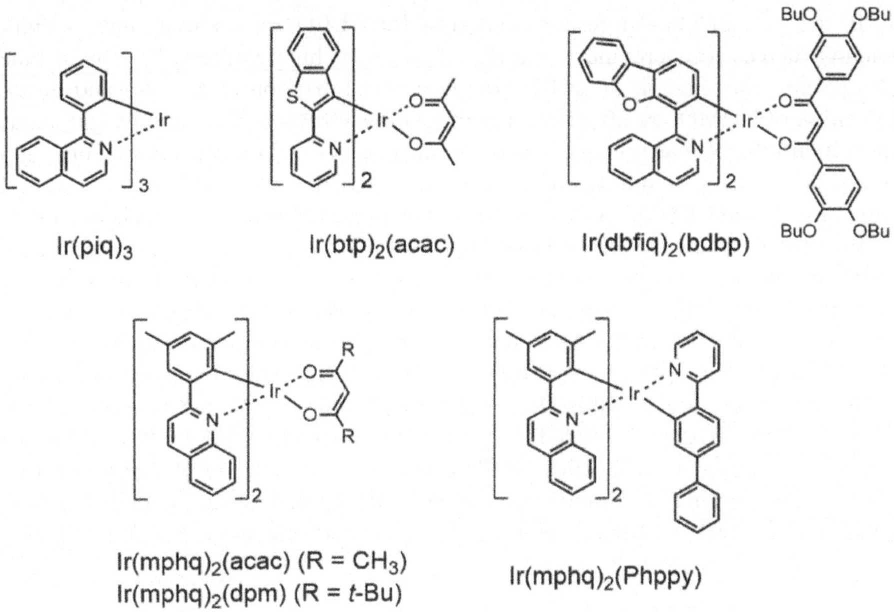

Fig. 16.15 Structures of red phosphorescent bis- and tris-cyclometalated iridium(III) complexes

(dbfiq) as C^N and 1,3-bis(3,4-dibutoxypheny)propan-1,3-dionate (bdbp) as O^O (Tsujimoto et al. 2010a). This complex Ir (dbfiq)$_2$(bdbp) exhibits intense red phosphorescence emission at 640 nm with Φ_{PL} of 0.61. It is noteworthy that the aromatic ancillary ligand is essential to obtain a high Φ_{PL} in a solid-state matrix: The Φ_{PL} of Ir(dbfiq)$_2$(bdbp) in poly(methyl metacrylate) (Φ_{PL}; 0.17) is larger than those of the analogous complexes with aliphatic ancillary ligands such as Ir(dbfiq)$_2$(dpm) and Ir(dbfiq)$_2$(acac) (Φ_{PL}s; 0.08 and 0.06, respectively). In the OLED with a PVCz-based EML, Ir(dbfiq)$_2$(bdbp) afforded pure red EL with the CIE coordinate of (0.68, 0.31), the maximum η_{ext} was 6.4%. The complexes based on a Ir(phq)$_2$(O^O) (phq; 2-phenylquinolinate) are practical red phosphorescent emitters, which were reported by the group of Kwon and Suh (Kim et al. 2011). Especially, Ir(mphp)$_2$(acac) and Ir(mphp)$_2$(dpm) (mphq; 2-(3,5-dimethylphenyl)quinolinate) are very emissive in the devices, although they are much less emissive in dichloromethane (Φ_{PL}s; *ca.* 0.1) with blue-shifted λ_{PL}s (*ca.* 595 nm). The devices based on Ir(mphp)$_2$(acac) and Ir(mphp)$_2$(dpm) exhibited red EL with good CIE coordinates of (0.654, 0.343) and (0.665, 0.343), respectively, accompanied by excellent η_{ext}s (19.8 and 21.9%, respectively). The heteroleptic tris-cyclometalated complex Ir(mphq)$_2$(Phppy) (Phppy; 2-(4-(phenyl)phenyl)quinolinate) exhibiting red emission at 615 nm in solution (Φ_{PL}; 0.43) was also reported as an excellent red emitter, and the device using this complex as an emitting dopant also afforded efficient red EL with a maximum η_{ext} of 19.2% (CIE (0.62, 0.38)) (Kim et al. 2015).

In the field of phosphorescent materials for OLED applications, one of challenging themes to overcome is the development of blue emitters. To obtain blue phosphorescent organoiridium(III) complexes, stabilization of HOMO and destabilization of LUMO are effective, and the wide HOMO–LUMO energy gap leads to a high-lying triplet energy level. In the case of typical ppy-based bis- and tris-cyclometalated complexes, HOMO is located at the metal center and the phenyl ring in C^N, and LUMO is localized at the pyridine moiety in C^N, as shown in Fig. 16.16. Thus, stabilization of HOMO is brought about by introduction of electron-withdrawing substituent(s) to the phenyl moiety. Replacement of the phenyl with an electron-deficient heteroaryl group such as pyridine is also effective. On the other hand, to destabilize LUMO, introduction of electron-donating group(s) to the phenyl moiety and replacement of the pyridyl with an electron-rich heteroaryl group such as pyrazole are possible. Representative examples of blue phosphorescent organoiridium(III) complexes so far reported are shown in Fig. 16.17. A well-known blue phosphorescent organoiridium(III) emitter is $Ir(dfppy)_2(pic)$ (dfppy and pic; 2-(4,6-difluorophenyl)pyridinate and picolinate, respectively), so-called FIrpic (Adachi et al. 2001b). FIrpic exhibits efficient sky blue emission at 471 nm with Φ_{PL}

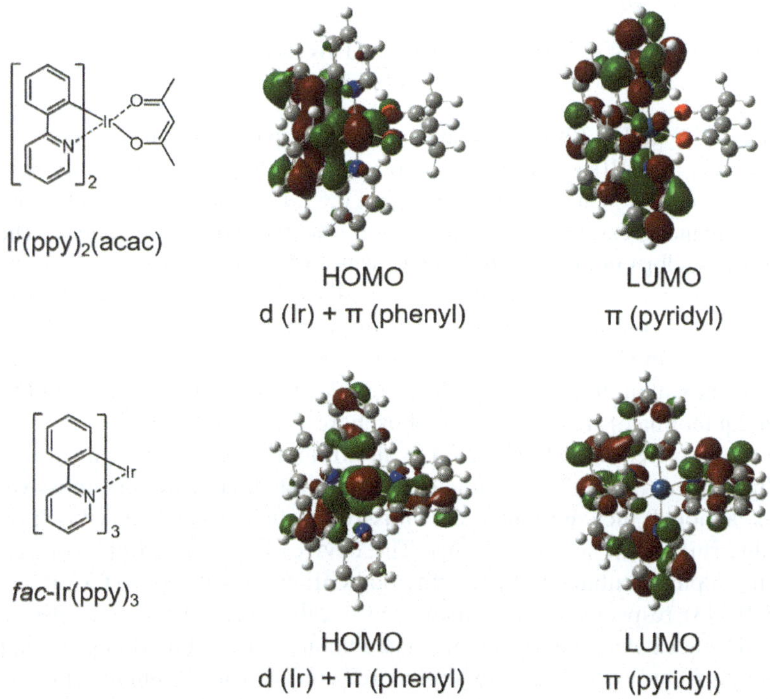

Fig. 16.16 Frontier molecular orbitals of $Ir(ppy)_2(acac)$ and *fac*-$Ir(ppy)_3$ calculated using the density functional theory with B3LYP/LANL2DZ (Ir) and 6–31G(d) (other atoms). For the former complex, the calculation for the *N,N-trans, C,C-cis* configuration was carried out. In both cases, the HOMO–LUMO transition mainly contributes to the excitation to the triplet state

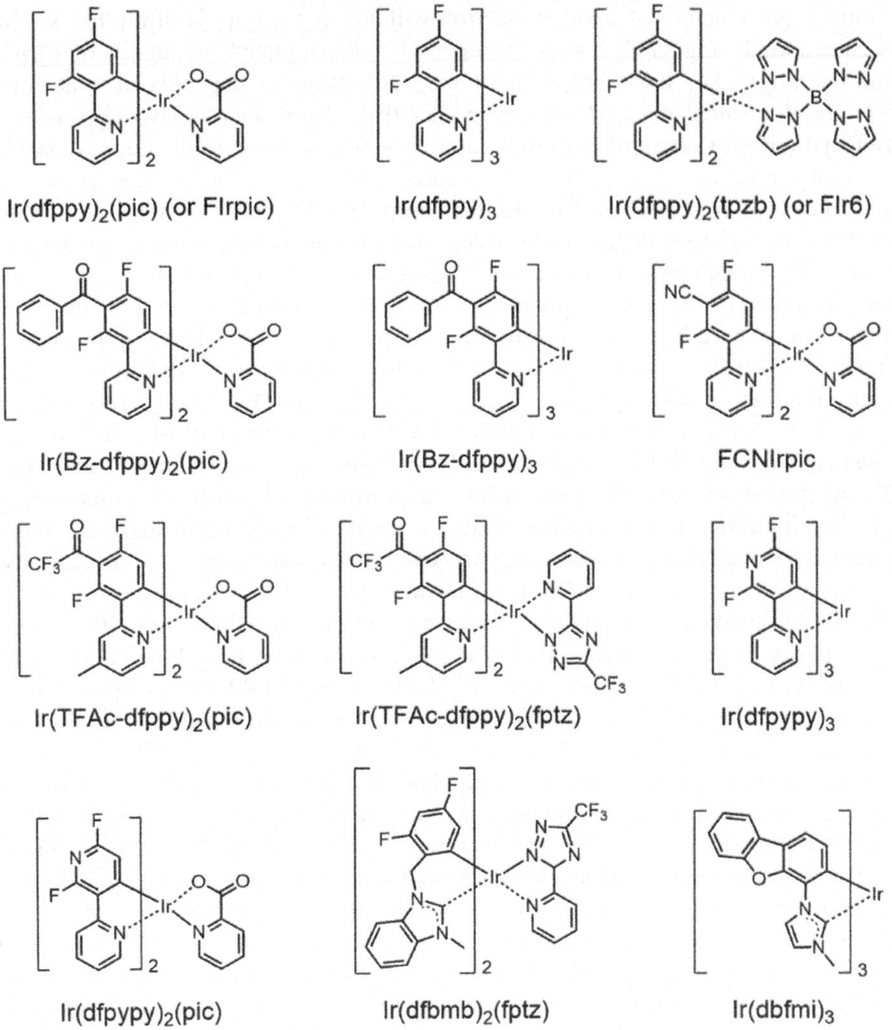

Fig. 16.17 Structures of blue phosphorescent bis- and tris-cyclometalated iridium(III) complexes

of 0.89 in solution (Endo et al. 2008). In this complex, pic is more effective than O^O such as acac to obtain blue-shifted emission: the PL maximum of Ir(dfppy)$_2$(acac) is *ca.* 20 nm red-shifted in comparison with that of FIrpic (Adachi et al. 2001b). The tris-cyclometalated complex *fac*-Ir(dfppy)$_3$ also exhibit sky blue emission at 467 nm with Φ_{PL} of 0.97 (Endo et al. 2008), and thus the fluorinated ppy is a practical cyclometalated ligand. Replacement of the ancillary ligand of FIrpic with tetrakis(1-pyrazolyl)borate (tpzb) leads to further blue-shifted emission: Ir(dfppy)$_2$(tpzb) (or FIr6) exhibits efficient blue emission at 458 nm with Φ_{PL} of 0.96 (Endo et al. 2008). Aimed at development of blue phosphorescent organoiridium(III) emitters, introduction of another electron-withdrawing group to dfppy has so far been reported. Yagi and coworkers reported 5′-banzoylated analogues of FIrpic and $Ir(dfppy)_3$, i.e., $Ir(Bz\text{-}dfppy)_2(pic)$ and $Ir(Bz\text{-}dfppy)_3$, respectively, to achieve 3–4 nm blue shifts of λ_{PL} (Okamura et al. 2016). The Φ_{PL}s of $Ir(Bz\text{-}dfppy)_2(pic)$ and $Ir(Bz\text{-}dfppy)_3$ are still high (0.82 and 0.90, respectively, in dichloromethane). A similar type of FIrpic analogue, so-called FCNIrpic, exhibits blue emission, and the FCNIrpic-based OLED afforded deep blue EL with the CIE coordinate of (0.14, 0.17) (Jeon et al. 2011). Deep blue phosphorescent bis-cyclometalated iridium(III) complexes consisting of trifluoromethylcarbonyl-substituted 2-(4′,6′-difluororphenyl)-3-methylpyridinate (TFAc-dfppy) cyclometalated ligands were also reported by a Korean group; $Ir(TFAc\text{-}dfppy)_2(pic)$ and $Ir(TFAc\text{-}dfppy)_2(fptz)$ (fptz; 2-(3-(trifluoromethyl)-1*H*-1,2,4-triazol-5-yl)pyridinate), exhibiting deep blue emission at 453 nm (Φ_{PL}; 0.74) and 447 nm (Φ_{PL}; 0.63), respectively (Lee et al. 2013). Deep blue EL with the CIE coordinate of (0.147, 0.116) was obtained from the OLED containing $Ir(TFAc\text{-}dfppy)_2(fptz)$ as an emitting dopant. To develop deep blue phosphorescent organoiridium(III) emitters, employment of 2,3′-bipyridine as a preligand is also a good strategy. Kang and coworkers reported the deep blue phosphorescent tris-cyclometalated complex, *fac*-$Ir(dfpypy)_3$ (dfpypy; 2′,6′-difluoro-2,3′-bipyridinate), which shows the emission spectrum with comparable peaks at 438 and 463 nm in dichloromethane (Lee et al. 2009). The color purity of the emission, CIE (0.14, 0.12), is excellent for a blue emitter, and Φ_{PL} of 0.71 is also fairly high. However, the low-lying HOMO of $Ir(dfpypy)_3$ (*ca.* –6.5 eV) prevents the hole injection from a host material, when it is used as an emitting dopant for OLED. Then, Kang and coworkers developed the deep blue phosphorescent bis-cyclometalated complex, $Ir(dfpypy)_2(pic)$, which shows the emission spectrum with a peak at 445 nm (Kang et al. 2013). This complex is highly emissive both in solution and polymer film: Φ_{PL}s in dichloromethane solution and poly(methyl methacrylate) film are 0.90 and 0.95, respectively. The HOMO level of $Ir(dfpypy)_2(pic)$ lies at –6.14 eV, destabilized in comparison with that of $Ir(dfpypy)_3$. As other examples of blue phosphorescent organoiridium(III) materials, the complexes with carbene-based cyclometalated ligands, such as $Ir(dfbmb)_2(fptz)$ (dfbmb; 1-(2,4-difluorobenzyl)-3-methylbenzimidazole) (Chang et al. 2008) and *mer*-$Ir(dbfmi)_3$ (dbfmi; 1-(dibenzofuran-4-yl)-3-methylimidazole) (Sasabe et al. 2010), have so far been reported. The OLED containing the latter complex as an emitter afforded highly efficient blue EL with CIE (0.15, 0.19), where high device performance ($\eta_{j\,max}$ of 28.6 cd A^{-1}, $\eta_{p\,max}$ of 35.9 lm W^{-1}, and $\eta_{ext\,max}$ of 18.6%) was obtained (Sasabe et al. 2010).

Generally, deep blue phosphorescent emitters possess high triplet energies and deep HOMO, and thus, the confinement of the excitons generated at EML is very difficult. This is one of the main reasons why the development of excellent blue phosphorescent emitters for OLED is still a challenging theme.

16.3.2 *Phosphorescent Emitters Based on Organoplatinum(II) Complexes*

Along with organoiridium(III) complexes, platinum complexes exhibiting efficient room-temperature phosphorescent emission have been frequently reported for OLED applications. In most cases, practical emitters are organoplatinum(II) complexes bearing cyclometalated ligands such as bidentate C^N and C^C, tridentate C^N^N and N^C^N, and a tetradentate N^C^C^N. The d^8 electronic configuration of platinum allows for obtaining stable square planar structures such as Pt(C^N)(L^X) (Brooks et al. 2002d), Pt(C^C)(L^X) (Unger et al. 2010), Pt(N^C^N)X (Williams et al. 2003), Pt(C^N^N)X (Lu et al. 2004), and Pt(N^C^C^N) (Fukagawa et al. 2012) (the general structures, shown in Fig. 16.18). As is the case with phosphorescent organoiridium(III) emitters, strong spin–orbit coupling is induced by the platinum atom to make the mixing of the singlet and triplet excited state efficient to reduce the spin-forbidden nature. So, effective perturbation between the metal center and the cyclometalated ligand through the metal–sp^2-carbon bonding plays a significant role in obtaining efficient phosphorescent emission, usually through ^{3}MLCT and metal-perturbed $^3\pi$–π^* transitions. In this term, as the cyclometalated ligands join in the electronic transition, they are also important to determine the emission color. In this section, phosphorescent organoplatinum(II) complexes for OLED applications are reviewed, focusing the attention on their structures. One of the characteristic features of organoplatinum(II) complexes is excimer or aggregate formation through metal–metal and/or π-based intermolecular stacking interactions caused by

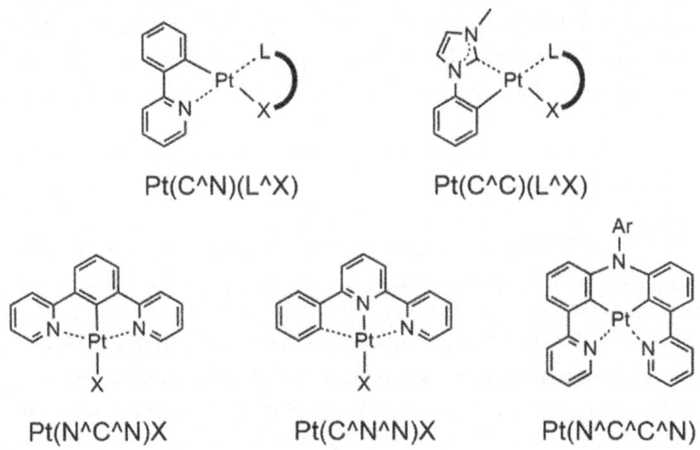

Fig. 16.18 General structures of organoplatinum(II) complexes bearing bidentate (C^N and C^C), tridentate (N^C^N and C^N^N), and tetradentate (N^C^C^N) ligands. L^X and X represent an anionic bidentate ligand and an anionic monodentate one, respectively. The substituent Ar represents an aryl group

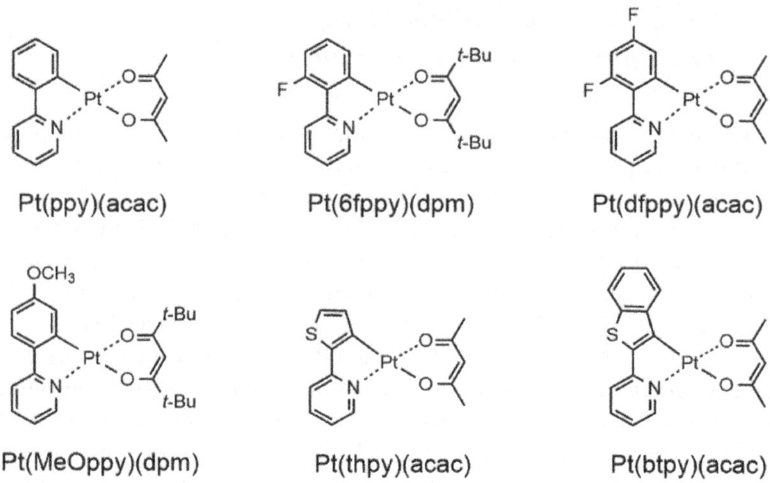

Fig. 16.19 Structures of phosphorescent heteroleptic cyclometalated platinum(II) complexes

the square planar structures, considerably affecting the optical and photophysical properties. This is also referred to in the later part of this section.

Cyclometalated platinum(II) complexes bearing a general structure of Pt(C^N)(O^O) are one of the most representative Pt(II)-based phosphorescent emitters. The complexes bearing a wide range of C^N were reported by Thompson and coworkers, where acac and dpm were employed as ancillary ligands, as shown in Fig. 16.19 (Brooks et al. 2002d). The complex Pt(ppy)(acac) exhibits its emission at 486 nm in degassed 2-methyltetrahydrofuran at room temperature. In the case of Pt(ppy)(O^O)-type complexes, the spectroscopic features are sensitive to the structure of C^N, although they are basically affected little by the ancillary ligand. The effect of C^N on the PL properties obeys those seen in $Ir(C^\wedge N)_3$ and $Ir(C^\wedge N)_2(O^\wedge O)$, because the phenyl and pyridine moieties in C^N mainly contribute to HOMO and LUMO, respectively, as shown in Fig. 16.20. Introduction of electron-withdrawing substituent(s) such as F to the phenyl moiety of C^N leads to a blue shift of the PL spectrum; for example, Pt(6fppy)(dpm) (6fppy, 2-(6-fluorophenyl)pyridinate; λ_{PL}, 476 nm; Φ_{PL}, 0.06) and Pt(dfppy)(acac) (λ_{PL}, 466 nm; Φ_{PL}, 0.02). On the other hand, introduction of an electron-donating group such as methoxy to the phenyl of C^N or replacement of the phenyl with an electron-rich heteroaromatic component such as thienyl and benzothienyl leads to a red shift of the PL spectrum, as seen in Pt(MeOppy)(dpm) (MeOppy, 2-(4-methoxyphenyl)pyridinate; λ_{PL}, 490 nm; Φ_{PL}, 0.20), Pt(thpy)(acac) (thpy, 2-(thiophen-2-yl)pyridinate; λ_{PL}, 575 nm; Φ_{PL}, 0.11), and Pt(btpy)(acac) (btpy, 2-(benzo[*b*]thiophen-2-yl)pyridinate; λ_{PL}, 612 nm; Φ_{PL}, 0.08). As mentioned above, the ancillary ligand of Pt(C^N)(O^O) has few effects on the PL properties in comparison with the cyclometalated ligand. However, it often improves the solubility in organic solvents. Although the complexes bearing acac as O^O often suffer from the insolubility due to strong intermolecular interaction

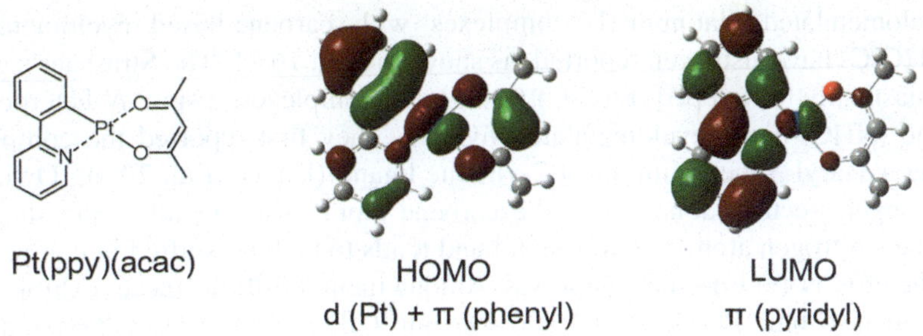

Fig. 16.20 Frontier molecular orbitals of Pt(ppy)(acac) calculated using the density functional theory with B3LYP/LANL2DZ (Pt) and 6–31G(d) (other atoms). The HOMO–LUMO transition mainly contributes to the excitation to the triplet state

caused by the highly planar structure, employing dpm in place of acac overcomes this problem. As reported by Yagi and coworkers (Table 16.1), employment of an aromatic 1,3-diketonate ancillary ligand with alkoxyl groups is also effective to improve the solubility (Tsujimoto et al. 2010b). For example, Pt(dbfp)(bdbp) (dbfp, 2-(dibenzo[*b*,*d*]furan-4-yl)pyridinate; bdbp, 1,3-bis(3,4-dibutoxyphenyl)propane-1,3-dionate) is dissolved up to 0.18 M in chloroform. It is noteworthy that some complexes with a bdbp ancillary ligand show higher Φ_{PL}s than the corresponding dpm-based complexes, as summarized in Table 16.1.

Table 16.1 Structures and photoluminescence properties (in deaerated $CHCl_3$, 298 K) of Pt(C^N)(bdbp) and Pt(C^N)(dpm)

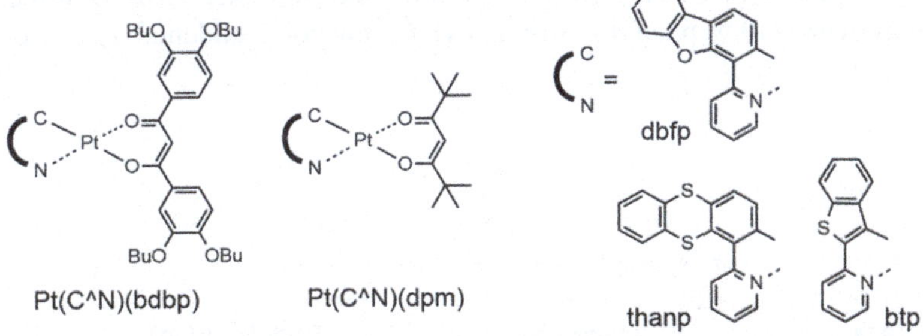

O^O	C^N		
	dbfp	thanp	btp
bdbp	Pt(dbfp)(bdbp)	Pt(thanp)(bdbp)	Pt(btp)(bdbp)
	λ_{PL} = 517, 552 nm Φ_{PL} = 0.59	λ_{PL} = 517, 538 nm Φ_{PL} = 0.43	λ_{PL} = 613, 667 nm Φ_{PL} = 0.10
dpm	Pt(dbfp)(dpm)	Pt(thanp)(dpm)	Pt(btp)(dpm)
	λ_{PL} = 515, 551 nm Φ_{PL} = 0.42	λ_{PL} = 513, 535 nm Φ_{PL} = 0.26	λ_{PL} = 610, 665 nm Φ_{PL} = 0.07

Cyclometalated platinum(II) complexes with carbene-based cyclometalated ligand (C^C) have also been reported, as shown in Fig. 16.21. The Strassner's group enthusiastically developed Pt(C^C)(O^O)-type complexes, using *N*-heterocyclic carbene (NHC)-based cyclometalated ligands. They first reported the complexes bearing a 1-aryl-3-methylimidazol-2-ylidene ligand (Unger et al. 2010). Owing to the stronger electron donation of the carbene atom to the metal center than the pyridine's nitrogen atom, the emission band tends to be blue-shifted in comparison with the Pt(C^N)(O^O)-type complexes. Among them, Pt(dbfmi)(acac) exhibits efficient greenish blue PL (λ_{PL}; 463 and 497 nm, CIE (0.162, 0.314), 2 wt% doped PMMA film) with Φ_{PL} of 0.90. Upon application to OLED, the device containing Pt(dbfmi)(acac) as an emitting dopant exhibited the EL whose spectrum corresponds to that of PL. The ancillary ligand significantly affects the PL properties. Strassner and coworkers investigated the effects of O^O on the PL properties of Pt(dbfmi)(O^O) (Tronnier et al. 2014). Replacement of acac of Pt(dbfmi)(acac) with 1,1,1,5,5,5-hexafluoropentane-2,4-dionate (hfacac) gives rise to no emission. The employment of 1,3-diphenylpropane-1,3-dionate (dpp) affords red-shifted emission (λ_{PL}; 530 nm) with a reduced Φ_{PL} of 0.51 in 2 wt% PMMA due to a considerable contribution from the ancillary ligand to the triplet exciton generation. Interestingly, 1,3-dimesitylpropane-1,3-dione (dmsp) allows for obtaining a highly emissive complex with the comparable PL characteristics to Pt(dbfmi)(acac) (λ_{PL}, 466 nm; Φ_{PL}, 0.91, 2 wt% doped PMMA film). Using other NHC-based ligands, the emission color is tuned to some extent. Employment of 1,3,4-triphenyl-4,5-dihydro-1*H*-1,2,4-triazol-5-ylidene, so-called "Enders triazole" (Enders et al. 1995), allows for preparation of the sky-blue-emitting complex Pt(tptz)(btmpp) (btmpp; 1,3-bis(2,3,5,6-tetramethylphenyl)propane-1,3-dionate) (λ_{PL}, 467 nm; Φ_{PL}, 0.87, CIE (0.175, 0.244), 2 wt% doped PMMA film) (Soellner and Strassner 2018). On the other hand, employment of *N*-phenyl-4,5-dimethyl-1,3-thiazol-2-ylidene-based C^C

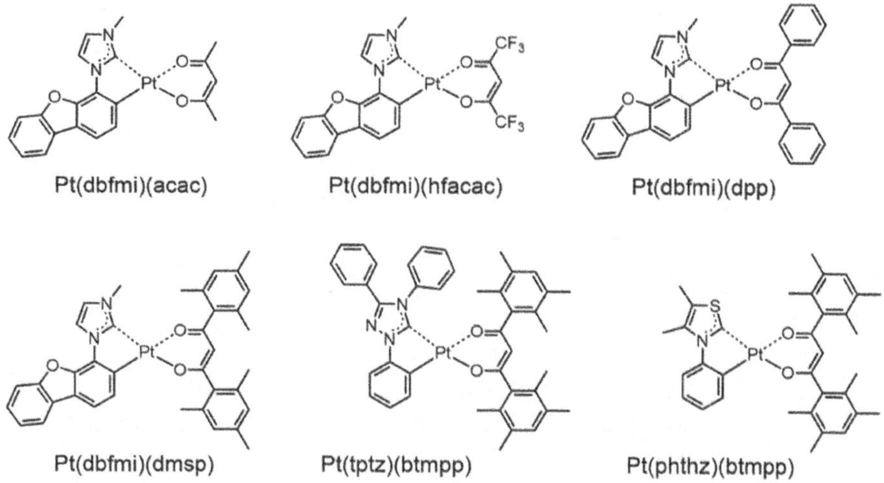

Fig. 16.21 Structures of phosphorescent platinum(II) complexes bearing an NHC-based cyclometalated ligand

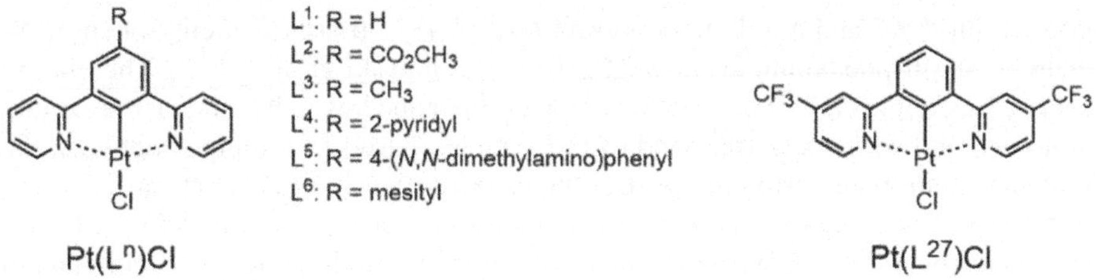

Fig. 16.22 Structures of phosphorescent organoplatinum(II) complexes with an N^C^N tridentate ligand

(pdmthz) affords the green-emitting complex Pt(phthz)(dmsp) (λ_{PL}, 508 nm; Φ_{PL}, 0.64, CIE (0.26, 0.52), 2 wt% doped PMMA film) (Leopold et al. 2017).

The representative examples of platinum(II) complexes with tridentate cyclometalated ligands are shown in Fig. 16.22. The Pt(N^C^N)X type complexes have been eagerly developed by Williams and coworkers. In the case of Pt(L^n)Cl, they emit bluish green to yellow in dichloromethane with the relatively high Φ_{PL} of 0.46–0.60, the spectral feature of which depends on the substituent R (Cocchi et al. 2007a). The OLED performance for these complexes was investigated for the following device; ITO (anode)/blend of 75 wt% *N*,*N'*-bis(3-methylphenyl)-*N*,*N'*-diphenylbenzidine (TPD) and 25 wt% polycarbonate (80 nm)/CBP (20 nm)/blend of 47 wt% CBP, 47 wt% 5-bis[5-(4-*tert*-butylphenyl)-1,3,4-oxadiazol-2-yl]benzene (OXD-7), and 6 wt% Pt(L^n)Cl (80 nm)/OXD-7 (20 nm)/Ca (cathode). The maximum η_{ext}s of 4–16%, the maximum current efficiencies of 15–40 cd A^{-1}, and the maximum luminances of 3500–12,100 cd m^{-2} were obtained, and EL corresponding to PL was observed for each complex: the CIE coordinates are (0.23, 0.57), (0.14, 0.46), (0.31, 0.61), (0.20, 0.76), (0.48, 0.51), and (0.24, 0.63) for Pt(L^1)Cl–Pt(L^6)Cl, respectively.

One of the characteristic features of this type of complex is excimer formation. Excimer is a dimeric species consisting of excited and ground-state monomers. Platinum complexes adopt square planar geometries, and thus they can interact with each other through interligand π–π stacking and bimetallic interactions to form an excimer (Adamovich et al. 2002). PL and EL from the excimer emerge in the longer wavelength region relative to the monomer emission. Williams and coworkers fabricated a single dopant white OLED by utilizing the excimer of Pt(L^2)Cl (Cocchi et al. 2007b). The complex Pt(L^2)Cl exhibited monomer- and excimer-based EL with structured and broadened spectral bands in the blue-to-green (λ_{EL}; *ca.* 490 nm) and red-to-deep red (λ_{EL}; *ca.* 690 nm) regions, respectively. The excimer-based EL was enhanced with the increase in the doping level of Pt(L^2)Cl. An appropriate combination of the monomer- and excimer-based EL provided pseudo-white emission with the CIE coordinate of (0.43, 0.43), where the maximum luminance of *ca.* 7000 cd m^{-2} and the maximum η_{ext} of 15.5% were obtained. To improve the color rendering index of this device, utilization of the exciplex (i.e., an excited heterodimeric species consisting of one excited monomer and another ground-state monomer) between Pt(L^2)Cl and a hole-transporting host (4,4′,4″-tris(*N*-(3-methylphenyl)-*N*-phenylamino)triphenylamine, m-MTDATA) (Kalinowski et al. 2007). The energy level of the exciplex is higher than that of the excimer and lower than that of the excited monomer, and so the spectral band of the exciplex-based EL emerges between the monomer- and excimer-based spectral bands, affording the EL spectrum covering the whole visible region. As a result, the color rendering index of white EL was improved from 60 to 90. Williams and coworkers also modified the N^C^N ligand to obtain near-infrared electrophosphorescence based on the excimer formation (Rossi et al. 2011). Introduction of a trifluoromethyl group to each pyridine ring of Pt(L^1)Cl led to red shifts of the monomer and excimer emission bands. Employing a neat film of Pt(L^{27})Cl (Fig. 16.22) as EML allowed for fabrication of OLED emitting in the near-infrared region of 750–800 nm.

As for organoplatinum(II) complexes with a tetradentate ligand, excellent red and blue phosphorescent emitters were reported. A research group in Science and Technology Research Laboratories of Japan Broadcasting Corporation (NHK) reported OLEDs employing organoplatinum(II) complexes with N^C^C^N-type ligands (TLEC-025 and TLEC-027, Fig. 16.23) (Fukagawa et al. 2012). In these ligands, two 2-phenylpyridinates are bridged by a nitrogen atom. Various types of multilayer OLEDs were fabricated to optimize the device performance. The NHK group confirmed that bis(benzo[*h*]quinolin-10-olato-κ*N*,κ*O*)beryllium(II) ($Bebq_2$) as a host material is useful to obtain excellent device efficiencies. With the help of exciton transfer from the $Bebq_2$ host, OLEDs with TLEC-025 and TLEC-027 exhibited highly efficient red EL with the CIE coordinates of (0.662, 0.337) and (0.657, 0.342), respectively (@100 cd m^{-2}). In terms of the device performance, the maximum η_{ext}s of 19.5 and 19.3% and power efficiencies of 25.5 and 30.3 lm W^{-1} were obtained for TLEC-025 and TLEC-027, respectively. Such excellent device

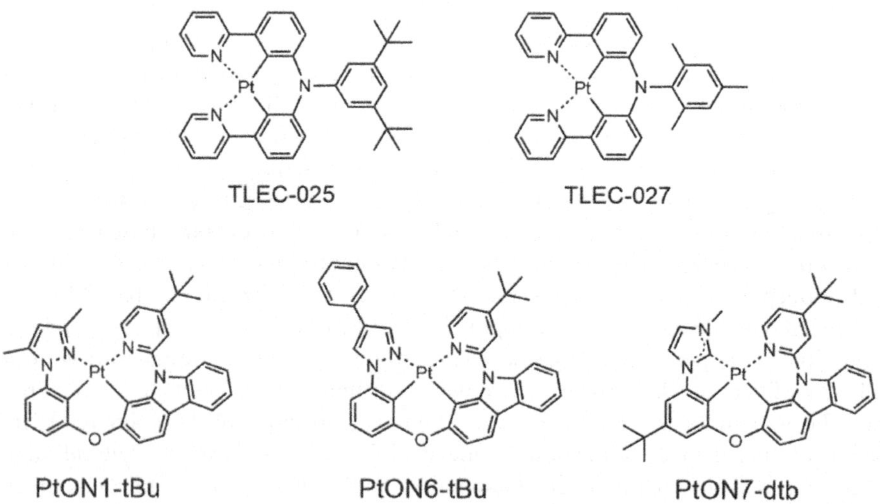

Fig. 16.23 Structures of phosphorescent organoplatinum(II) complexes with an tetradentae ligand

performances are comparable to those obtained from the red phosphorescent OLEDs based on the organoiridium(III) emitters (Kim et al. 2011). Li and coworkers reported excellent pure blue phosphorescent emitters by using unsymmetrical tetradentate ligands, namely, PON1-tBu, PtON6-tBu, and PtON7-dtb (Fleetham et al. 2014). These complexes exhibit blue emission at 444–448 nm with extremely narrow half bandwidths pf 19–20 nm. It is also interesting that they are still emissive in a solid matrix: the Φ_{PL}s of PON1-tBu, PtON6-tBu, and PtON7-dtb in PMMA films) are more than 0.7. The PL characteristics were reflected in the device performance, and OLEDs employing these complexes exhibited pure blue EL with spectral half bandwidths less than 30 nm. Optimizing the device structure, the OLED with PtON7-dtb recorded the maximum η_{ext} of 24.8%, comparable to the theoretical limit. Interestingly, the CIE coordinate of (0.148, 0.079) was very close to the NTSC standard for blue, defined as (0.14, 0.08).

16.4 TADF Materials: Third Generation Emitters

16.4.1 Advantage of TADF Emitters in OLED Applications

TADF is the exciton generation and light-emitting process in which the fluorescent emission is obtained through up-conversion from the low-lying triplet state to the high-lying singlet excited state, as shown in Fig. 16.3. This up-conversion process can theoretically convert the whole of triplet exciton into the singlet, different from the triplet–triplet annihilation process where one singlet exciton is produced from two triplet excitons. In this context, OLEDs utilizing TADF emitters allow us to overcome the problem of intrinsically low internal quantum efficiencies of fluorescent OLEDs: in TADF OLEDs, the whole of the first generated singlet and triplet excitons can be used for fluorescent emission, and thus, they are comparable to phosphorescent OLEDs in terms of η_{ext} of EL. Since TADF emitters are obtained as pure organic molecules, TADF OLEDs are also beneficial in terms of reducing the device fabrication cost and protecting the resource of precious metals such as iridium and platinum used for production of phosphorescent emitters. Thus, TADF emitting materials are often called the third generation emitters.

At the early stage of developing TADF OLEDs, eosin and tin(IV) porphyrins were known as TADF emitters. Using the latter compounds, the first TADF OLED was demonstrated by Adachi and coworkers in 2009 (Endo et al. 2009). EL from the device with SnF_2-OEP (Fig. 16.24) was exclusively based on the prompt and delayed fluorescence at relatively high temperatures (350 and 400 K), although phosphorescent emission was also included at 300 K. Anyway, they demonstrated utilization of both of singlet and triplet excitons to obtain EL. Just after this report, TADF from a cupper(I) dinuclear complex $\{Cu(PNP\text{-}tBu)\}_2$ (Fig. 16.24) was reported by Deaton and coworkers (Deaton et al. 2010). The OLED including this complex as an emitting dopant exhibited green EL. Although the red-shifted phosphorescent EL

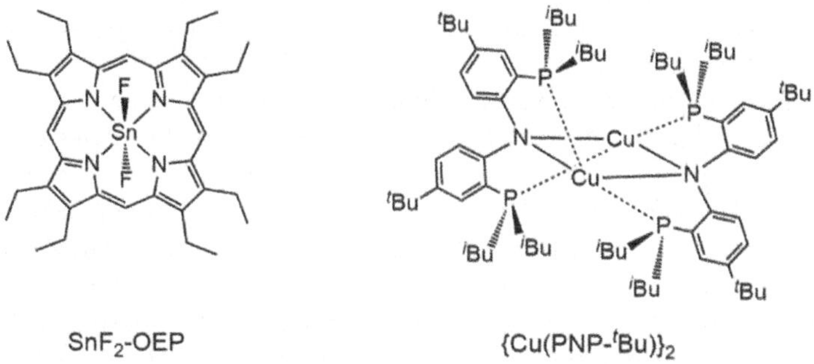

Fig. 16.24 Structures of SnF_2-OEP and $\{Cu(PNP\text{-}^tBu)\}_2$

was observed at 77 K, the increase in temperature led to considerable emergence of green fluorescent EL based on prompt and delayed fluorescence. The maximum η_{ext} was 16.1%, exceeding the limit of η_{ext} for traditional fluorescent OLEDs. This result obviously indicates that the $\{Cu(PNP\text{-}tBu)\}_2$-based device harvests singlet and triplet excitons. Therefore, a TADF emitter is one of the key categories for emitting materials to fabricate highly efficient OLEDs.

16.4.2 TADF Emitters Based on Donor–Acceptor Structures

A practical TADF emitter based on a pure organic molecule PIC-TRZ (Fig. 16.25) was reported by Adachi and coworkers in 2011 (Endo et al. 2011). TADF is realized by a narrow gap between the singlet and triplet excited states ΔE_{ST}. In the case of PIC-TRZ, peaks of the fluorescence and phosphorescence spectra are observed at 466 and 483 nm, and ΔE_{ST} of PIC-TRZ is just 0.11 eV. To obtain a small ΔE_{ST}, it is well known that separation of HOMO and LUMO with an appropriate overlap is essential: in other words, TADF fluorophores are obtained by achieving optimized balance between minimization of the exchange integral of two electrons occupying HOMO

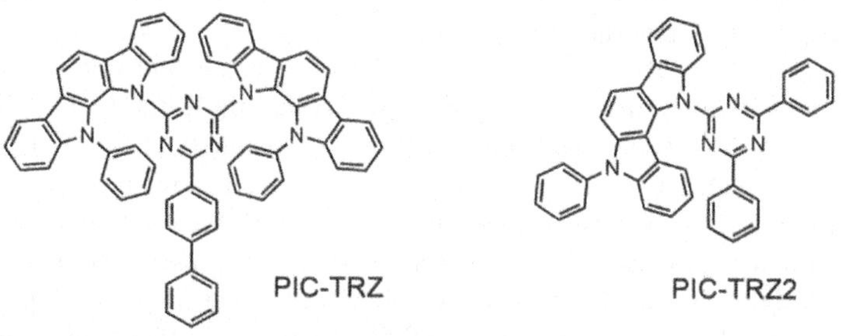

Fig. 16.25 Structures of TADF emitters PIC-TRZ and PIC-TRZ2

and LUMO and maximization of the transition dipole moment of the HOMO–LUMO transition. The molecule of PIC-TRZ consists of a (biphenyl-4-yl)-1,3,5-triazine unit as an acceptor and two 12-phenyl-11,12-dihydroindolo[2,3-*a*]carbazol-11-yl units as donors. The donor units are heavily distorted from the acceptor unit through steric hindrance, and thus the donor–acceptor π-conjugation system is partially broken. From the theoretical calculations, HOMO and LUMO are localized at the donor and acceptor units, respectively, with a small overlap at the triazine unit. As a result, a small ΔE_{ST} is achieved to obtain TADF. Indeed, although a toluene solution of PIC-TRZ without nitrogen gas bubbling showed a Φ_{PL} of 0.10, the solution filled with nitrogen gas did an improved Φ_{PL} of 0.35 with an increased TADF component. Using PIC-TRZ as an emitting dopant, an OLED was fabricated, and the maximum η_{ext} reached 5.3%, comparable to the theoretical limit. It is worthy to note that η_{ext} of 14% over the theoretical limit for fluorescent EL was achieved in 2013, where PIC-TRZ2 (Fig. 16.25), having an analogous structure to that of PIC-TRZ, was used as a TADF emitter (Sato et al. 2013). In the case of PIC-TRZ2, the HOMO and LUMO are completely separated, and indeed, a quite small ΔE_{ST} of 0.02 eV was achieved.

Starting with the development of PIC-TRZ, lots of TADF emitters having donor–acceptor ICT-type structures were developed to improve the device efficiencies of OLEDs. In 2012, Adachi et al. reported a series of TADF emitters based on a cyano-substituted benzene acceptor and carbazole donors, as shown in Fig. 16.26a (Uoyama et al. 2012). Various combination of cyano and carbazolyl groups on the benzene core allowed for a wide range of color variation of PL from sky blue to orange (Fig. 16.26b). Among the TADF emitters, 4CzIPN, 4CzTPN-Ph, and 2CzPN were employed to fabricate OLEDs. The device structure for 4CzIPN and 4CzTPN-Ph is ITO/α-NPD (35 nm)/5 wt% emitter-doped 4,4′-bis(*N*-carbazolyl)-1,1′-biphenyl (CBP, 15 nm)/TPBi (65 nm)/LiF (0.8 nm)/Al, and that for 2CzPN is ITO (100 nm)/α-NPD (40 nm)/1,3-bis(9-carbazolyl)benzene (mCP, 10 nm)/5 wt% emitter-doped 2,8-bis(diphenylphosphoryl)dibenzo[*b*,*d*]thiophene (PPT, 40 nm)/LiF (0.8 nm)/Al. In the case of the device with the green emitter 4CzIPN, an extremely high maximum η_{ext} of 19.3% was obtained, much higher than the theoretical limit for emitters exhibiting prompt fluorescent emission. This value is comparable to the internal quantum efficiency of 64.3–96.5% when light out-coupling efficiency of 20–30% is assumed. Thereafter, using TCz-TRZ (Hirata et al. 2015) and DACT-II (Kaji et al. 2015) (Fig. 16.27) as TADF emitters, the Adachi's group achieved η_{ext} more than 20%. In the case of the device with TCz-TRZ, the maximum η_{ext} of 20.6% was obtained when the device structure was ITO/α-NPD (30 nm)/3,3′-di(9*H*-carbazol-9-yl)-1,1′-biphenyl (m-CBP, 10 nm)/6 wt% emitter-doped (oxybis(2,1-phenylene))bis(diphenylphosphine oxide) (DPEPO, 15 nm)/DPEPO (10 nm)/TPBi (30 nm)/LiF (0.8 nm)/Al. The CIE chromaticity coordinate of the obtained EL was (0.19, 0.35), corresponding to sky blue. In the case of the device with the green emitter DACT-II, the maximum η_{ext} of 29.6% was achieved when the following device structure was employed; ITO/1,1-bis[4-(di-4-tolylamino)phenyl]cyclohexane (TAPC, 100 nm)/9 wt% emitter-doped CBP (40 nm)/([1,1′-biphenyl]-4-yloxy)bis((2-methylquinolin-8-yl)oxy)aluminum (BAlq, 30 nm)/LiF/Al. The extremely high η_{ext}

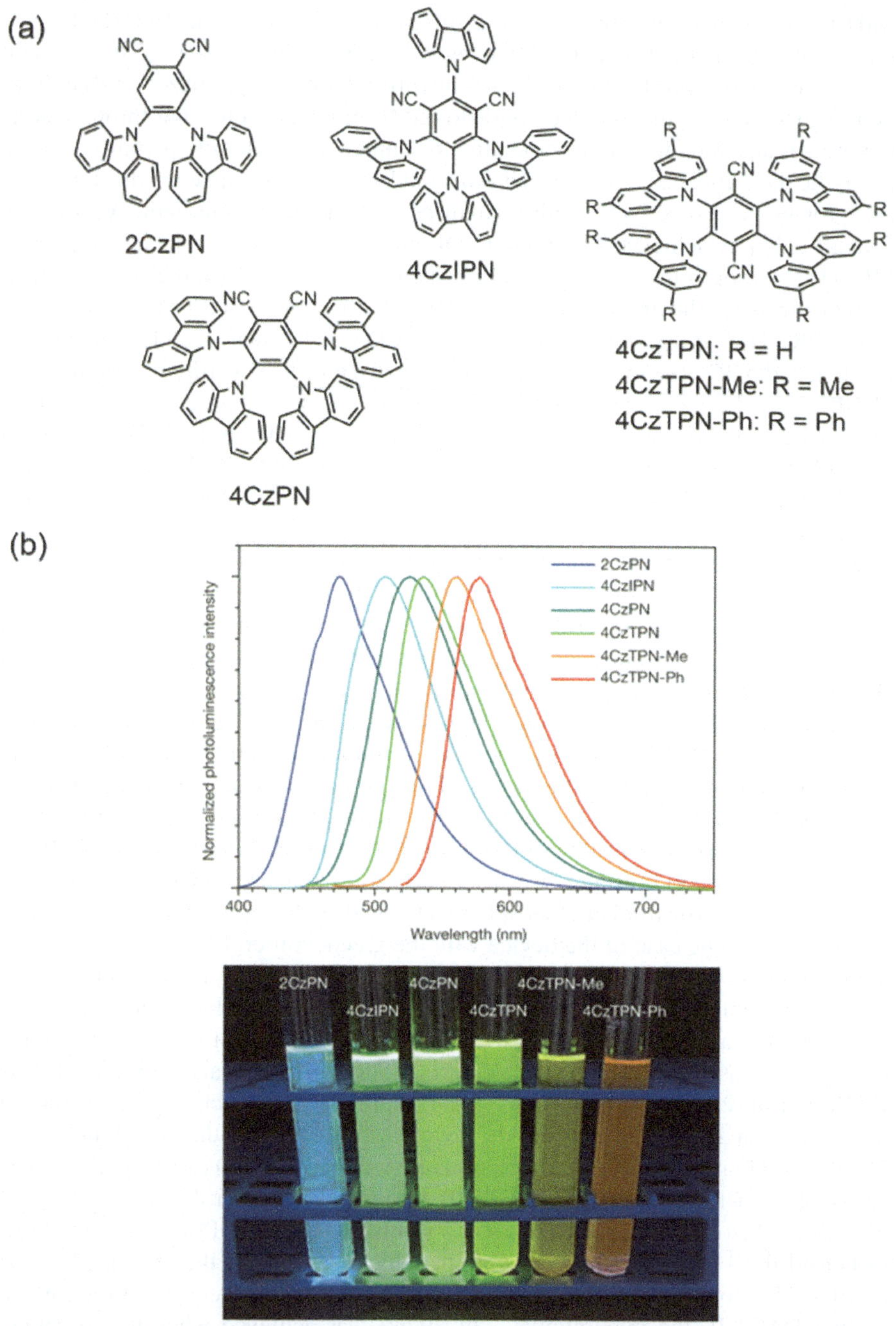

Fig. 16.26 **a** Structures of TADF emitters consisting of a cyano-substituted benzene core and carbazole moieties. **b** Photoluminescence spectra of the emitters in toluene and a photograph of their solutions upon exposure of 365 nm light. Adapted from Uoyama et al. (2012) with permission from Springer Nature

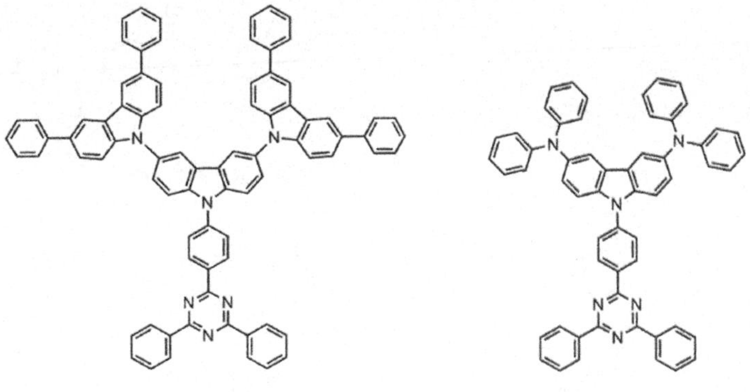

Fig. 16.27 Structures of TADF emitters TCz-TRZ and DACT-II

indicates that an internal quantum efficiency of 100% is achieved. Indeed, a CBP film doped with DACT-II emits with Φ_{PL} of approximately 100% under deaerated conditions, and thus approximately 100% of up-conversion for the triplet to the singlet excited stare was achieved. Using an out-coupling sheet, the maximum η_{ext} was further improved to 41.5%.

Except for the reports by Adachi's group, various unique TADF emitters with donor–acceptor combination were reported. For example, Congrave and coworkers developed a near-infrared emitting TADF material, the structure of which has a simple donor–acceptor combination consisting of 4-(*N*,*N*-diphenylamino)phenyl and acenaphtho[1,2-*b*]pyrazine-8,9-dicarbonitrile moieties (CAT-1, Fig. 16.28) (Congrave et al. 2019). Employing an evaporated neat film of CAT-1 as EML, the OLED was fabricated with the device structure of ITO (anode)/TAPC (40 nm)/CAT-1 (20 nm)/TPBi (40 nm)/LiF (1 nm)/Al (cathode, 100 nm). Interestingly near-infrared EL was observed at 904 nm, the emission band of which emerged in the region over 700 nm. As other examples, Tang and coworkers enthusiastically developed a series of AIE-TADF materials. As a typical example, DBT-BZ-DMAC (Fig. 16.28) (Guo et al. 2017) shows weak PL at 532 nm in THF (Φ_{PL}; 0.083), whereas the aggregate obtained by addition of a large amount of water to the THF solution exhibits intense green PL at 509 nm, showing AIE behavior. The neat film of DBT-BZ-DMAC is also highly emissive (λ_{PL}, 505 nm; Φ_{PL}, 0.802), and the observed emission includes *ca.* 50% of TADF (Φ_{PL} of TADF; 0.517). Using a neat film of this emitter as EML, non-doped OLED was fabricated, and green EL with the CIE coordinate of (0.261, 0.550) was obtained, where the maximum luminance, the maximum power efficiency, and the maximum η_{ext} were 27,270 cd m^{-2}, 35.7 lm W^{-1}, and 14.2%, respectively. Tang and coworkers further developed new AIE-TADF materials useful to fabricate high-efficiency non-doped OLEDs, namely, DCB-BP-PXZ, CBP-BP-PXZ, mCP-BP-PXZ, and mCBP-BP-PXZ (Fig. 16.28) (Liu et al. 2018). Using these emitters as a non-doped EML, OLEDs were fabricated, the device structure of which

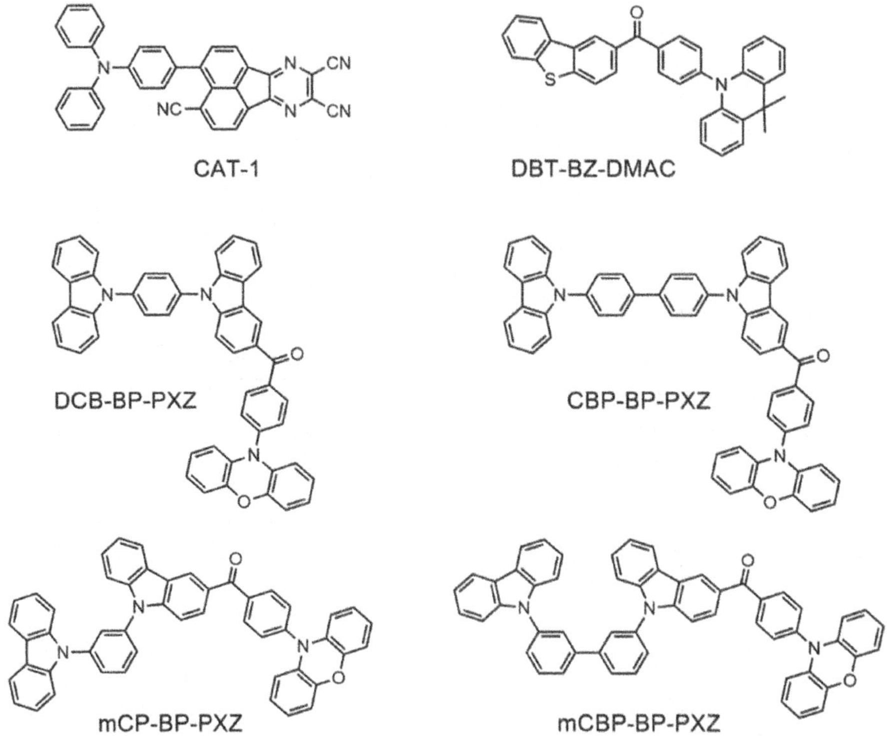

Fig. 16.28 Structures of CAT-1, DBT-BZ-DMAC, DCB-BP-PXZ, CBP-BP-PXZ, mCP-BP-PXZ, and mCBP-BP-PXZ

was ITO (anode)/TAPC (25 nm)/AIE-TADF emitter (35 nm)/1,3,5-tri(*m*-pyrid-3-yl-phenyl)benzene (TmPyPB, 55 nm)/LiF (1 nm)/Al (cathode). The devices exhibited yellow EL with the CIE coordinate of (0.38–0.39, 0.57), and showed excellent device performances: the maximum luminance, the maximum current efficiency, the maximum power efficiency, and the maximum η_{ext} were 95,577–100,126 cd m^{-2}, 69.0–72.9 cd A^{-1}, 75.0–81.8 lm W^{-1}, and 21.4–22.6%, respectively. Especially, the maximum η_{ext} is almost corresponding to the theoretical limit.

16.4.3 TADF Emitters Designed Through the Multiple Resonance Effect

As described above, TADF emitters based on donor–acceptor type fluorophores have so far been eagerly developed to overcome the theoretical limit for η_{ext} of traditional fluorescent emitters. In the case of such fluorophores, significant structural relaxation often occurs when they are excited, which gives rise to broadened spectral band shapes of PL with large half bandwidths of 70–100 nm. From the viewpoint of application to flat displays, this spectral feature is disadvantageous to tune precisely chromaticity coordinates of EL. Thus, another type of molecular design to achieve ΔE_{ST} has been required.

In 2016, Hatakeyama and coworkers have proposed a new approach to a TADF emitter with a narrow half bandwidth on the basis of efficient separation of HOMO and LUMO by the multiple resonance effect of a polycyclic aromatic compound. They employed triphenylboron with two nitrogen atoms bridging neighboring phenyl rings as a basic structural framework, i.e., 5,9-dihydro-5,9-diaza-13b-boranaphtho[3,2,1-*de*]anthracene (DABNA) (Fig. 16.29a) (Hatakeyama

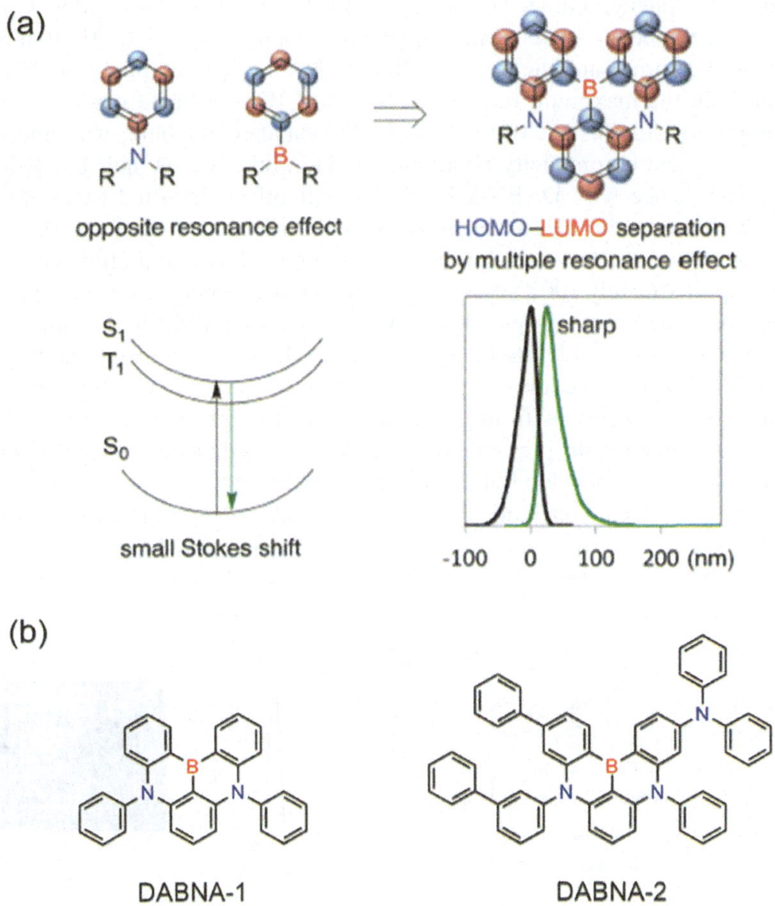

Fig. 16.29 **a** Molecular design of a TADF-active DABNA based on multiple resonance effect and **b** structures of DABNA-1 and DABNA-2. Adapted from Hatakeyama et al. 2016 with permission from Wiley

et al. 2016). This aromatic system contains nitrogen and boron atoms showing electron donating and withdrawing mesomeric effects, respectively, and shows effective HOMO–LUMO separation. Indeed, the ΔE_{ST} of 0.20 eV is achieved, enough small for triplet-to-singlet up-conversion. In addition, DABNA possesses rigid structural framework which suppresses structural relaxation upon electronic excitation, and the PL spectrum with a narrow bandwidth was expected. The 1 wt% doped m-CBP films of DABNA-1 and DABNA-2 (Fig. 16.29b) exhibit highly efficient blue PL at 460 and 469 nm with Φ_{PL}s of 0.88 and 0.90, respectively, containing prompt and delayed fluorescent emissions. The half bandwidths of their PL spectra are 28–30 nm, much smaller than those of the donor–acceptor type TADF emitters. The OLEDs with these TADF emitters were fabricated (device structure; ITO/α-NPD (40 nm)/tris(4-carbazolyl-9-ylphenyl)amine (TCTA, 15 nm)/mCP (15 nm)/1 wt% emitter-doped m-CBP (20 nm)/diphenyl-4-triphenylsilylphenylphosphine oxide (TSPO1, 40 nm)/LiF (1 nm)/Al (100 nm)), and the device with DABNA-1 exhibited pure blue EL (λ_{PL}; 459 nm) with the maximum η_{ext} of 13.5%. The CIE coordinate was (0.13, 0.09), comparable to CIE (0.14, 0.08) for the NTSC standard for blue. It is undoubted that this excellent chromaticity coordinate is due to the narrow half bandwidth of 28 nm. The device with DABNA-2 exhibited slightly red-shifted EL at 467 nm, but the half bandwidth of 28 nm is still narrow, affording CIE (0.12, 0.11). The maximum η_{ext} of 20.2% was achieved, and the internal quantum efficiency is estimated as approximately 100%, assuming the light out-coupling efficiency of 20%. In general, donor–acceptor type pure blue TADF emitters are difficult to obtain because the molecular design to elevate the singlet and triplet charge-transfer excited states (^{1}CT and ^{3}CT) brings the low-lying triplet locally excited state (^{3}LE) giving rise to trapping the triplet excitons from ^{1}CT and ^{3}CT to ^{3}LE (Zhang et al. 2014). Taking this into consideration, the present molecular design based on the multiple resonance effect is beneficial to develop blue TADF emitters.

Recently, Hatakeyama and coworkers realized deep-blue TADF emitter (ν-DABNA, Fig. 16.30a) with an extremely narrow half bandwidth by designing the new polycyclic aromatic framework in which two DABNA skeleton are fused

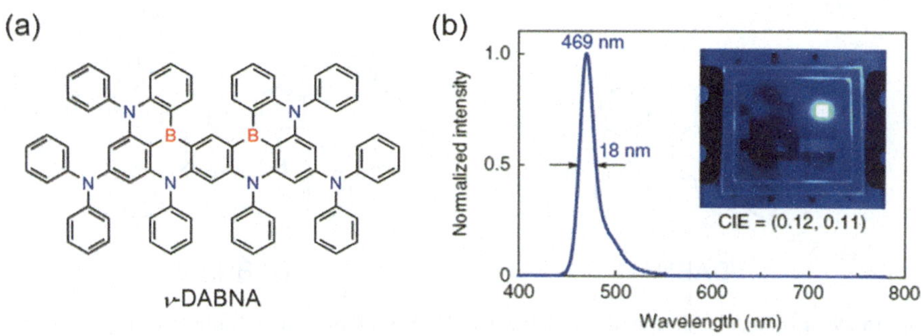

Fig. 16.30 **a** Structure of the blue TADF emitter ν-DABNA. **b** The electroluminescence spectrum of the OLED with ν-DABNA as an emitting dopant, with a photograph of the emitting device inset. Adapted from Kondo et al. (2019) with permission from Springer Nature

(Kondo et al. 2019). This TADF emitter exhibits blue emission at 457 nm with a quite narrow half bandwidth of 14 nm in toluene, the Φ_{PL} of which is fairly excellent (0.74). Thus, through the multiple resonance effect of the constituent boron and nitrogen atoms in the fused DABNA chromophore, the suppressed vibronic coupling between S_0 and S_1 caused the narrow half bandwidth of the PL spectrum. The OLED with ν-DABNA as an emitting dopant afforded efficient blue electroluminescence at 469 nm with a half bandwidth of 18 nm (Fig. 16.30b). The CIE coordinate was (0.12, 0.11), close to the NTSC standard for blue. The fabricated device showed the extremely high maximum η_{ext} of 34.4% at 15 cd m^{-2}, accompanied by a small device efficiency roll-off (η_{ext}; 26.0% at 1000 cd m^{-2}).

16.5 Summary

In this chapter, emitting materials for OLED were introduced, focusing on the exciton generation mechanism. First, fluorescent materials were reviewed according to their emission colors. For low-mass fluorescent emitters, various types of structural frameworks are available, allowing for tuning the luminescence properties such as the emission color and the quantum efficiencies. Fluorescent emitters based on polymer and dendrimer structures are applied to fabrication of solution-processed OLEDs, and their excellent film-forming properties are beneficial to fabricate flexible light-emitting devices. The η_{ext} of a fluorescent OLED is fatally low to be *ca.* 5% at most, because electric current excitation of an emitter provides the singlet and triplet excitons according to spin statics. However, fluorescent materials are still popular at the level of industrial implementation, especially as blue emitters.

Second, phosphorescent materials were reviewed, especially focusing on organoiridium(III) and organoplatinum(II) complexes. In these complexes, strong spin–orbit interactions induced by the constituent heavy metals facilitate effective mixing of the singlet and triplet states, and thus the triplet excitons are efficiently obtained even at ambient temperature. In OLEDs, triplet excitons are electrically generated directly and via the intersystem crossing from S_1 to T_1, and thus the η_{ext} of a phosphorescent OLED is as high as 100%, four times more than that of a fluorescent OLED. Thus, phosphorescent materials are widely used to fabricate high-efficiency devices, especially as red and green emitters. At this point, blue phosphorescent OLEDs do not meet the demands for industrial implementation, in terms of color purity and device lifetime.

In the last part, a new type of fluorescent material is introduced, that is, a TADF emitter. This emitter utilizes both singlet and triplet excitons to produce photons. The key process to achieve TADF is the reverse intersystem crossing from T_1 to S_1. To obtain a small ΔE_{ST}, a minimized HOMO–LUMO overlap and a miximized transition dipole moment are required. So, the molecular design for a TADF emitter has been based on construction of a distorted donor–acceptor π-conjugation system showing ICT electronic transition. So far, lots of TADF emitters achieving η_{ext}s over the theoretical upper limit of conventional fluorescent emitters have been reported.

Another type of TADF emitter is based on HOMO–LUMO separation by the multiple resonance effect. The ICT-type TADF emitters show PL and EL spectra with relatively large half bandwidths due to structural relaxation upon excitation. On the other hand, the structural relaxation is considerably suppressed in TADF emitters with the multiple resonance effect, and thus quite narrow half bandwidths of *ca.* 20 nm or less are achieved. It is noteworthy that pure blue OELDs with η_{ext}s almost reaching the theoretical limit have been developed using this type of TADF emitters.

In near future, OLEDs will be increasingly required to develop flat panel displays, illumination apparatuses, electronic papers, and other display devices. Toward widespread of OLEDs as smart devices, the device performance should be further improved. It is undoubted that emitting materials play an important role in the next breakthrough of the OLED technology.

References

Adachi C, Tsutsui T, Saito S (1990) Confinement of charge carriers and molecular excitons within 5-nm-thick emitter layer in organic electroluminescent devices with a double heterostructure. Appl Phys Lett 57:531–533

Adachi C, Baldo MA, Thompson ME, Forrest SR (2001a) Nearly 100% internal phosphorescence efficiency in an organic light-emitting device. J Appl Phys 90:5048–5051

Adachi C, Kwong RC, Djurovich P, Adamovich V, Baldo MA, Thompson ME, Forrest SR (2001b) Endothermic energy transfer: a mechanism for generating very efficient high-energy phosphorescent emission in organic materials 79:2082–2084

(a) Adamovich V,Brooks J, Tamayo A, Alexander AM, Djurovich PI, D'Andrade BW, Adachi C, Forrest SR, Thompson ME (2002) High efficiency single dopant white electrophosphorescent light emitting diodes. New J Chem 26:1171–1178; (b) Kim D, Brédas JL (2009) Triplet excimer formation in platinum-based phosphors: a theoretical study of the roles of Pt-Pt bimetallic interactions and interligand π–π interactions. J Am Chem Soc 131:11371–11380; (c) Shigehiro T, Yagi S, Maeda T, Nakazumi H, Fujiwara H, Sakurai Y (2013) Photo- and electroluminescence from 2-(Dibenzo[*b*,*d*]furan-4-yl)pyridine-based heteroleptic cyclometalated platinum(II) complexes: excimer formation drastically facilitated by an aromatic diketonate ancillary ligand. J Phys Chem C117:532–542; (d) Shigehiro T, Chen Q, Yagi S, Maeda T, Nakazumi H, Sakurai Y (2016) Substituent effect on photo- and electroluminescence properties of heteroleptic cyclometalated platinum(II) complexes based on a 2-(Dibenzo[b,d]furan-4-yl)pyridine ligand. Dyes Pigm 124:165–173; (e) Okamura N, Maeda T, Fujiwara H, Soman A, Unni KNN, Ajayaghosh A, Yagi S (2018) Photokinetic study on remarkable excimer phosphorescence from heteroleptic cyclometalated platinum(II) complexes bearing a benzoylated 2-phenylpyridinate ligand. Phys Chem Chem Phys 20:542–552

Auer-Berger M, Trattnig R, Qin T, Schlesinger R, Nardi MV, Ligorio G, Christodoulou C, Koch N, Baumgarten M, Müllen K, List-Kratochvil EJW (2016) All-solution-processed multilayer polymer/dendrimer light emitting diodes. Org Electron 35:164–170

Bolink HJ, Barea E, Costa RD, Coronado E, Sudhakar S, Zhen C, Sellinger A (2008) Efficient blue emitting organic light emitting diodes based on fluorescent solution processable cyclic phosphazenes. Org Electron 9:155–163

(a) Caspar JV, Kober EM, Sullivan BP, Meyer TJ (1982) Application of the energy gap low to the decay of charge-transfer excited states. J Am Chem Soc 104:630–632; (b) Sommer JR, Shelton AH, Parthasarathy A, Ghiviriga I, Reynords JR, Schanze KS (2011) Photophysical properties of near-infrared phosphorescent π-extended platinum porphyrins. Chem Mater 23:5296–5304

Chang CF, Cheng YM, Chi Y, Chiu YC, Lin CC, Lee, Pi-Tai Chou GH, Chen CC, Chang CH, Wu CC (2008) Highly efficient blue-emitting iridium(III) carbene complexes and phosphorescent OLEDs. Angew Chem Int Ed 47:4542–4545

Chen CH, Tang CW (2001) Efficient green organic light-emitting diodes with stericly hindered coumarin dopants. Appl Phys Lett 79:3711–3713

Chen CH, Tang CW, Shi J, Klubek KP (2000) Recent developments in the synthesis of red dopants for Alq3 hosted electroluminescence. Thin Solid Films 363:327–331

(a) Chen HF, Yang SJ, Tsai ZH, Hung WY, Wang TC, Wong KT (2009) 1,3,5-triazine derivatives as new electron transport–type host materials for highly efficient green phosphorescent OLEDs. J Mater Chem 19:8112–8118; (b) Sun C, Hudson ZM, Helander MG, Lu ZH, Wang S (2011) A polyboryl-functionalized triazine as an electron transport material for OLEDs. Organometallics 30:5552–5555; (c) Chen YM, Hung WY, You HW, Chaskar A, Ting HC, Chen HF, Wong KT, Liu YH (2011) Carbazole–benzimidazole hybrid bipolar host materials for highly efficient green and blue phosphorescent OLEDs. 21:14971–14978

Chen B, Jiang Y, Chen L, Nie H, He B, Lu P, Sung HHY, Williams ID, Kwok HS, Qin A, Zhao Z, Tang BZ (2013) 2,5-difuluorenyl-substituted siloles for the fabrication of high-performance yellow organic light-emitting diodes. Chem Eur J 20:1931–1939

Chuang TH, Yang CH, Kao PC (2009) Efficient red-emitting cyclometalated iridium(III) complex and applications of organic light-emitting diode. Inorg Chim Acta 362:5017–5022

Cocchi M, Virgili D, Rochester DL, Williams JAG (2007a) N^C^N-coordinated platinum(II) complexes as phosphorescent emitters in high-performance organic light-emitting devices. Adv Funct Mater 17:285–289

Cocchi M, Kalinowski J, Virgili D, Fattori V, Develay S, Williams JAG (2007b) Single-dopant organic white electrophosphorescent diodes with very high efficiency and its reduced current density roll-off. Appl Phys Lett 90:163508 (3 pp)

Congrave DG, Drummond BH, Conaghan PJ, Francis H, Jones STE, Grey CP, Greenham NC, Credgington D, Bronstein H (2019) A simple molecular design strategy for delayed fluorescence toward 1000 nm. J Am Chem Soc 141:18390–18394

Deaton JC, Switalski SC, Kondakov DY, Young RH, Pawlik TD, Giesen DJ, Harkins SB, Miller AJM, Mickenberg SF, Peters JC (2010) E-type delayed fluorescence of a phosphine-supported $Cu_2(\mu\text{-}NAr_2)_2$ diamond core: harvesting singlet and triplet excitons in OLEDs. J Am Chem Soc 132:9499–9508

Dedeian K, Shi J, Shepherd N, Forsythe E, Morton DC (2005) Photophysical and electrochemical properties of heteroleptic tris-cyclometalated iridium(III) complexes. Inorg Chem 44:4445–4447

Enders D, Breuer K, Raabe G, Runsink J, Teles JH, Melder JP, Ebel K, Brode S (1995) Preparation, structure, and reactivity of 1,3,4-Triphenyl-4,5-dihydro-1*H*-1,2,4-triazol-5-ylidene, a new stable carbene. Angew Chem Int Ed 34:1021–1023

Endo A, Suzuki K, Yoshihara T, Tobita S, Yahiro M, Adachi C (2008) Measurement of photoluminescence efficiency of Ir(III) phenylpyridine derivatives in solution and solid-state films. Chem Phys Lett 460:155–157

Endo A, Ogasawara A, Takahashi A, Yokoyama D, Kato Y, Adachi C (2009) Thermally activated delayed fluorescence from Sn^{4+}–porphyrin complexes and their application to organic light-emitting diodes—a novel mechanism for electroluminescence. Adv Mater 21:4802–4806

Endo A, Sato K, Yoshimura K, Kai T, Kawada A, Miyazaki H, Adachi C (2011) Efficient up-conversion of triplet excitons into a singlet state and its application for organic light emitting diodes. Appl Phys Lett 98:083302 (3 pp)

Fleetham T, Li G, Wen L, Li J (2014) Efficient "pure" blue OLEDs employing tetradentate Pt complexes with a narrow spectral bandwidth. Adv Mater 26:7116–7121

Fukagawa H, Shimizu T, Hanashima H, Osada Y, Suzuki M, Fujikake H (2012) Highly efficient and stable red phosphorescent organic light-emitting diodes using platinum complexes. Adv Mater 24:5099–5103

Giovanella U, Botta C, Galeotti F, Vercelli B, Battiato S, Pasinia M (2013) Perfluorinated polymer with unexpectedly efficient deep blue electroluminescence for full-colour OLED displays and light therapy applications. J Mater Chem C 1:5322–5329

Greenham NC, Moratti SC, Bradley DDC, Friend RH, Holmes AB (1993) Efficient light-emitting diodes based on polymers with high electron affinities. Nature 365:628–630

Grice AW, Bradley DDC (1998) High brightness and efficiency blue light-emitting polymer diodes. Appl Phys Lett 73:629–631

Guo J, Li XL, Nie H, Luo W, Gan S, Hu S, Hu R, Qin A, Zhao Z, Su SJ, Tang BZ (2017) Achieving high-performance nondoped OLEDs with extremely small efficiency roll-off by combining aggregation-induced emission and thermally activated delayed fluorescence. Adv Funct Mater 27:1606458 (9 pp)

Gustafsson G, Cao Y, Treacy GM, Klavetter F, Colaneri N, Heeger AJ (1992) Flexcible light-emitting diodes made from soluble conducting polymer. Nature 357:477–479

Hancock JM, Gifford AP, Zhu Y, Lou Y, Jenekhe SA (2006) n-type conjugated oligoquinoline and oligoquinoxaline with triphenylamine endgroups: efficient ambipolar light emitters for device applications. Chem Mater 18:4924–4932

Hancock JM, Gifford AP, Tonzola CJ, Jenekhe SA (2007) High-efficiency electroluminescence from new blue-emitting oligoquinolines bearing pyrenyl or triphenyl endgroups. J Phys Chem C 111:6875–6882

Hatakeyama T, Shiren K, Nakajima K, Nomura S, Nakatsuka S, Kinoshita K, Ni J, Ono Y, Ikuta T (2016) Ultrapure blue thermally activated delayed fluorescence molecules: efficient HOMO–LUMO separation by the multiple resonance effect 28:2777–2781

Helfrich W, Schneider WG (1965) Recombination radiation in anthracene crystals. Phys Rev Lett 14:229–231

Helfrich W, Schneider WG (1966) Transients of volume-controlled current and of recombination radiation in anthracene. J Chem Phys 44:2902–2909

He Y, Gong S, Hattori R, Kannicki J (1999) High performance organic polymer lightemitting heterostructure devices. Appl Phys Lett 74:2265–2267

Hirata S, Sakai Y, Masui K, Tanaka H, Lee SY, Nomura H, Nakamura N, Yasumatsu M, Nakanotani H, Zhang Q, Shizu K, Miyazaki H, Adachi C (2015) Highly effcient blue electroluminescence based on thermally activated delayed fluorescence. Nature Mater 14:330–336

Ikawa S, Yagi S, Maeda T, Nakazumi H, Fujiwara H, Sakurai Y (2012) Photoluminescence color tuning of phosphorescent bis-cyclometalated iridium(III) complexes by ancillary ligand replacement. Dyes Pigm 95:695–705

Jeon SO, Jang SE, Son HS, Lee JY (2011) External quantum efficiency above 20% in deep blue phosphorescent organic light-emitting diodes. Adv Mater 23:1436–1441

Kaji H, Suzuki H, Fukushima T, Shizu K, Suzuki K, Kubo S, Komino T, Oiwa H, Suzuki F, Wakamiya A, Murata Y, Adachi C (2015) Nature commun purely organic electroluminescent material realizing 100% conversion from electricity to light 6:8476 (8 pp)

Kalinowski J, Cocchi M, Virgili D, Fattori V, Williams JAG (2007) Mixing of excimer and exciplex emission: a new way to improve white light emitting organic electrophosphorescent diodes. Adv Mater 19:4000–4005

Kang Y, Chang YL, Lu JS, Ko SB, Rao Y, Varlan M, Lu ZH, Wang S (2013) Highly efficient blue phosphorescent and electroluminescent Ir(III) compounds. J Mater Chem C 1:441–450

Kim DH, Cho NS, Oh HY, Yang JH, Jeon WS, Park JS, Suh MC, Kwon JH (2011) Highly efficient red phosphorescent dopants in organic light-emitting devices. Adv Mater 23:2721–2726

Kim HU, Jang JH, Song W, Jung BJ, Lee JY, Hwang DH (2015) Improved luminance and external quantum efficiency of red and white organic light-emitting diodes with iridium(III) complexes with phenyl-substituted 2-phenylpyridine as a second cyclometalated ligand. J Mater Chem C 3:12107–12115

Kondo Y, Yoshiura K, Kitera S, Nishi H, Oda S, Gotoh H, Sasada Y, Yanai M, Hatakeyama T (2019) Narrowband deep-blue organic light-emitting diode featuring an organoboron-based emitter. Nat Photonics 13:678–682

Kwok CC, Wong MS (2001) Synthesis and light-emitting properties of difunctional dendritic dstyrylstilbenes. Macromolecules 34:6821–6830

(a) Lamansky S, Djurovich P, Murphy D, Abdel-Razzaq F, Kwong R, Tsyba I, Bortz M, Mui B, Bau R, Thompson ME (2001) Synthesis and characterization of phosphorescent cyclometalated iridium complexes. Inorg Chem 40:1704–1711; (b) Lamansky S, Djurovich P, Murphy D, Abdel-Razzaq F, Lee HE, Adachi A, Burrows PE, Forrest SR, Thompson ME (2001) Highly phosphorescent bis-cyclometalated iridium complexes: synthesis, photophysical characterization, and use in organic light emitting diodes. J Am Chem Soc 123:4304–4312; (c) Tamayo AB, Alleyne BD, Djurovich PI, Lamansky S, Tsyba I, Ho NN, Bau R, Thompson ME (2003) Synthesis and Characterization of Facial and Meridional Tris-cyclometalated Iridium(III) Complexes. J Am Chem Soc 125:7377–7387; (d) Brooks J, Babayan Y, Lamansky S, Djurovich PI, Tsyba I, Bau R, Thompson ME (2002) Synthesis and characterization of phosphorescent cyclometalated platinum complexes 41:3055–3066

Lee MT, Yen CK, Yang WP, Chen HH, Liao CH, Tsai CH, Chen CH (2004) Efficient green coumarin dopants for organic light-emitting devices. Org Lett 6:1241–1244

Lee SJ, Park KM, Yang K, Kang Y (2009) Blue phosphorescent Ir(III) complex with high color purity: *fac*-tris(2′,6′-difluoro-2,3′-bipyridinato-*N*, $^{C}4'$)iridium(III). Inorg Chem 48:1030–1037

Lee S, Kim SO, Shin H, Yun HJ, Yang K, Kwon SK, Kim JJ, Kim YH (2013) Deep-blue phosphorescence from perfluoro carbonyl-substituted iridium complexes. J Am Chem Soc 135:14321–14328

Leopold H, Heinemeyer U, Wagenblast G, Menster I, Strassner T (2017) Changing the emission properties of phosphorescent C^C*-cyclometalated thiazol-2-ylidene platinum(II) complexes by variation of the b-diketonate ligands 23:1118−1128

Liu H, Zeng J, Guo J, Nie H, Zhao Z, Tang BZ (2018) High-performance non-doped OLEDs with nearly 100% exciton use and negligible efficiency roll-off. Angew Chem Int Ed 57:9290–9294

Losurdo M, Giangregorio MM, Capezzuto P, Cardone A, Martinelli C, Farinola GM, Babudri F, Naso F, Büchel M, Bruno G (2009) Blue-gap poly(*p*-phenylene vinylene)s with fluorinated double bonds: interplay between supramolecular organization and optical properties in thin films. Adv Mater 21:1115–1120

Lu W, Mi BX, Chan MCW, Hui Z, Che CM, Zhu N, Lee ST (2004) Light-emitting tridentate cyclometalated platinum(II) complexes containing σ-alkynyl auxiliaries: tuning of photo- and electrophosphorescence. J Am Chem Soc 126:4958–4971

Mikami A, Koshiyama T, Tsubokawa T (2005) High-efficiency color and white organic light-emitting devices prepared on flexible plastic substrates. Jpn J Appl Phys 44:608–612

Mi BX, Gao ZQ, Lee CS, Lee ST, Kwong HL, Wong NB (1999) Reduction of molecular aggregation and its application to the high-performance blue perylene-doped organic electroluminescent device. Appl Phys Lett 75:4055–4057

Ohmori Y, Uchida M, Muro K, Yoshino K (1991) Blue electroluminescent diodes utilizing poly(alkylfluorene). Jpn J Appl Phys 30:L1941–L1943

Okamura N, Nakamura T, Yagi S, Maeda T, Nakazumi H, Fujiwara H, Koseki S (2016) Novel Bis-And Tris-Cyclometalated Iridium(III) complexes bearing a benzoyl group on each fluorinated 2-phenylpyridinate ligand aimed at development of blue phosphorescent materials for OLED. RSC Adv 6:51435–51445

Osken I, Gundogan AS, Tekin E, Eroglu MS, Ozturk T (2013) Fluorene−dithienothiophene-S, S-dioxide copolymers fine-tuning for OLED applications. Macromolecules 46:9202–9210

Pei J, Yu WL, Huang W, Heeger AJ (2000) The synthesis and characterization of an efficient green electroluminescent conjugated polymer: poly[2,7-bis(4-hexylthienyl)-9,9-dihexylfluorene]. Chem Commun 1631–1632

Qin W, Yang Z, Jiang Y, Lam JWY, Liang G, Kwok HS, Tang BZ (2015) Construction of efficient deep blue aggregation-induced emission luminogen from triphenylethene for nondoped organic light-emitting diodes. Chem Mater 27:3892–3901

(a) Reineke S, Walzer K, Leo K (2007) Triplet-exciton quenching in organic phosphorescent light-emitting diodes with Ir-based emitters. Phys Rev B 75:125328 (13 pp); (b) Reineke S, Schwartz

G, Walzer K, Leo K (2007) Reduced efficiency roll-off in phosphorescent organic light-emitting diodes by suppression of triplet-triplet annihilation. Appl Phys Lett 91:123508 (3 pp)

Rossi E, Murphy L, Brothwood PL, Colombo A, Dragonetti C, Roberto D, Ugo R, Cocchi M, Williams JAG (2011) Cyclometallated platinum(II) complexes of 1,3-Di(2-pyridyl)benzenes: tuning excimer emission from red to near-infrared for NIR-OLEDs 21:15501–15510

Sajoto T, Djurovich PI, Tamayo AB, Oxgaard J, Goddard WA III, Thompson ME (2009) Temperature dependence of blue phosphorescent cyclometalated Ir(III) complexes. J Am Chem Soc 131:9813–9822

Sasabe H, Takamatsu J, Motoyama T, Watanabe S, Wagenblast G, Langer N, Molt O, Fuchs E, Lennartz C, Kido J (2010) High-efficiency blue and white organic light-emitting devices incorporating a blue iridium carbene complex. Adv Mater 22:5003–5007

Sato K, Shizu K, Yoshimura K, Kawada A, Miyazaki H, Adachi C (2013) Organic luminescent molecule with energetically equivalent singlet and triplet excited states for organic light-emitting diodes. Phys Rev Lett 110:247401 (5 pp)

Schwartz G, Pfeiffer M, Reineke S, Walzer K, Leo K (2007) Harvesting triplet excitons from fluorescent blue emitters in white organic light-emitting diodes. Adv Mater 19:3672–3676

Shi J, Tang CW (1997) Doped organic electroluminescent devices with improved stability. Appl Phys Lett 70:1665–1667

Soellner J, Strassner T (2018) The "enders triazole" revisited: highly efficient, blue platinum(II) emitters. Organometallics 37:1821–1824

(a) Tanaka D, Sasabe H, Li Y-J, Su S-J, Takeda T, Kido J (2007) Ultra high efficiency green organic light-emitting devices. Jpn J Appl Phys 46:L10–L12; (b) Su S-J, Sasabe H, Pu Y-J, Nakayama K, Kido J (2010) Tuning energy levels of electron-transport materials by nitrogen orientation for electrophosphorescent devices with an 'ideal' operating voltage. Adv Mater 22:3311–3316; (c) Chang C-H, Kuo M-C, Lin W-C, Chen Y-T, Wong K-T, Chou S-H, Mondal E, Kwong RC, Xia S, Nakagawa T, Adachi C (2012) A dicarbazole–triazine hybrid bipolar host material for highly efficient green phosphorescent OLEDs. J Mater Chem 22:3832–3838; (d) Li W, Li J, Liu D, Li D, Zhang D (2016) Dual n-type units including pyridine and diphenylphosphine oxide: effective design strategy of host materials for high-performance organic light-emitting diodes. Chem Sci 7:6706–6714 (e) He G, Pfeiffer M, Leo K, Hofmann M, Birnstock J, Pudzich R, Salbeck J (2004) High-efficiency and low-voltage electrophosphorescent organic lightemitting diodes with double-emission layers. Appl Phys Lett 85:3911–3913

Tang CW, VanSlyke SA (1987) Organic electroluminescent diodes. Appl Phys Lett 51:913–915

Tang CW, VanSlyke SA (1989) Electroluminescence of doped organic thin films. J Appl Phys 65:3610–3616

Thomas KRJ, Lin JT, Velusamy M, Tao YT, Chuen CH (2004) Color tuning in benzo[1,2,5]thiadiazole-based small molecules by amino conjugation/deconjugation: bright red-light-emitting diodes. Adv Funct Mater 14:83–90

Tronnier A, Nischan N, Metz S, Wagenblast G, Münster I, Strassner T (2014) Phosphorescent C^C* cyclometalated PtII dibenzofuranyl-NHC complexes—an auxiliary ligand study. Eur J Inorg Chem 2014:256–264

Tsuboyama A, Iwawaki H, Furugori M, Mukaide T, Kamatani J, Igawa S, Moriyama T, Miura S, Takiguchi T, Okada S, Hoshino M, Ueno K (2003) Homoleptic cyclometalated iridium complexes with highly efficient red phosphorescence and application to organic light-emitting diode. J Am Chem Soc 125:12971–12979

Tsujimoto H, Yagi S, Asuka H, Inui Y, Ikawa S, Maeda T, Nakazumi H, Sakurai Y (2010a) Pure red electrophosphorescence from polymer light-emitting diodes doped with highly emissive bis-cyclometalated iridium(III) complexes. J Organomet Chem 695:1972–1978

Tsujimoto H, Yagi S, Honda Y, Terao H, Maeda T, Sakurai Y (2010b) Photoluminescent properties of heteroleptic cyclometalated platinum(II) complexes bearing 1,3-bis(3,4-dibutoxyphenyl)propane-1,3-dionate as an ancillary ligand. J Lumin 130:217–221

Unger Y, Meyer D, Molt O, Schildknecht C, Münster I, Wagenblast G, Strassner T (2010) Green-blue emitters: NHC-based cyclometalated [Pt(C^C*)(acac)] complexes. Angew Chem Int Ed 49:10214–10216

Uoyama H, Goushi K, Shizu K, Nomura H, Adachi C (2012) Highly efficient organic light-emitting diodes from delayed fluorescence. Nature 492:234–238

Williams JAG, Beeby A, Davies ES, Weinstein JA, Wilson C (2003) An alternative route to highly luminescent platinum(II) complexes: cyclometalation with N^C^N-coordinating dipyridylbenzene ligands. Inorg Chem 42:8609–8611

Yeh HC, Yeh SJ, Chen CT (2003) Readily synthesised arylamino fumaronitrile for non-doped red organic light-emitting diodes. Chem Commun 2632–2633

Yu J, Shirota Y (2002) A new class of high-performance red-fluorescent dyes for organic electroluminescent devices,[7-Diethylamino-3-(2-thienyl)chromen-2-ylidene]-2,2-dicyanovinylamine and {10-(2-Thienyl)-2,3,6,7-tetrahydro-1H,5H-chromeno[8,7,6-ij]quinolizin-11-ylideneg-2,2-dicyanovinylamine. Chem Lett 984

Yu YH, Huang CH, Yeh JM, Huang PT (2011) Effect of methyl substituents on the N-diaryl rings of anthracence-9,10-diamine derivatives for OLEDs applications. Org Electron 12:694–702

Zhang Q, Li B, Huang S, Nomura H, Tanaka H, Adachi C (2014) Efficient blue organic light-emitting diodes employing thermally activated delayed fluorescence. Nat Photonics 8:326–332

Zheng XY, Zhu WQ, Wu YZ Jiang XY, Sun RG, Zhang ZL, Xu SH (2003) A white OLED based on DPVBi blue light emitting host and dcjb red dopant. Displays 24:121–124

Zhou G, Wang Q, Ho CL, Wong WY, Ma D, Wang L, Lin Z (2008) Robust tris-cyclometalated iridium(III) phosphors with ligands for effective charge carrier injection/transport: synthesis, redox, photophysiscal, and electrophosphorescent behavior. Chem Asian J 3:1830–1841

Zhu YC, Zhou L, Li, HY, Xu QL, Teng MY, zheng YX, Zuo, JL, Zhang HJ, You XZ (2011a) Highly efficient green and blue-green phosphorescent OLEDs based on iridium complexes with the tetraphenylimidodiphosphinate ligand. Adv Mater 23:4041–4046

Zhu YC, Zhou L, Li HY, Xu QL, Teng MY, Zheng YX, Zuo JL, Zhang HJ, You XZ (2011b) Highly efficient green and blue-green phosphorescent OLEDs based on iridium complexes with the tetraphenylimidodiphosphinate ligand. Adv Mater 23:4041–4046

The manufacturer's authorised representative in the EU is Springer Nature Customer Service Centre GmbH, Europaplatz 3, 69115 Heidelberg, Germany. If you have any concerns regarding our products, please contact ProductSafety@springernature.com

Printed and bound by CPI Group (UK) Ltd, Croydon, CR0 4YY
30/09/2026
02225295-0017